Beginning and Intermediate Algebra

Beginning and Intermediate Algebra

John Tobey

North Shore Community College
Danvers, Massachusetts

Jeffrey Slater

North Shore Community College
Danvers, Massachusetts

Prentice Hall

Prentice Hall
Upper Saddle River, NJ 07458

Library of Congress Cataloging-in-Publication Data

Tobey, John
  Beginning and intermediate algebra / John Tobey, Jeffrey Slater.
    p. cm.
  Includes index.
  ISBN 0-13-090949-1
  1. Algebra.   I. Slater, Jeffrey – II. Title.
  QA152.3 .T64 2002
  512.9--dc21   2001052044

Executive Acquisition Editor: Karin E. Wagner
Editor in Chief: Christine Hoag
Project Manager: Mary Beckwith
Vice President/Director of Production and Manufacturing: David W. Riccardi
Executive Managing Editor: Kathleen Schiaparelli
Senior Managing Editor: Linda Mihatov Behrens
Production Management: Elm Street Publishing Services, Inc.
Manufacturing Buyer: Alan Fischer
Manufacturing Manager: Trudy Pisciotti
Executive Marketing Manager: Eilish Collins Main
Marketing Assistant: Annett Uebel
Development Editor: Kathy Sessa Federico
Editor in Chief, Development: Carol Trueheart
Media Project Manager, Developmental Math: Audra J. Walsh
Art Director: Maureen Eide
Assistant to the Art Director: John Christiana
Interior Designer: Studio Montage
Cover Designer: Studio Montage
Art Editor: Thomas Benfatti
Managing Editor, Audio/Video Assets: Grace Hazeldine
Creative Director: Carole Anson
Director of Creative Services: Paul Belfanti
Photo Researcher: Julie Tesser
Photo Editor: Beth Boyd
Cover Photo: Patrick Ingrand/Stone
Art Studio: Scientific Illustrators
Compositor: Preparé, Inc. / Emilcomp, Srl

Prentice Hall
© 2002 by Prentice-Hall, Inc.
Upper Saddle River, NJ 07458

Printed in the United States of America
10 9 8 7 6 5 4 3 2

Student ISBN (paperback) 0-13-090949-1
Student ISBN (case) 0-13-008540-5

Pearson Education Ltd.
Pearson Education Australia Pty., Limited
Pearson Education *Singapore*, Pte. Ltd
Pearson Education North Asia Ltd.
Pearson Education Canada, Ltd.
Pearson Educación de Mexico, S.A. de C.V.
Pearson Education—Japan
Pearson Education Malaysia, Pte. Ltd.

This book is dedicated to Emily B. Robinson,
a loving mother, grandmother, and great-grandmother.

Contents

Chapter 1

Real Numbers and Variables   1

Chapter 2

Equations, Inequalities, and Applications   63

Chapter **3**

Graphing and Functions   141

Chapter **4**

Systems of Linear Equations and Inequalities   211

Chapter **5**

Exponents and Polynomials   261

Chapter **6**

Factoring   319

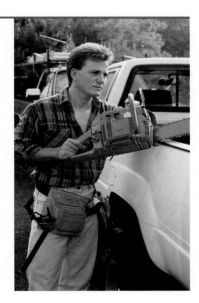

Chapter **7**

Rational Expressions and Equations 371

Chapter **8**

Rational Exponents and Radicals 423

Chapter **9**

Quadratic Equations, Inequalities, and Absolute Value   487

Chapter **10**

The Conic Sections   563

Chapter **11**

Chapter **12**

Preface

To the Instructor

We share a partnership with you. For over thirty years we have taught mathematics courses at North Shore Community College. Each semester we join you in the daily task of sharing the knowledge of mathematics with students who often struggle with this subject. We enjoy teaching and helping students—and we are confident that you share these joys with us.

Mathematics instructors and students face many challenges today. *Beginning and Intermediate Algebra* was written with these needs in mind. This textbook explains mathematics slowly, clearly, and in a way that is relevant to everyday life for the college student. Special attention has been given to problem solving. This text is written to help students organize the information in any problem-solving situation, to reduce anxiety, and to provide a guide that enables students to become confident problem solvers.

One of the hallmark characteristics of *Beginning and Intermediate Algebra* that makes the text easy to learn and teach from is the building-block organization. Each section is written to stand on its own, and each homework set is completely self-testing. Exercises are paired and graded and are of varying levels and types to ensure that all skills and concepts are covered. As a result, the text offers students an effective and proven learning program suitable for a variety of course formats—including lecture-based classes; discussion-oriented classes; distance learning centers; modular, self-paced courses; mathematics laboratories; and computer-supported centers.

Beginning and Intermediate Algebra is part of a series that includes the following:

Tobey/Slater, *Basic College Mathematics*, Fourth Edition

Blair/Tobey/Slater, *Prealgebra*, Second Edition

Tobey/Slater, *Beginning Algebra*, Fifth Edition

Tobey/Slater, *Intermediate Algebra*, Fourth Edition

We have visited and listened to teachers across the country and have incorporated a number of suggestions into this textbook to help you with the particular learning delivery system at your school. The following pages describe the key features of the text.

Key Features

Developing Problem-Solving Abilities

We are committed as authors to producing a textbook that emphasizes mathematical reasoning and problem-solving techniques as recommended by AMATYC, NCTM, AMS, NADE, MAA, and other bodies. To this end, the problem sets are built on a wealth of real-life and real-data applications. Unique problems have been developed and incorporated into the exercise sets that help train students in data interpretation, mental mathematics, estimation, geometry and graphing, number sense, critical thinking, and decision making.

Applied Problems

Numerous real-world and real-data application problems show students the relevance of the math they are learning. The applications relate to everyday life, global issues beyond the borders of the United States, and other academic disciplines. Many include source citations. Roughly 30 percent of the applications have been contributed by actual students based on scenarios they have encountered in their home or work lives.

Math in the Media

Math in the Media applications appear at the end of each chapter to offer students yet another opportunity to see why developing mastery of mathematical concepts enhances their understanding of the world around them. The applications are based on information from familiar media sources. The exercises may ask students to interpret or verify information, perform calculations, make decisions or predictions, or provide a rationale for their responses.

Putting Your Skills to Work Applications

The first edition incorporates this highly successful series feature. There are 19 Putting Your Skills to Work applications throughout the text. These nonroutine application problems challenge students to synthesize the knowledge they have gained and apply it to a totally new area. Each problem is specifically arranged for independent and cooperative learning or group investigation of mathematical problems that pique student interest. Students are given the opportunity to help one another discover mathematical solutions to extended problems. The investigations feature open-ended questions and extrapolation of data to areas beyond what is normally covered in such a course.

Internet Connections

As an integral part of each Putting Your Skills to Work problem, students are exposed to an interesting application of the Internet and encouraged to continue their investigations. This use of technology inspires students to have confidence in their abilities to successfully use mathematics.

The companion Web site (`http://www.prenhall.com/tobey_beg_int`) features annotated links to help students navigate the sites more efficiently and to provide a more user-friendly experience.

Increased Integration and Emphasis on Geometry

Due to the emphasis on geometry on many statewide exams, geometry problems are integrated throughout the text. Examples and exercises that incorporate a principle of geometry are marked with a triangle icon for easy identification.

Blueprint for Problem Solving

The successful Mathematics Blueprint for Problem Solving strengthens problem-solving skills by providing a consistent and interactive outline to help students organize their approach to problem solving. Once students fill in the blueprint, they can refer back to their plan as they do what is needed to solve the problem. Because of its flexibility, this feature can be used with single-step problems, multistep problems, applications, and nonroutine problems that require problem-solving strategies. Students will not need to use the blueprint to solve every problem. It is available for those faced with a problem with which they are not familiar, to alleviate anxiety, to show them where to begin, and to assist them in the steps of reasoning.

Developing Your Study Skills

The first edition incorporates this highly successful series feature. The boxed notes are integrated throughout the text to provide students with techniques for improving their study skills and succeeding in math courses.

Graphs, Charts, and Tables

When students encounter mathematics in real-world publications, they often encounter data represented in a graph, chart, or table and are asked to make a reasonable conclusion based on the data presented. This emphasis on graphical interpretation is a continuing trend with the expanding technology of our day. The first edition includes a significant number of mathematical problems based on charts, graphs, and tables. Students are asked to make simple interpretations, to solve medium-level problems, and to investigate challenging applied problems based on the data shown in a chart, graph, or table.

Design

Beginning and Intermediate Algebra has a design that enhances the accessible, student-friendly writing style. See the walkthrough of features in the preface on page xxii.

Mastering Mathematical Concepts

Text features that develop the mastery of concepts include the following:

Learning Objectives

Concise learning objectives listed at the beginning of each section allow students to preview the goals of that section.

Examples and Exercises

The examples and exercises in this text have been carefully chosen to guide students through *Beginning and Intermediate Algebra*. We have incorporated several different types of exercises and examples to assist your students in retaining the content of this course.

Chapter Pretests

Each chapter opens with a concise pretest to familiarize the students with the learning objectives for that particular chapter. The problems are keyed to appropriate sections of the chapter. All answers appear in the back of the book.

Practice Problems

Practice problems are found throughout the chapter, after the examples, and are designed to provide your students with immediate practice of the skills presented. The complete worked-out solution of each practice problem appears in the back of the book.

To Think About

These critical thinking questions follow some of the examples in the text and also appear in the exercise sets. They extend the concept being taught, providing the opportunity for all students to stretch their minds, to look for patterns, and to make conclusions based on their previous experience.

Exercise Sets

Exercise sets are paired and graded. This design helps ease the students into the problems, and the answers provide students with immediate feedback.

Cumulative Review Problems

Each exercise set concludes with a section of cumulative review problems. These problems review topics previously covered and are designed to assist students in retaining the material. Most sections include applied problems to continually reinforce problem solving abilities.

Graphing and Scientific Calculator Problems

Optional calculator boxes are placed in the margin of the text to alert students to a scientific or graphing calculator application. In the exercise section, icons indicate problems that are designed for solving with a graphing or scientific calculator.

Reviewing Mathematical Concepts

At the end of each chapter we have included problems and tests to provide students with several different formats to help them review and reinforce the ideas that they have learned. This not only assists them with the chapter, it reviews previously covered topics as well.

Chapter Organizers

The concepts and mathematical procedures covered in each chapter are reviewed at the end of the chapter in a unique chapter organizer. This device has been extremely popular with faculty and students alike. It not only lists concepts and methods, but provides a completely worked-out example for each type of problem. Students find that preparing a similar chapter organizer on their own in higher-level math courses becomes an invaluable way to master the content of a chapter.

Verbal and Writing Skills

These exercises provide students with the opportunity to extend a mathematical concept by allowing them to use their own words, to clarify their thinking, and to become familiar with mathematical terms.

Chapter Review Problems

These problems are grouped by section as a quick refresher at the end of the chapter. These problems can also be used by the student as a quiz of the chapter material.

Tests

Found at the end of the chapter, the chapter test is a representative review of the material from that particular chapter that simulates an actual testing format. This provides the students with a gauge to their preparedness for the actual examination.

Cumulative Tests

At the end of each chapter is a cumulative test. One-half of the content of each cumulative test is based on the math skills learned in previous chapters. By completing these tests for each chapter, the students build confidence that they have mastered not only the contents of the present chapter but the contents of the previous chapters as well.

Supplements Resource Package

Beginning and Intermediate Algebra is supported by a wealth of supplements designed for added effectiveness and efficiency. Highlights include the MathPro 4.0 Explorer tutorial software together with a unique video clip feature, MathPro 5—the online version of the popular tutorial program—providing online access anytime/anywhere and enhanced course management; a computerized testing system—TestGen-EQ with QuizMaster-EQ; lecture videos; lecture videos digitized on CD-ROM; Prentice Hall Tutoring Center; and options for online and distance learning courses. Please see the list of supplements and descriptions.

Options for Online and Distance Learning

For maximum convenience, Prentice Hall offers online interactivity and delivery options for a variety of distance learning needs. Instructors may access or adopt these in conjunction with this text, *Beginning and Intermediate Algebra*.

Companion Web Site

Visit `http://www.prenhall.com/tobey_beg_int`
The companion Web site includes basic distance learning access to provide links to the text's Internet Connections activities. The Web site features annotated links to facilitate student navigation of the sites associated with the exercises. Links to additional sites are also included.

This text-specific site offers students an online study guide via online self-quizzes. Questions are graded and students can e-mail their results. Syllabus Manager gives professors the option of creating their own online custom syllabus. Visit the Web site to learn more.

WebCT

Visit `http://www.prenhall.com/demo`
WebCT includes distance learning access to content found in the Tobey/Slater companion Web site plus more. WebCT provides tools to create, manage, and use online course materials. Save time and take advantage of items such as online help, communication tools, and access to instructor and student manuals. Your college may already have WebCT software installed on its server or you may choose to download it. Contact your local Prentice Hall sales representative for details.

BlackBoard

Visit `http://www.prenhall.com/demo`
For distance learning access to content and features from the Tobey/Slater companion Web site plus more. BlackBoard provides simple templates and tools to create, manage, and use online course materials. Take advantage of items such as online help, course management tools, communication tools, and access to instructor and student manuals. Contact your local Prentice Hall sales representative for details.

CourseCompass™ powered by BlackBoard

Visit `http://www.prenhall.com/demo`
For distance learning access to content and features from the Tobey/Slater companion Web site plus more. Prentice Hall content is preloaded in a customized version of BlackBoard 5. CourseCompass™ provides all of BlackBoard 5's powerful course management tools to create, manage, and use online course materials. Contact your local Prentice Hall sales representative for details.

Supplements for the Instructor

Printed Resources

Annotated Instructor's Edition (ISBN: 0-13-090940-8)

- Complete student text.
- Answers appear in place on the same text page as exercises.
- Teaching Tips placed in the margin at key points where students historically need extra help.
- Answers to all exercises in pretests, review problems, tests, cumulative tests, diagnostic pretest, and practice final.

Instructor's Solutions Manual (ISBN: 0-13-092413-X)

- Detailed step-by-step solutions to the even-numbered exercises.
- Solutions to every exercise (odd and even) in the diagnostic pretest, pretests, review problems, tests, cumulative tests, and practice final.
- Solution methods reflect those emphasized in the text.

Instructor's Resource Manual with Tests (ISBN: 0-13-092415-6)

- Nine test forms per chapter—6 free response, 3 multiple choice. Two of the free-response tests are cumulative in nature.
- Four forms of final examination.
- Answers to all items.

Media Resources

TestGen-EQ with QuizMaster-EQ (Windows/Macintosh) (ISBN: 0-13-092514-4)

- Algorithmically driven, text-specific testing program.
- Networkable for administering tests and capturing grades online.
- The built-in Question Editor allows you to edit or add your own questions to create a nearly unlimited number of tests and worksheets.
- Use the Function Plotter to create graphs.
- Side-by-side "Testbank" window and "Test" window show your test as you build it and as it will be printed.
- Extensive symbol palettes and expression templates assist professors in writing questions that include specialized tables and notation.
- Tests can be easily exported to HTML so they can be posted to the Web for student practice.
- QuizMaster-EQ tests can be used for practice and graded tests. Instructors can set preferences to determine test availability, time limits, and number of tries.
- QuizMaster-EQ provides detailed exam reports for individual students, classes, or the course.

MathPro Explorer 4.0
Network Version for Windows/Macintosh (ISBN: 0-13-009041-7)

- Enables instructors to create either customized or algorithmically generated practice tests from any section of a chapter, or a test of random items.
- Includes an e-mail function for network users, enabling instructors to send a message to a specific student or an entire group.
- Network-based reports and summaries for a class or student and for cumulative or selected scores are available.

MathPro 5 Anytime. Anywhere. With Assessment.

- The popular MathPro tutorial software available over the Internet.
- Online tutorial access—anytime/anywhere.
- Enhanced course management tools.

Companion Web Site

Visit `http://www.prenhall.com/tobey_beg_int`

- Annotated links facilitate student navigation of the sites associated with the Internet Connections exercises.
- Additional links provided to sites of interest or resources.
- Provides an online study guide via self-quizzes. Questions are graded and students can e-mail their results.
- Syllabus Manager gives professors the option of creating their own online custom syllabus. Visit the Web site to learn more.

Supplements for Students

Printed Resources

Student Solutions Manual (ISBN: 0-13-092414-8)

- Solutions to all odd-numbered exercises.
- Solutions to every (odd and even) exercise found in pretests, chapter tests, reviews, and cumulative reviews.
- Solution methods reflect those emphasized in the textbook.
- Ask your bookstore about ordering.

Media Resources

MathPro Explorer 4.0 CD-ROM (Student version: 0-13-067009-X)

- Keyed to each section of the text for text-specific tutorial exercises and instruction.
- Warm-up exercises and graded practice problems.
- Video clips, providing a problem similar to the one being attempted, explained and worked out on the board.
- Algorithmically generated exercises; includes bookmark, online help, glossary, and summary of scores for the exercises tried.
- Explorations enable students to explore concepts associated with each objective in more detail.

MathPro 5 Anytime. Anywhere. With Assessment.

- The popular MathPro tutorial software available over the Internet.
- Online tutorial access—anytime/anywhere.

Lecture Videos (ISBN: 0-13-035069-9)

- All-new videotapes.
- Keyed to each section of the text.
- Key concepts are explained step-by-step.

Digitized Lecture Videos on CD-ROM (ISBN: 0-13-092513-6)

- The entire set of *Beginning and Intermediate Algebra*, lecture videotapes in digital form.
- Convenient access anytime to video tutorial support from a computer at home or on campus.
- Available shrink-wrapped with the text or stand-alone.

Prentice Hall Tutoring Center

- Staffed with developmental math instructors and open 5 days a week, 7 hours per day.
- Obtain help for examples and exercises in Tobey/Slater, *Beginning and Intermediate Algebra*, First Edition via toll-free telephone, fax, or e-mail.
- The Prentice Hall Tutoring Center is accessed through a registration number that may be bundled with a new text or purchased separately with a used book.
- Contact your Prentice Hall sales representative for details, or visit http://www.prenhall.com/tutorcenter.

Companion Web Site

Visit http://www.prenhall.com/tobey_beg_int

- Annotated links facilitate navigation of the sites associated with the Internet Connections exercises.
- Additional links provided to sites of interest or resources.
- Provides an online study guide via self-quizzes. Questions are graded and students can e-mail their results to the instructor.

Additional Printed Material

Have your instructor contact the local Prentice Hall sales representative about the following resources:

- *How to Study Mathematics*
- *Math on the Internet: A Student's Guide*
- *Prentice Hall/New York Times, Theme of the Times Newspaper Supplement*

Acknowledgments

We want to thank the following focus group participants for their insights and suggestions for the series.

> Connie Buller, Metropolitan Community College
>
> Nelson Collins, Joliet Jr. College
>
> Doug Mace, Baker College
>
> Beverly Meyers, Jefferson College
>
> Jim Osborn, Baker College
>
> Linda Padilla, Joliet Jr. College
>
> Dennis Runde, Manatee Community College
>
> Margie Thrall, Manatee Community College

Special thanks to Jenny Crawford who spent many hours editing and managing the merging of *Beginning Algebra* and *Intermediate Algebra* into one combined book. Her diligent efforts and insight were invaluable.

We have been greatly helped by a supportive group of colleagues who not only teach at North Shore Community College but who have provided a number of ideas as well as extensive help on all of our mathematics books. Also, a special word of thanks to Hank Harmeling, Tom Rourke, Wally Hersey, Bob McDonald, Bob Campbell, Rick Ponticelli, Russ Sullivan, Kathy LeBlanc, Lora Connelly, Sharyn Sharaf, Donna Stefano, Elizabeth Lucas, and Nancy Tufo. Joan Peabody has done an excellent job of typing various materials for the manuscript and her help is gratefully acknowledged. Suellen Robinson and Judy Carter provided new problems, new ideas, new answer keys, and new perspective. Their excellent help was much appreciated. Sarah Street and Emily Keaton provided excellent help identifying geometry problems and incorporating accuracy suggestions.

We want to thank Louise Elton for providing several new applied problems and suggested applications. Error checking is a challenging task and few can do it well. So we especially want to thank Lauri Semarne and the staff of Laurel Technical Services for accuracy checking the content of the book at different stages of text preparation.

Additionally, Sherm Rosen researched the Internet Connections and provided splendid suggestions for improvements and helpful link annotations. Dave Nasby, Sherm Rosen, and others' work with Math in the Media is much appreciated.

Each textbook is a combination of ideas, writing, and revisions from the authors and wise editorial direction and assistance from the editors. We want to thank our Prentice Hall editor, Karin Wagner, for her helpful insight and perspective on each phase of the textbook. Her patience, her willingness to listen, and her flexibility to adapt to changing publishing decisions have been invaluable to the production of this book. Mary Beckwith, our project manager, provided daily support and encouragement as the book progressed. Her patient assistance with the art program and her attention to a variety of details was most appreciated. Kathy Sessa-Federico, our developmental editor, sifted through mountains of material and offered excellent suggestions for improvement and change. Gina Linko, our production director, kept things moving on schedule and cheerfully solved many crises.

Nancy Tobey retired from teaching and joined the team as our administrative assistant. Mailing, editing, photocopying, collating, and taping were cheerfully done each day. A special thanks to Nancy. We could not have finished the book without you.

Book writing is impossible for us without the loyal support of our families. Our deepest thanks and love to Nancy, Johnny, Melissa, Marcia, Shelley, Rusty, and Abby. Your understanding, your love and help, and your patience have been a source of great encouragement. Finally, we thank God for the strength and energy to write and the opportunity to help others through this textbook.

We have spent more than 30 years teaching mathematics. Each teaching day we find that our greatest joy is helping students learn. We take a personal interest in ensuring that each student has a good learning experience in taking this course. If you have some personal comments, suggestions, or ideas for future editions of this textbook, please write to us at:

Prof. John Tobey and Prof. Jeffrey Slater
Prentice Hall Publishing
Office of the College Mathematics Editor
One Lake Street
Upper Saddle River, NJ 07458
or e-mail us at
jtobey@nscc.mass.edu.

We wish you success in this course and in your future life!

John Tobey
Jeffrey Slater

Enhanced, Student-Friendly Pedagogy

The Tobey/Slater series is a comprehensive learning system that features several pedagogical tools designed for ease-of-use and student success.

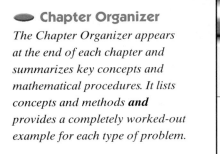 **Chapter Organizer**

*The Chapter Organizer appears at the end of each chapter and summarizes key concepts and mathematical procedures. It lists concepts and methods **and** provides a completely worked-out example for each type of problem.*

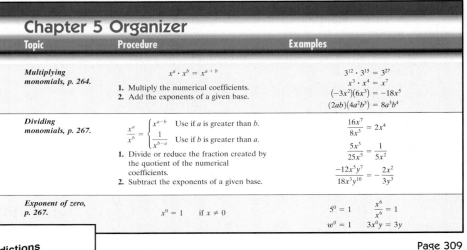

Chapter 5 Organizer

Topic	Procedure	Examples
Multiplying monomials, p. 264.	$x^a \cdot x^b = x^{a+b}$ 1. Multiply the numerical coefficients. 2. Add the exponents of a given base.	$3^{12} \cdot 3^{15} = 3^{27}$ $x^3 \cdot x^4 = x^7$ $(-3x^2)(6x^3) = -18x^5$ $(2ab)(4a^2b^3) = 8a^3b^4$
Dividing monomials, p. 267.	$\dfrac{x^a}{x^b} = \begin{cases} x^{a-b} & \text{Use if } a \text{ is greater than } b. \\ \dfrac{1}{x^{b-a}} & \text{Use if } b \text{ is greater than } a. \end{cases}$ 1. Divide or reduce the fraction created by the quotient of the numerical coefficients. 2. Subtract the exponents of a given base.	$\dfrac{16x^7}{8x^3} = 2x^4$ $\dfrac{5x^3}{25x^5} = \dfrac{1}{5x^2}$ $\dfrac{-12x^5y^7}{18x^3y^{10}} = -\dfrac{2x^2}{3y^3}$
Exponent of zero, p. 267.	$x^0 = 1 \qquad \text{if } x \neq 0$	$5^0 = 1 \qquad \dfrac{x^6}{x^6} = 1$ $w^0 = 1 \qquad 3x^0y = 3y$

Page 309

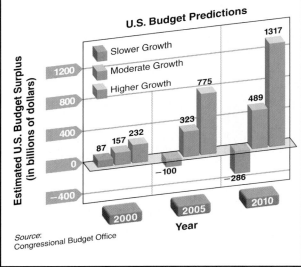

Page 392

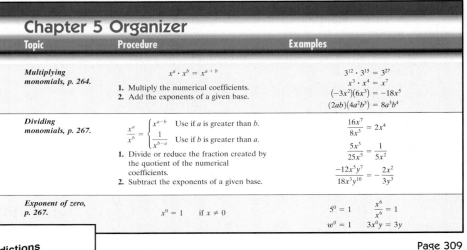 **Graphs, Charts, and Tables**

A significant number of problems based on charts, graphs, and tables are included. Students make simple interpretations, solve medium-level problems, and investigate challenging applied problems based on presented data.

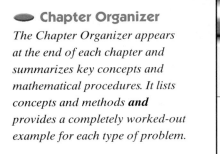 **Developing Your Study Skills**

Sprinkled throughout the text, these boxed notes provide students with techniques for improving their study skills and succeeding in math.

Developing Your Study Skills

Getting Help

Getting the right kind of help at the right time can be a key ingredient in being successful in mathematics. When you have gone to class on a regular basis, taken careful notes, methodically read your textbook, and diligently done your homework—all of which means making every effort possible to learn the mathematics—you may find that you are still having difficulty. If this is the case, then you need to seek help. Make an appointment with your instructor to find out what help is available to you. The instructor, tutoring services, a mathematics lab, videotapes, and computer software may be among the resources you can draw on.

Once you discover the resources available in your school, you need to take advantage of them. Do not put it off, or you will find yourself getting behind. You cannot afford that. When studying mathematics, you must keep up with your work.

Page 77

Math in the Media

Math Behind the Scenes

At close to 200 million dollars, *Titanic* is one of the most expensive movies ever made. For the movie, a replica of the original ship was constructed. *Entertainment Weekly Online* reported the replica to be 770 feet long—close to 90% to scale of the real ship.

Details of the original ship were followed closely to create the replica. A special studio was built for filming. The ship "sailed" in an enormous tank that held 17 million gallons of water.

Although movies frequently use models when filming, replicas tend to be on a much smaller scale than that used for the *Titanic*. In the questions that follow, you can use your own calculations to compare model and actual scales.

EXERCISES

1. The original Titanic was 880 feet long and 92 feet high. Use the exact length of the original Titanic and the exact length of the model to determine the percentage to scale of the real ship. Using this information, determine the height of the replica.

2. Does your calculation in Exercise 1 verify that the replica is 90% to scale of the actual ship? How

differ from the original estimate of 90%?

3. In the movie *The Hunt for Red October*, the story centers on the search for a Russian ballistic missile submarine that is reported to be 610 feet long and has a beam (width) of 46 feet. In filming some of the un[...] sequences, a m[...]

reported to be 14 feet wide. If that report is true what percent to scale is the model to the actual submarine in the story? If the entire model was built to that scale, what would be the expected length of the model submarine used in the filming?

Page 63

Math in the Media

Math in the Media exercises help students make connections between the real world and concepts learned in class. A brief news clip, graph, or scenario taken from the Web or print media appears at the end of each chapter. Related questions follow that may ask students to interpret the information, perform necessary calculations, or provide rationale for their decisions.

Chapter-Opening Application

A real-world application opens each chapter and links a specific situation to a Putting Your Skills to Work application that appears in that chapter to enhance students' awareness of the relevance of math.

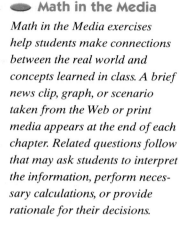

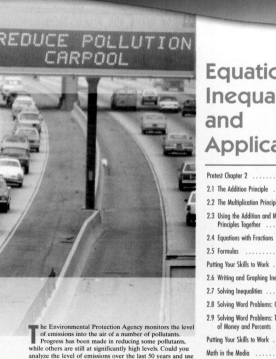

Chapter 2

Equations, Inequalities, and Applications

The Environmental Protection Agency monitors the level of emissions into the air of a number of pollutants. Progress has been made in reducing some pollutants, while others are still at significantly high levels. Could you analyze the level of emissions over the last 50 years and use your math skills to predict the level of emissions in the future? Turn to the Putting Your Skills to Work problems on page 98 to find out.

63

Integrated Problem Solving

● **Problem Solving**

Problem Solving is thorough and easy to follow; key steps are highlighted with the pedagogical use of color. A clear problem-solving process is defined and reinforced.

Procedure to Solve a Formula for a Specified Variable

1. Remove any parentheses.
2. If fractions exist, multiply all terms on both sides by the LCD of all the fractions.
3. Combine like terms on each side if possible.
4. Add or subtract terms on both sides of the equation to get all terms with the desired variable on one side of the equation.
5. Add or subtract the appropriate quantities to get all terms that do *not* have the desired variable on the other side of the equation.
6. Divide both sides of the equation by the coefficient of the desired variable.
7. Simplify if possible.

● **EXAMPLE 1** A business executive rented a car. The Supreme Car Rental Agency charged $39 per day and $0.28 per mile. The executive rented the car for two days and the total rental cost was computed to be $176. How many miles did the executive drive the rented car?

1. **Understand the problem.**

   How do you calculate the cost of renting a car?
   total cost = per-day cost + mileage cost
   What is known?
   It cost $176 to rent the car for two days.
   What do you need to find?
   The number of miles the car was driven.

   Choose a variable:
   Let m = the number of miles driven in the rented car.

2. **Write an equation.**
   Use the relationship for calculating the total cost.

per-day cost	+	mileage cost	=	total cost
$(39)(2)$	+	$(0.28)m$	=	176

3. **Solve and state the answer.**

   $(39)(2) + (0.28)(m) = 176$

   $78 + 0.28m = 176$ Simplify the equation.

   $0.28m = 98$ Subtract 78 from both sides.

   $\dfrac{0.28m}{0.28} = \dfrac{98}{0.28}$ Divide both sides by 0.28.

   $m = 350$ Simplify.

   The executive drove 350 miles.

4. **Check.**
   Does this seem reasonable? If he drove the car 350 miles in two days cost $176?

   (Cost of $39 per day for 2 days) + (cost of $0.28 per mile for 350 miles)
   $\overset{?}{=}$ total cost of $176

   $(\$39)(2) + (350)(\$0.28) \overset{?}{=} \$176$

   $\$78 + \$98 \overset{?}{=} \$176$

   $\$176 = \$176$ ✓

Page 118

● **Mathematics Blueprint for Problem Solving**

Students begin the problem-solving process and plan the steps to be taken along the way using an outline to organize their approach to problem solving. Once students fill in the blueprint, they can refer back to their plan to solve the problem.

Page 93

Page 121

Mathematics Blueprint For Problem Solving

Gather the Facts	Assign the Variable	Basic Formula or Equation	Key Points to Remember
$1250 is invested: part at 8% interest, part at 6% interest. The total interest for the year is $86.	Let x = the amount invested at 8%. $1250 - x$ = the amount invested at 6%. $0.08x$ = the amount of interest for x dollars at 8%. $0.06(1250 - x)$ = the amount of interest for $1250 - x$ dollars at 6%.	Interest earned at 8% + interest earned at 6% = total interest earned during the year, which is $86.	Be careful to write $1250 - x$ for the amount of money invested at 6%. The order is total - x. Do not use $x - 1250$.

interest earned at 8%	+	interest earned at 6%	=	total interest earned during the year
$0.08x$	+	$0.06(1250 - x)$	=	86

Note: Be sure you write $(1250 - x)$ for the amount of money invested at 6%. Students often write it backwards by mistake. It is *not* correct to use $(x - 1250)$ instead of $(1250 - x)$. The order of the terms is very important.

Solve and state the answer.

$0.08x + 75 - 0.06x = 86$ Remove parentheses.

$0.02x + 75 = 86$ Combine like terms.

$0.02x = 11$ Subtract 75 from both sides.

$\dfrac{0.02x}{0.02} = \dfrac{11}{0.02}$ Divide both sides by 0.02.

$x = 550$ The amount invested at 8% interest is $550.

$1250 - x = 1250 - 550 = 700$ The amount invested at 6% interest is $700.

Check.
Are these values reasonable? Yes. Do the amounts equal $1250?

$\$550 + \$700 \overset{?}{=} \$1250$

$\$1250 = \$1250$ ✓

Would these amounts earn $86 interest in one year invested at the specified rates?

$0.08(\$550) + 0.06(\$700) \overset{?}{=} \$86$

$\$44 + \$42 \overset{?}{=} \$86$

$\$86 = \$86$ ✓

xxiv

Putting Your Skills to Work

Controlling Emissions Levels

The U.S. Environmental Protection Agency monitors the emissions of a number of pollutants in the air in the United States. Each year, measurements are made to identify the number of tons of each type of pollutant. The following chart records in millions of tons the amount of emissions of nitrogen dioxide and volatile organic compounds.

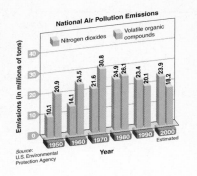

National Air Pollution Emissions

Source: U.S. Environmental Protection Agency

Problems for Individual Study and Investigation

1. How many more tons of nitrogen dioxide were emitted in 1980 than in 1970?

2. How many fewer tons of volatile organic compounds were emitted in 2000 than in 1990?

Problems for Cooperative Study and Investigation

Scientists sometimes use the equation $n = 23.9 + 0.05x$, where x represents the number of years since the year 2000, to predict the number of millions of tons of nitrogen dioxide that will be emitted per year.

3. Use the formula to estimate how many tons of nitrogen dioxide will be emitted in 2005.

4. If the actual amount of nitrogen dioxide emitted in 2010 turns out to be 24.7 million tons, by how many tons will the formula be in error?

Scientists sometimes use the equation $v = 18.2 - 0.19x$, where x represents the number of years since the year 2000, to predict the number of millions of tons of volatile organic compounds that will be emitted per year.

5. Solve the formula for the variable x. Leave the answer in the form of a fractional expression. Do not divide out the decimals.

6. Use the result from question 5 to find the year when 16.68 million tons of volatile organic compounds will be emitted.

Internet Connections

Netsite: http://www.prenhall.com/tobey_beg_int

Site: U.S. Environmental Protection Agency

7. Examine the three separate tables for carbon monoxide, nitrogen oxides, and sulfur dioxide emissions estimates 1989–1998. For each of these, by what percent did total emissions increase or decrease in 1998 as compared with 1989?

8. In what year did the emissions of carbon monoxide from transportation sources have the greatest percent decrease from the previous year? What was that percent decrease?

Page 98

Interesting and Diverse Exercises and Applications

Real-World Applications

Numerous real-world and real-data applications relate topics to everyday life, global issues, and other academic disciplines. An abundance of real-world application problems show students the relevance of math in their daily lives.

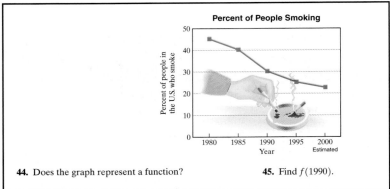

Percent of People Smoking

44. Does the graph represent a function?

45. Find $f(1990)$.

Page 198

EXAMPLE 7 The number of motor vehicle accidents in millions is recorded in the following table for the years 1980 to 2000.

(a) Plot points that represent this data on the given coordinate system.

(b) What trends are apparent from the plotted data?

Number of Years Since 1980	Number of Motor Vehicle Accidents (in Millions)
0	18
5	19
10	12
15	11
20	15

Source: U.S. National Highway Traffic Safety Administration

(a)

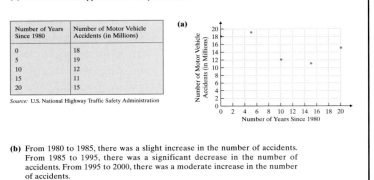

(b) From 1980 to 1985, there was a slight increase in the number of accidents. From 1985 to 1995, there was a significant decrease in the number of accidents. From 1995 to 2000, there was a moderate increase in the number of accidents.

Page 149

Verbal and Writing Skills Exercises

These exercises ask students to use their own words. Writing helps students clarify their thinking and become familiar with mathematical terms.

110 Chapter 2 Equations, Inequalities, and Applications

Verbal and Writing Skills

15. Add -2 to both sides of the inequality $5 > 3$. What is the result? Why is the direction of the inequality not reversed?

16. Divide -3 into both sides of the inequality $-21 > -29$. What is the result? Why is the direction of the inequality reversed?

Page 110

To Think About

Critical-thinking questions appear in the exercise sets to extend the concepts. They also give students the opportunity to look for patterns and draw conclusions.

To Think About

48. During the years from 1980 to 2000, the total income for the U.S. federal budget can be approximated by the equation $y = 14(4x + 35)$, where x is the number of years since 1980 and y is the amount of money in billions of dollars. (*Source:* U.S. Office of Management and Budget)
(a) Write the equation in slope–intercept form.

(b) Find the slope and the y-intercept.

(c) In this specific equation, what is the meaning of the slope? What does it indicate?

49. During the years from 1970 to 1990, the approximate number of civilians employed in the United States could be predicted by the equation $y = \frac{1}{10}(22x + 830)$, where x is the number of years since 1970 and y is the number of civilians employed, measured in millions. (*Source:* U.S. Bureau of Labor Statistics)
(a) Write the equation in slope–intercept form.

(b) Find the slope and the y-intercept.

(c) In this specific equation, what is the meaning of the slope? What does it indicate?

Page 174

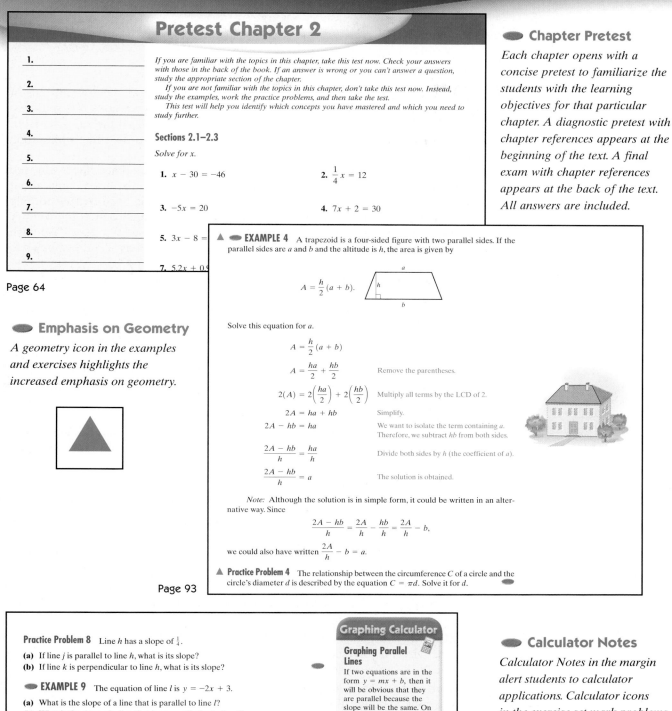

Pretest Chapter 2

Page 64

1. _____

2. _____

3. _____

4. _____

5. _____

6. _____

7. _____

8. _____

9. _____

If you are familiar with the topics in this chapter, take this test now. Check your answers with those in the back of the book. If an answer is wrong or you can't answer a question, study the appropriate section of the chapter.

If you are not familiar with the topics in this chapter, don't take this test now. Instead, study the examples, work the practice problems, and then take the test.

This test will help you identify which concepts you have mastered and which you need to study further.

Sections 2.1–2.3

Solve for x.

1. $x - 30 = -46$

2. $\frac{1}{4}x = 12$

3. $-5x = 20$

4. $7x + 2 = 30$

5. $3x - 8 =$

7. $5.2x + 0.9$

● Chapter Pretest

Each chapter opens with a concise pretest to familiarize the students with the learning objectives for that particular chapter. A diagnostic pretest with chapter references appears at the beginning of the text. A final exam with chapter references appears at the back of the text. All answers are included.

● Emphasis on Geometry

A geometry icon in the examples and exercises highlights the increased emphasis on geometry.

▲ ● **EXAMPLE 4** A trapezoid is a four-sided figure with two parallel sides. If the parallel sides are a and b and the altitude is h, the area is given by

$$A = \frac{h}{2}(a + b).$$

Solve this equation for a.

$$A = \frac{h}{2}(a + b)$$

$$A = \frac{ha}{2} + \frac{hb}{2} \qquad \text{Remove the parentheses.}$$

$$2(A) = 2\left(\frac{ha}{2}\right) + 2\left(\frac{hb}{2}\right) \qquad \text{Multiply all terms by the LCD of 2.}$$

$$2A = ha + hb \qquad \text{Simplify.}$$

$$2A - hb = ha \qquad \begin{array}{l}\text{We want to isolate the term containing } a.\\ \text{Therefore, we subtract } hb \text{ from both sides.}\end{array}$$

$$\frac{2A - hb}{h} = \frac{ha}{h} \qquad \text{Divide both sides by } h \text{ (the coefficient of } a\text{).}$$

$$\frac{2A - hb}{h} = a \qquad \text{The solution is obtained.}$$

Note: Although the solution is in simple form, it could be written in an alternative way. Since

$$\frac{2A - hb}{h} = \frac{2A}{h} - \frac{hb}{h} = \frac{2A}{h} - b,$$

we could also have written $\frac{2A}{h} - b = a$.

▲ **Practice Problem 4** The relationship between the circumference C of a circle and the circle's diameter d is described by the equation $C = \pi d$. Solve it for d. ●

Page 93

Practice Problem 8 Line h has a slope of $\frac{1}{4}$.

(a) If line j is parallel to line h, what is its slope?
(b) If line k is perpendicular to line h, what is its slope?

● **EXAMPLE 9** The equation of line l is $y = -2x + 3$.

(a) What is the slope of a line that is parallel to line l?
(b) What is the slope of a line that is perpendicular to line l?

(a) Looking at the equation, we can see that the slope of line l is -2. The slope of a line that is parallel to line l is -2.
(b) Perpendicular lines have slopes whose product is -1.

$$m_1 m_2 = -1$$

$$(-2)m_2 = -1 \qquad \text{Substitute } -2 \text{ for } m_1.$$

$$m_2 = \frac{1}{2} \qquad \text{Because } (-2)\left(\frac{1}{2}\right) = -1.$$

The slope of a line that is perpendicular to line l is $\frac{1}{2}$.

Practice Problem 9 The equation of line n is $y = \frac{1}{4}x - 1$.

(a) What is the slope of a line that is parallel to line n?
(b) What is the slope of a line that is perpendicular to line n?

Graphing Calculator

Graphing Parallel Lines

If two equations are in the form $y = mx + b$, then it will be obvious that they are parallel because the slope will be the same. On a graphing calculator graph both of these equations:

$$y = -2x + 6$$

$$y = -2x - 4$$

Use the window of -10 to 10 for both x and y. Display:

● Calculator Notes

Calculator Notes in the margin alert students to calculator applications. Calculator icons in the exercise set mark problems that can be solved using a scientific or graphing calculator.

Page 171

To the Student

This book was written with your needs and interests in mind. Several editions of *Beginning Algebra* and *Intermediate Algebra*, from which this textbook is based, have been class-tested with students all across the country. Based on the suggestions of many students, the book has been designed to maximize your learning while using this text.

We realize that students who enter college have sometimes never enjoyed math or never done well in a math course. You may find that you are anxious about taking this course. We want you to know that this book has been written to help you overcome those difficulties. Literally thousands of students across the country have found an amazing ability to learn math as they have used our series. We have incorporated several learning tools and various types of exercises and examples to assist you in learning this material.

It helps to know that learning mathematics is going to help you in life. Perhaps you have entered this course feeling that little or no mathematics will be necessary for you in your future job. However, elementary school teachers, bus drivers, laboratory technicians, nurses, telephone operators, cable TV repair personnel, photographers, pharmacists, salespeople, doctors, architects, inspectors, counselors, and custodians who once believed that they needed little if any mathematics are finding that the mathematical skills presented in this course can help them. Mathematics, you will find, can help you too. In this book, a great number of examples and problems come from everyday life. You will be amazed at the number of ways mathematics is used in the world each day.

Our greatest wish is that you will find success and personal satisfaction in your mathematics course. We have written this book to help you accomplish that very goal.

Suggestions for Students

1. Be sure to take the time to read the boxes marked **Developing Your Study Skills**. These ideas come from faculty and students throughout the United States. They have found ways to succeed in mathematics and they want to share those ideas with you.

2. **Read** through each section of the book that is assigned by your instructor. You will be amazed at how much information you will learn as you read. It will "fill in the pieces" of mathematical knowledge. Reading a math book is a key part of learning.

3. Study carefully each **sample example**. Then work out the related practice problem. Direct involvement in doing problems and not just thinking about them is one of the greatest guarantees of success in this course.

4. Work out all assigned **homework problems**. Verify your answers in the back of the book. Ask questions when you don't understand or when you are not sure of an answer.

5. Be sure to take advantage of **end-of-chapter helps**. Study the chapter organizers. Do the chapter review problems and the practice test problems.

6. If you need help, remember that teachers and tutors can assist you. There is a **Student Solutions Manual** for this textbook that you will find most valuable. This manual shows worked-out solutions for all the odd-numbered exercises as well as diagnostic pretest, chapter review, chapter test, and cumulative test problems in this book. (If your college bookstore does not carry the Student Solutions Manual, ask them to order a copy for you. ISBN 0-13-092414-8)

7. Watch the **videotapes or digitized videos on CD-ROM**. Every section of this text is explained in detail on videotape. The work is solved using the same methods as explained in the text.

8. Use **MathPro Explorer 4.0**, a tutorial software package that allows you to be tutored in any section in the book. It allows you to test yourself on your mastery of any section of any chapter of the text. Practice problem solving using the resources available to you.

9. Remember, learning mathematics takes time. However, the time spent is well worth the effort. **Take the time** to study, do homework, review, ask questions, and just reflect over what you have learned. The time you invest in learning mathematics will reap dividends in your future courses and your future life.

We encourage you to look over your textbook carefully. Many important features have been designed into the book to make learning mathematics a more enjoyable activity. There are some special features, unique to this textbook, that students throughout the country have told us that they found especially helpful.

Four Key Textbook Features to Help You

1. **Practice problems with worked-out solutions.** Immediately following every sample example is a similar problem called the practice problem. If you can work it out correctly by following the sample example as a general point of reference, then you will likely be able to do the homework exercises. If you encounter some difficulty, then you will find helpful the completely worked-out solutions to the practice problems that appear at the end of the text.

2. **Student-friendly application problems.** You will find in every chapter of this book application problems that are realistic and interesting. Many of the problems were actually written or suggested by students. As you develop your problem-solving and reasoning skills in this course, you will encounter a number of real-world situations that help you see how very helpful mathematical skills are in today's complex world.

3. **Putting Your Skills to Work problems.** In each chapter are unique problem sets that ask you to analyze in depth some mathematical aspect of daily life. You will be asked to extend your knowledge, do some creative thinking, and to work in small groups in a cooperative learning situation. You will even have a chance to explore some Web sites and see how the Internet can assist you in learning. These sections will awaken some new interests and help you to develop the critical thinking skills so necessary both in college and after you graduate.

4. **The Chapter Organizer.** Everything you need to learn in any one chapter of this book is readily available at your fingertips in the chapter organizer. This very popular chart summarizes all methods covered in the chapter, gives page references, and shows a sample example completely worked out for every major topic covered. Students have found this tool a most helpful way to master the content of any chapter of the book.

1. _____

2. _____

3. _____

4. _____

5. _____

6. _____

7. _____

8. _____

9. _____

10. _____

11. _____

12. _____

13. _____

14. _____

15. _____

16. _____

Follow the directions for each problem. Simplify each answer.

Chapter 1

1. Add. $-3 + (-4) + (+12)$

2. Subtract. $-20 - (-23)$

3. Combine. $5x - 6xy - 12x - 8xy$

4. Evaluate. $2x^2 - 3x - 4$ when $x = -3$.

5. Remove the grouping symbols. $2 - 3\{5 + 2[x - 4(3 - x)]\}$

6. Evaluate. $-3(2 - 6)^2 + (-12) \div (-4)$

Chapter 2

In questions 7–10, solve each equation for x.

7. $40 + 2x = 60 - 3x$

8. $7(3x - 1) = 5 + 4(x - 3)$

9. $\dfrac{2}{3}x - \dfrac{3}{4} = \dfrac{1}{6}x + \dfrac{21}{4}$

10. $\dfrac{4}{5}(3x + 4) = 20$

11. Solve for p. $A = \dfrac{1}{2}(3p - 4f)$

12. Solve for x and graph the result. $42 - 18x < 48x - 24$

13. The length of a rectangle is 7 meters longer than the width. The perimeter is 46 meters. Find the dimensions.

14. Marcia invested $6000 in two accounts. One earned 5% interest, while the other earned 7% interest. After one year she earned $394 in interest. How much did she invest in each account?

15. Melissa has three more dimes than nickels. She has twice as many quarters as nickels. The value of the coins is $4.20. How many of each coin does she have?

16. The drama club put on a play for Thursday, Friday, and Saturday nights. The total attendance for the three nights was 6210. Thursday night had 300 fewer people than Friday night. Saturday night had 510 more people than Friday night. How many people came each night?

Chapter 3

17. Graph $y = 2x - 4$.

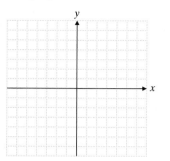

18. Graph $3x + 4y = -12$.

19. What is the slope of a line passing through $(6, -2)$ and $(-3, 4)$?

20. If $f(x) = 2x^2 - 3x + 1$, find $f(3)$.

21. Graph the region. $y \geq -\dfrac{1}{3}x + 2$

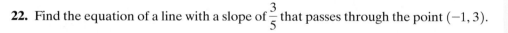

22. Find the equation of a line with a slope of $\dfrac{3}{5}$ that passes through the point $(-1, 3)$.

Chapter 4

Solve the following:

23. $3x + 5y = 30$
$5x + 3y = 34$

24. $2x - y + 2z = 8$
$x + y + z = 7$
$4x + y - 3z = -6$

25. A speedboat can travel 90 miles with the current in 2 hours. It can travel upstream 105 miles against the current in 3 hours. How fast is the boat in still water? How fast is the current?

26. Graph the system.

$x - y \leq -4$
$2x + y \leq 0$

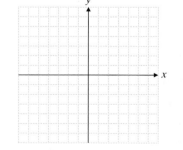

17. _____

18. _____

19. _____

20. _____

21. _____

22. _____

23. _____

24. _____

25. _____

26. _____

Chapter 5

27. Multiply. $(-2xy^2)(-4x^3y^4)$

28. Divide. $\dfrac{36x^5y^6}{-18x^3y^{10}}$

29. Raise to the indicated power. $(-2x^3y^4)^5$

30. Evaluate. $(-3)^{-4}$

31. Multiply. $(3x^2 + 2x - 5)(4x - 1)$

32. Divide. $(x^3 + 6x^2 - x - 30) \div (x - 2)$

Chapter 6

Factor completely.

33. $5x^2 - 5$

34. $x^2 - 12x + 32$

35. $8x^2 - 2x - 3$

36. $3ax - 8b - 6a + 4bx$

Solve for x.

37. $x^3 + 7x^2 + 12x = 0$

38. $16x^2 - 24x + 9 = 0$

Chapter 7

39. Simplify. $\dfrac{x^2 + 3x - 18}{2x - 6}$

40. Multiply.
$$\dfrac{6x^2 - 14x - 12}{6x + 4} \cdot \dfrac{x + 3}{2x^2 - 2x - 12}$$

41. Divide and simplify.
$$\dfrac{x^2}{x^2 - 4} \div \dfrac{x^2 - 3x}{x^2 - 5x + 6}$$

42. Add.
$$\dfrac{3}{x^2 - 7x + 12} + \dfrac{4}{x^2 - 9x + 20}$$

43. Solve for x. $2 - \dfrac{5}{2x} = \dfrac{2x}{x + 1}$

44. Simplify. $\dfrac{3 + \dfrac{1}{x}}{\dfrac{9}{x} + \dfrac{3}{x^2}}$

Chapter 8

Assume that all expressions under radicals represent nonnegative numbers.

45. Multiply and simplify.
$$(\sqrt{3} + \sqrt{2x})(\sqrt{7} - \sqrt{2x^3})$$

46. Rationalize the denominator.
$$\dfrac{3\sqrt{x} + \sqrt{y}}{\sqrt{x} - \sqrt{y}}$$

47. Solve and check your solutions. $2\sqrt{x - 1} = x - 4$

48. Solve for x. $\left|3\left(\dfrac{2}{3}x - 4\right)\right| \le 12$

49. Solve for y. $|3y - 2| + 5 = 8$

Chapter 9

50. Solve: $x^2 - 2x - 4 = 0$

51. $x^4 - 12x^2 + 20 = 0$

52. Graph $f(x) = (x - 2)^2 + 3$. Label the vertex.

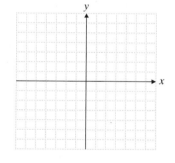

Chapter 10

53. Write in standard form the equation of a circle with center at $(5, -2)$ and a radius of 6.

54. Write in standard form the equation of an ellipse whose center is at $(0,0)$ and whose intercepts are at $(3, 0)$, $(-3, 0)$, $(0, 4)$ and $(0, -4)$.

55. Solve the following nonlinear system of equations.
$$x^2 + 4y^2 = 9$$
$$x + 2y = 3$$

Chapter 11

56. If $f(x) = 2x^2 - 3x + 4$, find $f(a + 2)$.

57. Graph on one axis $f(x) = |x + 3|$ and $g(x) = |x + 3| - 3$.

58. If $f(x) = \dfrac{3}{x + 2}$ and $g(x) = 3x^2 - 1$, find $g[f(x)]$.

59. If $f(x) = -\dfrac{1}{2}x - 5$, find $f^{-1}(x)$.

Chapter 12

60. Find y if $\log_5 125 = y$.

61. Find b if $\log_b 4 = \dfrac{2}{3}$.

62. What is $\log 10{,}000$?

63. Solve for x: $\log_6(5 + x) + \log_6 x = 2$

52. _____

53. _____

54. _____

55. _____

56. _____

57. _____

58. _____

59. _____

60. _____

61. _____

62. _____

63. _____

Real Numbers and Variables

Do you have a realistic sense of how much the cost of living has gone up in the last ten years? Do you know in what categories the increase has been the largest? If you moved to another city, do you know how you could determine the cost of living there? The federal government compiles a Consumer Price Index to keep a record of such information. Turn to the Putting Your Skills to Work problems on page 55 and see if you can answer these types of questions.

1. _____

2. _____

3. _____

4. _____

5. _____

6. _____

7. _____

8. _____

9. _____

10. _____

11. _____

12. _____

13. _____

14. _____

15. _____

16. _____

17. _____

18. _____

19. _____

20. _____

21. _____

22. _____

23. _____

24. _____

25. _____

26. _____

27. _____

If you are familiar with the topics in this chapter, take this test now. Check your answers with those in the back of the book. If an answer is wrong or you can't answer a question, study the appropriate section of the chapter.

If you are not familiar with the topics in this chapter, don't take this test now. Instead, study the examples, work the practice problems, and then take the test.

This test will help you identify which concepts you have mastered and which you need to study further.

Sections 1.1–1.2

Combine.

1. $-3 - (-6)$ **2.** $12 - 16 + 3 - 14$ **3.** $-7 + (-11)$ **4.** $7 - (-13)$

Section 1.3

Multiply or divide.

5. $-7(-2)(+3)(-1)$ **6.** $\dfrac{-\frac{2}{3}}{-\frac{1}{4}}$ **7.** $-2.3(-4)$ **8.** $20 \div (-5)$

Section 1.4

Evaluate.

9. $(-2)^4$ **10.** 4^3 **11.** $\left(-\dfrac{2}{3}\right)^3$ **12.** -4^4

Section 1.5

Simplify.

13. $6 \cdot 2 + 8 \cdot 3 - 4 \cdot 2$ **14.** $3(4)^3 + 18 \div 3 - 1$

15. $-\dfrac{1}{6}\left(\dfrac{1}{2}\right) + \dfrac{3}{4} \div \dfrac{1}{12}$ **16.** $1.22 - 4.1(1.4) + (-3.3)^2$

Section 1.6

Simplify.

17. $-2x(3x - 2xy + z)$ **18.** $-(3x - 4y - 12)$

Section 1.7

Combine like terms.

19. $5x^2 - 3xy - 6x^2y - 8xy$ **20.** $11x - 3y + 7 - 4x + y$

21. $3(2x - 5y) - (x - 8y)$ **22.** $2y(x^2 + y) - 3(x^2y - 5y^2)$

Section 1.8

23. Evaluate $3x^2 - 5x - 4$ for $x = -2$.

24. Evaluate $ab + 3a^2 - 5b$ for $a = 3$ and $b = -1$.

25. Determine the Celsius temperature when the Fahrenheit temperature is 77°. Use the formula $C = \dfrac{5}{9}(F - 32)$.

Section 1.9

Simplify.

26. $3\big[(x - y) - (2x - y)\big] - 3(2x + y)$ **27.** $2x^2 - 3x\big[2x - (x + 2y)\big]$

1.1 Addition of Real Numbers

① Different Types of Numbers

Student Learning Objectives

After studying this section, you will be able to:

① Understand the names of different types of numbers.

② Recognize real-life situations for real numbers.

③ Add real numbers with the same sign.

④ Add real numbers with opposite signs.

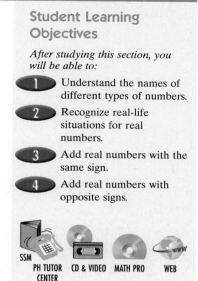

SSM PH TUTOR CENTER CD & VIDEO MATH PRO WEB

Let's review some of the basic terms we use to talk about numbers.

    Whole numbers are numbers such as $0, 1, 2, 3, 4, \ldots$.

    Integers are numbers such as $\ldots, -3, -2, -1, 0, 1, 2, 3, \ldots$.

    Rational numbers are numbers such as $\frac{3}{2}, \frac{5}{7}, -\frac{3}{8}, -\frac{4}{13}, \frac{6}{1}$, and $-\frac{8}{2}$.

    Rational numbers can be written as one integer divided by another integer (as long as the denominator is not zero!). Integers can be written as fractions ($3 = \frac{3}{1}$, for example), so we can see that all integers are rational numbers. Rational numbers can be expressed in decimal form. For example, $\frac{3}{2} = 1.5$, $-\frac{3}{8} = -0.375$, and $\frac{1}{3} = 0.333\ldots$ or $0.\overline{3}$. It is important to note that rational numbers in decimal form are either terminating decimals or repeating decimals.

    Irrational numbers are numbers that cannot be expressed as one integer divided by another integer. The numbers π, $\sqrt{2}$, and $\sqrt[3]{7}$ are irrational numbers.

    Irrational numbers can be expressed in decimal form. The decimal form of an irrational number is a nonterminating, nonrepeating decimal. For example, $\sqrt{2} = 1.414213\ldots$ can be carried out to an infinite number of decimal places with no repeating pattern of digits

    Finally, **real numbers** are all the rational numbers and all the irrational numbers.

EXAMPLE 1 Classify as an integer, a rational number, an irrational number, and/or a real number.

(a) 5   **(b)** $-\dfrac{1}{3}$ **(c)** 2.85   **(d)** $\sqrt{2}$ **(e)** $0.777\ldots$

Make a table. Check off the description of the number that applies.

	Number	Integer	Rational Number	Irrational Number	Real Number
(a)	5	✓	✓		✓
(b)	$-\frac{1}{3}$		✓		✓
(c)	2.85		✓		✓
(d)	$\sqrt{2}$			✓	✓
(e)	$0.777\ldots$		✓		✓

Practice Problem 1 Classify.

(a) $-\dfrac{2}{5}$ **(b)** $1.515151\ldots$ **(c)** -8 **(d)** π

    Any real number can be pictured on a **number line**.

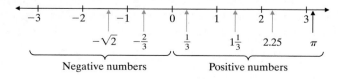

    Positive numbers are to the right of 0 on the number line.

    Negative numbers are to the left of 0 on the number line.

The **real numbers** include the positive numbers, the negative numbers, and zero.

2 Real-Life Situations for Real Numbers

We often encounter practical examples of number lines that include positive and negative rational numbers. For example, we can tell by reading the accompanying thermometer that the temperature is 20° below 0. From the stock market report, we see that the stock opened at 36 and closed at 34.5, and the net change for the day was −1.5.

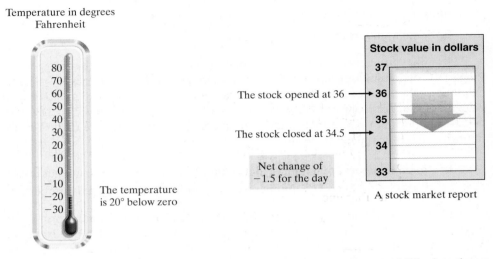

Temperature in degrees Fahrenheit

The temperature is 20° below zero

The stock opened at 36
The stock closed at 34.5
Net change of −1.5 for the day

Stock value in dollars

A stock market report

In the following example we use real numbers to represent real-life situations.

EXAMPLE 2 Use a real number to represent each situation.

(a) A temperature of 128.6°F below zero is recorded at Vostok, Antarctica.

(b) The Himalayan peak K2 rises 29,064 feet above sea level.

(c) The Dow gains 10.24 points.

(d) An oil drilling platform extends 328 feet below sea level.

A key word can help you to decide whether a number is positive or negative.

(a) 128.6°F *below* zero is −128.6.

(b) 29,064 feet *above* sea level is +29,064.

(c) A *gain* of 10.24 points is +10.24.

(d) 328 feet *below* sea level is −328.

Practice Problem 2 Use a real number to represent each situation.

(a) A population growth of 1,259

(b) A depreciation of $763

(c) A windchill factor of minus 10

In everyday life we consider positive numbers the opposite of negative numbers. For example, a gain of 3 yards in a football game is the opposite of a loss of 3 yards; a check written for $2.16 on a checking account is the opposite of a deposit of $2.16.

Each positive number has an opposite negative number. Similarly, each negative number has an opposite positive number. **Opposite numbers**, also called **additive inverses**, have the same magnitude but different signs and can be represented on the number line.

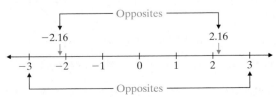

Opposites

−2.16   2.16

Opposites

EXAMPLE 3 Find the additive inverse (that is, the opposite). **(a)** −7 **(b)** $\frac{1}{4}$

(a) The opposite of −7 is +7. **(b)** The opposite of $\frac{1}{4}$ is $-\frac{1}{4}$.

Practice Problem 3 Find the additive inverse (the opposite).

(a) $+\frac{2}{5}$ **(b)** −1.92 **(c)** a loss of 12 yards on a football play

3 Adding Real Numbers with the Same Sign

To use a real number, we need to be clear about its sign. When we write the number three as +3, the sign indicates that it is a positive number. The positive sign can be omitted. If someone writes three (3), it is understood that it is a positive three (+3). To write a negative number such as negative three (−3), we must include the sign.

A concept that will help us add and subtract real numbers is the idea of absolute value. The **absolute value** of a number is the distance between that number and zero on the number line. The absolute value of 3 is written |3|.

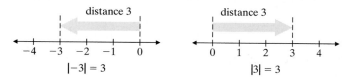

$|-3| = 3$ $|3| = 3$

Distance is always a positive number regardless of the direction we travel. This means that the absolute value of any number will be a positive value or zero. We place the symbols | and | around a number to mean the absolute value of the number.

The distance from 0 to 3 is 3, so |3| = 3. This is read "the absolute value of 3 is 3."

The distance from 0 to −3 is 3, so |−3| = 3. This is read "the absolute value of −3 is 3."

Some other examples are

$$|-22| = 22, \qquad |5.6| = 5.6, \qquad \text{and} \qquad |0| = 0.$$

Thus, the absolute value of a number can be thought of as the magnitude of the number, without regard to its sign.

EXAMPLE 4 Find the absolute value.

(a) $|-4.62|$ **(b)** $\left|\frac{3}{7}\right|$ **(c)** $|0|$

(a) $|-4.62| = 4.62$ **(b)** $\left|\frac{3}{7}\right| = \frac{3}{7}$ **(c)** $|0| = 0$

Practice Problem 4 Find the absolute value.

(a) $|-7.34|$ **(b)** $\left|\frac{5}{8}\right|$ **(c)** $\left|\frac{0}{2}\right|$

Now let's look at addition of real numbers when the two numbers have the same sign. Suppose that you are keeping track of your checking account at a local bank. When you make a deposit of 5 dollars, you record it as +5. When you write a check for 4 dollars, you record it as −4, as a debit. Consider two situations.

Situation 1

You made a deposit of 20 dollars on one day and a deposit of 15 dollars the next day. You want to know the total value of your deposits.

Your record for situation 1.

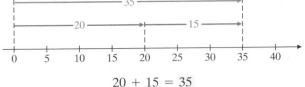

$$20 + 15 = 35$$

The amount of the deposit on the first day added to the amount of the deposit on the second day is the total of the deposits made over the two days.

Situation 2

You write a check for 25 dollars to pay one bill and two days later write a check for 5 dollars. You want to know the total value of debits to your account for the two checks.

Your record for situation 2.

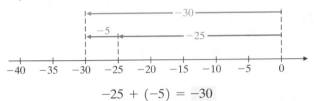

$$-25 + (-5) = -30$$

The value of the first check added to the value of the second check is the total debit to your account.

In each situation we found that we added the absolute value of each number. (That is, we added the numbers without regarding their sign.) The answer always contained the sign that was common to both numbers.

We will now state these results as a formal rule.

Addition Rule for Two Numbers with the Same Sign

To add two numbers with the same sign, add the absolute values of the numbers and use the common sign in the answer.

● **EXAMPLE 5**   Add.   **(a)** $14 + 16$ **(b)** $-8 + (-7)$

(a) $14 + 16$

$14 + 16 = 30$ Add the absolute values of the numbers.

$14 + 16 = +30$ Use the common sign in the answer. Here the common sign is the $+$ sign.

(b) $-8 + (-7)$

$8 + 7 = 15$ Add the absolute values of the numbers.

$-8 + (-7) = -15$ Use the common sign in the answer. Here the common sign is the $-$ sign.

Practice Problem 5   Add. $-23 + (-35)$

● **EXAMPLE 6**   Add. $\dfrac{2}{3} + \dfrac{1}{7}$

$\dfrac{2}{3} + \dfrac{1}{7}$

$\dfrac{14}{21} + \dfrac{3}{21}$ Change each fraction to an equivalent fraction with a common denominator of 21.

$\dfrac{14}{21} + \dfrac{3}{21} = +\dfrac{17}{21}$ or $\dfrac{17}{21}$ Add the absolute values of the numbers. Use the common sign in the answer. Note that if no sign is written, the number is understood to be positive.

Practice Problem 6 Add. $-\dfrac{3}{5} + \left(-\dfrac{4}{7}\right)$

EXAMPLE 7 Add. $-4.2 + (-3.9)$

$-4.2 + (-3.9)$

$4.2 + 3.9 = 8.1$ Add the absolute values of the numbers.

$-4.2 + (-3.9) = -8.1$ Use the common sign in the answer.

Practice Problem 7 Add. $-12.7 + (-9.38)$

The rule for adding two numbers with the same signs can be extended to more than two numbers. If we add more than two numbers with the same sign, the answer will have the sign common to all.

EXAMPLE 8 Add. $-7 + (-2) + (-5)$

$-7 + (-2) + (-5)$ We are adding three real numbers all with the same sign. We begin by adding the first two numbers.

$= -9 + (-5)$ Add $-7 + (-2) = -9$.

$= -14$ Add $-9 + (-5) = -14$.

Of course, this can be shortened by adding the three numbers without regard to sign and then using the common sign for the answer.

Practice Problem 8 Add. $-7 + (-11) + (-33)$

4 *Adding Real Numbers with Opposite Signs*

What if the signs of the numbers you are adding are different? Let's consider our checking account again to see how such a situation might occur.

Situation 3

You made a deposit of 30 dollars on one day. On the next day you write a check for 25 dollars. You want to know the result of your two transactions.

Your record for situation 3.

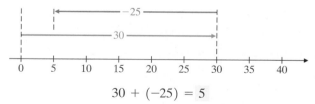

$$30 + (-25) = 5$$

A positive 30 for the deposit added to a negative 25 for the check, which is a debit, gives a net increase of 5 dollars in the account.

Situation 4

You made a deposit of 10 dollars on one day. The next day you write a check for 40 dollars. You want to know the result of your two transactions.

Your record for situation 4.

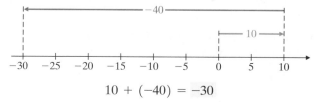

$$10 + (-40) = -30$$

A positive 10 for the deposit added to a negative 40 for the check, which is a debit, gives a net decrease of 30 dollars in the account.

The result is a negative thirty (−30) because the check was larger than the deposit. If you do not have at least 30 dollars in your account at the start of situation 4, you have overdrawn your account.

What do we observe from situations 3 and 4? In each case, first we found the difference of the absolute values of the two numbers. Then the sign of the result was always the sign of the number with the greater absolute value. Thus, in situation 3, 30 is larger than 25. The sign of 30 is positive. The sign of the answer (5) is positive. In situation 4, 40 is larger than 10. The sign of 40 is negative. The sign of the answer (−30) is negative.

We will now state these results as a formal rule.

Addition Rule for Two Numbers with Different Signs

1. **Find the difference between the larger absolute value and the smaller one.**
2. **Give the answer the sign of the number having the larger absolute value.**

EXAMPLE 9   Add. $8 + (-7)$

$8 + (-7)$	We are to add two numbers with opposite signs.
$8 - 7 = 1$	Find the difference between the two absolute values, which is 1.
$+8 + (-7) = +1$ or 1	The answer will have the sign of the number with the larger absolute value. That number is +8. Its sign is **positive**, so the answer will be +1.

Practice Problem 9   Add. $-9 + 15$

It is useful to know the following three properties of real numbers.

1. *Addition is commutative.*
   This property states that if two numbers are added, the result is the same no matter which number is written first. The order of the numbers does not affect the result.

   $$3 + 6 = 6 + 3 = 9$$
   $$-7 + (-8) = (-8) + (-7) = -15$$
   $$-15 + 3 = 3 + (-15) = -12$$

2. *Addition of zero to any given number will result in that given number again.*

   $$0 + 5 = 5$$
   $$-8 + 0 = -8$$

3. *Addition is associative.*
   This property states that if three numbers are added, it does not matter which two numbers are grouped by parentheses and added first.

$3 + (5 + 7) = (3 + 5) + 7$	First combine numbers inside parentheses; then combine the remaining numbers. The results are the same no matter which numbers are grouped first.
$3 + (12) = (8) + 7$	
$15 = 15$	

We can use these properties along with the rules we have for adding real numbers to add three or more numbers. We go from left to right, adding two numbers at a time.

EXAMPLE 10   Add. $\dfrac{3}{17} + \left(-\dfrac{8}{17}\right) + \dfrac{4}{17}$

$-\dfrac{5}{17} + \dfrac{4}{17}$ Add $\frac{3}{17} + \left(-\frac{8}{17}\right) = -\frac{5}{17}$.
The answer is negative since the larger of the two absolute values is negative.

$= -\dfrac{1}{17}$ Add $-\frac{5}{17} + \frac{4}{17} = -\frac{1}{17}$.
The answer is negative since the larger of the two absolute values is negative.

Practice Problem 10   Add. $-\dfrac{5}{12} + \dfrac{7}{12} + \left(-\dfrac{11}{12}\right)$

Sometimes the numbers being added have the same signs; sometimes the signs are different. When adding three or more numbers, you may encounter both situations.

EXAMPLE 11   Add. $-1.8 + 1.4 + (-2.6)$

$-0.4 + (-2.6)$ We take the difference of 1.8 and 1.4 and use the sign of the number with the larger absolute value.

$= -3.0$ Add $-0.4 + (-2.6) = -3.0$. The signs are the same; we add the absolute values of the numbers and use the common sign.

Practice Problem 11   Add. $-6.3 + (-8.0) + 3.5$

If many real numbers are added, it is often easier to add numbers with like signs in a column format. Remember that addition is commutative; therefore, real numbers can be added *in any order*. You do *not* need to combine the first two numbers as your first step.

EXAMPLE 12   Add. $-8 + 3 + (-5) + (-2) + 6 + 5$

$\begin{array}{r} -8 \\ -5 \\ -2 \\ \hline -15 \end{array}$ All the signs are the same. Add the three negative numbers to obtain -15.

$\begin{array}{r} +3 \\ +6 \\ +5 \\ \hline +14 \end{array}$ All the signs are the same. Add the three positive numbers to obtain $+14$.

Add the two results.

$$-15 + 14 = -1$$

The answer is negative because the number with the larger absolute value is negative.

Practice Problem 12   Add. $-6 + 5 + (-7) + (-2) + 5 + 3$

A word about notation: The only time we really need to show the sign of a number is when the number is negative—for example, -3. The only time we need to show parentheses when we add real numbers is when we have two different signs preceding a number. For example, $-5 + (-6)$.

EXAMPLE 13   Add.

(a) $2.8 + (-1.3)$ **(b)** $-\dfrac{2}{5} + \left(-\dfrac{3}{4}\right)$

(a) $2.8 + (-1.3) = 1.5$

(b) $-\dfrac{2}{5} + \left(-\dfrac{3}{4}\right) = -\dfrac{8}{20} + \left(-\dfrac{15}{20}\right) = -\dfrac{23}{20}$ or $-1\dfrac{3}{20}$

Practice Problem 13   Add.

(a) $-2.9 + (-5.7)$ **(b)** $\dfrac{2}{3} + \left(-\dfrac{1}{4}\right)$ **(c)** $-10 + (-3) + 15 + 4$

Verbal and Writing Skills

Check off any description of the number that applies.

Number	Whole Number	Rational Number	Irrational Number	Real Number
1. 23				
2. $-\frac{4}{5}$				
3. π				
4. 2.34				
5. $-6.666\ldots$				

Number	Whole Number	Rational Number	Irrational Number	Real Number
6. $-\frac{7}{9}$				
7. $-2.3434\ldots$				
8. 14				
9. $\sqrt{2}$				
10. $3.232232223\ldots$				

Use a real number to represent each situation.

11. Jules Verne wrote a book with the title *20,000 Leagues under the Sea*.

12. The value of the dollar is up $0.07 with respect to the yen.

13. The Dow Jones average is down by $2\frac{3}{8}$.

14. Jon withdraws $102 from his account.

15. The temperature rises $7°$F.

16. Maya won the game by 12 points.

Find the additive inverse (opposite).

17. $\frac{3}{4}$

18. -2

19. -2.73

20. 85.4

Find the absolute value.

21. $|-1.3|$

22. $|-5.9|$

23. $\left|\frac{5}{6}\right|$

24. $\left|\frac{7}{12}\right|$

Add.

25. $-6 + (-5)$

26. $-13 + (-3)$

27. $-\frac{1}{3} + \frac{2}{3}$

28. $-\frac{1}{5} + \left(-\frac{3}{5}\right)$

29. $-17 + (-14)$

30. $-12 + (-19)$

31. $-\frac{2}{13} + \left(-\frac{5}{13}\right)$

32. $-\frac{5}{14} + \frac{2}{14}$

33. $-1.5 + (-2.3)$

34. $-1.8 + (-1.4)$

35. $0.6 + (-0.2)$

36. $-0.8 + 0.5$

37. $-12 + (-13)$

38. $-17 + (-21)$

39. $-\frac{2}{5} + \frac{3}{7}$

40. $-\frac{2}{7} + \frac{3}{14}$

41. $-8 + 5 + (-3)$

42. $7 + (-8) + (-4)$

43. $-3 + 8 + 5 + (-7)$

Mixed Practice

Add.

44. $2 + (-17)$ **45.** $21 + (-4)$ **46.** $-83 + 42$ **47.** $-114 + 86$

48. $-\dfrac{4}{9} + \dfrac{5}{6}$ **49.** $-\dfrac{3}{5} + \dfrac{2}{3}$ **50.** $-\dfrac{1}{10} + \dfrac{1}{2}$ **51.** $-\dfrac{2}{3} + \left(-\dfrac{1}{4}\right)$

52. $4.3 + (-3.6)$ **53.** $5.7 + (-9.1)$ **54.** $4 + (-8) + 16$ **55.** $27 + (-11) + (-4)$

56. $-27 + 9 + (-54) + 30$ **57.** $18 + (-39) + 25 + (-3)$

58. $17.85 + (-2.06) + 0.15$ **59.** $23.17 + 5.03 + (-11.81)$

Applications

60. Hallie paid $47 for a vase at an estate auction. She resold it to an antiques dealer for $214. What was her profit or loss?

61. When we skied at Jackson Hole, Wyoming, yesterday, the temperature at the summit was $-12°$F. Today when we called the ski report, the temperature had risen $7°$F. What is the temperature at the summit today?

62. Donna is studying rain forest preservation in Central America. She stands on a ridge 126 feet below sea level. Then she and her team hike down to a gully 43 feet lower. Represent her distance below sea level as a real number.

63. Oceanographers studying effects of light on sea creatures are in a submarine 85 feet below sea level. Then they dive to a point 180 feet lower. Represent the submarine's distance below sea level as a real number.

64. On three successive football running plays, Jon gained 9 yards, lost 11 yards, and gained 5 yards. What was his total gain or loss?

65. Ureji's financial aid account at school held $643.85. She withdrew $185.50 to buy books for the semester. Does she have enough left in her account to pay the $475.00 registration fee for the next semester? If so, how much extra money does she have? If not, how much is she short?

66. The population of a particular butterfly species was 8000. Twenty years later there were 3000 fewer. Today, there are 1500 fewer. What is the new population?

67. Aaron owes $258 to a credit card company. He makes a purchase of $32 with the card and then makes a payment of $150 on the account. How much does he still owe?

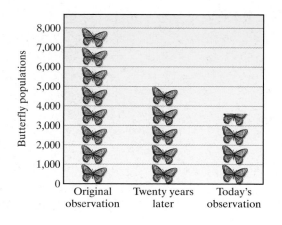

During the first five months of 2000, a regional Midwest airline posted profit and loss figures for each month of operation, as shown in the accompanying bar graph.

68. For the first three months of 2000, what were the total earnings of the airline?

69. For the first five months of 2000, what were the total earnings for the airline?

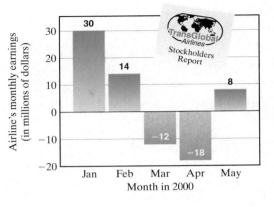

To Think About

70. What number must be added to −13 to get 5?

71. What number must be added to −18 to get 10?

72. Glen makes a deposit of $50 into his checking account and then writes out a check for $89. What is the least amount of money he needs in his checking account to cover the transactions?

73. The Roanoke Animal Shelter received 800 kittens from families whose cats had not been spayed. Only 328 kittens were placed with new families. Express as a percent how many kittens were able to find a home this year. If next year the same number of kittens are received, but 56% are placed in new homes, how many more kittens will find a home next year than did this year?

1.2 Subtraction of Real Numbers

Student Learning Objectives

After studying this section, you will be able to:

1. Subtract real numbers.

SSM PH TUTOR CD & VIDEO MATH PRO WEB
CENTER

1 Subtracting Real Numbers

So far we have developed the rules for adding real numbers. We can use these rules to subtract real numbers. Let's look at a checkbook situation to see how.

Situation 5

You have a balance of 20 dollars in your checking account. The bank calls you and says that a deposit of 5 dollars that belongs to another account was erroneously added to your account. They say they will correct the account balance to 15 dollars. The bank tells you that since they cannot take away the erroneous credit, they will add a debit to your account. You want to keep track of what's happening to your account.

Your record for situation 5.

$$20 - (+5) = 15$$

From your present balance subtract the deposit to give the new balance. This equation shows what needs to be done to your account. The bank tells you that because the error happened in the past they cannot "take it away." However, they can add to your account a debit of 5 dollars. Here is the equivalent addition.

$$20 + (-5) = 15$$

To your present balance add a debit to give the new balance. Subtracting a positive 5 has the same effect as adding a negative 5.

Subtraction of Real Numbers

To subtract real numbers, add the opposite of the second number (that is, the number you are subtracting) to the first.

The rule tells us to do three things when we subtract real numbers. First, change subtraction to addition. Second, replace the second number by its opposite. Third, add the two numbers using the rules for addition of real numbers.

EXAMPLE 1 Subtract. $6 - (-2)$

$$6 \quad - \quad (-2)$$

Change subtraction to addition. | Write the opposite of the second number.

$$= \quad 6 \quad + \quad (+2)$$

Add the two real numbers with the same sign.

$$= \quad 8$$

Practice Problem 1 Subtract. $9 - (-3)$

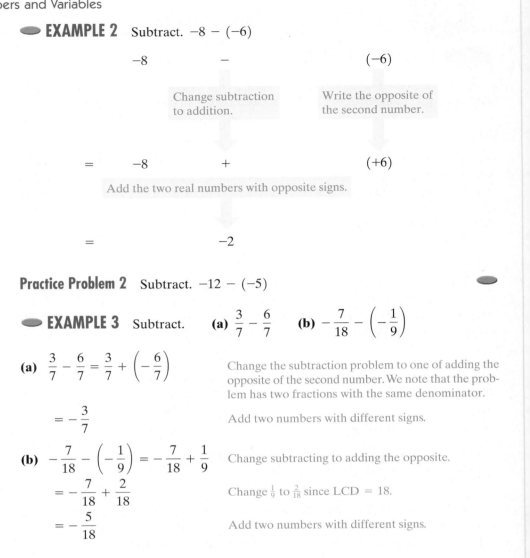

EXAMPLE 2 Subtract. $-8 - (-6)$

$$-8 \qquad\qquad - \qquad\qquad (-6)$$

Change subtraction to addition.

Write the opposite of the second number.

$$= \qquad -8 \qquad\qquad + \qquad\qquad (+6)$$

Add the two real numbers with opposite signs.

$$= \qquad\qquad\qquad -2$$

Practice Problem 2 Subtract. $-12 - (-5)$

EXAMPLE 3 Subtract. **(a)** $\dfrac{3}{7} - \dfrac{6}{7}$ **(b)** $-\dfrac{7}{18} - \left(-\dfrac{1}{9}\right)$

(a) $\dfrac{3}{7} - \dfrac{6}{7} = \dfrac{3}{7} + \left(-\dfrac{6}{7}\right)$ Change the subtraction problem to one of adding the opposite of the second number. We note that the problem has two fractions with the same denominator.

$\qquad = -\dfrac{3}{7}$ Add two numbers with different signs.

(b) $-\dfrac{7}{18} - \left(-\dfrac{1}{9}\right) = -\dfrac{7}{18} + \dfrac{1}{9}$ Change subtracting to adding the opposite.

$\qquad = -\dfrac{7}{18} + \dfrac{2}{18}$ Change $\frac{1}{9}$ to $\frac{2}{18}$ since LCD = 18.

$\qquad = -\dfrac{5}{18}$ Add two numbers with different signs.

Practice Problem 3 Subtract. $-\dfrac{1}{5} - \dfrac{1}{4}$

EXAMPLE 4 Subtract. $-5.2 - (-5.2)$

$-5.2 - (-5.2) = -5.2 + 5.2$ Change the subtraction problem to one of adding the opposite of the second number.

$\qquad = 0$ Add two numbers with different signs.

Example 4 illustrates what is sometimes called the **additive inverse property**. When you add two real numbers that are opposites of each other, you will obtain zero. Examples of this are the following:

$$5 + (-5) = 0 \qquad -186 + 186 = 0 \qquad -\dfrac{1}{8} + \dfrac{1}{8} = 0$$

Practice Problem 4 Subtract. $-17.3 - (-17.3)$

EXAMPLE 5 Calculate.

(a) $-8 - 2$ **(b)** $23 - 28$ **(c)** $5 - (-3)$ **(d)** $\dfrac{1}{4} - 8$

(a) $-8 - 2 = -8 + (-2)$ Notice that we are subtracting a positive 2.
                               Change to addition.
            $= -10$ Add.

In a similar fashion we have

(b) $23 - 28 = 23 + (-28) = -5$

(c) $5 - (-3) = 5 + 3 = 8$

(d) $\dfrac{1}{4} - 8 = \dfrac{1}{4} + (-8) = \dfrac{1}{4} + \left(-\dfrac{32}{4} \right) = -\dfrac{31}{4}$

Practice Problem 5   Calculate.   **(a)** $-21 - 9$ **(b)** $17 - 36$

EXAMPLE 6   A satellite is recording radioactive emissions from nuclear waste buried 3 miles below sea level. The satellite orbits the Earth at 98 miles above sea level. How far is the satellite from the nuclear waste?

     We want to find the difference between +98 miles and −3 miles.

$$98 - (-3) = 98 + 3$$
$$= 101$$

The satellite is 101 miles from the nuclear waste.

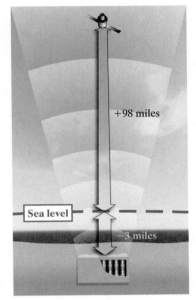

Practice Problem 6   A helicopter is directly over a sunken vessel. The helicopter is 350 feet above sea level. The vessel lies 186 feet below sea level. How far is the helicopter from the sunken vessel?

Verbal and Writing Skills

1. Explain in your own words how you would perform the necessary steps to find $-8 - (-3)$.

2. Explain in your own words how you would perform the necessary steps to find $-10 - (-15)$.

Subtract by adding the opposite.

3. $20 - 46$

4. $15 - 28$

5. $-14 - (-3)$

6. $-24 - (-7)$

7. $-52 - (-60)$

8. $-48 - (-80)$

9. $0 - (-5)$

10. $0 - (-7)$

11. $15 - 20$

12. $18 - 24$

13. $-18 - (-18)$

14. $24 - (-24)$

15. $-11 - (-8)$

16. $-35 - (-10)$

17. $-0.6 - 0.3$

18. $-0.9 - 0.5$

19. $2.64 - (-1.83)$

20. $-0.03 - 0.06$

21. $2.3 - (-1.4)$

22. $\dfrac{1}{3} - \left(-\dfrac{2}{5}\right)$

23. $\dfrac{3}{4} - \left(-\dfrac{3}{5}\right)$

24. $-\dfrac{2}{3} - \dfrac{1}{4}$

25. $-\dfrac{3}{4} - \dfrac{5}{6}$

26. $-\dfrac{7}{10} - \dfrac{10}{15}$

Mixed Practice

Calculate.

27. $34 - 87$

28. $19 - 76$

29. $-67 - 32$

30. $-98 - 34$

31. $2.3 - (-4.8)$

32. $8.4 - (-2.7)$

33. $8 - \left(-\dfrac{3}{4}\right)$

34. $\dfrac{2}{3} - (-6)$

35. $\dfrac{3}{5} - (-8)$

36. $\dfrac{5}{6} - 4$

37. $-\dfrac{3}{10} - \dfrac{3}{4}$

38. $-\dfrac{11}{12} - \dfrac{5}{18}$

39. $-135 - (-126.5)$

40. $-97.6 - (-146)$

41. $0.0067 - (-0.0432)$

42. $0.0762 - (-0.0094)$

43. $\dfrac{1}{5} - 6$

44. $\dfrac{2}{7} - (-3)$

45. $-0.0023 - 6$

46. $-2 - 0.071$ **47.** Subtract -9 from -2.   **48.** Subtract -12 from 20.   **49.** Subtract 13 from -35.

Change each subtraction operation to "adding the opposite." Then combine the numbers.

50. $9 + 6 - (-5)$ **51.** $7 + (-6) - 3$ **52.** $8 + (-4) - 10$ **53.** $-10 + 6 - (-15)$

54. $18 - (-15) - 3$ **55.** $7 + (-42) - 27$ **56.** $-37 - (-18) + 5$ **57.** $-21 - (-36) - 8$

58. $-3 - (-12) + 18 + 15 - (-6)$ **59.** $42 - (-30) - 65 - (-11) + 20$

Applications

60. A rescue helicopter is 300 feet above sea level. The captain has located an ailing submarine directly below it that is 126 feet below sea level. How far is the helicopter from the submarine?

61. Yesterday Jackie had $112 in her checking account. Today her account reads "balance $-$ \$37." Find the difference in these two amounts.

62. On January 6, 1971, Hawley Lake, Arizona, had a record low temperature of $-23°$F. The next day the temperature at the same place was $-40°$F. What was the change in temperature from January 6 to January 7, 1971?

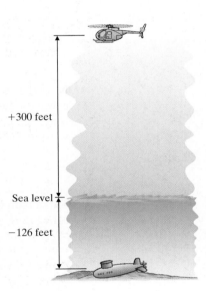

+300 feet

Sea level

−126 feet

63. On January 19, 1937, the temperature at Boca, California, was $-29°$F. On January 20, the temperature at the same place was $-45°$F. What was the change in temperature from January 19 to January 20, 1937?

64. In 2000, Rachel had \$1815 withheld from her paycheck in federal taxes. When she filed her tax return, she received a \$265 refund. How much did she actually pay in taxes?

Cumulative Review Problems

In exercises 65–67, perform the indicated operations.

65. $-37 + 16$ **66.** $-37 + (-14)$ **67.** $-3 + (-6) + (-10)$

68. What is the temperature after a rise of $13°$C from a start of $-21°$C?

69. One morning the temperature was $-15°$F. The forecast says the temperature should reach $8°$F. How many degrees will the temperature rise?

1.3 Multiplication and Division of Real Numbers

Student Learning Objectives

After studying this section, you will be able to:

1 Multiply real numbers.

2 Divide real numbers.

SSM
PH TUTOR   CD & VIDEO   MATH PRO   WEB
CENTER

1 *Multiplying Real Numbers*

We are familiar with the meaning of multiplication for positive numbers. For example, $5 \times 40 = 200$ might mean that you receive five weekly checks of 40 dollars each and you gain $200. Let's look at a situation that corresponds to $5 \times (-40)$. What might that mean?

Situation 6

You write a check for five weeks in a row to pay your weekly room rent of 40 dollars. You want to know the total impact on your checking account balance.

Your record for situation 6.

(+5)	×	(−40)	=	−200
The number of checks you have written	times	negative 40, the value of each check that was a debit to your account,	gives	negative 200 dollars, a net debit to your account.

Note that a multiplication symbol is not needed between the (+5) and the (−40) because the two sets of parentheses indicate multiplication. The multiplication (5)(−40) is the same as repeated addition of five (−40)'s. Note that 5 multiplied by −40 can be written as 5(−40) or (5)(−40).

$$\underbrace{(-40) + (-40) + (-40) + (-40) + (-40)}_{\text{repeated addition of five } (-40)\text{'s}} = -200$$

This example seems to show that a positive number multiplied by a negative number is negative.

What if the negative number is the one that is written first? If $(5)(-40) = -200$, then $(-40)(5) = -200$, by the commutative property of multiplication. This is an example showing that *when two numbers with opposite signs* (one positive, one negative) *are multiplied, the result is negative.*

But what if both numbers are negative? Consider the following situation.

Situation 7

Last year at college you rented a room at 40 dollars per week for 36 weeks, which included two semesters and summer school. This year you will not attend the summer session, so you will be renting the room for only 30 weeks. Thus the number of weekly rental checks will be six less than last year. You are making out your budget for this year. You want to know the financial impact of renting the room for six fewer weeks.

Your record for situation 7.

(−6)	×	(−40)	=	240
The difference in the number of checks this year compared to last is −6, which is negative to show a decrease,	times	−40, the value of each check paid out,	gives	+240 dollars. The product is positive because your financial situation will be 240 dollars better this year.

You could check that the answer is positive by calculating the total rental expenses.

Dollars in rent last year	$(36)(40) =$	1440
(subtract) Dollars in rent this year	$-(30)(40) =$	−1200
Extra dollars available this year	=	+240

This agrees with our previous answer: $(-6)(-40) = +240$.

18

In this situation it seems reasonable that a negative number times a negative number yields a positive answer. We already know from arithmetic that a positive number times a positive number yields a positive answer. Thus we might see the general rule that *when two numbers with the same sign* (both positive or both negative) *are multiplied, the result is positive.*

We will now state our rule.

Multiplication of Real Numbers

To multiply two real numbers with **the same sign**, multiply the absolute values. The sign of the result is **positive**.
To multiply two real numbers with **opposite signs**, multiply the absolute values. The sign of the result is **negative**.

Note that negative 6 times -40 can be written as $-6(-40)$ or $(-6)(-40)$.

EXAMPLE 1   Multiply.

(a) $(3)(6)$ **(b)** $\left(-\dfrac{5}{7}\right)\left(-\dfrac{2}{9}\right)$ **(c)** $-4(8)$ **(d)** $\left(\dfrac{2}{7}\right)(-3)$

(a) $(3)(6) = 18$

When multiplying two numbers with the same sign, the result is a positive number.

(b) $\left(-\dfrac{5}{7}\right)\left(-\dfrac{2}{9}\right) = \dfrac{10}{63}$

(c) $-4(8) = -32$

When multiplying two numbers with opposite signs, the result is a negative number.

(d) $\left(\dfrac{2}{7}\right)(-3) = \left(\dfrac{2}{7}\right)\left(-\dfrac{3}{1}\right) = -\dfrac{6}{7}$

Practice Problem 1   Multiply.

(a) $(-6)(-2)$ **(b)** $(7)(9)$ **(c)** $\left(-\dfrac{3}{5}\right)\left(\dfrac{2}{7}\right)$ **(d)** $40(-20)$

To multiply more than two numbers, multiply two numbers at a time.

EXAMPLE 2   Multiply. $(-4)(-3)(-2)$

$(-4)(-3)(-2) = (+12)(-2)$ We begin by multiplying the first two numbers, (-4) and (-3). The signs are the same. The answer is positive 12.

$= -24$ Now we multiply $(+12)$ and (-2). The signs are different. The answer is negative 24.

Practice Problem 2   Multiply. $(-5)(-2)(-6)$

EXAMPLE 3   Multiply.   **(a)** $-3(-8)$ **(b)** $\left(-\dfrac{1}{2}\right)(-1)(-4)$

(c) $-2(-2)(-2)(-2)$

Multiply two numbers at a time. See if you find a pattern.

(a) $-3(-8) = +24$ or 24

(b) $\left(-\dfrac{1}{2}\right)(-1)(-4) = +\dfrac{1}{2}(-4) = -2$

(c) $-2(-2)(-2)(-2) = +4(-2)(-2) = -8(-2) = +16$ or 16

What kind of answer would we obtain if we multiplied five negative numbers? If you guessed "negative," you probably see the pattern.

Practice Problem 3 Determine the sign of the product. Then multiply to check.

(a) $-2(-3)(-4)(-1)$ **(b)** $(-1)(-3)(-2)$ **(c)** $-4\left(-\dfrac{1}{4}\right)(-2)(-6)$

When you multiply two or more real numbers:
1. The result is always **positive** if there are an **even** number of negative signs.
2. The result is always **negative** if there are an **odd** number of negative signs.

For convenience, we will list the properties of multiplication.

---

1. *Multiplication is commutative.*
   This property states that if two real numbers are multiplied, the order of the numbers does not affect the result. The result is the same no matter which number is written first.

$$(5)(7) = (7)(5) = 35, \qquad \left(\frac{1}{3}\right)\left(\frac{2}{7}\right) = \left(\frac{2}{7}\right)\left(\frac{1}{3}\right) = \frac{2}{21}$$

2. *Multiplication of any real number by zero will result in zero.*

$$(5)(0) = 0, \qquad (-5)(0) = 0, \qquad (0)\left(\frac{3}{8}\right) = 0, \qquad (0)(0) = 0$$

3. *Multiplication of any real number by 1 will result in that same number.*

$$(5)(1) = 5, \qquad (1)(-7) = -7, \qquad (1)\left(-\frac{5}{3}\right) = -\frac{5}{3}$$

4. *Multiplication is associative.*
   This property states that if three real numbers are multiplied, it does not matter which two numbers are grouped by parentheses and multiplied first.

$$2 \times (3 \times 4) = (2 \times 3) \times 4 \qquad \text{First multiply the numbers in parentheses.}$$
$$\text{Then multiply the remaining numbers.}$$

$$2 \times (12) = (6) \times 4 \qquad \text{The results are the same no matter which}$$
$$\text{numbers are grouped and multiplied first.}$$

$$24 = 24$$

 Dividing Real Numbers

What about division? Any division problem can be rewritten as a multiplication problem.

We know that $20 \div 4 = 5$ because $4(5) = 20$.
Similarly, $-20 \div (-4) = 5$ because $-4(5) = -20$.

In both division problems the answer is positive 5. Thus we see that *when you divide two numbers with the same sign* (both positive or both negative), *the answer is positive*. What if the signs are different?

We know that $-20 \div 4 = -5$ because $4(-5) = -20$.
Similarly, $20 \div (-4) = -5$ because $-4(-5) = 20$.

In these two problems the answer is negative 5. So we have reasonable evidence to see that *when you divide two numbers with different signs* (one positive and one negative), *the answer is negative.*

We will now state our rule for division.

Division of Real Numbers

To divide two real numbers with **the same sign**, divide the absolute values. The sign of the result is **positive**.
To divide two real numbers with **different signs**, divide the absolute values. The sign of the result is **negative**.

● EXAMPLE 4   Divide.

(a) $12 \div 4$ **(b)** $(-25) \div (-5)$ **(c)** $\dfrac{-36}{18}$ **(d)** $\dfrac{42}{-7}$

(a) $12 \div 4 = 3$

> When dividing two numbers with the same sign, the result is a positive number.

(b) $(-25) \div (-5) = 5$

(c) $\dfrac{-36}{18} = -2$

> When dividing two numbers with different signs, the result is a negative number.

(d) $\dfrac{42}{-7} = -6$

Practice Problem 4   Divide.

(a) $-36 \div (-2)$ **(b)** $\dfrac{50}{-10}$ **(c)** $-49 \div 7$

● EXAMPLE 5   Divide.   **(a)** $-36 \div 0.12$ **(b)** $-2.4 \div (-0.6)$

(a) $-36 \div 0.12$ Look at the problem to determine the sign. When dividing two numbers with different signs, the result will be a negative number.

We then divide the absolute values.

$$0.12_{\wedge} \overline{)36.00_{\wedge}} \quad \begin{array}{r} 3\,00. \\ \underline{36} \\ 00 \end{array}$$

Thus $-36 \div 0.12 = -300$. The answer is a negative number.

(b) $-2.4 \div (-0.6)$ Look at the problem to determine the sign. When dividing two numbers with the same sign, the result will be positive.

We then divide the absolute values.

$$0.6_{\wedge} \overline{)2.4_{\wedge}} \quad \begin{array}{r} 4. \\ \underline{2\,4} \\ \end{array}$$

Thus $-2.4 \div (-0.6) = 4$. The answer is a positive number.

Practice Problem 5   Divide.

(a) $-1.242 \div (-1.8)$ **(b)** $0.235 \div (-0.0025)$

Note how similar the rules for multiplication and division are. When you **multiply** or **divide** two numbers with the **same** sign, you obtain a **positive** number. When you **multiply** or **divide** two numbers with **different** signs, you obtain a **negative** number.

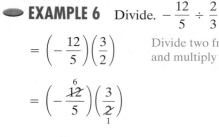

EXAMPLE 6 Divide. $-\dfrac{12}{5} \div \dfrac{2}{3}$

$= \left(-\dfrac{12}{5}\right)\left(\dfrac{3}{2}\right)$ Divide two fractions. We invert the second fraction and multiply by the first fraction.

$= \left(-\dfrac{\overset{6}{\cancel{12}}}{5}\right)\left(\dfrac{3}{\underset{1}{\cancel{2}}}\right)$

$= -\dfrac{18}{5}$ or $-3\dfrac{3}{5}$ The answer is negative since the two numbers divided have different signs.

Practice Problem 6 Divide. $-\dfrac{5}{16} \div \left(-\dfrac{10}{13}\right)$

Note that division can be indicated by the symbol $\div$ or by the fraction bar $-$. $\frac{2}{3}$ means $2 \div 3$.

EXAMPLE 7 Divide. **(a)** $\dfrac{\frac{7}{8}}{-21}$ **(b)** $\dfrac{-\frac{2}{3}}{-\frac{7}{13}}$

(a) $\dfrac{\frac{7}{8}}{-21}$

$= \dfrac{7}{8} \div \left(-\dfrac{21}{1}\right)$ Change -21 to a fraction. $-21 = -\frac{21}{1}$

$= \dfrac{\overset{1}{\cancel{7}}}{8}\left(-\dfrac{1}{\underset{3}{\cancel{21}}}\right)$ Change the division to multiplication. Cancel where possible.

$= -\dfrac{1}{24}$ Simplify.

(b) $\dfrac{-\frac{2}{3}}{-\frac{7}{13}} = -\dfrac{2}{3} \div \left(-\dfrac{7}{13}\right) = -\dfrac{2}{3}\left(-\dfrac{13}{7}\right) = \dfrac{26}{21}$ or $1\dfrac{5}{21}$

Practice Problem 7 Divide. **(a)** $\dfrac{-12}{-\frac{4}{5}}$ **(b)** $\dfrac{-\frac{2}{9}}{\frac{8}{13}}$

1. *Division of 0 by any nonzero real number gives 0 as a result.*

$$0 \div 5 = 0, \qquad 0 \div \dfrac{2}{3} = 0, \qquad \dfrac{0}{5.6} = 0, \qquad \dfrac{0}{1000} = 0$$

You can divide zero by $5, \frac{2}{3}, 5.6, 1000$, or any number (except 0).

2. *Division of any real number by 0 is* **undefined**.

$$7 \div 0 \qquad\qquad \dfrac{64}{0} \qquad\qquad \dfrac{0}{0}$$
$$\uparrow \qquad\qquad\quad \uparrow \qquad\qquad \uparrow$$

None of these operations is possible. **Division by zero is undefined.**

You may be wondering why division by zero is undefined. Let us think about it for a minute. We said that $7 \div 0$ is undefined. Suppose there were an answer. Let us call the answer a. So we assume for a minute that $7 \div 0 = a$. Then it would have to follow that $7 = 0(a)$. But this is impossible. Zero times any number is zero. So we see that if there were such a number, it would contradict known mathematical facts. Therefore there is no number a such that $7 \div 0 = a$. Thus we conclude that division by zero is undefined.

When combining two numbers, it is important to be sure you know which rule applies. Think about the concepts in the following chart. See if you agree with each example.

Operation	Two Real Numbers with the Same Sign	Two Real Numbers with Different Signs
Addition	Result may be positive or negative. $9 + 2 = 11$ $-5 + (-6) = -11$	Result may be positive or negative. $-3 + 7 = 4$ $4 + (-12) = -8$
Subtraction	Result may be positive or negative. $15 - 6 = 15 + (-6) = 9$ $-12 - (-3) = -12 + 3 = -9$	Result may be positive or negative. $-12 - 3 = -12 + (-3) = -15$ $5 - (-6) = 5 + 6 = 11$
Multiplication	Result is always positive. $9(3) = 27$ $-8(-5) = 40$	Result is always negative. $-6(12) = -72$ $8(-3) = -24$
Division	Result is always positive. $150 \div 6 = 25$ $-72 \div (-2) = 36$	Result is always negative. $-60 \div 10 = -6$ $30 \div (-6) = -5$

EXAMPLE 8 The Hamilton-Wenham Generals recently analyzed the 48 plays their team made while in the possession of the football during their last game. The following bar graph illustrates the number of plays made in each category. The team statistician prepared the following chart indicating the average number of yards gained or lost during each type of play.

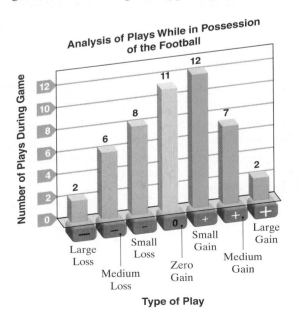

Type of Play	Average Yards Gained or Lost for Play
Large gain	+25
Medium gain	+15
Small gain	+5
Zero gain	0
Small loss	−5
Medium loss	−10
Large loss	−15

(a) How many yards were lost by the Generals in the plays that were considered small losses?

(b) How many yards were gained by the Generals in the plays that were considered small gains?

(c) If the total yards gained in small gains were combined with the total yards lost in small losses, what would be the result?

(a) We multiply the number of small losses by the average number of total yards lost on each small loss:

$$8(-5) = -40$$

The team lost approximately 40 yards with plays that were considered small losses.

(b) We multiply the number of small gains by the average number of yards gained on each small gain:

$$12(5) = 60$$

The team gained approximately 60 yards with plays that were considered small gains.

(c) We combine the results for (a) and (b):

$$-40 + 60 = 20$$

A total of 20 yards was gained during the plays that were small losses and small gains.

Practice Problem 8 Using the information provided in Example 8, answer the following:

(a) How many yards were lost by the Generals in the plays that were considered medium losses?

(b) How many yards were gained by the Generals in the plays that were considered medium gains?

(c) If the total yards gained in medium gains were combined with the total yards lost in medium losses, what would be the result?

Developing Your Study Skills

Reading the Textbook

Begin reading your textbook with a paper and pencil in hand. As you come across a new definition or concept, underline it in the text and/or write it down in your notebook. Whenever you encounter an unfamiliar term, look it up and make a note of it. When you come to an example, work through it step-by-step. Be sure to read each word and follow directions carefully.

Notice the helpful hints the author provides. They guide you to correct solutions and prevent you from making errors. Take advantage of these pieces of expert advice.

Be sure that you understand what you are reading. Make a note of any of those things that you do not understand and ask your instructor about them. Do not hurry through the material. Learning mathematics takes time.

Exercises

Verbal and Writing Skills

1. Explain in your own words the rule for determining the correct sign when multiplying two real numbers.

2. Explain in your own words the rule for determining the correct sign when multiplying three or more real numbers.

Multiply or divide. Be sure to write your answer in the simplest form.

3. $-8(3)$

4. $-5(11)$

5. $0(-12)$

6. $-3(150)$

7. $16(1.5)$

8. $24(2.5)$

9. $(-1.32)(-0.2)$

10. $(-2.3)(-0.11)$

11. $0.7(-2.5)$

12. $-6\left(\dfrac{3}{10}\right)$

13. $\left(\dfrac{12}{5}\right)(-10)$

14. $\left(\dfrac{3}{5}\right)\left(-\dfrac{15}{11}\right)$

15. $\left(-\dfrac{4}{9}\right)\left(-\dfrac{3}{5}\right)$

16. $\left(-\dfrac{5}{21}\right)\left(\dfrac{3}{10}\right)$

17. $12 \div (-4)$

18. $-36 \div (-9)$

19. $0 \div (-15)$

20. $-48 \div (-8)$

21. $-45 \div (9)$

22. $-220 \div (-11)$

23. $240 \div (-15)$

24. $156 \div (-13)$

25. $-0.6 \div 0.3$

26. $1.2 \div (-0.03)$

27. $-6.3 \div (0.7)$

28. $0.54 \div (-0.9)$

29. $-7.2 \div (8)$

30. $\dfrac{2}{7} \div \left(-\dfrac{3}{5}\right)$

31. $\left(-\dfrac{1}{5}\right) \div \left(\dfrac{2}{3}\right)$

32. $\left(-\dfrac{5}{6}\right) \div \left(-\dfrac{7}{18}\right)$

33. $\dfrac{5}{7} \div \left(-\dfrac{3}{28}\right)$

34. $\dfrac{4}{9} \div \left(-\dfrac{8}{15}\right)$

35. $\dfrac{12}{-\dfrac{2}{5}}$

36. $\dfrac{-\dfrac{3}{7}}{-6}$

37. $\dfrac{-\dfrac{3}{8}}{-\dfrac{2}{3}}$

38. $\dfrac{-\dfrac{1}{4}}{\dfrac{3}{8}}$

39. $\dfrac{\dfrac{7}{15}}{-\dfrac{3}{5}}$

Multiply. You may want to determine the sign of the product before you multiply.

40. $-6(2)(-3)(4)$

41. $-1(-2)(-3)(4)$

42. $-2(-1)(3)(-1)(-4)$

43. $-2(-2)(2)(-1)(-3)$

44. $-3(2)(-4)(0)(-2)$

45. $-3(-2)\left(\dfrac{1}{3}\right)(-4)(2)$

46. $60(-0.6)(-20)(0.5)$

47. $-3(-0.03)(100)(-2)$

48. $\left(\dfrac{3}{8}\right)\left(\dfrac{1}{2}\right)\left(-\dfrac{5}{6}\right)$

49. $\left(-\dfrac{4}{5}\right)\left(-\dfrac{6}{7}\right)\left(-\dfrac{1}{3}\right)$

50. $\left(-\dfrac{1}{2}\right)\left(\dfrac{4}{5}\right)\left(-\dfrac{7}{8}\right)\left(-\dfrac{2}{3}\right)$

51. $\left(-\dfrac{3}{4}\right)\left(-\dfrac{7}{15}\right)\left(-\dfrac{8}{21}\right)\left(-\dfrac{5}{9}\right)$

Mixed Practice

Take a minute to review the chart before Example 8. Be sure that you can remember the sign rules for each operation. Then do exercises 52–61. Perform the indicated calculations.

52. $-5-(-2)$

53. $-20 \div 5$

54. $-3(-9)$

55. $5+(-7)$

56. $-32 \div (-4)$

57. $8-(-9)$

58. $-6+(-3)$

59. $6(-12)$

60. $18 \div (-18)$

61. $-37 \div 37$

Applications

62. Debbie's monthly car loan payment is $250. If she pays by check, what is the net debit to her account after six months?

63. Phil pays a doctor bill for $100 but forgets to check the balance in his account. When he goes to the ATM, his balance reads −$70. How much was in his account before he wrote the check?

64. Ramon owes $6500 on his student loan. If $180 is automatically deducted from his bank account each month to pay the loan off, how much does he still owe after one year?

65. Keith's car loan of $15,768 is to be paid off in 48 equal monthly installments. What is his monthly bill?

The Beverly Panthers recently analyzed the 37 plays their team made while in the possession of the football during their last game. The team statistician prepared the following chart indicating the number of plays in each category and the average number of yards gained or lost during each type of play. Use this chart to answer exercises 66–73.

Type of Play	Number of Plays	Average Yards Gained or Lost per Play
Large gain	1	+25
Medium gain	6	+15
Small gain	4	+5
Zero gain	5	0
Small loss	10	−5
Medium loss	7	−10
Large loss	4	−15

66. How many yards were lost by the Panthers in the plays that were considered small losses?

67. How many yards were gained by the Panthers in the plays that were considered small gains?

68. If the total yards gained in small gains were combined with the total yards lost in small losses, what would be the result?

69. How many yards were lost by the Panthers in the plays that were considered medium losses?

70. How many yards were gained by the Panthers in the plays that were considered medium gains?

71. If the total yards gained in medium gains were combined with the total yards lost in medium losses, what would be the result?

72. The game being studied was lost by the Panthers. The coach said that if two of the large-loss plays had been avoided and the number of small-gain plays had been doubled, they would have won the game. If those actions listed by the coach had happened, how many additional yards would the Panthers have gained? Assume each play would reflect the average yards gained or lost per play.

73. The game being studied was lost by the Panthers. The coach of the opposing team said that they could have scored two more touchdowns against the Panthers if they had caused three more large-loss plays and avoided four medium-gain plays. If those actions had happened, what effect would this have had on total yards gained by the Panthers? Assume each play would reflect the average yards gained or lost per play.

Cumulative Review Problems

74. $-17.4 + 8.31 + 2.40$

75. $-13 + (-39) + (-20)$

76. $-3.7 - (-8.33)$

77. $-37 - 51$

▲ **78.** A very famous sneaker company needs $104\frac{1}{2}$ square yards of white leather, $88\frac{2}{3}$ square yards of neon-yellow leather, and $72\frac{5}{6}$ square yards of neon-red leather to make today's batch of designer sneakers. What is today's required total square yardage of leather?

Developing Your Study Skills

Steps Toward Success in Mathematics

Mathematics is a building process, mastered one step at a time. The foundation of this process consists of a few basic requirements. Those who are successful in mathematics realize the absolute necessity for building a study of mathematics on the firm foundation of these six minimum requirements.

1. Attend class every day.
2. Read the textbook.
3. Take notes in class.
4. Do assigned homework every day.
5. Get help immediately when needed.
6. Review regularly.

1.4 Exponents

1 *Writing Numbers in Exponent Form*

In mathematics, we use exponents as a way to abbreviate repeated multiplication.

Long Notation		**Exponent Form**
$2 \cdot 2 \cdot 2 \cdot 2 \cdot 2 \cdot 2$	$=$	2^6

There are two parts to exponent notation: (1) the **base** and (2) the **exponent**. The **base** tells you what number is being multiplied and the **exponent** tells you how many times this number is used as a factor. (A *factor*, you recall, is a number being multiplied.)

$$2 \cdot 2 \cdot 2 \cdot 2 \cdot 2 \cdot 2 = 2^6$$

The *base* is 2 (the number being multiplied)

The *exponent* is 6 (the number of times 2 is used as a factor)

If the base is a *positive* real number, the exponent appears to the right and slightly above the level of the number as in, for example, 5^6 and 8^3. If the base is a *negative* real number, then parentheses are used around the number and the exponent appears outside the parentheses. For example, $(-2)(-2)(-2) = (-2)^3$.

In algebra, if we do not know the value of a number, we use a letter to represent the unknown number. We call the letter a **variable**. This is quite useful in the case of exponents. Suppose we do not know the value of a number, but we know the number is multiplied by itself several times. We can represent this with a variable base and a whole-number exponent. For example, when we have an unknown number, represented by the variable x, and this number occurs as a factor four times, we have

$$(x)(x)(x)(x) = x^4.$$

Likewise if an unknown number, represented by the variable w, occurs as a factor five times, we have

$$(w)(w)(w)(w)(w) = w^5.$$

EXAMPLE 1 Write in exponent form.

(a) $9(9)(9)$ **(b)** $13(13)(13)(13)$ **(c)** $(-7)(-7)(-7)(-7)(-7)$
(d) $(-4)(-4)(-4)(-4)(-4)(-4)$ **(e)** $(x)(x)$ **(f)** $(y)(y)(y)$

(a) $9(9)(9) = 9^3$ **(b)** $13(13)(13)(13) = 13^4$

(c) The -7 is used as a factor five times. The answer must contain parentheses. Thus $(-7)(-7)(-7)(-7)(-7) = (-7)^5$.

(d) $(-4)(-4)(-4)(-4)(-4)(-4) = (-4)^6$ **(e)** $(x)(x) = x^2$ **(f)** $(y)(y)(y) = y^3$

Practice Problem 1 Write in exponent form.

(a) $6(6)(6)(6)$ **(b)** $(-2)(-2)(-2)(-2)(-2)$ **(c)** $108(108)(108)$
(d) $(-11)(-11)(-11)(-11)(-11)(-11)$ **(e)** $(w)(w)(w)$ **(f)** $(z)(z)(z)(z)$

If the base has an exponent of 2, we say the base is **squared**.

If the base has an exponent of 3, we say the base is **cubed**.

If the base has an exponent greater than 3, we say the base is raised **to the (exponent)-th power**.

x^2 is read "x squared."

y^3 is read "y cubed."

3^6 is read "three to the sixth power" or simply "three to the sixth."

② *Evaluating Numerical Expressions That Contain Exponents*

EXAMPLE 2 Evaluate. **(a)** 2^5 **(b)** $2^3 + 4^4$

(a) $2^5 = (2)(2)(2)(2)(2) = 32$

(b) First we evaluate each power.
$2^3 = 8 \qquad 4^4 = 256$
Then we add. $8 + 256 = 264$

Practice Problem 2 Evaluate. **(a)** 3^5 **(b)** $2^2 + 3^3$

If the base is negative, be especially careful in determining the sign. Notice the following:

$$(-3)^2 = (-3)(-3) = +9 \qquad (-3)^3 = (-3)(-3)(-3) = -27$$

From Section 1.3 we know that when you multiply two or more real numbers, first you multiply their absolute values.

- The result is positive if there are an even number of negative signs.
- The result is negative if there are an odd number of negative signs.

> **Sign Rule for Exponents**
>
> Suppose that a number is written in exponent form and the base is negative. The result is **positive** if the exponent is **even**. The result is **negative** if the exponent is **odd**.

Be careful how you read expressions with exponents and negative signs.

$$(-3)^4 \text{ means } (-3)(-3)(-3)(-3) \text{ or } +81.$$
$$-3^4 \text{ means } -(3)(3)(3)(3) \text{ or } -81.$$

> ## Calculator
>
> **Exponents**
> You can use a calculator to evaluate 3^5. Press the following keys:
>
> | 3 | y^x | 5 | = |
>
> The display should read
>
> | 243 |
>
> Try the following.
> **(a)** 4^6 **(b)** $(0.2)^5$
> **(c)** 18^6 **(d)** 3^{12}

EXAMPLE 3 Evaluate. **(a)** $(-2)^3$ **(b)** $(-4)^6$ **(c)** -3^6 **(d)** $-(5^4)$

(a) $(-2)^3 = -8$ — The answer is negative since the base is negative and the exponent 3 is odd.

(b) $(-4)^6 = +4096$ — The answer is positive since the exponent 6 is even.

(c) $-3^6 = -729$ — The negative sign is not contained in parentheses. Thus we find 3 raised to the sixth power and then take the negative of that value.

(d) $-(5^4) = -625$ — The negative sign is outside the parentheses.

Practice Problem 3 Evaluate. **(a)** $(-3)^3$ **(b)** $(-2)^6$ **(c)** -2^4 **(d)** $-(3^6)$

EXAMPLE 4 Evaluate.

(a) $\left(\dfrac{1}{2}\right)^4$ **(b)** $(0.2)^4$ **(c)** $\left(\dfrac{2}{5}\right)^3$ **(d)** $(3)^3(2)^5$ **(e)** $2^3 - 3^4$

(a) $\left(\dfrac{1}{2}\right)^4 = \left(\dfrac{1}{2}\right)\left(\dfrac{1}{2}\right)\left(\dfrac{1}{2}\right)\left(\dfrac{1}{2}\right) = \dfrac{1}{16}$ **(b)** $(0.2)^4 = (0.2)(0.2)(0.2)(0.2) = 0.0016$

(c) $\left(\dfrac{2}{5}\right)^3 = \left(\dfrac{2}{5}\right)\left(\dfrac{2}{5}\right)\left(\dfrac{2}{5}\right) = \dfrac{8}{125}$ **(d)** First we evaluate each power.
$3^3 = 27 \qquad 2^5 = 32$
Then we multiply. $(27)(32) = 864$

(e) $2^3 - 3^4 = 8 - 81 = -73$

Practice Problem 4 Evaluate.

(a) $\left(\dfrac{1}{3}\right)^3$ **(b)** $(0.3)^4$ **(c)** $\left(\dfrac{3}{2}\right)^4$ **(d)** $(3)^4(4)^2$ **(e)** $4^2 - 2^4$

1.4 Exercises

Verbal and Writing Skills

1. Explain in your own words how to evaluate 4^4.

2. Explain in your own words how to evaluate 9^2.

3. Explain how you would determine whether $(-5)^3$ is negative or positive.

4. Explain how you would determine whether $(-2)^5$ is negative or positive.

5. Explain the difference between $(-2)^4$ and -2^4. What answers do you obtain when you evaluate the expressions?

6. Explain the difference between $(-3)^4$ and -3^4. What answers do you obtain when you evaluate the expressions?

Evaluate.

7. 3^3

8. 4^2

9. 3^4

10. 8^3

11. 7^3

12. 5^4

13. $(-3)^3$

14. $(-2)^3$

15. $(-2)^6$

16. $(-3)^4$

17. -5^3

18. -4^3

19. $\left(\dfrac{1}{4}\right)^2$

20. $\left(\dfrac{1}{2}\right)^3$

21. $\left(\dfrac{2}{5}\right)^3$

22. $\left(\dfrac{2}{3}\right)^4$

23. $(0.9)^2$

24. $(0.4)^2$

25. $(0.2)^4$

26. $(0.7)^3$

27. $(-16)^2$

28. $(-7)^4$

29. -16^2

30. -7^4

Write in exponent form.

31. $(6)(6)(6)(6)(6)$

32. $(8)(8)(8)(8)(8)(8)$

33. $(w)(w)$

34. $(w)(w)(w)$

35. $(x)(x)(x)(x)$

36. $(x)(x)(x)(x)(x)$

37. $(3q)(3q)(3q)$

38. $(6x)(6x)(6x)(6x)$

Evaluate.

39. $5^3 + 6^2$

40. $7^2 + 6^3$

41. $(-3)^3 - (8)^2$

42. $(-2)^3 - (-5)^4$

43. $5^3 - (-3)^3$

44. $9^2 - (-4)^2$

45. $(-4)^3(-3)^2$

46. $(-7)^3(-2)^4$

47. $8^2(-2)^3$

48. $9^2(-3)^3$

49. 4^{12}

50. 6^{11}

To Think About

51. What number to the seventh power equals 128?

52. What number to the fifth power equals -243?

Cumulative Review Problems

Evaluate.

53. $(-11) + (-13) + 6 + (-9) + 8$

54. $\dfrac{3}{4} \div \left(-\dfrac{9}{20}\right)$

55. $-17 - (-9)$

56. $(-2.1)(-1.2)$

57. $\left(\dfrac{2}{5}\right)\left(-\dfrac{1}{4}\right)\left(\dfrac{10}{11}\right)\left(-\dfrac{3}{4}\right)$

1.5 Order of Arithmetic Operations

1 Using the Order of Operations

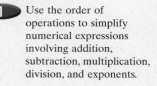

It is important to know *when* to do certain operations as well as how to do them. For example, to simplify the expression $2 - 4 \cdot 3$, should we subtract first or multiply first?

The following list will assist you. It tells which operations to do first: the correct **order of operations**. You might think of it as a *list of priorities*.

Order of Operations for Numbers

Follow this order of operations:

Do first
1. Combine numbers inside parentheses.
2. Raise numbers to a power.
3. Multiply and divide numbers from left to right.

Do last
4. Add and subtract numbers from left to right.

Let's return to the problem $2 - 4 \cdot 3$. There are no parentheses or numbers raised to a power, so multiplication comes next. We do that first. Then we subtract since this comes last on our list.

$$2 - 4 \cdot 3 = 2 - 12 \quad \text{Follow the order of operations by first multiplying } 4 \cdot 3 = 12.$$
$$= -10 \quad \text{Combine } 2 - 12 = -10.$$

EXAMPLE 1 Evaluate. $8 \div 2 \cdot 3 + 4^2$

$$8 \div 2 \cdot 3 + 4^2 = 8 \div 2 \cdot 3 + 16 \quad \text{Evaluate } 4^2 = 16 \text{ because the highest priority in this problem is raising to a power.}$$
$$= 4 \cdot 3 + 16 \quad \text{Next multiply and divide from left to right. So } 8 \div 2 = 4 \text{ and } 4 \cdot 3 = 12.$$
$$= 12 + 16$$
$$= 28 \quad \text{Finally, add.}$$

Practice Problem 1 Evaluate. $25 \div 5 \cdot 6 + 2^3$

Note: Multiplication and division have equal priority. We do not do multiplication first. Rather, we work from left to right, doing any multiplication or division that we encounter. Similarly, addition and subtraction have equal priority.

EXAMPLE 2 Evaluate. $(-3)^3 - 2^4$

The highest priority is to raise the expressions to the appropriate powers.

$$(-3)^3 - 2^4 = -27 - 16 \quad \text{In } (-3)^3 \text{ we are cubing the number } -3 \text{ to obtain } -27. \text{ Be careful; } -2^4 \text{ is not } (-2)^4! \text{ Raise 2 to the fourth power and take the negative of the result.}$$
$$= -43 \quad \text{The last step is to add and subtract from left to right.}$$

Practice Problem 2 Evaluate. $(-4)^3 - 2^6$

EXAMPLE 3 Evaluate. $2 \cdot (2 - 3)^3 + 6 \div 3 + (8 - 5)^2$

$2 \cdot (2 - 3)^3 + 6 \div 3 + (8 - 5)^2$ Combine the numbers inside the parentheses.
$= 2 \cdot (-1)^3 + 6 \div 3 + 3^2$
$= 2 \cdot (-1) + 6 \div 3 + 9$ Next, raise to a power. Note that we need parentheses for -1 because of the negative sign, but they are not needed for 3.

$= -2 + 2 + 9$ Next, multiply and divide from left to right.
$= 9$ Finally, add and subtract from left to right.

Practice Problem 3 Evaluate. $6 - (8 - 12)^2 + 8 \div 2$

EXAMPLE 4 Evaluate. $\left(-\frac{1}{5}\right)\left(\frac{1}{2}\right) - \left(\frac{3}{2}\right)^2$

The highest priority is to raise $\frac{3}{2}$ to the second power.

$$\left(\frac{3}{2}\right)^2 = \left(\frac{3}{2}\right)\left(\frac{3}{2}\right) = \frac{9}{4}$$ Next we multiply.

$$\left(-\frac{1}{5}\right)\left(\frac{1}{2}\right) - \left(\frac{3}{2}\right)^2 = \left(-\frac{1}{5}\right)\left(\frac{1}{2}\right) - \frac{9}{4}$$

$$= -\frac{1}{10} - \frac{9}{4}$$

$$= -\frac{1 \cdot 2}{10 \cdot 2} - \frac{9 \cdot 5}{4 \cdot 5}$$ We need to write each fraction as an equivalent fraction with the LCD of 20.

$$= -\frac{2}{20} - \frac{45}{20}$$

$$= -\frac{47}{20}$$ Add.

Practice Problem 4 Evaluate. $\left(-\frac{1}{7}\right)\left(-\frac{14}{5}\right) + \left(-\frac{1}{2}\right) \div \left(\frac{3}{4}\right)$

Calculator

Order of Operations
Use your calculator to evaluate $3 + 4 \cdot 5$. Enter

3 $+$ 4 $\times$ 5 $=$

If the display is ⟨ 23 ⟩, the correct order of operations is built in. If the display is not 23, you will need to modify the way you enter the problem. You should use

4 $\times$ 5 $+$ 3 $=$

Try $6 + 3 \cdot 4 - 8 \div 2$.

Developing Your Study Skills

Previewing New Material

Part of your study time each day should consist of looking over the sections in your text that are to be covered the following day. You do not necessarily need to study and learn the material on your own, but a survey of the concepts, terminology, diagrams, and examples will help the new ideas seem more familiar as the instructor presents them. You can look for concepts that appear confusing or difficult and be ready to listen carefully for your instructor's explanations. You can be prepared to ask the questions that will increase your understanding. Previewing new material enables you to see what is coming and prepares you to be ready to absorb it.

1.5 Exercises

Verbal and Writing Skills

You have lost a game of UNO and are counting the points left in your hand. You announce that you have three fours and six fives.

1. Write this as a number expression.

2. How many points have you in your hand?

3. What answer would you get for the number expression if you simplified it by
(a) performing the operations from left to right?

(b) following the order of operations?

4. Which procedure in exercise 3 gives the correct number of total points?

Evaluate.

5. $(2 - 5)^2 \div 3 \times 4$

6. $2(3 - 5 + 6) + 5$

7. $8 - 2^3 \cdot 5 + 3$

8. $-14 \div (-7) - 8 \cdot 2 + 3^3$

9. $4 + 27 \div 3 \cdot 2 - 8$

10. $3 \cdot 5 + 7 \cdot 3 - 5 \cdot 3$

11. $8 - 5(2)^3 \div (-8)$

12. $11 - 3(4)^2 \div (-6)$

13. $3(5 - 7)^2 - 6(3)$

14. $-2(3 - 6)^2 - (-2)$

15. $5 \cdot 6 - (3 - 5)^2 + 8 \cdot 2$

16. $(-3)^2 \cdot 6 \div 9 + 4 \cdot 2$

17. $\dfrac{1}{2} \div \dfrac{2}{3} + 6 \cdot \dfrac{1}{4}$

18. $\dfrac{5}{6} \div \dfrac{2}{3} - 6 \cdot \left(\dfrac{1}{2}\right)^2$

19. $0.8 + 0.3(0.6 - 0.2)^2$

20. $0.05 + 1.4 - (0.5 - 0.7)^3$

21. $\dfrac{3}{4}\left(-\dfrac{2}{5}\right) - \left(-\dfrac{3}{5}\right)$

22. $-\dfrac{2}{3}\left(\dfrac{3}{5}\right) + \dfrac{5}{7} \div \dfrac{5}{3}$

23. $-6.3 - (-2.7)(1.1) + (3.3)^2$

24. $4.35 + 8.06 \div (-2.6) - (2.1)^2$

25. $\left(\dfrac{1}{2}\right)^3 + \dfrac{1}{4} - \left(\dfrac{1}{6} - \dfrac{1}{12}\right) - \dfrac{2}{3} \cdot \left(\dfrac{1}{4}\right)^2$

26. $(2.4 \cdot 1.2)^2 - 1.6 \cdot 2.2 \div 4.0 - 3.6$

Cumulative Review Problems *Simplify.*

27. $(0.5)^3$

28. $-\dfrac{3}{4} - \dfrac{5}{6}$

29. -1^{20}

30. $\dfrac{18}{5} \div \dfrac{3}{5}$

31. An Olympic weight lifter has been told by his trainer to increase his daily consumption of protein by 15 grams. The trainer suggested a health drink that provides 2 grams of protein for every 6 ounces consumed. How many ounces of this health drink would the weight lifter need to consume daily to reach his goal of 15 additional grams per day?

Using the Distributive Property to Simplify Expressions

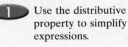
1 *Using the Distributive Property to Simplify Expressions*

As we learned previously, we use letters called *variables* to represent unknown numbers. If a number is multiplied by a variable we do not need any symbol between the number and variable. Thus, to indicate $(2)(x)$, we write $2x$. To indicate $3 \cdot y$, we write $3y$. If one variable is multiplied by another variable, we place the variables next to each other. Thus, $(a)(b)$ is written ab. We use exponent form if an unknown number (a variable) is used several times as a factor. Thus, $x \cdot x \cdot x = x^3$. Similarly, $(y)(y)(y)(y) = y^4$.

In algebra, we need to be familiar with several definitions. We will use them throughout the remainder of this book. Take some time to think through how each of these definitions is used.

An **algebraic expression** is a quantity that contains numbers and variables, such as $a + b$, $2x - 3$, and $5ab^2$. In this chapter we will be learning rules about adding and multiplying algebraic expressions. A **term** is a number, a variable, or a product of numbers and variables. 17, x, $5xy$, and $22xy^3$ are all examples of terms. We will refer to terms when we discuss the distributive property.

An important property of algebra is the **distributive property**. We can state it in an equation as follows:

Distributive Property

For all real numbers a, b, and c,

$$a(b + c) = ab + ac.$$

A numerical example shows that it does seem reasonable.

$$5(3 + 6) = 5(3) + 5(6)$$
$$5(9) = 15 + 30$$
$$45 = 45$$

We can use the distributive property to multiply any term by the sum of two or more terms. In Section 1.4, we defined the word *factor*. Two or more algebraic expressions that are multiplied are called **factors**. Consider the following examples of multiplying algebraic expressions.

EXAMPLE 1 Multiply. **(a)** $5(a + b)$ **(b)** $-1(3x + 2y)$

(a) $5(a + b) = 5a + 5b$ Multiply the factor $(a + b)$ by the factor 5.

(b) $-1(3x + 2y) = -1(3x) + (-1)(2y)$ Multiply the factor $(3x + 2y)$ by the factor -1.
$$= -3x - 2y$$

Practice Problem 1 Multiply.

(a) $-3(x + 2y)$ **(b)** $-a(a - 3b)$

If the parentheses are preceded by a negative sign, we consider this to be the product of (-1) and the expression inside the parentheses.

EXAMPLE 2 Multiply. $-(a - 2b)$

$$-(a - 2b) = (-1)(a - 2b) = (-1)(a) + (-1)(-2b) = -a + 2b$$

Practice Problem 2 Multiply. $-(-3x + y)$

In general, we see that in all these examples we have multiplied each term of the expression in the parentheses by the expression in front of the parentheses.

EXAMPLE 3 Multiply. **(a)** $\frac{2}{3}(x^2 - 6x + 8)$ **(b)** $1.4(a^2 + 2.5a + 1.8)$

(a) $\frac{2}{3}(x^2 - 6x + 8) = \left(\frac{2}{3}\right)(1x^2) + \left(\frac{2}{3}\right)(-6x) + \left(\frac{2}{3}\right)(8)$

$$= \frac{2}{3}x^2 + (-4x) + \frac{16}{3}$$

$$= \frac{2}{3}x^2 - 4x + \frac{16}{3}$$

(b) $1.4(a^2 + 2.5a + 1.8) = 1.4(1a^2) + (1.4)(2.5a) + (1.4)(1.8)$

$$= 1.4a^2 + 3.5a + 2.52$$

Practice Problem 3 Multiply.

(a) $\frac{3}{5}(a^2 - 5a + 25)$ **(b)** $2.5(x^2 - 3.5x + 1.2)$

There are times we multiply a variable by itself and use exponent notation. For example, $(x)(x) = x^2$ and $(x)(x)(x) = x^3$. In other cases there will be numbers and variables multiplied at the same time.

We will see problems like $(2x)(x) = (2)(x)(x) = 2x^2$. Some expressions will involve multiplication of more than one variable. We will see problems like $(3x)(xy) = (3)(x)(x)(y) = 3x^2y$. There will be times when we use the distributive property and all of these methods will be used. For example,

$$2x(x - 3y + 2) = 2x(x) + (2x)(-3y) + (2x)(2)$$
$$= 2x^2 + (-6)(xy) + 4(x)$$
$$= 2x^2 - 6xy + 4x.$$

We will discuss this type of multiplication of variables with exponents in more detail in Section 5.1. At that point we will expand these examples and other similar examples to develop the general rule for multiplication $(x^a)(x^b) = x^{a+b}$.

EXAMPLE 4 Multiply. $-2x(3x + y - 4)$

$$-2x(3x + y - 4) = -2(x)(3)(x) + (-2)(x)(y) + (-2)(x)(-4)$$
$$= -2(3)(x)(x) + (-2)(xy) + (-2)(-4)(x)$$
$$= -6x^2 - 2xy + 8x$$

Practice Problem 4 Multiply. $-4x(x - 2y + 3)$

The distributive property can also be presented with the a on the right.

$$(b + c)a = ba + ca$$

The a is "distributed" over the b and c inside the parentheses.

EXAMPLE 5   Multiply. $(2x^2 - x)(-3)$

$$(2x^2 - x)(-3) = 2x^2(-3) + (-x)(-3)$$
$$= -6x^2 + 3x$$

Practice Problem 5   Multiply.

$$(3x^2 - 2x)(-4)$$

EXAMPLE 6   A farmer has a rectangular field that is 300 feet wide. One portion of the field is $2x$ feet long. The other portion of the field is $3y$ feet long. Use the distributive property to find an expression for the area of this field.

First we draw a picture of a field that is 300 feet wide and $2x + 3y$ feet long.

To find the area of the field, we multiply the width times the length.

$$300(2x + 3y) = 300(2x) + 300(3y) = 600x + 900y$$

Thus the area of the field in square feet is $600x + 900y$.

Practice Problem 6   A farmer has a rectangular field that is 400 feet wide. One portion of the field is $6x$ feet long. The other portion of the field is $9y$ feet long. Use the distributive property to find an expression for the area of this field.

Developing Your Study Skills

How to Do Homework

As you begin your homework assignments, read the directions carefully. You need to understand what is being asked for. Concentrate on each exercise, taking time to solve it accurately. Rushing through your work usually causes errors. Check your answers with those given in the back of the textbook. If your answer is incorrect, check to see that you are doing the right exercise. Redo the exercise, watching for little errors. If it is still wrong, check with a friend. Perhaps the two of you can figure it out.

Verbal and Writing Skills

In exercises 1 and 2, complete each sentence by filling in the blank.

1. A _____ is a symbol used to represent an unknown number.

2. When we write an expression with numbers and variables such as $7x$, it indicates that we are _____ 7 by x.

3. Explain in your own words how we multiply a problem like $(4x)(x)$.

4. Explain why you think the property $a(b + c) = ab + ac$ is called the distributive property. What does distribute mean?

5. Does the following distributive property work? $a(b - c) = ab - ac$ Why or why not? Give an example.

Multiply. Use the distributive property.

6. $-(a - 2b)$

7. $-(x + 4y)$

8. $-2(4a - 3b)$

9. $-3(2a - 5b)$

10. $3(3x - y + 5)$

11. $2(4x + y - 2)$

12. $-2b(a - 3b + c)$

13. $-3x(-2x + 3y - z)$

14. $3x(x - 3y - 7)$

15. $2x(4x - y - 6)$

16. $-9(9x - 5y + 8)$

17. $-5(3x + 9 - 7y)$

18. $\frac{1}{3}(3x^2 + 2x - 1)$

19. $\frac{1}{4}(x^2 + 2x - 8)$

20. $\frac{5}{6}(12x^2 - 24x + 18)$

21. $\frac{2}{3}(-27a^4 + 9a^2 - 21)$

22. $\frac{x}{5}(x + 10y - 4)$

23. $\frac{y}{3}(3y - 4x - 6)$

24. $5x(x + 2y + z - 1)$

25. $3a(2a + b - c - 4)$

26. $(2x - 3)(-2)$

27. $(5x + 1)(-4)$

28. $2x(3x + y - 4)$

29. $3x(4x - 5y - 6)$

30. $(3x + 2y - 1)(-xy)$

31. $(4a - 2b - 1)(-ab)$

32. $(2x + 3y - 2)3xy$

33. $(-2x + y - 3)4xy$

34. $1.5(2.8x^2 + 3.0x - 2.5)$

35. $2.5(1.5a^2 - 3.5a + 2.0)$

36. $-0.3x(-1.2x^2 - 0.3x + 0.5)$

37. $-0.9q(2.1q - 0.2r - 0.8s)$

38. $0.5x(0.6x + 0.8y - 5)$

Applications

▲ **39.** Blaine Johnson has a rectangular field that is 700 feet wide. One portion of the field is $12x$ feet long. The other portion of the field is $8y$ feet long. Use the distributive property to find an expression for the area of this field.

▲ **40.** Kathy DesMaris has a rectangular field that is 800 feet wide. One portion of the field is $5x$ feet long. The other portion of the field is $14y$ feet long. Use the distributive property to find an expression for the area of this field

To Think About

▲ **41.** The athletic field at Baxford College is $3x$ feet wide. It used to be 1500 feet long. However, due to the construction of a new dormitory, the field length was decreased by $4y$ feet. Use the distributive property to find an expression for the area of the new field after construction of the dormitory.

▲ **42.** The Beverly Airport runway is $4x$ feet wide. The airport was supposed to have a 3000-foot-long runway. However, some of the land was wetland, so a runway could not be built on all of it. Therefore, the length of the runway was decreased by $2y$ feet. Use the distributive property to find an expression for the area of the final runway.

Cumulative Review Problems

In exercises 43–47, evaluate.

43. $-18 + (-20) + 36 + (-14)$

44. $(-2)^6$

45. $-27 - (-41)$

46. $4 - (-2)^2$

47. $3^3 \div (2 - 11) - 2 \cdot 5$

1 Identifying Like Terms

Student Learning Objectives

After studying this section, you will be able to:

1 Identify like terms.

2 Combine like terms.

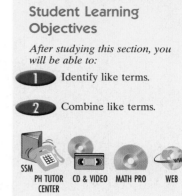

SSM
PH TUTOR CD & VIDEO MATH PRO WEB
CENTER

We can add or subtract quantities that are *like quantities*. This is called **combining** like quantities.

$$5 \text{ inches} + 6 \text{ inches} = 11 \text{ inches}$$
$$20 \text{ square inches} - 16 \text{ square inches} = 4 \text{ square inches}$$

However, we cannot combine things that are not the same.

$$16 \text{ square inches} - 4 \text{ inches} \text{ (Cannot be done!)}$$

Similarly, in algebra we can **combine like terms**. This means to add or subtract like terms. Remember, we cannot combine terms that are not the same. Recall that a *term* is a number, a variable, or a product of numbers and variables. **Like terms** are terms that have identical variables and exponents. In other words, like terms must have exactly the same letter parts.

EXAMPLE 1 List the like terms of each expression.

(a) $5x - 2y + 6x$ **(b)** $2x^2 - 3x - 5x^2 - 8x$

(a) $5x$ and $6x$ are like terms. These are the only like terms.

(b) $2x^2$ and $-5x^2$ are like terms.
$-3x$ and $-8x$ are like terms.
Note that x^2 and x are not like terms.

Practice Problem 1 List the like terms of each expression.

(a) $5a + 2b + 8a - 4b$ **(b)** $x^2 + y^2 + 3x - 7y^2$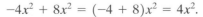

Do you really understand what a term is? A term is a number, a variable, or a product of numbers and variables. Terms are the parts of an algebraic expression separated by plus signs. The sign in front of the term is considered part of the term.

2 Combining Like Terms

It is important to know how to combine like terms. Since

$$4 \text{ inches} + 5 \text{ inches} = 9 \text{ inches},$$

we would expect in algebra that

$$4x + 5x = 9x.$$

Why is this true? Let's take a look at the distributive property.

> Like terms may be added or subtracted by using the distributive property:
> $$ab + ac = a(b + c) \quad \text{and} \quad ba + ca = (b + c)a$$

For example,

$$-7x + 9x = (-7 + 9)x = 2x$$
$$5x^2 + 12x^2 = (5 + 12)x^2 = 17x^2.$$

EXAMPLE 2 Combine like terms. **(a)** $-4x^2 + 8x^2$ **(b)** $5x + 3x + 2x$

(a) Notice that each term contains the factor x^2. Using the distributive property, we have

$$-4x^2 + 8x^2 = (-4 + 8)x^2 = 4x^2.$$

(b) Note that each term contains the factor x. Using the distributive property, we have

$$5x + 3x + 2x = (5 + 3 + 2)x = 10x.$$

Practice Problem 2   Combine like terms.

(a) $5a + 7a + 4a$ **(b)** $16y^3 + 9y^3$

In this section, the direction *simplify* means to remove parentheses and/or combine like terms.

EXAMPLE 3   Simplify. $5a^2 - 2a^2 + 6a^2$

$$5a^2 - 2a^2 + 6a^2 = (5 - 2 + 6)a^2 = 9a^2$$

Practice Problem 3   Simplify. $-8y^2 - 9y^2 + 4y^2$

After doing a few problems, you will find that it is not necessary to write out the step of using the distributive property. We will omit this step for the remaining examples in this section.

EXAMPLE 4   Simplify.   **(a)** $5a + 2b + 7a - 6b$
(b) $3x^2y - 2xy^2 + 6x^2y$ **(c)** $2a^2b + 3ab^2 - 6a^2b^2 - 8ab$

(a) $5a + 2b + 7a - 6b = 12a - 4b$ We combine the a terms and the b terms separately.

(b) $3x^2y - 2xy^2 + 6x^2y = 9x^2y - 2xy^2$ **Note:** x^2y and xy^2 are not like terms because of different powers.

(c) $2a^2b + 3ab^2 - 6a^2b^2 - 8ab$ These terms cannot be combined; there are no like terms in this expression.

Practice Problem 4   Simplify.   **(a)** $-x + 3a - 9x + 2a$
(b) $5ab - 2ab^2 - 3a^2b + 6ab$ **(c)** $7x^2y - 2xy^2 - 3x^2y - 4xy^2 + 5x^2y$

The two skills in this section that a student must practice are identifying like terms and correctly adding and subtracting like terms. If a problem involves many terms, you may find it helpful to rearrange the terms so that like terms are together.

EXAMPLE 5   Simplify. $3a - 2b + 5a^2 + 6a - 8b - 12a^2$

There are three pairs of like terms.

$$3a + 6a - 2b - 8b + 5a^2 - 12a^2$$ You can rearrange the terms so that like
$$\underbrace{\qquad}_{a \text{ terms}} \quad \underbrace{\qquad}_{b \text{ terms}} \quad \underbrace{\qquad}_{a^2 \text{ terms}}$$ terms are together, making it easier to combine them.
$$= 9a - 10b - 7a^2$$ Combine like terms.

The order of terms in an answer to this problem is not significant. These three terms can be rearranged in a different order. $-10b + 9a - 7a^2$ and $-7a^2 + 9a - 10b$ are also correct. However, we usually write polynomials in order of descending exponents. $-7a^2 + 9a - 10b$ would be the preferred way to write the answer.

Practice Problem 5   Simplify. $5xy - 2x^2y + 6xy^2 - xy - 3xy^2 - 7x^2y$

EXAMPLE 6 Simplify. $6(2x + 3xy) - 8x(3 - 4y)$

First remove the parentheses; then combine like terms.

$$6(2x + 3xy) - 8x(3 - 4y) = 12x + 18xy - 24x + 32xy$$ Use the distributive property.

$$= -12x + 50xy$$ Combine like terms.

Practice Problem 6 Simplify. $5a(2 - 3b) - 4(6a + 2ab)$

Use extra care with fractional values.

EXAMPLE 7 Simplify. $\dfrac{3}{4}x^2 - 5y - \dfrac{1}{8}x^2 + \dfrac{1}{3}y$

We need a least common denominator for the x^2 terms, which is 8.

Change $\dfrac{3}{4}$ to eighths by multiplying the numerator and denominator by 2.

$$\frac{3}{4}x^2 - \frac{1}{8}x^2 = \frac{3 \cdot 2}{4 \cdot 2}x^2 - \frac{1}{8}x^2 = \frac{6}{8}x^2 - \frac{1}{8}x^2 = \frac{5}{8}x^2$$

The least common denominator for the y terms is 3. Change 5 to thirds.

$$-\frac{5}{1}y + \frac{1}{3}y = \frac{-5 \cdot 3}{1 \cdot 3}y + \frac{1}{3}y = \frac{-15}{3}y + \frac{1}{3}y = -\frac{14}{3}y$$

Thus, our solution is $\dfrac{5}{8}x^2 - \dfrac{14}{3}y$.

Practice Problem 7 Simplify. $\dfrac{1}{7}a^2 - \dfrac{5}{12}b + 2a^2 - \dfrac{1}{3}b$

Verbal and Writing Skills

1. Explain in your own words the mathematical meaning of the word *term*.

2. Explain in your own words the mathematical meaning of the phrase *like terms*.

3. Explain which terms are like terms in the expression $5x - 7y - 8x$.

4. Explain which terms are like terms in the expression $12a - 3b - 9a$.

5. Explain which terms are like terms in the expression $7xy - 9x^2y - 15xy^2 - 14xy$.

6. Explain which terms are like terms in the expression $-3a^2b - 12ab + 5ab^2 + 9ab$.

Simplify.

7. $-14b^2 - 11b^2$

8. $-17x^5 + 3x^5$

9. $10x^4 + 8x^4 + 7x^2$

10. $3a^3 - 6a^2 + 5a^3$

11. $2ab + 1 - 6ab - 8$

12. $2x^2 + 3x^2 - 7 - 5x^2$

13. $1.3x - 2.6y + 5.8x - 0.9y$

14. $3.1ab - 0.2b - 0.8ab + 5.3b$

15. $1.6x - 2.8y - 3.6x - 5.9y$

16. $1.9x - 2.4b - 3.8x - 8.2b$

17. $\dfrac{1}{2}x^2 - 3y - \dfrac{1}{3}y + \dfrac{1}{4}x^2$

18. $\dfrac{1}{5}a^2 - 2b - \dfrac{1}{2}a^2 - 3b$

19. $\dfrac{1}{3}x - \dfrac{2}{3}y - \dfrac{2}{5}x + \dfrac{4}{7}y$

20. $\dfrac{2}{5}s - \dfrac{3}{8}t - \dfrac{4}{15}s - \dfrac{5}{12}t$

21. $3p - 4q + 2p + 3 + 5q - 21$

22. $6x - 5y - 3y + 7 - 11x - 5$

23. $5x^2y - 10xy^6 + 6xy^2 - 7xy^2$

24. $5bcd - 8cd - 12bcd + cd$

25. $2ab + 5bc - 6ac - 2ab$

26. $5x^2y + 12xy^2 - 8x^2 - 12xy^2$

27. $2x^2 - 3x - 5 - 7x + 8 - x^2$

28. $5x + 7 - 6x^3 + 6 - 11x + 4x^3$

29. $2y^2 - 8y + 9 - 12y^2 - 8y + 3$

30. $5 - 2y^2 + 3y - 8y - 9y^2 - 12$

31. $ab + 3a - 4ab + 2a - 8b$

Simplify. Use the distributive property to remove parentheses; then combine like terms.

32. $5(a - 3b) + 2(-b - 4a)$

33. $3(x + y) - 5(-2y + 3x)$

34. $-3b(5a - 3b) + 4(-3ab - 5b^2)$

35. $2x(x - 3y) - 4(-3x^2 - 2xy)$

36. $-3(x^2 + 3y) + 5(-6y - x^2)$

37. $-3(7xy - 11y^2) - 2y(-2x + 3y)$

38. $4(2 - x) - 3(-5 - 12x)$

39. $7(3 - x) - 6(8 - 13x)$

To Think About

▲ **40.** A triangle has sides of length $2a$ centimeters, $7b$ centimeters, and $5a + 3$ centimeters. What is the perimeter of the triangle?

▲ **41.** A rectangle has sides of length $7x - 2$ meters and $3x + 4$ meters. What is the perimeter of the rectangle?

▲ **42.** A square has a side of length $9x - 2$ inches. Each side is shortened by 3 inches. What is the perimeter of the new smaller square?

▲ **43.** A triangle has sides of length $4a - 5$ feet, $3a + 8$ feet, and $9a + 2$ feet. Each side is doubled in length. What is the perimeter of the new enlarged triangle?

Cumulative Review Problems

Evaluate.

44. $-\dfrac{1}{3} - \left(-\dfrac{1}{5}\right)$

45. $\left(-\dfrac{5}{3}\right)\left(\dfrac{1}{2}\right)$

46. $\dfrac{4}{5} + \left(-\dfrac{1}{25}\right) + \left(-\dfrac{3}{10}\right)$

47. $\left(\dfrac{5}{7}\right) \div \left(-\dfrac{14}{3}\right)$

48. A 2-quart container of orange juice produces 9.5 average servings. Two quarts is approximately 1.9 liters. How many liters are in one serving?

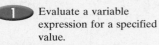 **Evaluating a Variable Expression for a Specified Value**

You will use the order of operations to **evaluate** variable expressions. Suppose we are asked to evaluate

$$6 + 3x \text{ for } x = -4.$$

In general, x represents some unknown number. Here we are told x has the value -4. We can replace x with -4. Use parentheses around -4. Note that we always put replacement values in parentheses.

$$6 + 3(-4) = 6 + (-12) = -6$$

When we replace a variable by a particular value, we say we have **substituted** the value for the variable. We then evaluate the expression (that is, find a value for it).

EXAMPLE 1 Evaluate $\frac{2}{3}x - 5$ for $x = -6$.

$$\frac{2}{3}x - 5 = \frac{2}{3}(-6) - 5 \quad \text{Substitute } x = -6. \text{ Be sure to enclose the } -6 \text{ in parentheses.}$$

$$= -4 - 5 \quad \text{Multiply } \left(\frac{2}{3}\right)\left(-\frac{6}{1}\right) = -4.$$

$$= -9 \quad \text{Combine.}$$

Practice Problem 1 Evaluate $4 - \frac{1}{2}x$ for $x = -8$.

Compare parts (a) and (b) in the next example. The two parts illustrate that you must be careful what value you raise to a power. *Note:* In part (b) we will need parentheses within parentheses. To avoid confusion, we use brackets [ ] to represent the outside parentheses.

EXAMPLE 2 **(a)** Evaluate $2x^2$ for $x = -3$. **(b)** Evaluate $(2x)^2$ for $x = -3$.

(a) Here the value x is squared.

$$2x^2 = 2(-3)^2$$
$$= 2(9) \quad \text{First square } -3.$$
$$= 18 \quad \text{Then multiply.}$$

(b) Here the value $(2x)$ is squared.

$$(2x)^2 = [(2)(-3)]^2$$
$$= (-6)^2 \quad \text{First multiply the numbers inside the parentheses.}$$
$$= 36 \quad \text{Then square } -6.$$

Practice Problem 2 Evaluate for $x = -3$. **(a)** $-x^4$ **(b)** $(-x)^4$

Carefully study the solutions to Example 2(a) and Example 2(b). You will find that taking the time to see *how* and *why* they are different is a good investment of study time.

EXAMPLE 3 Evaluate $x^2 + x$ for $x = -4$.

$$x^2 + x = (-4)^2 + (-4) \quad \text{Replace } x \text{ by } -4 \text{ in the original expression.}$$
$$= 16 + (-4) \quad \text{Raise to a power.}$$
$$= 12 \quad \text{Finally, add.}$$

Practice Problem 3 Evaluate $(5x)^3 + 2x$ for $x = -2$.

2   *Evaluating a Formula by Substitution*

We can *evaluate a formula* by substituting values for the variables. For example, the area of a triangle can be found using the formula $A = \frac{1}{2}ab$, where b is the length of the base of the triangle and a is the altitude of the triangle (see figure). If we know values for a and b, we can substitute those values into the formula to find the area. The units for area are *square units*.

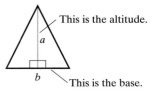

This is the altitude.

This is the base.

Because some of the examples and exercises in this section involve geometry, it may be helpful to review this topic.

However, before we proceed to do any examples, it may be helpful to review a little geometry. The following information is very important. If you have forgotten some of this material (or if you have never learned it), please take the time to learn it completely now. Throughout the entire book we will be using this information in solving applied problems.

Perimeter is the distance around a plane figure. Perimeter is measured in linear units (inches, feet, centimeters, miles). **Area** is a measure of the amount of surface in a region. Area is measured in square units (square inches, square feet, square centimeters).

In our sketches we will show angles of 90° by using a small square ($\llcorner$ $\lrcorner$). This indicates that the two lines are at right angles. All angles that measure 90° are called **right angles**. An **altitude** is perpendicular to the base of a figure. That is, the altitude forms right angles with the base. The small corner square in a sketch helps us identify the altitude of the figure.

The following box provides a handy guide to some facts and formulas you will need to know. Use it as a reference when solving word problems involving geometric figures.

Geometric Formulas: Two-Dimensional Figures

A **parallelogram** is a four-sided figure with opposite sides parallel. In a parallelogram, opposite sides are equal and opposite angles are equal.

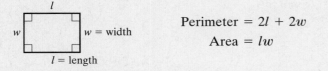

Perimeter = the sum of all four sides

Area = ab

A **rectangle** is a parallelogram with all interior angles measuring 90°.

Perimeter = $2l + 2w$

Area = lw

A **square** is a rectangle with all four sides equal.

Perimeter = $4s$

Area = s^2

A **trapezoid** is a four-sided figure with two sides parallel. The parallel sides are called the *bases* of the trapezoid.

Perimeter = the sum of all four sides

Area = $\dfrac{1}{2}a(b_1 + b_2)$

A **triangle** is a closed plane figure with three sides.

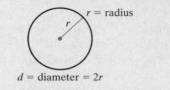

Perimeter = the sum of the three sides

$$\text{Area} = \frac{1}{2}ab$$

A **circle** is a plane curve consisting of all points at an equal distance from a given point called the center.

Circumference is the distance around a circle.

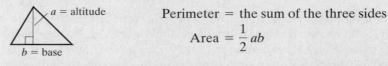

r = radius

d = diameter = $2r$

Circumference = $2\pi r$

Area = πr^2

π (the number *pi*) is a constant associated with circles. It is an irrational number that is approximately 3.141592654. We usually use 3.14 as a sufficiently accurate approximation. Thus we write $\pi \approx 3.14$ for most of our calculations involving π.

▲ ⬭ **EXAMPLE 4**   Find the area of a triangle with a base of 16 centimeters (cm) and a height of 12 centimeters (cm).

Use the formula

$$A = \frac{1}{2}ab.$$

Substitute 12 centimeters for a and 16 centimeters for b.

$$A = \frac{1}{2}(12 \text{ centimeters})(16 \text{ centimeters})$$

$$= \frac{1}{2}(12)(16)(\text{cm})(\text{cm}) \quad \text{If you take } \tfrac{1}{2} \text{ of 12 first, it will make your calculation easier.}$$

$$= (6)(16)(\text{cm})^2 = 96 \text{ square centimeters}$$

The area of the triangle is 96 square centimeters.

▲ **Practice Problem 4**   Find the area of a triangle with an altitude of 3 meters and a base of 7 meters.

The area of a circle is given by

$$A = \pi r^2.$$

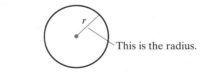

This is the radius.

We will use 3.14 as an approximation for the *irrational number π*.

▲ ⬭ **EXAMPLE 5**   Find the area of a circle if the radius is 2 inches.

$$A = \pi r^2 = (3.14)(2 \text{ inches})^2 \quad \text{Write the formula and substitute the given values for the letters.}$$

$$= (3.14)(4)(\text{in})^2 \quad \text{Raise to a power. Then multiply.}$$

$$= 12.56 \text{ square inches}$$

▲ **Practice Problem 5**   Find the area of a circle if the radius is 3 meters.

The formula $C = \frac{5}{9}(F - 32)$ allows us to find the Celsius temperature if we know the Fahrenheit temperature. That is, we can substitute a value for F in degrees Fahrenheit into the formula to obtain a temperature C in degrees Celsius.

EXAMPLE 6   What is the Celsius temperature when the Fahrenheit temperature is $F = -22°$?
Use the formula.

$$C = \frac{5}{9}(F - 32)$$

$$= \frac{5}{9}((-22) - 32) \quad \text{Substitute} -22 \text{ for } F \text{ in the formula.}$$

$$= \frac{5}{9}(-54) \quad\quad \text{Combine the numbers inside the parentheses.}$$

$$= (5)(-6) \quad\quad \text{Simplify.}$$

$$= -30 \quad\quad\quad \text{Multiply.}$$

The temperature is $-30°$ Celsius.

Practice Problem 6   What is the Celsius temperature when the Fahrenheit temperature is $F = 68°$? Use the formula $C = \frac{5}{9}(F - 32)$.

When driving in Canada or Mexico, we must observe speed limits posted in kilometers per hour. A formula that converts r (miles per hour) to k (kilometers per hour) is $k \approx 1.61r$. Note that this is an approximation.

EXAMPLE 7   You are driving on a highway in Mexico. It has a posted maximum speed of 100 kilometers per hour. You are driving at 61 miles per hour. Are you exceeding the speed limit?
Use the formula.

$$k \approx 1.61r$$

$$= (1.61)(61) \quad \text{Replace } r \text{ by 61.}$$

$$= 98.21 \quad\quad \text{Multiply the numbers.}$$

You are driving at approximately 98 kilometers per hour. You are not exceeding the speed limit.

Practice Problem 7   You are driving behind a heavily loaded truck on a Canadian highway. The highway has a posted minimum speed of 65 kilometers per hour. When you travel at exactly the same speed as the truck ahead of you, you observe that the speedometer reads 35 miles per hour. Assuming that your speedometer is accurate, determine whether the truck is violating the minimum speed law.

Evaluate.

1. $-2x + 1$ for $x = 3$ **2.** $-4x - 2$ for $x = 5$ **3.** $\frac{2}{3}x - 5$ for $x = -9$ **4.** $\frac{3}{4}x + 8$ for $x = -8$

5. $5x + 10$ for $x = \frac{1}{2}$ **6.** $7x + 20$ for $x = -\frac{1}{2}$ **7.** $2 - 4x$ for $x = 7$ **8.** $3 - 5x$ for $x = 8$

9. $x^2 - 3x$ for $x = -2$ **10.** $x^2 + 3x$ for $x = 4$ **11.** $3x^2$ for $x = -1$ **12.** $4x^2$ for $x = -1$

13. $-3x^3$ for $x = 2$ **14.** $-7x^2$ for $x = 5$ **15.** $9x + 13$ for $x = -\frac{3}{4}$ **16.** $5x + 7$ for $x = -\frac{2}{3}$

17. $2x^2 + 3x$ for $x = -3$ **18.** $18 - 5x$ for $x = -3$ **19.** $(2x)^2 + x$ for $x = 3$ **20.** $2 - x^2$ for $x = -2$

21. $2 - (-x)^2$ for $x = -2$ **22.** $2x - 3x^2$ for $x = -4$ **23.** $7x + (2x)^2$ for $x = -3$ **24.** $5x + (3x)^2$ for $x = -2$

25. $x^2 - 7x + 3$ for $x = 3$ **26.** $4x^2 - 3x + 9$ for $x = 2$ **27.** $\frac{1}{2}x^2 - 3x + 9$ for $x = -4$

28. $x^2 - 2y + 3y^2$ for $x = -3$ and $y = 4$ **29.** $2x^2 - 3xy + 2y$ for $x = 4$ and $y = -1$

30. $a^3 + 2abc - 3c^2$ for $a = 5$, $b = 9$, and $c = -1$ **31.** $a^2 - 2ab + 2c^2$ for $a = 3$, $b = 2$, and $c = -4$

Applications

▲ **32.** A sign is made in the shape of a parallelogram. The base measures 22 feet. The altitude measures 16 feet. What is the area of the sign?

▲ **33.** A field is shaped like a parallelogram. The base measures 92 feet. The altitude measures 54 feet. What is the area of the field?

▲ **34.** A square support unit in a television is made with a side measuring 3 centimeters. A new model being designed for next year will have a larger square with a side measuring 3.2 centimeters. By how much will the area of the square be increased?

▲ **35.** A square computer chip for last year's computer had a side measuring 23 millimeters. This year the computer chip has been reduced in size. The new square chip has a side of 20 millimeters. By how much has the area of the chip decreased?

▲ **36.** A carpenter cut out a small trapezoid as a wooden support for the front step. It has an altitude of 4 inches. One base of the trapezoid measures 9 inches and the other base measures 7 inches. What is the area of this support?

▲ **37.** The Media One signal tower has a small trapezoid frame on the top of the tower. The frame has an altitude of 9 inches. One base of the trapezoid is 20 inches and the other base measures 17 inches. What is the area of this small trapezoidal frame?

38. Bradley Palmer State Park has a triangular piece of land on the border. The altitude of the triangle is 400 feet. The base of the triangle is 280 feet. What is the area of this piece of land?

39. The ceiling in the Madisons' house has a leak. The roofer exposed a triangular region that needs to be sealed and then reroofed. The region has an altitude of 14 feet. The base of the region is 19 feet. What is the area of the region that needs to be reroofed?

40. The radius of a circular opening of a chemistry flask is 4 cm. What is the area of the opening?

41. An ancient outdoor sundial has a radius of 5 meters. What is its area?

For exercises 42 and 43, use the formula $F = \frac{9}{5}C + 32$ to find the Fahrenheit temperature.

42. Find the Fahrenheit temperature F when the Celsius temperature is 25°C.

43. It is not uncommon for parts of North Dakota to have a temperature of −10°C. Find the Fahrenheit temperature.

Solve.

44. Find the total cost of making a triangular sail that has a base dimension of 12 feet and a height of 20 feet if the price for making the sail is $19.50 per square foot.

45. A semicircular window of radius 15 inches is to be laminated with a sunblock coating that costs $0.85 per square inch to apply. What is the total cost of coating the window, to the nearest cent? (Use $\pi \approx 3.14$.)

46. Some new computers can be exposed to extreme temperatures (as high as 60°C and as low as −50°C). What is the temperature range in Fahrenheit that these computers can be exposed to? (Use the formula $F = \frac{9}{5}C + 32$.)

47. To deal with extreme temperatures while doing research at the South Pole, scientists have developed accommodations that can comfortably withstand an outside temperature of −60°C with no wind blowing, or −30°C with wind gusts of up to 50 miles per hour. What is the corresponding Fahrenheit temperature range? (Use the formula $F = \frac{9}{5}C + 32$.)

48. Bruce becomes exhausted while on a bicycle trip in Canada. He reads on the map that his present elevation is 2.3 kilometers above sea level. How many miles to the nearest tenth above sea level is he? Why is he so tired? Use the formula $r = 0.62k$ where r is the number of miles and k is the number of kilometers.

49. While biking down the Pacific coast of Mexico, you see on the map that it is 20 kilometers to the nearest town. Approximately how many miles is it to the nearest town? Use the formula $r = 0.62k$ where r is the number of miles and k is the number of kilometers.

Cumulative Review Problems

In exercises 50 and 51, simplify.

50. $(-2)^4 - 4 \div 2 - (-2)$

51. $3(x - 2y) - (x^2 - y) - (x - y)$

52. A 93-minute-long recordable compact disc is used to record 15 songs. Express in decimal form the average number of minutes available per song.

Putting Your Skills to Work

Measuring Your Level of Fitness

In recent years there has been a greater awareness of fitness, exercise, and weight control in our country. Often people would like a simple mathematical way to measure whether they are physically fit and of the proper weight.

A growing number of doctors and health fitness professionals refer to the government studies from the National Center for Health Statistics as supporting the BMI, or body mass index, as a helpful measure to determine whether a person is at the proper weight, overweight, or underweight.

To determine your BMI, multiply your weight in pounds by 0.45 to get your weight in kilograms. Next, convert your height to inches. Then multiply your height in inches by 0.0254 to get your height in meters. Now multiply that number by itself. Your weight in kilograms is divided by that number. (Your weight in kilograms is divided by the square of your height in meters.)

The result is your BMI. For most people this will be a number that ranges from the 20s to the low 30s. Federal guidelines suggest that you should keep your BMI under 25 in order not to be overweight. However, there is a limit to how low your BMI should be. Many doctors recommend that a person's BMI should not be lower than 18.

Problems for Individual Investigation and Study

Use the procedure for calculating BMI to answer the following questions. Round all measurements of BMI to the nearest tenth.

1. Find the BMI for a woman who weighs 125 pounds and is 5 feet, 4 inches tall.

2. Is a man who weighs 185 pounds and is 5 feet, 11 inches tall considered overweight by the government study? (Is his BMI 25 or higher?) What is his BMI?

Problems for Cooperative Group Activity

Together with other members of your class, see if you can complete the following.

3. Is a woman who weighs 95 pounds and is 1.6 meters tall considered underweight by the government study? (Is her BMI under 18?) What is her BMI?

4. Write a formula to find the BMI of a person if the height of the person is measured in meters (m) and the weight of the person is measured in pounds (p).

Internet Connections

www | Netsite: | http://www.prenhall.com/tobey_beg_int

Site: BMI Chart (Shape Up America!) or a related site

5. This site gives a shortcut method for calculating BMI. Write a formula based on this method to find the BMI of a person whose height is measured in inches (i) and weight is measured in pounds (p). Then use your formula to find the BMI of a person who is 5 feet, 10 inches tall and weighs 145 pounds.

1.9 Grouping Symbols

1 Simplifying Variable Expressions

Many expressions in algebra use **grouping symbols** such as parentheses, brackets, and braces. Sometimes expressions are inside other expressions. Because it can be confusing to have more than one set of parentheses, brackets and braces are also used. How do we know what to do first when we see an expression like $2[5 - 4(a + b)]$?

To simplify the expression, we start with the innermost grouping symbols. Here is a set of parentheses. We first use the distributive law to multiply.

$$2[5 - 4(a + b)] = 2[5 - 4a - 4b]$$

We use the distributive law again.

$$= 10 - 8a - 8b$$

There are no like terms, so this is our final answer.

Notice that we started with two sets of grouping symbols, but our final answer has none. So we can say we *removed* the grouping symbols. Of course, we didn't just take them away; we used the distributive law and the rules for real numbers to simplify as much as possible. Although simplifying expressions like this involves many steps, we sometimes say "remove parentheses" as a shorthand direction. Sometimes we say "simplify."

Remember to remove the innermost parentheses first. Keep working from the inside out.

EXAMPLE 1   Simplify. $3[6 - 2(x + y)]$

We want to remove the innermost parentheses first. Therefore, we first use the distributive property to simplify $-2(x + y)$.

$$3[6 - 2(x + y)] = 3[6 - 2x - 2y] \qquad \text{Use the distributive property.}$$

$$= 18 - 6x - 6y \qquad \text{Use the distributive property again.}$$

Practice Problem 1   Simplify. $5[4x - 3(y - 2)]$

You recall that a negative sign in front of parentheses is equivalent to having a coefficient of negative 1. You can write the -1 and then multiply by -1 using the distributive property.

$$-(x + 2y) = -1(x + 2y) = -x - 2y$$

Notice that this has the effect of removing the parentheses. Each term in the result now has its sign changed.

Similarly, a positive sign in front of parentheses can be viewed as multiplication by $+1$.

$$+(5x - 6y) = +1(5x - 6y) = 5x - 6y$$

If a grouping symbol has a positive or negative sign in front, we mentally multiply by $+1$ or -1, respectively.

Fraction bars are also considered grouping symbols. Later in this book we will encounter problems where our first step will be to simplify expressions above and below fraction bars. This type of operation will have some similarities to the operation of removing parentheses.

EXAMPLE 2   Simplify. $-2[3a - (b + 2c) + (d - 3e)]$

$$= -2[3a - b - 2c + d - 3e] \qquad \text{Remove the two innermost sets of parentheses. Since one is not inside the other, we remove both sets at once.}$$

$$= -6a + 2b + 4c - 2d + 6e \qquad \text{Now we remove the brackets by multiplying each term by } -2.$$

Practice Problem 2   Simplify. $3ab - [2ab - (2 - a)]$

Student Learning Objectives

After studying this section, you will be able to:

1 Simplify variable expressions with several grouping symbols.

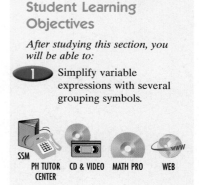

SSM

PH TUTOR CENTER   CD & VIDEO   MATH PRO   WEB

EXAMPLE 3 Simplify. $2[3x - (y + w)] - 3[2x + 2(3y - 2w)]$

$= 2[3x - y - w] - 3[2x + 6y - 4w]$ In each set of brackets, remove the inner parentheses.

$= 6x - 2y - 2w - 6x - 18y + 12w$ Remove each set of brackets by multiplying by the appropriate number.

$= -20y + 10w$ or $10w - 20y$ Combine like terms.
(Note that $6x - 6x = 0x = 0$.)

Practice Problem 3   Simplify. $3[4x - 2(1 - x)] - [3x + (x - 2)]$

You can always simplify problems with many sets of grouping symbols by the method shown. Essentially, you just keep removing one level of grouping symbols at each step. Finally, at the end you add up the like terms if possible.

Sometimes it is possible to combine like terms at each step.

EXAMPLE 4   Simplify. $-3\{7x - 2[x - (2x - 1)]\}$

$= -3\{7x - 2[x - 2x + 1]\}$ Remove the inner parentheses by multiplying each term within the parentheses by -1.

$= -3\{7x - 2[-x + 1]\}$ Combine like terms by combining $+x - 2x$.

$= -3\{7x + 2x - 2\}$ Remove the brackets by multiplying each term within them by -2.

$= -3\{9x - 2\}$ Combine the x-terms.

$= -27x + 6$ Remove the braces by multiplying each term by -3.

Practice Problem 4   Simplify. $-2\{5x - 3x[2x - (x^2 - 4x)]\}$

Developing Your Study Skills

Exam Time: The Night Before

With adequate preparation, you can spend the night before an exam pulling together the final details.

1. Look over each section to be covered in the exam. Review the steps needed to solve each type of problem.

2. Review your list of terms, rules, and formulas that you are expected to know for the exam.

3. Take the Practice Test at the end of the chapter just as though you were taking the actual exam. Do not look in your text or get help in any way. Time yourself so that you know how long it takes you to complete the test.

4. Check the Practice Test. Redo the problems you missed.

5. Be sure you have ready the necessary supplies for taking your exam.

Verbal and Writing Skills

1. Rewrite the expression $-3x - 2y$ using a negative sign and parentheses.

2. Rewrite the expression $-x + 5y$ using a negative sign and parentheses.

3. To simplify expressions with grouping symbols, we use the _____ property.

4. When an expression contains many grouping symbols, remove the _____ parentheses first.

Simplify. Remove grouping symbols and combine like terms.

5. $6x - 3(x - 2y)$

6. $-4x - 2(y - 3x)$

7. $2(a + 3b) - 3(b - a)$

8. $4(x - y) - 2(3x + y)$

9. $5[3 + 2(x - 26) + 3x]$

10. $-4[-(x + 3y) - 2(y - x)]$

11. $2x[4x^2 - 2(x - 3)]$

12. $4y[-3y^2 + 2(4 - y)]$

13. $2(x - 2y) - [3 - 2(x - y)]$

14. $3[x - y(3x + y) + y^2]$

15. $5[3a - 2a(3a + 6b) + 6a^2]$

16. $3(x + 2y) - [4 - 2(x + y)]$

17. $x(x^2 + 2x - 3) - 2(x^3 + 6)$

18. $5x^2(x + 6) - 2[x - 2(1 + 2x^2)]$

19. $3a^2 - 4[2b - 3b(b + 2)]$

20. $2a - \{6b - 4[a - (b - 3a)]\}$

21. $6b - \{5a - 2[a + (b - 2a)]\}$

22. $2b^2 - 3[5b + 2b(2 - b)]$

23. $-4\{3a^2 - 2[4a^2 - (b + a^2)]\}$

24. $-2\{x^2 - 3[x - (x - 2x^2)]\}$

To Think About

25. The dedicated job of a four-wheeled robot called Nomad, built by the Robotics Institute at Carnegie Mellon University, is to find samples of meteorites in Antarctica. If Nomad is successful in finding meteorites 2.5% of the time, out of four tries per day for six years, estimate the number of successful and unsuccessful meteorite search attempts.

26. Grandma and Grandpa Tobey had a tradition of eating out once a week. The average cost of the meal was $20. In Massachusetts there is a 5% state sales tax that is added to the cost of the meal. The Tobeys always left a 15% tip. They continued this pattern for ten years. Mrs. Tobey felt that they should calculate the tip on the total of the cost of the meal including the sales tax. Mr. Tobey felt they should calculate the tip on the cost of the meal alone. How much difference would this make over the ten-year period?

Cumulative Review Problems

27. Use $F = 1.8C + 32$ to find the Fahrenheit temperature equivalent to 36.4° Celsius.

▲ **28.** Use 3.14 as an approximation for π to compute the area covered by a circular irrigation system with radial arm of length 380 feet. $A = \pi r^2$.

▲ **29.** The base of an office building is in the shape of a trapezoid. The altitude of the trapezoid is 400 feet. The bases of the trapezoid are 700 feet and 800 feet, respectively. What is the area of the base of the office building? If the base has a marble floor that cost $55 per square foot, what was the cost of the marble floor?

▲ **30.** The Global Media Tower has a triangular signal tester at the top of the tower. The altitude of the triangle is 3.5 feet and the base is 6.5 feet. What is the area of the triangular signal tester? If the signal tester is coated with a special metallic surface paint that costs $122 per square foot, what was the cost of the amount of paint needed to coat one side of the triangle?

Putting Your Skills to Work

Consumer Price Index

The federal government produces a statistic that is helpful in determining how much the cost of living is increasing each year. It also produces different values for different regions of the country to track how the increase varies from city to city. This measure is called the Consumer Price Index (CPI). The CPI measures the average change in the price of goods and services purchased by all urban consumers. The following table shows the CPI for several years.

Type of Item	1975	1982–1984	1990	1995	2000*
Cost of all items	53.8	100	130.7	152.4	169.6
Food	59.8	100	132.4	148.4	188.4
Housing	50.7	100	128.5	148.5	173.2
Medical care	47.5	100	162.8	220.5	265.7
Entertainment	62.0	100	132.4	153.9	167.6

Source: U.S. Department of Labor
*estimated

The period of 1982–1984 is used as the focus of comparison, and the index for that period is assigned a value of 100. For example if the cost of food for a family was $100 in the period 1982–1984, then the cost of purchasing the same type of food in 1995 was $148.40. However, a similar purchase of the same type of food in 1975 would have amounted to only $59.80. Now use the table to answer the following questions.

Problems for Individual Investigation and Study

1. If a family spent $200 annually for entertainment during the 1982–1984 period, how much would the same entertainment have cost in 2000?

2. If an inpatient hospital operation cost $10,000 in the 1982–1984 period, how much would a similar operation have cost in 1995?

3. If a family found that three months' worth of food cost $1884.00 in 2000, how much would similar food have cost the family in 1975?

4. If a family found that renting an apartment for two months in 1995 cost $1485.00, how much would a similar apartment have cost for two months in 1975?

The CPI has increased more dramatically in certain parts of the country than in others. The following table records the CPI for five cities in 1998. For each city the index for each category is 100 for the 1982–1984 period.

CPI for Various Cities in 1998 (The CPI for each city in each category is 100 for 1982–1984.)					
Type of Item	Anchorage	Boston	Houston	New York	Chicago
Cost of all items	146.9	171.7	146.8	173.6	165.0
Food	147.5	166.3	150.3	165.3	164.3
Housing	131.0	165.8	129.6	175.9	164.2
Medical care	255.7	313.9	235.4	255.0	244.3

Use the preceding table to answer the following questions.

Problems for Cooperative Group Activity and Investigation

5. What category for what city saw the biggest increase in the CPI since 1982–1984?

6. How many times greater was the increase in the cost of housing in Boston than that in Anchorage from the 1982–1984 period to 1998?

Internet Connections   *Site: Consumer Price Index provided by the Bureau of Labor Statistics*

Use this Web site to obtain a CPI table for your city or region. Then answer the following questions.

7. Suppose you earned $22,000 during the 1982–1984 period. If your income increased at the same rate as the CPI for your city or region, how much would you have earned in 2000?

8. Suppose you spent $7000 on rent in 2000 in your city or region. How much would you have spent renting a similar location in your city or region during the 1982–1984 period?

Math in the Media

Readers Can Face Off with Pros and Darts

WSJ.com Staff Reporter

November 16, 2000. © 2000 Dow Jones & Company Inc. Reprinted by permission.

The *Wall Street Journal Interactive Edition* runs an investment contest, **Investment Dartboard**. Individual investors are invited to submit their best stock picks for the contest beginning in a given month and running for six months. Stock picks submitted by four readers will be selected in a drawing, and these picks will then be matched against the stocks selected by the pros and the darts.

There's no prize—just the glory, or the embarrassment, of publicly pitting your stock-picking skills against the investment professionals and the forces of chance.

The table shows Stock, Purchase Price, Latest Price, Day's Change, and % Gain/Loss.

Stock	Purchase Price	Latest Price	Day's Change	% Gain/Loss to Date
The Pros—professional investors				
A	38	$37\frac{3}{8}$	$-\frac{5}{8}$	−1.64%
B	$26\frac{9}{16}$	$26\frac{7}{8}$	$+\frac{5}{16}$	
C	$67\frac{3}{8}$	$65\frac{7}{8}$		−2.23%
D	$17\frac{3}{16}$	$17\frac{1}{8}$	$-\frac{1}{16}$	−0.36%
The Darts' Picks—Selected by Journal staffers flinging darts at the paper's stock tables.				
E	572	572	unch.	0
F	$5\frac{13}{16}$	$5\frac{1}{2}$	$-\frac{5}{16}$	−5.38%
G	$11\frac{1}{4}$	$10\frac{7}{8}$	$-\frac{3}{8}$	
H	$23\frac{1}{4}$	$22\frac{7}{8}$	$-\frac{3}{8}$	−1.61%
The Amateurs' Picks				
I	$14\frac{1}{2}$	$12\frac{11}{16}$	$-1\frac{13}{16}$	−12.50%
J	$29\frac{5}{16}$	$27\frac{29}{64}$	$-1\frac{55}{64}$	−6.34%
K	$41\frac{5}{16}$	40		−3.18%
L	$43\frac{5}{8}$	$43\frac{7}{8}$	$+\frac{1}{4}$	0.57%

EXERCISES

Investing can be risky business. Look at the table provided in the article. Test your skills. Try to answer questions 1–3 that follow.

1. How are the values in the Day's Change column achieved? Complete the tables for Stock C and Stock K.

2. How are the values in the % Gain/Loss column achieved? Complete the tables for Stock B and Stock G.

3. Compute the average % Gain/Loss for the Pros, Darts' Picks, and Amateurs. Which was the best?

Chapter 1 Organizer

Topic	Procedure	Examples
Absolute value, p. 5.	The absolute value of a number is the distance between that number and zero on the number line. The absolute value of any number will be positive or zero.	$\|3\| = 3$ $\|-2\| = 2$ $\|0\| = 0$ $\left\|-\dfrac{5}{6}\right\| = \dfrac{5}{6}$ $\|-1.38\| = 1.38$
Adding real numbers with the same sign, p. 6.	If the signs are the same, add the absolute values of the numbers. Use the common sign in the answer.	$-3 + (-7) = -10$
Adding real numbers with opposite signs, p. 8.	If the signs are different: **1.** Find the difference between the larger and the smaller absolute value. **2.** Give the answer the sign of the number having the larger absolute value.	$(-7) + 13 = 6$ $7 + (-13) = -6$
Adding several real numbers, p. 9.	When adding several real numbers, separate them into two groups by sign. Find the sum of all the positive numbers and the sum of all the negative numbers. Combine these two subtotals by the method described above.	$-7 + 6 + 8 + (-11) + (-13) + 22$ $\begin{array}{rr} -7 & +6 \\ -11 & +8 \\ -13 & +22 \\ \hline -31 & +36 \end{array}$ $-31 + 36 = 5$ The answer is positive since 36 is positive.
Subtracting real numbers, p. 13.	Change the sign of the second number (the number you are subtracting) and then add.	$-3 - (-13) = -3 + (+13) = 10$
Multiplying and dividing real numbers, p. 19 and p. 21.	**1.** If the two numbers have the same sign, multiply (or divide) the absolute values. The result is positive. **2.** If the two numbers have different signs, multiply (or divide) the absolute values. The result is negative.	$-5(-3) = +15$ $-36 \div (-4) = +9$ $28 \div (-7) = -4$ $-6(3) = -18$
Exponent form, p. 28.	The base tells you what number is being multiplied. The exponent tells you how many times this number is used as a factor.	$2^5 = 2 \cdot 2 \cdot 2 \cdot 2 \cdot 2 = 32$ $4^3 = 4 \cdot 4 \cdot 4 = 64$ $(-3)^4 = (-3)(-3)(-3)(-3) = 81$
Raising a negative number to a power, p. 29.	When the base is negative, the result is positive for even exponents, and negative for odd exponents.	$(-3)^3 = -27$ but $(-2)^4 = 16$
Order of operations, p. 31.	Remember the proper order of operations: **1.** Perform operations inside parentheses. **2.** Raise to powers. **3.** Multiply and divide from left to right. **4.** Add and subtract from left to right.	$3(5 + 4)^2 - 2^2 \cdot 3 \div (9 - 2^3)$ $= 3 \cdot 9^2 - 4 \cdot 3 \div (9 - 8)$ $= 3 \cdot 81 - 12 \div 1$ $= 243 - 12 = 231$
Removing parentheses, p. 34.	Use the distributive property to remove parentheses. $a(b + c) = ab + ac$	$3(5x + 2) = 15x + 6$

Topic	Procedure	Examples
Combining like terms, p. 39.	Combine terms that have identical letters and exponents.	$7x^2 - 3x + 4y + 2x^2 - 8x - 9y = 9x^2 - 11x - 5y$
Substituting into variable expressions, p. 44.	1. Replace each letter by the numerical value given for it. 2. Follow the order of operations in evaluating the expression.	Evaluate $2x^3 + 3xy + 4y^2$ for $x = -3$ and $y = 2$. $2(-3)^3 + 3(-3)(2) + 4(2)^2$ $\quad = 2(-27) + 3(-3)(2) + 4(4)$ $\quad = -54 - 18 + 16$ $\quad = -56$
Using formulas, p. 45.	1. Replace the variables in the formula by the given values. 2. Evaluate the expression. 3. Label units carefully.	Find the area of a circle with radius 4 feet. Use $A = \pi r^2$, with π as approximately 3.14. $A = (3.14)(4 \text{ feet})^2$ $\quad = (3.14)(16 \text{ feet}^2)$ $\quad = 50.24 \text{ feet}^2$ The area of the circle is approximately 50.24 square feet.
Removing grouping symbols, p. 51.	1. Remove innermost grouping symbols first. 2. Then remove remaining innermost grouping symbols. 3. Continue until all grouping symbols are removed. 4. Combine like terms.	$5\{3x - 2[4 + 3(x - 1)]\}$ $\quad = 5\{3x - 2[4 + 3x - 3]\}$ $\quad = 5\{3x - 8 - 6x + 6\}$ $\quad = 15x - 40 - 30x + 30$ $\quad = -15x - 10$

Developing Your Study Skills

Problems with Accuracy

Strive for accuracy. Mistakes are often made as a result of human error rather than from lack of understanding. Such mistakes are frustrating. A simple arithmetic or sign error can lead to an incorrect answer.

These five steps will help you cut down on errors.

1. Work carefully and take your time. Do not rush through a problem just to get it done.

2. Concentrate on one problem at a time. Sometimes problems become mechanical, and your mind begins to wander. You can become careless and make a mistake.

3. Check your problem. Be sure that you copied it correctly from the book.

4. Check your computations from step to step. Check the solution in the problem. Does it work? Does it make sense?

5. Keep practicing new skills. Remember the old saying, "Practice makes perfect." An increase in practice results in an increase in accuracy. Many errors are due simply to lack of practice.

There is no magic formula for eliminating all errors, but these five steps will be a tremendous help in reducing them.

Chapter 1 Review Problems

1.1 *Add.*

1. $-6 + (-2)$

2. $-12 + 7.8$

3. $5 + (-2) + (-12)$

4. $3.7 + (-1.8)$

5. $\frac{1}{2} + \left(-\frac{5}{6}\right)$

6. $-\frac{3}{11} + \left(-\frac{1}{22}\right)$

7. $\frac{3}{4} + \left(-\frac{1}{12}\right) + \left(-\frac{1}{2}\right)$

8. $-\frac{4}{15} + \frac{12}{5} + \left(-\frac{2}{3}\right)$

1.2 *Add or subtract.*

9. $5 - (-3)$

10. $-2 - (-15)$

11. $-30 - (+3)$

12. $8 - (-1.2)$

13. $-\frac{7}{8} + \left(-\frac{3}{4}\right)$

14. $-\frac{3}{14} + \frac{5}{7}$

15. $-20.8 - 1.9$

16. $-151 - (-63)$

Mixed Review

1.1–1.2 *Perform the operations indicated. Simplify all answers.*

17. $-5 + (-2) - (-3)$

18. $6 - (-4) + (-2) + 8$

19. $-16 + (-13)$

20. $-11 - (-12)$

1.3

21. $87 \div (-29)$

22. $-5(-6) + 4(-3)$

23. $\dfrac{-24}{-\frac{3}{4}}$

24. $-\frac{1}{2} \div \left(\frac{3}{4}\right)$

25. $\frac{5}{7} \div \left(-\frac{5}{25}\right)$

26. $-6(3)(4)$

27. $-1(-2)(-3)(-5)$

28. $(-5)\left(-\frac{1}{2}\right)(4)(-3)$

Mixed Review

1.1–1.3 *Perform the operations indicated. Simplify all answers.*

29. $-\frac{4}{3} + \frac{2}{3} + \frac{1}{6}$

30. $-\frac{6}{7} + \frac{1}{2} + \left(-\frac{3}{14}\right)$

31. $-3(-2)(-5)$

32. $-6 + (-2) - (-3)$

33. $3.5(-2.6)$

34. $-5.4 \div (-6)$

35. $5 - (-3.5) + 1.6$

36. $-8 + 2 - (-4.8)$

37. $17 + 3.4 + (-16) + (-2.5)$

38. $37 + (-44) + 12.5 + (-6.8)$

Solve.

39. The Dallas Cowboys football team had three plays in which they lost 8 yards each time. What was the total yardage lost?

40. The low temperature in Anchorage, Alaska, last night was $-34°$F. During the day the temperature rose $12°$F. What was the temperature during the day?

41. A mountain peak is 6895 feet above sea level. A location in Death Valley is 282 feet below sea level. What is the difference in height between these two locations?

42. During January 2000, IBM stock rose $1\frac{1}{2}$ points on Monday, dropped $3\frac{1}{4}$ points on Tuesday, rose 2 points on Wednesday, and dropped $2\frac{1}{2}$ points on Thursday. What was the total gain or loss on the value of the stock over this four-day period?

1.4   *Evaluate.*

43. $(-3)^5$

44. $(-2)^7$

45. $(-5)^4$

46. $\left(\dfrac{2}{3}\right)^3$

47. -9^2

48. $(0.6)^2$

49. $\left(\dfrac{5}{6}\right)^2$

50. $\left(\dfrac{3}{4}\right)^3$

1.5   *Simplify using the order of operations.*

51. $5(-4) + 3(-2)^3$

52. $20 - (-10) - (-6) + (-5) - 1$

53. $(7 - 9)^3 + -6(-2) + (-3)$

1.6   *Use the distributive property to multiply.*

54. $5(3x - 7y)$

55. $2x(3x - 7y + 4)$

56. $-(7x^2 - 3x + 11)$

57. $(2xy + x - y)(-3y)$

1.7   *Combine like terms.*

58. $3a^2b - 2bc + 6bc^2 - 8a^2b - 6bc^2 + 5bc$

59. $9x + 11y - 12x - 15y$

60. $4x^2 - 13x + 7 - 9x^2 - 22x - 16$

61. $-x + \dfrac{1}{2} + 14x^2 - 7x - 1 - 4x^2$

1.8   *Evaluate for the given value of the variable.*

62. $7x - 6$ for $x = -7$

63. $7 - \dfrac{3}{4}x$ for $x = 8$

64. $x^2 + 3x - 4$ for $x = -3$

65. $-3x^2 - 4x + 5$ for $x = 2$

66. $-3x^3 - 4x^2 + 2x + 6$ for $x = -2$

67. $vt - \dfrac{1}{2}at^2$ for $v = 24$, $t = 2$, and $a = 32$

68. $\dfrac{nRT}{V}$ for $n = 16$, $R = -2$, $T = 4$, and $V = -20$

Solve.

69. Find the simple interest on a loan of $6000 at an annual interest rate of 18% per year for $\frac{3}{4}$ of a year. Use $I = prt$, where p = principal, r = rate per year, and t = time in years.

70. Find the Fahrenheit temperature if a radio announcer in Mexico City says that the high temperature today was 30°C. Use the formula
$$F = \frac{9C + 160}{5}.$$

▲ **71.** How much will it cost to paint a circular sign with a radius of 15 meters if the painter charges $3 per square meter? Use $A = \pi r^2$, where π is approximately 3.14.

72. Find the daily profit P at a furniture factory if the initial cost of setting up the factory C = $1200, rent R = $300, and sales of furniture S = $56. Use the profit formula $P = 180S - R - C$.

▲ **73.** A parking lot is in the shape of a trapezoid. The altitude of the trapezoid is 200 feet, and the bases of the trapezoid are 300 feet and 700 feet. What is the area of the parking lot? If the parking lot had a sealer applied that costs $2 per square foot, what was the cost of the amount of sealer needed for the entire parking lot?

▲ **74.** The Green Mountain Telephone Company has a triangular signal tester at the top of a communications tower. The altitude of the triangle is 3.8 feet and the base is 5.5 feet. What is the area of the triangular signal tester? If the signal tester is painted with a special metallic surface paint that costs $66 per square foot, what was the cost of the amount of paint needed to paint one side of the triangle?

1.9   *Simplify.*

75. $5x - 7(x - 6)$

76. $3(x - 2) - 4(5x + 3)$

77. $2[3 - (4 - 5x)]$

78. $-3x[x + 3(x - 7)]$

79. $2xy^3 - 6x^3y - 4x^2y^2 + 3(xy^3 - 2x^2y - 3x^2y^2)$

80. $-5(x + 2y - 7) + 3x(2 - 5y)$

81. $2\{x - 3(y - 2) + 4[x - 2(y + 3)]\}$

82. $-5\{2a - b[5a - b(3 + 2a)]\}$

83. $-3\{2x - [x - 3y(x - 2y)]\}$

84. $2\{3x + 2[x + 2y(x - 4)]\}$

Chapter 1 Test

1. _____

2. _____

3. _____

4. _____

5. _____

6. _____

7. _____

8. _____

9. _____

10. _____

11. _____

12. _____

13. _____

14. _____

15. _____

16. _____

17. _____

18. _____

19. _____

20. _____

21. _____

22. _____

23. _____

24. _____

25. _____

26. _____

Simplify.

1. $-3 + (-4) + 9 + 2$

2. $-0.6 - (-0.8)$

3. $-8(-12)$

4. $-5(-2)(7)(-1)$

5. $-12 \div (-3)$

6. $-1.8 \div (0.6)$

7. $(-4)^3$

8. $(1.3)^2$

9. $\left(\dfrac{2}{3}\right)^4$

10. $7 - (6 - 9)^2 + 5(2)$

11. $3(4 - 6)^3 + 12 \div (-4) + 2$

12. $-5x(x + 2y - 7)$

13. $-2ab^2(-3a - 2b + 7ab)$

14. $6ab - \dfrac{1}{2}a^2b + \dfrac{3}{2}ab + \dfrac{5}{2}a^2b$

15. $12a(a + b) - 4(a^2 - 2ab)$

16. $3(2 - a) - 4(-6 - 2a)$

17. $5(3x - 2y) - (x + 6y)$

In questions 18–20, evaluate for the value of the variable indicated.

18. $x^3 - 3x^2y + 2y - 5$ for $x = 3$ and $y = -4$

19. $3x^2 - 7x - 11$ for $x = -3$

20. $2a - 3b$ for $a = \dfrac{1}{3}$ and $b = -\dfrac{1}{2}$

21. If you are traveling 60 miles per hour on a highway in Canada, how fast are you traveling in kilometers per hour? (Use $k = 1.61r$, where r = rate in miles per hour and k = rate in kilometers per hour.)

▲ **22.** A field is in the shape of a trapezoid. The altitude of the trapezoid is 120 feet and the bases of the trapezoid are 180 feet and 200 feet. What is the area of the field?

▲ **23.** Jeff Slater's garage has a triangular roof support beam. The support beam is covered with a sheet of plywood. The altitude of the triangular region is 6.8 feet and the base is 8.5 feet. If the triangular piece of plywood was painted with paint that cost $0.80 per square foot, what was the cost of the amount of paint needed to coat one side of the triangle?

▲ **24.** The front lawn of Westwood High School is watered by a sprinkler that waters a circular region. The radius of the circle is 12 feet. What is the area of this region of the lawn? Use $\pi \approx 3.14$.

Simplify.

25. $-3\{a + b[3a - b(1 - a)]\}$

26. $3\{x - (5 - 2y) - 4[3 + (6x - 7y)]\}$

Equations, Inequalities, and Applications

The Environmental Protection Agency monitors the level of emissions into the air of a number of pollutants. Progress has been made in reducing some pollutants, while others are still at significantly high levels. Could you analyze the level of emissions over the last 50 years and use your math skills to predict the level of emissions in the future? Turn to the Putting Your Skills to Work problems on page 98 to find out.

Pretest Chapter 2

1. _____

2. _____

3. _____

4. _____

5. _____

6. _____

7. _____

8. _____

9. _____

10. _____

11. _____

12. _____

13. (a) _____

(b) _____

14. (a) _____

(b) _____

15. _____

16. _____

17. _____

18. _____

If you are familiar with the topics in this chapter, take this test now. Check your answers with those in the back of the book. If an answer is wrong or you can't answer a question, study the appropriate section of the chapter.

If you are not familiar with the topics in this chapter, don't take this test now. Instead, study the examples, work the practice problems, and then take the test.

This test will help you identify which concepts you have mastered and which you need to study further.

Sections 2.1–2.3

Solve for x.

1. $x - 30 = -46$

2. $\frac{1}{4}x = 12$

3. $-5x = 20$

4. $7x + 2 = 30$

5. $3x - 8 = 7x + 6$

6. $5x + 3 - 6x = 4 - 8x - 2$

7. $5.2x + 0.9 = 2.8x + 3.3$

8. $2(x - 4) - 2 = 3(x - 2)$

Section 2.4

Solve for x.

9. $\frac{2}{5}x + \frac{1}{4} = \frac{1}{2}x$

10. $\frac{2}{3}(x - 2) + \frac{1}{2} = 5 - (3 + x)$

11. $x - \frac{3}{2} = -\frac{4}{3} - \frac{5}{6}$

12. $x - 2 = \frac{5}{7}(x + 4)$

Section 2.5

13. (a) Solve for F. $C = \frac{5}{9}(F - 32)$

(b) Find the temperature in degrees Fahrenheit if the Celsius temperature is $-15°$.

14. (a) Solve for r. $I = Prt$

(b) Find the simple interest rate r if $P = \$2000$, $t = 2$ years, and $I = \$240$.

Sections 2.6–2.7

Replace the question mark with $<$ or $>$.

15. $1\ ?\ 10$ **16.** $2\ ?\ -3$ **17.** $-12\ ?\ -13$ **18.** $0\ ?\ 0.9$

Solve and graph.

19. $-2x + 5 \leq 4 - x + 3$

20. $\frac{1}{2}(x + 2) - 1 > \frac{x}{3} - 5 + x$

Section 2.8

▲ **21.** A triangular piece of insulation is placed in the peak of a roof of a house. One side of the triangle is 3 feet shorter than triple the second side. The third side is five feet shorter than the length of the first side. The perimeter of this piece of insulation is 38 feet. Find the length of each side of the piece of insulation.

22. Charlene took three packages to the post office to mail. The second package was $3\frac{1}{2}$ pounds less than the first package. The third package was 2 pounds less than the first package. The total weight of the three packages was 17 pounds. How much did each package weigh?

▲ **23.** A rectangular storage room in the school has a perimeter of 88 feet. The length of the room is 11 feet longer than twice the width. Find the dimensions of the room.

Section 2.9

24. Two investments totaling $1000 were made. The investments gained interest once each year. The first investment yielded 12% interest after one year. The second investment yielded 9% interest. The total interest for both investments was $102. How much was invested at each interest rate?

25. The population of Springville has grown 11% in the last five years. The town now has a population of 24,420 people. What was the population five years ago?

26. Enrique has $3.35 in change in his pocket. He has only nickels, dimes, and quarters. He has four more dimes than nickels. He has three more quarters than dimes. How many coins of each type does he have?

19. _____

20. _____

21. _____

22. _____

23. _____

24. _____

25. _____

26. _____

2.1 The Addition Principle

Student Learning Objectives

After studying this section, you will be able to:

1 Use the addition principle to solve equations of the form $x + b = c$.

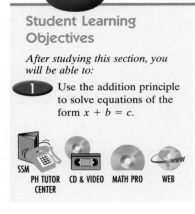

SSM PH TUTOR CD & VIDEO MATH PRO WEB
       CENTER

1 Using the Addition Principle

When we use an equal sign ($=$), we are indicating that two expressions are equal in value. Such a statement is called an **equation.** For example, $x + 5 = 23$ is an equation. A **solution** of an equation is a number that when substituted for the variable makes the equation true. Thus 18 is a solution of $x + 5 = 23$ because $18 + 5 = 23$. Equations that have exactly the same solutions are called **equivalent equations.** By following certain procedures, we can often transform an equation to a simpler equivalent one that has the form $x = $ some number. Then this number is a solution of the equation. The process of finding all solutions of an equation is called **solving the equation.**

One of the first procedures used in solving equations has an application in our everyday world. Suppose that we place a 10-kilogram box on one side of a seesaw and a 10-kilogram stone on the other side. If the center of the box is the same distance from the balance point as the center of the stone, we would expect the seesaw to balance. The box and the stone do not look the same, but their weights are equal. If we add a 2-kilogram lead weight to the center of weight of each object at the same time, the seesaw should still balance. The weights are still equal.

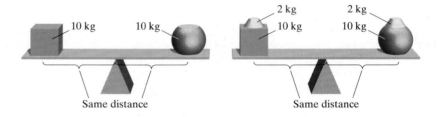

There is a similar principle in mathematics. We can state it in words as follows.

The Addition Principle

If the same number is added to both sides of an equation, the results on both sides are equal in value.

We can restate it in symbols this way.

For real numbers a, b, and c, if $a = b$, then $a + c = b + c$.

Here is an example.

$$\text{If } 3 = \frac{6}{2}, \quad \text{then } 3 + 5 = \frac{6}{2} + 5.$$

Since we added the same amount, 5, to both sides, the sides remain equal to each other.

$$3 + 5 = \frac{6}{2} + 5$$

$$8 = \frac{6}{2} + \frac{10}{2}$$

$$8 = \frac{16}{2}$$

$$8 = 8$$

We can use the addition principle to solve certain equations.

EXAMPLE 1   Solve for x. $x + 16 = 20$

$$x + 16 + (-16) = 20 + (-16) \qquad \text{Use the addition principle to add} \\ \qquad\qquad\qquad\qquad\qquad\qquad\qquad -16 \text{ to both sides.}$$

$$x + 0 = 4 \qquad\qquad\qquad \text{Simplify.}$$

$$x = 4 \qquad\qquad\qquad\quad \text{The value of } x \text{ is 4.}$$

We have just found a solution of the equation. A **solution** is a value for the variable that makes the equation true. We then say that the value 4 in our example **satisfies** the equation. We can easily verify that 4 is a solution by substituting this value into the original equation. This step is called **checking** the solution.

$$\text{Check.} \qquad\qquad x + 16 = 20$$
$$4 + 16 \overset{?}{=} 20$$
$$20 = 20 \;\; \checkmark$$

When the same value appears on both sides of the equal sign, we call the equation an **identity.** Because the two sides of the equation in our check have the same value, we know that the original equation has been solved correctly. We have found a solution, and since no other number makes the equation true, it is the only solution.

Practice Problem 1   Solve for x and check your solution. $x + 0.3 = 1.2$

Notice that when you are trying to solve these types of equations, you must add a particular number to both sides of the equation. What is the number to choose? Look at the number that is on the same side of the equation with x, that is, the number added to x. Then think of the number that is **opposite in sign.** This is called the **additive inverse** of the number. The additive inverse of 16 is -16. The additive inverse of -3 is 3. The number to add to both sides of the equation is precisely this additive inverse.

It does not matter which side of the equation contains the variable. The x-term may be on the right or left. In the next example the x-term will be on the right.

EXAMPLE 2   Solve for x. $14 = x - 3$

$$14 + 3 = x - 3 + 3 \qquad \text{Notice that } -3 \text{ is being added to } x \text{ in the original} \\ \qquad\qquad\qquad\qquad\qquad\quad \text{equation. Add 3 to both sides, since 3 is the} \\ \qquad\qquad\qquad\qquad\qquad\quad \text{additive inverse of } -3. \text{ This will eliminate the } -3 \\ \qquad\qquad\qquad\qquad\qquad\quad \text{on the right and isolate } x.$$

$$17 = x + 0 \qquad\qquad \text{Simplify.}$$
$$17 = x \qquad\qquad\quad\;\; \text{The value of } x \text{ is 17.}$$
$$\text{Check.} \qquad 14 = x - 3$$
$$14 \overset{?}{=} 17 - 3 \qquad\qquad \text{Replace } x \text{ by 17.}$$
$$14 = 14 \;\; \checkmark \qquad\qquad \text{Simplify. It checks. The solution is 17.}$$

Practice Problem 2   Solve for x and check your solution. $17 = x - 5$

Before you add a number to both sides, you should always simplify the equation. The following example shows how combining numbers by addition—separately, on both sides of the equation—simplifies the equation.

EXAMPLE 3   Solve for x. $15 + 2 = 3 + x + 2$

$$17 = x + 5 \qquad\qquad \text{Simplify by adding.}$$
$$17 + (-5) = x + 5 + (-5) \qquad \text{Add the value } -5 \text{ to both sides, since } -5 \\ \qquad\qquad\qquad\qquad\qquad\qquad\quad \text{is the additive inverse of 5.}$$
$$12 = x \qquad\qquad\qquad\qquad \text{Simplify. The value of } x \text{ is 12.}$$

Check. $15 + 2 = 3 + x + 2$

$15 + 2 \overset{?}{=} 3 + 12 + 2$ Replace x by 12 in the original equation.

$17 = 17$ ✓   It checks.

Practice Problem 3   Solve for x and check your solution. $5 - 12 = x - 3$ ●

In Example 3 we added -5 to each side. You could subtract 5 from each side and get the same result. In Chapter 1 we discussed how subtracting a 5 is the same as adding a negative 5. Do you see why?

Just as it is possible to add the same number to both sides of an equation, it is also possible to subtract the same number from both sides of an equation. This is so because any subtraction problem can be rewritten as an addition problem. For example, $17 - 5 = 17 + (-5)$. Thus the addition principle tells us that we can subtract the same number from both sides of the equation.

We can determine whether a value is the solution to an equation by following the same steps used to check an answer. Substitute the value to be tested for the variable in the original equation. We will obtain an identity if the value is the solution.

● **EXAMPLE 4**   Is 10 the solution to the equation $-15 + 2 = x - 3$? If it is not, find the solution.

We substitute 10 for x in the equation and see if we obtain an identity.

$-15 + 2 = x - 3$

$-15 + 2 \overset{?}{=} 10 - 3$

$-13 \neq 7$ The values are not equal. The statement is not an identity.

Thus, 10 is not the solution. Now we take the original equation and solve to find the solution.

$-15 + 2 = x - 3$

$-13 = x - 3$ Simplify by adding.

$-13 + 3 = x - 3 + 3$ Add 3 to both sides. 3 is the additive inverse of -3.

$-10 = x$

Check to see if -10 is the solution. The value 10 was incorrect because of a sign error. We must be especially careful to write the correct sign for each number when solving equations.

Practice Problem 4   Is -2 the solution to the equation $x + 8 = -22 + 6$? If it is not, find the solution.   ●

● **EXAMPLE 5**   Find the value of x that satisfies the equation $\dfrac{1}{5} + x = -\dfrac{1}{10} + \dfrac{1}{2}$.

To be combined, the fractions must have common denominators. The least common denominator (LCD) of the fractions is 10.

$\dfrac{1 \cdot 2}{5 \cdot 2} + x = -\dfrac{1}{10} + \dfrac{1 \cdot 5}{2 \cdot 5}$ Change each fraction to an equivalent fraction with a denominator of 10.

$\dfrac{2}{10} + x = -\dfrac{1}{10} + \dfrac{5}{10}$ This is an equivalent equation.

$\dfrac{2}{10} + x = \dfrac{4}{10}$ Simplify by adding.

$$\frac{2}{10} + \left(-\frac{2}{10}\right) + x = \frac{4}{10} + \left(-\frac{2}{10}\right)$$

Add the additive inverse of $\frac{2}{10}$ to each side. You could also say that you are subtracting $\frac{2}{10}$ from each side.

$$x = \frac{2}{10}$$

Add the fractions.

$$x = \frac{1}{5}$$

Simplify the answer.

Check.   We substitute $\frac{1}{5}$ for x in the original equation and see if we obtain an identity.

$$\frac{1}{5} + x = -\frac{1}{10} + \frac{1}{2}$$

$$\frac{1}{5} + \frac{1}{5} \overset{?}{=} -\frac{1}{10} + \frac{1}{2}$$

Substitute $\frac{1}{5}$ for x.

$$\frac{2}{5} \overset{?}{=} -\frac{1}{10} + \frac{5}{10}$$

$$\frac{2}{5} \overset{?}{=} \frac{4}{10}$$

$$\frac{2}{5} = \frac{2}{5} \checkmark$$

It checks.

Practice Problem 5   Find the value of x that satisfies the equation $\dfrac{1}{20} - \dfrac{1}{2} = x + \dfrac{3}{5}$.

Developing Your Study Skills

Why Study Mathematics?

In our present-day, technological world, it is easy to see mathematics at work. Many vocational and professional areas—such as the fields of business, statistics, economics, psychology, finance, computer science, chemistry, physics, engineering, electronics, nuclear energy, banking, quality control, and teaching—require a certain level of expertise in mathematics. Those who want to work in these fields must be able to function at a given mathematical level. Those who cannot will not make it. So if your field of study requires you to take higher-level mathematics courses, be sure to master the basics of this course. Then you will be ready for the next one.

Verbal and Writing Skills

In exercises 1–3, fill in the blank with the appropriate word.

1. When we use the _____ sign, we indicate two expressions are _____ in value.

2. If the _____ _____ is added to both sides of an equation, the results on each side are equal in value.

3. The _____ of an equation is a value of the variable that makes the equation true.

4. What is the additive inverse of -20?

5. Why do we add the additive inverse of a to each side of $x + a = b$ to solve for x?

Find the value of x that satisfies each equation. Check your answers.

6. $x + 11 = 15$

7. $x + 12 = 18$

8. $13 = 4 + x$

9. $15 = x + 9$

10. $x - 3 = 14$

11. $x - 11 = 5$

12. $0 = x + 5$

13. $0 = x - 7$

14. $3 + 5 = x - 7$

15. $8 - 2 = x + 5$

16. $7 + 3 + x = 5 + 5$

17. $18 - 2 + 3 = x + 19$

18. $x - 6 = -19$

19. $x - 11 = -13$

20. $-12 + x = 50$

21. $-18 + x = 48$

22. $18 - 11 = x - 5$

23. $23 - 8 = x - 12$

24. $8 - 23 + 7 = 1 + x - 2$

25. $3 - 17 + 8 = 8 + x - 3$

In exercises 26–33, determine whether the given solution is correct. If it is not, find the solution.

26. Is 8 the solution to $-12 + x = 4$?

27. Is 12 the solution to $-19 + x = 7$?

28. Is -3 the solution to $-18 - 2 = x - 7$?

29. Is -5 the solution to $-16 + 5 = x - 6$?

30. Is -33 the solution to $x - 23 = -56$?

31. Is -8 the solution to $-39 = x - 47$?

32. Is 35 the solution to $15 - 3 + 20 = x - 3$?

33. Is -12 the solution to $x + 8 = 12 - 19 + 3$?

Find the value of x that satisfies each equation.

34. $-3 = x - 8$

35. $-11 + x = -7$

36. $-16 = x + 25$

37. $27 = 5 + x$

38. $1.3 + x + 1.8 = 0.2$

39. $3.6 + 1.2 = x + 1.3$

40. $2.5 + x = 0.7$

41. $4.2 + x = 1.3$

42. $x - \dfrac{1}{4} = \dfrac{3}{4}$

43. $x + \dfrac{1}{3} = \dfrac{2}{3}$

44. $\dfrac{2}{3} + x = \dfrac{1}{6} + \dfrac{1}{4}$

45. $\dfrac{2}{5} + x = \dfrac{1}{2} - \dfrac{3}{10}$

46. $\dfrac{1}{18} - \dfrac{5}{9} = x - \dfrac{1}{2}$

47. $\dfrac{7}{12} - \dfrac{2}{3} = x - \dfrac{5}{4}$

48. $5\dfrac{1}{6} + x = 8$

49. $7\dfrac{1}{8} = -20 + x$

50. $\dfrac{1}{2} + 3x = \dfrac{1}{4} + 2x - 3 - \dfrac{1}{2}$

51. $1.6 - 5x - 3.2 = -2x + 5.6 + 4x - 8x$

52. $x + 0.7513 = 2.2419$

53. $x - 0.2314 = -4.0144$

Cumulative Review Problems

Simplify by adding like terms.

54. $x + 3y - 5x - 7y + 2x$

55. $y^2 + y - 12 - 3y^2 - 5y + 16$

56. A 90-meter-wide radar picture is taken of a swamp in northern Australia. The radar detects a rock outcrop that is 90 feet above sea level, and a vein of opal (a semiprecious stone) 27 feet below sea level. How far is the top of the rock from the location of the opal?

57. Trevor pays his monthly computer lease bill for $49.99, but forgets to look at his checking account balance before doing so. When he gets his checking account statement at the local ATM, his balance reads −$35.07. How much was in his account before he wrote the check?

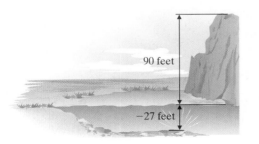

90 feet

−27 feet

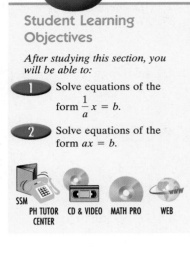

1 ***Solving Equations of the Form $\frac{1}{a}x = b$***

The addition principle allows us to add the same number to both sides of an equation. What would happen if we multiplied each side of an equation by the same number? For example, what would happen if we multiplied each side of an equation by 3?

To answer this question, let's return to our simple example of the box and the stone on a balanced seesaw. If we triple the weight on each side (that is, multiply the weight on each side by 3), the seesaw should still balance. The weight values of both sides remain equal.

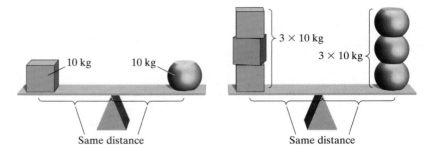

In words we can state this principle thus.

> **Multiplication Principle**
>
> If both sides of an equation are multiplied by the same nonzero number, the results on both sides are equal in value.

In symbols we can restate the multiplication principle this way.

> For real numbers a, b, and c with $c \neq 0$,   if $a = b$,   then $ca = cb$.

It is important that we say $c \neq 0$. We will explore this idea in the To Think About exercises.

Let us look at an equation where it would be helpful to multiply each side by 3.

EXAMPLE 1   Solve for x. $\frac{1}{3}x = -15$

We know that $(3)(\frac{1}{3}) = 1$. We will multiply each side of the equation by 3 because we want to isolate the variable x.

$$3\left(\frac{1}{3}x\right) = 3(-15)$$ Multiply each side of the equation by 3 since $(3)(\frac{1}{3}) = 1$.

$$\left(\frac{3}{1}\right)\left(\frac{1}{3}\right)(x) = -45$$

$$1x = -45$$ Simplify.

$$x = -45$$ The solution is -45.

Check. $\qquad \frac{1}{3}(-45) \overset{?}{=} -15$ Substitute -45 for x in the original equation.

$$-15 = -15 \quad \checkmark \quad \text{It checks.}$$

Practice Problem 1   Solve for x. $\dfrac{1}{8}x = -2$

Note that $\frac{1}{5}x$ can be written as $\frac{x}{5}$. To solve the equation $\frac{x}{5} = 3$, we could multiply each side of the equation by 5. Try it. Then check your solution.

2   Solving Equations of the Form $ax = b$

We can see that using the multiplication principle to multiply each side of an equation by $\frac{1}{2}$ is the same as dividing each side of the equation by 2. Thus, it would seem that the multiplication principle would allow us to divide each side of the equation by any nonzero real number. Is there a real-life example of this idea?

Let's return to our simple example of the box and the stone on a balanced seesaw. Suppose that we were to cut the two objects in half (so that the amount of weight of each was divided by 2). We then return the objects to the same places on the seesaw. The seesaw would still balance. The weight values of both sides remain equal.

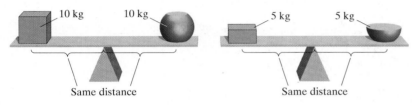

In words we can state this principle thus.

> ### Division Principle
> If both sides of an equation are divided by the same nonzero number, the results on both sides are equal in value.

Note: We put a restriction on the number by which we are dividing. We cannot divide by zero. We say that expressions like $\frac{2}{0}$ are not defined. Thus we restrict our divisor to *nonzero* numbers. We can restate the division principle this way.

> For real numbers a, b, and c where $c \neq 0$,   if $a = b$,   then $\dfrac{a}{c} = \dfrac{b}{c}$.

EXAMPLE 2   Solve for x. $5x = 125$

$$\frac{5x}{5} = \frac{125}{5} \qquad \text{Divide both sides by 5.}$$

$$x = 25 \qquad \text{Simplify. The solution is 25.}$$

Check. $5x = 125$

$5(25) \overset{?}{=} 125$ Replace x by 25.

$125 = 125$ ✓   It checks.

Practice Problem 2   Solve for x. $9x = 72$

For equations of the form $ax = b$ (a number multiplied by x equals another number), we solve the equation by choosing to divide both sides by a particular number. What is the number to choose? We look at the side of the equation that contains x. We notice the number that is multiplied by x. We divide by that number. The division principle tells us that we can still have a true equation provided that we divide by that number *on both sides* of the equation.

The solution to an equation may be a proper fraction or an improper fraction.

EXAMPLE 3 Solve for x. $4x = 38$

$$\frac{4x}{4} = \frac{38}{4}$$ Divide both sides by 4.

$$x = \frac{19}{2}$$ Simplify. The solution is $\frac{19}{2}$.

If you leave the solution as a fraction, it will be easier to check that solution in the original equation.

Check. $4x = 38$

$$\overset{2}{\cancel{4}}\left(\frac{19}{\cancel{2}}\right) \overset{?}{=} 38$$ Replace x by $\frac{19}{2}$.

$$38 = 38 \checkmark$$ It checks.

Practice Problem 3 Solve for x. $6x = 50$

In Examples 2 and 3 we *divided by the number multiplied by* x. This procedure is followed regardless of whether the sign of that number is positive or negative. In equations of the form $ax = b$ the **coefficient** of x is a. A coefficient is a multiplier.

EXAMPLE 4 Solve for x. $-3x = 48$

$$\frac{-3x}{-3} = \frac{48}{-3}$$ Divide both sides by -3.

$$x = -16$$ The solution is -16.

Check. **Can you check this solution?**

Practice Problem 4 Solve for x. $-27x = 54$

The coefficient of x may be 1 or -1. You may have to rewrite the equation so that the coefficient of 1 or -1 is obvious. With practice you may be able to recognize the coefficient without actually rewriting the equation.

EXAMPLE 5 Solve for x. $-x = -24$

$$-1x = -24$$ Rewrite the equation. $-1x$ is the same as $-x$.
   Now the coefficient of -1 is obvious.

$$\frac{-1x}{-1} = \frac{-24}{-1}$$ Divide both sides by -1.

$$x = 24$$ The solution is 24.

Check. **Can you check this solution?**

Practice Problem 5 Solve for x. $-x = 36$

The variable can be on either side of the equation. The equation $-78 = -3x$ can be solved in exactly the same way as $-3x = -78$.

EXAMPLE 6 Solve for x. $-78 = -3x$

$$\frac{-78}{-3} = \frac{-3x}{-3}$$ Divide both sides by -3.

$$26 = x$$ The solution is 26.

Check. $-78 = -3x$

$$-78 \overset{?}{=} -3(26)$$ Replace x by 26.

$$-78 = -78 \quad \checkmark \quad \text{It checks.}$$

Practice Problem 6 Solve for x. $-51 = 6x$

There is a mathematical concept that unites what we have learned in this section. The concept uses the idea of a multiplicative inverse. For any nonzero number a, the **multiplicative inverse** of a is $1/a$. Likewise, for any nonzero number a, the multiplicative inverse of $1/a$ is a. So to solve an equation of the form $ax = b$, we say that we need to multiply each side by the multiplicative inverse of a. Thus to solve $5x = 45$, we would multiply each side of the equation by the multiplicative inverse of 5, which is $1/5$. In similar fashion, if we wanted to solve the equation $(1/6)x = 4$, we would multiply each side of the equation by the multiplicative inverse of $1/6$, which is 6. In general, all the problems we have covered so far in this section can be solved by multiplying both sides of the equation by the multiplicative inverse of the coefficient of x.

EXAMPLE 7 Solve for x. $31.2 = 6.0x - 0.8x$

$31.2 = 6.0x - 0.8x$ There are like terms on the right side.

$31.2 = 5.2x$ Collect like terms.

$$\frac{31.2}{5.2} = \frac{5.2x}{5.2}$$ Divide both sides by 5.2 (which is the same as multiplying both sides by the multiplicative inverse of 5.2).

$6 = x$ The solution is 6.

Note: Be sure to place the decimal point in the quotient directly above the caret $(\wedge)$ when performing the division.

$$
\begin{array}{r}
6. \\
5.2_{\wedge}\overline{)31.2_{\wedge}} \\
\underline{31\ 2} \\
0
\end{array}
$$

Check. The check is up to you.

Practice Problem 7 Solve for x. $21 = 4.2x$

2.2 Exercises

Verbal and Writing Skills

1. To solve the equation $6x = -24$, divide each side of the equation by _____.

2. To solve the equation $-7x = 56$, divide each side of the equation by _____.

3. To solve the equation $\frac{1}{7}x = -2$, multiply each side of the equation by _____.

4. To solve the equation $\frac{1}{9}x = 5$, multiply each side of the equation by _____.

Solve for x. Be sure to reduce your answer. Check your solution.

5. $\frac{1}{7}x = 5$

6. $\frac{1}{5}x = 6$

7. $\frac{1}{3}x = -9$

8. $\frac{1}{4}x = -20$

9. $\frac{x}{5} = 16$

10. $\frac{x}{10} = 8$

11. $-3 = \frac{x}{5}$

12. $\frac{x}{3} = -12$

13. $12x = 48$

14. $8x = 72$

15. $-16 = 6x$

16. $-35 = 21x$

17. $1.5x = 75$

18. $2x = 0.36$

19. $-15 = -x$

20. $32 = -x$

21. $-84 = 12x$

22. $-72 = -9x$

23. $0.4x = 0.08$

24. $2.1x = 0.3$

Determine whether the given solution is correct. If it is not, find the correct solution.

25. Is 7 the solution for $-3x = 21$?

26. Is 8 the solution for $5x = -40$?

27. Is -6 the solution for $-11x = 66$?

28. Is -20 the solution for $-x = 20$?

Find the value of the variable that satisfies the equation.

29. $-3y = 2.4$

30. $5z = -1.8$

31. $-56 = -21t$

32. $34 = -51q$

33. $4.6y = -3.22$

34. $-2.8y = -3.08$

35. $4x + 3x = 21$

36. $5x + 4x = 36$

37. $2x - 7x = 20$

38. $3x - 9x = 18$

39. $-6x - 3x = -7$

40. $y - 11y = 7$

41. $-\frac{2}{3} = -\frac{4}{7}x$

42. $5.6 = -2.7x$

43. $3.6172x = -19.026472$

44. $-4.0518x = 14.505444$

To Think About

45. We have said that if $a = b$ and $c \neq 0$, then $ac = bc$. Why is it important that $c \neq 0$? What would happen if we tried to solve an equation by multiplying both sides by zero?

46. We have said that if $a = b$ and $c \neq 0$ then $\frac{a}{c} = \frac{b}{c}$. Why is it important that $c \neq 0$? What would happen if we tried to solve an equation by dividing both sides by zero?

Cumulative Review Problems

Evaluate using the correct order of operations. (Be careful to avoid sign errors.)

47. $(-6)(-8) + (-3)(2)$

48. $(-3)^3 + (-20) \div 2$

49. $5 + (2 - 6)^2$

50. In 1995, the humpback whale calf population at Stellwagen Bank, near Gloucester, Massachusetts, was estimated at 12 calves. The population grew by 21 calves in 1996, 18 calves in 1997, and 51 calves in 1998. In 1999, the number of whale calves decreased by 4, and in 2000 it increased by 6. What was the whale calf population at the end of 2000?

51. In January, Keiko invested $600 in a certain stock. In February, the stock gained $82.00. In March, the stock lost $47.00. In April, the stock gained $103.00. In May, the stock lost $106.00. What was Keiko's stock holding worth after the May loss?

Developing Your Study Skills

Getting Help

Getting the right kind of help at the right time can be a key ingredient in being successful in mathematics. When you have gone to class on a regular basis, taken careful notes, methodically read your textbook, and diligently done your homework—all of which means making every effort possible to learn the mathematics—you may find that you are still having difficulty. If this is the case, then you need to seek help. Make an appointment with your instructor to find out what help is available to you. The instructor, tutoring services, a mathematics lab, videotapes, and computer software may be among the resources you can draw on.

Once you discover the resources available in your school, you need to take advantage of them. Do not put it off, or you will find yourself getting behind. You cannot afford that. When studying mathematics, you must keep up with your work.

Student Learning Objectives

After studying this section, you will be able to:

1 Solve equations of the form $ax + b = c$.

2 Solve equations with the variable on both sides of the equation.

3 Solve equations with parentheses.

SSM

PH TUTOR CENTER   CD & VIDEO   MATH PRO   WEB

1 **Solving Equations of the Form** $ax + b = c$

Jenny Crawford scored several goals in field hockey during April. Her teammates scored three more than five times the number she scored. Together the team scored 18 goals in April. How many did Jenny score? To solve this problem, we need to solve the equation $5x + 3 = 18$.

To solve an equation of the form $ax + b = c$, we must use both the addition principle and the multiplication principle.

EXAMPLE 1 Solve for x to determine how many goals Jenny scored and check your solution.

$$5x + 3 = 18$$

We first want to isolate the variable term.

$5x + 3 + (-3) = 18 + (-3)$	Use the addition principle to add -3 to both sides.
$5x = 15$	Simplify.
$\dfrac{5x}{5} = \dfrac{15}{5}$	Use the division principle to divide both sides by 5.
$x = 3$	The solution is 3. Thus Jenny scored 3 goals.

Check.
$$5(3) + 3 \stackrel{?}{=} 18$$
$$15 + 3 \stackrel{?}{=} 18$$
$$18 = 18 \checkmark \qquad \text{It checks.}$$

Practice Problem 1 Solve for x and check your solution. $9x + 2 = 38$

2 **Solving Equations with the Variable on Both Sides of the Equation**

In some cases the variable appears on both sides of the equation. We would like to rewrite the equation so that all the terms containing the variable appear on one side. To do this, we apply the addition principle to the variable term.

EXAMPLE 2 Solve for x. $9x = 6x + 15$

$9x + (-6x) = 6x + (-6x) + 15$	Add $-6x$ to both sides. Notice $6x + (-6x)$ eliminates the variable on the right side.
$3x = 15$	Combine like terms.
$\dfrac{3x}{3} = \dfrac{15}{3}$	Divide both sides by 3.
$x = 5$	The solution is 5.

Check. The check is left to the student.

Practice Problem 2 Solve for x. $13x = 2x - 66$

In many problems the variable terms and constant terms appear on both sides of the equations. You will want to get all the variable terms on one side and all the constant terms on the other side.

EXAMPLE 3   Solve for x and check your solution. $9x + 3 = 7x - 2$

$$9x + (-7x) + 3 = 7x + (-7x) - 2 \qquad \text{Add } -7x \text{ to both sides of the equation.}$$
$$2x + 3 = -2 \qquad \text{Combine like terms.}$$
$$2x + 3 + (-3) = -2 + (-3) \qquad \text{Add } -3 \text{ to both sides.}$$
$$2x = -5 \qquad \text{Simplify.}$$
$$\frac{2x}{2} = \frac{-5}{2} \qquad \text{Divide both sides by 2.}$$
$$x = -\frac{5}{2} \qquad \text{The solution is } -\tfrac{5}{2}.$$

Check.
$$9x + 3 = 7x - 2$$
$$9\left(-\frac{5}{2}\right) + 3 \overset{?}{=} 7\left(-\frac{5}{2}\right) - 2 \qquad \text{Replace } x \text{ by } -\frac{5}{2}.$$
$$-\frac{45}{2} + 3 \overset{?}{=} -\frac{35}{2} - 2 \qquad \text{Simplify.}$$
$$-\frac{45}{2} + \frac{6}{2} \overset{?}{=} -\frac{35}{2} - \frac{4}{2} \qquad \begin{array}{l}\text{Change to equivalent fractions} \\ \text{with a common denominator.}\end{array}$$
$$-\frac{39}{2} = -\frac{39}{2} \ \checkmark \qquad \text{It checks. The solution is } -\tfrac{5}{2}.$$

Practice Problem 3   Solve for x and check your solution. $3x + 2 = 5x + 2$

In our next example we will study equations that need simplifying before any other steps are taken. Where it is possible, you should first collect like terms on one or both sides of the equation. The variable terms can be collected on the right side or the left side. In this example we will collect all the x-terms on the right side.

EXAMPLE 4   Solve for x. $5x + 26 - 6 = 9x + 12x$

$$5x + 20 = 21x \qquad \text{Combine like terms.}$$
$$5x + (-5x) + 20 = 21x + (-5x) \qquad \text{Add } -5x \text{ to both sides.}$$
$$20 = 16x \qquad \text{Combine like terms.}$$
$$\frac{20}{16} = \frac{16x}{16} \qquad \text{Divide both sides by 16.}$$
$$\frac{5}{4} = x \qquad \text{Don't forget to reduce the resulting fraction.}$$

Check.   The check is left to the student.

Practice Problem 4   Solve for z. $-z + 8 - z = 3z + 10 - 3$

Do you really need all these steps? No. As you become more proficient, you will be able to combine or eliminate some of these steps. However, it is best to write each step in its entirety until you are consistently obtaining the correct solution. It is much better to show every step than to take a lot of shortcuts and possibly obtain a wrong answer. This is a section of the algebra course where working neatly and accurately will help you—both now and as you progress through the course.

All the equations we have been studying so far are called **first-degree equations**. This means the variable terms are not squared (such as x^2 or y^2) or raised to some higher power. It is possible to solve equations with x^2 and y^2 terms by the same methods we have used so far. If x^2 or y^2 terms appear, try to collect them on one side of the equation. If the squared term drops out, you may solve the equation as a first-degree equation using the methods discussed in this section.

EXAMPLE 5 Solve for y. $5y^2 + 6y - 2 = -y + 5y^2 + 12$

$5y^2 - 5y^2 + 6y - 2 = -y + 5y^2 - 5y^2 + 12$	Subtract $5y^2$ from both sides.
$6y - 2 = -y + 12$	Combine, since $5y^2 - 5y^2 = 0$.
$6y + y - 2 = -y + y + 12$	Add y to each side.
$7y - 2 = 12$	Simplify.
$7y - 2 + 2 = 12 + 2$	Add 2 to each side.
$7y = 14$	Simplify.
$\dfrac{7y}{7} = \dfrac{14}{7}$	Divide each side by 7.
$y = 2$	Simplify. The solution is 2.

Check. The check is left to the student.

Practice Problem 5 Solve for x. $2x^2 - 6x + 3 = -4x - 7 + 2x^2$

 Solving Equations with Parentheses

The equations that you just solved are simpler versions of equations that we will now discuss. These equations contain parentheses. If the parentheses are first removed, the problems then become just like those encountered previously. We use the distributive property to remove the parentheses.

EXAMPLE 6 Solve for x and check your solution. $4(x + 1) - 3(x - 3) = 25$

$4(x + 1) - 3(x - 3) = 25$	
$4x + 4 - 3x + 9 = 25$	Multiply by 4 and -3 to remove parentheses. Be careful of the signs. Remember that $(-3)(-3) = 9$.

After removing the parentheses, it is important to collect like terms on each side of the equation. Do this before going on to isolate the variable.

$x + 13 = 25$	Collect like terms.
$x + 13 - 13 = 25 - 13$	Add -13 to both sides to isolate the variable.
$x = 12$	The solution is 12.
Check. $4(12 + 1) - 3(12 - 3) \overset{?}{=} 25$	Replace x by 12.
$4(13) - 3(9) \overset{?}{=} 25$	Combine numbers inside parentheses.
$52 - 27 \overset{?}{=} 25$	Multiply.
$25 = 25$ ✓	Simplify. It checks.

Practice Problem 6 Solve for x and check your solution.

$$4x - (x + 3) = 12 - 3(x - 2)$$

EXAMPLE 7 Solve for x. $3(-x - 7) = -2(2x + 5)$

$-3x - 21 = -4x - 10$	Remove parentheses. Watch the signs carefully.
$-3x + 4x - 21 = -4x + 4x - 10$	Add $4x$ to both sides.
$x - 21 = -10$	Simplify.
$x - 21 + 21 = -10 + 21$	Add 21 to both sides.
$x = 11$	The solution is 11.

Check.   The check is left to the student.

Practice Problem 7   Solve for x. $4(-2x - 3) = -5(x - 2) + 2$

In problems that involve decimals, great care should be taken. In some steps you will be multiplying decimal quantities, and in other steps you will be adding them.

EXAMPLE 8   Solve for x. $0.3(1.2x - 3.6) = 4.2x - 16.44$

$0.36x - 1.08 = 4.2x - 16.44$	Remove parentheses.
$0.36x - 0.36x - 1.08 = 4.2x - 0.36x - 16.44$	Subtract $0.36x$ from both sides.
$-1.08 = 3.84x - 16.44$	Collect like terms.
$-1.08 + 16.44 = 3.84x - 16.44 + 16.44$	Add 16.44 to both sides.
$15.36 = 3.84x$	Simplify.
$\dfrac{15.36}{3.84} = \dfrac{3.84x}{3.84}$	Divide both sides by 3.84.
$4 = x$	The solution is 4.

Check.   The check is left to the student.

Practice Problem 8   Solve for x. $0.3x - 2(x + 0.1) = 0.4(x - 3) - 1.1$

EXAMPLE 9   Solve for z and check. $2(3z - 5) + 2 = 4z - 3(2z + 8)$

$6z - 10 + 2 = 4z - 6z - 24$	Remove parentheses.
$6z - 8 = -2z - 24$	Collect like terms.
$6z - 8 + 2z = -2z + 2z - 24$	Add $2z$ to each side.
$8z - 8 = -24$	Simplify.
$8z - 8 + 8 = -24 + 8$	Add 8 to each side.
$8z = -16$	Simplify.
$\dfrac{8z}{8} = \dfrac{-16}{8}$	Divide each side by 8.
$z = -2$	Simplify. The solution is -2.

Check. $2[3(-2) - 5] + 2 \overset{?}{=} 4(-2) - 3[2(-2) + 8]$	Replace z by -2.
$2[-6 - 5] + 2 \overset{?}{=} -8 - 3[-4 + 8]$	Multiply.
$2[-11] + 2 \overset{?}{=} -8 - 3[4]$	Simplify.
$-22 + 2 \overset{?}{=} -8 - 12$	
$-20 = -20$ ✓	It checks.

Practice Problem 9   Solve for z and check. $5(2z - 1) + 7 = 7z - 4(z + 3)$

Find the value of the variable that satisfies the equation. Check your solution. Answers that are not integers may be left in fractional form or decimal form.

1. $4x + 13 = 21$

2. $7x - 4 = 59$

3. $4x - 11 = 13$

4. $5x - 11 = 39$

5. $4x - 21 = -91$

6. $9x - 13 = -76$

7. $-4x + 17 = -35$

8. $-6x + 25 = -83$

9. $-15 + 2x = 15$

10. $-8 + 4x = 8$

11. $\frac{1}{5}x - 2 = 6$

12. $\frac{1}{3}x - 7 = 4$

13. $\frac{1}{6}x + 2 = -4$

14. $\frac{1}{5}x + 6 = -24$

15. $8x = 48 + 2x$

16. $5x = 22 + 3x$

17. $-6x = -27 + 3x$

18. $-7x = -26 + 6x$

19. $63 - x = 8x$

20. $56 - 3x = 5x$

21. $54 - 2x = -8x$

To Think About

22. Is 2 the solution for $2y + 3y = 12 - y$?

23. Is 4 the solution for $5y + 2 = 6y - 6 + y$?

24. Is 11 a solution for $7x + 6 - 3x = 2x - 5 + x$?

25. Is -12 a solution for $9x + 2 - 5x = -8 + 5x - 2$?

Solve for y by getting all the y-terms on the left. Then solve for y by getting all the y-terms on the right. Which approach is better?

26. $-3 + 10y + 6 = 15 + 12y - 18$

27. $7y + 21 - 5y = 5y - 7 + y$

Solve for the variable. You may move the variable terms to the right or to the left.

28. $14 - 2x = -5x + 11$

29. $8 - 3x = 7x + 8$

30. $x - 6 = 8 - x$

31. $2x + 5 = 4x - 5$

32. $6y - 5 = 8y - 7$

33. $11y - 8 = 9y - 16$

34. $5x - 9 + 2x = 3x + 23 - 4x$

35. $9x - 5 + 4x = 7x + 43 - 2x$

Remove the parentheses and solve for the variable. Check your solution. Answers that are not integers may be left in fractional form or decimal form.

36. $5(x + 3) = 35$

37. $6(x + 2) = 42$

38. $6(3x + 2) - 8 = -2$ **39.** $4(2x + 1) - 7 = 6 - 5$

40. $7x - 3(5 - x) = 10$

41. $6(3 - 4x) + 17 = 8x - 3(2 - 3x)$

42. $0.7x - 0.2(x + 1) = 0.16$

43. $3(x + 0.2) = 2(x - 0.3) + 4.3$

44. $5(x - 3) + 5 = 3(x + 2) - 4$

45. $3(x - 2) + 2 = 2(x - 4)$

46. $0.2(x + 3) - (x - 1.5) = 0.3(x + 2) - 2.9$

47. $3(x + 0.2) - (2x + 0.5) = 2(x + 0.3) - 0.5$

48. $-3(y - 3y) + 4 = -4(3y - y) + 6 + 13y$

49. $2(4x - x) + 6 = 2(2x + x) + 8 - x$

Mixed Practice

Solve for the variable.

50. $5.7x + 3 = 4.2x - 3$

51. $4x - 3.1 = 5.3 - 3x$

52. $5z + 7 - 2z = 32 - 2z$

53. $8 - 7z + 2z = 20 + 5z$

54. $-4w - 28 = -7 - w$

55. $-6w - 7 = -3 - 8w$

56. $6x + 8 - 3x = 11 - 12x - 13$

57. $4 - 7x - 13 = 8x - 3 - 5x$

58. $2x^2 - 3x - 8 = 2x^2 + 5x - 6$

59. $3x^2 + 4x - 7 = 3x^2 - 5x + 2$

60. $-3.5x + 1.3 = -2.7x + 1.5$

61. $2.8x - 0.9 = 5.2x - 3.3$

62. $5(4 + x) = 3(3x - 1) - 9$ **63.** $x - 0.8x + 4 = 2.6$ **64.** $17(y + 3) - 4(y - 10) = 13$

65. $3x + 2 - 1.7x = 0.6x + 31.4$ **66.** $3(x + 4) - 5(3x - 2) = 8$ **67.** $3(2z - 4) - 4(z + 5) = 5(z - 4)$

Solve for x. Round your answer to the nearest hundredth.

68. $1.63x - 9.23 = 5.71x + 8.04$

69. $-2.21x + 8.65 = 3.69x - 7.78$

Cumulative Review Problems

Simplify.

70. $2x(3x - y) + 4(2x^2 - 3xy)$

71. $2\{x - 3[4 + 2(3 + x)]\}$

72. On March 30, 2000, William owned three different stocks: ABC, RTC, and JTJ. His portfolio contained the following:

4.0 shares of ABC stock valued at $42.25,

3.2 shares of RTC stock valued at $161.50, and

5.2 shares of JTJ stock valued at $102.

Find the market value of William's stock holdings on March 30, 2000.

73. A rectangle has a width of $\frac{1}{4}x + 5$ feet and a length of $\frac{5}{8}x - 1$ feet. Write an expression for the perimeter of the rectangle.

Equations with Fractions

1 Solving Equations with Fractions

Student Learning
Objectives

After studying this section, you
will be able to:

1 Solve equations with
fractions.

SSM   PH TUTOR   CD & VIDEO   MATH PRO   WEB
        CENTER

Equations with fractions can be rather difficult to solve. This difficulty is simply due to the extra care we usually have to use when computing with fractions. The actual equation-solving procedures are the same, with fractions or without. To avoid unnecessary work, we transform the given equation with fractions to an equivalent equation that does not contain fractions. How do we do this? We multiply each side of the equation by the least common denominator of all the fractions contained in the equation. We then use the distributive property so that the LCD is multiplied by each term of the equation.

EXAMPLE 1 Solve for x. $\frac{1}{4}x - \frac{2}{3} = \frac{5}{12}x$

First we find that the LCD $= 12$.

$$12\left(\frac{1}{4}x - \frac{2}{3}\right) = 12\left(\frac{5}{12}x\right) \qquad \text{Multiply each side by 12.}$$

$$\left(\frac{12}{1}\right)\left(\frac{1}{4}\right)(x) - \left(\frac{12}{1}\right)\left(\frac{2}{3}\right) = \left(\frac{12}{1}\right)\left(\frac{5}{12}\right)(x) \qquad \text{Use the distributive property.}$$

$$3x - 8 = 5x \qquad \text{Simplify.}$$

$$3x + (-3x) - 8 = 5x + (-3x) \qquad \text{Add } -3x \text{ to each side.}$$

$$-8 = 2x \qquad \text{Simplify.}$$

$$\frac{-8}{2} = \frac{2x}{2} \qquad \text{Divide each side by 2.}$$

$$-4 = x \qquad \text{Simplify.}$$

Check.

$$\frac{1}{4}(-4) - \frac{2}{3} \overset{?}{=} \frac{5}{12}(-4)$$

$$-1 - \frac{2}{3} \overset{?}{=} -\frac{5}{3}$$

$$-\frac{3}{3} - \frac{2}{3} \overset{?}{=} -\frac{5}{3}$$

$$-\frac{5}{3} = -\frac{5}{3} \quad \checkmark \qquad \text{It checks.}$$

Practice Problem 1 Solve for x. $\frac{3}{8}x - \frac{3}{2} = \frac{1}{4}x$

In Example 1 we multiplied each side of the equation by the LCD. However, most students prefer to go immediately to the second step and multiply each term by the LCD. This avoids having to write out a separate step using the distributive property.

EXAMPLE 2 Solve for x and check your solution. $\frac{x}{3} + 3 = \frac{x}{5} - \frac{1}{3}$

$$15\left(\frac{x}{3}\right) + 15(3) = 15\left(\frac{x}{5}\right) - 15\left(\frac{1}{3}\right) \qquad \text{The LCD is 15. Use the multiplication principle to multiply each term by 15.}$$

$$5x + 45 = 3x - 5 \qquad \text{Simplify.}$$

$$5x - 3x + 45 = 3x - 3x - 5 \qquad \text{Add } -3x \text{ to both sides.}$$

$$2x + 45 = -5 \qquad \text{Combine like terms.}$$

$$2x + 45 - 45 = -5 - 45 \qquad \text{Add } -45 \text{ to both sides.}$$

$$2x = -50 \qquad \text{Simplify.}$$

85

$$\frac{2x}{2} = \frac{-50}{2}$$ Divide both sides by 2.

$$x = -25$$ The solution is −25.

Check.

$$\frac{-25}{3} + 3 \overset{?}{=} \frac{-25}{5} - \frac{1}{3}$$

$$-\frac{25}{3} + \frac{9}{3} \overset{?}{=} -\frac{5}{1} - \frac{1}{3}$$

$$-\frac{16}{3} \overset{?}{=} -\frac{15}{3} - \frac{1}{3}$$

$$-\frac{16}{3} = -\frac{16}{3} \checkmark$$

Practice Problem 2 Solve for x and check your solution. $\frac{5x}{4} - 1 = \frac{3x}{4} + \frac{1}{2}$

EXAMPLE 3 Solve for x. $\frac{x + 5}{7} = \frac{x}{4} + \frac{1}{2}$

$$\frac{x}{7} + \frac{5}{7} = \frac{x}{4} + \frac{1}{2}$$ First we rewrite the left side as two fractions. This is actually multiplying $\frac{1}{7}(x + 5) = \frac{x}{7} + \frac{5}{7}$.

$$28\left(\frac{x}{7}\right) + 28\left(\frac{5}{7}\right) = 28\left(\frac{x}{4}\right) + 28\left(\frac{1}{2}\right)$$ We observe that the LCD is 28, so we multiply each term by 28.

$$4x + 20 = 7x + 14$$ Simplify.

$$4x - 4x + 20 = 7x - 4x + 14$$ Add $-4x$ to both sides.

$$20 = 3x + 14$$ Combine like terms.

$$20 - 14 = 3x + 14 - 14$$ Add -14 to both sides.

$$6 = 3x$$ Combine like terms.

$$\frac{6}{3} = \frac{3x}{3}$$ Divide both sides by 3.

$$2 = x$$ The solution is 2.

Check.   The check is left to the student.

Practice Problem 3 Solve for x. $\frac{5x}{6} - \frac{5}{8} = \frac{3x}{4} - \frac{1}{3}$

If a problem contains both parentheses and fractions, it is best to remove the parentheses first. Many students find it is helpful to have a written procedure to follow in solving these more involved equations.

Procedure to Solve Equations

1. Remove any parentheses.
2. If fractions exist, multiply all terms on both sides by the least common denominator of all the fractions.
3. Combine like terms if possible.
4. Add or subtract terms on both sides of the equation to get all terms with the variable on one side of the equation.
5. Add or subtract a constant value on both sides of the equation to get all terms not containing the variable on the other side of the equation.
6. Divide both sides of the equation by the coefficient of the variable.
7. Simplify the solution (if possible).
8. Check your solution.

Let's use each step in solving the next example.

EXAMPLE 4 Solve for x and check your solution. $\frac{1}{3}(x - 2) = \frac{1}{5}(x + 4) + 2$

Step 1 $\dfrac{x}{3} - \dfrac{2}{3} = \dfrac{x}{5} + \dfrac{4}{5} + 2$ Remove parentheses.

Step 2 $15\left(\dfrac{x}{3}\right) - 15\left(\dfrac{2}{3}\right) = 15\left(\dfrac{x}{5}\right) + 15\left(\dfrac{4}{5}\right) + 15(2)$ Multiply by the LCD, 15.

$5x - 10 = 3x + 12 + 30$ Simplify.

Step 3 $5x - 10 = 3x + 42$ Combine like terms on each side.

Step 4 $5x - 3x - 10 = 3x - 3x + 42$ Add $-3x$ to both sides.

$2x - 10 = 42$ Simplify.

Step 5 $2x - 10 + 10 = 42 + 10$ Add 10 to both sides.

$2x = 52$ Simplify.

Step 6 $\dfrac{2x}{2} = \dfrac{52}{2}$ Divide both sides by 2.

Step 7 $x = 26$ Simplify the solution.

Step 8 Check. $\dfrac{1}{3}(26 - 2) \stackrel{?}{=} \dfrac{1}{5}(26 + 4) + 2$ Replace x by 26.

$\dfrac{1}{3}(24) \stackrel{?}{=} \dfrac{1}{5}(30) + 2$ Combine values within parentheses.

$8 \stackrel{?}{=} 6 + 2$ Simplify.

$8 = 8 \checkmark$ The solution is 26.

Practice Problem 4 Solve for x and check your solution.

$$\frac{1}{3}(x - 2) = \frac{1}{4}(x + 5) - \frac{5}{3}$$

Remember that not every step will be needed in each problem. You can combine some steps as well, *as long as you are consistently obtaining the correct solution.* However, you are encouraged to write out every step as a way of helping you to avoid careless errors.

It is important to remember that when we write decimals, these numbers are really fractions written in a special way. Thus, $0.3 = \frac{3}{10}$ and $0.07 = \frac{7}{100}$. It is possible to take an equation containing decimals and to multiply each term by the appropriate value to obtain integer coefficients.

EXAMPLE 5 Solve for x. $0.2(1 - 8x) + 1.1 = -5(0.4x - 0.3)$

$0.2 - 1.6x + 1.1 = -2.0x + 1.5$ Remove parentheses.

$10(0.2) - 10(1.6x) + 10(1.1) = 10(-2.0x) + 10(1.5)$ Multiply each term by 10.

$2 - 16x + 11 = -20x + 15$ Multiplying by 10 moves the decimal point one place to the right.

$-16x + 13 = -20x + 15$ Simplify.

$-16x + 20x + 13 = -20x + 20x + 15$ Add $20x$ to each side.

$4x + 13 = 15$ Simplify.

$4x + 13 + (-13) = 15 + (-13)$ Add -13 to each side.

$4x = 2$ Simplify.

$\dfrac{4x}{4} = \dfrac{2}{4}$ Divide each side by 4.

$x = \dfrac{1}{2}$ or 0.5 Simplify.

Check. $0.2[1 - 8(0.5)] + 1.1 \overset{?}{=} -5[0.4(0.5) - 0.3]$

$0.2[1 - 4] + 1.1 \overset{?}{=} -5[0.2 - 0.3]$

$0.2[-3] + 1.1 \overset{?}{=} -5[-0.1]$

$-0.6 + 1.1 \overset{?}{=} 0.5$

$0.5 = 0.5$ ✓

Practice Problem 5   Solve for x. $2.8 = 0.3(x - 2) + 2(0.1x - 0.3)$

To Think About   Does every equation have one solution? Actually, no. There are some rare cases where an equation has no solution at all. Suppose we try to solve the equation

$$5(x + 3) = 2x - 8 + 3x.$$

If we remove the parentheses and collect like terms we have

$$5x + 15 = 5x - 8.$$

If we add $-5x$ to each side, we obtain

$$15 = -8.$$

Clearly this is impossible. There is no value of x for which these two numbers are equal. We would say this equation has **no solution.**

One additional surprise may happen. An equation may have an infinite number of solutions. Suppose we try to solve the equation

$$9x - 8x - 7 = 3 + x - 10.$$

If we combine like terms on each side, we have the equation

$$x - 7 = x - 7.$$

If we add $-x$ to each side, we obtain

$$-7 = -7.$$

Now this statement is always true, no matter what the value of x. We would say this equation has **an infinite number of solutions.**

In the To Think About exercises in this section, we will encounter some equations that have no solution or an infinite number of solutions.

Developing Your Study Skills

Taking Notes in Class

An important part of studying mathematics is taking notes. To take meaningful notes, you must be an active listener. Keep your mind on what the instructor is saying, and be ready with questions whenever you do not understand something.

If you have previewed the lesson material, you will be prepared to take good notes. The important concepts will seem somewhat familiar. If you frantically try to write all that the instructor says or copy all the examples done in class, you may find your notes nearly worthless when you are home alone. Write down *important* ideas and examples as the instructor lectures, making sure that you are listening and following the logic. Include any helpful hints or suggestions that your instructor gives you or refers to in your text.

In exercises 1–16, solve for the variable and check your answer. Noninteger answers may be left in fractional form or decimal form.

1. $\dfrac{1}{5}x + \dfrac{1}{10} = \dfrac{1}{2}$

2. $\dfrac{1}{3}x - \dfrac{1}{9} = \dfrac{8}{9}$

3. $\dfrac{3}{4}x = \dfrac{1}{2}x + \dfrac{5}{8}$

4. $\dfrac{1}{2} = \dfrac{3}{10} - \dfrac{2}{5}x$

5. $\dfrac{x}{2} + \dfrac{x}{5} = \dfrac{7}{10}$

6. $\dfrac{x}{5} - \dfrac{x}{3} = \dfrac{8}{15}$

7. $20 - \dfrac{1}{3}x = \dfrac{1}{2}x$

8. $\dfrac{1}{5}x - \dfrac{1}{2} = \dfrac{1}{6}x$

9. $2 + \dfrac{y}{2} = \dfrac{3y}{4} - 3$

10. $\dfrac{x}{3} - 1 = -\dfrac{1}{2} - x$

11. $\dfrac{y-1}{2} = 4 - \dfrac{y}{7}$

12. $\dfrac{x-7}{6} = -\dfrac{1}{2}$

13. $0.3x - 2.2 = 3.2$

14. $2.8 - 0.4x = 8$

15. $0.6x + 5.9 = 3.8$

16. $1.2x - 2.2 = 5.6$

17. Is 4 a solution to $\dfrac{1}{2}(y - 2) + 2 = \dfrac{3}{8}(3y - 4)$?

18. Is 2 a solution to $\dfrac{1}{5}(y + 2) = \dfrac{1}{10}y + \dfrac{3}{5}$?

19. Is $\dfrac{5}{8}$ a solution to $\dfrac{1}{2}\left(y - \dfrac{1}{5}\right) = \dfrac{1}{5}(y + 2)$?

20. Is $\dfrac{13}{3}$ a solution to $\dfrac{y}{2} - \dfrac{7}{9} = \dfrac{y}{6} + \dfrac{2}{3}$?

Solve for the variable. Noninteger answers may be left in fractional form or decimal form.

21. $\dfrac{3}{4}(3x + 1) = 2(3 - 2x) + 1$

22. $2(x - 2) = \dfrac{2}{5}(3x + 1) + 2$

23. $0.7x - 3.3 = 2.5 - 0.2x - 5.8$

24. $0.5 - 2.1x = 6.4 + 0.3x - 5.9$

25. $0.3x - 0.2(3 - 5x) = -0.5(x - 6)$

26. $0.3(x - 2) + 0.4x = -0.2(x - 6)$

27. $-5(0.2x + 0.1) - 0.6 = 1.9$

28. $0.3x + 1.7 = 0.2x - 0.4(5x + 1)$

Mixed Practice

Solve. Noninteger answers may be left in fractional form or decimal form.

29. $\frac{1}{3}(y + 2) = 3y - 5(y - 2)$

30. $\frac{2}{5}(y + 3) - \frac{1}{2} = \frac{1}{3}(y - 2) + \frac{1}{2}$

31. $\frac{1 + 2x}{5} + \frac{4 - x}{3} = \frac{1}{15}$

32. $\frac{x + 3}{4} = 4x - 2(x - 3)$

33. $\frac{1}{5}(x + 3) = 2x - 3(2 - x) - 3$

34. $\frac{2}{3}(x + 4) = 6 - \frac{1}{4}(3x - 2) - 1$

35. $\frac{1}{3}(x - 2) = 3x - 2(x - 1) + \frac{16}{3}$

36. $\frac{3}{4}(x - 2) + \frac{3}{5} = \frac{1}{5}(x + 1)$

37. $\frac{3}{2}x + \frac{1}{3} = \frac{2x - 3}{4}$

38. $\frac{5}{3} - \frac{1}{6}x = \frac{3x + 5}{4}$

39. $0.8(x - 3) = -5(2.1x + 0.4)$

40. $0.7(2x + 3) = -4(6.1x + 0.2)$

41. $\frac{1}{5}x + \frac{2}{3} + \frac{1}{15} = \frac{4}{3}x - \frac{7}{15}$

42. $\frac{1}{20}x - \frac{1}{4} + \frac{3}{5} = -\frac{1}{2}x + \frac{1}{4}x$

To Think About

Solve. Be careful to examine your work to see if the equation may have no solution or an infinite number of solutions.

43. $-1 + 5(x - 2) = 12x + 3 - 7x$

44. $x + 3x - 2 + 3x = -11 + 7(x + 2)$

45. $9(x + 3) - 6 = 24 - 2x - 3 + 11x$

46. $7(x + 4) - 10 = 3x + 20 + 4x - 2$

Cumulative Review Problems

47. Add. $\dfrac{3}{7} + 1\dfrac{5}{10}$

48. Subtract. $\dfrac{3}{8} - 9$

49. Multiply. $\left(2\dfrac{1}{5}\right)(6)$

50. Divide. $15 \div 1\dfrac{1}{4}$

51. Derek is training for a 30-kilometer running race. How many miles is the race? Round your answer to the nearest tenth of a mile.

▲ **52.** Tom Rourke needs to replace the sail on his sailboat. It is in the shape of a triangle with an altitude of 9 feet and a base of 8 feet. The material to make the sail costs $3 per square foot. How much will the material cost to make a new sail for the boat?

▲ **53.** The Newbury Elementary School needs a new air vent drilled into the wall of the maintenance room. The circular hole that is necessary will have a radius of 6 inches. The stainless steel grill that will cover the vent costs $2 per square inch. How much will the vent cost? Use $\pi \approx 3.14$.

▲ **54.** Sally constructed a new countertop for her kitchen. It is in the shape of a parallelogram. The altitude of the parallelogram is 19 inches. The base of the parallelogram is 24 inches. If the laminate used to make the countertop costs $0.60 per square inch, how much will the laminate cost?

2.5 Formulas

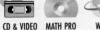
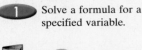
1 Solving a Formula for a Specified Variable

Formulas are equations with one or more variables that are used to describe real-life situations. The formula describes the relationship that exists among the variables. For example, in the formula $d = rt$, distance (d) is equal to the rate of speed (r) multiplied by the time (t). We can use this formula to find distance if we know the rate and time. Sometimes, however, we are given the distance and the rate, and we are asked to find the time.

EXAMPLE 1 Prior to a tragic accident in Paris in 2000, the Concorde was used extensively for business travel between New York and London. The British Airways Concorde had a range of approximately 3640 miles. The cruising speed for this plane was approximately 1300 miles per hour. How many hours did it take the Concorde to fly 3640 miles while traveling at cruising speed?

$$d = rt \qquad \text{Use the distance formula.}$$

$$3640 = 1300t \qquad \text{Substitute the known values for the variables.}$$

$$\frac{3640}{1300} = \frac{1300t}{1300} \qquad \text{Divide both sides of the equation by 1300 to solve for } t.$$

$$2.8 = t \qquad \text{Simplify.}$$

It took the Concorde 2.8 hours (2 hours and 48 minutes) to travel 3640 miles at cruising speed.

Practice Problem 1 On a flight in December 1999, the Concorde flew 3525 miles in 2.5 hours. Find the average rate of speed for the flight.

If we have many problems that ask us to find the time given the distance and rate, it may be worthwhile to rewrite the formula in terms of time.

EXAMPLE 2 Solve for t. $d = rt$

$$\frac{d}{r} = \frac{rt}{r} \qquad \text{We want to isolate } t. \text{ Therefore we divide both sides of the equation by the coefficient of } t, \text{ which is } r.$$

$$\frac{d}{r} = t \qquad \text{We have solved for the variable indicated.}$$

Practice Problem 2 Einstein's equation relating energy E to mass m and the speed of light c is $E = mc^2$. Solve it for m.

A straight line can be described by an equation of the form $Ax + By = C$ where A, B, and C are real numbers and A and B are not both zero. We will study this in later chapters. Often it is useful to solve such an equation for the variable y in order to make graphing the equation easier.

EXAMPLE 3 Solve for y. $3x - 2y = 6$

$$-2y = 6 - 3x \qquad \text{We want to isolate the term containing } y, \text{ so we subtract } 3x \text{ from both sides.}$$

$$\frac{-2y}{-2} = \frac{6 - 3x}{-2} \qquad \text{Divide both sides by the coefficient of } y.$$

$$y = \frac{6}{-2} + \frac{-3x}{-2} \qquad \text{Rewrite the fraction on the right side as two fractions.}$$

$$y = \frac{3}{2}x - 3 \qquad \text{Simplify and reorder the terms on the right.}$$

This is known as the slope–intercept form of the equation of a line.

Practice Problem 3 Solve for y. $8 - 2y + 3x = 0$

Our procedure for solving an equation can be rewritten to give us a procedure for solving a formula for a specified variable.

> ### Procedure to Solve a Formula for a Specified Variable
>
> **1.** Remove any parentheses.
> **2.** If fractions exist, multiply all terms on both sides by the LCD of all the fractions.
> **3.** Combine like terms on each side if possible.
> **4.** Add or subtract terms on both sides of the equation to get all terms with the desired variable on one side of the equation.
> **5.** Add or subtract the appropriate quantities to get all terms that do *not* have the desired variable on the other side of the equation.
> **6.** Divide both sides of the equation by the coefficient of the desired variable.
> **7.** Simplify if possible.

▲ ● **EXAMPLE 4**   A trapezoid is a four-sided figure with two parallel sides. If the parallel sides are a and b and the altitude is h, the area is given by

$$A = \frac{h}{2}(a + b).$$

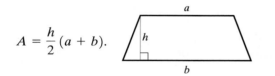

Solve this equation for a.

$$A = \frac{h}{2}(a + b)$$

$$A = \frac{ha}{2} + \frac{hb}{2} \qquad \text{Remove the parentheses.}$$

$$2(A) = 2\left(\frac{ha}{2}\right) + 2\left(\frac{hb}{2}\right) \qquad \text{Multiply all terms by the LCD of 2.}$$

$$2A = ha + hb \qquad \text{Simplify.}$$

$$2A - hb = ha \qquad \begin{array}{l}\text{We want to isolate the term containing } a. \\ \text{Therefore, we subtract } hb \text{ from both sides.}\end{array}$$

$$\frac{2A - hb}{h} = \frac{ha}{h} \qquad \text{Divide both sides by } h \text{ (the coefficient of } a).$$

$$\frac{2A - hb}{h} = a \qquad \text{The solution is obtained.}$$

Note: Although the solution is in simple form, it could be written in an alternative way. Since

$$\frac{2A - hb}{h} = \frac{2A}{h} - \frac{hb}{h} = \frac{2A}{h} - b,$$

we could also have written $\dfrac{2A}{h} - b = a$.

▲ **Practice Problem 4**   The relationship between the circumference C of a circle and the circle's diameter d is described by the equation $C = \pi d$. Solve it for d.   ●

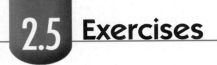

Verbal and Writing Skills

1. The formula for calculating the temperature in degrees Fahrenheit when you know the temperature in degrees Celsius is $F = \frac{9}{5}C + 32$. Explain in your own words how you would solve this equation for C.

▲ 2. The formula for finding the area of a trapezoid with an altitude of 9 meters and bases of b meters and c meters is given by the equation $A = \frac{9}{2}(b + c)$. Explain in your own words how you would solve this equation for b.

Applications

▲ 3. The formula for the area of a triangle is $A = \frac{1}{2}ab$, where b is the *base* of the triangle and a is the *altitude* of the triangle.
   (a) Use this formula to find the base of a triangle that has an area of 60 square meters and an altitude of 12 meters.

   (b) Use this formula to find the altitude of a triangle that has an area of 88 square meters and a base of 11 meters.

4. The formula for calculating simple interest is $I = Prt$, where P is the *principal* (amount of money invested), r is the *rate* at which the money is invested, and t is the *time*.
   (a) Use this formula to find how long it would take to earn \$720 in interest on an investment of \$3000 at the rate of 6%.

   (b) Use this formula to find the rate of interest if \$5000 earns \$400 interest in 2 years.

   (c) Use this formula to find the amount of money invested if the interest earned was \$120 and the rate of interest was 5% over 3 years.

5. The equation $3x - 5y = 15$ describes a line and is written in standard form.
   (a) Solve for the variable y.

   (b) Use this result to find y with $x = -5$.

6. The equation $6x + 3y = -14$ describes a line and is written in standard form.
   (a) Solve for the variable x.

   (b) Use this result to find x with $y = -2$.

In each formula or equation, solve for the variable indicated.

Area of a triangle

▲ **7.** $A = \frac{1}{2}bh$ Solve for b.

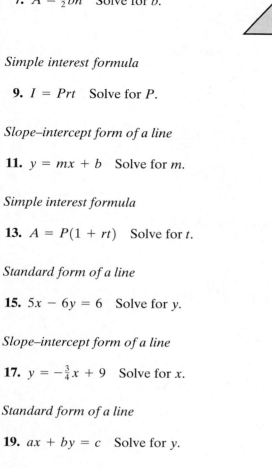

▲ **8.** $A = \frac{1}{2}bh$ Solve for h.

Simple interest formula

9. $I = Prt$ Solve for P.

10. $I = Prt$ Solve for r.

Slope–intercept form of a line

11. $y = mx + b$ Solve for m.

12. $y = mx + b$ Solve for b.

Simple interest formula

13. $A = P(1 + rt)$ Solve for t.

Area of a trapezoid

▲ **14.** $A = \frac{1}{2}a(b_1 + b_2)$ Solve for b_1.

Standard form of a line

15. $5x - 6y = 6$ Solve for y.

16. $5x + 9y = -18$ Solve for y.

Slope–intercept form of a line

17. $y = -\frac{3}{4}x + 9$ Solve for x.

18. $y = \frac{6}{7}x - 12$ Solve for x.

Standard form of a line

19. $ax + by = c$ Solve for y.

20. $ax + by = c$ Solve for x.

Area of a circle

▲ **21.** $A = \pi r^2$ Solve for r^2.

Surface area of a sphere

▲ **22.** $s = 4\pi r^2$ Solve for r^2.

Distance of a falling object

23. $S = \frac{1}{2}gt^2$ Solve for g.

24. $S = \frac{1}{2}gt^2$ Solve for t^2.

Surface area of a right circular cylinder

▲ **25.** $S = 2\pi rh + 2\pi r^2$ Solve for h.

Perimeter of a square

▲ **26.** $P = 4s$ Solve for s.

Volume of a right circular cylinder

▲ **27.** $V = \pi r^2 h$ Solve for h.

▲ **28.** $V = \pi r^2 h$ Solve for r^2.

Volume of a rectangular prism

▲ **29.** $V = LWH$ Solve for L.

▲ **30.** $V = LWH$ Solve for H.

Volume of a cone

▲ **31.** $V = \frac{1}{3}\pi r^2 h$ Solve for r^2.

▲ **32.** $V = \frac{1}{3}\pi r^2 h$ Solve for h.

Perimeter of a rectangle

▲ **33.** $P = 2L + 2W$ Solve for W.

▲ **34.** $P = 2L + 2W$ Solve for L.

Pythagorean theorem

▲ **35.** $c^2 = a^2 + b^2$ Solve for a^2.

▲ **36.** $c^2 = a^2 + b^2$ Solve for b^2.

Temperature conversion formulas

37. $F = \frac{9}{5}C + 32$ Solve for C.

38. $C = \frac{5}{9}(F - 32)$ Solve for F.

Boyle's law for gases

39. $P = k\left(\dfrac{T}{V}\right)$ Solve for T.

40. $P = k\left(\dfrac{T}{V}\right)$ Solve for V.

Area of a sector of a circle

▲ **41.** $A = \dfrac{\pi r^2 S}{360}$ Solve for S.

▲ **42.** $A = \dfrac{\pi r^2 S}{360}$ Solve for r^2.

Applications

▲ **43.** Use the result you obtained in exercise 33 to solve the following problem. A farmer has a rectangular field with a perimeter of 5.8 miles and a length of 2.1 miles. Find the width of the field.

▲ **44.** Use the result you obtained in exercise 34 to solve the following problem. Smithfield High School has a rectangular athletic field with a perimeter of 1260 yards and a width of 280 yards. Find the length of the field.

▲ **45.** Use the result you obtained in exercise 29 to solve the following problem. The foundation of a house is in the shape of a rectangular solid. The volume held by the foundation is 5940 cubic feet. The height of the foundation is 9 feet and the width is 22 feet. What is the length of the foundation?

▲ **46.** Use the result you obtained in exercise 30 to solve the following problem. The fish tank at the Mandarin Danvers Restaurant is in the shape of a rectangular solid. The volume held by the tank is 3024 cubic inches. The length of the tank is 18 inches while the width of the tank is 14 inches. What is the height of the tank?

47. The number of foreign visitors measured in thousands (V) admitted to the United States for a pleasure trip for any given year can be predicted by the equation $V = 1100x + 7050$, where x is the number of years since 1985. For example, if $x = 3$ (this would be the year 1988), the predicted number of visitors in thousands would be $1100(3) + 7050 = 10,350$. Thus we would predict that in 1988, a total of 10,350,000 visitors came to the United States for a pleasure trip. (Source: U.S. Immigration and Naturalization Service.)

(a) Solve this equation for x.

(b) Use the result of your answer in (a) to find the year in which the number of visitors will be predicted to be 25,750,000. (*Hint:* Let $V = 25,750$ in your answer for (a).)

48. The number of foreign visitors from Europe measured in thousands (E) admitted to the United States from Europe for a pleasure trip for any given year can be predicted by the equation $E = 480x + 2400$, where x is the number of years since 1985. For example, if $x = 7$ (this would be the year 1992), the predicted number of visitors in thousands would be $480(7) + 2400 = 5760$. Thus we would predict that in 1992, a total of 5,760,000 visitors from Europe came to the United States for a pleasure trip. (Source: U.S. Immigration and Naturalization Service.)

(a) Solve this equation for x.

(b) Use the result of your answer in (a) to find the year in which the number of visitors from Europe will be predicted to be 11,520,000. (*Hint:* Let $E = 11,520$ in your answer for (a).)

To Think About

49. In $I = Prt$, if t doubles, what is the effect on I?

50. In $I = Prt$, if both r and t double, what is the effect on I?

▲ **51.** In $A = \pi r^2$, if r doubles, what is the effect on A?

▲ **52.** In $A = \pi r^2$, if r is halved, what is the effect on A?

Cumulative Review Problems

53. $-8 + 7 - 3 - (-12) + 2$

54. $-\dfrac{12}{35} \div \left(-\dfrac{8}{7}\right)$

55. $3[2 - 2\{5 + 3(2 - 5)\}]$

56. Evaluate $2x^2 - 3x$ for $x = -5$

▲ **57.** A very popular handheld electronic game requires 3.25 square feet of a certain type of durable plastic in the manufacturing process. How many square feet of durable plastic does this company need to make 12,000 handheld games?

58. The Superstar Lighting Company rents out giant spotlights that shine up in the sky to mark the location of special events, such as the opening of a movie, a major sports play-off game, or a huge sales event at an auto dealership. A giant spotlight was used for $4\frac{1}{3}$ hours on Saturday, $2\frac{3}{4}$ hours on Tuesday, and $3\frac{1}{2}$ hours on Wednesday. What was the total number of hours that the spotlight was in use?

Putting Your Skills to Work

Controlling Emissions Levels

The U.S. Environmental Protection Agency monitors the emissions of a number of pollutants in the air in the United States. Each year, measurements are made to identify the number of tons of each type of pollutant. The following chart records in millions of tons the amount of emissions of nitrogen dioxide and volatile organic compounds.

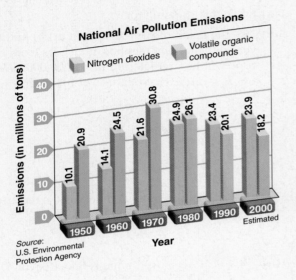

Problems for Individual Study and Investigation

1. How many more tons of nitrogen dioxide were emitted in 1980 than in 1970?

2. How many fewer tons of volatile organic compounds were emitted in 2000 than in 1990?

Problems for Cooperative Study and Investigation

Scientists sometimes use the equation $n = 23.9 + 0.05x$, where x represents the number of years since the year 2000, to predict the number of millions of tons of nitrogen dioxide that will be emitted per year.

3. Use the formula to estimate how many tons of nitrogen dioxide will be emitted in 2005.

4. If the actual amount of nitrogen dioxide emitted in 2010 turns out to be 24.7 million tons, by how many tons will the formula be in error?

Scientists sometimes use the equation $v = 18.2 - 0.19x$, where x represents the number of years since the year 2000, to predict the number of millions of tons of volatile organic compounds that will be emitted per year.

5. Solve the formula for the variable x. Leave the answer in the form of a fractional expression. Do not divide out the decimals.

6. Use the result from question 5 to find the year when 16.68 million tons of volatile organic compounds will be emitted.

Internet Connections

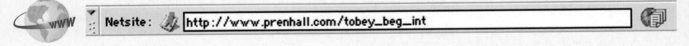

Site: U.S. Environmental Protection Agency

7. Examine the three separate tables for carbon monoxide, nitrogen oxides, and sulfur dioxide emissions estimates 1989–1998. For each of these, by what percent did total emissions increase or decrease in 1998 as compared with 1989?

8. In what year did the emissions of carbon monoxide from transportation sources have the greatest percent decrease from the previous year? What was that percent decrease?

2.6 Writing and Graphing Inequalities

1 Interpreting Inequality Statements

Student Learning Objectives

After studying this section, you will be able to:

1 Interpret inequality statements.

2 Graph an inequality on a number line.

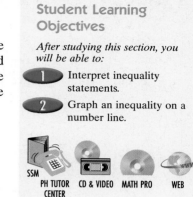

SSM PH TUTOR CD & VIDEO MATH PRO WEB
CENTER

We frequently speak of one value being greater than or less than another value. We say that "5 is less than 7" or "9 is greater than 4." These relationships are called **inequalities.** We can write inequalities in mathematics by using symbols. We use the symbol **<** to represent the words "**is less than.**" We use the symbol **>** to represent the words "**is greater than.**"

Statement in Words	Statement in Algebra
5 is less than 7.	$5 < 7$
9 is greater than 4.	$9 > 4$

Note: "5 is less than 7" and "7 is greater than 5" have the same meaning. Similarly, $5 < 7$ and $7 > 5$ have the same meaning. They represent two equivalent ways of describing the same relationship between the two numbers 5 and 7.

We can better understand the concept of inequality if we examine a number line.

We say that one number is greater than another if it is to the right of the other on the number line. Thus $7 > 5$, since 7 is to the right of 5.

What about negative numbers? We can say "-1 is greater than -3" and write it in symbols as $-1 > -3$ because we know that -1 lies to the right of -3 on the number line.

EXAMPLE 1   In each statement, replace the question mark with the symbol $<$ or $>$.

(a) $3 \,?\, -1$ **(b)** $-2 \,?\, 1$ **(c)** $-3 \,?\, -4$ **(d)** $0 \,?\, 3$ **(e)** $-3 \,?\, 0$

(a) $3 > -1$ Use $>$, since 3 is to the right of -1 on the number line.

(b) $-2 < 1$ Use $<$, since -2 is to the left of 1.
    (Or equivalently, we could say that 1 is to the right of -2.)

(c) $-3 > -4$ Note that -3 is to the right of -4.

(d) $0 < 3$

(e) $-3 < 0$

Practice Problem 1   In each statement, replace the question mark with the symbol $<$ or $>$.

(a) $7 \,?\, 2$ **(b)** $-3 \,?\, -4$ **(c)** $-1 \,?\, 2$ **(d)** $-8 \,?\, -5$ **(e)** $0 \,?\, -2$ **(f)** $\dfrac{2}{5} \,?\, \dfrac{3}{8}$

2 Graphing an Inequality on a Number Line

Sometimes we will use an inequality to express the relationship between a variable and a number. $x > 3$ means that x could have the value of *any number* greater than 3.

Any number that makes an inequality true is called a **solution** of the inequality. The set of all numbers that make the inequality true is called the **solution set.** A picture that represents all of the solutions of an inequality is called a **graph** of the inequality. The inequality $x > 3$ can be graphed on the number line as follows:

Note that all of the points to the right of 3 are shaded. The open circle at 3 indicates that we do not include the point for the number 3.

Similarly, we can graph $x < -2$ as follows:

Note that all of the points to the left of -2 are shaded.

Sometimes a variable will be either greater than or equal to a certain number. In the statement "x is greater than or equal to 3," we are implying that x could have the value of 3 or any number greater than 3. We write this as $x \geq 3$. We graph it as follows:

Note that the closed circle at 3 indicates that we *do* include the point for the number 3.

Similarly, we can graph $x \leq -2$ as follows:

EXAMPLE 2   State each mathematical relationship in words and then graph it.

(a) $x < -2$ **(b)** $-3 < x$ **(c)** $x \geq -2$ **(d)** $x \leq -6$

(a) We state that "x is less than -2."

$$x < -2$$

(b) We can state that "-3 is less than x" or, equivalently, that "x is greater than -3." Be sure you see that $-3 < x$ is equivalent to $x > -3$. Although both statements are correct, we *usually write the variable first* in a simple inequality containing a variable and a numerical value.

$$x > -3$$

(c) We state that "x is greater than or equal to -2."

$$x \geq -2$$

(d) We state that "x is less than or equal to -6."

$$x \leq -6$$

Practice Problem 2   State each mathematical relationship in words and then graph it on a number line in the margin.

(a) $x > 5$

(b) $x \leq -2$

(c) $3 > x$

(d) $x \geq -\dfrac{3}{2}$

We can translate many everyday situations into algebraic statements with an unknown value and an inequality symbol. This is the first step in solving word problems using inequalities.

EXAMPLE 3   Translate each English statement into an algebraic statement.

(a) The police on the scene said that the car was traveling more than 80 miles per hour. (Use the variable s for speed.)

(b) The owner of the trucking company said that the payload of a truck must never exceed 4500 pounds. (Use the variable p for payload.)

(a) Since the speed must be greater than 80, we have $s > 80$.

(b) If the payload of the truck can never exceed 4500 pounds, then the payload must be always less than or equal to 4500 pounds. Thus we write $p \leq 4500$.

Practice Problem 3   Translate each English statement into an inequality.

(a) During the drying cycle, the temperature inside the clothes dryer must never exceed 180 degrees Fahrenheit. (Use the variable t for temperature.)

(b) The bank loan officer said that the total consumer debt incurred by Wally and Mary must be less than $15,000 if they want to qualify for a mortgage to buy their first home. (Use the variable d for debt.)

Developing Your Study Skills

Keep Trying

You may be one of those students who have had much difficulty with mathematics in the past and who are sure that you cannot do well in this course. Perhaps you are thinking, "I have never been any good at mathematics," or "I have always hated mathematics," or "Math always scares me," or "I have not had any math for so long that I have forgotten it all." You may have even picked up the label "math anxiety" and attached it to yourself. That is most unfortunate, and it is time for you to reprogram your thinking. Replace those negative thoughts with more positive ones. You need to say things like, "I will give this math class my best shot," or "I can learn mathematics if I work at it," or "I will try to do better than I have done in previous math classes." You will be pleasantly surprised at the difference a positive attitude makes!

We live in a highly technical world, and you cannot afford to give up on the study of mathematics. Dropping mathematics may prevent you from entering certain career fields that you may find interesting. You may not have to take math courses as high-level as calculus, but such courses as intermediate algebra, finite math, college algebra, and trigonometry may be necessary. Learning mathematics can open new doors for you.

Learning mathematics is a process that takes time and effort. You will find that regular study and daily practice are necessary to strengthen your skills and to help you to grow academically. This process will lead you toward success in mathematics. Then, as you become more successful, your confidence in your ability to do mathematics will grow.

Verbal and Writing Skills

1. Is the statement $5 > -6$ equivalent to the statement $-6 < 5$? Why?

2. Is the statement $-8 < -3$ equivalent to the statement $-3 > -8$? Why?

Replace the ? by $<$ or $>$.

3. $9 ? -3$

4. $-2 ? 5$

5. $-4 ? -2$

6. $-3 ? -6$

7. $\dfrac{3}{5} ? \dfrac{4}{7}$

8. $\dfrac{4}{6} ? \dfrac{7}{9}$

9. $-1.2 ? 2.1$

10. $-3.6 ? 2.4$

11. $-\dfrac{13}{3} ? -4$

12. $-3 ? -\dfrac{15}{4}$

13. $-\dfrac{5}{8} ? -\dfrac{3}{5}$

14. $-\dfrac{2}{3} ? -\dfrac{3}{4}$

Which is greater?

15. $\dfrac{123}{4986}$ or 0.0247?

16. $\dfrac{997}{6384}$ or 0.15613?

Graph each inequality on the number line.

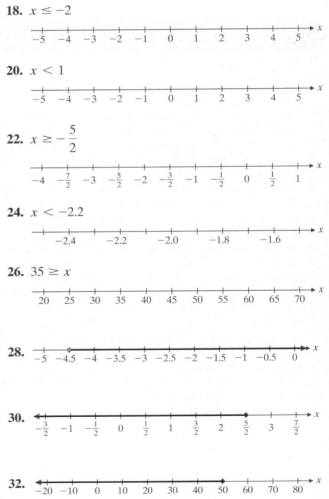

17. $x \geq -6$

18. $x \leq -2$

19. $x > 7$

20. $x < 1$

21. $x > \dfrac{3}{4}$

22. $x \geq -\dfrac{5}{2}$

23. $x \leq -3.6$

24. $x < -2.2$

25. $25 < x$

26. $35 \geq x$

Translate each graph to an inequality using the variable x.

27.

28.

29.

30.

31.

32.

Translate each English statement into an inequality.

33. The speed of the rocket was greater than 580 kilometers per hour. (Use the variable V for speed.)

34. The cost of the hiking boots must be less than $56. (Use the variable c for cost.)

35. The number of hours for a full-time position at this company cannot be less than 37 in order to receive full-time benefits. (Use the variable h for hours.)

36. The number of nurses on duty on the floor can never exceed 6. (Use the variable n for the number of nurses.)

37. In order for you to be allowed to ride the roller coaster at the theme park, your height must be at least 48″. (Use h for height.)

38. In order for you to avoid paying extra tuition for a semester, the number of credits you are taking must not exceed 18. (Use C for credits.)

To Think About

39. Suppose that the variable x must satisfy *all* of these conditions.

$$x \le 2, \quad x > -3, \quad x < \frac{5}{2}, \quad x \ge -\frac{5}{2}$$

Graph on a number line the region that satisfies all of the conditions.

$$\xrightarrow{\quad}$$
$$-5 \quad -4 \quad -3 \quad -2 \quad -1 \quad 0 \quad 1 \quad 2 \quad 3 \quad 4 \quad 5 \qquad x$$

40. Suppose that the variable x must satisfy *all* of these conditions.

$$x < 4, \quad x > -4, \quad x \le \frac{7}{2}, \quad x \ge -\frac{9}{2}$$

Graph on a number line the region that satisfies all of the conditions.

$$\xrightarrow{\quad}$$
$$-5 \quad -4 \quad -3 \quad -2 \quad -1 \quad 0 \quad 1 \quad 2 \quad 3 \quad 4 \quad 5 \qquad x$$

Cumulative Review Problems

Simplify.

41. $\left(-\dfrac{3}{4}\right)^2$

42. $-6(x + 3) + 4(5 - x)$

43. Solve. $\dfrac{x}{6} = 13$

44. Is -5 the solution for $2 + 3x = 2x - 8 - x$?

45. Find the simple interest on a loan of $3500 at an annual interest rate of 9% for 5 years. Use $I = prt$, where p = principal, r = rate per year, and t = time in years.

46. A radar picture is taken of a portion of the Amazon rain forest. The radar detects that a cliff is 414 feet above sea level and that a giant tree trunk is directly below it in a body of water. The tree trunk is 81 feet below sea level. How far is the top of the cliff from the tree trunk?

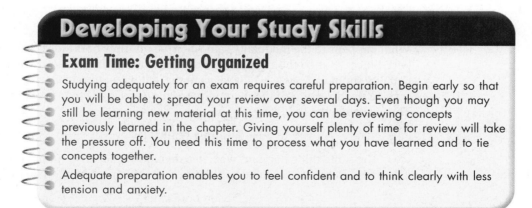

Developing Your Study Skills

Exam Time: Getting Organized

Studying adequately for an exam requires careful preparation. Begin early so that you will be able to spread your review over several days. Even though you may still be learning new material at this time, you can be reviewing concepts previously learned in the chapter. Giving yourself plenty of time for review will take the pressure off. You need this time to process what you have learned and to tie concepts together.

Adequate preparation enables you to feel confident and to think clearly with less tension and anxiety.

2.7 Solving Inequalities

Student Learning Objectives

After studying this section, you will be able to:

  Solve an inequality.

SSM PH TUTOR CD & VIDEO MATH PRO WEB
CENTER

1 Solving an Inequality

As stated in Section 2.6, the possible values that make an inequality true are called its **solutions.** Thus, when we **solve an inequality,** we are finding *all* the values that make it true. To solve an inequality, we simplify it to the point where we can clearly see all possible values for the variable. We've solved equations by adding, subtracting, multiplying by, and dividing by a particular value on both sides of the equation. Here we perform similar operations with inequalities with one important exception. We'll show some examples so that you can see how these operations can be used with inequalities just as with equations.

We will first examine the pattern that occurs when we perform these operations *with a positive value* on both sides of an inequality.

⬤ EXAMPLE 1

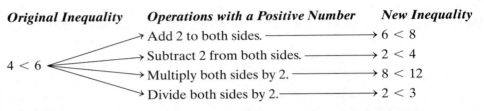

Original Inequality **Operations with a Positive Number** **New Inequality**

4 < 6
- Add 2 to both sides. ⟶ 6 < 8
- Subtract 2 from both sides. ⟶ 2 < 4
- Multiply both sides by 2. ⟶ 8 < 12
- Divide both sides by 2. ⟶ 2 < 3

Notice that the inequality symbol remains the same when these operations are performed.

Practice Problem 1   Perform the given operation and write a new inequality.

(a) 9 > 6   Add 4 to each side.
(b) −2 < 5   Subtract 3 from both sides.
(c) 1 > −3   Multiply both sides by 2.
(d) 10 < 15   Divide both sides by 5.

Now let us examine what happens when we perform these operations *with a negative value.*

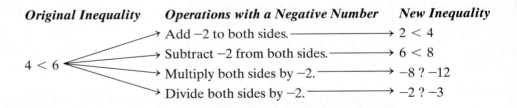

Original Inequality **Operations with a Negative Number** **New Inequality**

4 < 6
- Add −2 to both sides. ⟶ 2 < 4
- Subtract −2 from both sides. ⟶ 6 < 8
- Multiply both sides by −2. ⟶ −8 ? −12
- Divide both sides by −2. ⟶ −2 ? −3

What happens to the inequality sign when we multiply both sides by a negative number? Since −8 is to the right of −12 on the number line, we know that the new inequality should be −8 > −12 if we want the statement to remain true. Notice how we reverse the direction of the inequality from < (less than) to > (greater than). Thus we have the following.

$$4 < 6 \longrightarrow \text{Multiply both sides by } -2. \longrightarrow -8 > -12$$

The same thing happens when we divide by a negative number. The inequality is reversed from < to >. We know this since −2 is to the right of −3 on the number line.

$$4 < 6 \longrightarrow \text{Divide both sides by } -2. \longrightarrow -2 > -3$$

Similar reversals take place in the next example.

⬤ EXAMPLE 2

Original Inequality			*New Inequality*
(a) $-2 < -1$	$\longrightarrow$ Multiply both sides by -3.	$\longrightarrow$	$6 > 3$
(b) $0 > -4$	$\longrightarrow$ Divide both sides by -2.	$\longrightarrow$	$0 < 2$
(c) $8 \geq 4$	$\longrightarrow$ Divide both sides by -4.	$\longrightarrow$	$-2 \leq -1$

Notice that we perform the arithmetic with signed numbers just as we always do. But the new inequality signs reversed (from those of the original inequalities). *Whenever both sides of an inequality are multiplied or divided by a negative quantity, the direction of the inequality is reversed.*

Practice Problem 2

(a) $7 > 2$ Multiply each side by -2.

(b) $-3 < -1$ Multiply each side by -1.

(c) $-10 \geq -20$ Divide each side by -10.

(d) $-15 \leq -5$ Divide each side by -5.   ⬤

Procedure for Solving Inequalities

You may use the same procedures to solve inequalities that you did to solve equations *except* that the direction of an inequality is *reversed* if you *multiply* or *divide* both sides *by a negative number.*

It may be helpful to think over quickly what we have discussed here. The inequalities remain the same when we add a number to both sides or subtract a number from both sides of the equation. The inequalities remain the same when we multiply both sides by a positive number or divide both sides by a positive number.

Rules When Inequalities Remain the Same

For all real numbers a, b, and c:

1. If $a > b$, then $a + c > b + c$.

2. If $a > b$, then $a - c > b - c$.

3. If $a > b$ and c is a **positive number** ($c > 0$), then $ac > bc$.

4. If $a > b$ and c is a **positive number** ($c > 0$), then $a/c > b/c$.

However, if we multiply both sides of an inequality by a negative number or if we divide both sides of an inequality by a negative number, then the inequality is reversed.

Rules When Inequalities Are Reversed

For all real numbers a, b, and c:

1. If $a > b$, and c is a **negative number** ($c < 0$), then $ac < bc$.

2. If $a > b$, and c is a **negative number** ($c < 0$), then $a/c < b/c$.

The pattern is fairly simple. We could also make similar rule boxes for the cases where $a \geq b$, $a < b$, and $a \leq b$, but they are really not necessary. We simply remember that the inequality is reversed when we multiply or divide by a negative number. Otherwise the inequality remains unchanged.

EXAMPLE 3 Solve and graph $3x + 7 \geq 13$.

$$3x + 7 - 7 \geq 13 - 7 \qquad \text{Subtract 7 from both sides.}$$

$$3x \geq 6 \qquad \text{Simplify.}$$

$$\frac{3x}{3} \geq \frac{6}{3} \qquad \text{Divide both sides by 3.}$$

$$x \geq 2 \qquad \text{Simplify. Note that the direction of the inequality is not changed since we have divided by a positive number.}$$

The graph is as follows:

Practice Problem 3 Solve and graph $8x - 2 < 3$.

EXAMPLE 4 Solve and graph $5 - 3x > 7$.

$$5 - 5 - 3x > 7 - 5 \qquad \text{Subtract 5 from both sides.}$$

$$-3x > 2 \qquad \text{Simplify.}$$

$$\frac{-3x}{-3} < \frac{2}{-3} \qquad \text{Divide by } -3 \text{ and \textbf{reverse the inequality} since you are dividing by a negative number.}$$

$$x < -\frac{2}{3} \qquad \text{Note the direction of the inequality.}$$

The graph is as follows:

Practice Problem 4 Solve and graph $4 - 5x > 7$.

Just like equations, some inequalities contain parentheses and fractions. The initial steps to solve these inequalities will be the same as those used to solve equations with parentheses and fractions. When the variable appears on both sides of the inequality, it is advisable to collect the x-terms on the left side of the inequality symbol.

EXAMPLE 5 Solve and graph $-\dfrac{13x}{2} \leq \dfrac{x}{2} - \dfrac{15}{8}$.

$$8\left(\frac{-13x}{2}\right) \leq 8\left(\frac{x}{2}\right) - 8\left(\frac{15}{8}\right) \qquad \text{Multiply all terms by LCD} = 8. \text{ We do \textbf{not} reverse the direction of the inequality symbol since we are multiplying by a positive number.}$$

$$-52x \leq 4x - 15 \qquad \text{Simplify.}$$

$$-52x - 4x \leq 4x - 15 - 4x \qquad \text{Subtract } 4x \text{ from both sides.}$$

$$-56x \leq -15 \qquad \text{Combine like terms.}$$

$$\frac{-56x}{-56} \geq \frac{-15}{-56} \qquad \text{Divide both sides by } -56. \text{ We \textbf{reverse} the direction of the inequality when we divide both sides by a negative number.}$$

$$x \geq \frac{15}{56}$$

The graph is as follows:

Practice Problem 5 Solve and graph $\frac{1}{2}x + 3 < \frac{2}{3}x$.

$$\xrightarrow{\hspace{1cm}} x$$
$$16 \quad 17 \quad 18 \quad 19 \quad 20 \quad 21$$

EXAMPLE 6 Solve and graph $\frac{1}{3}(3 - 2x) \le -4(x + 1)$.

$$1 - \frac{2x}{3} \le -4x - 4 \qquad \text{Remove parentheses.}$$

$$3(1) - 3\left(\frac{2x}{3}\right) \le 3(-4x) - 3(4) \qquad \text{Multiply all terms by LCD} = 3.$$

$$3 - 2x \le -12x - 12 \qquad \text{Simplify.}$$

$$3 - 2x + 12x \le -12x + 12x - 12 \qquad \text{Add } 12x \text{ to both sides.}$$

$$3 + 10x \le -12 \qquad \text{Combine like terms.}$$

$$3 - 3 + 10x \le -12 - 3 \qquad \text{Subtract 3 from both sides.}$$

$$10x \le -15 \qquad \text{Simplify.}$$

$$\frac{10x}{10} \le \frac{-15}{10} \qquad \text{Divide both sides by 10. Since we are dividing by a **positive** number, the inequality is **not** reversed.}$$

$$x \le -\frac{3}{2}$$

The graph is as follows:

$$\xleftarrow{\hspace{1cm}}\xrightarrow{\hspace{1cm}} x$$
$$-\frac{9}{2} \quad -4 \quad -\frac{7}{2} \quad -3 \quad -\frac{5}{2} \quad -2 \quad -\frac{3}{2} \quad -1 \quad -\frac{1}{2} \quad 0 \quad \frac{1}{2}$$

Practice Problem 6 Solve and graph $\frac{1}{2}(3 - x) \le 2x + 5$.

$$\xrightarrow{\hspace{1cm}} x$$
$$-2 \quad -\frac{9}{5} \quad -\frac{8}{5} \quad -\frac{7}{5} \quad -\frac{6}{5} \quad -1$$

⊘ **CAUTION** The most common error students make in solving inequalities is forgetting to reverse the direction of the inequality symbol when multiplying or dividing both sides of the inequality by a negative number.

EXAMPLE 7   A hospital director has determined that the costs of operating one floor of the hospital for an eight-hour shift must never exceed $2370. An expression for the cost of operating one floor of the hospital is $130n + 1200$, where n is the number of nurses. This expression is based on an estimate of $1200 in fixed costs and a cost of $130 per nurse for an eight-hour shift. Solve the inequality $130n + 1200 \leq 2370$ to determine the number of nurses that may be on duty on this floor during an eight-hour shift if the director's cost control measure is to be followed.

$130n + 1200 \leq 2370$	The inequality we must solve.
$130n + 1200 - 1200 \leq 2370 - 1200$	Subtract 1200 from each side.
$130n \leq 1170$	Simplify.
$\dfrac{130n}{130} \leq \dfrac{1170}{130}$	Divide each side by 130.
$n \leq 9$	

The number of nurses on duty on this floor during an eight-hour shift must always be less than or equal to nine.

Practice Problem 7   The company president of Staywell, Inc., wants the monthly profits never to be less than $2,500,000. He has determined that an expression for monthly profit for the company is $2000n - 700,000$. In the expression, n is the number of exercise machines manufactured each month. The profit on each machine is $2000, and the −$700,000 in the expression represents the fixed costs of running the manufacturing division. Solve the inequality $2000n - 700,000 \geq 2,500,000$ to find how many machines must be made and sold each month to satisfy these financial goals.

Solve and graph the result.

1. $x + 7 \leq 4$

2. $x - 5 < -3$

3. $-2x < 18$

4. $5x \leq 25$

5. $\frac{1}{2}x \geq 4$

6. $-\frac{1}{5}x < 10$

7. $2x - 3 < 4$

8. $3 - 3x > 12$

9. $-4 - 14x < 6 - 6x$

10. $7 - 8x \leq -6x - 5$

11. $\frac{5x}{6} - 5 > \frac{x}{6} - 9$

12. $\frac{x}{4} - 2 < \frac{3x}{4} + 5$

13. $2(3x + 4) > 3(x + 3)$

14. $5(x - 3) \leq 2(x - 3)$

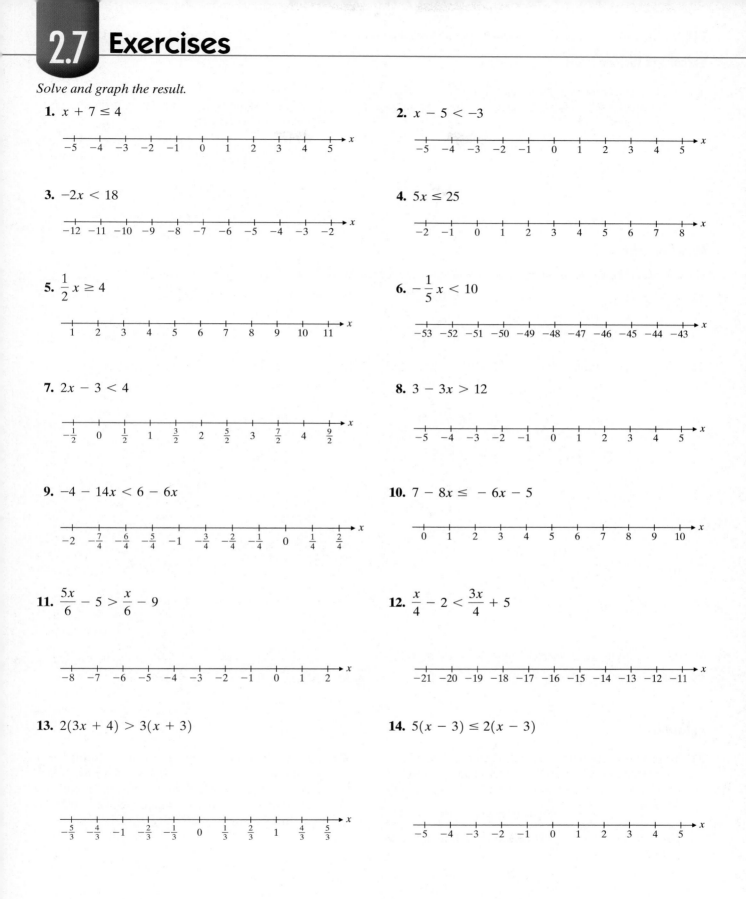

Verbal and Writing Skills

15. Add -2 to both sides of the inequality $5 > 3$. What is the result? Why is the direction of the inequality not reversed?

16. Divide -3 into both sides of the inequality $-21 > -29$. What is the result? Why is the direction of the inequality reversed?

Mixed Practice

Solve. Collect the variable terms on the left side of the inequality.

17. $2x - 5 < 5x - 11$

18. $4x - 7 > 9x - 2$

19. $6x - 2 \geq 4x + 6$

20. $5x - 5 \leq 2x + 10$

21. $0.3(x - 1) < 0.1x - 0.5$

22. $0.2(3 - x) + 0.1 > 0.1(x - 2)$

23. $3 + 5(2 - x) \geq -3(x + 5)$

24. $7 - 2(x - 4) \leq 7(x - 3)$

25. $\dfrac{x + 6}{7} - \dfrac{6}{14} > \dfrac{x + 3}{2}$

26. $\dfrac{3x + 5}{4} + \dfrac{7}{12} > -\dfrac{x}{6}$

27. $\dfrac{1}{6} - \dfrac{1}{2}(3x + 2) < \dfrac{1}{3}\left(x - \dfrac{1}{2}\right)$

28. $\dfrac{2}{3}(2x - 5) + 3 \geq \dfrac{1}{4}(3x + 1) - 5$

Solve for x. Round your answer to the nearest hundredth.

29. $1.96x - 2.58 < 9.36x + 8.21$

30. $3.5(1.7x - 2.8) \leq 7.96x - 5.38$

Applications

31. To pass a course with a B grade, a student must have an average of 80 or greater. A student's grades on three tests are 75, 83, and 86. Solve the inequality $\dfrac{75 + 83 + 86 + x}{4} \geq 80$ to find what score the student must get on the next test to get a B average or better.

32. Sharon sells very expensive European sports cars. She may choose to receive $10,000.00 or 8% of her sales as payment for her work. Solve the inequality $0.08x > 10,000$ to find how much she needs to sell to make the 8% offer a better deal.

33. The average African elephant weighs 268 pounds at birth. During the first three weeks of life the baby elephant will usually gain about 4 pounds per day. Assuming that growth rate, solve the inequality $268 + 4x \geq 300$ to find how many days it will be until a baby elephant weighs at least 300 pounds.

34. Tess supervises a computer chip manufacturing facility. She has determined that her monthly profit factor is given by the expression $12.5n - 300,000$. Here n represents the number of chips manufactured each month. Each finished chip produces a profit of \$12.50. The fixed costs (overhead) of the factory are \$300,000 per month. Solve the inequality $12.5n - 300,000 \geq 650,000$ to find how many chips must be manufactured monthly to ensure a profit of \$650,000.

Cumulative Review Problems

▲ **35.** A rectangular tennis court measures 36 feet wide and 78 feet long. Robert is building a fence to surround the tennis court. He wants the fence to be 4 feet from each side of the court. How many feet of fence will he need?

▲ **36.** Maria is looking at an enlargement of a rectangular photograph. The photograph has a perimeter of 29 inches. The length of the photograph is 10 inches. What is the width of the photograph?

▲ **37.** Melinda is making a poster for her college basketball team. On the poster she is placing a life-sized picture of a basketball. When inflated properly, a basketball has a diameter of 9 inches. What is the area of the basketball on her poster? Use $\pi \approx 3.14$. Round your answer to the nearest tenth.

▲ **38.** The seats of an outdoor amphitheater are arranged in the shape of a trapezoid. The altitude of the trapezoid is 120 feet. The bases of the trapezoid are 90 feet and 170 feet. What is the area of this seating area?

1 Solving Word Problems Involving Comparisons

Many real-life problems involve comparisons. We often compare quantities such as length, height, or income. Sometimes not all the information is known about the quantities that are being compared. You need to identify each quantity and write an algebraic expression that describes the situation in the word problem.

EXAMPLE 1 The Center City Animal Hospital treated a total of 18,360 dogs and cats last year. The hospital treated 1376 more dogs than cats. How many dogs were treated last year? How many cats were treated last year?

1. Understand the problem.

What information is given?   The combined number of dogs and cats is 18,360.

What is being compared?   There were 1376 more dogs than cats.

If you compare one quantity to another, usually the second quantity is represented by the variable. Since we are comparing the number of dogs to the number of cats, we start with the number of cats.

Let c = the number of cats treated at the hospital.

Then $c + 1376$ = the number of dogs treated at the hospital.

2. Write an equation.

The number of cats   plus   the number of dogs   is   18,360.

$$c + (c + 1376) = 18{,}360$$

3. Solve and state the answer.

$$c + c + 1376 = 18{,}360$$

$$2c + 1376 = 18{,}360 \quad \text{Combine like terms.}$$

$$2c + 1376 - 1376 = 18{,}360 - 1376 \quad \text{Subtract 1376 from both sides.}$$

$$2c = 16{,}984$$

$$\frac{2c}{2} = \frac{16{,}984}{2} \quad \text{Divide both sides by 2.}$$

$$c = 8492 \quad \text{The number of cats treated is 8492.}$$

$$c + 1376 = 8492 + 1376 = 9868 \quad \text{The number of dogs treated is 9868.}$$

4. Check.

The number of dogs treated plus the number of cats treated should total 18,360.

$$8492 + 9868 \overset{?}{=} 18{,}360$$

$$18{,}360 = 18{,}360 \quad ✓$$

Practice Problem 1 A deck hand on a fishing boat is working with a rope that measures 89 feet. He needs to cut it into two pieces. The long piece must be 17 feet longer than the short piece. Find the length of each piece of rope.

If the word problem contains three unknown quantities, determine the basis of comparison for two of the quantities.

EXAMPLE 2 An airport filed a report showing the number of plane departures that took off from the airport during each month last year. The number of departures in March was 50 more than the number of departures in January. In July, the number of departures was 150 less than triple the number of departures in January. In those three months, the airport had 2250 departures. How many departures were recorded for each month?

1. ***Understand the problem.***
   What is the basis of comparison?

   The number of departures in March is compared to the number in January.

   The number of departures in July is compared to the number in January.

   Express this algebraically. It may help to underline the key phrases.

   Let j = the departures in January.

   March was 50 more than January

   Then $j + 50$ = the departures in March.

   July was 150 less than triple January

   And $3j - 150$ = the departures in July.

2. ***Write an equation.***

number of departures in January	+	number of departures in March	+	number of departures in July	=	three months' total departures
j	+	$(j + 50)$	+	$(3j - 150)$	=	2250

3. ***Solve and state the answer.***

   $$j + (j + 50) + (3j - 150) = 2250$$
   $$5j - 100 = 2250 \quad \text{Collect like terms.}$$
   $$5j = 2350 \quad \text{Add 100 to each side.}$$
   $$j = 470 \quad \text{Divide both sides by 5.}$$

   Now, if $j = 470$, then

   $$j + 50 = 470 + 50 = 520$$

   and

   $$3j - 150 = 3(470) - 150 = 1410 - 150 = 1260.$$

   The number of departures in January was 470; the number of departures in March was 520; the number of departures in July was 1260.

4. ***Check.***
   Do these answers seem reasonable? Yes. Do these answers agree with all the statements in the word problem?
   Is the number of departures in March 50 more than those in January?

   $$520 \stackrel{?}{=} 50 + 470$$
   $$520 = 520 \quad \checkmark$$

   Is the number of departures in July 150 less than triple those in January?

   $$1260 \stackrel{?}{=} 3(470) - 150$$
   $$1260 \stackrel{?}{=} 1410 - 150$$
   $$1260 = 1260 \quad \checkmark$$

   Is the total number of departures in the three months equal to 2250?

   $$470 + 520 + 1260 \stackrel{?}{=} 2250$$
   $$2250 = 2250 \quad \checkmark$$

   Yes, all conditions are satisfied. The three answers are correct.

Practice Problem 2   A social services worker was comparing the cost incurred by three families in heating their homes for the year. The first family had an annual heating bill that was $360 more than that of the second family. The third family had a heating bill that was $200 less than double the heating bill of the second family. The total annual heating bill for the three families was $3960. What was the annual heating bill for each family?

▲ ◢▬ **EXAMPLE 3**   A small plot of land is in the shape of a rectangle. The length is 7 meters longer than the width. The perimeter of the rectangle is 86 meters. Find the dimensions of the rectangle.

1. *Understand the problem.*
   Read the problem: What information is given?

   > *The perimeter of a rectangle is 86 meters.*
   > What is being compared?
   > *The length is being compared to the width.*

   Express this algebraically and draw a picture.

   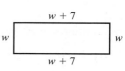

   > Let w = the width.
   > Then $w + 7$ = the length.

   Reread the problem: What are you being asked to do?
   > *Find the dimensions of the rectangle. The dimensions of the rectangle are the length and the width of the rectangle.*

2. *Write an equation.*
   The perimeter is the total distance around the rectangle.

   $$w + (w + 7) + w + (w + 7) = 86$$

3. *Solve and state the answer.*

   $$w + (w + 7) + w + (w + 7) = 86$$
   $$4w + 14 = 86 \quad \text{Combine like terms.}$$
   $$4w = 72 \quad \text{Subtract 14 from both sides.}$$
   $$w = 18 \quad \text{Divide both sides by 4.}$$

   The width of the rectangle is 18 meters. What is the length?

   $$w + 7 = \text{the length}$$
   $$18 + 7 = 25$$

   The length of the rectangle is 25 meters.

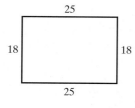

4. *Check.*
   Put the actual dimensions in your drawing and add the lengths of the sides. Is the sum 86 meters? ✓

▲ **Practice Problem 3**   A farmer purchased 720 meters of wire fencing to enclose a pasture. The pasture is in the shape of a triangle. The first side of the triangle is 30 meters less than the second side. The third side is one-half as long as the second side. Find the dimensions of the triangle.

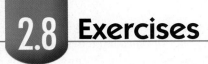

2.8 Exercises

Applications

Solve. Check to see if your answer is reasonable. Have you answered the question that was asked?

1. For their homecoming parade, the students of Wheaton College have created a colorful banner, 47 meters in length, that is made of two pieces of parachute material. The short piece is 17 meters shorter than the long piece. Find the length of each piece.

2. A copper conducting wire measures 84 centimeters in length. It is cut into two pieces. The shorter piece is 5 centimeters shorter than the long piece. Find the length of each piece.

3. Three siblings—David, Kate, and Sarah—committed themselves to volunteer 100 hours for Habitat for Humanity to build new housing. Sarah worked 15 hours more than David. Kate worked 5 hours fewer than David. How many hours did each of these siblings volunteer?

4. Jerome counted 24 pieces of fruit in a fruit basket. There were twice as many apples as pears. There were four fewer peaches than pears. How many of each fruit were there?

5. Mt. McKinley in Alaska is 5826 feet taller than Mt. Whitney in California. Mt. Oxford in Colorado is 341 feet shorter than Mt. Whitney. The combined height of the three mountains is 48,967 feet. What is the height of each of the mountains? (*Source:* U.S. Department of the Interior)

6. The Sears Tower in Chicago is 200 feet taller than the Empire State Building in New York City. The Nations Bank Tower in Atlanta is 227 feet shorter than the Empire State Building. The total height of all three buildings is 3723 feet. What is the height of each of the three buildings? (*Source:* U.S. Department of the Interior)

▲ 7. The length of a polo field is 50 meters more than twice its width. If the perimeter is 700 meters, find the dimensions of the polo field.

8. The part of a diving board that is over the water is one meter longer than the part that is over the deck around the pool. The total length of the diving board is two meters less than three times the part that is over the deck. Find the length of the board.

▲ 9. A solid-gold jewelry box was found in the underwater palace of Cleopatra just off the shore of Alexandria. The length of the rectangular box is 35 centimeters less than triple the width. The perimeter of the box is 190 centimeters. Find the length and width of Cleopatra's solid-gold jewelry box.

▲ 10. A giant rectangular chocolate bar was made for a special promotion. The length was 6 meters more than half the width. The perimeter of the chocolate bar was 24 meters. Find the length and width of the giant chocolate bar.

11. The top running speed of a cheetah is double the top running speed of a jackal. The top running speed of an elk is 10 miles per hour faster than that of a jackal. If each of these three animals could run at top speed for an hour (which of course is not possible), they could run a combined distance of 150 miles. What is the top running speed of each of these three animals? (*Source:* American Museum of Natural History)

12. The Missouri River is 149 miles longer than double the length of the Snake River. The Potomac River is 796 miles shorter than the Snake River. The combined lengths of these three rivers is 3685 miles. What is the length of each of the three rivers? (*Source:* U.S. Department of the Interior)

▲ **13.** A small triangular piece of metal is welded onto the hull of a Gloucester whale watch boat. The perimeter of the metal piece is 46 inches. The shortest side of the triangle is four inches longer than one-half of the longest side. The second-longest side is 3 inches shorter than the longest side. Find the length of each side.

▲ **14.** Carmelina's uncle owns a triangular piece of land in Maryland. The perimeter fence that surrounds the land measures 378 yards. The shortest side is 30 yards longer than one-half of the longest side. The second-longest side is 2 yards shorter than the longest side. Find the length of each side.

15. A balloonist is trying to complete a nonstop trip around the world. During the course of his trip, he notes that he took 18 hours to travel 684 miles over land from Rockford, Illinois, to Washington, D.C. He took 21 hours to travel 1197 miles over water from Washington, D.C., to San Juan, Puerto Rico.
   (a) How fast did he travel over land?
   (b) How fast did he travel over water?
   (c) How much faster did he travel over water than over land?

16. In a typical year in New York City, 600 waterline breaks are reported along the 6181 miles of water pipes that twist and turn underground. Water breaks occur five times more often during the period from October to March than they do during the period from April to September.
   (a) On average, how many waterline breaks would be expected during the period from October to March?
   (b) On average, how many waterline breaks would be expected during the period from April to September?
   (c) A water department official once complained to city hall that he expects a break in the waterline every five miles during the course of a year. Is his comment correct? Why?

To Think About

▲ **17.** A small square is constructed. Then a new square is made by increasing each side by 2 meters. The perimeter of the new square is 3 meters shorter than five times the length of one side of the original square. Find the dimensions of the original square.

▲ **18.** A rectangle is constructed. The length of this rectangle is double the width. Then a new rectangle is made by increasing each side by 3 meters. The perimeter of the new rectangle is 2 meters greater than four times the length of the old rectangle. Find the dimensions of the original rectangle.

Cumulative Review Problems

Simplify.

19. $-4x(2x^2 - 3x + 8)$

20. $5a(ab + 6b - 2a)$

21. $-7x + 10y - 12x - 8y - 2$

22. $3x^2y - 6xy^2 + 7xy + 6x^2y$

23. Solve $\dfrac{2}{3}x - \dfrac{1}{6} = 2x + \dfrac{1}{2}$.

24. Solve for h. $y = \dfrac{3h - 2a}{b}$

2.9 Solving Word Problems: The Value of Money and Percents

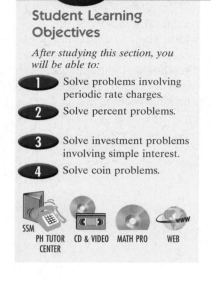
The problems we now present are frequently encountered in business. They deal with money: buying, selling, and renting items; earning and borrowing money; and the value of collections of stamps or coins. Many applications require an understanding of the use of percents and decimals.

1 Solving Problems Involving Periodic Rate Charges

EXAMPLE 1 A business executive rented a car. The Supreme Car Rental Agency charged $39 per day and $0.28 per mile. The executive rented the car for two days and the total rental cost was computed to be $176. How many miles did the executive drive the rented car?

1. **Understand the problem.**

   How do you calculate the cost of renting a car?

   total cost = per-day cost + mileage cost

   What is known?

   It cost $176 to rent the car for two days.

   What do you need to find?

   The number of miles the car was driven.

   Choose a variable:

   Let m = the number of miles driven in the rented car.

2. **Write an equation.**

   Use the relationship for calculating the total cost.

per-day cost	+	mileage cost	=	total cost
(39)(2)	+	(0.28)m	=	176

3. **Solve and state the answer.**

$$(39)(2) + (0.28)(m) = 176$$
$$78 + 0.28m = 176 \qquad \text{Simplify the equation.}$$
$$0.28m = 98 \qquad \text{Subtract 78 from both sides.}$$
$$\frac{0.28m}{0.28} = \frac{98}{0.28} \qquad \text{Divide both sides by 0.28.}$$
$$m = 350 \qquad \text{Simplify.}$$

   The executive drove 350 miles.

4. **Check.**

   Does this seem reasonable? If he drove the car 350 miles in two days, would it cost $176?

   (Cost of $39 per day for 2 days) + (cost of $0.28 per mile for 350 miles)
   $$\overset{?}{=} \text{total cost of } \$176$$
   $$(\$39)(2) + (350)(\$0.28) \overset{?}{=} \$176$$
   $$\$78 + \$98 \overset{?}{=} \$176$$
   $$\$176 = \$176 \quad ✓$$

Practice Problem 1 Alfredo wants to rent a truck to move to Florida. He has determined that the cheapest rental rates for a truck of the correct size are from a local company that will charge him $25 per day and $0.20 per mile. He has not yet completed an estimate of the mileage of the trip, but he knows that he will need the truck for three days. He has allowed $350 in his moving budget for the truck. How far can he travel for a rental cost of exactly $350?

2 Solving Percent Problems

Many applied situations require finding a percent of an unknown number. If we want to find 23% of $400, we multiply 0.23 by 400: $0.23(400) = 92$. If we want to find 23% of an unknown number, we can express this using algebra by writing $0.23n$, where n represents the unknown number.

EXAMPLE 2   A sofa was marked with the following sign: "The price of this sofa has been reduced by 23%. You can save $138 if you buy now." What was the original price of the sofa?

1. **Understand the problem.**

$$\text{Let } s = \text{the original price of the sofa.}$$
$$\text{Then } 0.23s = \text{the amount of the price reduction, which is \$138.}$$

2. **Write an equation and solve.**

$0.23s = 138$ Write the equation.

$\dfrac{0.23s}{0.23} = \dfrac{138}{0.23}$ Divide each side of the equation by 0.23.

$s = 600$ Simplify.

The original price of the sofa was $600.

3. **Check:**

Is $600 a reasonable answer? ✓   Does 23% of $600 = $138? ✓

Practice Problem 2   John earns a commission of 38% of the cost of every set of encyclopedias that he sells. Last year he earned $4560 in commissions. What was the cost of the encyclopedias that he sold last year?

EXAMPLE 3   Hector received a pay raise this year. The raise was 6% of last year's salary. This year he will earn $15,900. What was his salary last year before the raise?

1. **Understand the problem.**

What do we need to find?

Hector's salary last year.

What do we know?

Hector received a 6% pay raise and now earns $15,900.

What does this mean?

Reword the problem: *This year's salary of $15,900 is 6% more than last year's salary.*

Choose a variable:

$$\text{Let } x = \text{Hector's salary last year.}$$
$$\text{Then } 0.06x = \text{the amount of the raise.}$$

2. **Write an equation and solve.**

Last year's salary	+	the amount of his raise	=	this year's salary	
x	+	$0.06x$	=	15,900	Write the equation.
$1.00x$	+	$0.06x$	=	15,900	Rewrite x as $1.00x$.
		$1.06x$	=	15,900	Combine like terms.
x			=	$\dfrac{15,900}{1.06}$	Divide by 1.06.
x			=	15,000	Simplify.

Thus Hector's salary was $15,000 last year before the raise.

3. *Check.*

Does it seem reasonable that Hector's salary last year was $15,000? The check is up to you.

Practice Problem 3   The price of Betsy's new car is 7% more than the price of a similar model last year. She paid $13,910 for her car this year. What would a similar model have cost last year?

3 Solving Investment Problems Involving Simple Interest

Interest is a charge for borrowing money or an income from investing money. Interest rates affect our lives. They affect the national economy and they affect a consumer's ability to borrow money for big purchases. For these reasons, a student of mathematics should be able to solve problems involving interest.

There are two basic types of interest: simple and compound. **Simple interest** is computed by multiplying the amount of money borrowed or invested (which is called the *principal*) times the rate of interest times the period of time over which it is borrowed or invested (usually measured in years unless otherwise stated).

$$\text{Interest} = \text{principal} \times \text{rate} \times \text{time}$$
$$I = prt$$

You often hear of banks offering a certain interest rate *compounded* quarterly, monthly, weekly, or daily. In **compound interest** the amount of interest is added to the amount of the original principal at the end of each time period, so future interest is based on the sum of both principal and previous interest. Most financial institutions use compound interest in their transactions.

Problems involving compound interest may be solved by:

1. Repeated calculations using the simple interest formula.
2. Using a compound interest table.
3. Using exponential functions, a topic that is usually covered in a higher-level college algebra course.

*All examples and exercises in this chapter will involve **simple interest**.*

EXAMPLE 4   Find the interest on $3000 borrowed at a simple interest rate of 18% for one year.

$I = prt$ The simple interest formula.

$I = (3000)(0.18)(1)$ Substitute the values of the variables: principal $= 3000$, the rate $= 18\% = 0.18$, the time $=$ one year.

$I = 540$

Thus the interest charge for borrowing $3000 for one year at a simple interest rate of 18% is $540.

Practice Problem 4   Find the interest on $7000 borrowed at a simple interest rate of 12% for one year.

Now we apply this concept to a word problem about investments.

EXAMPLE 5   A woman invested an amount of money in two accounts for one year. She invested some at 8% simple interest and the rest at 6% simple interest. Her total amount invested was $1250. At the end of the year she had earned $86 in interest. How much money had she invested in each account?

We will use the Mathematics Blueprint For Problem Solving in this example. To see more problems solved using the blueprint, see Appendix B.

Mathematics Blueprint For Problem Solving

Gather the Facts	Assign the Variable	Basic Formula or Equation	Key Points to Remember
$1250 is invested: part at 8% interest, part at 6% interest. The total interest for the year is $86.	Let x = the amount invested at 8%. $1250 - x$ = the amount invested at 6%. $0.08x$ = the amount of interest for x dollars at 8%. $0.06(1250 - x)$ = the amount of interest for $1250 - x$ dollars at 6%.	Interest earned at 8% + interest earned at 6% = total interest earned during the year, which is $86.	Be careful to write $1250 - x$ for the amount of money invested at 6%. The order is total $- x$. Do not use $x - 1250$.

$$\begin{array}{ccccc} \text{interest} & & \text{interest} & & \text{total interest} \\ \text{earned} & + & \text{earned} & = & \text{earned during} \\ \text{at 8\%} & & \text{at 6\%} & & \text{the year} \\ 0.08x & + & 0.06(1250 - x) & = & 86 \end{array}$$

Note: Be sure you write $(1250 - x)$ for the amount of money invested at 6%. Students often write it backwards by mistake. It is *not* correct to use $(x - 1250)$ instead of $(1250 - x)$. The order of the terms is very important.

Solve and state the answer.

$0.08x + 75 - 0.06x = 86$	Remove parentheses.
$0.02x + 75 = 86$	Combine like terms.
$0.02x = 11$	Subtract 75 from both sides.
$\dfrac{0.02x}{0.02} = \dfrac{11}{0.02}$	Divide both sides by 0.02.
$x = 550$	The amount invested at 8% interest is $550.
$1250 - x = 1250 - 550 = 700$	The amount invested at 6% interest is $700.

Check.
Are these values reasonable? Yes. Do the amounts equal $1250?

$$\$550 + \$700 \stackrel{?}{=} \$1250$$
$$\$1250 = \$1250 \quad \checkmark$$

Would these amounts earn $86 interest in one year invested at the specified rates?
$$0.08(\$550) + 0.06(\$700) \stackrel{?}{=} \$86$$
$$\$44 + \$42 \stackrel{?}{=} \$86$$
$$\$86 = \$86 \quad \checkmark$$

Practice Problem 5 A woman invested her savings of $8000 in two accounts that each calculate interest only once per year. She placed one amount in a special notice account that yields 9% annual interest. The remainder she placed in a tax-free All-Savers account that yields 7% annual interest. At the end of the year, she had earned $630 in interest from the two accounts together. How much had she invested in each account?

 **4** *Solving Coin Problems*

Coin problems provide an unmatched opportunity to use the concept of *value*. We must make a distinction between how many coins there are and the *value* of the coins.

Consider the next example. Here we know *the value* of some coins, but do not know *how many* we have.

EXAMPLE 6   When Bob got out of math class, he had to make a long-distance call. He had exactly enough dimes and quarters to make a phone call that would cost $2.55. He had one less quarter than he had dimes. How many coins of each type did he have?

$$\text{Let } d = \text{ the number of dimes.}$$
$$\text{Then } d - 1 = \text{ the number of quarters.}$$

The total value of the coins was $2.55. How can we represent the value of the dimes and the value of the quarters? Think.

Each dime is worth $0.10.	Each quarter is worth $0.25.
5 dimes are worth $(5)(0.10) = 0.50$.	8 quarters are worth $(8)(0.25) = 2.00$.
d dimes are worth $(d)(0.10) = 0.10d$.	$(d - 1)$ quarters are worth $(d - 1)(0.25) = 0.25(d - 1)$.

Now we can write an equation for the total value.

$$\text{(value of dimes)} + \text{(value of quarters)} = \$2.55$$
$$0.10d \quad + \quad 0.25(d - 1) \quad = \quad 2.55$$

$$0.10d + 0.25d - 0.25 = 2.55 \quad \text{Remove parentheses.}$$
$$0.35d - 0.25 = 2.55 \quad \text{Combine like terms.}$$
$$0.35d = 2.80 \quad \text{Add 0.25 to both sides.}$$
$$\frac{0.35d}{0.35} = \frac{2.80}{0.35} \quad \text{Divide both sides by 0.35.}$$
$$d = 8 \quad \text{Simplify.}$$
$$d - 1 = 7$$

Thus Bob had eight dimes and seven quarters.
Check.
Is this answer reasonable? Yes. Does Bob have one less quarter than he has dimes?

$$8 - 7 \overset{?}{=} 1$$
$$1 = 1 \quad \checkmark$$

Are eight dimes and seven quarters worth $2.55?

$$8(\$0.10) + 7(\$0.25) \overset{?}{=} \$2.55$$
$$\$0.80 + \$1.75 \overset{?}{=} \$2.55$$
$$\$2.55 = \$2.55 \quad \checkmark$$

Practice Problem 6   Ginger has five more quarters than dimes. She has $5.10 in change. If she has only quarters and dimes, how many coins of each type does she have?

EXAMPLE 7   Michele and her two children returned from the grocery store with only $2.80 in change. She had twice as many quarters as nickels. She had two more dimes than nickels. How many nickels, dimes, and quarters did she have?

Mathematics Blueprint For Problem Solving

Gather the Facts	Assign the Variable	Basic Formula or Equation	Key Points to Remember
Michele had $2.80 in change. She had twice as many quarters as nickels. She had two more dimes than nickels.	Let x = the number of nickels. $2x$ = the number of quarters. $x + 2$ = the number of dimes. $0.05x$ = the value of the nickels. $0.25(2x)$ = the value of the quarters. $0.10(x + 2)$ = the value of the dimes.	The value of the nickels + the value of the dimes + the value of the quarters = $2.80.	Don't add the number of coins to get $2.80. You must add the value of the coins!

$$\text{(value of nickels)} + \text{(value of dimes)} + \text{(value of quarters)} = \$2.80$$
$$0.05x \quad + \quad 0.10(x + 2) \quad + \quad 0.25(2x) \quad = \quad 2.80$$

Solve.

$0.05x + 0.10x + 0.20 + 0.50x = 2.80$	Remove parentheses.
$0.65x + 0.20 = 2.80$	Combine like terms.
$0.65x = 2.60$	Subtract 0.20 from both sides.
$\dfrac{0.65x}{0.65} = \dfrac{2.60}{0.65}$	Divide both sides by 0.65.
$x = 4$	Simplify. Michele had four nickels.
$2x = 8$	She had eight quarters.
$x + 2 = 6$	She had six dimes.

FOOD MARKET

When Michele left the grocery store she had four nickels, eight quarters, and six dimes.

Check.
Is the answer reasonable? Yes. Did Michele have twice as many quarters as nickels?

$$(4)(2) \overset{?}{=} 8 \qquad 8 = 8 \ \checkmark$$

Did she have two more dimes than nickels?

$$4 + 2 \overset{?}{=} 6 \qquad 6 = 6 \ \checkmark$$

Do four nickels, eight quarters, and six dimes have a value of $2.80?

$$4(\$0.05) + 8(\$0.25) + 6(\$0.10) \overset{?}{=} \$2.80$$
$$\$0.20 + \$2.00 + \$0.60 \overset{?}{=} \$2.80$$
$$\$2.80 = \$2.80 \ \checkmark$$

Practice Problem 7   A young boy told his friend that he had twice as many nickels as dimes in his pocket. He also said that he had four more quarters than dimes. He said that he had $2.35 in change in his pocket. Can you determine how many nickels, dimes, and quarters he had?

Developing Your Study Skills

Applications or Word Problems

Applications or word problems are the very life of mathematics! They are the reason for doing mathematics because they teach you how to put into use the mathematical skills you have developed. Learning mathematics without ever doing word problems is similar to learning all the skills of a sport without ever playing a game or learning all the notes on an instrument without ever playing a song.

The key to success is practice. Make yourself do as many problems as you can. If you need help organizing your facts, use the Mathematics Blueprint. You may not be able to do all problems correctly at first, but keep trying. Do not give up whenever you reach a difficult one. If you cannot solve it, just try another one. Then come back and try it again later.

A misconception among students when they begin studying word problems is that each problem is different. At first the problems may seem this way, but as you practice more and more, you will begin to see the similarities, the different "types." You will see patterns in solving problems, which will enable you to solve them more easily.

Applications

Solve. All problems involving interest refer to simple interest.

1. Paul has a job raking leaves for a neighbor. He makes $6.50 an hour, plus $0.75 for each bag he fills. Last Saturday he worked three hours and made a total of $27.75. How many bags of leaves did he fill?

2. Marybelle is contemplating a job as a waitress. She would be paid $4 per hour plus tips. The other waitresses have told her that an average tip at that restaurant is $3 per table served. If she works 20 hours per week, how many tables would she have to serve in order to make $191 during the week?

3. Ramon has a summer job as a lifeguard to earn his college tuition. He gets paid $6.00 per hour for the first 40 hours and $9.00 per hour for each hour in the week worked more than 40 hours. This summer his goal is to earn at least $303.00 per week. How many hours of overtime per week will he need to achieve his goal?

4. Maria has a summer job as a tennis instructor to earn her college tuition. She gets paid $6.50 per hour for the first 40 hours and $9.75 for each hour in the week worked more than 40 hours. This summer her goal is to earn at least $338.00 per week. How many hours of overtime per week will she need to achieve her goal?

5. The camera Melissa wanted for her birthday is on sale at 28% off the usual price. The amount of the discount is $100.80. What was the original price of the camera?

6. The number of women working full-time in Springfield has risen 7% this year. This means 126 more women have full-time jobs. What was the number of women working full-time last year?

7. In the bullish market at the end of 1999, two partners, Jim Brown and Walter Payton, split a profit of $9540 from the sale of stock. Jim had invested more, so he received 12% more of the profit than Walter did. How much money did each partner make?

8. A speculator bought stocks and later sold them for $5136, making a profit of 7%. How much did the stocks cost him?

9. Don Williams invested some money at 9% simple interest. At the end of the year, the total amount of his original principal and the interest was $6540. How much did he originally invest?

10. The cost of living last year went up 7%. Fortunately, Alice Swanson got a 7% raise in her salary from last year. This year she is earning $21,400. How much did she make last year?

11. Mr. and Mrs. Wright set up a trust fund for their children last year. Part of it is earning 7% simple interest per year while the rest of it is earning 5% simple interest per year. They placed $5000 in the trust fund. In one year the trust fund has earned $310. How much did they invest at each interest rate?

12. Anne and Michael invested $8000 last year in tax-free bonds. Some of the bonds earned 8% simple interest while the rest earned 6% simple interest. At the end of the year, they had earned $580 in interest. How much did they invest at each interest rate?

13. Plymouth Rock Bank invested $400,000 last year in mutual funds. The conservative fund earned 8% simple interest. The growth fund earned 12% simple interest. At the end of the year, the bank had earned $38,000 from these mutual funds. How much did it invest in each fund?

14. Millennium Securities last year invested $600,000 in mutual funds. The international fund earned 11% simple interest. The high-tech fund earned 7% simple interest. At the end of the year, the company had earned $50,000 in interest. How much did it invest in each fund?

15. William Tell invested half of his money at 5%, one-third of his money at 4%, and the rest of his money at 3.5%. If his total annual investment income was $530, how much had he invested?

16. Last year Pete Pfeffer decided to invest half of his money in a credit union paying 4.5% interest, one-third of his money in a mutual fund paying 5% interest, and the rest of his money in a bank CD paying 4% interest. If his annual investment income was $357.50 last year, how much money had he invested?

17. Little Melinda has nickels and quarters in her bank. She has four fewer nickels than quarters. She has $3.70 in the bank. How many coins of each type does she have?

18. Fred's younger brother had several coins when he returned from his paper route. He had a total of $5.35 in dimes and quarters. He had six more quarters than he had dimes. How many of each coin did he have?

19. A newspaper carrier has $3.75 in change. He has three more quarters than dimes but twice as many nickels as quarters. How many coins of each type does he have?

20. A baseball club bought 70 balls for $120. The balls used for practice cost $1.50 each and the balls used for games cost $2.25 each. How many of each kind were bought?

21. Charlie Saulnier cashed his paycheck and came home from the bank with $100 bills, $20 bills, and $10 bills. He has twice as many $20 bills as he has $10 bills. He has three more $100 bills than he has $10 bills. He is carrying $1500 in bills. How many of each denomination does he have?

22. Roberta Burgess came home with $325 in tips from two nights on her job as a waitress. She had $20 bills, $10 bills, and $5 bills. She discovered that she had three times as many $5 bills as she had $10 bills. She also found that she had 4 fewer $20 bills than she had $10 bills. How many of each denomination did she have?

▲ **23.** The sum of the measures of the three interior angles of a triangle is always 180°. In a certain triangle the measure of one angle is double the measure of the second angle but is 5 degrees less than the measure of the third angle. What is the measure of each angle?

24. In a citywide veterinary study of a group of cats over age 5 with no regular exercise and playtime, 136 of the cats, or 4%, had no health problems.
  (a) How many cats were in the original study?
  (b) 28% of the cats had minor health problems, and the rest had major health problems. How many cats had major health problems?

25. Walter invested $2969 for one year and earned $273.47 in interest. He had invested part of the $2969 at 7% and the remainder at 11%. How much had he invested at each rate?

26. Last year the town of Waterbury paid an interest payment of $52,396.08 for a one-year note. This represented 11% of the amount the town borrowed. How much had the town borrowed?

To Think About

27. The West Suburban Car Rental Agency will rent a compact car for $35 per day and an additional charge of $0.24 per mile. The Golden Gate Car Rental Agency charges only $0.16 per mile but charges $41 per day. If a salesperson wanted to rent a car for three days, how many miles would that person have to drive to make the Golden Gate Car Rental Agency car a better bargain?

28. A Peabody pumping station pumps 2000 gallons of water per hour into an empty reservoir tank for a town's drinking supply. The station pumps for three hours. Then a leak in the reservoir tank is created by a large crack. Some water flows out of the reservoir tank at a constant rate. The pumping station continues pumping for six more hours while the leak is undetected. At the end of nine hours the reservoir contains 17,640 gallons of water. During the last six hours, how many gallons per hour are leaking from the reservoir?

Cumulative Review Problems

Perform the operations in the proper order.

29. $5(3) + 6 \div (-2)$

30. $2 - (8 - 10)^3 + 5$

Evaluate for x = −2 and y = 3.

31. $2x^2 + 3xy - 2y^2$

32. $x^3 - 5x^2 + 3y - 6$

Brian sells high-tech parts to satellite communications companies. In his negotiations he originally offers to sell one company 200 parts for a total of $22,400. However, after negotiations, he offers to sell that company the same parts at a 15% discount if the company agrees to sign a purchasing contract for 200 additional parts at some future date.

33. What is the average cost per part if the parts are sold at the discounted price?

34. How much will the total bill be for the 200 parts at the discounted price?

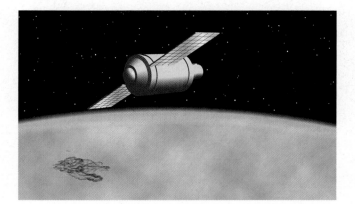

Putting Your Skills to Work

Automobile Loans and Installment Loans

The following table provides the amount of the monthly payment of a loan of $1000 at various interest rates for a period of 2 to 5 years. To find the monthly payment for a loan of a larger amount of money, merely multiply this payment by the number of thousands of the loan. For example, to borrow $14,000 at 8% interest for 3 years, you multiply $31.34 by 14 to obtain a payment of $438.76.

LOAN AMORTIZATION TABLE FOR REPRESENTATIVE INTEREST RATES
Monthly Payment per $1000 to Pay Principal and Interest

Period of Loan in Years	7%	8%	9%	10%	12%	16%
2	44.77	45.23	45.68	46.14	47.07	48.96
3	30.88	31.34	31.80	32.27	33.21	35.16
4	23.95	24.41	24.89	25.36	26.33	28.34
5	19.80	20.28	20.76	21.25	22.24	24.32

Problems for Individual Study and Investigation

Use the preceding table to solve the following problems.

1. Ricardo Sanchez wants to buy a new car. He will not make a down payment, but he will need to borrow $10,000 at a 9% interest rate for four years. What will his monthly payments be? How much interest will he pay over the four-year period? (That is, how much more than $10,000 will he pay over four years?)

2. If Alicia Wong wants to purchase a new stereo system for her apartment and borrow $3000 at 12% interest for two years, what will her monthly payments be? How much interest will she pay over the two-year period? (That is, how much more than $3000 will she pay over the two years?)

Problems for Cooperative Group Investigation

With some other members of your class, determine answers to the following.

3. If Walter Swenson plans to borrow $15,000 to purchase a new car at an interest rate of 7%, how much more in interest will he pay if he borrows the money for five years as opposed to a period of four years?

4. Ray and Shirley Peterson are planning to build a garage and a connecting family room for their house. It will cost $35,000. They have $17,000 in savings, and they plan to borrow the rest at 16% interest. They can afford a maximum monthly payment of $640. What time period for the loan will they have to select to meet that requirement? What will their monthly payment be?

Internet Connections

Netsite: http://www.prenhall.com/tobey_beg_int

Site: Mortgage Amortization Calculator or a related site

Alternate Site: Amortization Calculator

Kimberly Jones is buying a $135,000 house. She will make a down payment of 10%. The rest will be financed with a 30-year loan at an interest rate of 8.5%.

5. What will her monthly payment be? How much interest will she pay over the life of the loan?

6. Suppose she decides to pay an extra $200 every month. How much interest will she save by doing this? In how many years will the loan be paid off?

Math in the Media

Fastest Land Bird

Source: Guinness World Records.com

According to the *Guinness Book of World Records*®, the ostrich holds the world record as the fastest land bird. Its maximum running speed is 45 miles per hour. The ostrich lives in Africa. Despite its fast running speed, this bird cannot fly. Remarkably, the volume of just one ostrich egg is equal to the volume of approximately 25–40 chicken eggs.

EXERCISES

Can you use your skills from this chapter to perform some calculations in the following questions about this speedy bird?

1. An ostrich ran 20 miles per hour and traveled 10 miles. How long did it take?

2. If an ostrich ran for 45 minutes at a speed of 15 miles per hour, how far could it travel?

3. Could you arrive at a formula for calculating the maximum expected volume of one ostrich egg?

4. A man makes a huge order of scrambled eggs using 3 ostrich eggs. If he had to make this large of a quantity of scrambled eggs using chicken eggs, what is the smallest number of chicken eggs that he would need?

Chapter 2 Organizer

Topic	Procedure	Examples
Solving equations without parentheses or fractions, p. 79.	1. On each side of the equation, combine like terms if possible. 2. Add or subtract terms on both sides of the equation in order to get all terms with the variable on one side of the equation. 3. Add or subtract a value on both sides of the equation to get all terms not containing the variable on the other side of the equation. 4. Divide both sides of the equation by the coefficient of the variable. 5. If possible, simplify the solution. 6. Check your solution by substituting the obtained value into the original equation.	Solve for x. $$5x + 2 + 2x = -10 + 4x + 3$$ $$7x + 2 = -7 + 4x$$ $$7x - 4x + 2 = -7 + 4x - 4x$$ $$3x + 2 = -7$$ $$3x + 2 - 2 = -7 - 2$$ $$3x = -9$$ $$\frac{3x}{3} = \frac{-9}{3}$$ $$x = -3$$ *Check:* Is -3 the solution of $$5x + 2 + 2x = -10 + 4x + 3?$$ $$5(-3) + 2 + 2(-3) \overset{?}{=} -10 + 4(-3) + 3$$ $$-15 + 2 - 6 \overset{?}{=} -10 + (-12) + 3$$ $$-13 - 6 \overset{?}{=} -22 + 3$$ $$-19 = -19 \checkmark$$
Solving equations with parentheses and/or fractions, p. 80 and p. 85.	1. Remove any parentheses. 2. Simplify, if possible. 3. If fractions exist, multiply all terms on both sides by the least common denominator of all the fractions. 4. Now follow the remaining steps for solving an equation without parentheses or fractions.	Solve for y. $$5(3y - 4) = \frac{1}{4}(6y + 4) - 48$$ $$15y - 20 = \frac{3}{2}y + 1 - 48$$ $$15y - 20 = \frac{3}{2}y - 47$$ $$2(15y) - 2(20) = 2\left(\frac{3}{2}y\right) - 2(47)$$ $$30y - 40 = 3y - 94$$ $$30y - 3y - 40 = 3y - 3y - 94$$ $$27y - 40 = -94$$ $$27y - 40 + 40 = -94 + 40$$ $$27y = -54$$ $$\frac{27y}{27} = \frac{-54}{27}$$ $$y = -2$$

Topic	Procedure	Examples
Solving formulas, p. 93.	1. Remove any parentheses. 2. If fractions exist, multiply all terms on both sides by the LCD, which may be a variable. 3. Add or subtract terms on both sides of the equation in order to get all terms containing the *desired variable* on one side of the equation. 4. Add or subtract terms on both sides of the equation in order to get all other terms on the opposite side of the equation. 5. Divide both sides of the equation by the coefficient of the desired variable. This division may involve other variables. 6. Simplify, if possible. 7. (Optional) Check your solution by substituting the obtained expression into the original equation.	Solve for z. $B = \frac{1}{3}(hx + hz)$ First we remove parentheses. $$B = \frac{1}{3}hx + \frac{1}{3}hz$$ Now we multiply each term by 3. $$3(B) = 3\left(\frac{1}{3}hx\right) + 3\left(\frac{1}{3}hz\right)$$ $$3B = hx + hz$$ $$3B - hx = hx - hx + hz$$ $$3B - hx = hz$$ The coefficient of z is h, so we divide each side by h. $$\frac{3B - hx}{h} = z$$
Solving inequalities, p. 105.	1. Follow the steps for solving an equation up until the division step. 2. If you divide both sides of the inequality by a *positive number*, the direction of the inequality is not reversed. 3. If you divide both sides of the inequality by a *negative number*, the direction of the inequality is reversed.	Solve for x and graph your solution. $$\frac{1}{2}(3x - 2) \le -5 + 5x - 3$$ First remove parentheses and simplify. $$\frac{3}{2}x - 1 \le -8 + 5x$$ Now multiply each term by 2. $$2\left(\frac{3}{2}x\right) - 2(1) \le 2(-8) + 2(5x)$$ $$3x - 2 \le -16 + 10x$$ $$3x - 10x - 2 \le -16 + 10x - 10x$$ $$-7x - 2 \le -16$$ $$-7x - 2 + 2 \le -16 + 2$$ $$-7x \le -14$$ When we divide both sides by a negative number, the inequality is reversed. $$\frac{-7x}{-7} \ge \frac{-14}{-7}$$ $$x \ge 2$$ Graphical solution:

Topic	Procedure

Solving word problems, p. 112 and p. 118.	**1. Understand the problem.**
	  (a) Read the word problem carefully to get an overview.
	  (b) Determine what information you will need to solve the problem.
	  (c) Draw a sketch. Label it with the known information. Determine what needs to be found.
	  (d) Choose a variable to represent one unknown quantity.
	  (e) If necessary, represent other unknown quantities in terms of that same variable.
	2. Write an equation.
	  (a) Look for key words to help you to translate the words into algebraic symbols.
	  (b) Use a given relationship in the problem or an appropriate formula in order to write an equation.
	3. Solve and state the answer.
	4. Check.
	  (a) Check the solution in the original equation. Is the answer reasonable?
	  (b) Be sure the solution to the equation answers the question in the word problem. You may need to do some additional calculations if it does not.

Example A

The perimeter of a rectangle is 126 meters. The length of the rectangle is 6 meters less than double the width. Find the dimensions of the rectangle.

1. **Understand the problem.**
   We want to find the length and the width of a rectangle whose perimeter is 126 meters.

   The length is *compared to* the width, so we start with the width.

   Let $w =$ width.

   The length is 6 meters less than double the width.

   Then $l = 2w - 6$.

2. **Write an equation.**
   The perimeter of a rectangle is $P = 2w + 2l$.

   $$126 = 2w + 2(2w - 6)$$

3. **Solve.**

   $$126 = 2w + 4w - 12$$
   $$126 = 6w - 12$$
   $$138 = 6w$$
   $$23 = w \qquad \text{The width is 23 meters.}$$
   $$2w - 6 = 2(23) - 6 =$$
   $$46 - 6 = 40$$
   The length is 40 meters.

4. **Check:**
   Is this reasonable?
   Yes. A rectangle 23 meters wide and 40 meters long seems to be about right for the perimeter to be 126 meters.

   $$2(23) + 2(40) \stackrel{?}{=} 126$$

   Is the perimeter exactly 126 meters?

   $$46 + 80 \stackrel{?}{=} 126$$
   $$126 = 126 \checkmark$$

   Is the length exactly 6 meters less than double the width?

   $$40 \stackrel{?}{=} 2(23) - 6$$
   $$40 \stackrel{?}{=} 46 - 6$$
   $$40 = 40 \checkmark$$

Example B

Gina saved some money for college. She invested $2400 for one year and earned $225 in simple interest. She invested part of it at 12% and the rest of it at 9%. How much did she invest at each rate?

1. **Understand the problem.**
   We want to find each amount that was invested.

   $$\begin{array}{c}\text{interest earned} \\ \text{at 12%}\end{array} + \begin{array}{c}\text{interest} \\ \text{earned} \\ \text{at 9%}\end{array} = \begin{array}{c}\text{total} \\ \text{interest} \\ \text{of \$225}\end{array}$$

   We let x represent one quantity of money.

   Let $x =$ amount of money invested at 12%.

   We started with $2400. If we invest x at 12%, we still have $(2400 - x)$ left.

   Then $2400 - x =$ the amount of money invested at 9%.

   Interest $= prt$
   Interest at 12% $I_1 = 0.12x$
   Interest at 9% $I_2 = 0.09(2400 - x)$

2. **Write an equation.**

   $I_1 + I_2 = 225.$ $0.12x + 0.09(2400 - x) = 225$

3. **Solve.**

   $$0.12x + 216 - 0.09x = 225$$
   $$0.03x + 216 = 225$$
   $$0.03x = 9$$
   $$x = \frac{9}{0.03}$$
   $$x = 300$$

   $300 was invested at 12%.
   $$2400 - x = 2400 - 300 = 2100$$
   $2100 was invested at 9%.

4. **Check:**
   Is this reasonable? Yes.
   Are the conditions of the problem satisfied? Does the total amount invested equal $2400?

   $$2100 + 300 \stackrel{?}{=} 2400$$
   $$2400 = 2400 \checkmark$$

   Will $2100 at 9% and $300 at 12% yield $225 in interest?

   $$0.09(2100) + 0.12(300) \stackrel{?}{=} 225$$
   $$189 + 36 \stackrel{?}{=} 225$$
   $$225 = 225 \checkmark$$

Chapter 2 Review Problems

2.1–2.3 *Solve for the variable. Noninteger answers may be left in fractional form or decimal form.*

1. $5x + 20 = 3x$

2. $7x + 3 = 4x$

3. $6 - 18x = 4 - 17x$

4. $18 - 10x = 63 + 5x$

5. $6x - 2(x + 3) = 5$

6. $1 - 2(6 - x) = 3x + 2$

7. $x - (0.5x + 2.6) = 17.6$

8. $-0.2(x + 1) = 0.3(x + 11)$

9. $3(x - 2) = -4(5 + x)$

10. $\frac{1}{4} y = -16$

11. $y + 37 = 26$

12. $6(8x + 3) = 5(9x + 8)$

13. $3(x - 3) = 13x + 21$

14. $9x + 10 = 3x + 4$

15. $24 - 3x = 4(x - 1)$

16. $36 = 9x - (3x - 18)$

17. $2(3 - x) = 1 - (x - 2)$

18. $4(x + 5) - 7 = 2(x + 3)$

19. $0.9y + 3 = 0.4y + 1.5$

20. $7y - 3.4 = 11.3$

21. $8(3x + 5) - 10 = 9(x - 2) + 13$

22. $3 = 2x + 5 - 3(x - 1)$

23. $-2(x - 3) = -4x + 3(3x + 2)$

24. $2(5x - 1) - 7 = 3(x - 1) + 5 - 4x$

2.4 *Solve for the variable. Noninteger answers may be left in fractional form or decimal form.*

25. $1 = \frac{5x}{6} + \frac{2x}{3}$

26. $\frac{7x}{5} = 5 + \frac{2x}{5}$

27. $\frac{7x - 3}{2} - 4 = \frac{5x + 1}{3}$

28. $\frac{3x - 2}{2} + \frac{x}{4} = 2 + x$

29. $\frac{-3}{2}(x + 5) = 1 - x$

30. $\frac{-4}{3}(2x + 1) = -x - 2$

31. $\frac{1}{3}(x - 2) = \frac{x}{4} + 2$

32. $\frac{1}{5}(x - 3) = 2 - \frac{x}{2}$

33. $\frac{4}{5} + \frac{1}{2} x = \frac{1}{5} x + \frac{1}{2}$

34. $3x + \frac{6}{5} - x = \frac{6}{5} x - \frac{4}{5}$

35. $\frac{10}{3} - \frac{5}{3} x + x = \frac{2}{9} + \frac{1}{9} x$

36. $-\frac{8}{3} x - 8 + 2x - 5 = -\frac{5}{3}$

37. $\frac{1}{2} + \frac{5}{4} x = \frac{2}{5} x - \frac{1}{10} + 4$

38. $\frac{1}{6} x - \frac{2}{3} = \frac{1}{3}(x - 4)$

39. $\frac{1}{2}(x - 3) = \frac{1}{4}(3x - 1)$

40. $\frac{7}{12}(x - 3) = \frac{1}{3} x + 4$

41. $\dfrac{1}{6} + \dfrac{1}{3}(x - 3) = \dfrac{1}{2}(x + 9)$

42. $\dfrac{1}{7}(x + 5) - \dfrac{6}{14} = \dfrac{1}{2}(x + 3)$

43. $-\dfrac{1}{3}(2x - 6) = \dfrac{2}{3}(3 + x)$

44. $\dfrac{7}{9}x + \dfrac{2}{3} = 5 + \dfrac{1}{3}x$

2.5   *Solve for the variable indicated.*

45. Solve for y. $3x - y = 10$

46. Solve for y. $5x + 2y + 7 = 0$

47. Solve for r. $A = P(1 + rt)$

48. Solve for h. $A = 4\pi r^2 + 2\pi rh$

49. Solve for p. $H = \dfrac{1}{3}(a + 2p + 3)$

50. Solve for y. $ax + by = c$

51. Solve for d. $H + 2d = 6c - 3d$

52. Solve for b. $H = \dfrac{3c + 2b}{4}$

53. **(a)** Solve for T. $C = \dfrac{WRT}{1000}$.

     (b) Use your result to find T if $C = 0.36$, $W = 30$, and $R = 0.002$.

54. **(a)** Solve for y. $5x - 3y = 12$

     (b) Use your result to find y if $x = 9$.

55. **(a)** Solve for R. $I = \dfrac{E}{R}$

     (b) Use your result to find R if $E = 100$ and $I = 20$.

2.6–2.7   *Solve each inequality and graph the result.*

56. $2 - 3x \le -5 + 4x$

57. $2x - 3 + x > 5(x + 1)$

58. $-x + 4 < 3x + 16$

59. $4x \geq 2(12 - 2x)$ $\overset{\xrightarrow{\hspace{3cm}} x}{\underset{-5\ -4\ -3\ -2\ -1\ \ 0\ \ 1\ \ 2\ \ 3\ \ 4\ \ 5}{}}$

60. $5 - \dfrac{1}{2}x > 4$ $\overset{\xrightarrow{\hspace{3cm}} x}{\underset{-5\ -4\ -3\ -2\ -1\ \ 0\ \ 1\ \ 2\ \ 3\ \ 4\ \ 5}{}}$

61. $2(x - 1) \geq 3(2 + x)$ $\overset{\xrightarrow{\hspace{3cm}} x}{\underset{-10\ -9\ -8\ -7\ -6\ -5\ -4\ -3\ -2\ -1\ \ 0}{}}$

62. $3x + 5 - 7x \leq -2x - 1$ $\overset{\xrightarrow{\hspace{3cm}} x}{\underset{-5\ -4\ -3\ -2\ -1\ \ 0\ \ 1\ \ 2\ \ 3\ \ 4\ \ 5}{}}$

63. $-4x - 14 < 4 - 2(3x - 1)$ $\overset{\xrightarrow{\hspace{3cm}} x}{\underset{7\ \ 8\ \ 9\ \ 10\ \ 11\ \ 12\ \ 13\ \ 14\ \ 15\ \ 16\ \ 17}{}}$

64. $\dfrac{1}{2}(2x + 3) > 10$ $\overset{\xrightarrow{\hspace{3cm}} x}{\underset{5\ \ \frac{11}{2}\ \ 6\ \ \frac{13}{2}\ \ 7\ \ \frac{15}{2}\ \ 8\ \ \frac{17}{2}\ \ 9\ \ \frac{19}{2}\ \ 10}{}}$

65. $\dfrac{1}{3}(x + 2) \leq \dfrac{1}{2}(3x - 5)$ $\overset{\xrightarrow{\hspace{3cm}} x}{\underset{\frac{12}{7}\ \ \frac{13}{7}\ \ 2\ \ \frac{15}{7}\ \ \frac{16}{7}\ \ \frac{17}{7}\ \ \frac{18}{7}\ \ \frac{19}{7}\ \ \frac{20}{7}\ \ 3\ \ \frac{22}{7}}{}}$

66. $4(2 - x) - (-5x + 1) \geq -8$ $\overset{\xrightarrow{\hspace{3cm}} x}{\underset{-18\ -17\ -16\ -15\ -14\ -13\ -12\ -11\ -10\ -9\ -8}{}}$

67. $5(1 - x) < 3(x - 1) - 2(3 - x)$ $\overset{\xrightarrow{\hspace{3cm}} x}{\underset{1\ \ \frac{6}{5}\ \ \frac{7}{5}\ \ \frac{8}{5}\ \ \frac{9}{5}\ \ 2\ \ \frac{11}{5}\ \ \frac{12}{5}\ \ \frac{13}{5}\ \ \frac{14}{5}\ \ 3}{}}$

Use an inequality to solve.

68. The cost of a substitute teacher for one day at Central High School is $70. Let n = the number of substitute teachers. Set up an inequality to determine how many times a substitute teacher may be hired if the monthly budget for substitute teachers is $3220. What is the maximum number of substitute teachers that may be hired during the month? (*Hint:* Use $70n \leq 3220$.)

69. The cost of hiring a temporary secretary for a day is $85. Let n = the number of temporary secretaries. Set up an inequality to determine how many times a temporary secretary may be hired if the company budget for temporary secretaries is $1445 per month. What is the maximum number of days a temporary secretary may be hired during the month? (*Hint:* Use $85n \leq 1445$.)

2.8

▲ **70.** The perimeter of a triangle is 40 yards. The second side is 1 yard shorter than double the length of the first side. The third side is 9 yards longer than the first side. Find the length of each side.

▲ **71.** The measure of the second angle of a triangle is three times the measure of the first angle. The measure of the third angle is 12 degrees less than twice the measure of the first. Find the measure of each of the three angles.

72. A piece of rope 50 yards long is cut into two pieces. One piece is three-fifths as long as the other. Find the length of each piece.

73. Jon and Lauren have a combined income of $48,000. Lauren earns $3000 more than half of what Jon earns. How much does each earn?

2.9

74. The electric bill at Jane's house this month was $71.50. The charge is based on a flat rate of $25 per month plus a charge of $0.15 per kilowatt-hour of electricity used. How many kilowatt-hours of electricity were used?

75. Abel rented a car from Sunshine Car Rentals. He was charged $39 per day and $0.25 per mile. He rented the car for three days and paid a rental cost of $187. How many miles did he drive the car?

76. Alfred has worked for his company for 10 years as a store manager. Last year he received a 9% raise in salary. He now earns $24,525 per year. What was his salary last year before the raise?

77. Jamie bought a new tape deck for her car. When she bought it, she found that the price had been decreased by 18%. She was able to buy the tape deck for $36 less than the original price. What was the original price?

78. Peter and Shelly invested $9000 for one year. Part of it was invested at 12% and the remainder was invested at 8%. At the end of one year the couple had earned exactly $1000 in simple interest. How much did they invest at each rate?

79. A man invests $5000 in a savings bank. He places part of it in a checking account that earns 4.5% and the rest in a regular savings account that earns 6%. His total annual income from this investment in simple interest is $270. How much was invested in each account?

80. Mary has $3.75 in nickels, dimes, and quarters. She has three more quarters than dimes. She has twice as many nickels as quarters. How many of each coin does she have?

81. Tim brought in $3.65 in coins. He had two more quarters than he had dimes. He had one fewer nickel than the number of dimes. How many coins of each type did he have?

Chapter 2 Test

Solve for the variable. Noninteger answers may be left in fractional form or decimal form.

1. $3x + 5.6 = 11.6$

2. $9x - 8 = -6x - 3$

3. $2(2y - 3) = 4(2y + 2)$

4. $\frac{1}{7}y + 3 = \frac{1}{2}y$

5. $5(8 - 2x) = -6x + 8$

6. $0.8x + 0.18 - 0.4x = 0.3(x + 0.2)$

7. $\frac{2y}{3} + \frac{1}{5} - \frac{3y}{5} + \frac{1}{3} = 1$

8. $3 - 2y = 2(3y - 2) - 5y$

9. $5(20 - x) + 10x = 165$

10. $2(x + 75) + 5x = 1025$

11. $-2(2 - 3x) = 76 - 2x$

12. $20 - (2x + 6) = 5(2 - x) + 2x$

In questions 13–17, solve for x.

13. $2x - 3 = 12 - 6x + 3(2x + 3)$

14. $\frac{1}{3}x - \frac{3}{4}x = \frac{1}{12}$

15. $\frac{3}{5}x + \frac{7}{10} = \frac{1}{3}x + \frac{3}{2}$

16. $\frac{15x - 2}{28} = \frac{5x - 3}{7}$

17. $\frac{1}{3}(7x - 1) + \frac{1}{4}(2 - 5x) = \frac{1}{3}(5 + 3x)$

18. Solve for w. $A = 3w + 2P$

19. Solve for w. $\frac{2w}{3} = 4 - \frac{1}{2}(x + 6)$

20. Solve for a. $A = \frac{1}{2}h(a + b)$

21. Solve for y. $5ax(2 - y) = 3axy + 5$

22. Solve for W. $P = 2L + 2W$

▲ **23.** Use your result from question 22 to find the width of a field if the perimeter is 120 feet and the length is 42 feet.

1.
2.
3.
4.
5.
6.
7.
8.
9.
10.
11.
12.
13.
14.
15.
16.
17.
18.
19.
20.
21.
22.
23.

24. _____

25. _____

26. _____

27. _____

Solve and graph the inequality.

24. $4(x - 1) \geq 12x$

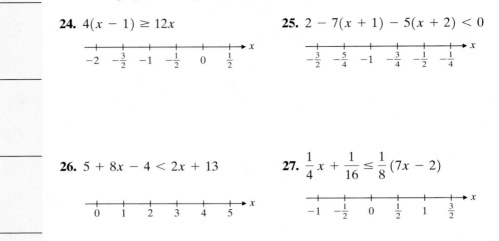

25. $2 - 7(x + 1) - 5(x + 2) < 0$

26. $5 + 8x - 4 < 2x + 13$

27. $\dfrac{1}{4} x + \dfrac{1}{16} \leq \dfrac{1}{8} (7x - 2)$

28. _____

▲ **28.** A triangular region has a perimeter of 66 meters. The first side is two-thirds of the second side. The third side is 14 meters shorter than the second side. What are the lengths of the three sides of the triangular region?

29. _____

▲ **29.** A rectangle has a length 7 meters longer than double the width. The perimeter is 134 meters. Find the dimensions of the rectangle.

30. _____

30. Three harmful pollutants were measured by a consumer group in the city. The total sample contained 15 parts per million of three harmful pollutants. The amount of the first pollutant was double the second. The amount of the third pollutant was 75% of the second. How many parts per million of each pollutant were found?

31. _____

31. Raymond has a budget of $1940 to rent a computer for his company office. The computer company he wants to rent from charges $200 for installation and service as a one-time fee. Then they charge $116 per month rental for the computer. How many months will Raymond be able to rent a computer with this budget?

32. _____

32. Franco invested $4000 in money market funds. Part was invested at 14% interest, the rest at 11% interest. At the end of each year the fund company pays interest. After one year he earned $482 in simple interest. How much was invested at each interest rate?

33. _____

33. Mary has $3.50 in change. She has twice as many nickels as quarters. She has one less dime than she has quarters. How many of each coin does she have?

Approximately one-half of this test covers the content of Chapter 1. The remainder covers the content of Chapter 2. In questions 1–4, simplify.

1. Add. $-\dfrac{2}{3} + -\dfrac{2}{7}$

2. Multiply. $(-3)(-5)(-1)(2)(-1)$

3. Divide. $1.025 \div 2.5$

4. Collect like terms. $5ab - 7ab^2 - 3ab - 12ab^2 + 10ab - 9ab^2$

5. Simplify. $(5x)^2$

6. Simplify. $(3 - 5)^2 - 9 \div (2 - 11)$

7. Simplify. $2\{x + y[3 - 2x(1 - 4y)]\}$

8. Solve for x. $4(7 - 2x) = 3x - 12$

9. Solve for x. $\dfrac{1}{3}(x + 5) = 2x - 5$

10. Solve for y. $\dfrac{2y}{3} - \dfrac{1}{4} = \dfrac{1}{6} + \dfrac{y}{4}$

11. Solve for b. $H = \dfrac{2}{3}(b + 4a)$

12. Solve for t. $I = Prt$

13. Solve for a. $A = \dfrac{ha}{2} + \dfrac{hb}{2}$

In questions 14–17, solve and graph the inequality.

14. $\dfrac{1}{2}(x - 5) \geq x - 4$

15. $4(2 - x) > 1 - 5x - 8$

16. $x + \dfrac{5}{9} \leq \dfrac{1}{3} + \dfrac{7}{9}x$

17. $4 - 16x \leq 6 - 5(3x - 2)$

1. _____

2. _____

3. _____

4. _____

5. _____

6. _____

7. _____

8. _____

9. _____

10. _____

11. _____

12. _____

13. _____

14. _____

15. _____

16. _____

17. _____

18. _____

18. The football team will not let Chuck play unless he passes biology with a C (70 or better) average. There are five tests in the semester, and he has failed (0) the first one. However, he found a tutor and received an 82, an 89, and an 87 on the next three tests. Solve the inequality $\dfrac{0 + 82 + 89 + 87 + x}{5} \geq 70$ to find what his minimum score must be on the last test in order to pass the course and play football.

19. _____

▲ **19.** A rectangle has a perimeter of 78 centimeters. The length of the rectangle is 11 centimeters longer than triple the width. Find the dimensions of the rectangle.

20. _____

20. Hassan invested $7000 for one year. He invested part of it in a high-risk fund that pays 15% interest. He placed the rest in a safe, low-risk fund that pays 7% interest. At the end of one year he earned $730 in simple interest. How much did he invest in each of the two funds?

21. _____

21. Linda has some dimes, nickels, and quarters in her purse. The total value of these coins is $2.55. She has three times as many dimes as quarters. She has three more nickels than quarters. How many coins of each type does she have?

Graphing and Functions

A rainforest is one of the most amazing ecosystems in the entire world. It is a huge region with its own climate, an amazing variety of animals and plant life, and home to many species that do not exist anywhere else in the world. However, all is not well. The amount of land covered by rainforests is decreasing. Can you predict how rapidly this decrease is taking place? Turn to the Putting Your Skills to Work problems on page 199 to find out.

1. _____

2. _____

3. (a) _____

(b) _____

4. _____

5. _____

6. _____

7. _____

8. (a) _____

(b) _____

If you are familiar with the topics in this chapter, take this test now. Check your answers with those in the back of the book. If an answer is wrong or you can't answer a question, study the appropriate section of the chapter.

If you are not familiar with the topics in this chapter, don't take this test now. Instead, study the examples, work the practice exercises, and then take the test.

This test will help you to identify which concepts you have mastered and which you need to study further.

Section 3.1

1. Graph the points.
$(-3, 1), (4, -4), (-2, -6)$

2. Write the coordinates of each point.

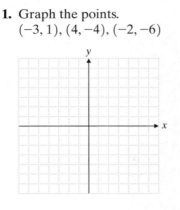

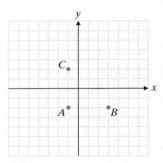

3. Complete each ordered pair so that it is a solution to the equation $y = -3x + 7$.
(a) $(-1,)$ **(b)** $(, 1)$

Section 3.2

4. Graph $5x - 2y = -10$.

5. Graph $y = 4$.

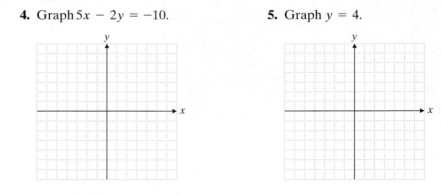

Sections 3.3 and 3.4

6. What is the slope of the line passing through $(4, -2)$ and $(-4, 6)$?

7. Find the slope of the line $4x - 3y - 7 = 0$.

8. Find the equation of the line with slope 4 and y-intercept $(0, -5)$.
(a) Write the equation in slope–intercept form.
(b) Write the equation in the form $Ax + By = C$.

9. Graph the equation $y = 2x + 1$.

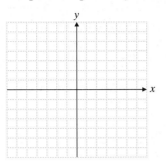

10. Write the equation of the graph.

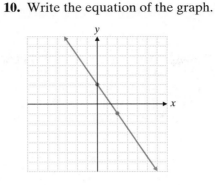

11. Line p has a slope of $\frac{3}{5}$.
   (a) What is the slope of a line parallel to line p?
   (b) What is the slope of a line perpendicular to line p?

12. Find the equation of a line with slope $\frac{2}{3}$ that passes through the point $(-3, -3)$.

13. Find the equation of a line that passes through $(-1, 6)$ and $(2, 3)$.

Section 3.5

14. Graph $y \geq 3x - 2$.

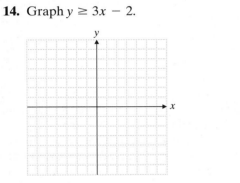

15. Graph $4x + 2y < -12$.

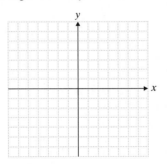

Section 3.6

Determine whether each of the following is a function.

16. Circle with radius r and circumference C

r	1	5	10	15
C	6.28	31.4	62.8	94.2

17.

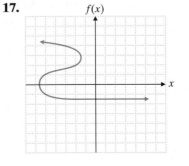

Find the indicated values for the function.

18. $f(x) = 3x - 7$
   (a) $f(-2)$ **(b)** $f(7)$

19. $g(x) = 2x^2$
   (a) $g(-6)$ **(b)** $g\left(\frac{1}{2}\right)$

9. _____

10. _____

11. (a) _____

(b) _____

12. _____

13. _____

14. _____

15. _____

16. _____

17. _____

18. (a) _____

(b) _____

19. (a) _____

(b) _____

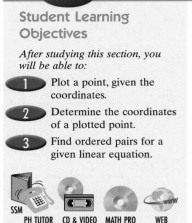
1 Plotting a Point, Given the Coordinates

Oftentimes we can better understand an idea if we see a picture. This is the case with many mathematical concepts, including those relating to algebra. We can illustrate algebraic relationships with drawings called **graphs**. Before we can draw a graph, however, we need a frame of reference.

In Chapter 1 we showed that any real number can be represented on a number line. Look at the following number line. The arrow indicates the positive direction.

To form a **rectangular coordinate system**, we draw a second number line vertically. We construct it so that the 0 point on each number line is exactly at the same place. We refer to this location as the **origin**. The horizontal number line is often called the **x-axis**. The vertical number line is often called the **y-axis**. Arrows show the positive direction for each axis.

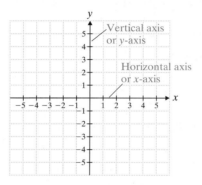

We can represent a point in this rectangular coordinate system by using an **ordered pair** of numbers. For example, $(5, 2)$ is an ordered pair that represents a point in the rectangular coordinate system. The numbers in an ordered pair are often referred to as the **coordinates** of the point. The first number is called the **x-coordinate** and it represents the distance from the origin measured along the horizontal or x-axis. If the x-coordinate is positive, we count the proper number of squares to the right (that is, in the positive direction). If the x-coordinate is negative, we count to the left. The second number in the pair is called the **y-coordinate** and it represents the distance from the origin measured along the y-axis. If the y-coordinate is positive, we count the proper number of squares upward (that is, in the positive direction). If the y-coordinate is negative, we count downward.

$$(5, 2)$$
x-coordinate ⤴ ⤴ y-coordinate

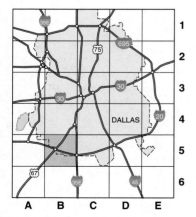

Suppose the directory for the map on the left indicated that you would find a certain street in the region B5. To find the street you would first scan across the horizontal scale until you found section B; from there you would scan up the map until you hit section 5 along the vertical scale. As we will see in the next example, plotting a point in the rectangular coordinate system is much like finding a street on a map with grids.

EXAMPLE 1 Plot the point $(5, 2)$ on a rectangular coordinate system. Label this as point A.

Since the x-coordinate is 5, we first count 5 units to the right on the x-axis. Then, because the y-coordinate is 2, we count 2 units up from the point where we stopped on the x-axis. This locates the point corresponding to $(5, 2)$. We mark this point with a dot and label it A.

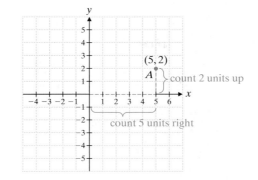

Practice Problem 1   Plot the point $(3, 4)$ on the preceding coordinate system. Label this as point B.

It is important to remember that the first number in an ordered pair is the x-coordinate and the second number is the y-coordinate. The ordered pairs $(5, 2)$ and $(2, 5)$ represent different points.

EXAMPLE 2   Plot each point on the following coordinate system. Label the points F, G, and H, respectively.

(a) $(-5, 3)$ **(b)** $(2, -6)$ **(c)** $(-4, -5)$

(a) Notice that the x-coordinate, -5, is negative. On the coordinate grid, negative x-values appear to the left of the origin. Thus, we will begin by counting 5 squares to the left starting at the origin. Since the y-coordinate, 3, is positive, we will count 3 units up from the point where we stopped on the x-axis.

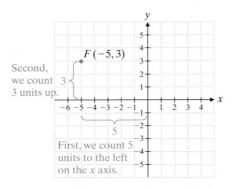

(b) The x-coordinate is positive. Begin by counting 2 squares to the right of the origin. Then count down because the y-coordinate is negative.

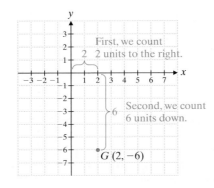

(c) The x-coordinate is negative. Begin by counting 4 squares to the left of the origin. Then count down because the y-coordinate is negative.

PRACTICE PROBLEM 2

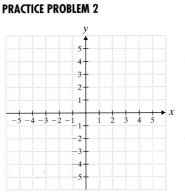

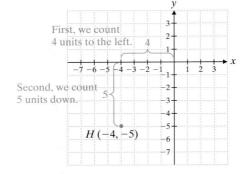

Practice Problem 2 Use the coordinate system in the margin to plot each point. Label the points I, J, and K, respectively.

(a) $(-2, -4)$ **(b)** $(-4, 5)$ **(c)** $(4, -2)$

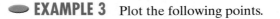

EXAMPLE 3 Plot the following points.

$F: (0, 5)$ $G: \left(3, \frac{3}{2}\right)$ $H: (-6, 4)$ $I: (-3, -4)$
$J: (-4, 0)$ $K: (2, -3)$ $L: (6.5, -7.2)$

These points are plotted in the figure.

PRACTICE PROBLEM 3

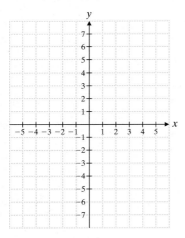

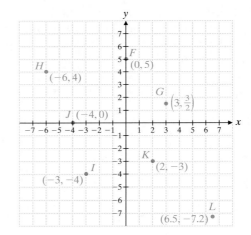

Note: When you are plotting decimal values like $(6.5, -7.2)$, plot the point halfway between 6 and 7 in the x-direction (for the 6.5) and at your best approximation of -7.2 in the y-direction.

Practice Problem 3 Plot the following points. Label each point with both the letter and the ordered pair. Use the coordinate system provided in the margin.

$A: (3, 7)$ $B: (0, -6)$ $C: (3, -4.5)$ $D: \left(-\frac{7}{2}, 2\right)$

2 *Determining the Coordinates of a Plotted Point*

Sometimes we need to find the coordinates of a point that has been plotted. First, we count the units we need on the x-axis to get as close as possible to the point. Next we count the units up or down we need to go from the x-axis to reach the point.

EXAMPLE 4   What ordered pair of numbers represents point A in the graph below?

   If we move along the x-axis until we get as close as possible to A, we end up at the number 5. Thus we obtain 5 as the first number of the ordered pair. Then we count 4 units upward on a line parallel to the y-axis to reach A. So we obtain 4 as the second number of the ordered pair. Thus point A is represented by the ordered pair $(5, 4)$.

Practice Problem 4   What ordered pair of numbers represents point B in the graph below?

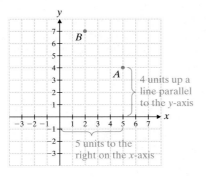

EXAMPLE 5   Write the coordinates of each point plotted in the following graph.

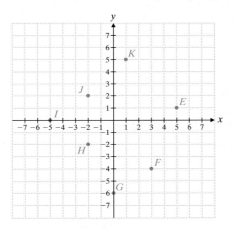

The coordinates of each point are as follows.

$E = (5, 1)$	$I = (-5, 0)$
$F = (3, -4)$	$J = (-2, 2)$
$G = (0, -6)$	$K = (1, 5)$
$H = (-2, -2)$	

Be very careful that you put the x-coordinate first and the y-coordinate second. Be careful that each sign is correct.

Practice Problem 5   Give the coordinates of each point plotted in the graph in the margin.

PRACTICE PROBLEM 5

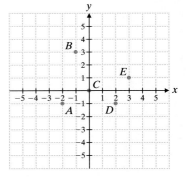

3   Finding Ordered Pairs for a Given Linear Equation

Equations such as $3x + 2y = 5$ and $6x + y = 3$ are called linear equations in two variables.

> A **linear equation in two variables** is an equation that can be written in the form $Ax + By = C$ where A, B, and C are real numbers but A and B are not *both* zero.

   Replacement values for x and y that make *true mathematical statements* of the equation are called *truth values*, and an ordered pair of these truth values is called a **solution**.

Consider the equation $3x + 2y = 5$. The ordered pair $(1, 1)$ is a solution to the equation because when we replace x by 1 and y by 1 in the equation, we obtain a true statement.

$$3(1) + 2(1) = 5 \quad \text{or} \quad 3 + 2 = 5$$

Likewise $(-1, 4)$, $(3, -2)$, $(5, -5)$, and $(7, -8)$ are also solutions to the equation. In fact, there are an infinite number of solutions for any given linear equation in two variables.

If one value of an ordered-pair solution to a linear equation is known, the other can be quickly obtained. To do so, we replace the proper variable in the equation by the known value. Then using the methods learned in Chapter 2, we solve the resulting equation for the other variable.

EXAMPLE 6 Find the missing coordinate to complete the following ordered-pair solutions for the equation $2x + 3y = 15$.

(a) $(0, ?)$ **(b)** $(?, 1)$

(a) For the ordered pair $(0, ?)$ we know that $x = 0$. Replace x by 0 in the equation.

$$2x + 3y = 15$$
$$2(0) + 3y = 15$$
$$0 + 3y = 15$$
$$y = 5$$

Thus we have the ordered pair $(0, 5)$.

(b) For the ordered pair $(?, 1)$, we *do not know* the value of x. However, we do know that $y = 1$. So we start by replacing the variable y by 1. We will end up with an equation with one variable, x. We can then solve for x.

$$2x + 3y = 15$$
$$2x + 3(1) = 15$$
$$2x + 3 = 15$$
$$2x = 12$$
$$x = 6$$

Thus we have the ordered pair $(6, 1)$.

Practice Problem 6 Find the missing coordinate to complete the following ordered-pair solutions for the equation $3x - 4y = 12$.

(a) $(0, ?)$ **(b)** $(?, 3)$ **(c)** $(?, -6)$

The linear equations that we work with are not always written in the form $Ax + By = C$, but are sometimes solved for y, as in $y = -6x + 3$. Consider the equation $y = -6x + 3$. The ordered pair $(2, -9)$ is a solution to the equation. When we replace x by 2 and y by -9 we obtain a true mathematical statement:

$$(-9) = -6(2) + 3 \quad \text{or} \quad -9 = -12 + 3.$$

In examining data from real-world situations, we often find that plotting data points shows useful trends. In such cases, it is often necessary to use a different scale, one that displays only positive values.

EXAMPLE 7   The number of motor vehicle accidents in millions is recorded in the following table for the years 1980 to 2000.

(a) Plot points that represent this data on the given coordinate system.

(b) What trends are apparent from the plotted data?

(a)

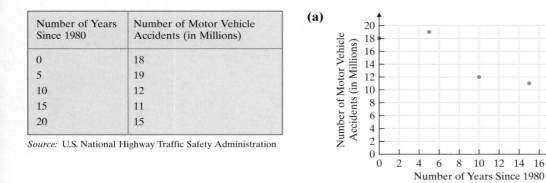

Number of Years Since 1980	Number of Motor Vehicle Accidents (in Millions)
0	18
5	19
10	12
15	11
20	15

Source: U.S. National Highway Traffic Safety Administration

(b) From 1980 to 1985, there was a slight increase in the number of accidents. From 1985 to 1995, there was a significant decrease in the number of accidents. From 1995 to 2000, there was a moderate increase in the number of accidents.

Practice Problem 7   The number of motor vehicle deaths in thousands is recorded in the following table for the years 1980 to 2000.

(a) Plot points that represent this data on the given coordinate system.

(b) What trends are apparent from the plotted data?

Number of Years Since 1980	Number of Motor Vehicle Deaths (in Thousands)
0	51
5	44
10	45
15	42
20	43

Source: U.S. National Highway Traffic Safety Administration

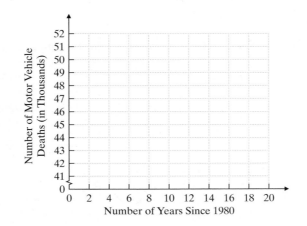

1. Plot the following points.

$J: (-4, 3.5)$ $K: (6, 0)$
$L: (5, -6)$ $M: (0, -4)$

2. Plot the following points.

$R: (-3, 0)$ $S: (3.5, 4)$
$T: (-2, -2.5)$ $V: (0, 5)$

Consider the points plotted in the graph at right.

3. Give the coordinates for points R, S, X, and Y.

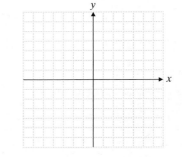

4. Give the coordinates for points T, V, W, and Z.

Verbal and Writing Skills

5. What is the x-coordinate of the origin?

6. What is the y-coordinate of the origin?

7. Explain why $(5, 1)$ is referred to as an *ordered* pair of numbers.

In exercises 8 and 9, 10 points are plotted in the figure. List all the ordered pairs needed to represent the points.

8.

9.

Complete the ordered pairs so that each is a solution to the given linear equation.

10. $y = 2x + 5$
   (a) $(0, \quad)$ **(b)** $(3, \quad)$

11. $y = 3x + 8$
   (a) $(0, \quad)$ **(b)** $(4, \quad)$

12. $y = -4x + 2$
   (a) $(-5, \quad)$ **(b)** $(4, \quad)$

13. $y = -2x + 3$
   (a) $(-6, \quad)$ **(b)** $(3, \quad)$

14. $3x - 4y = 11$
   (a) $(-3, \quad)$ **(b)** $(\quad, 1)$

15. $5x - 2y = 9$
   (a) $(7, \quad)$ **(b)** $(\quad, -7)$

16. $2y + 3x = -6$
   (a) $(-2, \quad)$ **(b)** $(\quad, 3)$

17. $-4x + 5y = -20$
   (a) $(10, \quad)$ **(b)** $(\quad, -8)$

18. $3x + \dfrac{1}{2}y = 2$
   (a) $(\quad, 16)$ **(b)** $\left(\dfrac{5}{2}, \quad\right)$

19. $4x + \dfrac{1}{3}y = 8$ **(a)** $(\quad, -12)$ **(b)** $\left(\dfrac{3}{2}, \quad\right)$

The preceding map shows a portion of New York, Connecticut, and Massachusetts. Like many maps used in driving or flying, it has horizontal and vertical grid markers for ease of use. For example, Newburgh, New York, is located in grid B3. Use the grid labels to indicate the locations of the following cities.

20. Lynbrook, New York

21. Hampton Bays, New York

22. Athol, Massachusetts

23. Pittsfield, Massachusetts

24. Middletown, Connecticut

25. Waterbury, Connecticut

26. The number of farms in the United States for selected years starting in 1940 is recorded in the following table. The number of farms is measured in hundred thousands. For example, 61 in the second column means 61 hundred thousand or 6,100,000.

(a) Plot points that represent this data on the given rectangular coordinate system.

(b) What trends are apparent from the plotted data?

Number of Years Since 1940	Number of Farms (in Hundred Thousands)
0	61
10	54
20	40
30	30
40	24
50	21
60	22

Source: U.S. Department of Agriculture

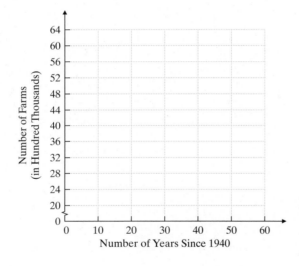

27. The number of barrels of oil produced in the United States for selected years starting in 1940 is recorded in the following table. The number of barrels of oil is measured in hundred thousands. For example, 14 in the second column means 14 hundred thousand, or 1,400,000.

(a) Plot points on the given rectangular coordinate system that represent this data.

(b) What trends are apparent from the plotted data?

Number of Years Since 1940	Number of Barrels of Crude Oil (in Hundred Thousands)
0	14
10	20
20	26
30	35
40	31
50	27
60	21

Source: U.S. Department of Energy

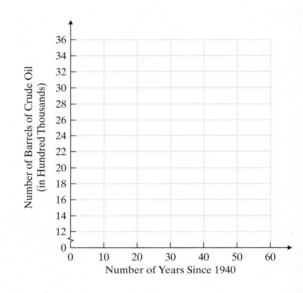

Cumulative Review Problems

28. Solve. $\dfrac{1}{4}(x + 2) = \dfrac{1}{2} + 3x$

29. Solve and graph. $x - 3 \le -5(x - 6)$

▲ **30.** The circular pool at the hotel where Bob and Linda stayed in Orlando, Florida, has a radius of 19 yards. What is the area of the pool? (Use $\pi \approx 3.14$).

31. The membership at David's gym has increased by 16% over the last 2 years. It used to have 795 members. How many members does it have now?

32. A major Russian newspaper called *Izvestia* had a circulation of 6,109,005 before the dissipation of the Soviet Union. Currently the circulation is 262,045. What is the percent of decrease of readership of *Izvestia*?

▲ **33.** An expensive Persian rug that measures 30 feet by 22 feet is priced at $44,020. A customer negotiates with the owner of the rug, and they agree upon a price of $36,300.
  (a) What is the cost per square foot of the rug at the discounted price?
  (b) The rug dealer has a smaller matching Persian rug measuring 14 feet by 8 feet. He says he will sell it to the customer at the same cost per square foot as the larger one. How much will the small rug cost?

1 Graphing a Linear Equation by Plotting Three Ordered Pairs

We have seen that a solution to a linear equation in two variables is an ordered pair. The graph of an ordered pair is a point. Thus we can graph an equation by graphing the points corresponding to its ordered-pair solutions.

A linear equation in two variables has an infinite number of ordered-pair solutions. We can see that this is true by noting that we can substitute any number for x in the equation and solve it to obtain a y-value. For example, if we substitute $x = 0$, $1, 2, 3, \ldots$ into the equation $y = -x + 3$ and solve for y, we obtain the ordered-pair solutions $(0, 3)$, $(1, 2)$, $(2, 1)$, $(3, 0), \ldots$ (Substitute these values into the equation to convince yourself.) If we plot these points on a rectangular coordinate system, we notice that they form a straight line.

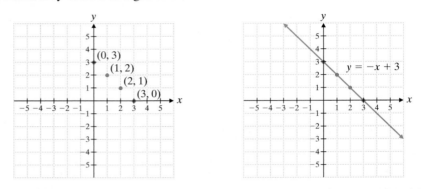

It turns out that all of the points corresponding to the ordered-pair solutions of $y = -x + 3$ lie on this line, and the line extends forever in both directions. A similar statement can be made about any linear equation in two variables.

> The graph of any linear equation in two variables is a straight line.

From geometry we know that two points determine a line. Thus to graph a linear equation in two variables, we need only graph two ordered-pair solutions of the equation and then draw the line that passes through them. Having said this, we recommend that you use three points to graph a line. Two points will determine where the line is. The third point verifies that you have drawn the line correctly. For ease in plotting, it is better if the ordered pairs contain integers.

To Graph a Linear Equation

1. Look for three ordered pairs that are solutions to the equation.
2. Plot the points.
3. Draw a line through the points.

● EXAMPLE 1 Find three ordered pairs that satisfy $2x + y = 4$. Then graph the resulting straight line.

Since we can choose any value for x, choose numbers that are convenient. To organize the results, we will make a table of values. We will let $x = 0$, $x = 1$, and $x = 3$, respectively. We write these numbers under x in our table of values. For each of these x-values, we find the corresponding y-value in the equation $2x + y = 4$.

Table of Values	
x	y
0	4
1	2
3	−2

$$2x + y = 4 \qquad\qquad 2x + y = 4 \qquad\qquad 2x + y = 4$$
$$2(0) + y = 4 \qquad\qquad 2(1) + y = 4 \qquad\qquad 2(3) + y = 4$$

154

$$0 + y = 4 \qquad 2 + y = 4 \qquad 6 + y = 4$$
$$y = 4 \qquad\qquad y = 2 \qquad\qquad y = -2$$

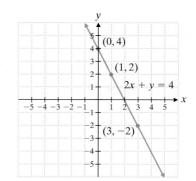

We record these results by placing each y-value in the table next to its corresponding x-value. Keep in mind that these values represent ordered pairs, each of which is a solution to the equation. To make calculating and graphing easier, we choose integer values whenever possible. If we plot these ordered pairs and connect the three points, we get a straight line that is the graph of the equation $2x + y = 4$. The graph of the equation is shown in the figure at the right.

Practice Problem 1   Graph $x + y = 10$ on the given coordinate system.

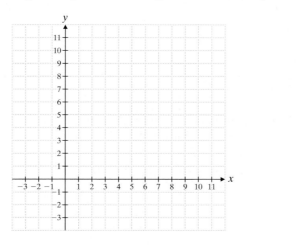

EXAMPLE 2   Graph $5x - 4y + 2 = 2$.

First, we simplify the equation $5x - 4y + 2 = 2$ by adding -2 to each side.

$$5x - 4y + 2 - 2 = 2 - 2$$
$$5x - 4y = 0$$

Since we are free to choose any value of x, $x = 0$ is a natural choice. Calculate the value of y when $x = 0$.

$$5(0) - 4y = 0$$
$$-4y = 0$$
$$y = 0 \quad \text{Remember: Any number times 0 is 0.}$$
$$\text{Since } -4y = 0, y \text{ must equal } 0.$$

Now let's see what happens when $x = 1$.

$$5(1) - 4y = 0$$
$$5 - 4y = 0$$
$$-4y = -5$$
$$y = \frac{-5}{-4} \quad \text{or} \quad \frac{5}{4} \quad \text{This is not an easy number to graph.}$$

A better choice for a replacement of x is a number that is divisible by 4. Let's see why. Let $x = 4$ and let $x = -4$.

$$5(4) - 4y = 0 \qquad\qquad 5(-4) - 4y = 0$$
$$20 - 4y = 0 \qquad\qquad -20 - 4y = 0$$
$$-4y = -20 \qquad\qquad -4y = 20$$
$$y = \frac{-20}{-4} \quad \text{or} \quad 5 \qquad y = \frac{20}{-4} \quad \text{or} \quad -5$$

Now we can put these numbers into our table of values and graph the line.

PRACTICE PROBLEM 2

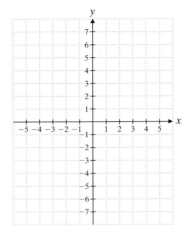

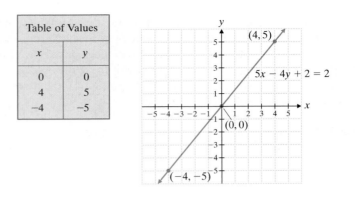

Table of Values	
x	y
0	0
4	5
−4	−5

Practice Problem 2 Graph $7x + 3 = -2y + 3$ on the coordinate system in the margin.

To Think About In Example 2, we picked values of x and found the corresponding values for y. An alternative approach is to solve the equation for the variable y first.

$$5x - 4y = 0$$
$$-4y = -5x \quad \text{Add } -5x \text{ to each side.}$$
$$\frac{-4y}{-4} = \frac{-5x}{-4} \quad \text{Divide each side by } -4.$$
$$y = \frac{5}{4}x$$

Now let $x = -4$, $x = 0$, and $x = 4$, and find the corresponding values of y. Explain why you would choose multiples of 4 as replacements of x in this equation. Graph the equation and compare it to the graph in Example 2.

In the previous two examples we began by picking values for x. We could just as easily have chosen values for y.

2 *Graphing a Straight Line by Plotting Its Intercepts*

What values should we pick for x and y? Which points should we use for plotting? For many straight lines it is easiest to pick the two *intercepts*. Some lines have only one intercept. We will discuss these separately.

> The *x-intercept* of a line is the point where the line crosses the x-axis; it has the form $(a, 0)$. The *y-intercept* of a line is the point where the line crosses the y-axis; it has the form $(0, b)$.

Intercept Method of Graphing

To graph an equation using intercepts, we:

1. Find the x-intercept by letting $y = 0$ and solving for x.
2. Find the y-intercept by letting $x = 0$ and solving for y.
3. Find one additional ordered pair so that we have three points with which to plot the line.

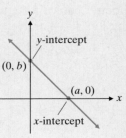

EXAMPLE 3 Use the intercept method to graph $5y - 3x = 15$.

Let $y = 0$. $5(0) - 3x = 15$ Replace y by 0.

$-3x = 15$ Divide both sides by -3.

$x = -5$

The ordered pair $(-5, 0)$ is the x-intercept.

Let $x = 0$. $5y - 3x = 15$

$5y - 3(0) = 15$ Replace x by 0.

$5y = 15$ Divide both sides by 5.

$y = 3$

The ordered pair $(0, 3)$ is the y-intercept.
We find another ordered pair to have a third point.

Let $y = 6$. $5(6) - 3x = 15$ Replace y by 6.

$30 - 3x = 15$ Simplify.

$-3x = -15$ Subtract 30 from both sides.

$x = \dfrac{-15}{-3}$ or   5

The ordered pair is $(5, 6)$.
Our table of values is

x	y
0	3
-5	0
5	6

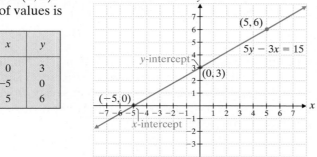

Practice Problem 3 Use the intercept method to graph $2y - x = 6$. Use the given coordinate system.

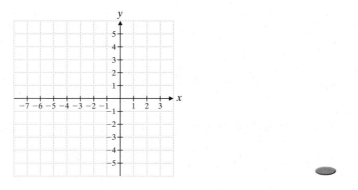

Sidelight

Can you draw all straight lines by the intercept method? Not really. Some straight lines may go through the origin and have only one intercept. If a line goes through the origin, it will have an equation of the form $Ax + By = 0$, where $A \neq 0$ or $B \neq 0$ or both. Examples are $3x + 4y = 0$ and $5x - 2y = 0$. In such cases you should plot two additional points besides the origin. Be sure to simplify each equation before attempting to graph it.

3 *Graphing Horizontal and Vertical Lines*

You will notice that the *x*-axis is a horizontal line. It is the line $y = 0$, since for any value of *x*, the value of *y* is 0. Try a few points. The points $(1, 0)$, $(3, 0)$, and $(-2, 0)$ all lie on the *x*-axis. Any horizontal line will be parallel to the *x*-axis. Lines such as $y = 5$ and $y = -2$ are horizontal lines. What does $y = 5$ mean? It means that for any value of *x*, *y* is 5. Likewise $y = -2$ means that for any value of *x*, $y = -2$.

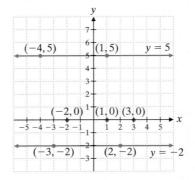

How can we recognize the equation of a line that is horizontal, that is, parallel to the *x*-axis?

> If the graph of an equation is a straight line that is parallel to the *x*-axis (that is, a horizontal line), the equation will be of the form $y = b$, where *b* is some real number.

EXAMPLE 4   Graph $y = -3$.

You could write the equation as $0x + y = -3$. Then it is clear that for any value of *x* that you substitute, you will always obtain $y = -3$. Thus, as shown in the figure, $(4, -3)$, $(0, -3)$, and $(-3, -3)$ are all ordered pairs that satisfy the equation $y = -3$. Since the *y*-coordinate of every point on this line is -3, it is easy to see that the horizontal line will be 3 units below the *x*-axis.

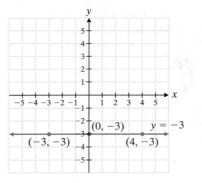

Practice Problem 4   Graph $2y - 3 = 0$ on the given coordinate system.

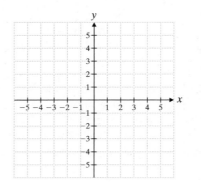

Notice that the y-axis is a vertical line. This is the line $x = 0$, since for any y, x is 0. Try a few points. The points $(0, 2)$, $(0, -3)$, and $\left(0, \frac{1}{2}\right)$ all lie on the y-axis. Any vertical line will be parallel to the y-axis. Lines such as $x = 2$ and $x = -3$ are vertical lines. Think of what $x = 2$ means. It means that for any value of y, x is 2. The graph of $x = 2$ is a vertical line two units to the right of the y-axis.

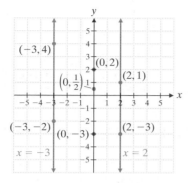

How can we recognize the equation of a line that is vertical, that is, parallel to the y-axis?

If the graph of an equation is a straight line that is parallel to the y-axis (that is, a vertical line), the equation will be of the form $x = a$, where a is some real number.

EXAMPLE 5   Graph $x = 5$.

This can be done immediately by drawing a vertical line 5 units to the right of the origin. The x-coordinate of every point on this line is 5.

The equation $x - 5 = 0$ can be rewritten as $x = 5$ and graphed as shown.

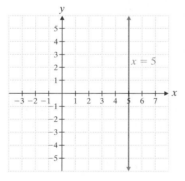

Practice Problem 5   Graph $x + 3 = 0$ on the following coordinate system.

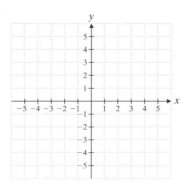

3.2 Exercises

Verbal and Writing Skills

1. Is the point $(-2, 5)$ a solution to the equation $2x + 5y = 0$? Why or why not?

2. The graph of a linear equation in two variables is a _____.

3. The x-intercept of a line is the point where the line crosses the _____ .

4. The graph of the equation $y = b$ is a _____ line.

Complete the ordered pairs so that each is a solution of the given linear equation. Then plot each solution and graph the equation by connecting the points by a straight line.

5. $y = -2x + 1$
(0,)
(−2,)
(1,)

6. $y = -3x - 4$
(−2,)
(−1,)
(0,)

7. $y = x - 4$
(0,)
(2,)
(4,)

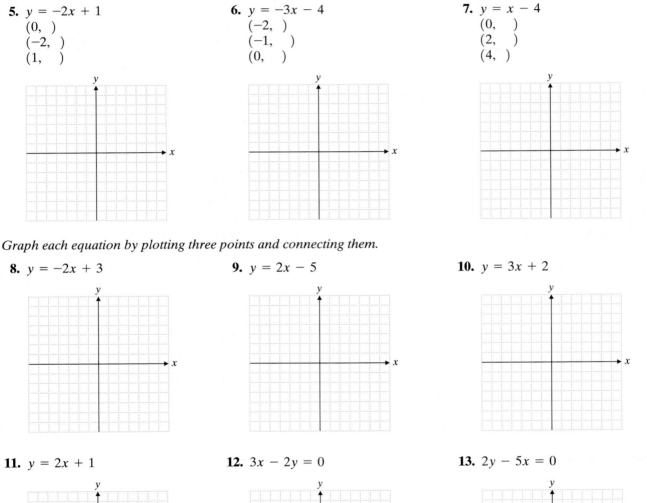

Graph each equation by plotting three points and connecting them.

8. $y = -2x + 3$

9. $y = 2x - 5$

10. $y = 3x + 2$

11. $y = 2x + 1$

12. $3x - 2y = 0$

13. $2y - 5x = 0$

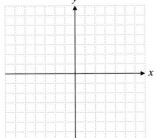
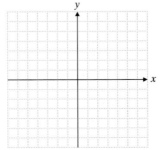

Graph each equation by plotting the intercepts and one other point.

14. $y = -\frac{1}{2}x + 3$

15. $y = -\frac{2}{3}x + 2$

16. $4x + 3y = 12$

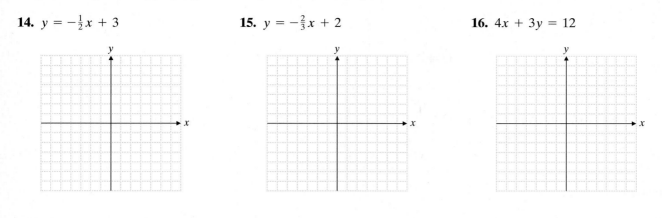

17. $y = 6 - 2x$

18. $y = 4 - 2x$

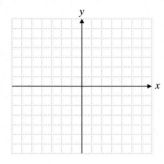

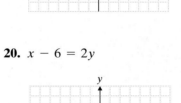

19. $x + 3 = 6y$

20. $x - 6 = 2y$

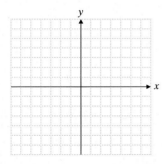

Graph the equation. Be sure to simplify the equation before graphing it.

21. $3x + 2y = 6$

22. $y - 2 = 3y$

23. $2x + 5y - 2 = -12$

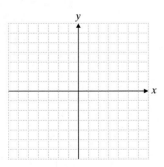

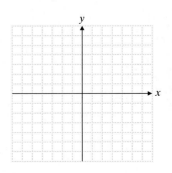

24. $3x - 4y - 5 = -17$

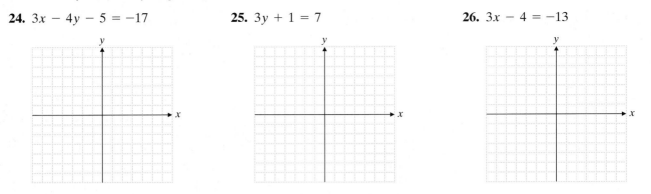

25. $3y + 1 = 7$

26. $3x - 4 = -13$

Applications

27. The number of calories burned by an average person while cross-country skiing is given by the equation $C = 8m$, where m is the number of minutes. (*Source: National Center for Health Statistics.*) Find the values of C for $m = 0, 15, 30, 45, 60,$ and 75. Graph your result.

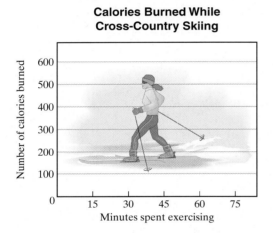

Calories Burned While Cross-Country Skiing

28. The number of calories burned by an average person while jogging is given by the equation $C = \dfrac{28}{3} m$, where m is the number of minutes. (*Source: National Center for Health Statistics.*) Find the values of C for $m = 0, 15, 30, 45, 60,$ and 75. Graph your result.

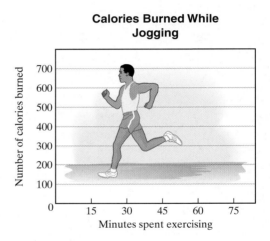

Calories Burned While Jogging

29. The maximum amount of carbon dioxide in the air in Mauna Loa, Hawaii, is approximated by the equation $C = 1.6t + 343$, where C is the concentration of CO_2 measured in parts per million and t is the number of years since 1982. (*Source: Environmental Science: The Way the World Works*, Seventh Edition, B. Nebel and R. Wright, Upper Saddle River, NJ, Prentice-Hall.) Graph the equation for $t = 0, 4, 8,$ and 18.

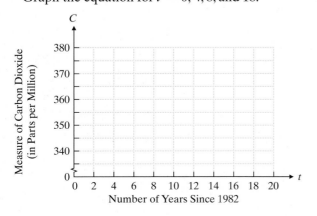

30. The approximate population of the United States is given by the equation $P = 2.3t + 179$ where P is the population in millions and t is the number of years since 1960. (*Source: Bureau of the Census, U.S. Department of Commerce.*) Graph the equation for $t = 0, 10, 30,$ and 40.

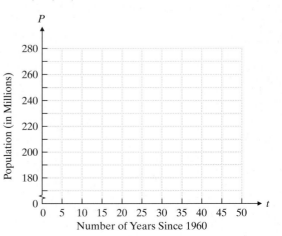

Cumulative Review Problems

31. Evaluate. $6 + (-2)^3 - (5 - 9) \div 2$

32. Solve and graph on a number line. $4 - 3x \le 18$

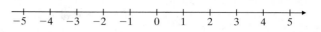

33. A rectangle's width is 1 meter more than half of its length. Find the dimensions if the perimeter measures 53 meters.

34. Last semester there were 29 students for every 2 faculty members at Skyline University. At that time the school had 3074 students. How many faculty members were there last semester?

35. Hank is working for six months in Japan while his wife is back in California with their two children. Each of them sends the other a letter each week. The cost of mailing a letter from Japan and the cost of mailing a letter from the United States totals $1.84. The cost of mailing a letter from the United States is two cents less than half the cost of mailing a letter from Japan. What is the cost of mailing a letter from Japan?

36. Each day Melinda has a cup of instant coffee at home in the morning and a cup of percolated coffee at the office in the afternoon. She consumes 194 milligrams of caffeine each day with these two cups. If the instant coffee has seventeen more milligrams than half the number of milligrams of caffeine in the percolated coffee, how many milligrams of caffeine does a cup of the instant coffee contain?

3.3 Slope of a Line

Student Learning Objectives

After studying this section, you will be able to:

 1 Find the slope of a line given two points on the line.

2 Find the slope and *y*-intercept of a line given the equation of the line.

3 Write the equation of a line given the slope and *y*-intercept.

4 Graph a line using the slope and *y*-intercept.

5 Find the slopes of lines that are parallel or perpendicular.

SSM · PH TUTOR CENTER · CD & VIDEO · MATH PRO · WEB

1 Finding the Slope of a Line Given Two Points on the Line

We often use the word *slope* to describe the incline (the steepness) of a hill. A carpenter or a builder will refer to the *pitch* or *slope* of a roof. The slope is the change in the vertical distance compared with the change in the horizontal distance as you go from one point to another point along the roof. If the change in the vertical distance is greater than the change in the horizontal distance, the slope will be steep. If the change in the horizontal distance is greater than the change in the vertical distance, the slope will be gentle.

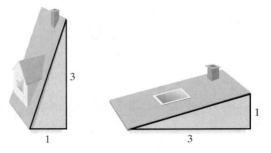

In a coordinate plane, the **slope** of a straight line is defined by the change in *y* divided by the change in *x*.

$$\text{Slope} = \frac{\text{change in } y}{\text{change in } x}$$

Consider the line drawn through points *A* and *B* in the figure. If we measure the change from point *A* to point *B* in the *x*-direction and the *y*-direction, we will have an idea of the steepness (or the slope) of the line. From point *A* to point *B* the change in *y* values is from 2 to 4, a *change of* 2. From point *A* to point *B* the change in *x* values is from 1 to 5, a *change of* 4. Thus

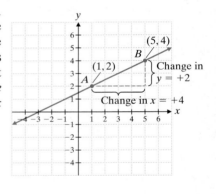

$$\text{slope} = \frac{\text{change in } y}{\text{change in } x} = \frac{2}{4} = \frac{1}{2}$$

Informally, we can describe this move as the rise over the run: $\text{slope} = \dfrac{\text{rise}}{\text{run}}$.

We now state a more formal (and more frequently used) definition.

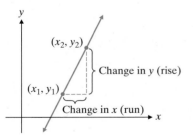

Definition of Slope of a Line

The **slope** of any *nonvertical* straight line that contains the points with coordinates (x_1, y_1) and (x_2, y_2) has a slope defined by the difference ratio

$$\text{slope} = m = \frac{y_2 - y_1}{x_2 - x_1} \qquad \text{where } x_2 \neq x_1.$$

The use of subscripted terms such as x_1, x_2, and so on, is just a way of indicating that the first *x*-value is x_1 and the second *x*-value is x_2. Thus (x_1, y_1) are the coordinates of the first point and (x_2, y_2) are the coordinates of the second point. The letter *m* is commonly used for the slope.

● **EXAMPLE 1** Find the slope of the line that passes through $(2, 0)$ and $(4, 2)$. Let $(2, 0)$ be the first point (x_1, y_1) and $(4, 2)$ be the second point (x_2, y_2).

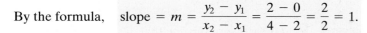

$$(2, 0) \qquad (4, 2) \qquad m = \frac{2}{2} = 1$$

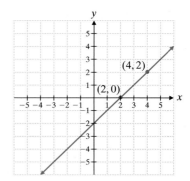

By the formula, $\quad$ slope $= m = \dfrac{y_2 - y_1}{x_2 - x_1} = \dfrac{2 - 0}{4 - 2} = \dfrac{2}{2} = 1.$

The sketch of the line is shown in the figure at the right.

Note that the slope of the line will be the same if we let $(4, 2)$ be the first point (x_1, y_1) and $(2, 0)$ be the second point (x_2, y_2).

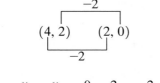

$$(4, 2) \qquad (2, 0)$$

$$m = \frac{y_2 - y_1}{x_2 - x_1} = \frac{0 - 2}{2 - 4} = \frac{-2}{-2} = 1$$

Thus, given two points, it does not matter which you call (x_1, y_1) and which you call (x_2, y_2).

🚫 **WARNING** Be careful, however, not to put the x's in one order and the y's in another order when finding the slope from two points on a line.

Practice Problem 1 Find the slope of the line that passes through $(6, 1)$ and $(-4, -1)$.

It is a good idea to have some concept of the values of slopes. In downhill skiing, a very gentle slope used for teaching beginning skiers might drop one foot vertically for each 10 feet horizontally. The slope would be $\frac{1}{10}$. The speed of a skier on a hill with such a gentle slope would be only about 6 miles per hour.

A triple diamond slope for experts might drop 11 feet vertically for each 10 feet horizontally. The slope would be $\frac{11}{10}$. The speed of a skier on such an expert trail would be in the range of 60 miles per hour.

It is important to see how positive and negative slopes affect the graphs of lines.

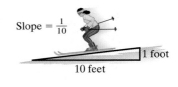

Slope $= \frac{1}{10}$

1 foot

10 feet

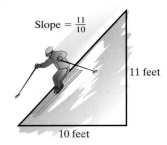

Slope $= \frac{11}{10}$

11 feet

10 feet

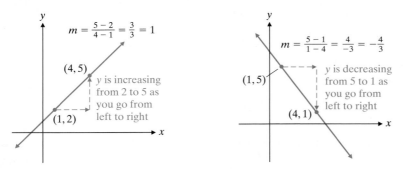

$$m = \frac{5 - 2}{4 - 1} = \frac{3}{3} = 1$$

$(4, 5)$

y is increasing from 2 to 5 as you go from left to right

$(1, 2)$

$$m = \frac{5 - 1}{1 - 4} = \frac{4}{-3} = -\frac{4}{3}$$

$(1, 5)$

y is decreasing from 5 to 1 as you go from left to right

$(4, 1)$

1. If the y-values increase as you go from left to right, the slope of the line is positive.

2. If the y-values decrease as you go from left to right, the slope of the line is negative.

EXAMPLE 2 Find the slope of the line that passes through $(-3, 2)$ and $(2, -4)$.
Let $(-3, 2)$ be (x_1, y_1) and $(2, -4)$ be (x_2, y_2).

$$m = \frac{y_2 - y_1}{x_2 - x_1} = \frac{-4 - 2}{2 - (-3)} = \frac{-4 - 2}{2 + 3} = \frac{-6}{5} = -\frac{6}{5}$$

The slope of this line is negative. We would expect this since the y-value decreased from 2 to -4 as the x-value increased. What does the graph of this line look like? Plot the points and draw the line to verify.

Practice Problem 2   Find the slope of the line that passes through $(2, 0)$ and $(-1, 1)$.

To Think About   Describe the line in Practice Problem 2 by looking at its slope. Then verify by drawing the graph.

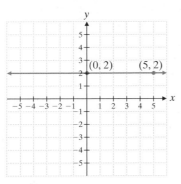

EXAMPLE 3   Find the slope of the line that passes through the given points.

(a) $(0, 2)$ and $(5, 2)$ **(b)** $(-4, 0)$ and $(-4, -4)$

(a) Take a moment to look at the y-values. What do you notice? What does this tell you about the line? Now calculate the slope.

$$m = \frac{2 - 2}{5 - 0} = \frac{0}{5} = 0$$

Since any two points on a horizontal line will have the same y-value, the slope of a horizontal line is 0.

(b) Take a moment to look at the x-values. What do you notice? What does this tell you about the line? Now calculate the slope.

$$m = \frac{-4 - 0}{-4 - (-4)} = \frac{-4}{0}$$

Recall that division by 0 is undefined.

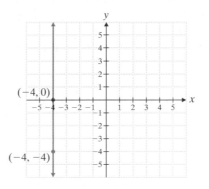

The slope of a vertical line is undefined. We say that a vertical line has **no slope**. Notice in our definition of slope that $x_2 \neq x_1$. Thus it is not appropriate to use the formula for slope for the points in (b). We did so to illustrate what would happen if $x_2 = x_1$. We get an impossible situation, $\dfrac{y_2 - y_1}{0}$. Now you can see why we include the restriction $x_2 \neq x_1$ in our definition.

Practice Problem 3   Find the slope of the line that passes through the given points.

(a) $(-5, 6)$ and $(-5, 3)$ **(b)** $(-7, -11)$ and $(3, -11)$

Slope of a Straight Line

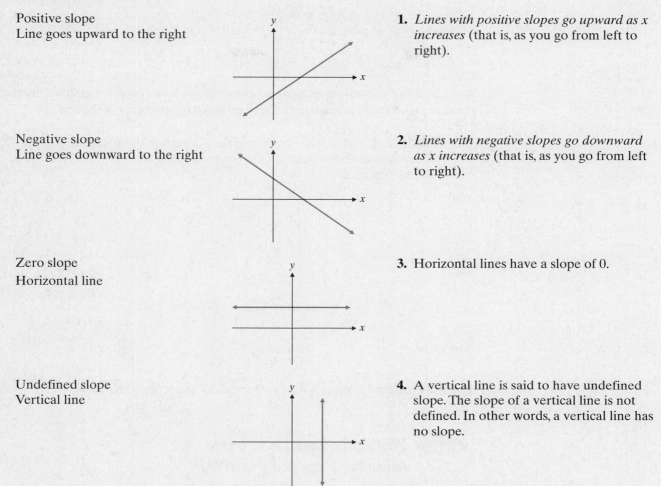

Positive slope
Line goes upward to the right

1. *Lines with positive slopes go upward as x increases* (that is, as you go from left to right).

Negative slope
Line goes downward to the right

2. *Lines with negative slopes go downward as x increases* (that is, as you go from left to right).

Zero slope
Horizontal line

3. Horizontal lines have a slope of 0.

Undefined slope
Vertical line

4. A vertical line is said to have undefined slope. The slope of a vertical line is not defined. In other words, a vertical line has no slope.

2 *Finding the Slope and y-Intercept of a Line Given the Equation of the Line*

Recall that the equation of a line is a linear equation in two variables. This equation can be written in several different ways. A very useful form of the equation of a straight line is the slope–intercept form. This form can be derived in the following way. Suppose that a straight line with slope m crosses the y-axis at a point $(0, b)$. Consider any other point on the line and label the point (x, y). Then we have the following.

$$\frac{y_2 - y_1}{x_2 - x_1} = m$$ Definition of slope.

$$\frac{y - b}{x - 0} = m$$ Substitute $(0, b)$ for (x_1, y_1) and (x, y) for (x_2, y_2).

$$\frac{y - b}{x} = m$$ Simplify.

$$y - b = mx$$ Multiply both sides by x.

$$y = mx + b$$ Add b to both sides.

y-intercept *b*
Line of slope *m*
(x, y)
$(0, b)$

This form of a linear equation immediately reveals the slope of the line, m, and the y-coordinate of the point where the line intercepts (crosses) the y-axis, b.

> **Slope–Intercept Form of a Line**
>
> The slope–intercept form of the equation of the line that has slope m and the y-intercept $(0, b)$ is given by
>
> $$y = mx + b.$$

By using algebraic operations, we can write any linear equation in slope–intercept form and use this form to identify the slope and the y-intercept of the line.

EXAMPLE 4   What is the slope and the y-intercept of the line $5x + 3y = 2$?

We want to solve for y and get the equation in the form $y = mx + b$. We need to isolate the y-variable.

$5x + 3y = 2$

$3y = -5x + 2$ Subtract $5x$ from both sides.

$y = \dfrac{-5x + 2}{3}$ Divide both sides by 3.

$y = -\dfrac{5}{3}x + \dfrac{2}{3}$ Using the property $\dfrac{a + b}{c} = \dfrac{a}{c} + \dfrac{b}{c}$, write the right-hand side as two fractions.

The *slope* is $-\dfrac{5}{3}$.   The y-intercept is $\left(0, \dfrac{2}{3}\right)$.

Practice Problem 4   What is the slope and the y-intercept of the line $4x - 2y = -5$?

3 Writing the Equation of a Line Given the Slope and y-Intercept

If we know the slope of a line and the y-intercept, we can write the equation of the line, $y = mx + b$.

EXAMPLE 5   Find the equation of the line with slope $\frac{2}{5}$ and y-intercept $(0, -3)$.

(a) Write the equation in slope–intercept form, $y = mx + b$.

(b) Write the equation in the form $Ax + By = C$.

(a) We are given that $m = \frac{2}{5}$ and $b = -3$. Thus we have the following.

$$y = mx + b$$

$$y = \frac{2}{5}x + (-3)$$

$$y = \frac{2}{5}x - 3$$

(b) Recall, for the form $Ax + By = C$, that A, B, and C are integers. We first clear the equation of fractions. Then we move the x-term to the left side.

$5y = 5\left(\dfrac{2x}{5}\right) - 5(3)$ Multiply each term by 5.

$5y = 2x - 15$ Simplify.

$-2x + 5y = -15$ Subtract $2x$ from each side.

$2x - 5y = 15$ Multiply each term by -1. The form $Ax + By = C$ is usually written with A as a positive integer.

Practice Problem 5   Find the equation of the line with slope $-\frac{3}{7}$ and y-intercept $\left(0, \frac{2}{7}\right)$.

(a) Write the equation in slope–intercept form.

(b) Write the equation in the form $Ax + By = C$.

---

4   *Graphing a Line Using the Slope and y-Intercept*

If we know the slope of a line and the y-intercept, we can draw the graph of the line.

EXAMPLE 6   Graph the line with slope $m = \frac{2}{3}$ and y-intercept $(0, -3)$. Use the given coordinate system.

Recall that the y-intercept is the point where the line crosses the y-axis. We plot the point $(0, -3)$ on the y-axis.

Recall that slope $= \dfrac{\text{rise}}{\text{run}}$. Since the slope for this line is $\frac{2}{3}$, we will go up (rise) 2 units and go over (run) to the right 3 units from the point $(0, -3)$. Look at the figure below. This is the point $(3, -1)$. Plot the point. Draw a line that connects the two points $(0, -3)$ and $(3, -1)$.

This is the graph of the line with slope $\frac{2}{3}$ and y-intercept $(0, -3)$.

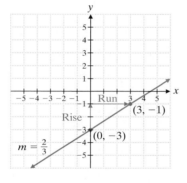

PRACTICE PROBLEM 6

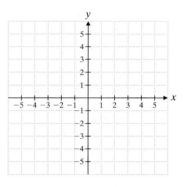

Practice Problem 6   Graph the line with slope $= \frac{3}{4}$ and y-intercept $(0, -1)$. Use the coordinate system in the margin.

EXAMPLE 7   Graph the equation $y = -\frac{1}{2}x + 4$. Use the following coordinate system.

Begin with the y-intercept. Since $b = 4$, plot the point $(0, 4)$. Now look at the slope $-\frac{1}{2}$. This can be written as $\frac{-1}{2}$. Begin at $(0, 4)$ and go *down* 1 unit and to the right 2 units. This is the point $(2, 3)$. Plot the point. Draw the line that connects the points $(0, 4)$ and $(2, 3)$.

This is the graph of the equation $y = -\frac{1}{2}x + 4$.

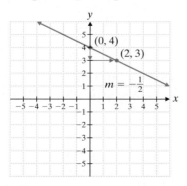

PRACTICE PROBLEM 7

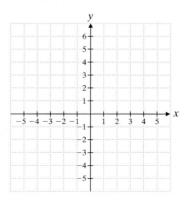

Practice Problem 7   Graph the equation $y = -\frac{2}{3}x + 5$. Use the coordinate system in the margin.

To Think About Explain why you count down 1 unit and move to the right 2 units to represent the slope $-\frac{1}{2}$. Could you have done this in another way? Try it. Verify that this is the same line.

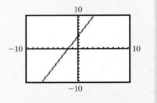

Graphing Calculator

Graphing Lines
You can graph a line given in the form $y = mx + b$ using a graphing calculator. For example, to graph $y = 2x + 4$, enter the right-hand side of the equation in the Y = editor of your calculator and graph. Choose an appropriate window to show all the intercepts. The following window is −10 to 10 by −10 to 10.
Display:

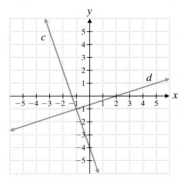

Try graphing other equations given in slope–intercept form.

5 *Finding the Slopes of Lines That Are Parallel or Perpendicular*

Parallel lines are two straight lines that never touch. Look at the parallel lines in the figure below. Notice that the slope of line a is −3 and the slope of line b is also −3. Why do you think the slopes must be equal? What would happen if the slope of line b were −1? Graph it and see.

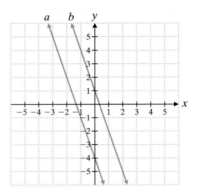

Parallel Lines

Parallel lines are two straight lines that never touch.
Parallel lines have the same slope.

$$m_1 = m_2$$

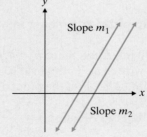

Slope m_1

Slope m_2

Perpendicular lines are two lines that meet at a 90° angle. Look at the perpendicular lines in the figure at left. The slope of line c is −3. The slope of line d is $\frac{1}{3}$. Notice that

$$(-3)\left(\frac{1}{3}\right) = \left(-\frac{3}{1}\right)\left(\frac{1}{3}\right) = -1.$$

You may wish to draw several pairs of perpendicular lines to determine whether the product of their slopes is always −1.

Perpendicular Lines

Perpendicular lines are two lines that meet at a 90° angle.
Perpendicular lines have slopes whose product is –1. If m_1 and m_2 are slopes of perpendicular lines, then

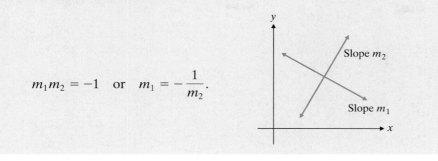

$$m_1 m_2 = -1 \quad \text{or} \quad m_1 = -\frac{1}{m_2}.$$

EXAMPLE 8   Line e has a slope of $-\frac{2}{3}$.

(a) If line f is parallel to line e, what is its slope?

(b) If line g is perpendicular to line e, what is its slope?

(a) Parallel lines have the same slope. Line f has a slope of $-\frac{2}{3}$.

(b) Perpendicular lines have slopes whose product is -1.

$$m_1 m_2 = -1$$

$$-\frac{2}{3} m_2 = -1 \qquad \text{Substitute } -\frac{2}{3} \text{ for } m_1.$$

$$\left(-\frac{3}{2}\right)\left(-\frac{2}{3}\right) m_2 = -1\left(-\frac{3}{2}\right) \qquad \text{Multiply both sides by } -\frac{3}{2}.$$

$$m_2 = \frac{3}{2}$$

Thus line g has a slope of $\frac{3}{2}$.

Practice Problem 8   Line h has a slope of $\frac{1}{4}$.

(a) If line j is parallel to line h, what is its slope?

(b) If line k is perpendicular to line h, what is its slope?

EXAMPLE 9   The equation of line l is $y = -2x + 3$.

(a) What is the slope of a line that is parallel to line l?

(b) What is the slope of a line that is perpendicular to line l?

(a) Looking at the equation, we can see that the slope of line l is -2. The slope of a line that is parallel to line l is -2.

(b) Perpendicular lines have slopes whose product is -1.

$$m_1 m_2 = -1$$
$$(-2)m_2 = -1 \quad \text{Substitute } -2 \text{ for } m_1.$$
$$m_2 = \frac{1}{2} \quad \text{Because } (-2)\left(\frac{1}{2}\right) = -1.$$

The slope of a line that is perpendicular to line l is $\frac{1}{2}$.

Practice Problem 9   The equation of line n is $y = \frac{1}{4}x - 1$.

(a) What is the slope of a line that is parallel to line n?

(b) What is the slope of a line that is perpendicular to line n?

Graphing Calculator

Graphing Parallel Lines

If two equations are in the form $y = mx + b$, then it will be obvious that they are parallel because the slope will be the same. On a graphing calculator graph both of these equations:

$$y = -2x + 6$$
$$y = -2x - 4$$

Use the window of -10 to 10 for both x and y. Display:

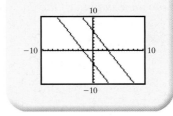

3.3 Exercises

Find the slope of a straight line that passes through the given pair of points.

1. $(6, 6)$ and $(9, 3)$

2. $(4, 1)$ and $(6, 7)$

3. $(-2, 1)$ and $(3, 4)$

4. $(5, 6)$ and $(-3, 1)$

5. $(-7, -4)$ and $(3, -8)$

6. $(-4, -6)$ and $(5, -9)$

7. $(-3, 0)$ and $(0, -4)$

8. $(0, 5)$ and $(5, 3)$

9. $\left(\dfrac{3}{4}, -4\right)$ and $(2, -8)$

10. $\left(\dfrac{5}{3}, -2\right)$ and $(3, 6)$

Verbal and Writing Skills

11. Can you find the slope of the line passing through $(5, -12)$ and $(5, -6)$? Why or why not?

12. Can you find the slope of the line passing through $(6, -2)$ and $(-8, -2)$? Why or why not?

Find the slope and the y-intercept.

13. $y = 8x + 9$

14. $y = 2x + 10$

15. $y = -3x + 4$

16. $y = -8x - 7$

17. $y = \dfrac{5}{6}x - \dfrac{2}{9}$

18. $y = -\dfrac{3}{4}x + \dfrac{5}{6}$

19. $y = -6x$

20. $y = -2$

21. $6x + y = \dfrac{4}{5}$

22. $2x + y = -\dfrac{3}{4}$

23. $5x + 2y = 3$

24. $7x + 3y = 4$

25. $7x - 3y = 4$

26. $9x - 4y = 18$

Write the equation of the line (a) in slope–intercept form and (b) in the form $Ax + By = C$.

27. $m = \dfrac{3}{4}, b = 2$

28. $m = \dfrac{4}{5}, b = 3$

29. $m = 6, b = -3$

30. $m = 5, b = -6$

31. $m = -\dfrac{5}{4}, b = -\dfrac{3}{4}$

32. $m = -4, b = \dfrac{1}{2}$

Graph the line $y = mx + b$ for the given values.

33. $m = \dfrac{1}{2}, b = -3$

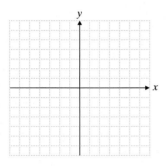

34. $m = \dfrac{2}{3}, b = -4$

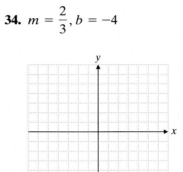

35. $m = -\dfrac{5}{3}, b = 2$

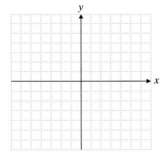

36. $m = -\dfrac{3}{2}, b = 4$

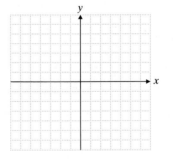

In exercises 37–41, graph the line.

37. $y = \dfrac{3}{4}x - 5$

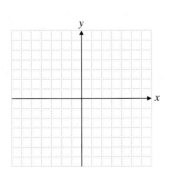

38. $y = \dfrac{2}{3}x - 6$

39. $y + 2x = 3$

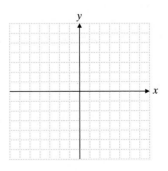

40. $y + 4x = 5$

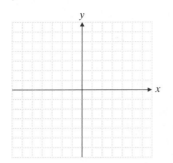

41. $y = 2x$

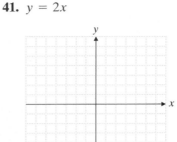

42. A line has a slope of $\dfrac{2}{3}$.

   (a) What is the slope of a line parallel to it?

   (b) What is the slope of a line perpendicular to it?

43. A line has a slope of $\dfrac{13}{5}$.

   (a) What is the slope of a line parallel to it?

   (b) What is the slope of a line perpendicular to it?

44. A line has a slope of 6.

   (a) What is the slope of a line parallel to it?

   (b) What is the slope of a line perpendicular to it?

45. A line has a slope of $-\dfrac{5}{8}$.

   (a) What is the slope of a line parallel to it?

   (b) What is the slope of a line perpendicular to it?

46. The equation of a line is $y = \dfrac{1}{3}x + 2$.

   (a) What is the slope of a line parallel to it?

   (b) What is the slope of a line perpendicular to it?

47. Do the points $(3, -4)$, $(18, 6)$, and $(9, 0)$ all lie on the same line? If so, what is the equation of the line?

To Think About

48. During the years from 1980 to 2000, the total income for the U.S. federal budget can be approximated by the equation $y = 14(4x + 35)$, where x is the number of years since 1980 and y is the amount of money in billions of dollars. (*Source:* U.S. Office of Management and Budget)

   (a) Write the equation in slope–intercept form.

   (b) Find the slope and the y-intercept.

   (c) In this specific equation, what is the meaning of the slope? What does it indicate?

49. During the years from 1970 to 1990, the approximate number of civilians employed in the United States could be predicted by the equation $y = \frac{1}{10}(22x + 830)$, where x is the number of years since 1970 and y is the number of civilians employed, measured in millions. (*Source:* U.S. Bureau of Labor Statistics)

   (a) Write the equation in slope–intercept form.

   (b) Find the slope and the y-intercept.

   (c) In this specific equation, what is the meaning of the slope? What does it indicate?

Cumulative Review Problems

Solve for x and graph the solution.

50. $3x + 8 > 2x + 12$

51. $\frac{1}{4}x + 3 > \frac{2}{3}x + 2$

52. $\frac{1}{2}(x + 2) \leq \frac{1}{3}x + 5$

53. $7x - 2(x + 3) \leq 4(x - 7)$

54. One angle of a triangle is seven degrees less than double the second angle. The third angle is twelve degrees more than double the second angle. What is the measure of each angle?

55. New England Camp Cherith tries to maintain a ratio of 3 counselors for every 25 campers. One week last summer there were 189 campers. What is the minimum number of counselors who should have been there that week?

56. New York City subway ridership rose from 206 million a year in 1990 to 240 million in 1999. (*Source:* Bureau of Transportation Statistics.) The mayor predicts it will rise to 288 million by 2008. How much larger is the projected percent of increase from 1999 to 2008 than the observed percent of increase from 1990 to 1999?

Putting Your Skills to Work

Underwater Pressure

The pressure on a submarine or other submersible object deep under the ocean's surface is very significant. In 1960 a record was set for a manned submersible by a special research vessel named the *Trieste*, which descended to a depth of more than 35,800 feet. The pressure on the hull of the *Trieste* at that depth was more than 8 tons per square inch!

The underwater pressure in pounds per square inch on an object submerged in the ocean is given by the equation $p = \frac{5}{11}d + 15$, where d is the depth in feet below the surface.

Problems for Individual Study and Investigation

1. Find the underwater pressure at a depth of 22 feet.

2. At a certain location, the underwater pressure was found to be 35 pounds per square inch. What was the depth at which this measurement was taken?

3. Using the coordinates of the two points obtained in questions 1 and 2, calculate the slope of the line.

4. From the linear equation, determine the coordinates of a third point. Plot this point and the points obtained in questions 1 and 2. Draw a line.

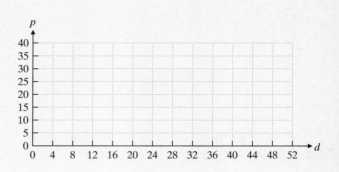

5. From the graph, determine the slope of the line. Does this agree with the slope in question 3? Does this agree with the equation? Why? Why not?

6. Complete the following table of values.

d	0	11	22	33	44
p					

(a) What is the difference between successive p values?

(b) What is the difference between successive d values?

(c) How is this related to the slope? Why?

Problems For Cooperative Group Activity and Analysis

Together with members of your class, see if you can determine the following.

Scientists are working to find a pressure equation similar to $p = \frac{5}{11}d + 15$ for a liquid that has different properties than water. They have compiled the following table of values in which p is the pressure in pounds per square inch and d is the depth in feet below the surface of this liquid.

d	0	5	12
p	18	22	27.6

7. Using the values $(0, 18)$ and $(5, 22)$, find the slope of the line.

8. Using the values $(5, 22)$ and $(12, 27.6)$, find the slope of the line. How does this compare with your answer from question 7? What does this imply?

9. Write the equation of the pressure for this new liquid in the form $p = md + b$ by determining the slope m and the p-intercept $(0, b)$.

10. What is the pressure on an object if it is submerged in this liquid at a depth of 44 feet? How does this compare with the pressure on an object submerged in water at a depth of 44 feet?

11. Is this new liquid more or less dense than water? What evidence do you have to support your conclusions?

Internet Connections

wwW   Netsite: http://www.prenhall.com/tobey_beg_int

Site: Boat Dives (Kona Coast Divers) or a related site

This site provides information about diving locations in Hawaii. Use the information provided to find the underwater pressure that you would experience in a dive to the maximum possible depth at each of the following locations.

12. Robert's Reef

13. Garden Eel Cove

14. Fish Rock (Ka'iwi Point)

15. Long Lava Tube

16. Amphitheatre

3.4 Obtaining the Equation of a Line

Student Learning Objectives

After studying this section, you will be able to:

1 Write the equation of a line given a point and a slope.

2 Write the equation of a line given two points.

3 Write the equation of a line given a graph of the line.

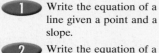

SSM
PH TUTOR CENTER   CD & VIDEO   MATH PRO   WEB

---

1 Writing the Equation of a Line Given a Point and a Slope

If we know the slope of a line and the *y*-intercept, we can write the equation of the line in slope–intercept form. Sometimes we are given the slope and a point on the line. We use the information to find the *y*-intercept. Then we can write the equation of the line.

It may be helpful to summarize our approach.

> **To Find the Equation of a Line Given a Point and a Slope**
>
> **1.** Substitute the given values of *x*, *y*, and *m* into the equation $y = mx + b$.
> **2.** Solve for *b*.
> **3.** Use the values of *b* and *m* to write the equation in the form $y = mx + b$.

EXAMPLE 1 Find an equation of the line that passes through $(-3, 6)$ with slope $-\frac{2}{3}$.

We are given the values $m = -\frac{2}{3}$, $x = -3$, and $y = 6$.

$$y = mx + b$$

$$6 = \left(-\frac{2}{3}\right)(-3) + b \quad \text{Substitute known values.}$$

$$6 = 2 + b$$

$$4 = b$$

The equation of the line is $y = -\frac{2}{3}x + 4$.

Practice Problem 1 Find an equation of the line that passes through $(-8, 12)$ with slope $-\frac{3}{4}$.

2 Writing the Equation of a Line Given Two Points

Our procedure can be extended to the case for which two points are given.

EXAMPLE 2 Find an equation of the line that passes through $(2, 5)$ and $(6, 3)$.

We first find the slope of the line. Then we proceed as in Example 1.

$$m = \frac{y_2 - y_1}{x_2 - x_1}$$

$$m = \frac{3 - 5}{6 - 2} \quad \text{Substitute } (x_1, y_1) = (2, 5) \text{ and } (x_2, y_2) = (6, 3) \text{ into the formula.}$$

$$= \frac{-2}{4} = -\frac{1}{2}$$

Choose either point, say $(2, 5)$, to substitute into $y = mx + b$ as in Example 1.

$$5 = -\frac{1}{2}(2) + b$$
$$5 = -1 + b$$
$$6 = b$$

The equation of the line is $y = -\frac{1}{2}x + 6$.

    Note: We could have substituted the slope and the other point, $(6, 3)$, into the slope–intercept form and arrived at the same answer. Try it.

Practice Problem 2 Find an equation of the line that passes through $(3, 5)$ and $(-1, 1)$.

---

3 *Writing the Equation of a Line Given a Graph of the Line*

EXAMPLE 3 What is the equation of the line in the figure at right?

    First, look for the y-intercept. The line crosses the y-axis at $(0, 4)$. Thus $b = 4$.

    Second, find the slope.

$$m = \frac{\text{change in } y}{\text{change in } x}$$

Look for another point on the line. We choose $(5, -2)$. Count the number of vertical units from 4 to -2 (rise). Count the number of horizontal units from 0 to 5 (run).

$$m = \frac{-6}{5}$$

    Now using $m = -\frac{6}{5}$ and $b = 4$, we can write the equation of the line.

$$y = mx + b$$
$$y = -\frac{6}{5}x + 4$$

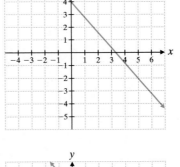

PRACTICE PROBLEM 3

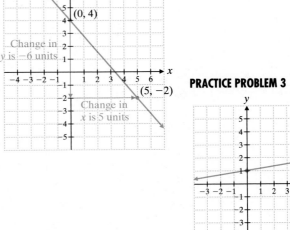

Practice Problem 3 What is the equation of the line in the figure at the right?

Find an equation of the line that has the given slope and passes through the given point.

1. $m = 4, (-3, 0)$ **2.** $m = 3, (2, -2)$ **3.** $m = -2, (4, 3)$ **4.** $m = -4, (5, 7)$

5. $m = -3, \left(\dfrac{1}{2}, 2\right)$ **6.** $m = -2, \left(3, \dfrac{1}{3}\right)$ **7.** $m = -\dfrac{2}{5}, (5, -3)$ **8.** $m = \dfrac{2}{3}, (3, -2)$

Write an equation of the line passing through the given points.

9. $(3, -12)$ and $(-4, 2)$ **10.** $(-3, 9)$ and $(2, -11)$ **11.** $(2, -3)$ and $(-1, 6)$ **12.** $(1, -8)$ and $(2, -14)$

13. $(3, 5)$ and $(-1, -15)$ **14.** $(-1, -19)$ and $(2, 2)$ **15.** $\left(1, \dfrac{5}{6}\right)$ and $\left(3, \dfrac{3}{2}\right)$ **16.** $(2, 0)$ and $\left(\dfrac{3}{2}, \dfrac{1}{2}\right)$

Write an equation of each line.

17.

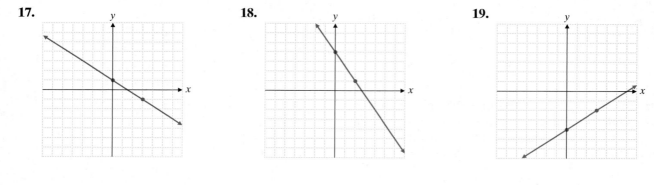

18.

19.

20.

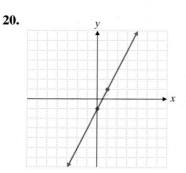

21.

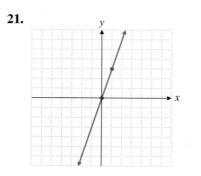

22.

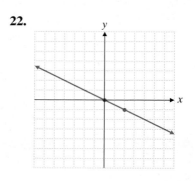

23.

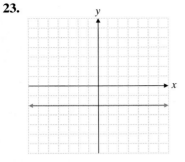

24.

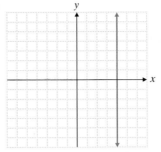

To Think About

Find an equation of the line that fits each description.

25. Passes through $(7, -2)$ and has zero slope

26. Passes through $(4, 6)$ and has undefined slope

27. Passes through $(0, 2)$ and is perpendicular to the *y*-axis

28. Passes through $(-1, 4)$ and is perpendicular to the *x*-axis

29. Passes through $(0, -4)$ and is parallel to $y = \frac{3}{4}x + 2$

30. Passes through $(0, -4)$ and is perpendicular to $y = \frac{3}{4}x + 2$

31. Passes through $(2, 3)$ and is perpendicular to $y = 2x - 9$

32. Passes through $(2, 9)$ and is parallel to $y = 5x - 3$

33. The growth of the population of the United States during the period from 1980 to 2000 can be approximated by an equation of the form $y = mx + b$, where *x* is the number of years since 1980 and *y* is the population measured in millions. (*Source:* U.S. Census Bureau.) Find the equation if two ordered pairs that satisfy it are $(0, 227)$ and $(10, 251)$.

34. The amount of debt outstanding on home equity loans in the United States during the period from 1993 to 2000 can be approximated by an equation of the form $y = mx + b$, where *x* is the number of years since 1993 and *y* is the debt measured in billions of dollars. (*Source:* Board of Governors of the Federal Reserve System.) Find the equation if two ordered pairs that satisfy it are $(1, 280)$ and $(6, 500)$.

Cumulative Review Problems

35. Is $x = 5$ the solution to $\dfrac{x}{2} + \dfrac{1}{3} = 4\left(\dfrac{1}{2} + \dfrac{x}{24}\right)$?

36. Solve. $0.5(x + 2) = -4(1.8x + 2.2)$

37. A pair of basketball sneakers sells for $80. The next week the sneakers go on sale for 15% off. The third week there is a coupon in the newspaper offering a 10% discount off the second week's price. How much would you have paid for the sneakers during the third week if you had used the coupon?

38. Dave and Jane Wells have a cell phone. The plan they subscribe to costs $50 per month. It includes 200 free minutes of calling time. Each minute after the 200 is charged at the rate of $0.21 per minute. Last month their cell phone bill was $68.90. How many total minutes did they use their cell phone during the month?

39. A Scandinavian furniture wholesaler had a shipment of oak entertainment centers shipped by boat across the Atlantic. There were 35 cargo containers, each containing 60 entertainment centers. The distributor guarantees that no more than 5% of the shipment will be defective. What is the maximum number of entertainment centers the furniture wholesaler should expect to be defective?

40. Jackie earned $2122.40 in interest after investing $24,000 for one year. She placed part of the money in an account earning 7% simple interest. She took a risk on some stock options with the rest of the savings, which yielded her 10.5%. How much was invested in the stock options?

3.5 Graphing Linear Inequalities

In Sections 2.6 and 2.7 we discussed inequalities in one variable. Look at the inequality $x < -2$ (x is less than -2). Some of the solutions to the inequality are $-3, -5,$ and $-5\frac{1}{2}$. In fact all numbers to the left of -2 on the number line are solutions. The graph of the inequality is given in the following figure. Notice that the open circle at -2 indicates that -2 is *not* a solution.

$$x < -2$$

Student Learning Objectives

After studying this section, you will be able to:

① Graph linear inequalities in two variables.

SSM PH TUTOR CD & VIDEO MATH PRO WEB
CENTER

Now we will extend our discussion to consider linear inequalities in two variables.

① Graphing Linear Inequalities in Two Variables

Consider the inequality $y \geq x$. The solution of the inequality is the set of all possible ordered pairs that when substituted into the inequality will yield a true statement. Which ordered pairs will make the statement $y \geq x$ true? Let's try some.

$(0, 6)$	$(-2, 1)$	$(1, -2)$	$(3, 5)$	$(4, 4)$
$6 \geq 0,$ true	$1 \geq -2,$ true	$-2 \geq 1,$ false	$5 \geq 3,$ true	$4 \geq 4,$ true

$(0, 6), (-2, 1), (3, 5),$ and $(4, 4)$ are solutions to the inequality $y \geq x$. In fact, every point at which the y-coordinate is greater than or equal to the x-coordinate is a solution to the inequality. This is shown by the solid line and the shaded region in the graph at the right.

Is there an easier way to graph a linear inequality in two variables? It turns out that we can graph such an inequality by first graphing the associated linear equation and then testing one point that is not on that line. That is, we can change the inequality symbol to an equal sign and graph the equation. If the inequality symbol is $\geq$ or $\leq$, we use a solid line to indicate that the points on the line are included in the solution of the inequality. If the inequality symbol is $>$ or $<$, we use a dashed line to indicate that the points on the line are not included in the solution of the inequality. Then we test one point that is not on the line. If the point is a solution to the inequality, we shade the region on the side of the line that includes the point. If the point is not a solution, we shade the region on the other side of the line.

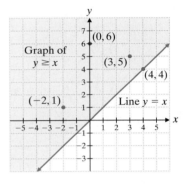

EXAMPLE 1 Graph $5x + 3y > 15$. Use the given coordinate system.

We begin by graphing the line $5x + 3y = 15$. You may use any method discussed previously to graph the line. Since there is no equal sign in the inequality, we will draw a dashed line to indicate that the line is *not* part of the solution set.

Look for a test point. The easiest point to test is $(0, 0)$. Substitute $(0, 0)$ for (x, y) in the inequality.

$$5x + 3y > 15$$
$$5(0) + 3(0) > 15$$
$$0 > 15 \quad \text{false}$$

$(0, 0)$ is not a solution. Shade the side of the line that does *not* include $(0, 0)$.

Practice Problem 1 Graph $x - y \geq -10$. Use the coordinate system in the margin.

PRACTICE PROBLEM 1

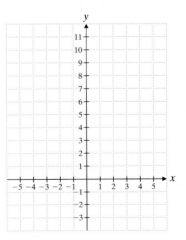

Graphing Linear Inequalities

1. Replace the inequality symbol by an equality symbol. Graph the line.
   (a) The line will be solid if the inequality is $\geq$ or $\leq$.
   (b) The line will be dashed if the inequality is $>$ or $<$.
2. Test the point $(0, 0)$ in the inequality if $(0, 0)$ does not lie on the graphed line in step 1.
   (a) If the inequality is true, shade the side of the line that includes $(0, 0)$.
   (b) If the inequality is false, shade the side of the line that does not include $(0, 0)$.
3. If the point $(0, 0)$ is a point on the line, choose another test point and proceed accordingly.

PRACTICE PROBLEM 2

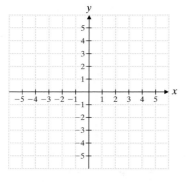

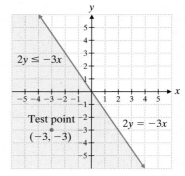

EXAMPLE 2   Graph $2y \leq -3x$.

Step 1   Graph $2y = -3x$. Since $\leq$ is used, the line should be a solid line.

Step 2   We see that the line passes through $(0, 0)$.

Step 3   Choose another test point. We will choose $(-3, -3)$.

$$2y \leq -3x$$
$$2(-3) \leq -3(-3)$$
$$-6 \leq 9 \quad \text{true}$$

Shade the region that includes $(-3, -3)$, that is, the region below the line.

Practice Problem 2   Graph $y > \frac{1}{2}x$ on the coordinate system in the margin.

If we are graphing the inequality $x < -2$ on the coordinate plane, the solution will be a region. Notice that this is very different from the solution $x < -2$ on the number line discussed earlier.

PRACTICE PROBLEM 3

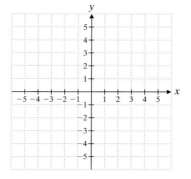

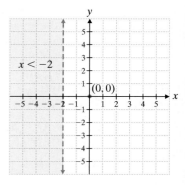

EXAMPLE 3   Graph $x < -2$.

Step 1   Graph $x = -2$. Since $<$ is used, the line should be dashed.

Step 2   Test $(0, 0)$ in the inequality.

$$x < -2$$
$$0 < -2 \quad \text{false}$$

Shade the region that does not include $(0, 0)$, that is, the region to the left of the line $x = -2$. Observe that every point in the shaded region has an x-value that is less than -2.

Practice Problem 3   Graph $y \geq -3$ on the coordinate system in the margin.

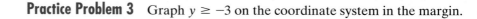

3.5 Exercises

Verbal and Writing Skills

1. Does it matter what point you use as your test point? Justify your response.

2. Explain when to use a solid line or a dashed line when graphing linear inequalities in two variables.

Graph the region described by the inequality.

3. $y > 2 - 3x$

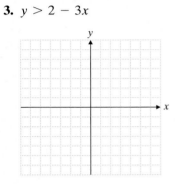

4. $y > 3x - 1$

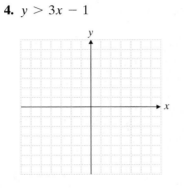

5. $2x - 3y < 6$

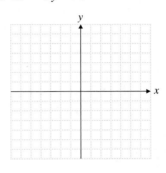

6. $3x + 2y < -6$

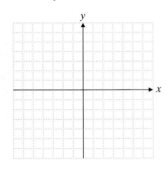

7. $2x - y \geq 3$

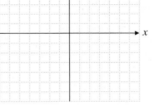

8. $3x - y \geq 4$

9. $y \geq 3x$

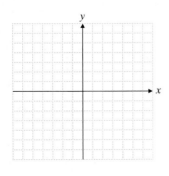

10. $y \geq -4x$

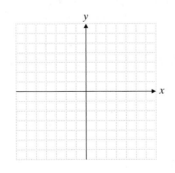

11. $y < -\dfrac{1}{2}x$

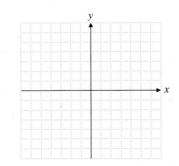

12. $y > \dfrac{1}{5}x$

13. $x \geq 2$

14. $y \leq -2$

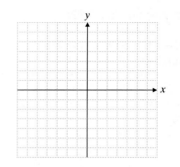

15. $3x - y + 1 \geq 0$

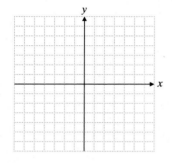

16. $2x + y - 5 \leq 0$

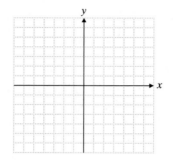

17. $2x > -3y$

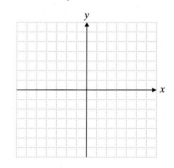

18. $3x \leq -2y$

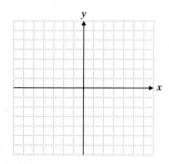

19. $2x > 3 - y$

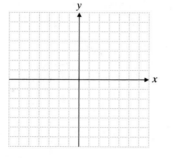

20. $3x > 1 + y$

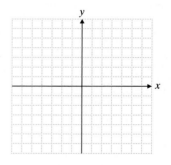

21. $x > -2y$

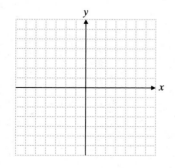

22. $x < -3y$

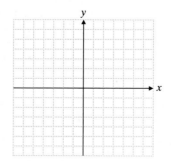

To Think About

23. Graph the inequality $3x + 3 > 5x - 3$ by graphing the lines $y = 3x + 3$ and $y = 5x - 3$ on the same coordinate plane.

(a) For which values of x does the graph of $y = 3x + 3$ lie above the graph of $y = 5x - 3$?

(b) For which values of x is $3x + 3 > 5x - 3$ true?

(c) Explain how the answer to (a) can help you to find the answer to (b).

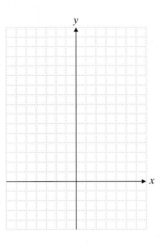

Cumulative Review Problems

24. Simplify. $-3\{5a^2 + a[3a - b(2 + a)]\}$

25. What is the slope–intercept form of the equation for the line through $(3, 0)$ and $(6, -7)$?

26. Angela has a new sales job that allows her to lease a car. Her lease budget is $8400 for a 3-year period. She wants to lease a new Chrysler Cirrus at $195 per month for the next 36 months. However, all mileage in excess of 36,000 miles driven over this 3-year period will be charged at $0.15 per mile. How many miles above the 36,000-mile limit can Angela drive and still not exceed the leasing budget?

27. Samantha is a great telemarketer. She has great success in selling subscriptions to a travel club. She has an average of 0.492 after having made 1000 telephone calls. (She sells a subscription 49.2% of the time.) Assuming that she makes a sale with each of her next telephone calls, what is the minimum number of calls she needs to make in order to raise her average to 0.500?

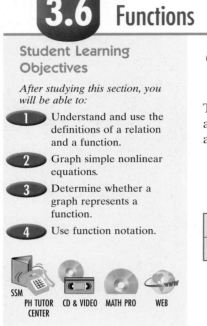
1 Understanding and Using the Definitions of a Relation and a Function

Thus far you have studied linear equations in two variables. You have seen that such an equation can be represented by a table of values, by the algebraic equation itself, and by a graph.

x	0	1	3
y	4	1	−5

$$y = -3x + 4$$

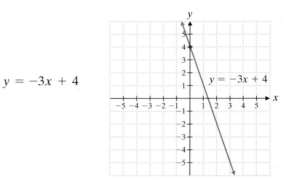

$y = -3x + 4$

The solutions to the linear equation are all the ordered pairs that satisfy the equation (make the equation true). They are all the points that lie on the graph of the line. These ordered pairs can be represented in a table of values. Notice the relationship between the ordered pairs. We can choose any value for x. But once we have chosen a value for x, the value of y is determined. For the preceding equation, if x is 0, then y must be 4. We say that x is the **independent variable** and that y is the **dependent variable**.

Mathematicians call such a pairing of two values a *relation*.

Definition of a Relation

A **relation** is any set of ordered pairs.

All the first coordinates in all of the ordered pairs of the relation make up the **domain** of the relation. All the second coordinates in all of the ordered pairs make up the **range** of the relation. Notice that the definition of a relation is very broad. Some relations cannot be described by an equation. These relations may simply be a set of discrete ordered pairs.

EXAMPLE 1 State the domain and range of the relation.

$$\{(5, 7), (9, 11), (10, 7), (12, 14)\}$$

The domain consists of all the first coordinates in the ordered pairs.

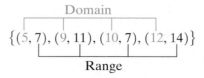

Domain

$$\{(5, 7), (9, 11), (10, 7), (12, 14)\}$$

Range

The range consists of all of the second coordinates in the ordered pairs.

The domain is $\{5, 9, 10, 12\}$.

The range is $\{7, 11, 14\}$.   We list 7 only once.

Practice Problem 1 State the domain and range of the relation.

$$\{(-3, -5), (3, 5), (0, -5), (20, 5)\}$$

Some relations have the special property that no two different ordered pairs have the same first coordinate. Such relations are called **functions**. The relation $y = -3x + 4$ is a function. If we substitute a value for x, we get just one value for y. Thus no two ordered pairs will have the same x-coordinate and different y-coordinates.

> ### Definition of a Function
>
> A **function** is a relation in which no two different ordered pairs have the same first coordinate.

EXAMPLE 2 Determine whether the relation is a function.

(a) $\{(3, 9), (4, 16), (5, 9), (6, 36)\}$ **(b)** $\{(7, 8), (9, 10), (12, 13), (7, 14)\}$

(a) Look at the ordered pairs. No two ordered pairs have the same first coordinate. Thus this set of ordered pairs defines a function. Note that the ordered pairs $(3, 9)$ and $(5, 9)$ have the same second coordinate, but the relation is still a function. It is the first coordinates that cannot be the same.

(b) Look at the ordered pairs. Two different ordered pairs, $(7, 8)$ and $(7, 14)$, have the same first coordinate. Thus this relation is *not* a function.

Practice Problem 2 Determine whether the relation is a function.

(a) $\{(-5, -6), (9, 30), (-3, -3), (8, 30)\}$

(b) $\{(60, 30), (40, 20), (20, 10), (60, 120)\}$

A functional relationship is often what we find when we analyze two sets of data. Look at the following table of values, which compares Celsius temperature with Fahrenheit temperature. Is there a relationship between degrees Fahrenheit and degrees Celsius? Is the relation a function?

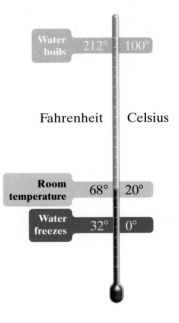

Temperature

°F	23	32	41	50
°C	−5	0	5	10

Since every Fahrenheit temperature produces a unique Celsius temperature, we would expect this to be a function. We can verify our assumption by looking at the formula $C = \frac{5}{9}(F - 32)$ and its graph. The formula is a linear equation, and its graph is a line with slope $\frac{5}{9}$ and y-intercept at about -17.8. The relation is a function. In the equation given here, notice that the *dependent variable* is C, since the value of C depends on the value of F. We say that F is the *independent variable*. The *domain* can be described as the set of possible values of the independent variable. The *range* is the set of corresponding values of the dependent variable. Scientists believe that the coldest temperature possible is approximately $-273°C$. They call this temperature **absolute zero**. Thus,

Domain = {all possible Fahrenheit temperatures from absolute zero to infinity}

Range = {all corresponding Celsius temperatures from $-273°$ C to infinity}

EXAMPLE 3 Each of the following tables contains some data pertaining to a relation. Determine whether the relation suggested by the table is a function. If it is a function, identify the domain and range.

(a)   Circle

Radius	1	2	3	4	5
Area	3.14	12.56	28.26	50.24	78.5

(b) $4000 Loan at 8% for a Minimum of One Year

Time (yr)	1	2	3	4	5
Interest	$320	$665.60	$1038.85	$1441.96	$1877.31

(a) Looking at the table, we see that no two different ordered pairs have the same first coordinate. The area of a circle is a function of the length of the radius.

Next we need to identify the independent variable to determine the domain. Sometimes it is easier to identify the dependent variable. Here we notice that the area of the circle depends on the length of the radius. Thus radius is the independent variable. Since a negative length does not make sense, the radius cannot be a negative number. Although only integer radius values are listed in the table, the radius of a circle can be any nonnegative real number.

Domain = {all nonnegative real numbers}

Range = {all nonnegative real numbers}

(b) No two different ordered pairs have the same first coordinate. Interest is a function of time.

Since the amount of interest paid on a loan depends on the number of years (term of the loan), interest is the dependent variable and time is the independent variable. Negative numbers do not apply in this situation. Although the table includes only integer values for the time, the length of a loan in years can be any real number that is greater than or equal to 1.

Domain = {all real numbers greater than or equal to 1}

Range = {all positive real numbers greater than or equal to $320}

Practice Problem 3 Determine whether the relation suggested by the table is a function. If it is a function, identify the domain and the range.

(a)   28 Mpg at $1.20 per Gallon

Distance	0	28	42	56	70
Cost	0	1.20	1.80	2.40	3.00

(b)   Store's Inventory of Shirts

Number of Shirts	5	10	5	2	8
Price of Shirt	$20	$25	$30	$45	$50

To Think About Look at the following bus schedule. Determine whether the relation is a function. Which is the independent variable? Explain your choice.

                              Bus Schedule

Bus Stop	Main St.	8th Ave.	42nd St.	Sunset Blvd.	Cedar Lane
Time	7:00	7:10	7:15	7:30	7:39

2   *Graphing Simple Nonlinear Equations*

Thus far in this chapter we have graphed linear equations in two variables. We now turn to graphing a few nonlinear equations. We will need to plot more than three points to get a good idea of what the graph of a nonlinear equation will look like.

● **EXAMPLE 4**   Graph $y = x^2$.

Begin by constructing a table of values. We select values for x and then determine by the equation the corresponding values of y. We will include negative values for x as well as positive values. We then plot the ordered pairs and connect the points with a smooth curve.

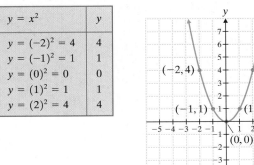

x	$y = x^2$	y
-2	$y = (-2)^2 = 4$	4
-1	$y = (-1)^2 = 1$	1
0	$y = (0)^2 = 0$	0
1	$y = (1)^2 = 1$	1
2	$y = (2)^2 = 4$	4

This type of curve is called a *parabola*. We will study this type of curve more extensively in Chapters 9 and 10.

Practice Problem 4   Graph $y = x^2 - 2$ on the coordinate system in the margin. ●

Some equations are solved for x. Usually, in those cases we pick values of y and then obtain the corresponding values of x from the equation.

● **EXAMPLE 5**   Graph $x = y^2 + 2$.

We will select a value of y and then substitute it into the equation to obtain x. For convenience in graphing, we will repeat the y column at the end so that it is easy to write the ordered pairs (x, y).

y	$x = y^2 + 2$	x	y
-2	$x = (-2)^2 + 2 = 4 + 2 = 6$	6	-2
-1	$x = (-1)^2 + 2 = 1 + 2 = 3$	3	-1
0	$x = (0)^2 + 2 = 0 + 2 = 2$	2	0
1	$x = (1)^2 + 2 = 1 + 2 = 3$	3	1
2	$x = (2)^2 + 2 = 4 + 2 = 6$	6	2

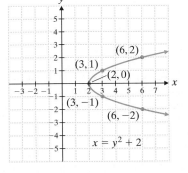

Practice Problem 5   Graph $x = y^2 - 1$ on the coordinate system in the margin. ●

If the equation involves fractions with variables in the denominator, we must use extra caution. Remember that you may never divide by zero.

Graphing Nonlinear Equations

You can graph nonlinear equations solved for y using a graphing calculator. For example, graph $y = x^2 - 2$ on a graphing calculator using an appropriate window. Display:

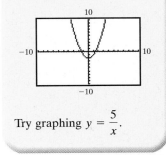

Try graphing $y = \dfrac{5}{x}$.

PRACTICE PROBLEM 4

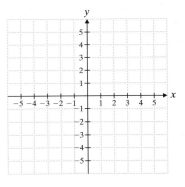

PRACTICE PROBLEM 5

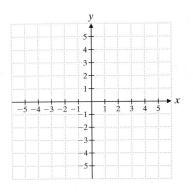

EXAMPLE 6 Graph $y = \dfrac{4}{x}$.

It is important to note that x cannot be zero because division by zero is not defined. $y = \dfrac{4}{0}$ is not allowed! Observe that when we draw the graph we get two separate branches that do not touch.

x	$y = \dfrac{4}{x}$	y
-4	$y = \dfrac{4}{-4} = -1$	-1
-2	$y = \dfrac{4}{-2} = -2$	-2
-1	$y = \dfrac{4}{-1} = -4$	-4
0	We cannot divide by zero.	There is no value.
1	$y = \dfrac{4}{1} = 4$	4
2	$y = \dfrac{4}{2} = 2$	2
4	$y = \dfrac{4}{4} = 1$	1

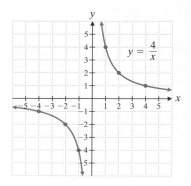

PRACTICE PROBLEM 6

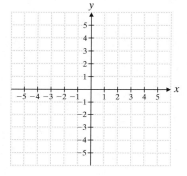

Practice Problem 6 Graph $y = \dfrac{6}{x}$ on the coordinate system in the margin.

3   *Determining Whether a Graph Represents a Function*

Can we tell whether a graph represents a function? Recall that a function cannot have two different ordered pairs with the same first coordinate. That is, each value of x must have a separate unique value of y. Look at the graph of the function $y = x^2$ in Example 4. Each x-value has a unique y-value. Look at the graph of $x = y^2 + 2$ in Example 5. At $x = 3$ there are two y-values, 1 and -1. In fact, for every x-value greater than 2 there are two y-values. $x = y^2 + 2$ is not a function.

Observe that we can draw a vertical line through $(6, 2)$ and $(6, -2)$. Any graph that is not a function will have at least one region in which a vertical line will cross the graph more than once.

Vertical Line Test

If a vertical line can intersect the graph of a relation more than once, the relation is not a function. If no such line can be drawn, then the relation is a function.

EXAMPLE 7   Determine whether each of the following is the graph of a function.

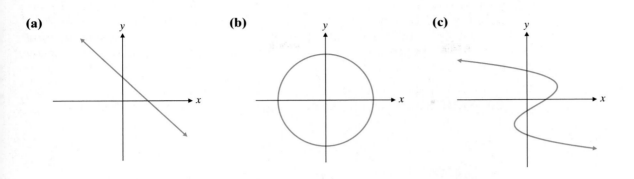

(a) The graph of the straight line is a function. Any vertical line will cross this straight line in only one location.

(b) and **(c)** Each of these graphs is not the graph of a function. In each case there exists a vertical line that will cross the curve in more than one place.

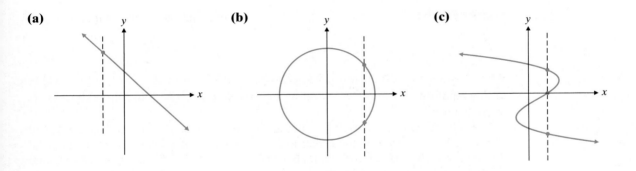

Practice Problem 7   Determine whether each of the following is the graph of a function.

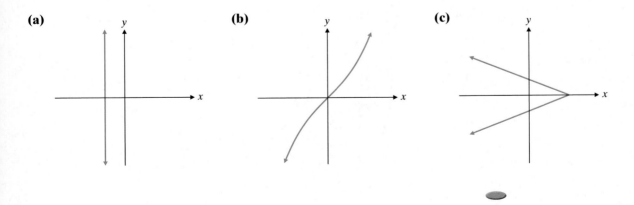

4 *Using Function Notation*

We have seen that an equation like $y = 2x + 7$ is a function. For each value of x, the equation assigns a unique value to y. We could say "y is a function of x." If we name the function f, this statement can be symbolized by using the **function notation** $y = f(x)$. Many times we avoid using the y-variable completely and write the function as $f(x) = 2x + 7$.

⊘ **WARNING** Be careful. The notation $f(x)$ does not mean f multiplied by x.

EXAMPLE 8 If $f(x) = 3x^2 - 4x + 5$, find each of the following.

(a) $f(-2)$ **(b)** $f(4)$ **(c)** $f(0)$

(a) $f(-2) = 3(-2)^2 - 4(-2) + 5 = 3(4) - 4(-2) + 5 = 12 + 8 + 5 = 25$

(b) $f(4) = 3(4)^2 - 4(4) + 5 = 3(16) - 4(4) + 5 = 48 - 16 + 5 = 37$

(c) $f(0) = 3(0)^2 - 4(0) + 5 = 3(0) - 4(0) + 5 = 0 - 0 + 5 = 5$

Practice Problem 8 If $f(x) = -2x^2 + 3x - 8$, find each of the following.

(a) $f(2)$ **(b)** $f(-3)$ **(c)** $f(0)$

When evaluating a function, it is helpful to place parentheses around the value that is being substituted for x. Taking the time to do this will minimize sign errors in your work.

Some functions are useful in medicine. For example, the approximate width of a man's elbow is given by the function $e(x) = 0.03x + 0.6$, where x is the height of the man in inches. If a man is 68 inches tall, $e(68) = 0.03(68) + 0.6 = 2.64$. A man 68 inches tall would have an elbow width of approximately 2.64 inches.

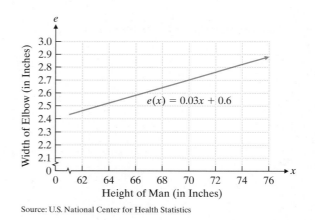

Source: U.S. National Center for Health Statistics

3.6 Exercises

Verbal and Writing Skills

1. What are the three ways you can describe a function?

2. What is the difference between a function and a relation?

3. The domain of a function is the set of _____ _____ of the _____ variable.

4. The range of a function is the set of _____ _____ of the _____ variable.

5. How can we tell whether a graph is the graph of a function?

Find the domain and range of the relation. Determine whether the relation is a function.

6. $\left\{\left(\frac{1}{2}, \frac{1}{2}\right), \left(-10, \frac{1}{2}\right), \left(7, \frac{1}{4}\right), \left(\frac{1}{2}, \frac{1}{4}\right)\right\}$

7. $\left\{\left(\frac{1}{2}, 5\right), \left(\frac{1}{4}, 10\right), \left(\frac{3}{4}, 6\right), \left(\frac{1}{2}, 6\right)\right\}$

8. $\left\{(7, 3.1), (5, 0), (7, 2.3)\right\}$

9. $\left\{(7.3, 1), (0, 8), (2, 1)\right\}$

10. $\left\{(12, 1), (14, 3), (1, 12), (9, 12)\right\}$

11. $\left\{(5.6, 8), (5.8, 6), (6, 5.8), (5, 6)\right\}$

12. $\left\{(3, 75), (5, 95), (3, 85), (7, 100)\right\}$

13. $\left\{(85, 3), (95, 11), (110, 15), (110, 20)\right\}$

Graph the equation.

14. $y = x^2 + 3$

15. $y = x^2 - 1$

16. $y = 2x^2$

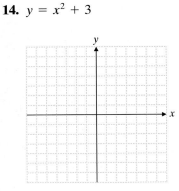

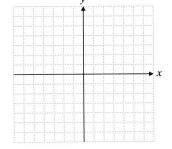

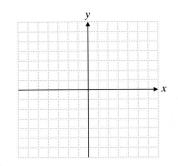

17. $y = \dfrac{1}{2}x^2$

18. $x = -2y^2$

19. $x = \dfrac{1}{2}y^2$

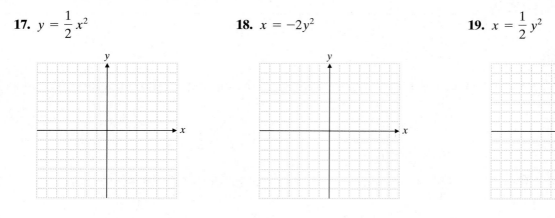

20. $x = y^2 + 1$

21. $x = 2y^2$

22. $y = \dfrac{2}{x}$

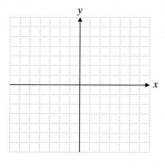
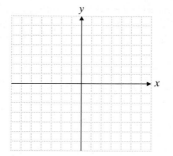
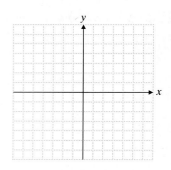

23. $y = -\dfrac{2}{x}$

24. $y = \dfrac{4}{x^2}$

25. $y = -\dfrac{6}{x^2}$

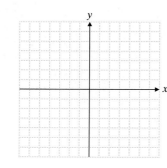
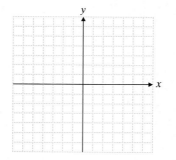
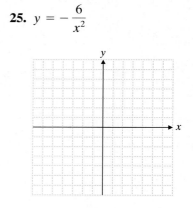

26. $y = (x + 1)^2$

27. $x = (y - 2)^2$

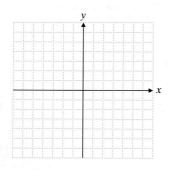
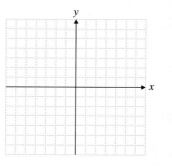

28. $y = \dfrac{4}{x - 2}$

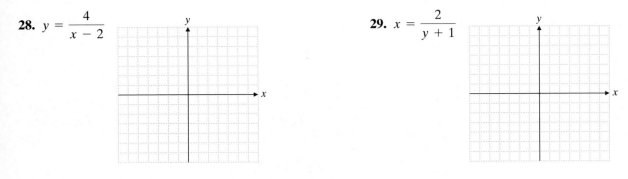

29. $x = \dfrac{2}{y + 1}$

Determine whether each relation is a function.

30.

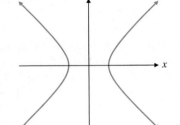

31.

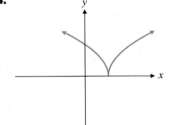

32.

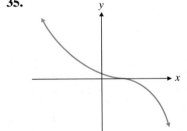

33.

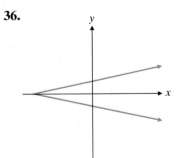

34.

35.

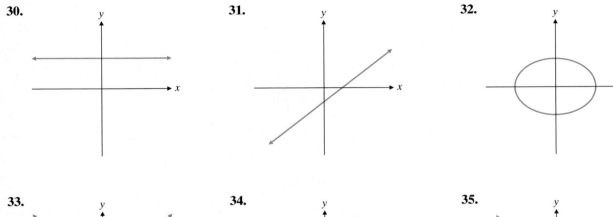

36.

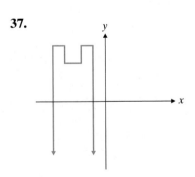

37.

Given the following functions, find the indicated values.

38. $f(x) = 2 - 3x$
   (a) $f(-8)$ **(b)** $f(0)$ **(c)** $f(2)$

39. $f(x) = 3 - 5x$
   (a) $f(-1)$ **(b)** $f(1)$ **(c)** $f(2)$

40. $f(x) = x^2 + 3x + 4$
   (a) $f(2)$ **(b)** $f(0)$ **(c)** $f(-3)$

41. $f(x) = x^2 + 2x - 3$
   (a) $f(-1)$ **(b)** $f(0)$ **(c)** $f(3)$

42. $f(x) = 2x + 3 - \dfrac{12}{x}$

    (a) $f(-2)$ **(b)** $f(1)$ **(c)** $f(4)$

43. $f(x) = x - 2 + \dfrac{8}{x}$

    (a) $f(-2)$ **(b)** $f(1)$ **(c)** $f(4)$

Applications

The accompanying graph represents the approximate percent of people in the United States who smoked during selected years from 1980 to 2000. Source: National Center for Health Statistics.

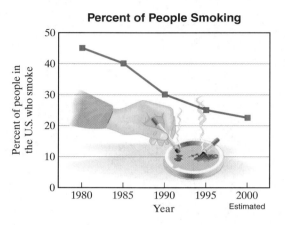

Percent of People Smoking

44. Does the graph represent a function?

45. Find $f(1990)$.

46. If $f(x) = 45\%$, what year is represented by x?

47. Between what two years is the difference in function values equal to 10%?

48. During a recent population growth period in Nevada from 1980 to 1990, the approximate population of the state measured in thousands could be predicted by the function $f(x) = 2x^2 + 20x + 800$, where x is the number of years since 1980. (*Source:* U.S. Census Bureau.) Find $f(0), f(3), f(6)$, and $f(10)$. Graph the function. What pattern do you observe?

49. During a recent population growth period in Alaska from 1980 to 2000, the approximate population of the state measured in thousands could be predicted by the function $f(x) = -0.6x^2 + 22x + 411$, where x is the number of years since 1980. (*Source:* U.S. Census Bureau.) Find $f(0), f(4), f(10)$, and $f(20)$. Graph the function. What pattern do you observe?

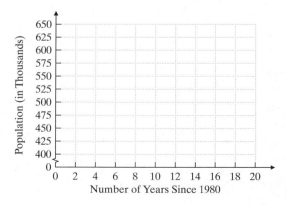

Cumulative Review Problems

50. Evaluate. $-3\dfrac{1}{2} \div 2\dfrac{1}{4}$

51. Evaluate. $x^2 + xy - 5$ when $x = 2$ and $y = -3$

52. Simplify. $3(5 - x) + 4(x + 3)$

53. Solve for y. $6x - 2y = -10$

Putting Your Skills to Work

Graphing the Size of Rainforests

Most of us are aware that the world's rainforests are disappearing, but do you realize how quickly? Rainforests cover only 2% of the Earth's surface but are home to almost half of all life forms (plants and animals) on the planet. As deforestation continues, many species are becoming extinct and the world's climate is being negatively affected (a phenomenon known as the greenhouse effect).

In 1990, there were 1756 million hectares of rainforest on the planet. Since that time, rainforest destruction has occurred at 2.47 acres per second. (This area is about as large as two football fields.)

Problems for Individual Investigation and Study

Round to the nearest million.

1. How many acres of rainforest are destroyed each day? Each year?

2. Approximately how many acres of rainforest were there in 1990? (1 hectare = 2.47 acres)

3. Use your answers from questions 1 and 2 to determine how many acres of rainforest remained in 1992. How many acres of rainforest were there in 1995?

Problems for Cooperative Group Activity

Suppose that the current rate of rainforest destruction continues without intervention. How many acres of rainforest will there be 20 years from now? 35 years from now? In what year will there be no rainforests left in the world? Answer the following questions to find out.

4. Write a function that gives the number of acres of rainforest *r* (in millions of acres) that remain *t* years after 1990.

5. Complete the following table.

Number of Years after 1990	Amount of Rainforest Remaining on Earth (in Millions of Acres)
0	
5	
10	
15	
20	

6. Graph the information from question 5. What is the dependent variable? What is the independent variable? Be sure to use appropriate scales on the axes.

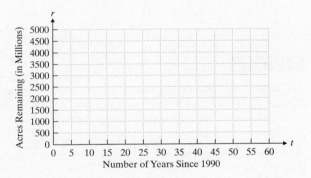

7. Using the graph, determine how many acres of rainforest will remain in the year 2035.

8. Using the graph, determine when there will be no rainforests left on Earth.

Internet Connections

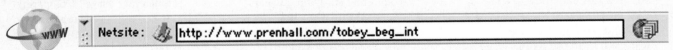

Site: Food and Agricultural Organization of the United Nations

9. Comparing Asia, Latin America, and Africa, which of these is projected to have the greatest absolute growth in population in the period 2000–2050? Can you tell just by looking at the slopes of the three plots which of these areas will experience the highest *rate* of growth in that period?

10. What is the relative growth of urban population as a percent of total population between 1950 and 2030 in Asia, Latin America/Caribbean, and Africa, respectively?

199

Math in the Media

Automobiles — Costs Beyond the Purchase Price

By Jim Mateja

An article in the *Chicago Tribune-Northern Light*, August 25, 2000, reported, "The Saturn sedan and Chevy Metro 2000 models cost less than half as much to own and operate as the Cadillac DeVille or Lincoln Town Car. They were less than one-third as much as the Mercedes-Benz 500S, according to Runzheiner International, the Wisconsin-based management consulting firm."

The analysis considered the Total Annual Costs of each model.

The Total Annual Costs = annual operating costs + fixed costs.

It is important to consider all the costs associated with a car when making a decision to purchase. The questions that follow give you the chance to compare costs in a similar way using four different vehicle models.

EXERCISES

The chart below shows the Total Annual Costs of operating four different motor vehicles, including the costs associated with gasoline and oil, maintenance, tires, insurance, license, registration, taxes, depreciation, and finance charges.

Total Annual Costs in Dollars

Miles Driven per Year	Vehicle #1	Vehicle #2	Vehicle #3	Vehicle #4
10,000 miles per year	3,800	5,500	5,800	6,700
15,000 miles per year	5,100	7,500	6,800	7,600
20,000 miles per year	7,300	9,200	8,700	9,900

1. Draw a graph with *Miles per Year* on the *x*-axis and *Total Annual Cost* on the *y*-axis. Use a scale of 0 to 20,000 on the *x*-axis and a scale of 0 to 10,000 on the *y*-axis. Plot three points for each of the vehicle types and, for each vehicle, connect each pair of ordered pairs with straight lines.
   a. Based on your graph, which vehicle type had the highest dollar increase in cost per year comparing 10,000 miles/year usage with 15,000 miles/year usage? Comparing 15,000 miles/year usage with 20,000 miles/year usage?
   b. Based on your graph, which vehicle type had the lowest percent increase in cost per mile comparing 10,000 miles/year usage with 15,000 miles/year usage? Comparing 15,000 miles/year usage with 20,000 miles/year usage?

2. What is being measured by the slope between the points plotted for any given vehicle on the graph you constructed?

Chapter 3 Organizer

Topic	Procedure	Examples
Plotting points, p. 144.	To plot (x, y): 1. Begin at the origin. 2. If x is positive, move to the right along the x-axis. If x is negative, move to the left along the x-axis. 3. If y is positive, move up. If y is negative, move down.	To plot $(-2, 3)$:
Graphing straight lines, p. 154.	An equation of the form $$Ax + By = C$$ has a graph that is a straight line. To graph such an equation, plot any three points; two give the line and the third checks it. (Where possible, use the x- and y-intercepts.)	Graph $3x + 2y = 6$. <table><tr><td>x</td><td>y</td></tr><tr><td>0</td><td>3</td></tr><tr><td>4</td><td>-3</td></tr><tr><td>2</td><td>0</td></tr></table>
Finding the slope given two points, p. 164.	Nonvertical lines passing through distinct points (x_1, y_1) and (x_2, y_2) have slope $$m = \frac{y_2 - y_1}{x_2 - x_1}.$$ The slope of a horizontal line is 0. The slope of a vertical line is undefined.	What is the slope of the line through $(2, 8)$ and $(5, 1)$? $$m = \frac{1 - 8}{5 - 2} = -\frac{7}{3}$$
Finding the slope and y-intercept of a line given the equation, p. 168.	1. Rewrite the equation in the form $y = mx + b$. 2. The slope is m. 3. The y-intercept is $(0, b)$.	Find the slope and y-intercept. $$3x - 4y = 8$$ $$-4y = -3x + 8$$ $$y = \tfrac{3}{4}x - 2$$ The slope is $\frac{3}{4}$. The y-intercept is $(0, -2)$.
Finding the equation of a line given the slope and y-intercept, p. 168.	The slope–intercept form of the equation of a line is $$y = mx + b$$ The slope is m and the y-intercept is $(0, b)$.	Find the equation of the line with y-intercept $(0, 7)$ and with slope $m = 3$. $$y = 3x + 7$$
Graphing a line using slope and y-intercept, p. 169.	1. Plot the y-intercept. 2. Starting from $(0, b)$, plot a second point using the slope. $$\text{slope} = \frac{\text{rise}}{\text{run}}$$ 3. Draw a line that connects the two points.	Graph $y = -4x + 1$. First plot the y-intercept at $(0, 1)$. Slope $= -4$ or $\frac{-4}{1}$
Finding the slope of parallel and perpendicular lines, p. 170.	Parallel lines have the same slope. Perpendicular lines have slopes whose product is -1.	Line q has a slope of 2. The slope of a line parallel to q is 2. The slope of a line perpendicular to q is $-\frac{1}{2}$.

Topic	Procedure	Examples
Finding the equation of a line through a point with a given slope, p. 178.	1. Substitute the known values in the equation $y = mx + b$. 2. Solve for b. 3. Use the values of m and b to write the general equation.	Find the equation of the line through $(3, 2)$ with slope $m = \frac{4}{5}$. $$y = mx + b$$ $$2 = \frac{4}{5}(3) + b$$ $$2 = \frac{12}{5} + b$$ $$-\frac{2}{5} = b$$ The equation is $y = \frac{4}{5}x - \frac{2}{5}$.
Finding the equation of a line through two points, p. 178.	1. Find the slope. 2. Use the procedure when given a point and the slope.	Find the equation of the line through $(3, 2)$ and $(13, 10)$. $$m = \frac{y_2 - y_1}{x_2 - x_1} = \frac{10 - 2}{13 - 3} = \frac{8}{10} = \frac{4}{5}$$ We choose the point $(3, 2)$. $$y = mx + b$$ $$2 = \frac{4}{5}(3) + b$$ $$2 = \frac{12}{5} + b$$ $$-\frac{2}{5} = b$$ The equation is $y = \frac{4}{5}x - \frac{2}{5}$.
Writing the equation of a line given the graph, p. 179.	1. Identify the y-intercept. 2. Find the slope. $$\text{slope} = \frac{\text{change in } y}{\text{change in } x}$$	 The equation is $y = -\frac{5}{4}x - 2$.
Graphing linear inequalities, p. 183.	1. Graph as if it were an equation. If the inequality symbol is $>$ or $<$, use a dashed line. If the inequality symbol is $\geq$ or $\leq$, use a solid line. 2. Look for a test point. The easiest test point is $(0, 0)$, unless the line passes through $(0, 0)$. In that case, choose another test point. 3. Substitute the coordinates of the test point into the inequality. 4. If it is a true statement, shade the side of the line containing the test point. If it is a false statement, shade the side of the line that does *not* contain the test point.	Graph $y \geq 3x + 2$. Graph the line $y = 3x + 2$. Use a solid line. $$\text{Test } (0, 0). \quad 0 \geq 3(0) + 2$$ $$0 \geq 2 \quad \text{false}$$ Shade the side of the line that does not contain $(0, 0)$.

Topic	Procedure	Examples
Determining whether a relation is a function, p. 189.	A function is a relation in which no two different ordered pairs have the same first coordinate.	Is this relation a function? $\{(5, 7), (3, 8), (5, 10)\}$ It is *not* a function since $(5, 7)$ and $(5, 10)$ are two different ordered pairs with the same x-coordinate, 5.
Determining whether a graph represents a function, p. 192.	If a vertical line can intersect the graph of a relation more than once, the relation is not a function. If no such line exists, the relation is a function.	Is this graph a function? 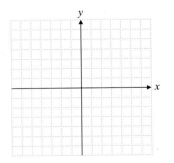 Yes. Any vertical line will intersect it at most once.

Chapter 3 Review Problems

3.1

1. Plot and label the following points.
   $A: (2, -3)$ $B: (-1, 0)$ $C: (3, 2)$ $D: (-2, -3)$

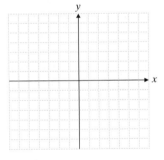

2. Give the coordinates of each point.

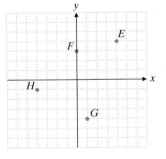

Complete the ordered pairs so that each is a solution to the given equation.

3. $y = 3x - 5$
   (a) $(0, \quad)$ **(b)** $(3, \quad)$

4. $2x + 5y = 12$
   (a) $(1, \quad)$ **(b)** $(\quad, 4)$

5. $x = 6$
   (a) $(\quad, -1)$ **(b)** $(\quad, 3)$

3.2

6. Graph $3y = 2x + 6$.

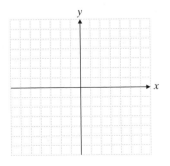

7. Graph $5y + x = -15$.

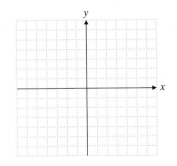

8. Graph $3x + 5y = 15 + 3x$.

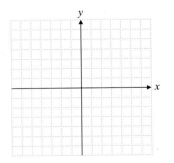

3.3

9. Find the slope of the line passing through $(5, -3)$ and $\left(2, -\frac{1}{2}\right)$.

10. Find the slope and y-intercept of the line $9x - 11y + 15 = 0$.

11. Write an equation of the line with slope $-\frac{1}{2}$ and y-intercept $(0, 3)$.

12. A line has a slope of $-\frac{2}{3}$. What is the slope of a line perpendicular to that line?

13. Graph $y = -\dfrac{1}{2}x + 3$.

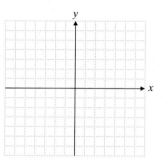

14. Graph $2x - 3y = -12$.

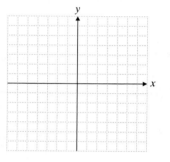

15. Graph $5x + 2y = 20 + 2y$.

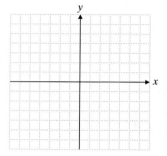

3.4

16. Write an equation of the line passing through $(5, 6)$ having a slope of 2.

17. Write an equation of the line passing through $(3, -4)$ having a slope of -6.

18. Write an equation of the line passing through $(6, -3)$ having a slope of $\frac{1}{3}$.

19. Write an equation of the line passing through $(3, 7)$ and $(-6, 7)$.

Write an equation of the graph.

20.

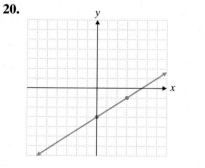

21.

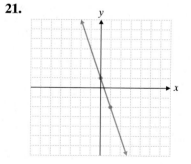

22.

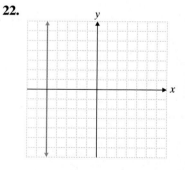

3.5

23. Graph $y < \frac{1}{3}x + 2$.

24. Graph $3y + 2x \geq 12$.

25. Graph $y \geq -2$.

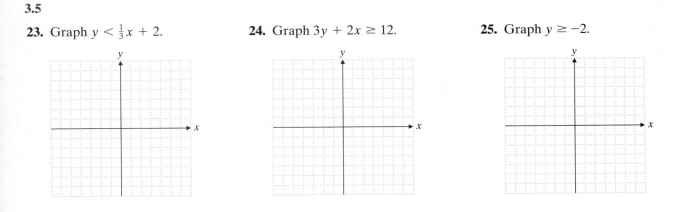

3.6

Determine the domain and range of the relation. Determine whether the relation is a function.

26. $\{(5, -6), (-6, 5), (-5, 5), (-6, -6)\}$

27. $\{(3, -7), (-7, 3), (-3, 7), (7, -3)\}$

In exercises 28–30, determine whether the graph represent a function.

28.

29.

30.

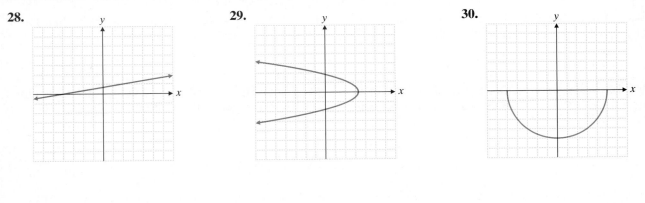

31. Graph $y = x^2 - 5$.

32. Graph $x = y^2 + 3$.

33. Graph $y = (x - 3)^2$.

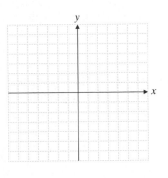
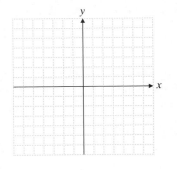
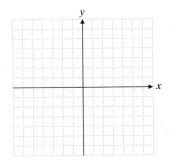

Given the following functions, find the indicated values.

34. $f(x) = 7 - 6x$ **(a)** $f(0)$ **(b)** $f(-4)$

35. $g(x) = -2x^2 + 3x + 4$ **(a)** $g(-1)$ **(b)** $g(3)$

36. $p(x) = \dfrac{-3}{x}$ **(a)** $p(-3)$ **(b)** $p(0.5)$

37. $f(x) = \dfrac{2}{x + 4}$ **(a)** $f(-2)$ **(b)** $f(6)$

38. $f(x) = \dfrac{x + 6}{x - 5}$ **(a)** $f(0)$ **(b)** $f(-2)$

Applications

Bob and Evelyn Hanson have found that when they travel across the country, they can estimate the cost of their trip with the equation $y = 150 + 0.15x$. In this equation, y represents the cost in dollars, and x represents the number of miles traveled.

39. What is the approximate cost to travel 1000 miles?

40. What is the approximate cost to travel 3000 miles?

41. Write the equation in the form $y = mx + b$, and determine the numerical value of the slope.

42. What is the significance of the slope? What does it tell us about the cross-country costs?

43. Bob Hanson estimated that the total budget for a trip was $540. With that limit, how many miles can they travel?

44. Evelyn Hanson estimated that the total budget for a trip was $750. With that limit, how many miles can they travel?

Russ and Norma Camp found that their monthly electric bill could be calculated by the equation $y = 30 + 0.09x$. In this equation, y represents the amount of the monthly bill in dollars, and x represents the number of kilowatt-hours used during the month.

45. What would be their monthly bill if they used 2000 kilowatt-hours of electricity?

46. What would be their monthly bill if they used 1600 kilowatt-hours of electricity?

47. Write the equation in the form $y = mx + b$, and determine the numerical value of the y-intercept. What is the significance of this y-intercept? What does it tell us?

48. If the equation is placed in the form $y = mx + b$, what is the numerical value of the slope? What is the significance of this slope? What does it tell us?

49. If Russ and Norma have a monthly bill of $147, how many kilowatt-hours of electricity did they use?

50. If Russ and Norma have a monthly bill of $246, how many kilowatt-hours of electricity did they use?

Chapter 3 Test

1. Plot and label the following points.
   B: $(6, 1)$ C: $(-4, -3)$
   D: $(-3, 0)$ E: $(5, -2)$

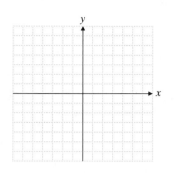

2. Graph the line $6x - 3 = 5x - 2y$.

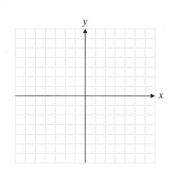

3. Graph the line $12x - 3y = 6$.

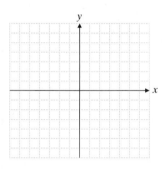

4. Graph $y = \frac{2}{3}x - 4$.

5. What is the slope and the y-intercept of the line $3x + 2y - 5 = 0$?

6. Find the slope of the line that passes through $(8, 6)$ and $(-3, -5)$.

7. Write an equation for the line that passes through $(4, -2)$ and has a slope of $\frac{1}{2}$.

8. Find the slope of the line through $(-3, 11)$ and $(6, 11)$.

9. Find an equation for the line passing through $(2, 5)$ and $(8, 3)$.

10. Write an equation for the line through $(2, 7)$ and $(2, -2)$. What is the slope of this line?

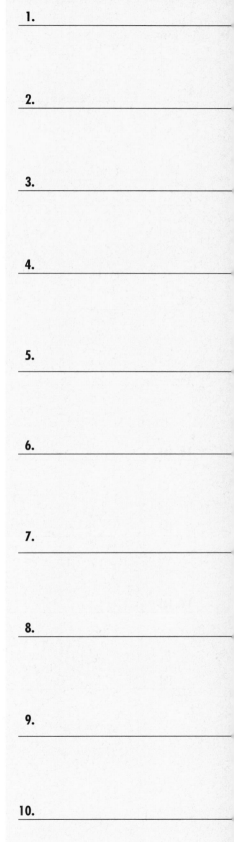

1. _____

2. _____

3. _____

4. _____

5. _____

6. _____

7. _____

8. _____

9. _____

10. _____

11. Graph the region described by $4y \le 3x$.

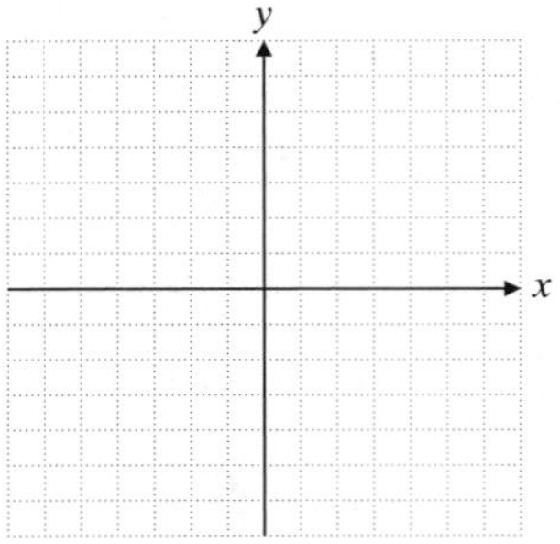

12. Graph the region described by $-3x - 2y > 10$.

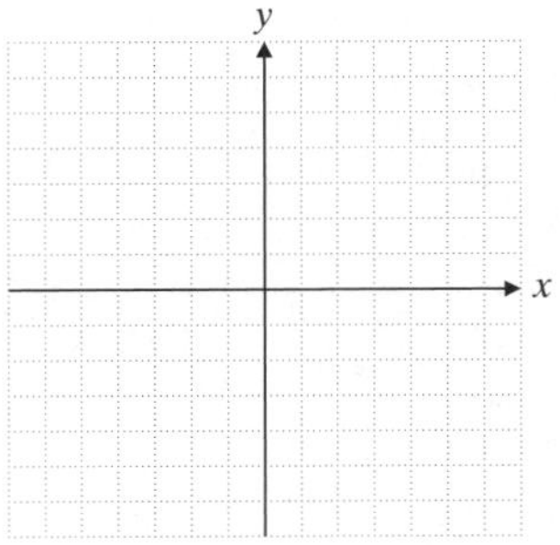

13. Is this relation a function? $\{(-8, 2), (3, 2), (5, -2)\}$ Why?

14. Look at the relation graphed below. Is this relation a function? Why?

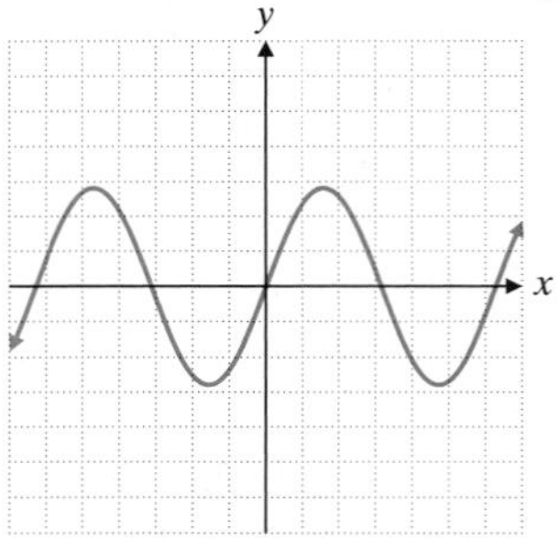

15. Graph $y = 2x^2 - 3$.

x	y
-2	
-1	
0	
1	
2	

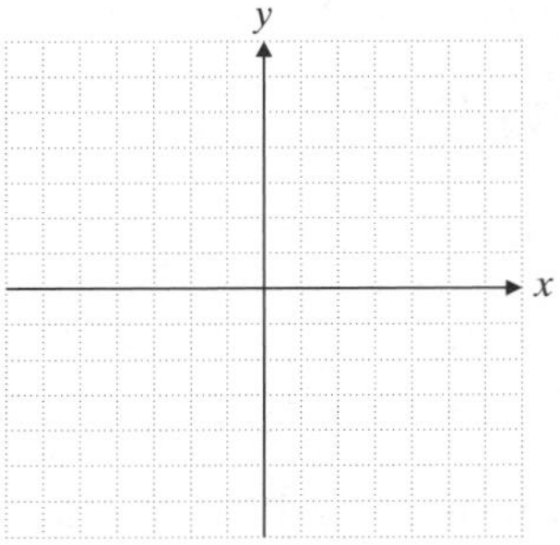

16. For $f(x) = 2x^2 + 3x - 4$:

   (a) Find $f(0)$.

   (b) Find $f(-2)$.

17. For $g(x) = \dfrac{3}{x - 4}$:

   (a) Find $g(3)$.

   (b) Find $g(-11)$.

11. _____

12. _____

13. _____

14. _____

15. _____

16. (a) _____

(b) _____

17. (a) _____

(b) _____

Approximately one-half of this test covers the content of Chapters 1–2. The remainder covers the content of Chapter 3.

1. Evaluate. $\left(-\dfrac{1}{3}\right)\left(\dfrac{2}{5}\right) - \left(\dfrac{2}{5}\right)^2$

2. Solve. $x = 2^3 - 6 \div 3 - 1$

3. Simplify. $-5x[3x + 2(x - 6)]$

4. Evaluate. $I = prt$ for $p = \$3500$, $r = 7\%, t = 5$ years

5. Solve. $2 - 3(4 - x) = x - (3 - x)$

6. Solve. $\dfrac{3x - 2}{4} - \dfrac{x}{3} = 2$

7. Solve. $A = lw + lh$ for w

8. Solve and graph.
$3(x - 2) \le 4x - 7$

▲ **9.** Ricardo was hired by a new fitness center to paint a mural on one of the gym walls. The length of the mural is to be 3 feet less than triple the height. The perimeter of the mural will be 90 feet. What is the length and height of the mural?

10. Find an equation of the line through $(6, 8)$ and $(7, 11)$.

In questions 11 and 12, give the equation of the line that fits each description.

11. Passes through $(7, -4)$ and is vertical.

12. Passes through $(-2, 3)$ with slope $\frac{1}{3}$.

13. Find the slope of the line through $(-8, -3)$ and $(11, -3)$.

14. What is the slope of the line $3x - 7y = -2$?

1. _____

2. _____

3. _____

4. _____

5. _____

6. _____

7. _____

8. _____

9. _____

10. _____

11. _____

12. _____

13. _____

14. _____

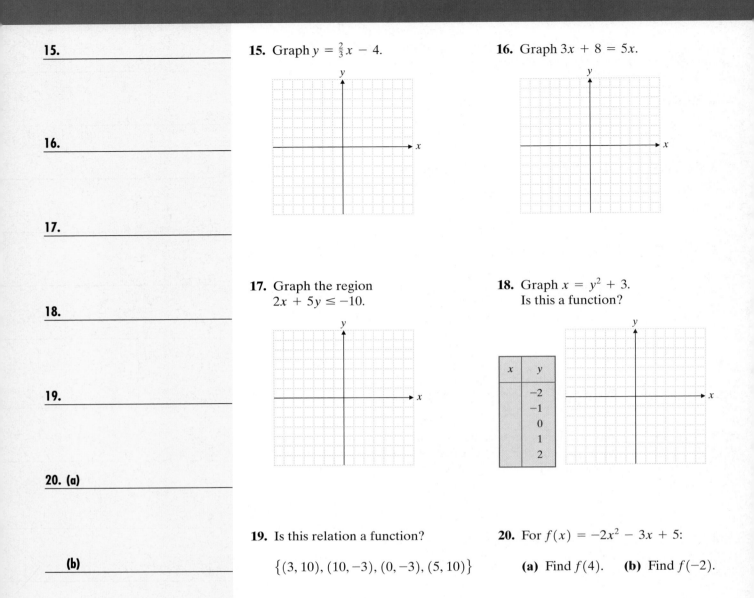

15. _____

16. _____

17. _____

18. _____

19. _____

20. (a) _____

_**(b)**_ _____

15. Graph $y = \frac{2}{3}x - 4$.

16. Graph $3x + 8 = 5x$.

17. Graph the region $2x + 5y \leq -10$.

18. Graph $x = y^2 + 3$. Is this a function?

x	y
	−2
	−1
	0
	1
	2

19. Is this relation a function?

$\{(3, 10), (10, -3), (0, -3), (5, 10)\}$

20. For $f(x) = -2x^2 - 3x + 5$:

(a) Find $f(4)$. **(b)** Find $f(-2)$.

Systems of Linear Equations and Inequalities

The cheetah is an amazing animal. Beautiful, powerful, and dangerous, these fast creatures were once plentiful in many African countries. However, they are now on the endangered species list. From the estimated 100,000 that roamed free in 1900, fewer than twelve thousand exist at the present time. What can be done about this dangerous trend? Exactly how fast is their population decreasing? What predictions can be made about the future? Please turn to the Putting Your Skills to Work exercises on page 225 to see whether you can answer these questions.

Pretest Chapter 4

1. _____

2. _____

If you are familiar with the topics in this chapter, take this test now. Check your answers with those in the back of the book. If an answer is wrong or you can't do an exercise, study the appropriate section of the chapter.

If you are not familiar with the topics in this chapter, don't take this test now. Instead, study the examples, work the practice exercises, and then take the test.

This test will help you identify those concepts that you have mastered and those that need more study.

Section 4.1

3. _____

Find the solution to each system of equations. If there is no single solution to a system, state the reason.

1. $5x - 2y = 27$
$3x - 5y = -18$

2. $7x + 3y = 15$
$\dfrac{1}{3}x - \dfrac{1}{2}y = 2$

3. $2x = 3 + y$
$3y = 6x - 9$

4. $2x + 7y = -10$
$5x + 6y = -2$

4. _____

Section 4.2

Find the solution to each system of equations.

5. _____

5. $5x - 2y + z = -1$
$3x + y - 2z = 6$
$-2x + 3y - 5z = 7$

6. $x + y + 2z = 9$
$3x + 2y + 4z = 16$
$2y + z = 10$

7. $x - 2z = -5$
$y - 3z = -3$
$2x - z = -4$

Section 4.3

6. _____

Use a system of linear equations to solve each exercise.

8. A coach purchased two shirts and three pairs of pants for his team and paid $75. His assistant purchased three shirts and five pairs of pants at the same store for $121. What was the cost for a shirt? What was the cost for a pair of pants?

7. _____

9. A biologist needs to use 21 milligrams of iron, 22 milligrams of vitamin B_{12}, and 26 milligrams of niacin for an experiment. She has available packets *A*, *B*, and *C* to meet these requirements. Packet *A* contains 3 milligrams of iron, 2 milligrams of vitamin B_{12}, and 4 milligrams of niacin. Packet *B* contains 2 milligrams of iron, 4 milligrams of vitamin B_{12}, and 5 milligrams of niacin. Packet *C* contains 2 milligrams of iron, 2 milligrams of vitamin B_{12}, and 1 milligram of niacin. How many of each packet should she use?

8. _____

10. You can solve ten math exercises in 82 minutes. Easy exercises take 7 minutes. Difficult exercises take 10 minutes. How many exercises of each type can you solve in 82 minutes?

9. _____

Section 4.4

Solve the following systems of inequalities by graphing.

11. $2x + 2y \geq -4$
$-3x + y \leq 2$

12. $x - 3y < -6$
$x < 3$

10. _____

11. _____

12. _____

4.1 Systems of Equations in Two Variables

1 **Determining Whether an Ordered Pair Is a Solution to a System of Two Equations**

In Chapter 3 we found that a linear equation containing two variables, such as $4x + 3y = 12$, has an unlimited number of ordered pairs (x, y) that satisfy it. For example, $(3, 0), (0, 4)$, and $(-3, 8)$ all satisfy the equation $4x + 3y = 12$. We call *two* linear equations in two unknowns a **system of two linear equations in two variables**. Many such systems have exactly one solution. A **solution to a system** of two linear equations in two variables is an *ordered pair* that is a solution to *each* equation.

EXAMPLE 1 Determine whether $(3, -2)$ is a solution to the following system.

$$x + 3y = -3$$
$$4x + 3y = 6$$

We will begin by substituting $(3, -2)$ into the first equation to see whether the ordered pair is a solution to the first equation.

$$3 + 3(-2) \stackrel{?}{=} -3$$
$$3 - 6 \stackrel{?}{=} -3$$
$$-3 = -3 \checkmark$$

Likewise, we will determine whether $(3, -2)$ is a solution to the second equation.

$$4(3) + 3(-2) \stackrel{?}{=} 6$$
$$12 - 6 \stackrel{?}{=} 6$$
$$6 = 6 \checkmark$$

Since $(3, -2)$ is a solution to each equation in the system, it is a solution to the system itself.

It is important to remember that we cannot confirm that a particular ordered pair is in fact the solution to a system of two equations unless we have checked to see whether the solution satisfies both equations. Merely checking one equation is not sufficient. Determining whether an ordered pair is a solution to a system of equations requires that we verify that the solution satisfies *both* equations.

Practice Problem 1 Determine whether $(-3, 4)$ is a solution to the following system.

$$2x + 3y = 6$$
$$3x - 4y = 7$$

2 **Solving a System of Two Linear Equations by the Graphing Method**

We can verify the solution to a system of linear equations by graphing each equation. If the lines intersect, the system has a unique solution. The point of intersection lies on both lines. Thus, it is a solution to each equation and the solution to the system. We will illustrate this by graphing the equations in Example 1. Notice that the coordinates of the point of intersection are $(3, -2)$. The solution to the system is $(3, -2)$.

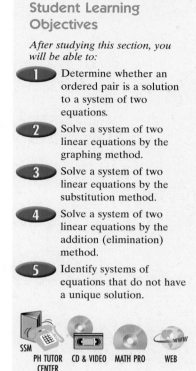

Student Learning Objectives

After studying this section, you will be able to:

1 Determine whether an ordered pair is a solution to a system of two equations.

2 Solve a system of two linear equations by the graphing method.

3 Solve a system of two linear equations by the substitution method.

4 Solve a system of two linear equations by the addition (elimination) method.

5 Identify systems of equations that do not have a unique solution.

SSM PH TUTOR CD & VIDEO MATH PRO WEB
CENTER

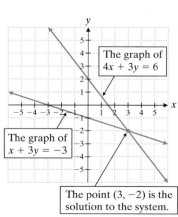

The graph of $4x + 3y = 6$

The graph of $x + 3y = -3$

The point $(3, -2)$ is the solution to the system.

This example shows that we can find the solution to a system of linear equations by graphing each line and determining the point of intersection.

EXAMPLE 2   Solve this system of equations by graphing.

$$2x + 3y = 12$$
$$x - y = 1$$

Using the methods that we developed in Chapter 3, we graph each line and determine the point at which the two lines intersect.

Finding the solution by the graphing method does not always lead to an accurate result, however, because it involves visual estimation of the point of intersection. Also, our plotting of one or more of the lines could be off slightly. Thus, we verify that our answer is correct by substituting $x = 3$ and $y = 2$ into the system of equations.

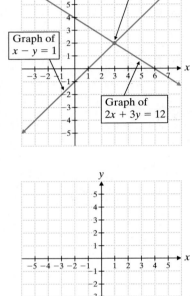

The point $(3, 2)$ is the solution.

Graph of $x - y = 1$

Graph of $2x + 3y = 12$

$$
\begin{array}{ll}
x - y = 1 & \qquad 2x + 3y = 12 \\
3 - 2 \overset{?}{=} 1 & \qquad 2(3) + 3(2) \overset{?}{=} 12 \\
\quad 1 = 1 \ \checkmark & \qquad \qquad 12 = 12 \ \checkmark
\end{array}
$$

Thus, we have verified that the solution to the system is $(3, 2)$.

Practice Problem 2   Solve this system of equations by graphing. Check your solution.

$$3x + 2y = 10$$
$$x - y = 5$$

Many times when we graph a system, we find that the two straight lines intersect at one point. However, it is possible for a given system to have as its graph two parallel lines. In such a case there is no solution because there is no point that lies on both lines (i.e., no ordered pair that satisfies both equations). Such a system of equations is said to be **inconsistent**. Another possibility is that when we graph each equation in the system, we obtain one line. In such a case there are an infinite number of solutions. Any point that lies on the first line will also lie on the second line (i.e., any ordered pair). A system of equations in two variables is said to have **dependent equations** if it has infinitely many solutions. We will discuss these situations in more detail after we have developed algebraic methods for solving a system of equations.

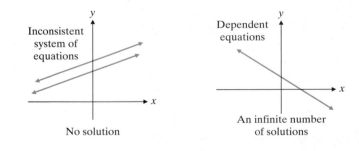

Inconsistent system of equations

No solution

Dependent equations

An infinite number of solutions

3   *Solving a System of Two Linear Equations by the Substitution Method*

An algebraic method of solving a system of linear equations in two variables is the **substitution method**. To use this method, we choose one equation and solve for one variable. It is usually best to solve for a variable that has a coefficient of $+1$ or -1. This will help us avoid introducing fractions. When we solve for one variable, we obtain an expression that contains the other variable. We *substitute* this expression into the second equation. Then we have one equation with one unknown, which we can easily solve. Once we know the value of this variable, we can substitute it into one of the original equations to find the value of the other variable.

⬭ **EXAMPLE 3**   Find the solution to the following system of equations. Use the substitution method.

$$x + 3y = -7 \quad \textbf{(1)}$$
$$4x + 3y = -1 \quad \textbf{(2)}$$

We can work with equation **(1)** or equation **(2)**. Let's choose equation **(1)** and solve for x. This gives us equation **(3)**.

$$x = -7 - 3y \quad \textbf{(3)}$$

Now we substitute this expression for x into equation **(2)** and solve the equation for y.

$$4x + 3y = -1 \quad \textbf{(2)}$$
$$4(-7 - 3y) + 3y = -1$$
$$-28 - 12y + 3y = -1$$
$$-28 - 9y = -1$$
$$-9y = -1 + 28$$
$$-9y = 27$$
$$y = -3$$

Now we substitute $y = -3$ into equation **(1)** or **(2)** to find x. Let's use **(1)**:

$$x + 3(-3) = -7$$
$$x - 9 = -7$$
$$x = -7 + 9$$
$$x = 2$$

Therefore, our solution is the ordered pair $(2, -3)$.
   Check: We must verify the solution in both of the *original* equations.

$$\begin{array}{ll}
x + 3y = -7 & 4x + 3y = -1 \\
2 + 3(-3) \overset{?}{=} -7 & 4(2) + 3(-3) \overset{?}{=} -1 \\
2 - 9 \overset{?}{=} -7 & 8 - 9 \overset{?}{=} -1 \\
-7 = -7 \ \checkmark & -1 = -1 \ \checkmark
\end{array}$$

Practice Problem 3   Use the substitution method to solve this system.

$$2x - y = 7$$
$$3x + 4y = -6$$

We summarize the substitution method here.

How to Solve a System of Two Linear Equations by the Substitution Method

1. Choose one of the two equations and solve for one variable in terms of the other variable.
2. Substitute this expression from step 1 into the *other* equation.
3. You now have one equation with one variable. Solve this equation for that variable.
4. Substitute this value for the variable into one of the original equations to obtain a value for the second variable.
5. Check the solution in both original equations.

Graphing Calculator

Solving Systems of Equations

We can solve systems of equations graphically by using a graphing calculator. For example, to solve the system of equations in Example 2, first rewrite each equation in slope–intercept form.

$$y = -\frac{2}{3}x + 4$$
$$y = x - 1$$

Then graph $y_1 = -\frac{2}{3}x + 4$ and $y_2 = x - 1$ on the same screen.

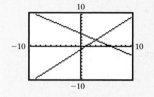

Next, use the Zoom and Trace features to find the intersection of the two lines.
   Some graphing calculators have a command to find and calculate the intersection.

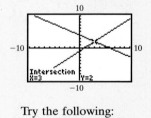

Try the following:

1. $y_1 = -3x + 9$
   $y_2 = 4x - 5$

2. $y_1 = -0.48x + 4.79$
   $y_2 = 1.52x - 2.98$

Optional Graphing Calculator Note

Before solving the system in Example 3 with a graphing calculator, you will first need to solve each equation for y. Equation **(1)** can be written as $y_1 = -\dfrac{1}{3}x - \dfrac{7}{3}$ or as $y_1 = \dfrac{-x-7}{3}$. Likewise, equation **(2)** can be written as $y_2 = -\dfrac{4}{3}x - \dfrac{1}{3}$ or as $y_2 = \dfrac{-4x-1}{3}$.

EXAMPLE 4   Solve the following system of equations.

$$\frac{1}{2}x - \frac{1}{4}y = -\frac{3}{4} \quad \textbf{(1)}$$

$$3x - 2y = -6 \quad \textbf{(2)}$$

First clear equation **(1)** of fractions by multiplying each term by 4.

$$4\left(\frac{1}{2}x\right) - 4\left(\frac{1}{4}y\right) = 4\left(-\frac{3}{4}\right)$$

$$2x - y = -3 \quad \textbf{(3)}$$

The new system is as follows:

$$2x - y = -3 \quad \textbf{(3)}$$

$$3x - 2y = -6 \quad \textbf{(2)}$$

Step 1   Let's solve equation **(3)** for y.

$$-y = -3 - 2x$$

$$y = 3 + 2x$$

Step 2   Substitute this expression for y into equation **(2)**.

$$3x - 2(3 + 2x) = -6$$

Step 3   Solve this equation for x.

$$3x - 6 - 4x = -6$$

$$-6 - x = -6$$

$$-x = -6 + 6$$

$$-x = 0$$

$$x = 0$$

Step 4   Substitute $x = 0$ into equation **(2)**.

$$3(0) - 2y = -6$$

$$-2y = -6$$

$$y = 3$$

So our solution is $(0, 3)$.

Step 5   We must verify the solution in both original equations.

$$\frac{1}{2}x - \frac{1}{4}y = -\frac{3}{4}$$

$$\frac{0}{2} - \frac{3}{4} \overset{?}{=} -\frac{3}{4}$$

$$-\frac{3}{4} = -\frac{3}{4} \quad \checkmark$$

$$3x - 2y = -6$$

$$3(0) - 2(3) \overset{?}{=} -6$$

$$-6 = -6 \quad \checkmark$$

Practice Problem 4 Use the substitution method to solve this system.

$$\frac{1}{2}x + \frac{2}{3}y = 1$$

$$\frac{1}{3}x + y = -1$$

4 *Solving a System of Equations by the Addition Method*

Another way to solve a system of two linear equations in two variables is to add the two equations so that a variable is eliminated. This technique is called the **addition method** or the **elimination method**. We usually have to multiply one or both of the equations by suitable factors so that we obtain opposite coefficients on one variable (either x or y) in the equations.

EXAMPLE 5 Solve the following system by the addition method.

$$5x + 8y = -1 \quad \textbf{(1)}$$
$$3x + y = 7 \quad \textbf{(2)}$$

We can eliminate either the x- or the y-variable. Let's choose y. We multiply equation **(2)** by -8.

$$-8(3x) + (-8)(y) = -8(7)$$
$$-24x - 8y = -56 \quad \textbf{(3)}$$

We now add equations **(1)** and **(3)**.

$$5x + 8y = -1 \quad \textbf{(1)}$$
$$\underline{-24x - 8y = -56 \quad \textbf{(3)}}$$
$$-19x = -57$$

We solve for x.

$$x = \frac{-57}{-19} = 3$$

Now we substitute $x = 3$ into equation **(2)** (or equation **(1)**).

$$3(3) + y = 7$$
$$9 + y = 7$$
$$y = -2$$

Our solution is $(3, -2)$.

Check:

$$5(3) + 8(-2) \overset{?}{=} -1$$
$$15 + (-16) \overset{?}{=} -1$$
$$-1 = -1 \ \checkmark$$
$$3(3) + (-2) \overset{?}{=} 7$$
$$9 + (-2) \overset{?}{=} 7$$
$$7 = 7 \ \checkmark$$

Practice Problem 5   Use the addition method to solve this system.

$$-3x + y = 5$$
$$2x + 3y = 4$$

For convenience, we summarize the addition method here.

How to Solve a System of Two Linear Equations by the Addition (Elimination) Method

1. Arrange each equation in the form $ax + by = c$. (Remember that a, b, and c can be any real numbers.)
2. Multiply one or both equations by appropriate numbers so that the coefficients of one of the variables are opposites.
3. Add the two equations from step 2 so that one variable is eliminated.
4. Solve the resulting equation for the remaining variable.
5. Substitute this value into one of the *original* equations and solve to find the value of the other variable.
6. Check the solution in both of the original equations.

EXAMPLE 6   Solve the following system by the addition method.

$$3x + 2y = -8 \quad \textbf{(1)}$$
$$2x + 5y = 2 \quad \textbf{(2)}$$

To eliminate the variable x, we multiply equation **(1)** by 2 and equation **(2)** by -3. We now have the following equivalent system.

$$6x + 4y = -16$$
$$\underline{-6x - 15y = -6}$$

$$-11y = -22 \qquad \text{Add the equations.}$$
$$y = 2 \qquad \text{Solve for } y.$$

Substitute $y = 2$ into equation **(1)**.

$$3x + 2(2) = -8$$
$$3x + 4 = -8$$
$$3x = -12$$
$$x = -4$$

The solution to the system is $(-4, 2)$.

Check: Verify that this solution is correct.

Note: We could have easily eliminated the variable y in Example 6 by multiplying equation **(1)** by 5 and equation **(2)** by -2. Try it. Is the solution the same? Why?

Practice Problem 6   Use the addition (elimination) method to solve this system.

$$5x + 4y = 23$$
$$7x - 3y = 15$$

5 *Identifying Systems of Equations That Do Not Have a Unique Solution*

So far we have examined only those systems that have one solution. But other systems must also be considered. These systems can best be illustrated with graphs. In general, the system of equations

$$ax + by = c$$
$$dx + ey = f$$

may have one solution, no solution, or an infinite number of solutions.

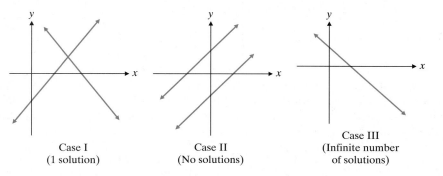

Case I
(1 solution)

Case II
(No solutions)

Case III
(Infinite number
of solutions)

Case I: *One solution.* The two graphs intersect at one point, which is the solution. We say that the equations are **independent**. It is a **consistent system** of equations. There is a point (an ordered pair) *consistent* with both equations.

Case II: *No solution.* The two graphs are parallel and so do not intersect. We say that the system of equations is **inconsistent** because there is no point consistent with both equations.

Case III: *An infinite number of solutions.* The graphs of each equation yield the same line. Every ordered pair on this line is a solution to both of the equations. We say that the equations are **dependent**.

● EXAMPLE 7   If possible, solve the system.

$$2x + 8y = 16 \quad \textbf{(1)}$$
$$4x + 16y = -8 \quad \textbf{(2)}$$

To eliminate the variable y, we'll multiply equation **(1)** by -2.

$$-2(2x) + (-2)(8y) = (-2)(16)$$
$$-4x - 16y = -32 \quad \textbf{(3)}$$

We now have the following equivalent system.

$$-4x - 16y = -32 \quad \textbf{(3)}$$
$$4x + 16y = -8 \quad \textbf{(2)}$$

When we add equations **(3)** and **(2)**, we get

$$0 = -40,$$

which, of course, is false. Thus, we conclude that this system of equations is inconsistent, and **there is no solution**. Therefore, equations **(1)** and **(2)** do not intersect as we can see on the graph to the right.

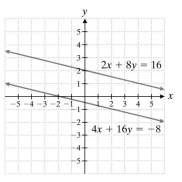

$2x + 8y = 16$

$4x + 16y = -8$

Graphing Calculator

Identifying Systems of Equations

If there is a concern as to whether or not a system of equations has no solution, how can this be determined quickly using a graphing calculator? What is the most important thing to look for on the graph?

If the equations are dependent, how can we be sure of this by looking at the display of a graphing calculator? Why? Determine whether there is one solution, no solution, or an infinite number of solutions for each of the following systems.

1. $y_1 = -2x + 1$
$3y_2 = -6x + 3$

2. $y_1 = 3x + 6$
$y_2 = 3x + 2$

3. $y_1 = x - 3$
$y_2 = -2x + 12$

If we had used the substitution method to solve this system, we still would have obtained a false statement. When you try to solve an inconsistent system of linear equations by any method, you will always obtain a mathematical equation that is not true.

Practice Problem 7 If possible, solve the system.

$$4x - 2y = 6$$
$$-6x + 3y = 9$$

EXAMPLE 8 If possible, solve the system.

$$0.5x - 0.2y = 1.3 \quad \textbf{(1)}$$
$$-1.0x + 0.4y = -2.6 \quad \textbf{(2)}$$

Although we could work directly with the decimals, it is easier to multiply each equation by the appropriate power of 10 (10, 100, and so on) so that the coefficients of the new system are integers. Therefore, we will multiply equations **(1)** and **(2)** by 10 to obtain the following equivalent system.

$$5x - 2y = 13 \quad \textbf{(3)}$$
$$-10x + 4y = -26 \quad \textbf{(4)}$$

We can eliminate the variable y by multiplying each term of equation **(3)** by 2.

$$10x - 4y = 26 \quad \textbf{(5)}$$
$$\underline{-10x + 4y = -26} \quad \textbf{(4)}$$
$$0 = 0 \quad \text{Add the equations.}$$

This statement is always true; it is an **identity**. Hence, the two equations are dependent, and there are an infinite number of solutions. Any solution satisfying equation **(1)** will also satisfy equation **(2)**. For example, $(3, 1)$ is a solution to equation **(3)**. (Prove this.) Hence, it must also be a solution to equation **(4)**. (Prove it.) Thus, the equations actually describe the same line, as you can see on the graph.

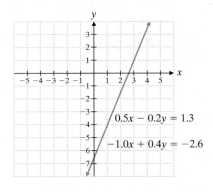

Practice Problem 8 If possible, solve the system.

$$0.3x - 0.9y = 1.8$$
$$-0.4x + 1.2y = -2.4$$

Exercises

Verbal and Writing Skills

1. Explain what happens when a system of two linear equations is inconsistent. What effect does it have in obtaining a solution? What would the graph of such a system look like?

2. Explain what happens when a system of two linear equations has dependent equations. What effect does it have in obtaining a solution? What would the graph of such a system look like?

Determine whether the given ordered pair is a solution to the system of equations.

3. $\left(\dfrac{3}{2}, -1\right)$ $4x + 1 = 6 - y$
              $2x - 5y = 8$

4. $\left(-4, \dfrac{2}{3}\right)$ $2x - 3(y - 5) = 5$
                   $6y = x + 8$

Solve the system of equations by graphing. Check your solution.

5. $3x + y = 2$
    $2x - y = 3$

6. $3x + y = 5$
    $2x - y = 5$

7. $2x + 3y = 6$
    $2x + y = -2$

8. $2x + 3y = -6$
     $x - 3y = 6$

9. $y = -x + 3$
    $3x + 3y = -2$

10. $y = \dfrac{1}{3}x - 2$
     $-x + 3y = 9$

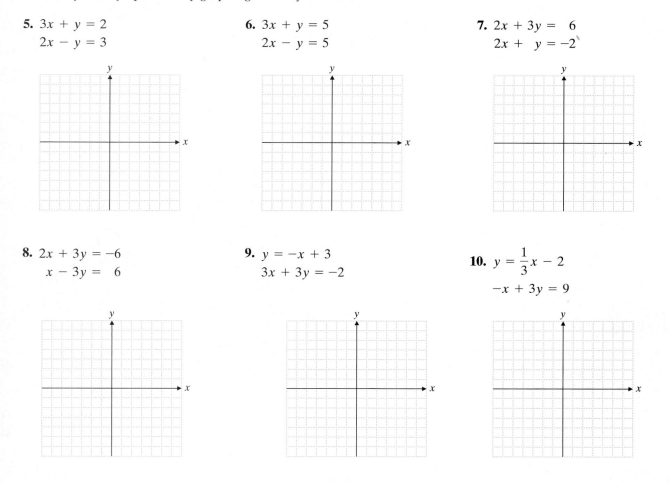

Find the solution to each system by the substitution method. Check your answers for Exercises 11–14.

11. $3x + 2y = -17$
$\quad\;\; 2x + y = 3$

12. $5x - 2y = 8$
$\quad\;\; 3x - y = 7$

13. $-x + 3y = -8$
$\quad\;\; 2x - y = 6$

14. $10x + 3y = 8$
$\quad\;\; 2x + y = 2$

15. $\dfrac{1}{5}x - \dfrac{1}{2}y = 1$
$\quad\;\; \dfrac{1}{5}x - 3y = -9$

16. $x - \dfrac{3}{2}y = 1$
$\quad\;\; 2x - 7y = 10$

17. $3(x - 1) + 2(y + 2) = 6$
$\quad\;\; 4(x - 2) + y = 2$

18. $2(x - 2) + 3(y + 1) = -2$
$\quad\;\; x + 2(y - 1) = -4$

Find the solution to each system by the addition (elimination) method. Check your answers for Exercises 23–26.

19. $9x + 2y = 2$
$\quad\;\; 3x + 5y = 5$

20. $12x - 5y = -7$
$\quad\;\; 4x + 2y = 5$

21. $6s - 3t = 1$
$\quad\;\; 5s + 6t = 15$

22. $2s + 3t = 5$
$\quad\;\; 3s - 6t = 18$

23. $\dfrac{7}{2}x + \dfrac{5}{2}y = -4$
$\quad\;\; 3x + \dfrac{2}{3}y = 1$

24. $\dfrac{4}{3}x - y = 4$
$\quad\;\; \dfrac{3}{4}x - y = \dfrac{1}{2}$

25. $1.6x + 1.5y = 1.8$
$\quad\;\; 0.4x + 0.3y = 0.6$

26. $2.5x + 0.6y = 0.2$
$\quad\;\; 0.5x - 1.2y = 0.7$

Mixed Practice

If possible, solve each system of equations. Use any method. If there is not a unique solution to a system, say why.

27. $2x + y = 4$
$\quad\;\; \dfrac{2}{3}x + \dfrac{1}{4}y = 2$

28. $2x + 3y = 16$
$\quad\;\; 5x - \dfrac{3}{4}y = 7$

29. $0.2x = 0.1y - 1.2$
$\quad\;\; 2x - y = 6$

30. $0.1x - 0.6 = 0.3y$
$\quad\;\; 0.3x + 0.1y + 2.2 = 0$

31. $\begin{aligned} 5x - 7y &= 12 \\ -10x + 14y &= -24 \end{aligned}$

32. $\begin{aligned} 3x - 11y &= 9 \\ -9x + 33y &= 18 \end{aligned}$

33. $\begin{aligned} 0.8x + 0.9y &= 1.3 \\ 0.6x - 0.5y &= 4.5 \end{aligned}$

34. $\begin{aligned} 0.6y &= 0.9x + 1 \\ 3x &= 2y - 4 \end{aligned}$

35. $\begin{aligned} \frac{4}{5}b &= \frac{1}{5} + a \\ 15a - 12b &= 4 \end{aligned}$

36. $\begin{aligned} 3a - 2b &= \frac{3}{2} \\ \frac{3a}{2} &= \frac{3}{4} + b \end{aligned}$

37. $\begin{aligned} \frac{3}{8}x + y &= 14 \\ 2x - \frac{7}{4}y &= 18 \end{aligned}$

38. $\begin{aligned} \frac{4}{5}x - y &= 7 \\ x - \frac{4}{3}y &= 8 \end{aligned}$

To Think About

39. Wayne Burton is having some tile replaced in his bathroom. He has obtained an estimate from two tile companies. Old World Tile gave an estimate of $200 to remove the old tile and $50 per hour to place new tile on the wall. Modern Bathroom Headquarters gave an estimate of $300 to remove the old tile and $30 per hour to place new tile on the wall.

(a) Create a cost equation for each company where y is the total cost of the tile work and x is the number of hours of labor. Write a system of equations.

(b) Graph the two equations using the values $x = 0, 4$, and 8.

(c) Determine from your graph how many hours of installing new tile will be required for the two companies to cost the same.

(d) Determine from your graph which company costs less to remove old tile and to install new tile if the time needed to install new tile is 6 hours.

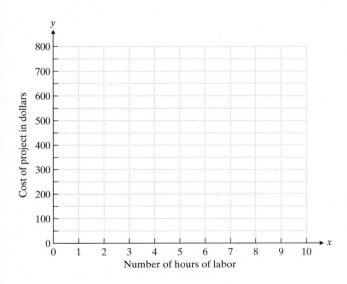

40. Jeff and Shelley are planning to move some furniture to their daughter's house. Seaside Movers quoted a price of $100 for the truck and $40 per hour for the movers. Beverly Rapid Mover quoted a price of $50 for the truck and $50 per hour for the movers.
   (a) Create a cost equation for each company where y is the total cost of the move and x is the number of hours of labor. Write a system of equations.
   (b) Graph the two equations using the values $x = 0, 4,$ and 7.
   (c) Determine from your graph how many hours of moving would be required for the two companies to cost the same.
   (d) Determine from your graph which company costs less to conduct the move if the total number of hours needed for moving is 3 hours.

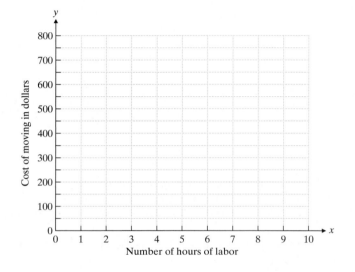

Optional Graphing Calculator Problems

On your graphing calculator, graph each system of equations on the same set of axes. Find the point of intersection to the nearest hundredth.

41. $y_1 = -1.7x + 3.8$
$y_2 = 0.7x - 2.1$

42. $y_1 = -0.81x + 2.3$
$y_2 = 1.6x + 0.8$

43. $0.5x + 1.1y = 5.5$
$-3.1x + 0.9y = 13.1$

44. $5.86x + 6.22y = -8.89$
$-2.33x + 4.72y = -10.61$

Cumulative Review Problems

45. Gina is covering the top of a small table with a stained-glass mosaic. The stained glass she will use costs $8 per square foot. If the table top measures 18 inches by 27 inches, how much will Gina spend on glass?

46. Eighty percent of the automobiles that enter the city of Boston during rush hour will have to park in private or municipal parking lots. If there are 273,511 private or municipal lot spaces filled by cars entering the city during rush hour every morning, how many cars enter the city during rush hour? Round your answer to the nearest car.

Putting Your Skills to Work

Endangered Species

Two of Africa's most endangered species are the cheetah and the black rhino. The cheetah faces a significant threat from human poachers, who kill them for their hides, while the black rhino is hunted for its horns. Do you realize how quickly these animal populations are decreasing?

Source: United Nations Statistics Division.

Problems for Individual Investigation

1. In 1900, the cheetah population in Africa was estimated to be 100,000. By 2000, their numbers had fallen to twelve thousand. Determine the average number of cheetahs lost per year during this time period. Write a function describing the cheetah population $c(t)$ in terms of the number of years t since 1900.

2. The black rhino population in Africa in 1970 was seventy thousand. By 2000, the population was 10,500. Determine the average number of black rhinos lost per year during this time period. (Round to the nearest ten.) Write a function describing the population $r(t)$ in terms of the number of years t since 1970.

3. Using your answer to question 1, write a function describing the cheetah population $c(t)$ in terms of the number of years t *since 1990.*

4. Using your answer to question 2, write a function describing the black rhino population $r(t)$ in terms of the number of years t *since 1990.*

Problems for Cooperative Group Investigation

5. Use the functions you found in question 3 and 4 to determine the populations for the cheetah and black rhino for the years 1990, 1995, and 2000. Use this information to graph the functions on the set of axes given. (Note: On the t-axis, 0 corresponds to the year 1990.)

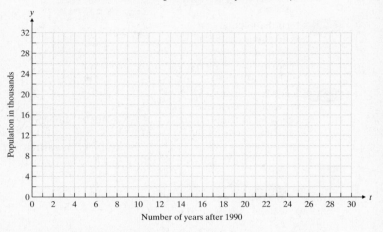

6. Interpret the slope of each line in terms of population and time. Which animal's population is decreasing at a faster rate?

7. In what year were the populations of the cheetah and black rhino approximately equal?

8. You can use the graphs to make predictions. If the populations of these two animals continue to decrease at these rates, when will the black rhino become extinct? When will the cheetah become extinct?

Internet Connections

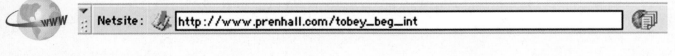

Site: The Endangered Species List

9. How many mountain gorillas are thought to be still living? How did the civil war in Rwanda of the early 1990s affect the gorilla population there?

10. Seven of the 23 species of crocodilians, which include crocodiles and alligators, are critically endangered. What are three types of measures that have been taken to protect crocodilian species? Have these measures been successful?

1 *Determining Whether an Ordered Triple Is the Solution to a System of Three Equations in Three Variables*

We are now going to study **systems of three linear equations in three variables** (unknowns). A **solution** to a system of three linear equations in three unknowns is an **ordered triple** of real numbers (x, y, z) that satisfies each equation in the system.

EXAMPLE 1 Determine whether $(2, -5, 1)$ is the solution to the following system.

$$3x + y + 2z = 3$$
$$4x + 2y - z = -3$$
$$x + y + 5z = 2$$

How can we prove that $(2, -5, 1)$ is a solution to this system? We will substitute $x = 2$, $y = -5$, and $z = 1$ into each equation. If a true statement occurs each time, $(2, -5, 1)$ is a solution to each equation and hence, a solution to the system. For the first equation:

$$3(2) + (-5) + 2(1) \stackrel{?}{=} 3$$
$$6 - 5 + 2 \stackrel{?}{=} 3$$
$$3 = 3 \checkmark$$

For the second equation:

$$4(2) + 2(-5) - 1 \stackrel{?}{=} -3$$
$$8 - 10 - 1 \stackrel{?}{=} -3$$
$$-3 = -3 \checkmark$$

For the third equation:

$$2 + (-5) + 5(1) \stackrel{?}{=} 2$$
$$2 - 5 + 5 \stackrel{?}{=} 2$$
$$2 = 2 \checkmark$$

Since we obtained three true statements, the ordered triple $(2, -5, 1)$ is a solution to the system.

Practice Problem 1 Determine whether $(3, -2, 2)$ is a solution to this system.

$$2x + 4y + z = 0$$
$$x - 2y + 5z = 17$$
$$3x - 4y + z = 19$$

To Think About Can we graph an equation in three variables? How? What would the graph look like? What would the graph of the system in Example 1 look like? Describe the graph of the solution.

2 *Finding the Solution to a System of Three Linear Equations in Three Variables If None of the Coefficients Is Zero*

One way to solve a system of three equations with three variables is to obtain from it a system of two equations in two variables; in other words, we eliminate one variable from both equations. We can then use the methods of Section 4.1 to solve the resulting system. You can find the third variable (the one that was eliminated) by substituting the two variables that you have found into one of the original equations.

EXAMPLE 2   Find the solution to (that is, solve) the following system of equations.

$$-2x + 5y + z = 8 \quad \textbf{(1)}$$
$$-x + 2y + 3z = 13 \quad \textbf{(2)}$$
$$x + 3y - z = 5 \quad \textbf{(3)}$$

Let's eliminate z because it can be done easily by adding equations **(1)** and **(3)**.

$$-2x + 5y + z = 8 \quad \textbf{(1)}$$
$$\underline{x + 3y - z = 5} \quad \textbf{(3)}$$
$$-x + 8y \quad\quad = 13 \quad \textbf{(4)}$$

Now we need to choose a *different pair* from the original equations and once again eliminate the same variable. In other words, we have to use equations **(1)** and **(2)** or equations **(2)** and **(3)** and eliminate z. Let's multiply each term of equation **(3)** by 3 (and call it equation **(6)**) and add the result to equation **(2)**.

$$-x + 2y + 3z = 13 \quad \textbf{(2)}$$
$$\underline{3x + 9y - 3z = 15} \quad \textbf{(6)}$$
$$2x + 11y \quad\quad = 28 \quad \textbf{(5)}$$

We now can solve the resulting system of two linear equations.

$$-x + 8y = 13 \quad \textbf{(4)}$$
$$2x + 11y = 28 \quad \textbf{(5)}$$

Multiply each term of equation **(4)** by 2.

$$-2x + 16y = 26$$
$$\underline{2x + 11y = 28}$$
$$27y = 54 \quad \text{Add the equations.}$$
$$y = 2 \quad \text{Solve for } y.$$

Substituting $y = 2$ into equation **(4)**, we have the following:

$$-x + 8(2) = 13$$
$$-x = -3$$
$$x = 3$$

Now substitute $x = 3$ and $y = 2$ into one of the original equations (any one will do) to solve for z. Let's use equation **(1)**.

$$-2x + 5y + z = 8$$
$$-2(3) + 5(2) + z = 8$$
$$-6 + 10 + z = 8$$
$$z = 4$$

The solution to the system is $(3, 2, 4)$.
Check: Verify that $(3, 2, 4)$ satisfies *each* of the three *original* equations.

Practice Problem 2   Solve this system.

$$x + 2y + 3z = 4$$
$$2x + y - 2z = 3$$
$$3x + 3y + 4z = 10$$

Here's a summary of the procedure that we just used.

> **How to Solve a System of Three Linear Equations in Three Unknowns**
>
> 1. Use the addition method to eliminate any variable from any pair of equations. (The choice of variable is arbitrary.)
> 2. Use appropriate steps to eliminate the *same variable* from a *different pair* of equations. (If you don't eliminate the same variable, you will still have three unknowns.)
> 3. Solve the resulting system of two equations in two variables.
> 4. Substitute the values obtained in step 3 into one of the three original equations. Solve for the remaining variable.
> 5. Check the solution in all of the original equations.

It is helpful to write all equations in the form $Ax + By + Cz = D$ before using this five-step method.

3   Finding the Solution to a System of Three Linear Equations in Three Variables If Some of the Coefficients Are Zero

If a system of three linear equations in three variables contains one or more equations of the form $Ax + By + Cz = 0$, where one of the values of A, B, or C is zero, then we will slightly modify our approach to solving the system. We will select one equation that contains only two variables. Then we will take the remaining system of two equations and eliminate the variable that was missing in the equation that we selected.

EXAMPLE 3   Solve the system.

$$\begin{cases} 4x + 3y + 3z = 4 & \textbf{(1)} \\ 3x + 0y + 2z = 2 & \textbf{(2)} \\ 2x - 5y + 0z = -4 & \textbf{(3)} \end{cases}$$

Note that equation **(2)** has no y-term and equation **(3)** has no z-term. Obviously, that makes our work easier. Let's work with equations **(2)** and **(1)** to obtain an equation that contains only x and y.

Step 1   Multiply equation **(1)** by 2 and equation **(2)** by -3 to obtain the following system.

$$\begin{array}{rcl} 8x + 6y + 6z = & 8 & \textbf{(4)} \\ -9x - 6y - 6z = & -6 & \textbf{(5)} \\ \hline -x + 6y = & 2 & \textbf{(6)} \end{array}$$

Step 2   This step is already done since equation **(3)** has no z-term.

Step 3 Now we can solve the system formed by equations **(3)** and **(6)**.

$$2x - 5y = -4 \quad \textbf{(3)}$$
$$-x + 6y = 2 \quad \textbf{(6)}$$

If we multiply each term of equation **(6)** by 2, we obtain the system

$$2x - 5y = -4$$
$$-2x + 12y = 4$$
$$\overline{7y = 0} \quad \text{Add.}$$
$$y = 0 \quad \text{Solve for } y.$$

Substituting $y = 0$ in equation **(6)**, we find the following:

$$-x + 6(0) = 2$$
$$-x = 2$$
$$x = -2$$

Step 4 To find z, we substitute $x = -2$ and $y = 0$ into one of the original equations containing z. Since equation **(2)** has only two variables, let's use it.

$$3x + 2z = 2$$
$$3(-2) + 2z = 2$$
$$2z = 8$$
$$z = 4$$

The solution to the system is $(-2, 0, 4)$.

Step 5 *Check:* Verify this solution by substituting these values into equations **(1)**, **(2)**, and **(3)**.

Practice Problem 3 Solve the system.

$$2x + y + z = 11$$
$$4y + 3z = -8$$
$$x - 5y = 2$$

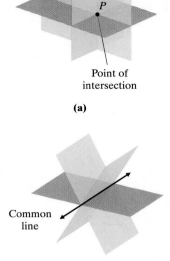

(a) Point of intersection

(b) Common line

(c)

(d)

(e)

A linear equation in three variables is a plane in three-dimensional space. A system of linear equations in three variables is three planes. The solution to the system is the set of points at which all three planes intersect. There are three possible results. The three planes may intersect at one point. (See figure **(a)** in the margin.)

This point is described by an ordered triple of the form (x, y, z) and lies in each plane. The three planes may intersect at a line. (See figure **(b)** in the margin.) In this case the system has an infinite number of solutions; that is, all the points on the line are solutions to the system.

Finally, all three planes may not intersect at any points. It may mean that all three planes never share any point of intersection, but that two planes intersect. (See figures **(c)**, **(d)**, and **(e)** in the margin.) In all such cases there is no solution to the system of equations.

1. Determine whether $(2, 1, -4)$ is a solution to the system.

$$2x - 3y + 2z = -7$$
$$x + 4y - z = 10$$
$$3x + 2y + z = 4$$

2. Determine whether $\left(5, 0, \dfrac{1}{2}\right)$ is a solution to the system.

$$x + y + 2z = 6$$
$$-x + 6y + 4z = -3$$
$$2x - 3y - 6z = 7$$

Solve each system.

3.
$$x + y + 2z = 0$$
$$2x - y - z = 1$$
$$x + 2y + 3z = 1$$

4.
$$2x + y + 3z = 2$$
$$x - y + 2z = -4$$
$$x + 3y - z = 1$$

5.
$$x + 2y - 3z = -11$$
$$-2x + y - z = -11$$
$$x + y + z = 6$$

6.
$$-5x + 3y + 2z = 1$$
$$x + y + z = 7$$
$$2x - y + z = 7$$

7.
$$-4x + y - 3z = 2$$
$$5x - 3y + 4z = 1$$
$$3x - 2y + 5z = 1$$

8.
$$8x - 5y + z = 15$$
$$3x + y - z = -7$$
$$x + 4y + z = -3$$

9.
$$x + 4y - z = -5$$
$$-2x - 3y + 2z = 5$$
$$x - \dfrac{2}{3}y + z = \dfrac{11}{3}$$

10.
$$x - 4y + 4z = -1$$
$$-x + \dfrac{y}{2} - \dfrac{5}{2}z = -3$$
$$-x + 3y - z = 5$$

11.
$$2x - 3y + 2z = -7$$
$$\dfrac{3}{2}x + y + \dfrac{1}{2}z = 2$$
$$x + 4y - z = 10$$

12. $\dfrac{x}{2} - y - 2z = 0$

$x + \dfrac{y}{2} - 3z = 3$

$2x + \dfrac{3y}{2} + z = 10$

13. $a = 8 + 3b - 2c$

$4a + 2b - 3c = 10$

$c = 10 + b - 2a$

14. $a = c - b$

$3a - 2b + 6c = 1$

$c = 4 - 3b - 7a$

15. $0.2a + 0.1b + 0.2c = 0.1$

$0.3a + 0.2b + 0.4c = -0.1$

$0.6a + 1.1b + 0.2c = 0.3$

16. $-0.1a + 0.2b + 0.3c = 0.1$

$0.2a - 0.6b + 0.3c = 0.5$

$0.3a - 1.2b - 0.4c = -0.4$

17. $-x + 3y - 2z = 11$

$2x - 4y + 3z + 15 = 0$

$3x - 5y - 4z = 5$

18. $-3x - 2y + 3z = 2$

$2x - 5y + 2z + 2 = 0$

$4x - 3y + 4z = 10$

Find the solution for each system of equations. Round your answers to five decimal places.

19. $x - 4y + 4z = -3.72186$

$-x + 3y - z = 5.98115$

$2x - y + 5z = 7.93645$

20. $4x + 2y + 3z = 9$

$9x + 3y + 2z = 3$

$2.987x + 5.027y + 3.867z = 18.642$

Solve each system.

21. $x + y = 1$

$y - z = -3$

$2x + 3y + z = 1$

22. $y - 2z = 5$

$2x + z = -1$

$3x + y - z = 4$

23. $-y + 2z = 1$

$x + y + z = 2$

$-x + 3z = 2$

24. $-2x + y - 3z = 0$

$-2y - z = -1$

$x + 2y - z = 5$

25. $x - 2y + z = 0$

$-3x - y = -6$

$y - 2z = -7$

26. $x + 2z = 0$

$3x + 3y + z = 6$

$6y + 5z = -3$

27. $\dfrac{a}{2} - b + c = 8$

$\dfrac{3}{2}a + b + 2c = 0$

$a + c = 2$

28. $2a + b + \dfrac{c}{3} = -2$

$\dfrac{a}{3} + \dfrac{b}{3} = -1$

$3b + c = 0$

Try to solve the system of equations. Explain your result in each case.

29.
$$2x + y \qquad = -3$$
$$2y + 16z = -18$$
$$-7x - 3y + 4z = 6$$

30.
$$6x - 2y + 2z = 2$$
$$4x + 8y - 2z = 5$$
$$-2x - 4y + z = -2$$

31.
$$3x + 3y - 3z = -1$$
$$4x + y - 2z = 1$$
$$-2x + 4y - 2z = -8$$

32.
$$-3x + 4y - z = -4$$
$$x + 2y + z = 4$$
$$-12x + 16y - 4z = -16$$

Cumulative Review Problems

33. Find the slope-intercept form of the equation of the line that passes through $(1, 4)$ and $(-2, 3)$.

34. Find the slope-intercept form of the equation of the line that is perpendicular to $y = -\dfrac{2}{3}x + 4$ and passes through $(-4, 2)$.

35. A rancher in Australia has 346 horses, 545 sheep, and 601 cattle. He wants to purchase more animals so that he has 80% more cattle than horses and 74% more sheep than horses. How many animals of each type will he have to buy? How many animals of each type will he have after his purchases? (Round all answers to the nearest whole number.)

36. The current in a river moves at a speed of 3.5 miles per hour. A boat travels with the current 48 miles downstream in a total of 3 hours. What would the speed of the boat have been if there had been no current?

1 *Solving an Applied Exercise Requiring the Use of a System of Two Linear Equations in Two Unknowns*

We will now examine how a system of linear equations can assist us in solving applied exercises.

EXAMPLE 1 For the paleontology lecture on campus, advance tickets cost $5 and tickets at the door cost $6. The ticket sales this year came to $4540. The department chairman wants to raise prices next year to $7 for advance tickets and $9 for tickets at the door. He said that if exactly the same number of people attend next year, the ticket sales at these new prices will total $6560. If he is correct, how many tickets were sold in advance this year? How many tickets were sold at the door?

1. **Understand the problem.**
   Since we are looking for the number of tickets sold, we let

   x = number of tickets bought in advance and
   y = number of tickets bought at the door.

2. **Write a system of two equations in two unknowns.**
   If advance tickets cost $5, then the total sales will be $5x$; similarly, total sales of door tickets will be $6y$. Since the total sales of both types of tickets was $4540, we have

   $$5x + 6y = 4540.$$

   By the same reasoning, we have

   $$7x + 9y = 6560.$$

   Thus, our system is as follows:

   $$5x + 6y = 4540 \quad \textbf{(1)}$$
   $$7x + 9y = 6560 \quad \textbf{(2)}$$

Allosaurus

3. **Solve the system of equations and state the answer.**
   We will multiply each term of equation **(1)** by -3 and each term of equation **(2)** by 2 to obtain the following equivalent system.

   $$-15x - 18y = -13,620 \quad \textbf{(3)}$$
   $$\underline{14x + 18y = 13,120} \quad \textbf{(4)}$$
   $$-x = -500$$

   Therefore, $x = 500$. Substituting $x = 500$ into equation **(1)**, we have the following:

   $$5(500) + 6y = 4540$$
   $$6y = 2040$$
   $$y = 340$$

   Thus, 500 advance tickets and 340 door tickets were sold.

4. **Check.**
   We need to check our answers. Do they seem reasonable?

Would 500 advance tickets at $5 and 340 door tickets at $6 yield $4540?	Would 500 advance tickets at $7 and 340 door tickets at $9 yield $6560?
$5(500) + 6(340) \overset{?}{=} 4540$	$7(500) + 9(340) \overset{?}{=} 6560$
$2500 + 2040 \overset{?}{=} 4540$	$3500 + 3060 \overset{?}{=} 6560$
$4540 = 4540$ ✓	$6560 = 6560$ ✓

Practice Problem 1   Coach Perez purchased baseballs at $6 each and bats at $21 each last week for the college baseball team. The total cost of the purchase was $318. This week he noticed that the same items are on sale. Baseballs are now $5 each and bats are $17. He found that if he made the same purchase this week, it would cost only $259. How many baseballs and how many bats did he buy last week?

EXAMPLE 2   An electronics firm makes two types of switching devices. Type A takes 4 minutes to make and requires $3 worth of materials. Type B takes 5 minutes to make and requires $5 worth of materials. When the production manager reviewed the latest batch, he found that it took 35 hours to make these switches with a materials cost of $1900. How many switches of each type were produced for this latest batch?

1. **Understand the problem.**
   We are given a lot of information, but the major concern is to find out how many of the type A devices and the type B devices were produced. This becomes our starting point to define the variables we will use.

   Let $A =$ the number of type A devices produced and
   $B =$ the number of type B devices produced.

2. **Write a system of two equations.**
   How should we construct the equations? What relationships exist between our variables (or unknowns)? According to the problem, the devices are related by time and by cost. So we set up one equation in terms of time (minutes in this case) and one in terms of cost (dollars). Each type A took 4 minutes to make; each type B took 5 minutes to make, and the total time was 2100 minutes. Each type A used $3 worth of materials; each type B used $5 worth of materials, and the total material cost was $1900. We can gather this information in a table. Making a table will help us form the equations.

	Type A Devices	Type B Devices	Total
Number of minutes	$4A$	$5B$	2100
Cost of materials	$3A$	$5B$	1900

$$4A + 5B = 2100$$
$$3A + 5B = 1900$$

   Therefore, we have the following system.

$$4A + 5B = 2100 \quad \textbf{(1)}$$
$$3A + 5B = 1900 \quad \textbf{(2)}$$

3. **Solve the system of equations and state the answers.**
   Multiplying equation **(2)** by -1 and adding the equations, we find the following:

$$4A + 5B = 2100$$
$$\underline{-3A - 5B = -1900}$$
$$A = 200$$

   Substituting $A = 200$ into equation **(1)**, we have the following:

$$800 + 5B = 2100$$
$$5B = 1300$$
$$B = 260$$

   Thus, 200 type A devices and 260 type B devices were produced.

4. *Check.*

If each type *A* requires 4 minutes and each type *B* requires 5 minutes, does this amount to a total time of 2100 minutes?

$$4A + 5B = 2100$$
$$4(200) + 5(260) \overset{?}{=} 2100$$
$$800 + 1300 \overset{?}{=} 2100$$
$$2100 = 2100 \checkmark$$

If each type *A* costs $3 and each type *B* costs $5, does this amount to a total cost of $1900?

$$3A + 5B = 1900$$
$$3(200) + 5(260) \overset{?}{=} 1900$$
$$600 + 1300 \overset{?}{=} 1900$$
$$1900 = 1900 \checkmark$$

Practice Problem 2   A furniture company makes both small and large chairs. It takes 30 minutes of machine time and 1 hour and 15 minutes of labor to build the small chair. The large chair requires 40 minutes of machine time and 1 hour and 20 minutes of labor. The company has 57 hours of labor time and 26 hours of machine time available each day. If all available time is used, how many chairs of each type can the company make?

When we encounter motion problems involving rate, time, or distance, it is useful to recall the formula $D = RT$ or distance = (rate)(time).

EXAMPLE 3   An airplane travels between two cities that are 1500 miles apart. The trip against the wind takes 3 hours. The return trip with the wind takes $2\frac{1}{2}$ hours. What is the speed of the plane in still air (in other words, how fast would the plane travel if there were no wind)? What is the speed of the wind?

1. *Understand the problem.*

Our unknowns are the speed of the plane in still air and the speed of the wind.

Let

$$x = \text{the speed of the plane in still air and}$$
$$y = \text{the speed of the wind.}$$

Let's make a sketch to help us see how these speeds are related to one another. When we travel against the wind, the wind is slowing us down. Since the wind speed opposes the plane's speed in still air, we must subtract: $x - y$.

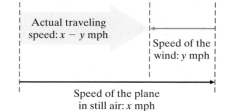

When we travel with the wind, the wind is helping us travel forward. Thus, the wind speed is added to the plane's speed in still air, and we add: $x + y$.

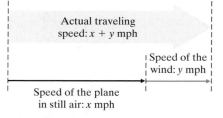

2. ***Write a system of two equations.***
   To help us write our equations, we organize the information in a chart. The chart will be based on the formula $RT = D$, which is (rate)(time) = distance.

	R	$\cdot$ T	$= D$
Flying against the wind	$x - y$	3	1500
Flying with the wind	$x + y$	2.5	1500

   Using the rows of the chart, we obtain a system of equations.

   $$(x - y)(3) = 1500$$
   $$(x + y)(2.5) = 1500$$

   If we remove the parentheses, we will obtain the following system.

   $$3x - 3y = 1500 \quad \textbf{(1)}$$
   $$2.5x + 2.5y = 1500 \quad \textbf{(2)}$$

3. ***Solve the system of equations and state the answer.***
   It will be helpful to clear equation **(2)** of decimal coefficients. Although we could multiply each term by 10, doing so will result in large coefficients on x and y. For this equation, multiplying by 2 is a better choice.

   $$3x - 3y = 1500 \quad \textbf{(1)}$$
   $$5x + 5y = 3000 \quad \textbf{(3)}$$

   If we multiply equation **(1)** by 5 and equation **(3)** by 3, we will obtain the following system.

   $$15x - 15y = 7500$$
   $$\underline{15x + 15y = 9000}$$
   $$30x \quad\quad = 16{,}500$$
   $$x = 550$$

Substituting this result in equation **(1)**, we obtain the following:

$$3(550) - 3y = 1500$$
$$1650 - 3y = 1500$$
$$-3y = -150$$
$$y = 50$$

Thus, the speed of the plane in still air is 550 miles per hour, and the speed of the wind is 50 miles per hour.

4. *Check.*
   The check is left to the student.

Practice Problem 3   An airplane travels west from city A to city B against the wind. It takes 3 hours to travel 1950 kilometers. On the return trip the plane travels east from city B to city C, a distance of 1600 kilometers in a time of 2 hours. On the return trip the plane travels with the wind. What is the speed of the plane in still air? What is the speed of the wind?

2 *Solving an Applied Exercise Requiring the Use of a System of Three Linear Equations with Three Unknowns*

EXAMPLE 4   A trucking firm has three sizes of trucks. The biggest truck holds 10 tons of gravel; the next size holds 6 tons, and the smallest holds 4 tons. The firm's manager has fifteen trucks available to haul 104 tons of gravel. However, to reduce fuel costs she wants to use two more of the fuel-efficient, 10-ton trucks than the 6-ton trucks. Her assistant tells her that she has two more 10-ton trucks than 6-ton trucks available. How many trucks of each type should she use?

1. *Understand the problem.*
   Since we need to find three things (the numbers of 10-ton trucks, 6-ton trucks, and 4-ton trucks), it would be helpful to have three variables. Let

   x = the number of 10-ton trucks used,

   y = the number of 6-ton trucks used, and

   z = the number of 4-ton trucks used.

2. *Write a system of three equations.*
   We know that fifteen trucks will be used; hence, we have the following:

   $$x + y + z = 15 \quad \textbf{(1)}$$

   How can we get our second equation? Well, we also know the *capacity* of each truck type, and we know the total tonnage to be hauled. The first type of truck hauls 10 tons, the second type 6 tons, and the third type 4 tons, and the total tonnage is 104 tons. Hence, we can write the following:

   $$10x + 6y + 4z = 104 \quad \textbf{(2)}$$

   We still need one more equation. What other given information can we use? The problem states that the manager wants to use two more 10-ton trucks than 6-ton trucks. Thus, we have the following:

   $$x = 2 + y \quad \textbf{(3)}$$

   (We could also have written $x - y = 2$.) Hence, our system of equations is as follows:

   $$x + y + z = 15 \quad \textbf{(1)}$$
   $$10x + 6y + 4z = 104 \quad \textbf{(2)}$$
   $$x - y = 2 \quad \textbf{(3)}$$

Graphing Calculator

Exploration
A visual interpretation of two equations in two unknowns is sometimes helpful. Study Example 3. Graph the two equations.

$$3x - 3y = 1500$$
$$2.5x + 2.5y = 1500$$

What is the significance of the point of intersection? If you were an air traffic controller, how would you interpret the linear equation $3x - 3y = 1500$? Why would this be useful? How would you interpret $2.5x + 2.5y = 1500$? How would this be useful?

3. *Solve the system of equations and state the answers.*

Equation **(3)** doesn't contain the variable z. Let's work with equations **(1)** and **(2)** to eliminate z. First, we multiply equation **(1)** by -4 and add it to equation **(2)**.

$$
\begin{array}{rl}
-4x - 4y - 4z = -60 & \textbf{(4)} \\
\underline{10x + 6y + 4z = 104} & \textbf{(2)} \\
6x + 2y = 44 & \textbf{(5)}
\end{array}
$$

Make sure you understand how we got equation **(5)**. Dividing each term of equation **(5)** by 2 and adding to equation **(3)** gives the following:

$$
\begin{array}{rl}
3x + y = 22 & \textbf{(6)} \\
\underline{x - y = 2} & \textbf{(3)} \\
4x = 24 & \\
x = 6 &
\end{array}
$$

For $x = 6$, equation **(3)** yields the following:

$$
\begin{array}{r}
6 - y = 2 \\
4 = y
\end{array}
$$

Now we substitute the known x- and y-values into equation **(1)**.

$$
\begin{array}{r}
6 + 4 + z = 15 \\
z = 5
\end{array}
$$

Thus, the manager needs six 10-ton trucks, four 6-ton trucks, and five 4-ton trucks.

4. *Check.*

The check is left to the student.

Practice Problem 4   A factory uses three machines to wrap boxes for shipment. Machines A, B, and C can wrap 260 boxes in 1 hour. If machine A runs 3 hours and machine B runs 2 hours, they can wrap 390 boxes. If machine B runs 3 hours and machine C runs 4 hours, 655 boxes can be wrapped. How many boxes per hour can each machine wrap?

Developing Your Study Skills

Applications or Word Problems

Applications, or word problems, are the very life of mathematics! They are the reason for doing mathematics because they teach you how to put into use the mathematical skills you have developed. Learning mathematics without ever doing word problems is similar to learning all the skills of a sport without ever playing a game or learning all the notes on an instrument without ever playing a song.

The key to success is practice. Make yourself work through as many exercises as you can. You may not be able to do them all correctly at first, but keep trying. Do not give up when you reach a difficult exercise. If you cannot solve it, just try another one. Then go back and try the "difficult" one again later.

A misconception among students when they begin studying word problems is that each one is different. At first the exercises may seem this way, but as you practice more and more, you will begin to see the similarities, the different "types." You will see patterns, which will enable you to solve exercises of a given type more easily.

Applications

Use a system of two linear equations to solve each exercise.

1. An employment agency specializing in temporary construction help pays heavy equipment operators $140 per day and general laborers $90 per day. If thirty-five people were hired and the payroll was $3950, how many heavy equipment operators were employed? How many laborers?

2. A Broadway performance of *The Phantom of the Opera* had a paid attendance of 320 people. Balcony tickets cost $42, and orchestra tickets cost $64. Ticket sales receipts totaled $16,630. How many tickets of each type were sold?

3. Ninety-eight passengers rode in an Amtrak train from Boston to Denver. Tickets for regular coach seats cost $120. Tickets for sleeper car seats cost $290. The receipts for the trip totaled $19,750. How many passengers purchased each type of ticket?

4. The Tupper Farm has 450 acres of land allotted for raising corn and wheat. The cost to cultivate corn is $42 per acre. The cost to cultivate wheat is $35 per acre. The Tuppers have $16,520 available to cultivate these crops. How many acres of each crop should the Tuppers plant?

5. A large company wants to train its managers to use new word processing and spreadsheet software. Managers with computer experience can learn the word processing software in 2 hours and the spreadsheet software in 3 hours. Managers without computer experience require 5 hours to learn the word processing software and 8 hours to learn the spreadsheet software. The company can afford to pay for 190 hours of word processing instruction and 295 hours of spreadsheet instruction. How many of each type of manager can the company train?

6. During a time study, a company that makes auto radar detectors found that its basic model requires 3 hours of manufacturing time for the inside components and 2 hours of manufacturing time for the housing and controls. The advanced model needs 5 hours of manufacturing time for the inside components and 3 hours manufacturing time for the housing and controls. The production division has 1050 hours available this week to manufacture inside components and 660 hours available to manufacture housing and controls. How many detectors of each type can be made?

7. A farmer has several packages of fertilizer for his new grain crop. The old packages contain 50 pounds of long-term-growth supplement and 60 pounds of weed killer. The new packages contain 65 pounds of long-term-growth supplement and 45 pounds of weed killer. Using past experience, the farmer estimates that he needs 3125 pounds of long-term-growth supplement and 2925 pounds of weed killer for the fields. How many old packages of fertilizer and how many new packages of fertilizer should he use?

8. A staff hospital dietician has two prepackaged mixtures of vitamin additives available for patients. Mixture 1 contains 5 grams of vitamin C and 3 grams of niacin; mixture 2 contains 6 grams of vitamin C and 5 grams of niacin. On an average day she needs 87 grams of niacin and 117 grams of vitamin C. How many packets of each mixture will she need?

9. On Monday, Harold picked up three doughnuts and four large coffees for the office staff. He paid $4.91. On Tuesday, Melinda picked up five doughnuts and six large coffees for the office staff. She paid $7.59. What is the cost of one doughnut? What is the cost of one large coffee?

10. A local department store is preparing four-color sales brochures to insert into the *Salem Evening News*. The printer has a fixed charge to set up the printing of the brochure and a specific per-copy amount for each brochure printed. He quoted a price of $1350 for printing five thousand brochures and a price of $1750 for printing seven thousand brochures. What is the fixed charge to set up the printing? What is the per-copy cost for printing a brochure?

11. Against the wind a small plane flew 210 miles in 1 hour and 10 minutes. The return trip took only 50 minutes. What was the speed of the wind? What was the speed of the plane in still air?

12. Don Williams uses his small motorboat to go 8 miles upstream to his favorite fishing spot. Against the current, the trip takes $\frac{2}{3}$ hour. With the current, the trip takes $\frac{1}{2}$ hour. How fast can the boat travel in still water? What is the speed of the current?

13. Against the wind a commercial airline in South America flew 630 miles in 3 hours and 30 minutes. With a tailwind the return trip took 3 hours. What was the speed of the wind? What was the speed of the plane in still air?

14. It took Linda and Alice 4 hours to travel 24 miles downstream by canoe on Indian River. The next day they traveled for 6 hours upstream for 18 miles. What was the rate of the current? What was their average speed in still water?

15. Kobe Bryant scored 32 points in an NBA basketball game without scoring any 3-point shots. He scored 21 times. He made several free throws worth 1 point each and several regular shots from the floor, which were worth 2 points each. How many free throws did he make? How many 2-point shots did he make?

16. Shaquille O'Neal scored 38 points in a recent basketball game. He scored no free throws, but he made a number of 2-point shots and 3-point shots. He scored 16 times. How many 2-point baskets did he make? How many 3-point baskets did he make?

17. Carlos has found that his new Neon gets 32 miles per gallon on the highway and 24 miles per gallon in the city. He recently drove 432 miles on 16 gallons of gasoline. How many miles did he drive on the highway? How many miles did he drive in the city?

18. Brenda has found that her new Ford Escort gets 35 miles per gallon on the highway but only 21 miles per gallon in the city. She recently drove 420 miles on 14 gallons of gasoline. How many miles did she drive on the highway? How many miles did she drive in the city?

19. This year the state highway department in Montana purchased 256 identical cars and 183 identical trucks for official use. The purchase price was $5,791,948. Because of a budget shortfall, next year the department plans to purchase only 64 cars and 107 trucks. It will be charged the same price for each car and for each truck. Next year it plans to spend $2,507,612. How much does the department pay for each car and for each truck?

20. A recent concert at Gordon College had a paid audience of 987 people. Advance tickets were $9.95 and tickets at the door were $12.95. A total of $10,738.65 was collected in ticket sales. How many of each type of ticket were sold?

Use a system of three linear equations to solve Exercises 21–30.

21. The White Mountain Ski Lodge has three types of vehicles. The large van holds 15 passengers. The Dodge minivan holds 7 passengers. The Ford Explorer holds 5 passengers. The lodge has a total of 14 vehicles, and all together they can carry 98 passengers. The total number of large vans and Dodge minivans is 6. How many of each type of vehicle does the White Mountain Ski Lodge have?

22. Doug Camp's trucking firm has three types of trucks. One holds 10 tons of gravel; one holds 5 tons of gravel, and one holds 3 tons of gravel. Today this firm has twelve trucks on the road. They are available to haul 78 tons of gravel today. The total number of 5-ton trucks and 3-ton trucks on the road today is eight. How many of each type of truck is on the road?

23. A total of three hundred people attended the high school play. The admission prices were $5 for adults, $3 for high school students, and $2 for any children not yet in high school. The ticket sales totaled $1010. The school principal suggested that next year they raise prices to $7 for adults, $4 for high school students, and $3 for children not yet in high school. He said that if exactly the same number of people attend next year, the ticket sales at the higher prices will total $1390. How many adults, high school students, and children not yet in high school attended this year?

24. The college conducted a CPR training class for students, faculty, and staff. Faculty were charged $10, staff were charged $8, and students were charged $2 to attend the class. A total of four hundred people came. The receipts for all who attended totaled $2130. The college president remarked that if he had charged faculty $15 and staff $10 and let students come free, the receipts this year would have been $2425. How many students, faculty, and staff came to the CPR training class?

25. A total of twelve thousand passengers normally ride the green line of the MBTA during the morning rush hour. The token prices for a ride are $0.25 for children under 12, $1 for adults, and $0.50 for senior citizens, and the revenue from these riders is $10,700. If the token prices were raised to $0.35 for children under 12 and $1.50 for adults, and the senior citizen price were unchanged, the expected revenue from these riders would be $15,820. How many riders in each category normally ride the green line during the morning rush hour?

26. The owner of Danvers Ford found that he sold a total of 520 cars, Windstars, and Explorers last year. He paid the sales staff a commission of $100 for every car, $200 for every Windstar, and $300 for every Explorer sold. The total of these commissions last year was $87,000. In the coming year he is contemplating an increase so that the commission will be $150 for every car and $250 for every Windstar, with no change in the commission for Explorer sales. If the sales are the same this year as they were last year, the commissions will total $106,500. How many vehicles in each category were sold last year?

27. One of the favorite meeting places for local college students is Nick's Roast Beef in Beverly, Massachusetts. Last night from 8 PM to 9 PM Nick served twenty-four roast beef sandwiches. He sliced 15 pounds 8 ounces of roast beef to make these sandwiches and collected $82 for them. The medium roast beef sandwich has 6 ounces of beef and costs $2.50. The large roast beef sandwich has 10 ounces of beef and costs $3. The extra large roast beef sandwich has 14 ounces of beef and costs $4.50. How many of each size of roast beef sandwich did Nick sell from 8 PM to 9 PM?

28. The Essex House of Pizza delivered twenty pepperoni pizzas to Gordon College on the first night of final exams. The cost of these pizzas totaled $181. A small pizza costs $5 and contains 3 ounces of pepperoni. A medium pizza costs $9 and contains 4 ounces of pepperoni. A large pizza costs $12 and contains 5 ounces of pepperoni. The owner of the pizza shop used 5 pounds 2 ounces of pepperoni in making these twenty pizzas. How many pizzas of each size were delivered to Gordon College?

29. Sunshine Fruit Company packs three types of gift boxes of oranges, pink grapefruit, and white grapefruit. Box *A* contains 10 oranges, 3 pink grapefruit, and 3 white grapefruit. Box *B* contains 5 oranges, 2 pink grapefruit, and 3 white grapefruit. Box *C* contains 4 oranges, 1 pink grapefruit, and 2 white grapefruit. The shipping manager has available 51 oranges, 16 pink grapefruit, and 23 white grapefruit. How many gift boxes of each type can she prepare?

30. A company packs three types of packages in a large shipping box. Package type *A* weighs 0.5 pound; package type *B* weighs 0.25 pound, and package type *C* weighs 1.0 pound. Seventy-five packages weighing a total of 42 pounds were placed into the box. The number of type *A* packages was three less than the total number of types *C* and *B* combined. How many packages of each type were placed into the box?

Solve each of the following exercises by using three equations and three unknowns. Recall that the sum of the measures of the angles of a triangle is 180°.

▲ **31.** An engineer constructed a triangular piece of metal for the wheel housing of a jet airliner. He measured each angle of the piece of metal. The sum of the measures of the first angle and the second angle is half the measure of the third angle. The sum of the measures of the second angle and the third angle is eight times the measure of the first angle. Find the measure of each angle.

▲ **32.** In Caribou, Maine, Nancy Alberto owns a triangular field. She and her husband measured each angle of the field. The sum of the measures of the first and second angles is 100° more than the measure of the third angle. If the measure of the third angle is subtracted from the measure of the second angle, the result is 30° greater than the measure of the first angle. Find the measure of each angle.

To Think About

Use a system of four linear equations and four unknowns to solve the following problem.

33. A scientist at the University of Chicago is performing an experiment to determine how to increase the life span of mice through a controlled diet. The mice need 134 grams of carbohydrates, 150 grams of protein, 178 grams of fat, and 405 grams of moisture during the length of the experiment. The food is available in four packets, as shown in the table. How many packets of each type should the scientist use?

	Packet			
Contents	*A*	*B*	*C*	*D*
Carbohydrates	42	20	0	10
Protein	20	10	20	0
Fat	34	0	10	20
Moisture	50	35	30	40

Cumulative Review Problems

Solve for the variable indicated.

34. $\dfrac{1}{3}(4 - 2x) = \dfrac{1}{2}x - 3$

35. $0.06x + 0.15(0.5 - x) = 0.04$

36. $2(y - 3) - (2y + 4) = -6y$

1. Graphing a System of Linear Inequalities

We learned how to graph a linear inequality in two variables in Section 3.5. We call two linear inequalities in two variables a **system of linear inequalities in two variables**. We now consider how to graph such a system. The solution to a system of inequalities is the intersection of the solution sets of individual inequalities.

EXAMPLE 1   Graph the solution of the system.

$$y \le -3x + 2$$
$$-2x + y \ge -1$$

In this example, we will first graph each inequality separately. The graph of $y \le -3x + 2$ is the region on or below the line $y = -3x + 2$.

The graph of $-2x + y \ge -1$ consists of the region on or above the line $-2x + y = -1$.

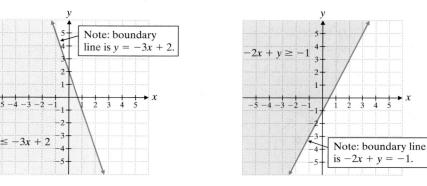

We will now place these graphs on one rectangular coordinate system. The darker shaded region is the intersection of the two graphs. Thus, the solution to the system of two inequalities is the darker shaded area and its boundary lines.

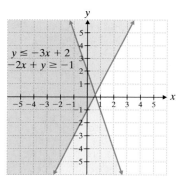

PRACTICE PROBLEM 1

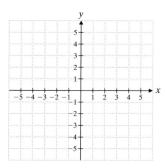

Practice Problem 1   Graph the solution of the system.

$$-2x + y \le -3$$
$$x + 2y \ge 4$$

Usually we sketch the graphs of the individual inequalities on one set of axes. We will illustrate that concept with the following example.

EXAMPLE 2   Graph the solution of the system.

$$y < 4$$
$$y > \frac{3}{2}x - 2$$

The region $y < 4$ is the area below the line $y = 4$. It does not include the line since we have the $<$ symbol. Thus, we use a dashed line to indicate that the boundary line is not part of the answer. The region $y > \frac{3}{2}x - 2$ is the area above the line $y = \frac{3}{2}x - 2$. Again, we use the dashed line to indicate that the boundary line is not part of the answer. The final solution is the darker shaded area. The solution does *not* include the dashed boundary lines.

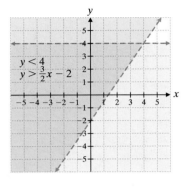

PRACTICE PROBLEM 2

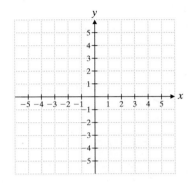

Practice Problem 2   Graph the solution of the system.

$$y > -1$$
$$y < -\tfrac{3}{4}x + 2$$

There are times when we require the exact location of the point where two boundary lines intersect. In these cases the boundary points are labeled on the final sketch of the solution.

EXAMPLE 3   Graph the solution to the following system of inequalities. Find the coordinates of any point where boundary lines intersect.

$$x + y \le 5$$
$$x + 2y \le 8$$
$$x \ge 0$$
$$y \ge 0$$

The region $x + y \le 5$ is the region on and under the line $x + y = 5$. The region $x + 2y \le 8$ is the region on and under the line $x + 2y = 8$. We solve the system containing the equations $x + y = 5$ and $x + 2y = 8$ to find that their point of intersection is $(2, 3)$. The region $x \ge 0$ is the y-axis and all the region to the right of the y-axis. The region $y \ge 0$ is the x-axis and all the region above the x-axis. Thus, the solution to the system is the shaded region and its boundary lines. The boundary lines intersect at four points. These points are called the **vertices** of the solution. Thus, the vertices of the solution are $(0, 0)$, $(0, 4)$, $(2, 3)$, and $(5, 0)$.

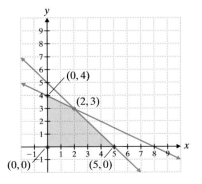

Graphing Systems of Linear Inequalities

On some graphing calculators, you can graph a system of linear inequalities. To graph the system in Example 1, first rewrite it as follows:

$$y \leq -3x + 2$$

$$y \geq 2x - 1$$

Enter each expression into the Y = editor of your graphing calculator and then select the appropriate direction for shading.

Display:

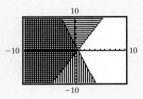

Notice that one inequality is shaded vertically and the other is shaded horizontally. The intersection is the solution. Note also that the graphing calculator will not indicate whether the boundary of the solution region is included in the solution.

Practice Problem 3   Graph the solution to the system of inequalities. Find the vertices of the solution.

$$x + y \leq 6$$
$$3x + y \leq 12$$
$$x \geq 0$$
$$y \geq 0$$

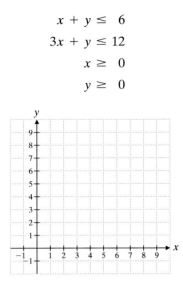

Developing Your Study Skills

Exam Time: How To Review

Reviewing adequately for an exam enables you to bring together the concepts you have learned over several sections. For your review, you will need to:

1. Reread your textbook. Make a list of any terms, rules, or formulas you need to know for the exam. Be sure you understand them all.

2. Reread your notes. Go over returned homework and quizzes. Redo the exercises you missed.

3. Practice some of each type of exercise covered in the chapter(s) you are to be tested on.

4. Use the end-of-chapter materials provided in your textbook. Read carefully through the Chapter Organizer. Take the Chapter Test. When you are finished, check your answers. Redo any exercises you missed.

5. Get help if any concepts give you difficulty.

Graph the solution for each of the following systems.

1. $y \geq 2x - 1$
    $x + y \leq 6$

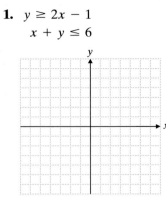

2. $y \geq x - 3$
    $x + y \geq 2$

3. $y \geq -4x$
    $y \geq 3x - 2$

4. $y \geq x$
    $y \geq -x$

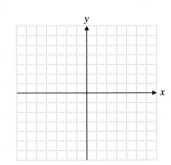

5. $y \geq 2x - 3$
    $y \leq \dfrac{2}{3} x$

6. $y \leq \dfrac{1}{2} x - 3$
    $y \geq -\dfrac{1}{2} x$

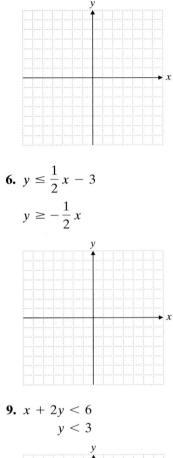

7. $x - y \geq -1$
    $-3x - y \leq \ 4$

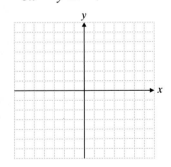

8. $2x + \ y \leq 3$
    $x - 2y \leq 4$

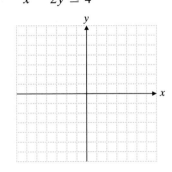

9. $x + 2y < 6$
    $y < 3$

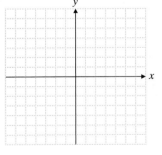

10. $x + 3y \leq 12$
    $y \geq \ 4$

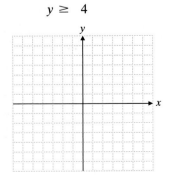

11. $y < \ 4$
    $x > -2$

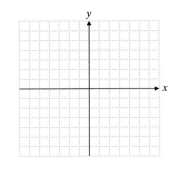

12. $y > -3$
    $x < \ 2$

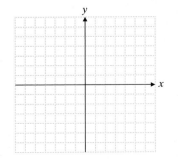

13. $3x + 2y < 6$
$3x + 2y > -6$

14. $2x - y < 2$
$2x - y > -2$

15. $x - 4y \geq -4$
$3x + y \leq 3$

16. $5x - 2y \leq 10$
$x - y \geq -1$

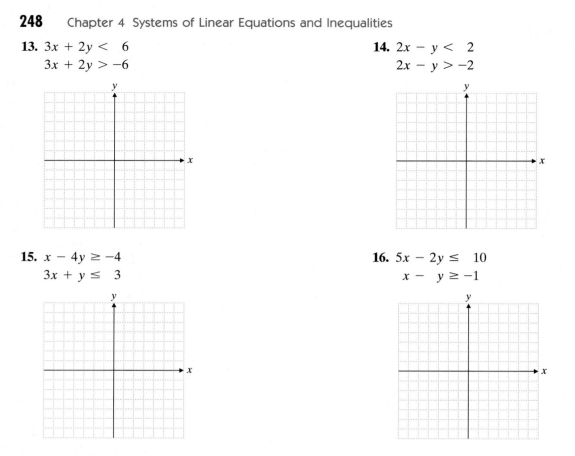

Graph the solution to the following systems of inequalities. Find the vertices of the solution.

17. $x + y \leq 5$
$2x - y \geq 1$

18. $x + y \geq 2$
$y + 4x \leq -1$

19. $x + 3y \leq 12$
$y < x$

20. $x + 2y \leq 4$
$y < -x$

21. $x + y \geq 1$
$x - y \geq 1$
$x \geq 3$

22. $x - y \leq 2$
$x + y \leq 2$
$x \geq -2$

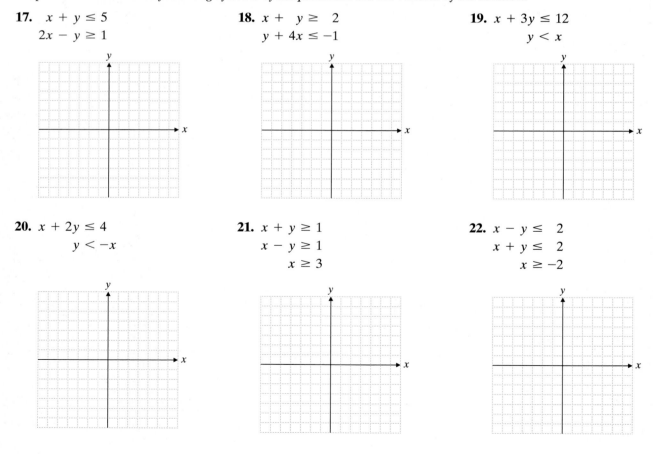

To Think About

Graph the region determined by each of the following systems.

23.
$$y \le 3x + 6$$
$$4y + 3x \le 3$$
$$x \ge -2$$
$$y \ge -3$$

24.
$$-x + y \le 100$$
$$x + 3y \le 150$$
$$x \ge 0$$
$$y \ge 20$$

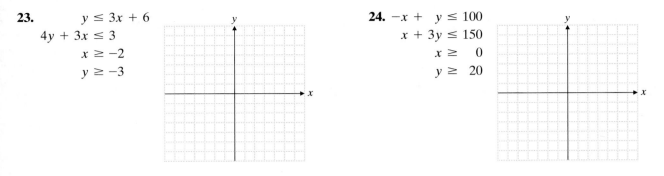

Applications

Hint: In Exercises 25 and 26, if the coordinates of a boundary point contain fractions, it is wise to obtain the point of intersection algebraically rather than graphically.

25. The equation that represents the proper level of medical staffing in the cardiac care unit of a local hospital is $N \le 2D$, where N is the number of nurses on duty and D is the number of doctors on duty. In order to control costs, the equation $4N + 3D \le 20$ is appropriate. The number of doctors and nurses on duty at any time cannot be negative, so $N \ge 0$ and $D \ge 0$.

(a) Graph the region satisfying all of the medical requirements for the cardiac care unit. Use the following special graph grids, where D is measured on the horizontal axis and N is measured on the vertical axis.

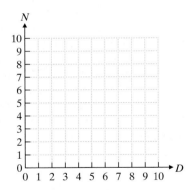

(b) If there are three doctors and two nurses on duty in the cardiac care unit, are all of the medical requirements satisfied?

(c) If there is one doctor and four nurses on duty in the cardiac care unit, are all of the medical requirements satisfied?

26. The equation that represents the proper traffic control and emergency vehicle response availability in the city of Salem is $P + 3F \le 18$, where P is the number of police cars on active duty and F is the number of fire trucks that have left the firehouse and are involved in a response to a call. In order to comply with staffing limitations, the equation $4P + F \le 28$ is appropriate. The number of police cars on active duty and the number of fire trucks that have left the firehouse cannot be negative, so $P \ge 0$ and $F \ge 0$.

(a) Graph the regions satisfying all of the availability and staffing limitation requirements for the city of Salem. Use the following special graph grids, where P is measured on the horizontal axis and F is measured on the vertical axis.

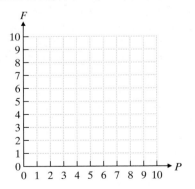

(b) If four police cars are on active duty and four fire trucks have left the firehouse in response to a call, are all of the requirements satisfied?

(c) If two police cars are on active duty and six fire trucks have left the firehouse in response to a call, are all of the requirements satisfied?

Cumulative Review

27. Find the slope of the line passing through $(3, -4)$ and $(-1, -2)$.

28. What is the slope and the y-intercept of the line defined by the equation $3x + 4y = -8$?

29. Find the equation of a line passing through $(2, 6)$ and $(-2, 1)$. Write the equation in slope-intercept form.

30. Find the equation of a line passing through $(5, 0)$ that is parallel to the line $y + 2x = 8$.

31. The Cape Cod Cinema took in $23,400 on two rainy days and five sunny days. The next week the cinema took in $25,800 on four rainy days and three sunny days. What is the average amount of money taken in when a film is shown at the Cape Cod Cinema on a rainy day? On a sunny day?

32. Illinois is establishing a number of new bicycle trails throughout the state. When a group of volunteers worked three days and a group of experienced professionals worked four days, they were able to establish 389 feet of bicycle trails. When the same group of volunteers worked five days and the same group of experienced professionals worked seven days, they were able to establish 670 feet of bicycle trails. How many feet per day are established by the group of volunteers? By the group of experienced professionals?

33. Hector sells televisions at a local department store. He earns $200 per week plus a commission of 5% on his total sales. His brother Fernando sells automobile tires at a tire store. His salary is $100 per week plus a commission of 8% on his total sales. After they had both worked several weeks, they discovered that they had sold the same dollar value in sales. However, their total earnings were quite different. Hector had earned $7400, and Fernando had earned $9200. How many weeks had each worked? What was the dollar value of the amount of sales that each had made?

34. Two weeks ago Larry went from his office to Nick's Roast Beef and bought 3 roast beef sandwiches, 2 orders of french fries, and 3 sodas for $13.85. One week ago it was Alice's turn to take the trip to Nick's. She bought 4 roast beef sandwiches, 3 orders of french fries, and 5 sodas for $20. This week Roberta made the trip. She purchased 3 roast beef sandwiches, 3 orders of french fries, and 4 sodas for $16.55. What is the cost of one roast beef sandwich? Of one order of french fries? Of one soda?

Math in the Media

Harvest Costs

Ideal weather, planting conditions, and improved technology resulted in a record-breaking potato harvest in 2000. This surplus resulted in record low prices, drastically affecting growers.

In the March 17, 2001, *New York Times Late Edition*, the reported price for a 100-pound sack of fresh potatoes had dropped to $1.00 from $8.00. Production costs for a 100-pound sack—typically $5.00—were anticipated to rise due to increasing energy costs.

Using the skills you learned in this chapter, try calculating cultivating costs in questions 1–3 below. Visit the text companion Web site for links to further information on the consequences of the surplus.

EXERCISES

1. A farmer sells two lots of potatoes totaling 7,000 100-pound sacks. The better grade of potato sells for $3.25 per sack and the lesser grade sells for $2.25 per sack. The two lots together sell for $20,000. How many sacks were in each lot?

2. A farmer has 400 acres on which he can raise rye and/or alfalfa. The cost to cultivate the alfalfa is $32 per acre. Because of the market situation, he wants to plant 150 acres in rye and the rest of the land in alfalfa. He wants to raise both crops on a budget of $14,600. He calculates that he is able to do this. What was the cost per acre of rye cultivation that he uses in the calculation?

3. If potatoes were selling in the supermarket for about 89 cents per pound in March of 2001, approximately what percent of that amount could be attributed to the actual cost of raising potatoes?

Chapter 4 Organizer

Topic	Procedure	Examples
Finding a solution to a system of equations by the graphing method, p. 213.	1. Graph the first equation. 2. Graph the second equation. 3. Approximate from your graph where the two lines intersect, if they intersect at one point. 4. If the lines are parallel, there is no solution. If the lines coincide, there are an infinite number of solutions.	Solve the system by graphing. $$x + y = 6$$ $$2x - y = 6$$ We graph each line and determine from our sketch that the lines intersect at $(4, 2)$. The solution is $(4, 2)$.
Solving a system of two linear equations by the substitution method, p. 215.	The substitution method is most appropriate when *at least one variable has a coefficient of 1 or −1.* 1. Solve for one variable in one of the equations. 2. In the other equation, replace that variable with the expression you obtained in step 1. 3. Solve the resulting equation. 4. Substitute the numerical value you obtain for a variable into one of the original equations and solve for the other variable. 5. Check the solution in both original equations.	Solve: $2x + y = 11$ **(1)** $x + 3y = 18$ **(2)** $y = 11 - 2x$ from equation **(1)**. Substitute this into equation **(2)**. $$x + 3(11 - 2x) = 18$$ $$x + 33 - 6x = 18$$ $$-5x = -15$$ $$x = 3$$ Substitute $x = 3$ into $2x + y = 11$. $$2(3) + y = 11$$ $$y = 5$$ The solution is $(3, 5)$.
Solving a system of two linear equations by the addition method, p. 217.	The addition method is most appropriate when the variables *all have coefficients other than 1 or −1.* 1. Arrange each equation in the form $ax + by = c$. 2. Multiply one or both equations by appropriate numerical values so that when the two resulting equations are added, one variable is eliminated. 3. Solve the resulting equation. 4. Substitute the numerical value you obtain for the variable into one of the original equations. 5. Solve this equation to find the other variable.	Solve: $2x + 3y = 5$ **(1)** $-3x - 4y = -2$ **(2)** Multiply equation **(1)** by 3 and equation **(2)** by 2. $$6x + 9y = 15$$ $$\underline{-6x - 8y = -4}$$ $$y = 11$$ Substitute $y = 11$ into equation **(1)**. $$2x + 3(11) = 5$$ $$2x + 33 = 5$$ $$2x = -28$$ $$x = -14$$ The solution is $(-14, 11)$.

Topic	Procedure	Examples
Solving a system of three linear equations by algebraic methods, p. 226.	If there is one solution to a system of three linear equations in three unknowns, it may be obtained in the following manner. 1. Choose two equations from the system. 2. Multiply one or both of the equations by the appropriate constants so that by adding the two equations together, one variable can be eliminated. 3. Choose a *different* pair of the three original equations and eliminate the *same* variable using the procedure of step 2. 4. Solve the system formed by the two equations resulting from steps 2 and 3 for both variables. 5. Substitute the two values obtained in step 4 into one of the original three equations to find the third variable.	Solve: $\quad 2x - y - 2z = -1 \quad$ **(1)** $\qquad\quad x - 2y - z = 1 \quad$ **(2)** $\qquad\quad x + y + z = 4 \quad$ **(3)** Add equations **(2)** and **(3)** together to eliminate z. $\qquad\qquad 2x - y = 5 \qquad$ **(4)** Multiply equation **(3)** by 2 and add to equation **(1)**. $\qquad\qquad 4x + y = 7 \qquad$ **(5)** Add equations **(4)** and **(5)**. $\qquad\qquad 2x - y = 5$ $\qquad\qquad \underline{4x + y = 7}$ $\qquad\qquad\quad 6x = 12$ $\qquad\qquad\quad\ x = 2$ Substitute $x = 2$ into equation **(5)**. $\qquad\qquad 4(2) + y = 7$ $\qquad\qquad\qquad\ \ y = -1$ Substitute $x = 2$, $y = -1$ into equation **(3)**. $\qquad\qquad 2 + (-1) + z = 4$ $\qquad\qquad\qquad\qquad\ z = 3$ The solution is $(2, -1, 3)$.
Inconsistent system of equations, p. 219.	If there is no solution to a system of linear equations, the system of equations is inconsistent. When you try to solve an inconsistent system, you obtain an equation that is not true, such as $0 = 5$.	Attempt to solve the system. $\qquad\qquad 4x + 3y = 10 \qquad$ **(1)** $\qquad\quad -8x - 6y = 5 \qquad$ **(2)** Multiply equation **(1)** by 2 and add to equation **(2)**. $\qquad\qquad 8x + 6y = 20$ $\qquad\quad \underline{-8x - 6y = \ \ 5}$ $\qquad\qquad\qquad\quad 0 = 25$ But $0 \neq 25$. Thus, there is no solution. The system of equations is inconsistent.
Dependent equations, p. 219.	If there are an *infinite number of solutions* to a system of linear equations, at least one pair of equations is dependent. When you try to solve a system that contains dependent equations, you will obtain an equation that is always true (such as $0 = 0$ or $3 = 3$). These equations are called *identities*.	Attempt to solve the system. $\qquad\qquad x - 2y = -5 \qquad$ **(1)** $\qquad\quad -3x + 6y = 15 \qquad$ **(2)** Multiply equation **(1)** by 3 and add to equation **(2)**. $\qquad\qquad 3x - 6y = -15$ $\qquad\quad \underline{-3x + 6y = \ \ 15}$ $\qquad\qquad\qquad 0 = \quad 0$ There are an infinite number of solutions. The equations are dependent.

Topic	Procedure	Examples
Graphing the solution to a system of inequalities in two variables, p. 244.	1. Determine the region that satisfies each individual inequality. 2. Shade the common region that satisfies all the inequalities.	Graph: $\quad 3x + 2y \le 10$ $\qquad\qquad -1x + 2y \ge 2$ 1. $3x + 2y \le 10$ can be graphed more easily as $y \le -\dfrac{3}{2}x + 5$. We draw a solid line and shade the region below it. $-1x + 2y \ge 2$ can be graphed more easily as $y \ge \dfrac{1}{2}x + 1$. We draw a solid line and shade the region above it. 2. The common region is shaded. 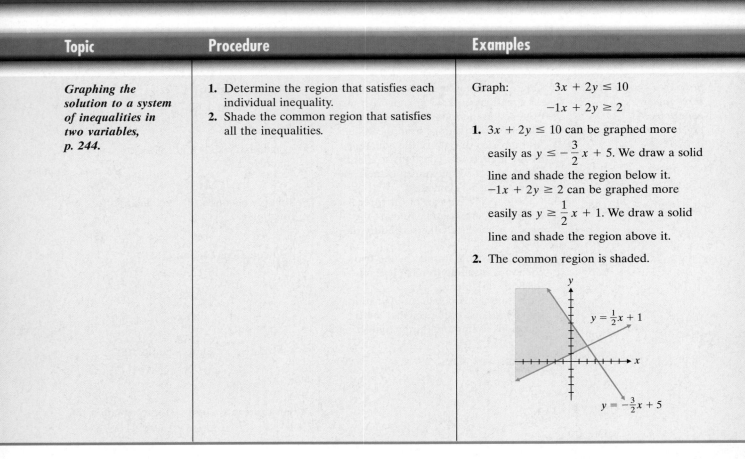

Chapter 4 Review Problems

Solve the following systems by graphing.

1. $x + 2y = 8$
$\quad x - \ y = 2$

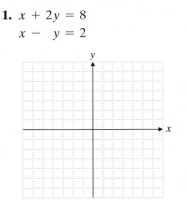

2. $\quad x + y = 2$
$\quad 3x - y = 6$

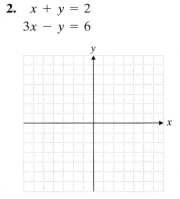

3. $2x + \ y = 6$
$\quad 3x + 4y = 4$

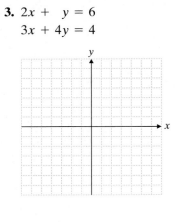

Solve the following systems by substitution.

4. $3x - 2y = -9$
$\quad 2x + \ y = \ \ 1$

5. $-6x - \ y = \ \ 1$
$\quad 3x - 4y = 31$

6. $4x + 3y = 10$
$\quad 5x - \ y = \ \ 3$

7. $-7x + \ y = -4$
$\quad 5x + 2y = 11$

Solve the following systems by addition.

8. $-2x + 5y = -12$
$\quad\;\; 3x + \;\; y = \quad 1$

9. $-4x + 2y = -16$
$\quad\;\; 3x + 5y = \;\; -1$

10. $7x - 4y = \quad 2$
$\quad\;\; 6x - 5y = -3$

11. $\quad 3x + 4y = \;\; -6$
$\quad -2x + 7y = -25$

Solve by any appropriate method. If there is no unique solution, state why.

12. $2x + 4y = 9$
$\quad\;\; 3x + 6y = 8$

13. $x + 5y = 10$
$\quad\;\; y = 2 - \dfrac{1}{5}x$

14. $7x + 6y = -10$
$\quad\;\; 2x + \;\; y = \quad 0$

15. $3x + 4y = \quad 1$
$\quad\;\; 9x - 2y = -4$

16. $\quad x + \dfrac{1}{3}y = \quad 1$
$\quad \dfrac{1}{4}x - \dfrac{3}{4}y = -\dfrac{9}{4}$

17. $\dfrac{2}{3}x + y = \quad \dfrac{14}{3}$
$\quad \dfrac{2}{3}x - y = -\dfrac{22}{3}$

18. $9a + 10b = \quad 7$
$\quad\;\; 6a - \;\; 4b = 10$

19. $3a + 5b = 8$
$\quad\;\; 2a + 4b = 3$

20. $x + 3 = 3y + 1$
$\quad\;\; 1 - 2(x - 2) = 6y + 1$

21. $10(x + 1) - 13 = -8y$
$\quad\;\; 4(2 - y) = 5(x + 1)$

22. $\quad 0.3x - 0.2y = 0.7$
$\quad -0.6x + 0.4y = 0.3$

23. $0.2x - 0.1y = 0.8$
$\quad\;\; 0.1x + 0.3y = 1.1$

Solve by an appropriate method.

24. $3x - 2y - z = \quad 3$
$\quad\;\; 2x + \;\; y + z = \quad 1$
$\quad -x - \;\; y + z = -4$

25. $-2x + \;\; y - \;\; z = -7$
$\quad\;\; x - 2y - \;\; z = \quad 2$
$\quad\;\; 6x + 4y + 2z = \quad 4$

26. $2x + 5y + \;\; z = \quad 3$
$\quad\;\; x + \;\; y + 5z = 42$
$\quad\;\; 2x + \;\; y \quad\;\;\; = \quad 7$

27. $x + 2y + z = 5$
$\quad\;\; 3x - 8y \quad\;\;\; = 17$
$\quad\;\; 2y + z = -2$

28. $2x - 4y + 3z = \quad 0$
$\quad\;\; x - 2y - 5z = 13$
$\quad\;\; 5x + 3y - 2z = 19$

29. $\quad 5x + 2y + 3z = 10$
$\quad\;\; 6x - 3y + 4z = 24$
$\quad -2x + \;\; y + 2z = \quad 2$

30. $3x + 2y \quad\quad\;\; = \quad 7$
$\quad\;\; 2x \quad\quad + 7z = -26$
$\quad\quad\quad\;\; 5y + \;\; z = \quad 6$

31. $\quad x - \;\; y \quad\quad\;\; = \quad 2$
$\quad\;\; 5x + 7y - 5z = \quad 2$
$\quad\;\; 3x - 5y + 2z = -2$

Use a system of linear equations to solve each of the following exercises.

32. A plane flies 720 miles against the wind in 3 hours. The return trip with the wind takes only $2\frac{1}{2}$ hours. Find the speed of the wind. Find the speed of the plane in still air.

33. A local company has computerized its production line. Company officials have found that it takes 25 hours to train new employees to use the computerized equipment and 8 hours to review mathematics skills. Previously laid-off employees can be trained in 10 hours to use the computerized equipment and require only 3 hours to review mathematics skills. This month the management team has 275 hours available to train people to use computerized equipment and 86 hours available to review necessary mathematics skills. How many new employees can be trained? How many previously laid-off employees can be trained this month?

34. When the circus came to town last year, they hired general laborers at $70 per day and mechanics at $90 per day. They paid $1950 for this temporary help for one day. This year they hired exactly the same number of people of each type, but they paid $80 for general laborers and $100 for mechanics for the one day. This year they paid $2200 for temporary help. How many general laborers did they hire? How many mechanics did they hire?

35. A total of 590 tickets were sold for the circus matinee performance. Children's admission tickets were $6, and adult tickets were $11. The ticket receipts for the matinee performance were $4790. How many children's tickets were sold? How many adult tickets were sold?

36. A baseball coach bought two hats, five shirts, and four pairs of pants for $129. His assistant purchased one hat, one shirt, and two pairs of pants for $42. The next week the coach bought two hats, three shirts, and one pair of pants for $63. What was the cost of each item?

37. A scientist needs three types of food packets for an experiment. Packet *A* has 2 grams of carbohydrates, 4 grams of protein, and 3 grams of fat. Packet *B* has 3 grams of carbohydrates, 1 gram of protein, and 1 gram of fat. Packet *C* has 4 grams of carbohydrates, 3 grams of protein, and 2 grams of fat. The experiment requires 29 grams of carbohydrates, 23 grams of protein, and 17 grams of fat. How many packets should she use?

38. Four jars of jelly, three jars of peanut butter, and five jars of honey cost $9.80. Two jars of jelly, two jars of peanut butter, and one jar of honey cost $4.20. Three jars of jelly, four jars of peanut butter, and two jars of honey cost $7.70. Find the cost for one jar of jelly, one jar of peanut butter, and one jar of honey.

39. The church youth group is planning a trip to Mount Washington. A total of 127 people need rides. The church has buses available that hold forty passengers, and several parents have volunteered station wagons that hold eight passengers or sedans that hold five passengers. The youth leader is planning to use nine vehicles to transport the people. One parent said that if they didn't use any buses, tripled the number of station wagons, and doubled the number of sedans, they would be able to transport 126 people. How many buses, station wagons, and sedans are they planning to use if they use nine vehicles?

Solve by any appropriate method.

40. $-x - 5z = -5$
$\quad\;\; 13x + 2z = \;\;\; 2$

41. $x + \;\; y = 10$
$\quad\;\; 6x + 9y = 70$

42. $2x + 5y = \;\;\;\; 4$
$\quad\;\; 5x - 7y = -29$

43. $\dfrac{x}{2} - 3y = -6$
$\quad\;\; \dfrac{4}{3}x + 2y = \;\;\; 4$

44. $\dfrac{3}{5}x - y = \;\; 6$
$\quad\;\; x + \dfrac{y}{3} = 10$

45. $\dfrac{x + 1}{5} = y + 2$
$\quad\;\; \dfrac{2y + 7}{3} = x - y$

46. $3(2 + x) = y + 1$
$\quad\;\; 5(x - y) = -7 - 3y$

47. $7(x + 3) = 2y + 25$
$\quad\;\; 3(x - 6) = -2(y + 1)$

48. $0.3x - 0.4y = 0.9$
$\quad\;\; 0.2x - 0.3y = 0.4$

49. $0.05x + 0.08y = -0.76$
$\quad\;\; 0.04x - 0.03y = \;\;\; 0.05$

50. $x - \dfrac{y}{2} + \dfrac{1}{2}z = -1$
$\quad\;\; 2x \quad\quad + \dfrac{5}{2}z = -1$
$\quad\;\; \dfrac{3}{2}y + \; 2z = \;\;\; 1$

51. $2x - 3y + 2z = \;\;\; 0$
$\quad\;\; x + 2y - \;\; z = \;\;\; 2$
$\quad\;\; 2x + \;\; y + 3z = -1$

52. $x - 4y + 4z = -1$
$\quad\;\; 2x - \;\; y + 5z = -3$
$\quad\;\; x - 3y + \;\; z = \;\;\; 4$

53. $x - 2y + z = \;\; -5$
$\quad\;\; 2x + \quad\quad z = -10$
$\quad\;\; y - z = \;\;\; 15$

Solve each of the following systems of linear inequalities by graphing.

54. $x - y \le 3$
$\quad\;\; y \le -\dfrac{1}{4}x + 2$

55. $-2x + 3y < \;\; 6$
$\quad\;\; y > -2$

56. $x + y > 1$
$\quad\;\; 2x - y < 5$

57. $x + y \ge 4$
$\quad\;\; y \le x$
$\quad\;\; x \le 6$

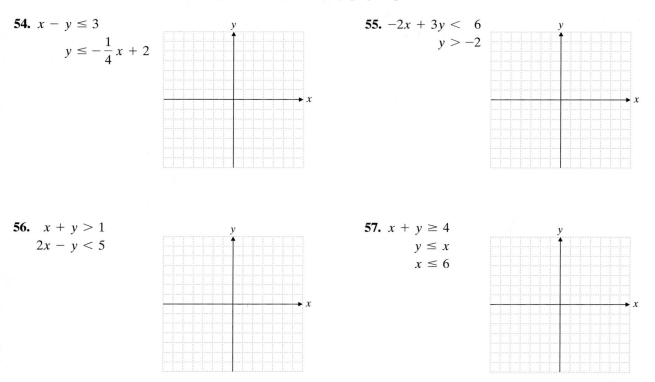

Chapter 4 Test

1. _____

2. _____

3. _____

4. _____

5. _____

6. _____

7. _____

8. _____

9. _____

10. _____

11. _____

12. _____

13 _____

14. _____

Solve each system of equations. If there is no solution to the system, give a reason.

1. $3x - 2y = -8$
$\quad x + 6y = 4$

2. $6x - 2y = -2$
$\quad 3x + 4y = 14$

3. $\frac{1}{4}a - \frac{3}{4}b = -1$
$\quad \frac{1}{3}a + b = \frac{5}{3}$

4. $7x - 1 = 3(1 + y)$
$\quad 1 - 6y = -7(2x + 1)$

5. $5x - 3y = 3$
$\quad 7x + y = 25$

6. $\frac{1}{3}x + \frac{5}{6}y = 2$
$\quad \frac{3}{5}x - y = -\frac{7}{5}$

7. $3x + 5y - 2z = -5$
$\quad 2x + 3y - z = -2$
$\quad 2x + 4y + 6z = 18$

8. $3x + 2y = 0$
$\quad 2x - y + 3z = 8$
$\quad 5x + 3y + z = 4$

9. $x + 5y + 4z = -3$
$\quad x - y - 2z = -3$
$\quad x + 2y + 3z = -5$

Use a system of linear equations to solve the following exercises.

10. A plane flew 1000 miles with a tailwind in 2 hours. The return trip against the wind took $2\frac{1}{2}$ hours. Find the speed of the wind and the speed of the plane in still air.

11. On an automobile assembly line, station wagons require 5 minutes for rustproofing, 4 minutes for painting, and 3 minutes for heat drying. Four-door sedans require 4 minutes for rustproofing, 3 minutes for painting, and 2 minutes for drying. Two-door sedans require 3 minutes for rustproofing, 3 minutes for painting, and 2 minutes for drying. The assembly line supervisor wants to find an assembly plan that uses 62 minutes of rustproofing time, 52 minutes of painting time, and 36 minutes of heat drying time. How many vehicles of each type should be sent down the assembly line?

12. Sue Miller had to move some supplies to Camp Cherith for the summer camp program. She rented a Portland Rent-A-Truck in April for 5 days and drove 150 miles. She paid $180 for the rental in April. Then in May she rented the same truck again for 7 days and drove 320 miles. She paid $274 for the rental in May. How much does Portland Rent-A-Truck charge for a daily rental of the truck? How much do they charge per mile?

Solve the following systems of linear inequalities by graphing.

13. $x + 2y \leq 6$
$\quad -2x + y \geq -2$

14. $3x + y \geq 8$
$\quad x - 2y \geq 5$

Approximately one-half of this test covers the content of Chapters 1–3.
The remainder covers the content of Chapter 4.

1. Subtract. $5\dfrac{1}{2} - 3\dfrac{7}{8}$

2. Evaluate. $(2 - 3)^3 + 20 \div (-10)\left(\dfrac{1}{5}\right)$

3. Evaluate. $(-2)^2 + (-3)^3$

4. Simplify. $2x - 4[x - 3(2x + 1)]$

5. Solve for P: $A = P(3 + 4rt)$

6. Solve for x: $\dfrac{1}{4}x + 5 = \dfrac{1}{3}(x - 2)$

7. Graph the line $4x - 8y = 10$. Plot at least three points.

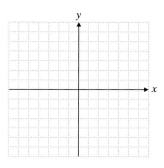

8. Find the slope of the line passing through $(6, -1)$ and $(-4, -2)$.

Solve the following linear inequalities and graph the solution on a number line.

9. $4x + 3 - 13x - 7 < 2(3 - 4x)$

10. $\dfrac{2x - 1}{3} \le 7$ and $2(x + 1) \ge 12$

11. Find the slope-intercept form of the equation of the line passing through $(2, -3)$ and perpendicular to $3x + 6y = -2$.

▲ **12.** A triangle has a perimeter of 69 meters. The second side is 7 meters longer than the first side. The third side is 6 meters shorter than double the length of the first side. Find the length of each side.

13. Victor invests $6000 in a bank. Part is invested at 7% simple interest and part at 9% simple interest. In 1 year Victor earns $510 in interest. How much did he invest at each amount?

1. _____

2. _____

3. _____

4. _____

5. _____

6. _____

7. _____

8. _____

9. _____

10. _____

11. _____

12. _____

13. _____

14. _____

15. _____

16. _____

17. _____

18. _____

19. _____

20. _____

14. Solve the system.

$$5x + 2y = 2$$
$$4x + 3y = -4$$

15. Solve the system.

$$2x + y - z = 4$$
$$x + 2y - 2z = 2$$
$$x - 3y + z = 4$$

16. Patricia bought five shirts and eight pairs of slacks for $345 at Super Discount Center. Joanna bought seven of the same shirts and three pairs of the same slacks at the same store, and her total was $237. How much did a shirt cost? How much did a pair of slacks cost?

17. Solve the system.

$$7x - 6y = 17$$
$$3x + y = 18$$

18. Solve the system.

$$x + 3y + z = 5$$
$$2x - 3y - 2z = 0$$
$$x - 2y + 3z = -9$$

19. What happens when you attempt to solve the system below? Why is this?

$$-5x + 6y = 2$$
$$10x - 12y = -4$$

20. Solve the following system of inequalities by graphing.

$$x - y \geq -4$$
$$x + 2y \geq 2$$

Exponents and Polynomials

Polynomials are an important tool of mathematics. They are used to create formulas that model many real-world phenomena. For example, using records of past student enrollments in the United States, one can create a polynomial that predicts future student enrollments. Do you think you could use your mathematical skills to predict the number of students who will be in school in the year 2009? Turn to the Putting Your Skills to Work problems on page 307 to find out.

1. _____

2. _____

3. _____

4. _____

5. _____

6. _____

7. _____

8. _____

9. _____

10. _____

11. _____

12. _____

13. _____

14. _____

15. _____

16. _____

If you are familiar with the topics in this chapter, take this test now. Check your answers with those in the back of the book. If an answer is wrong or you can't answer a question, study the appropriate section of the chapter.

If you are not familiar with the topics in this chapter, don't take this test now. Instead, study the examples, work the practice problems, and then take the test.

This test will help you to identify which concepts you have mastered and which you need to study further.

Section 5.1

Multiply. Leave your answer in exponent form.

1. $(3^6)(3^{10})$ **2.** $(-5x^3)(2x^4)$ **3.** $(-4ab^2)(2a^3b)(3ab)$

Simplify. Assume that all variables are nonzero.

4. $\dfrac{x^{26}}{x^3}$ **5.** $\dfrac{12xy^2}{-6x^3y^4}$ **6.** $\dfrac{-25a^0b^2c^3}{-15abc^2}$

7. $(x^5)^{10}$ **8.** $(-2x^2y)^3$ **9.** $\left(\dfrac{4a^3b}{c^2}\right)^3$

Section 5.2

In questions 10–12, express with positive exponents.

10. $3x^2y^{-3}z^{-4}$ **11.** $\dfrac{-2a^{-3}}{b^{-2}c^4}$ **12.** Evaluate. $(-3)^{-2}$

13. Write in scientific notation. 0.000638

14. Write in decimal notation. 1.894×10^{12}

Section 5.3

Combine.

15. $(3x^2 - 6x - 9) + (-5x^2 + 13x - 20)$

16. $(2x^3 - 5x^2 + 7x) - (x^3 - 4x + 12)$

Section 5.4

Multiply.

17. $-3x(2 - 7x)$

18. $2x^2(x^2 + 4x - 1)$

19. $(x^3 - 3xy + y^2)(5xy^2)$

20. $(x + 5)(x + 9)$

21. $(3x - 2y)(4x + 3y)$

22. $(2x^2 - y^2)(x^2 - 3y^2)$

Section 5.5

Multiply.

23. $(8x - 11y)(8x + 11y)$

24. $(5x - 3y)^2$

25. $(5ab - 6)^2$

26. $(x^2 - 3x + 2)(4x - 1)$

27. $(2x - 3)(2x + 3)(4x^2 + 9)$

Section 5.6

Divide.

28. $(42x^4 - 36x^3 + 66x^2) \div (6x)$

29. $(15x^3 - 28x^2 + 18x - 7) \div (3x - 2)$

17. _____

18. _____

19. _____

20. _____

21. _____

22. _____

23. _____

24. _____

25. _____

26. _____

27. _____

28. _____

29. _____

1 Multiplying Exponential Expressions with Like Bases

Recall that x^2 means $x \cdot x$. That is, x appears as a factor two times. The 2 is called the **exponent**. The **base** is the variable x. The expression x^2 is called an **exponential expression**. What happens when we multiply $x^2 \cdot x^2$? Is there a pattern that will help us form a general rule?

Notice that

$$(2^2)(2^3) = \overbrace{(2 \cdot 2)(2 \cdot 2 \cdot 2)}^{5 \text{ twos}} = 2^5$$

The exponent means 2 occurs 5 times as a factor.

$$(3^3)(3^4) = \overbrace{(3 \cdot 3 \cdot 3)(3 \cdot 3 \cdot 3 \cdot 3)}^{7 \text{ threes}} = 3^7$$

Notice that $3 + 4 = 7$.

$$(x^3)(x^5) = \overbrace{(x \cdot x \cdot x)(x \cdot x \cdot x \cdot x \cdot x)}^{8 \text{ } x\text{'s}} = x^8$$

The sum of the exponents is $3 + 5 = 8$.

$$(y^4)(y^2) = \overbrace{(y \cdot y \cdot y \cdot y)(y \cdot y)}^{6 \text{ } y\text{'s}} = y^6$$

The sum of the exponents is $4 + 2 = 6$.

We can state the pattern in words and then use variables.

> **The Product Rule**
>
> To multiply two exponential expressions that have the same base, keep the base and *add the exponents*.
>
> $$x^a \cdot x^b = x^{a+b}$$

Be sure to notice that this rule applies only to expressions that have the *same base*. Here x represents the base while the letters a and b represent the exponents that are added.

It is important that you apply this rule even when an exponent is 1. Every variable that does not have a written exponent is understood to have an exponent of 1. Thus $x^1 = x$, $y^1 = y$, and so on.

EXAMPLE 1 Multiply. **(a)** $x^3 \cdot x^6$ **(b)** $x \cdot x^5$

(a) $x^3 \cdot x^6 = x^{3+6} = x^9$

(b) $x \cdot x^5 = x^{1+5} = x^6$ Note that the exponent of the first x is 1.

Practice Problem 1 Multiply. **(a)** $a^7 \cdot a^5$ **(b)** $w^{10} \cdot w$

EXAMPLE 2 Simplify. **(a)** $y^5 \cdot y^{11}$ **(b)** $2^3 \cdot 2^5$ **(c)** $x^6 \cdot y^8$

(a) $y^5 \cdot y^{11} = y^{5+11} = y^{16}$

(b) $2^3 \cdot 2^5 = 2^{3+5} = 2^8$ Note that the base does not change! Only the exponent changes.

(c) $x^6 \cdot y^8$ The rule for multiplying exponential expressions does not apply since the bases are not the same. This cannot be simplified.

Practice Problem 2 Simplify, if possible.

(a) $x^3 \cdot x^9$ **(b)** $3^7 \cdot 3^4$ **(c)** $a^3 \cdot b^2$

We can now look at multiplying expressions such as

$$(2x^5)(3x^6).$$

The number 2 in $2x^5$ is called the **numerical coefficient**. Recall that a numerical coefficient is a number that is multiplied by a variable. When we multiply two expressions such as $2x^5$ and $3x^6$, we first multiply the numerical coefficients; we multiply the variables with exponents separately.

EXAMPLE 3 Multiply. **(a)** $(2x^5)(3x^6)$ **(b)** $(5x^3)(x^6)$ **(c)** $(-6x)(-4x^5)$

(a) $(2x^5)(3x^6) = (2 \cdot 3)(x^5 \cdot x^6)$ Multiply the numerical coefficients.

$= 6(x^5 \cdot x^6)$ Use the rule for multiplying expressions with exponents. Add the exponents.

$= 6x^{11}$

(b) Every variable that does not have a visible numerical coefficient is understood to have a numerical coefficient of 1. Thus x^6 has a numerical coefficient of 1.

$$(5x^3)(x^6) = (5 \cdot 1)(x^3 \cdot x^6) = 5x^9$$

(c) $(-6x)(-4x^5) = (-6)(-4)(x^1 \cdot x^5) = 24x^6$ Remember that x has an exponent of 1.

Practice Problem 3 Multiply.

(a) $(-a^8)(a^4)$ **(b)** $(3y^2)(-2y^3)$ **(c)** $(-4x^3)(-5x^2)$

Problems of this type may involve more than one variable or more than two factors.

EXAMPLE 4 Multiply. $(5ab)(-\frac{1}{3}a)(9b^2)$

$$(5ab)(-\tfrac{1}{3}a)(9b^2) = (5)(-\tfrac{1}{3})(9)(a \cdot a)(b \cdot b^2)$$
$$= -15a^2b^3$$

As you do the following problems, keep in mind the rule for multiplying numbers with exponents and the rules for multiplying signed numbers.

Practice Problem 4 Multiply. $(2xy)(-\frac{1}{4}x^2y)(6xy^3)$

2   *Dividing Exponential Expressions with Like Bases*

Frequently, we must divide exponential expressions. Since division by zero is undefined, in all problems in this chapter we assume that the denominator of any variable expression is not zero. We'll look at division in three separate parts.

Suppose that we want to simplify $x^5 \div x^2$. We could do the division the long way.

$$\frac{x^5}{x^2} = \frac{(x)(x)(x)\cancel{(x)}\cancel{(x)}}{\cancel{(x)}\cancel{(x)}} = x^3$$

Here we are using the arithmetical property of reducing fractions. When the same factor appears in both numerator and denominator, that factor can be removed.

A simpler way is to *subtract the exponents*. Notice that the base remains the same.

The Quotient Rule (Part 1)

$\frac{x^a}{x^b} = x^{a-b}$ Use this form if the larger exponent is in the numerator and $x \neq 0$.

EXAMPLE 5 Divide. **(a)** $\dfrac{2^{16}}{2^{11}}$ **(b)** $\dfrac{x^5}{x^3}$ **(c)** $\dfrac{y^{16}}{y^7}$

(a) $\dfrac{2^{16}}{2^{11}} = 2^{16-11} = 2^5$ Note that the base does *not* change.

(b) $\dfrac{x^5}{x^3} = x^{5-3} = x^2$ **(c)** $\dfrac{y^{16}}{y^7} = y^{16-7} = y^9$

Practice Problem 5 Divide. **(a)** $\dfrac{10^{13}}{10^7}$ **(b)** $\dfrac{x^{11}}{x}$

Now we consider the situation where the larger exponent is in the denominator. Suppose that we want to simplify $x^2 \div x^5$.

$$\frac{x^2}{x^5} = \frac{\cancel{(x)}\,\cancel{(x)}}{\cancel{(x)}\,\cancel{(x)}(x)(x)(x)} = \frac{1}{x^3}$$

The Quotient Rule (Part 2)

$\dfrac{x^a}{x^b} = \dfrac{1}{x^{b-a}}$ Use this form if the larger exponent is in the denominator and $x \neq 0$.

EXAMPLE 6 Divide. **(a)** $\dfrac{12^{17}}{12^{20}}$ **(b)** $\dfrac{b^7}{b^9}$ **(c)** $\dfrac{x^{20}}{x^{24}}$

(a) $\dfrac{12^{17}}{12^{20}} = \dfrac{1}{12^{20-17}} = \dfrac{1}{12^3}$ Note that the base does *not* change.

(b) $\dfrac{b^7}{b^9} = \dfrac{1}{b^{9-7}} = \dfrac{1}{b^2}$ **(c)** $\dfrac{x^{20}}{x^{24}} = \dfrac{1}{x^{24-20}} = \dfrac{1}{x^4}$

Practice Problem 6 Divide. **(a)** $\dfrac{c^3}{c^4}$ **(b)** $\dfrac{10^{31}}{10^{56}}$

When there are numerical coefficients, use the rules for dividing signed numbers to reduce fractions to lowest terms.

EXAMPLE 7 Divide. **(a)** $\dfrac{5x^5}{25x^7}$ **(b)** $\dfrac{-12x^8}{4x^3}$ **(c)** $\dfrac{-16x^7}{-24x^8}$

(a) $\dfrac{5x^5}{25x^7} = \dfrac{1}{5x^{7-5}} = \dfrac{1}{5x^2}$ **(b)** $\dfrac{-12x^8}{4x^3} = -3x^{8-3} = -3x^5$ **(c)** $\dfrac{-16x^7}{-24x^8} = \dfrac{2}{3x^{8-7}} = \dfrac{2}{3x}$

Practice Problem 7 Divide. **(a)** $\dfrac{-7x^7}{-21x^9}$ **(b)** $\dfrac{15x^{11}}{-3x^4}$

You have to work very carefully if two or more variables are involved. Treat the coefficients and each variable separately.

EXAMPLE 8 Divide. **(a)** $\dfrac{x^3 y^2}{5x y^6}$ **(b)** $\dfrac{-3x^2 y^5}{12x^6 y^8}$

(a) $\dfrac{x^3 y^2}{5x y^6} = \dfrac{x^2}{5y^4}$ **(b)** $\dfrac{-3x^2 y^5}{12x^6 y^8} = -\dfrac{1}{4x^4 y^3}$

Practice Problem 8 Divide. **(a)** $\dfrac{x^7y^9}{y^{10}}$ **(b)** $\dfrac{12x^5y^6}{-24x^3y^8}$

Suppose that a given base appears with the same exponent in the numerator and denominator of a fraction. In this case we can use the fact that *any nonzero number divided by itself is* 1.

EXAMPLE 9 Divide. **(a)** $\dfrac{x^6}{x^6}$ **(b)** $\dfrac{3x^5}{x^5}$

(a) $\dfrac{x^6}{x^6} = 1$ **(b)** $\dfrac{3x^5}{x^5} = 3\left(\dfrac{x^5}{x^5}\right) = 3(1) = 3$

Practice Problem 9 Divide. **(a)** $\dfrac{10^7}{10^7}$ **(b)** $\dfrac{12a^4}{15a^4}$

Do you see that if we had subtracted exponents when simplifying $\dfrac{x^6}{x^6}$ we would have obtained x^0 in Example 9? So we can surmise that any number (except 0) to the 0 power equals 1. We can write this fact as a separate rule.

> ### The Quotient Rule (Part 3)
>
> $$\dfrac{x^a}{x^a} = x^0 = 1 \qquad \text{if } x \neq 0 \qquad (0^0 \text{ remains undefined}).$$

To Think About What about 0^0? Why is it undefined? $0^0 = 0^{1-1}$. If we use the quotient rule, $0^{1-1} = \dfrac{0}{0}$. Since division by zero is undefined, we must agree that 0^0 is undefined.

EXAMPLE 10 Divide. **(a)** $\dfrac{4x^0y^2}{8^0y^5z^3}$ **(b)** $\dfrac{5x^2y}{10x^2y^3}$

(a) $\dfrac{4x^0y^2}{8^0y^5z^3} = \dfrac{4(1)y^2}{(1)y^5z^3} = \dfrac{4y^2}{y^5z^3} = \dfrac{4}{y^3z^3}$ **(b)** $\dfrac{5x^2y}{10x^2y^3} = \dfrac{1x^0}{2y^2} = \dfrac{(1)(1)}{2y^2} = \dfrac{1}{2y^2}$

Practice Problem 10 Divide. $\dfrac{-20a^3b^8c^4}{28a^3b^7c^5}$

We can combine all three parts of the quotient rule we have developed.

> ### The Quotient Rule
>
> $\dfrac{x^a}{x^b} = x^{a-b}$ Use this form if the larger exponent is in the numerator and $x \neq 0$.
>
> $\dfrac{x^a}{x^b} = \dfrac{1}{x^{b-a}}$ Use this form if the larger exponent is in the denominator and $x \neq 0$.
>
> $\dfrac{x^a}{x^a} = x^0 = 1$ if $x \neq 0$.

We can combine the product rule and the quotient rule to simplify algebraic expressions that involve both multiplication and division.

EXAMPLE 11   Simplify. $\dfrac{(8x^2 y)(-3x^3 y^2)}{-6x^4 y^3}$

$$\frac{(8x^2 y)(-3x^3 y^2)}{-6x^4 y^3} = \frac{-24x^5 y^3}{-6x^4 y^3} = 4x$$

Practice Problem 11   Simplify. $\dfrac{(-6ab^5)(3a^2 b^4)}{16a^5 b^7}$

3 Raising Exponential Expressions to a Power

How do we simplify an expression such as $(x^4)^3$? $(x^4)^3$ is x^4 raised to the third power. For this type of problem we say that we are raising a power to a power. A problem such as $(x^4)^3$ could be done by writing the following.

$$\begin{aligned} (x^4)^3 &= x^4 \cdot x^4 \cdot x^4 \quad \text{By definition} \\ &= x^{12} \quad\quad\quad\quad \text{By adding exponents} \end{aligned}$$

Notice that when we add the exponents we get $4 + 4 + 4 = 12$. This is the same as multiplying 4 by 3. That is, $4 \cdot 3 = 12$. This process can be summarized by the following rule.

> **Raising a Power to a Power**
>
> To raise a power to a power, keep the same base and multiply the exponents.
>
> $$(x^a)^b = x^{ab}$$

Recall what happens when you raise a negative number to a power. $(-1)^2 = 1$. $(-1)^3 = -1$. In general,

$$(-1)^n = \begin{cases} +1 & \text{if } n \text{ is even} \\ -1 & \text{if } n \text{ is odd.} \end{cases}$$

EXAMPLE 12   Simplify.   **(a)** $(x^3)^5$ **(b)** $(2^7)^3$ **(c)** $(-1)^8$

(a) $(x^3)^5 = x^{3\cdot5} = x^{15}$ **(b)** $(2^7)^3 = 2^{7\cdot3} = 2^{21}$ **(c)** $(-1)^8 = +1$

Note that in both parts (a) and (b) the base does not change.

Practice Problem 12   Simplify.

(a) $(a^4)^3$ **(b)** $(10^5)^2$ **(c)** $(-1)^{15}$

Here are two rules involving products and quotients that are very useful. We'll illustrate each with an example.

If a product in parentheses is raised to a power, the parentheses indicate that *each factor* must be raised to that power.

$$(xy)^2 = x^2 y^2 \qquad (xy)^3 = x^3 y^3$$

$$(xy)^a = x^a y^a$$

EXAMPLE 13   Simplify.   **(a)** $(ab)^8$ **(b)** $(3x)^4$ **(c)** $(-2x^2)^3$

(a) $(ab)^8 = a^8 b^8$ **(b)** $(3x)^4 = (3)^4 x^4 = 81x^4$ **(c)** $(-2x^2)^3 = (-2)^3 \cdot (x^2)^3 = -8x^6$

Practice Problem 13   Simplify.

(a) $(3xy)^3$ **(b)** $(yz)^{37}$

If a fractional expression within parentheses is raised to a power, the parentheses indicate that both numerator and denominator must be raised to that power.

$$\left(\frac{x}{y}\right)^5 = \frac{x^5}{y^5} \qquad \left(\frac{x}{y}\right)^2 = \frac{x^2}{y^2} \qquad \text{if } y \neq 0$$

$$\left(\frac{x}{y}\right)^a = \frac{x^a}{y^a} \qquad \text{if } y \neq 0.$$

EXAMPLE 14   Simplify.   **(a)** $\left(\dfrac{x}{y}\right)^5$ **(b)** $\left(\dfrac{7}{w}\right)^4$

(a) $\left(\dfrac{x}{y}\right)^5 = \dfrac{x^5}{y^5}$ **(b)** $\left(\dfrac{7}{w}\right)^4 = \dfrac{7^4}{w^4} = \dfrac{2401}{w^4}$

Practice Problem 14   Simplify. $\left(\dfrac{4a}{b}\right)^6$

Many expressions can be simplified by using the previous rules involving exponents. Be sure to take particular care to determine the correct sign, especially if there is a negative numerical coefficient.

EXAMPLE 15   Simplify. $\left(\dfrac{-3x^2z^0}{y^3}\right)^4$

$\left(\dfrac{-3x^2z^0}{y^3}\right)^4 = \left(\dfrac{-3x^2}{y^3}\right)^4$ Simplify inside the parentheses first. Note that $z^0 = 1$.

$\qquad = \dfrac{(-3)^4x^8}{y^{12}}$ Apply the rules for raising a power to a power. Notice that we wrote $(-3)^4$ and not -3^4. We are raising -3 to the fourth power.

$\qquad = \dfrac{81x^8}{y^{12}}$ Simplify the coefficient: $(-3)^4 = +81$.

Practice Problem 15   Simplify. $\left(\dfrac{-2x^3y^0z}{4xz^2}\right)^5$

Developing Your Study Skills

Why Is Review Necessary?

You master a course in mathematics by learning the concepts one step at a time. Thus the study of mathematics is built step-by-step, with each step a supporting foundation for the next. The process is a carefully designed procedure, so no steps can be skipped. A student of mathematics needs to realize the importance of this building process to succeed.

Because new concepts depend on those previously learned, students often need to take time to review. The reviewing process will strengthen understanding and skills, which may be weak due to a lack of mastery or the passage of time. Review at the right time on the right concepts can strengthen previously-learned skills and make progress possible.

Timely, periodic review of previously-learned mathematical concepts is absolutely necessary for mastery of new concepts. You may have forgotten a concept or grown a bit rusty in applying it. Reviewing is the answer. Make use of the cumulative review problems in your textbook, whether they are assigned or not. Look back to previous chapters whenever you have forgotten how to do something. Review the chapter organizers from previous chapters. Study the examples and practice some exercises to refresh your understanding.

Be sure that you understand and can perform the computations of each new concept. This will enable you to move successfully on to the next one.

Remember, mathematics is a step-by-step building process. Learn each concept and reinforce and strengthen with review whenever necessary.

Verbal and Writing Skills

1. Write in your own words the product rule for exponents.

2. To be able to use the rules of exponents, what must be true of the bases?

3. If the larger exponent is in the denominator, the quotient rule states that $\dfrac{x^a}{x^b} = \dfrac{1}{x^{b-a}}$. Provide an example to show why this is true.

In exercises 4 and 5, identify the numerical coefficient, the base(s), and the exponent(s).

4. $-8x^5y^2$

5. $6x^{11}y$

6. Evaluate **(a)** $3x^0$ and **(b)** $(3x)^0$. **(c)** Why are the results different?

Write in simplest exponent form.

7. $2 \cdot 2 \cdot a \cdot a \cdot a \cdot b$

8. $5 \cdot x \cdot x \cdot x \cdot y \cdot y$

9. $(-3)(a)(a)(b)(c)(b)(c)(c)$

10. $(-7)(x)(y)(z)(y)(x)$

Multiply. Leave your answer in exponent form.

11. $(3^8)(3^7)$ **12.** $(2^5)(2^8)$ **13.** $(5^{10})(5^{16})$ **14.** $(5^3)(2^6)$ **15.** $(3^5)(8^2)$

Multiply.

16. $-5x^4(4x)$ **17.** $6x^2(-9x^3)$ **18.** $(5x)(10x^2)$ **19.** $(-4x^2)(-3x^3)$

20. $(-2a^5)(-5a^3)$ **21.** $(4x^8)(-3x^3)$ **22.** $\left(\dfrac{2}{5}xy^3\right)\left(\dfrac{1}{3}x^2y^2\right)$ **23.** $\left(\dfrac{4}{5}x^5y\right)\left(\dfrac{15}{16}x^2y^4\right)$

24. $(1.1x^2z)(-2.5xy)$ **25.** $(2.3x^4w)(-3.5xy^4)$ **26.** $(-3x)(2y)(5x^3y)$ **27.** $(8a)(2a^3b)(0)$

28. $(-5ab)(2a^2)(0)$ **29.** $(-16x^2y^4)(-5xy^3)$ **30.** $(-12x^4y)(-7x^5y^3)$ **31.** $(14a^5)(-2b^6)$

32. $(5a^3b^2)(-ab)$ **33.** $(-2x^3y^2)(0)(-3x^4y)$ **34.** $(-4x^8y^2)(13y^3)(0)$ **35.** $(4x^3y)(0)(-12x^4y^2)$

36. $(5x^3y)(-2w^4z)$ **37.** $(6w^5z^6)(-4xy)$ **38.** $(3ab)(5a^2c)(-2b^2c^3)$

Divide. Leave your answer in exponent form. Assume that all variables in any denominator are nonzero.

39. $\dfrac{y^5}{y^8}$ **40.** $\dfrac{x^{13}}{x^3}$ **41.** $\dfrac{y^{12}}{y^5}$ **42.** $\dfrac{b^{20}}{b^{23}}$ **43.** $\dfrac{13^{20}}{13^{30}}$ **44.** $\dfrac{5^{19}}{5^{11}}$

45. $\dfrac{-3^{18}}{-3^{14}}$ **46.** $\dfrac{-5^{16}}{-5^{12}}$ **47.** $\dfrac{a^{13}}{4a^5}$ **48.** $\dfrac{4b^{16}}{b^{13}}$ **49.** $\dfrac{x^7}{y^9}$ **50.** $\dfrac{x^{20}}{y^3}$

51. $\dfrac{-12x^5y^3}{-24xy^3}$ **52.** $\dfrac{-45a^4b^3}{-15a^4b^2}$ **53.** $\dfrac{a^5b^6}{a^5b^6}$ **54.** $\dfrac{-36x^3y^7}{72x^5y}$ **55.** $\dfrac{48x^7y}{-24x^3y^6}$ **56.** $\dfrac{84x^5y^2}{4.2xy^3}$

57. $\dfrac{3.1s^5t^3}{62s^8t}$ **58.** $\dfrac{-51x^6y^8z^{12}}{17x^3y^8z^7}$ **59.** $\dfrac{30x^5y^4}{5x^3y^4}$ **60.** $\dfrac{27x^3y^2}{3xy^2}$ **61.** $\dfrac{8^0x^2y^3}{16x^5y}$ **62.** $\dfrac{3^2x^3y^7}{3^0x^5y^2}$

63. $\dfrac{18a^6b^3c^0}{24a^5b^3}$ **64.** $\dfrac{12a^7b^8}{16a^3b^8c^0}$ **65.** $\dfrac{25x^6}{35y^8}$ **66.** $\dfrac{24y^5}{16x^3}$

Simplify. Multiply the numerator first. Then divide the two expressions.

67. $\dfrac{(4x)(9x^4)}{12x^3}$ **68.** $\dfrac{(6x^7)(3x^2)}{9x^4}$ **69.** $\dfrac{(9a^2b)(2a^3b^6)}{-27a^8b^7}$

To Think About

70. What expression can be multiplied by $(-3x^3yz)$ to obtain $81x^8y^2z^4$?

71. $63a^5b^6$ is divided by an expression and the result is $-9a^4b$. What is this expression?

Choose a value for the variable to show that the two expressions are not equivalent.

72. $(x^2)(x^3);\ x^6$ **73.** $\dfrac{y^8}{y^4};\ y^2$ **74.** $\dfrac{z^3}{z^6};\ z^3$

Simplify.

75. $\dfrac{x^{5a}}{x^{2a}}$ **76.** $y^{7b} \cdot y^{2b}$ **77.** $\dfrac{(c^{3y})(c^4)}{c^{2y+2}}$

Simplify.

78. $(x^2)^6$ **79.** $(w^5)^8$ **80.** $(xy^2)^7$ **81.** $(a^3b)^4$

82. $\left(a^5b^2\right)^6$

83. $\left(m^3n^2p\right)^5$

84. $\left(3a^3b^2c\right)^3$

85. $\left(3^2xy^2\right)^4$

86. $\left(-3a^4\right)^2$

87. $\left(-2a^5\right)^4$

88. $\left(\dfrac{7}{w^2}\right)^8$

89. $\left(\dfrac{12x}{y^2}\right)^5$

90. $\left(\dfrac{5x}{7y^2}\right)^2$

91. $\left(\dfrac{2a^4}{3b^3}\right)^4$

92. $\left(-3a^2b^3c^0\right)^4$

93. $\left(-2a^5b^2c\right)^5$

94. $\left(-2x^3yz\right)^3$

95. $\left(-4xy^0z^4\right)^3$

96. $\dfrac{(2x)^4}{(2x)^5}$

97. $\dfrac{\left(4a^2b\right)^2}{\left(4ab^2\right)^3}$

98. $\left(3ab^2\right)^3(ab)$

99. $\left(-2a^2b^3\right)^3(ab^2)$

100. $\left(\dfrac{8}{y^5}\right)^2$

101. $\left(\dfrac{4}{x^6}\right)^3$

102. $\left(\dfrac{ab^2}{c^3d^4}\right)^4$

103. $\left(\dfrac{a^3b}{c^5d}\right)^5$

104. $\left(5xy^2\right)^3(xy^2)$

105. $\left(4x^3y\right)^2(x^3y)$

To Think About

106. What expression raised to the third power is $-27x^9y^{12}z^{21}$?

107. What expression raised to the fourth power is $16x^{20}y^{16}z^{28}$?

Cumulative Review Problems

Simplify.

108. $-3 - 8$

109. $-17 + (-32) + (-24) + 27$

110. $\left(\dfrac{2}{3}\right)\left(-\dfrac{21}{8}\right)$

111. $\dfrac{-3}{4} \div \dfrac{-12}{80}$

Solve each system.

112. $-3x + y = -3$
$\dfrac{1}{4}x - 2y = -\dfrac{11}{2}$

113. $3x + y - z = 1$
$x - 2y + 3z = -9$
$-x + 5y + 6z = 21$

5.2 Negative Exponents and Scientific Notation

1 Using Negative Exponents

If n is an integer, and $x \neq 0$, then x^{-n} is defined as follows:

Definition of a Negative Exponent

$$x^{-n} = \frac{1}{x^n}, \quad x \neq 0$$

Student Learning Objectives

After studying this section, you will be able to:

1 Use negative exponents.

2 Write numbers in scientific notation.

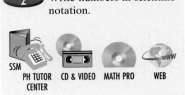

SSM   PH TUTOR   CD & VIDEO   MATH PRO   WEB
          CENTER

EXAMPLE 1 Write with positive exponents.

(a) y^{-3} **(b)** z^{-6} **(c)** w^{-1}

(a) $y^{-3} = \dfrac{1}{y^3}$ **(b)** $z^{-6} = \dfrac{1}{z^6}$ **(c)** $w^{-1} = \dfrac{1}{w^1} = \dfrac{1}{w}$

Practice Problem 1 Write with positive exponents.

(a) x^{-12} **(b)** w^{-5} **(c)** z^{-2}

To evaluate a numerical expression with a negative exponent, first write the expression with a positive exponent. Then simplify.

EXAMPLE 2 Evaluate. 2^{-5}

$$2^{-5} = \frac{1}{2^5} = \frac{1}{32}$$

Practice Problem 2 Evaluate. 4^{-3}

All the previously studied laws of exponents are true for any integer exponent. These laws are summarized in the following box. Assume that $x, y \neq 0$.

Laws of Exponents

The Product Rule

$$x^a \cdot x^b = x^{a+b}$$

The Quotient Rule

$$\frac{x^a}{x^b} = x^{a-b} \quad \text{Use if } a > b, \qquad \frac{x^a}{x^b} = \frac{1}{x^{b-a}} \quad \text{Use if } a < b.$$

Power Rules

$$(xy)^a = x^a y^a, \qquad (x^a)^b = x^{ab}, \qquad \left(\frac{x}{y}\right)^a = \frac{x^a}{y^a}$$

By using the definition of a negative exponent and the properties of fractions, we can derive two more helpful properties of exponents. Assume that $x, y \neq 0$.

Properties of Negative Exponents

$$\frac{1}{x^{-n}} = x^n \qquad \frac{x^{-m}}{y^{-n}} = \frac{y^n}{x^m}$$

273

EXAMPLE 3 Simplify. Write the expression with no negative exponents.

(a) $\dfrac{1}{x^{-6}}$ **(b)** $\dfrac{x^{-3}y^{-2}}{z^{-4}}$ **(c)** $x^{-2}y^3$

(a) $\dfrac{1}{x^{-6}} = x^6$ **(b)** $\dfrac{x^{-3}y^{-2}}{z^{-4}} = \dfrac{z^4}{x^3y^2}$ **(c)** $x^{-2}y^3 = \dfrac{y^3}{x^2}$

Practice Problem 3 Simplify. Write the expression with no negative exponents.

(a) $\dfrac{3}{w^{-4}}$ **(b)** $\dfrac{x^{-6}y^4}{z^{-2}}$ **(c)** $x^{-6}y^{-5}$

EXAMPLE 4 Simplify. Write the expression with no negative exponents.

(a) $(3x^{-4}y^2)^{-3}$ **(b)** $\dfrac{x^2y^{-4}}{x^{-5}y^3}$

(a) $(3x^{-4}y^2)^{-3} = 3^{-3}x^{12}y^{-6} = \dfrac{x^{12}}{3^3y^6} = \dfrac{x^{12}}{27y^6}$

(b) $\dfrac{x^2y^{-4}}{x^{-5}y^3} = \dfrac{x^2x^5}{y^4y^3} = \dfrac{x^7}{y^7}$ First rewrite the expression so that only positive exponents appear. Then simplify using the product rule.

Practice Problem 4 Simplify. Write the expression with no negative exponents.

(a) $(2x^4y^{-5})^{-2}$ **(b)** $\dfrac{y^{-3}z^{-4}}{y^2z^{-6}}$

2 *Writing Numbers in Scientific Notation*

One common use of negative exponents is in writing numbers in scientific notation. Scientific notation is most useful in expressing very large and very small numbers.

> **Scientific Notation**
>
> A positive number is written in **scientific notation** if it is in the form $a \times 10^n$, where $1 \le a < 10$ and n is an integer.

EXAMPLE 5 Write in scientific notation. **(a)** 4567 **(b)** 157,000,000

(a) $4567 = 4.567 \times 1000$ To change 4567 to a number that is greater than 1 but less than 10, we move the decimal point three places to the left. We must then multiply the number by a power of 10 so that we do not change the value of the number. Use 1000.

$= 4.567 \times 10^3$

(b) $157,000,000 = \underset{\text{8 places}}{\underleftarrow{1.57000000}} \times \underset{\text{8 zeros}}{\underbrace{100000000}}$

$= 1.57 \times 10^8$

Practice Problem 5 Write in scientific notation.

(a) 78,200 **(b)** 4,786,000

Numbers that are smaller than 1 will have a negative power of 10 if they are written in scientific notation.

● **EXAMPLE 6**   Write in scientific notation.

(a) 0.061   **(b)** 0.000052

(a) We need to write 0.061 as a number that is greater than 1 but less than 10. In which direction do we move the decimal point?

$0.061 = 6.1 \times 10^{-2}$ Move the decimal point 2 places to the right. Then multiply by 10^{-2}.

(b) $0.000052 = 5.2 \times 10^{-5}$ Why?

Practice Problem 6   Write in scientific notation.

(a) 0.98   **(b)** 0.000092   ●

The reverse procedure transforms scientific notation into ordinary decimal notation.

● **EXAMPLE 7**   Write in decimal notation.   **(a)** 1.568×10^2 **(b)** 7.432×10^{-3}

(a) $1.568 \times 10^2 = 1.568 \times 100$
$= 156.8$

Alternative Method

$1.568 \times 10^2 = 156.8$ The exponent 2 tells us to move the decimal point 2 places to the right.

(b) $7.432 \times 10^{-3} = 7.432 \times \dfrac{1}{1000}$
$= 0.007432$

Alternative Method

$7.432 \times 10^{-3} = 0.007432$ The exponent -3 tells us to move the decimal point 3 places to the left.

Practice Problem 7   Write in decimal notation.

(a) 2.96×10^3 **(b)** 1.93×10^6 **(c)** 5.43×10^{-2} **(d)** 8.562×10^{-5} ●

The distance light travels in one year is called a *light-year*. A light-year is a convenient unit of measure to use when investigating the distances between stars.

● **EXAMPLE 8**   A light-year is a distance of 9,460,000,000,000,000 meters. Write this in scientific notation.

$$9,460,000,000,000,000 = 9.46 \times 10^{15} \text{ meters}$$

Practice Problem 8   Astronomers measure distances to faraway galaxies in parsecs. A parsec is a distance of 30,900,000,000,000,000 meters. Write this in scientific notation.   ●

To perform a calculation involving very large or very small numbers, it is usually helpful to write the numbers in scientific notation and then use the laws of exponents to do the calculation.

Calculator

Scientific Notation

Most calculators can display only eight digits at one time. Numbers with more than eight digits are usually shown in scientific notation. 1.12 E 08 or 1.12 8 means 1.12×10^8. You can use a calculator to compute with large numbers by entering the numbers using scientific notation. For example,

$$(7.48 \times 10^{24}) \times (3.5 \times 10^8)$$

is entered as follows.

7.48 $\boxed{\text{EXP}}$

24 $\boxed{\times}$ 3.5

$\boxed{\text{EXP}}$ 8 $\boxed{=}$

Display: $\boxed{2.618 \text{ E } 33}$

or $\boxed{2.618 \qquad 33}$

Note: Some calculators have an $\boxed{\text{EE}}$ key instead of $\boxed{\text{EXP}}$.

Compute on a calculator.

1. 35,000,000,000 + 77,000,000,000

2. $(6.23 \times 10^{12}) \times (4.9 \times 10^5)$

3. $(2.5 \times 10^7)^5$

4. $3.3284 \times 10^{32} \div (6.28 \times 10^{24})$

5. How many seconds are there in 1000 years?

EXAMPLE 9   Use scientific notation and the laws of exponents to find the following. Leave your answer in scientific notation.

(a) $(32,000,000)(1,500,000,000,000)$

(b) $\dfrac{0.00063}{0.021}$

(a) $(32,000,000)(1,500,000,000,000)$

$\quad = (3.2 \times 10^7)(1.5 \times 10^{12})$ Write each number in scientific notation.

$\quad = 3.2 \times 1.5 \times 10^7 \times 10^{12}$ Rearrange the order. Remember that multiplication is commutative.

$\quad = 4.8 \times 10^{19}$ Multiply 3.2×1.5. Multiply $10^7 \times 10^{12}$.

(b) $\dfrac{0.00063}{0.021} = \dfrac{6.3 \times 10^{-4}}{2.1 \times 10^{-2}}$ Write each number in scientific notation.

$\quad = \dfrac{6.3}{2.1} \times \dfrac{10^{-4}}{10^{-2}}$ Rearrange the order. We are actually using the definition of multiplication of fractions.

$\quad = \dfrac{6.3}{2.1} \times \dfrac{10^2}{10^4}$ Rewrite with positive exponents.

$\quad = 3.0 \times 10^{-2}$

Practice Problem 9   Use scientific notation and the laws of exponents to find the following. Leave your answer in scientific notation.

(a) $(56,000)(1,400,000,000)$

(b) $\dfrac{0.000111}{0.00000037}$

When we use scientific notation, we are writing approximate numbers. We must include some zeros so that the decimal point can be properly located. However, all other digits except for these zeros are considered **significant digits**. The number 34.56 has four significant digits. The number 0.0049 has two significant digits. The zeros are considered placeholders. The number 634,000 has three significant digits. We sometimes round numbers to a specific number of significant digits. For example, 0.08746 rounded to two significant digits is 0.087. When we round 1,348,593 to three significant digits, we obtain 1,350,000.

EXAMPLE 10   The approximate distance from Earth to the star Polaris is 208 parsecs. A parsec is a distance of approximately 3.09×10^{13} kilometers. How long would it take a space probe traveling at 40,000 kilometers per hour to reach the star? Round to three significant digits.

1. ***Understand the problem.***
   Recall that the distance formula is

   $$\text{distance} = \text{rate} \times \text{time}.$$

   We are given the distance and the rate. We need to find the time.
   Let's take a look at the distance. The distance is given in parsecs, but the rate is given in kilometers per hour. We need to change the distance to kilometers. We are told that a parsec is approximately 3.09×10^{13} kilometers. That is, there are 3.09×10^{13} kilometers per parsec. We use this information to change 208 parsecs to kilometers.

$$208 \text{ parsecs} = \frac{(208 \text{ parsecs})(3.09 \times 10^{13} \text{ kilometers})}{1 \text{ parsec}} = 642.72 \times 10^{13} \text{ kilometers}$$

2. **Write an equation.**
   Use the distance formula.

$$d = r \times t$$

3. **Solve the equation and state the answer.**
   Substitute the known values into the formula and solve for the unknown, time.

$$642.72 \times 10^{13} \text{ km} = \frac{40,000 \text{ km}}{1 \text{ hr}} \times t$$

$$6.4272 \times 10^{15} \text{ km} = \frac{4 \times 10^4 \text{ km}}{1 \text{ hr}} \times t \qquad \text{Change the numbers to scientific notation.}$$

$$\frac{6.4272 \times 10^{15} \text{ km}}{\dfrac{4 \times 10^4 \text{ km}}{1 \text{ hr}}} = t \qquad \text{Divide both sides by } \frac{4 \times 10^4 \text{ km}}{1 \text{ hr}}.$$

$$\frac{(6.4272 \times 10^{15} \text{ km})(1 \text{ hr})}{4 \times 10^4 \text{ km}} = t$$

$$1.6068 \times 10^{11} \text{ hr} = t$$

1.6068×10^{11} is 160.68×10^9 or 160.68 billion hours. The space probe will take approximately 160.68 billion hours to reach the star.

Reread the problem. Are we finished? What is left to do? We need to round the answer to three significant digits. Rounding to three significant digits, we have

$$160.68 \times 10^9 \approx 161 \times 10^9.$$

This is approximately 161 billion hours or a little more than 18 million years.

4. **Check.**
   Unless you have had a great deal of experience working in astronomy, it would be difficult to determine whether this is a reasonable answer. You may wish to reread your analysis and redo your calculations as a check.

Practice Problem 10   The average distance from Earth to the star Betelgeuse is 159 parsecs. How many hours would it take a space probe to travel from Earth to Betelgeuse at a speed of 50,000 kilometers per hour? Round to three significant digits.

Simplify. Express your answer with positive exponents. Assume that all variables are nonzero.

1. $3x^{-2}$

2. $4xy^{-4}$

3. $(4x^2y)^{-2}$

4. $(2x^3y^5)^{-3}$

5. $\dfrac{3xy^{-2}}{z^{-3}}$

6. $\dfrac{4x^{-2}y^{-3}z^0}{y^4}$

7. $\dfrac{(3x)^{-2}}{(3x)^{-3}}$

8. $\dfrac{(2ab^2)^{-3}}{(2ab^2)^{-4}}$

9. $wx^{-5}y^3z^{-2}$

10. $a^5b^{-3}c^{-4}d^0$

11. $(8^{-2})(2^3)$

12. $(9^2)(3^{-3})$

13. $\left(\dfrac{3xy^2}{z^4}\right)^{-2}$

14. $\left(\dfrac{2a^3b^0}{c^2}\right)^{-3}$

15. $\dfrac{x^{-2}y^{-3}}{x^4y^{-2}}$

16. $\dfrac{a^{-6}b^3}{a^{-2}b^{-5}}$

Write in scientific notation.

17. 123,780

18. 0.063

19. 0.000742

20. 889,610,000,000

21. 7,652,000,000

22. 0.00000001963

In exercises 23–28, write in decimal notation.

23. 3.02×10^5

24. 8.137×10^7

25. 3.3×10^{-5}

26. 1.99×10^{-1}

27. 9.83×10^5

28. 3.5×10^{-8}

29. The mass of a proton is 0.000000000000000000000000000167 kilogram. Write this in scientific notation.

30. The speed of light is 3.00×10^8 meters per second. Write this in decimal notation.

31. A single human red blood cell is about 7×10^{-6} meters in diameter. Write this in decimal notation.

32. The average volume of an atom of gold is 0.00000000000000000000001695 cubic centimeters. Write this in scientific notation.

Evaluate by using scientific notation and the laws of exponents. Leave your answer in scientific notation.

33. $\dfrac{(5,000,000)(16,000)}{8,000,000,000}$

34. $(0.0075)(0.0000002)(0.001)$

35. $(0.0002)^5$

36. $(40,000,000)^3$

37. $(150,000,000)(0.00005)(0.002)(30,000)$

38. $\dfrac{(1,600,000)(0.00003)}{2400}$

Applications

For fiscal year 1999, the national debt was determined to be approximately 5.614×10^{12} *dollars.*

39. The census bureau estimates that in 1999, the entire population of the United States was 2.76×10^8 people. If the national debt were evenly divided among every person in the country, how much debt would be assigned to each individual? Round to three significant digits.

40. The census bureau estimates that in 1999, the number of people in the United States who were over age 18 was approximately 2.05×10^8 people. If the national debt were evenly divided among every person over age 18 in the country, how much debt would be assigned to each individual? Round to three significant digits.

A parsec is a distance of approximately 3.09×10^{13} *kilometers.*

41. How long would it take a space probe to travel from Earth to the star Rigel, which is 276 parsecs from Earth? Assume that the space probe travels at 45,000 kilometers per hour. Round to three significant digits.

42. How long would it take a space probe to travel from Earth to the star Hadar, which is 150 parsecs from Earth? Assume that the space probe travels at 55,000 kilometers per hour. Round to three significant digits.

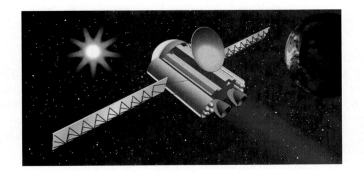

43. Antares, a supergiant star, is one of the brightest stars seen by the naked eye. It is 130 parsecs from Earth. How long would it take an experimental robotic probe to reach Antares if we assume that the probe travels at 35,000 kilometers per hour? Round to three significant digits.

44. The mass of a neutron is approximately 1.675×10^{-27} kilogram. Find the mass of 180,000 neutrons.

45. The sun radiates energy into space at the rate of 3.9×10^{26} joules per second. How many joules are emitted in two weeks?

46. Avogadro's number says that there are approximately 6.02×10^{23} molecules/mole. How many molecules can one expect in 0.00483 mole?

47. In 1990 the cost for construction of new private buildings was estimated at 3.61×10^{11}. By 2000 the estimated cost for construction of new private buildings had risen to 5.28×10^{11}. What was the percent of increase from 1990 to 2000? Round to the nearest tenth of a percent. (*Source:* U.S. Census Bureau)

48. In 1990 the cost for construction of new public buildings was estimated at 1.07×10^{11}. By 2000 the estimated cost for construction of new public buildings had risen to 1.53×10^{11}. What was the percent of increase from 1990 to 2000? Round to the nearest tenth of a percent. (*Source:* U.S. Census Bureau)

Cumulative Review Problems

Simplify.

49. $-2.7 - (-1.9)$

50. $(-1)^{33}$

51. $\dfrac{-3}{4} + \dfrac{5}{7}$

52. A recent debate between two political candidates attracted 3540 people to a local gymnasium. At the conclusion of the debate, each person went to one side of the gymnasium to shake the hand of candidate #1 or to the other side of the room to shake the hand of candidate #2. Candidate #1 shook 524 less than triple the number of hands shaken by candidate #2. How many hands did each candidate shake?

53. Gina has a bachelor's degree. She earns $12,460 more a year than Mario, who holds an associate's degree. Alfonso, who has not yet been able to attend college, earns $8742 a year less than Mario. The combined annual salaries of the three people is $112,000. What is the annual salary of each person?

54. Graph $y = 3x - 5$

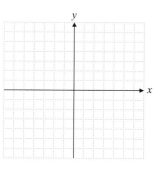

55. Graph $-x + 2y = 10 - x$

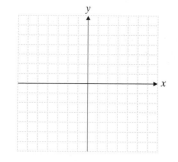

Putting Your Skills to Work

The Mathematics of Forests

One-third of the world's area that was originally covered by forests and woodlands no longer has either. The welfare of people and animals in future generations on Earth may well depend on the ability of people to stop the destruction of forests worldwide. Some countries have a very significant amount of the present total of 9.258×10^9 acres of forests in the world.

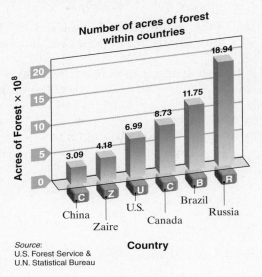

Number of acres of forest within countries

Acres of Forest $\times 10^8$

Source:
U.S. Forest Service &
U.N. Statistical Bureau

Country

Problems for Individual Investigation and Analysis

1. How many more acres of forests are in Brazil than are in Canada?

2. What is the combined acreage of forests in Russia and China?

Problems for Group Investigation and Cooperative Group Activity

Together with some members of your class, see if you can determine the answers to the following.

3. If the current population of China is 1.256×10^9 people, how many people are there for each acre of forest in China? Round to the nearest whole number.

4. For quite some time rainforests have been destroyed at the rate of 40 million acres per year. If we assume that this rate has been the same for the last 25 years, approximately how many acres of forests were in the world 25 years ago?

Internet Connections

Netsite: http://www.prenhall.com/tobey_beg_int

Site: Rates of Rainforest Loss (Rainforest Action Network) or a related site

This site presents information about rates of rainforest destruction. Use the information to answer the following questions. (There is some difference of opinion among scientists as to the rate of destruction.)

5. About how many acres of rainforest are destroyed in one hour?

6. About how many acres of Brazilian rainforest were destroyed during the last 12 years?

7. The current indigenous population of Brazil's forests is about what percent of the indigenous population in 1500?

5.3 Fundamental Polynomial Operations

Student Learning Objectives

After studying this section, you will be able to:

1 Recognize polynomials and determine their degrees.

2 Add polynomials.

3 Subtract polynomials.

4 Evaluate polynomials.

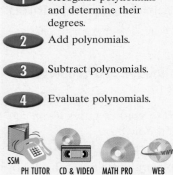

SSM · PH TUTOR CENTER · CD & VIDEO · MATH PRO · WEB

1 *Recognizing Polynomials and Determining Their Degrees*

A **polynomial** in x is the sum of a finite number of terms of the form ax^n, where a is any real number and n is a whole number. Usually these polynomials are written in descending powers of the variable, as in

$$5x^3 + 3x^2 - 2x - 5 \quad \text{and} \quad 3.2x^2 - 1.4x + 5.6.$$

A **multivariable polynomial** is a polynomial with more than one variable. The following are multivariable polynomials:

$$5xy + 8, \qquad 2x^2 - 7xy + 9y^2, \qquad 17x^3y^9$$

The **degree of a term** is the sum of the exponents of all of the variables in the term. For example, the degree of $7x^3$ is three. The degree of $4xy$ is two. The degree of $10x^4y^2$ is six.

The **degree of a polynomial** is the highest degree of all of the terms in the polynomial. For example, the degree of $5x^3 + 8x^2 - 20x - 2$ is three. The degree of $6xy - 4x^2y + 2xy^3$ is four.

The polynomial 0 is said to have **no degree**.

There are special names for polynomials with one, two, or three terms.

A **monomial** has *one* term:

$$5a, \qquad 3x^3yz^4, \qquad 12xy$$

A **binomial** has *two* terms:

$$7x + 9y, \qquad -6x - 4, \qquad 5x^4 + 2xy^2$$

A **trinomial** has *three* terms:

$$8x^2 - 7x + 4, \qquad 2ab^3 - 6ab^2 - 15ab, \qquad 2 + 5y + y^4$$

EXAMPLE 1 State the degree of the polynomial and whether it is a monomial, a binomial, or a trinomial.

(a) $5xy + 3x^3$ **(b)** $-7a^5b^2$ **(c)** $8x^4 - 9x - 15$

(a) This polynomial is of degree 3. It has two terms, so it is a binomial.

(b) The sum of the exponents is $5 + 2 = 7$. Therefore this polynomial is of degree 7. It has one term, so it is a monomial.

(c) This polynomial is of degree 4. It has three terms, so it is a trinomial.

Practice Problem 1 State the degree of the polynomial and whether it is a monomial, a binomial, or a trinomial.

(a) $-7x^5 - 3xy$ **(b)** $22a^3b^4$ **(c)** $-3x^3 + 5x^2 - 6x$

2 *Adding Polynomials*

We usually write a polynomial in x so that the exponents on x decrease from left to right. You can add, subtract, multiply, and divide polynomials. Let us take a look at addition. To add two polynomials, we add their like terms.

EXAMPLE 2 Add. $(5x^2 - 6x - 12) + (-3x^2 - 9x + 5)$

$$
\begin{aligned}
(5x^2 - 6x - 12) + (-3x^2 - 9x + 5) &= \left[5x^2 + (-3x^2)\right] + \left[-6x + (-9x)\right] + \left[-12 + 5\right] \\
&= \left[(5 - 3)x^2\right] + \left[(-6 - 9)x\right] + \left[-12 + 5\right] \\
&= 2x^2 + (-15x) + (-7) \\
&= 2x^2 - 15x - 7
\end{aligned}
$$

Practice Problem 2   Add. $(-8x^3 + 3x^2 + 6) + (2x^3 - 7x^2 - 3)$

The numerical coefficients of the polynomials may be any real number. Thus the polynomials may have numerical coefficients that are decimals or fractions.

 EXAMPLE 3   Add. $\left(\frac{1}{2}x^2 - 6x + \frac{1}{3}\right) + \left(\frac{1}{5}x^2 - 2x - \frac{1}{2}\right)$

$$\left(\tfrac{1}{2}x^2 - 6x + \tfrac{1}{3}\right) + \left(\tfrac{1}{5}x^2 - 2x - \tfrac{1}{2}\right) = \left[\tfrac{1}{2}x^2 + \tfrac{1}{5}x^2\right] + \left[-6x + (-2x)\right] + \left[\tfrac{1}{3} + \left(-\tfrac{1}{2}\right)\right]$$
$$= \left[\left(\tfrac{1}{2} + \tfrac{1}{5}\right)x^2\right] + \left[(-6 - 2)x\right] + \left[\tfrac{1}{3} + \left(-\tfrac{1}{2}\right)\right]$$
$$= \left[\left(\tfrac{5}{10} + \tfrac{2}{10}\right)x^2\right] + \left[-8x\right] + \left[\tfrac{2}{6} - \tfrac{3}{6}\right]$$
$$= \tfrac{7}{10}x^2 - 8x - \tfrac{1}{6}$$

Practice Problem 3   Add. $\left(-\tfrac{1}{3}x^2 - 6x - \tfrac{1}{12}\right) + \left(\tfrac{1}{4}x^2 + 5x - \tfrac{1}{3}\right)$

 EXAMPLE 4   Add. $\left(1.2x^3 - 5.6x^2 + 5\right) + \left(-3.4x^3 - 1.2x^2 + 4.5x - 7\right)$
Group like terms.

$$\left(1.2x^3 - 5.6x^2 + 5\right) + \left(-3.4x^3 - 1.2x^2 + 4.5x - 7\right) = (1.2 - 3.4)x^3 + (-5.6 - 1.2)x^2 + 4.5x + (5 - 7)$$
$$= -2.2x^3 - 6.8x^2 + 4.5x - 2$$

Practice Problem 4   Add.

$$\left(3.5x^3 - 0.02x^2 + 1.56x - 3.5\right) + \left(-0.08x^2 - 1.98x + 4\right)$$

As mentioned previously, polynomials may involve more than one variable.

3 ⬤ *Subtracting Polynomials*

Recall that subtraction of real numbers can be defined as adding the opposite of the second number. Thus $a - b = a + (-b)$. That is, $3 - 5 = 3 + (-5)$. A similar method is used to subtract two polynomials.

To subtract two polynomials, change the sign of each term in the second polynomial and then add.

 EXAMPLE 5   Subtract. $\left(7x^2 - 6x + 3\right) - \left(5x^2 - 8x - 12\right)$

We change the sign of each term in the second polynomial and then add.

$$\left(7x^2 - 6x + 3\right) - \left(5x^2 - 8x - 12\right) = \left(7x^2 - 6x + 3\right) + \left(-5x^2 + 8x + 12\right)$$
$$= (7 - 5)x^2 + (-6 + 8)x + (3 + 12)$$
$$= 2x^2 + 2x + 15$$

Practice Problem 5   Subtract.
$$\left(5x^3 - 15x^2 + 6x - 3\right) - \left(-4x^3 - 10x^2 + 5x + 13\right)$$

EXAMPLE 6 Subtract. $(-6x^2y - 3xy + 7xy^2) - (5x^2y - 8xy - 15x^2y^2)$

Change the sign of each term in the second polynomial and add. Look for like terms.

$$(-6x^2y - 3xy + 7xy^2) + (-5x^2y + 8xy + 15x^2y^2) = (-6 - 5)x^2y + (-3 + 8)xy + 7xy^2 + 15x^2y^2$$

$$= -11x^2y + 5xy + 7xy^2 + 15x^2y^2$$

Nothing further can be done to combine these four terms.

Practice Problem 6 Subtract.
$(x^3 - 7x^2y + 3xy^2 - 2y^3) - (2x^3 + 4xy - 6y^3)$

4 Evaluating Polynomials

Sometimes polynomials are used to predict a value. In such cases we need to **evaluate** the polynomial. We do this by substituting a known value for the variable and determining the value of the polynomial.

EXAMPLE 7 Automobiles sold in the United States have become more fuel efficient over the years due to regulations from Congress. The number of miles per gallon obtained by the average automobile in the United States can be described by the polynomial

$$0.3x + 12.9,$$

where x is the number of years since 1970. (*Source:* U.S. Federal Highway Administration.) Use this polynomial to estimate the number of miles per gallon obtained by the average automobile in   **(a)** 1972   **(b)** 2004.

(a) The year 1972 is two years later than 1970, so $x = 2$.
Thus the number of miles per gallon obtained by the average automobile in 1972 can be estimated by evaluating $0.3x + 12.9$ when $x = 2$.

$$0.3(2) + 12.9 = 0.6 + 12.9$$
$$= 13.5$$

We estimate that the average car in 1972 obtained 13.5 miles per gallon.

(b) The year 2004 will be 34 years after 1970, so $x = 34$.
Thus the estimated number of miles per gallon obtained by the average automobile in 2004 can be predicted by evaluating $0.3x + 12.9$ when $x = 34$.

$$0.3(34) + 12.9 = 10.2 + 12.9$$
$$= 23.1$$

We therefore predict that the average car in 2004 will obtain 23.1 miles per gallon.

Practice Problem 7 The number of miles per gallon obtained by the average truck in the United States can be described by the polynomial $0.03x + 5.4$, where x is the number of years since 1970. (*Source:* U.S. Federal Highway Administration.) Use this polynomial to estimate the number of miles per gallon obtained by the average truck in

(a) 1974   **(b)** 2006.

Verbal and Writing Skills

1. State in your own words a definition for a polynomial in x and give an example.

2. State in your own words a definition for a multivariable polynomial and give an example.

3. State in your own words how to determine the degree of a polynomial in x.

4. State in your own words how to determine the degree of a multivariable polynomial.

State the degree of the polynomial and whether it is a monomial, a binomial, or a trinomial.

5. $6x^3y$

6. $5xy^6$

7. $20x^5 + 6x^3 - 7x$

8. $13x^4 - 12x + 20$

9. $5xy^2 - 3x^2y^3$

10. $7x^3y + 5x^4y^4$

Add.

11. $(-3x + 15) + (8x - 43)$

12. $(5x - 11) + (-7x + 34)$

13. $(2x^2 - 8x + 7) + (-5x^2 - 2x + 3)$

14. $(x^2 - 6x - 3) + (-4x^2 + x - 4)$

15. $\left(\frac{1}{2}x^2 + \frac{1}{3}x - 4\right) + \left(\frac{1}{3}x^2 + \frac{1}{6}x - 5\right)$

16. $\left(\frac{1}{4}x^2 - \frac{2}{3}x - 10\right) + \left(-\frac{1}{3}x^2 + \frac{1}{9}x + 2\right)$

17. $(3.4x^3 - 5.6x^2 - 7.1x + 3.4) + (-1.7x^3 + 2.2x^2 - 6.1x - 8.8)$

18. $(-4.6x^3 - 2.9x^2 + 5.6x - 0.3) + (-2.6x^3 + 9.8x^2 + 4.5x - 1.7)$

Subtract.

19. $(2x - 19) - (-3x + 5)$

20. $(5x - 5) - (6x - 3)$

21. $\left(\frac{1}{2}x^2 + 3x - \frac{1}{5}\right) - \left(\frac{2}{3}x^2 - 4x + \frac{1}{10}\right)$

22. $\left(\frac{1}{8}x^2 - 5x + \frac{2}{7}\right) - \left(\frac{1}{4}x^2 - 6x + \frac{1}{14}\right)$

23. $(-3x^2 + 5x) - (2x^3 - 3x^2 + 10)$

24. $(7x^3 - 3x^2 - 5x + 4) - (x^3 - 7x + 3)$

25. $(0.5x^4 - 0.7x^2 + 8.3) - (5.2x^4 + 1.6x + 7.9)$

26. $(1.3x^4 - 3.1x^3 + 6.3x) - (x^4 - 5.2x^2 + 6.5x)$

Perform the indicated operations.

27. $(7x - 8) - (5x - 6) + (8x + 12)$

28. $(-8x + 3) + (-6x - 20) - (-3x + 9)$

29. $(5x^2y - 6xy^2 + 2) + (-8x^2y + 12xy^2 - 6)$

30. $(7x^2y^2 - 6xy + 5) + (-15x^2y^2 - 6xy + 18)$

31. $(3x^4 - 4x^2 - 18) - (2x^4 + 3x^3 + 6)$

32. $(2b^3 + 3b - 5) - (-3b^3 + 5b^2 + 7b)$

To Think About

Buses operated in this country have become more fuel efficient in recent years. The number of miles per gallon achieved by the average bus operated in the United States can be described by the polynomial $0.04x + 5.2$, where x is the number of years since 1970. (Source: U.S. Federal Highway Administration)

33. Estimate the number of miles per gallon obtained by the average bus in 1975.

34. Estimate the number of miles per gallon obtained by the average bus in 1995.

35. Estimate in what year the rating of the average bus will be 7.2 miles per gallon.

36. Estimate in what year the rating of the average bus will be 6.8 miles per gallon.

The average number of prisoners held in federal and state prisons increases each year. The number of prisoners measured in thousands can be described by the polynomial $1.8x^2 + 22.2x + 325$, where x is the number of years since 1980. (Source: U.S. Bureau of Justice Statistics)

37. Estimate the number of prisoners in 1990.

38. Estimate the number of prisoners in 1995.

39. According to the polynomial, by how much will the prison population increase from 2002 to 2006?

40. According to the polynomial, by how much will the prison population increase from 2004 to 2010?

Applications

▲ **41.** The lengths and the widths of the following three rectangles are labeled. Create a polynomial that describes the sum of the *area* of these three rectangles.

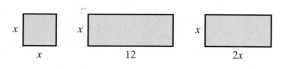

▲ **42.** The dimensions of the sides of the following figure are labeled. Create a polynomial that describes the *perimeter* of this figure.

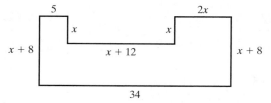

Cumulative Review Problems

43. Solve for x. $3y - 8x = 2$

44. Solve for d. $B = \dfrac{5xy}{d}$

45. The total approximate expenditure for health care in the United States increased by 90% over the years 1990 to 2000. In 2000 a total of 1.324 trillion dollars was spent. How much was spent in 1990? (*Source: U.S. Department of Health and Human Services*)

46. Find the slope of the line passing through $(-3, 4)$ and $(0, 2)$.

47. Write an equation of the line with slope -4 and y-intercept $(0, -3)$.

5.4 Multiplication of Polynomials

1 ➤ Multiplying a Monomial by a Polynomial

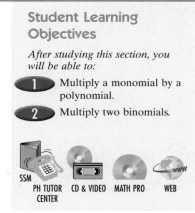

We use the distributive property to multiply a monomial by a polynomial. Remember, the distributive property states that for real numbers a, b, and c,

$$a(b + c) = ab + ac.$$

EXAMPLE 1 Multiply. $3x^2(5x - 2)$

$$
\begin{aligned}
3x^2(5x - 2) &= 3x^2(5x) + 3x^2(-2) \qquad \text{Use the distributive property.}\\
&= (3 \cdot 5)(x^2 \cdot x) + (3)(-2)x^2\\
&= 15x^3 - 6x^2
\end{aligned}
$$

Practice Problem 1 Multiply. $4x^3(-2x^2 + 3x)$

Try to do as much of the multiplication as you can mentally.

EXAMPLE 2 Multiply. **(a)** $2x(x^2 + 3x - 1)$ **(b)** $-2xy^2(x^2 - 2xy - 3y^2)$

(a) $2x(x^2 + 3x - 1) = 2x^3 + 6x^2 - 2x$

(b) $-2xy^2(x^2 - 2xy - 3y^2) = -2x^3y^2 + 4x^2y^3 + 6xy^4$

Notice in part (b) that you are multiplying each term by the negative term $-2xy^2$. This will affect the sign of each term in the product.

Practice Problem 2 Multiply. **(a)** $-3x(x^2 + 2x - 4)$ **(b)** $6xy(x^3 + 2x^2y - y^2)$

When we multiply by a monomial, the monomial may be on the right side.

EXAMPLE 3 Multiply. $(x^2 - 2x + 6)(-2xy)$

$$(x^2 - 2x + 6)(-2xy) = -2x^3y + 4x^2y - 12xy$$

Practice Problem 3 Multiply.

$$(-6x^3 + 4x^2 - 2x)(-3xy)$$

2 ➤ Multiplying Two Binomials

We can build on our knowledge of the distributive property and our experience with multiplying monomials to learn how to multiply two binomials. Let's suppose that we want to multiply $(x + 2)(3x + 1)$. We can use the distributive property. Since a can represent any number, let $a = x + 2$. Then let $b = 3x$ and $c = 1$. We now have the following.

$$
\begin{aligned}
a(b + c) &= ab + ac\\
(x + 2)(3x + 1) &= (x + 2)(3x) + (x + 2)(1)\\
&= 3x^2 + 6x + x + 2\\
&= 3x^2 + 7x + 2
\end{aligned}
$$

287

Let's take another look at the original problem, $(x + 2)(3x + 1)$. This time we will assign a letter to each term in the binomials. That is, let $a = x$, $b = 2$, $c = 3x$, and $d = 1$. Using substitution, we have the following.

$$(x + 2)(3x + 1) = (a + b)(c + d)$$
$$= (a + b)c + (a + b)d$$
$$= ac + bc + ad + bd$$
$$= (x)(3x) + (2)(3x) + (x)(1) + (2)(1) \quad \text{By substitution}$$
$$= 3x^2 + 6x + x + 2$$
$$= 3x^2 + 7x + 2$$

How does this compare with the preceding result?

The distributive property shows us *how* the problem can be done and *why* it can be done. In actual practice there is a somewhat easier approach to obtain the answer. It is often referred to as the FOIL method. The letters FOIL stand for the following.

F multiply the *F*irst terms

O multiply the *O*uter terms

I multiply the *I*nner terms

L multiply the *L*ast terms

The FOIL letters are simply a way to remember the four terms in the final product and how they are obtained. Let's return to our original problem.

$(x + 2)(3x + 1)$ F   Multiply the *first* terms to obtain $3x^2$.

$(x + 2)(3x + 1)$ O   Multiply the *outer* terms to obtain x.

$(x + 2)(3x + 1)$ I   Multiply the *inner* terms to obtain $6x$.

$(x + 2)(3x + 1)$ L   Multiply the *last* terms to obtain 2.

The result so far is $3x^2 + x + 6x + 2$. These four terms are the same four terms that we obtained when we multiplied using the distributive property. We can combine the like terms to obtain the final answer: $3x^2 + 7x + 2$. Now let's study the FOIL method in a few examples.

EXAMPLE 4   Multiply. $(2x - 1)(3x + 2)$

First   Last

$(2x - 1)(3x + 2)$

Inner

Outer

First	+	Outer	+	Inner	+	Last
F		O		I		L
$= 6x^2 +$		$4x$	$-$	$3x$	$-$	2
$= 6x^2 +$		$x - 2$				

Collect like terms.

Practice Problem 4   Multiply. $(5x - 1)(x - 2)$

EXAMPLE 5 Multiply. $(3a + 2b)(4a - b)$

First   Last

$(3a + 2b)(4a - b)$ $= 12a^2 - 3ab + 8ab - 2b^2$

                                  $= 12a^2 + 5ab + 2b^2$

Inner

    Outer

Practice Problem 5 Multiply. $(8a - 5b)(3a - b)$

After you have done several problems, you may be able to combine the outer and inner products mentally.

In some problems the inner and outer products cannot be combined.

EXAMPLE 6 Multiply. $(3x + 2y)(5x - 3z)$

First   Last

$(3x + 2y)(5x - 3z)$ $= 15x^2 - 9xz + 10xy - 6yz$

    Inner

      Outer   Since there are no like terms, we cannot combine any terms.

Practice Problem 6 Multiply. $(3a + 2b)(2a - 3c)$

EXAMPLE 7 Multiply. $(7x - 2y)^2$

$(7x - 2y)(7x - 2y)$ When we square a binomial, it is the same as multiplying the binomial by itself.

First   Last

$(7x - 2y)(7x - 2y)$ $= 49x^2 - 14xy - 14xy + 4y^2$

                                  $= 49x^2 - 28xy + 4y^2$

Inner

    Outer

Practice Problem 7 Multiply. $(3x - 2y)^2$

We can multiply binomials containing exponents that are greater than 1. That is, we can multiply binomials containing x^2 or y^3, and so on.

EXAMPLE 8 Multiply. $(3x^2 + 4y^3)(2x^2 + 5y^3)$

$$(3x^2 + 4y^3)(2x^2 + 5y^3) = 6x^4 + 15x^2y^3 + 8x^2y^3 + 20y^6$$
$$= 6x^4 + 23x^2y^3 + 20y^6$$

Practice Problem 8 Multiply. $(2x^2 + 3y^2)(5x^2 + 6y^2)$

▲ ⬭ **EXAMPLE 9**   The width of a living room is $(x + 4)$ feet. The length of the room is $(3x + 5)$ feet. What is the area of the room in square feet?

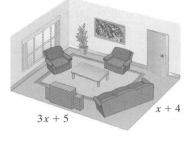

$3x + 5$ $x + 4$

$$A = (\text{length})(\text{width}) = (3x + 5)(x + 4)$$
$$= 3x^2 + 12x + 5x + 20$$
$$= 3x^2 + 17x + 20$$

There are $(3x^2 + 17x + 20)$ square feet in the room.

Practice Problem 9   What is the area in square feet of a room that is $(2x - 1)$ feet wide and $(7x + 3)$ feet long?

Developing Your Study Skills

Exam Time: How to Review

Reviewing adequately for an exam enables you to bring together the concepts you have learned over several sections. For your review, you will need to:

1. Reread your textbook. Make a list of any terms, rules, or formulas you need to know for the exam. Be sure you understand them all.

2. Reread your notes. Go over returned homework and quizzes. Redo the problems you missed.

3. Practice some of each type of problem covered in the chapter(s) you are to be tested on.

4. Use the end-of-chapter materials provided in your textbook. Read carefully through the chapter organizer. Do the review problems. Take the chapter test. When you are finished, check your answers. Redo any problems you missed.

5. Get help if any concepts give you difficulty.

Multiply.

1. $-2x(6x^3 - x)$

2. $5x(-3x^4 + 4x)$

3. $-5x(3x^2 - 2x + 5)$

4. $3x(-2x^2 + 7x - 11)$

5. $2x^3(3x^4 - 2x^3 + 5x - 1)$

6. $5x^2(8 - 4x + 7x^5 - 8x^7)$

7. $\frac{1}{2}(2x + 3x^2 + 5x^3)$

8. $\frac{2}{3}(4x + 6x^2 - 2x^3)$

9. $(5x^3 - 2x^2 + 6x)(-3xy^2)$

10. $(3b^2 - 6b + 8ab)(5b^3)$

11. $(2b^2 + 3b - 4)(2b^2)$

12. $(2x^3 + x^2 - 6x)(2xy)$

13. $(x^3 - 3x^2 + 5x - 2)(3x)$

14. $(-2x^3 + 4x^2 - 7x + 3)(2x)$

15. $(x^2y^2 - 6xy + 8)(-2xy)$

16. $(x^2y^2 + 5xy - 9)(-3xy)$

17. $(-7x^3 + 3x^2 + 2x - 1)(4x^2y)$

18. $(-5x^3 - 6x^2 + x - 1)(5xy^2)$

19. $(3d^4 - 4d^2 + 6)(-2c^2d)$

20. $(-4x^3 + 6x^2 - 5x)(-7xy^2)$

21. $6x^3(2x^4 - x^2 + 3x + 9)$

22. $8x^3(-2x^4 + 3x^2 - 5x - 14)$

23. $-4x^3(6x^4 - 3x^2 + 2)$

24. $-3x^5(-5x^4 + 2x^2 - 3)$

Multiply. Try to do most of the exercises mentally without writing down intermediate steps.

25. $(x + 10)(x + 3)$

26. $(x + 3)(x + 4)$

27. $(x + 6)(x + 2)$

28. $(x - 8)(x + 2)$

29. $(x + 3)(x - 6)$

30. $(x - 5)(x - 4)$

31. $(x - 6)(x - 5)$

32. $(5x - 2)(-x - 3)$

33. $(7x + 1)(-2x - 3)$

34. $(7x - 4)(x + 2y)$

35. $(x - 9)(3x + 4y)$

36. $(2y + 3)(5y - 2)$

37. $(3y + 2)(4y - 3)$

38. $(7y - 2)(3y - 1)$

39. $(5y - 3)(4y - 2)$

To Think About

40. What is wrong with this multiplication?
$(x - 2)(-3) = 3x - 6$

41. What is wrong with this answer?
$-(3x - 7) = -3x - 7$

42. What is the missing term?
$(5x + 2)(5x + 2) = 25x^2 +$ _____ $+ 4$

43. Multiply the binomials and write a brief description of what is special about the result.
$(5x - 1)(5x + 1)$

Multiply.

44. $(4x^2 - 3y)(5x^2 - 2y)$

45. $(3b^2 - 5c)(2b^2 - 7c)$

46. $(8x^2 - 3y^2)(8x^2 - 3y^2)$

47. $(8x - 2)^2$

48. $(5x - 3)^2$

49. $(5a^2 - 3b^2)^2$

50. $(6x^2 + 5y^3)^2$

51. $(0.2x + 3)(4x - 0.3)$

52. $(0.5x - 2)(6x - 0.2)$

53. $(5x + 8y)(6x - y)$

54. $\left(\dfrac{1}{2}x + \dfrac{1}{3}\right)\left(\dfrac{1}{2}x - \dfrac{1}{4}\right)$

55. $\left(\dfrac{1}{3}x + \dfrac{1}{5}\right)\left(\dfrac{1}{3}x - \dfrac{1}{2}\right)$

56. $(2a - 3b)(x - 5b)$

57. $(5b - 7c)(x - 2b)$

58. $(x - 3y)(z - 8y)$

Find the area of the rectangle.

▲ **59.**

5x + 2
2x − 3

▲ **60.**
7x + 3
4x − 6

Cumulative Review Problems

61. Solve for x. $3(x - 6) = -2(x + 4) + 6x$

62. Solve for w. $3(w - 7) - (4 - w) = 11w$

63. La Tanya has three more quarters than dimes, giving her $3.55. How many of each coin does she have? (Use d for the number of dimes.)

64. A number decreased by three is seven more than three times the number. Find the number.

65. Write an equation of the line passing through $(6, -2)$ and $(2, 0)$.

The polynomial $1.5x + 16$ can be used to describe the number of hours per week that young people age 12–18 spend playing video games, where x is the number of years since 1992. (Source: U.S. Department of Health and Human Services)

66. What was the average number of hours per week young people spent playing video games in 1994?

67. What was the average number of hours per week young people spent playing video games in 1996?

68. Predict the number of hours per week young people will spend playing video games in 2002.

69. Predict the number of hours per week young people will spend playing video games in 2003.

5.5 Multiplication: Special Cases

1 ▶ Multiplying Binomials of the Type $(a + b)(a - b)$

The case when you multiply $(x + y)(x - y)$ is interesting and deserves special consideration. Using the FOIL method, we find

$$(x + y)(x - y) = x^2 - xy + xy - y^2 = x^2 - y^2.$$

Notice that the sum of the inner product and the outer product is zero. We see that

$$(x + y)(x - y) = x^2 - y^2.$$

This works in all cases when the binomials are the sum and difference of the same two terms. That is, in one factor the terms are added while in the other factor the same two terms are subtracted.

$$(5a + 2b)(5a - 2b) = 25a^2 - 10ab + 10ab - 4b^2$$
$$= 25a^2 - 4b^2$$

The product is the difference of the squares of the terms. That is, $(5a)^2 - (2b)^2$ or $25a^2 - 4b^2$.

    Many students find it helpful to memorize this equation.

Student Learning Objectives

After studying this section, you will be able to:

1 Multiply binomials of the type $(a + b)(a - b)$.

2 Multiply binomials of the type $(a + b)^2$ and $(a - b)^2$.

3 Multiply polynomials with more than two terms.

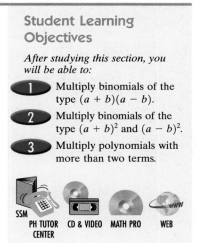

SSM   PH TUTOR   CD & VIDEO   MATH PRO   WEB
     CENTER

> ### Multiplying Binomials: A Sum and a Difference
>
> $$(a + b)(a - b) = a^2 - b^2$$

You may use this relationship to find the product quickly in cases where it applies. The terms must be the same and there must be a sum and a difference.

EXAMPLE 1   Multiply. $(7x + 2)(7x - 2)$

$$(7x + 2)(7x - 2) = (7x)^2 - (2)^2 = 49x^2 - 4$$

Practice Problem 1   Multiply. $(6x + 7)(6x - 7)$

EXAMPLE 2   Multiply. $(5x - 8y)(5x + 8y)$

$$(5x - 8y)(5x + 8y) = (5x)^2 - (8y)^2 = 25x^2 - 64y^2$$

Practice Problem 2   Multiply. $(3x + 5y)(3x - 5y)$

2 ▶ Multiplying Binomials of the Type $(a + b)^2$ and $(a - b)^2$

A second case that is worth special consideration is a binomial that is squared. Consider the following problem.

$$(3x + 2)^2 = (3x + 2)(3x + 2)$$
$$= 9x^2 + 6x + 6x + 4$$
$$= 9x^2 + 12x + 4$$

If you complete enough problems of this type, you will notice a pattern. The answer always contains the square of the first term added to double the product of the first and last terms added to the square of the last term.

$3x$ is the first term	2 is the last term	Square the first term: $(3x)^2$	Double the product of the first and last terms: $2(3x)(2)$	Square the last term: $(2)^2$
↓	↓	↓	↓	↓

$$(3x \quad + \quad 2)^2 \quad = \quad 9x^2 \quad + \quad 12x \quad + \quad 4$$

We can show the same steps using variables instead of words.

$$(a + b)^2 = a^2 + 2ab + b^2$$

There is a similar formula for the square of a difference:

$$(a - b)^2 = a^2 - 2ab + b^2$$

We can use this formula to simplify $(2x - 3)^2$.

$$(2x - 3)^2 = (2x)^2 - 2(2x)(3) + (3)^2$$
$$= 4x^2 - 12x + 9$$

You may wish to multiply this product using FOIL to verify.

These two types of products, the square of a sum and the square of a difference, can be summarized as follows.

A Binomial Squared

$$(a + b)^2 = a^2 + 2ab + b^2$$
$$(a - b)^2 = a^2 - 2ab + b^2$$

EXAMPLE 3 Multiply. **(a)** $(5y - 2)^2$ **(b)** $(8x + 9y)^2$

(a) $(5y - 2)^2 = (5y)^2 - (2)(5y)(2) + (2)^2$
$$= 25y^2 - 20y + 4$$

(b) $(8x + 9y)^2 = (8x)^2 + (2)(8x)(9y) + (9y)^2$
$$= 64x^2 + 144xy + 81y^2$$

Practice Problem 3 Multiply. **(a)** $(5x + 4)^2$ **(b)** $(4a - 9b)^2$

Warning

$(a + b)^2 \neq a^2 + b^2$! The two sides are not equal! Squaring the sum $(a + b)$ does not give $a^2 + b^2$! Beginning algebra students often make this error. Make sure you remember that when you square a binomial there is always a *middle term*.

$$(a + b)^2 = a^2 + 2ab + b^2$$

Sometimes a numerical example helps you to see this.

$$(3 + 4)^2 \neq 3^2 + 4^2$$
$$7^2 \neq 9 + 16$$
$$49 \neq 25$$

Notice that what is missing on the right is $2ab = 2 \cdot 3 \cdot 4 = 24$.

3   *Multiplying Polynomials with More Than Two Terms*

We used the distributive property to multiply two binomials $(a + b)(c + d)$, and we obtained $ac + ad + bc + bd$. We could also use the distributive property to multiply the polynomials $(a + b)$ and $(c + d + e)$, and we would obtain $ac + ad + ae + bc + bd + be$. Let us see if we can find a direct way to multiply products such as $(3x - 2)(x^2 - 2x + 3)$. It can be done quickly using an approach similar to that used in arithmetic for multiplying whole numbers. Consider the following arithmetic problem.

$$
\begin{array}{r}
128 \\
\times \quad 43 \\
\hline
384 \\
512 \\
\hline
5504
\end{array}
$$

$384 \leftarrow$ The product of 128 and 3
$512 \leftarrow$ The product of 128 and 4 moved one space to the left
$5504 \leftarrow$ The sum of the two partial products

Let us follow a similar format to multiply the two polynomials. For example, multiply $(x^2 - 2x + 3)$ and $(3x - 2)$.

$$
\begin{array}{r}
x^2 - \ 2x + 3 \\
3x - 2 \\
\hline
-2x^2 + \ 4x - 6 \\
3x^3 - 6x^2 + \ 9x \\
\hline
3x^3 - 8x^2 + 13x - 6
\end{array}
$$

$-2x^2 + 4x - 6 \leftarrow$ The product $(x^2 - 2x + 3)(-2)$ This is often called **vertical multiplication**.
$3x^3 - 6x^2 + 9x \leftarrow$ The product $(x^2 - 2x + 3)(3x)$ moved one space to the left so that like terms are underneath each other.
$3x^3 - 8x^2 + 13x - 6 \leftarrow$ The sum of the two partial products

 EXAMPLE 4   Multiply vertically. $(3x^3 + 2x^2 + x)(x^2 - 2x - 4)$

$$
\begin{array}{r}
3x^3 + \ 2x^2 + \ x \\
x^2 - \ 2x - 4 \\
\hline
-12x^3 - \ 8x^2 - 4x \\
-6x^4 - \ 4x^3 - \ 2x^2 \\
3x^5 + 2x^4 + \ x^3 \\
\hline
3x^5 - 4x^4 - 15x^3 - 10x^2 - 4x
\end{array}
$$

We place one polynomial over the other.
$-12x^3 - 8x^2 - 4x \leftarrow$ The product $(3x^3 + 2x^2 + x)(-4)$
$-6x^4 - 4x^3 - 2x^2 \leftarrow$ The product $(3x^3 + 2x^2 + x)(-2x)$
$3x^5 + 2x^4 + x^3 \leftarrow$ The product $(3x^3 + 2x^2 + x)(x^2)$
$3x^5 - 4x^4 - 15x^3 - 10x^2 - 4x \leftarrow$ The sum of the three partial products

Note that the answers for each partial product are placed so that like terms are underneath each other.

Practice Problem 4   Multiply vertically. $(3x^2 - 2xy + 4y^2)(x - 2y)$

Alternative Method

Some students prefer to do this type of multiplication using a horizontal format similar to the FOIL method. The following example illustrates this approach.

 EXAMPLE 5   Multiply horizontally. $(x^2 + 3x + 5)(x^2 - 2x - 6)$
We will use the distributive property repeatedly.

$$(x^2 + 3x + 5)(x^2 - 2x - 6) = x^2(x^2 - 2x - 6) + 3x(x^2 - 2x - 6) + 5(x^2 - 2x - 6)$$
$$= x^4 - 2x^3 - 6x^2 + 3x^3 - 6x^2 - 18x + 5x^2 - 10x - 30$$
$$= x^4 + x^3 - 7x^2 - 28x - 30$$

Practice Problem 5   Multiply horizontally. $(2x^2 - 5x + 3)(x^2 + 3x - 4)$

Some problems may need to be done in two or more separate steps.

EXAMPLE 6   Multiply. $(2x - 3y)(x + 2y)(x + y)$

We first need to multiply any two of the binomials. Let us select the first pair.

$$\underbrace{(2x - 3y)(x + 2y)}_{\text{Find this product first.}}(x + y)$$

$$(2x - 3y)(x + 2y) = 2x^2 + 4xy - 3xy - 6y^2$$
$$= 2x^2 + xy - 6y^2$$

Now we replace the first two factors with their resulting product.

$$\underbrace{(2x^2 + xy - 6y^2)}_{\text{Result of first product}}(x + y)$$

We then multiply again.

$$(2x^2 + xy - 6y^2)(x + y) = (2x^2 + xy - 6y^2)x + (2x^2 + xy - 6y^2)y$$
$$= 2x^3 + x^2y - 6xy^2 + 2x^2y + xy^2 - 6y^3$$
$$= 2x^3 + 3x^2y - 5xy^2 - 6y^3$$

The vertical format of Example 4 is an alternative method for this type of problem.

$$
\begin{array}{r}
2x^2 + xy - 6y^2 \\
x + y \\
\hline
2x^2y + xy^2 - 6y^3 \\
2x^3 + x^2y - 6xy^2 \\
\hline
2x^3 + 3x^2y - 5xy^2 - 6y^3
\end{array}
$$

$\Bigg\}$ Be sure to use special care in writing the exponents correctly in problems with more than one variable.

Thus we have

$$(2x - 3y)(x + 2y)(x + y) = 2x^3 + 3x^2y - 5xy^2 - 6y^3.$$

Note that it does not matter which two binomials are multiplied first. For example, you could first multiply $(2x - 3y)(x + y)$ to obtain $2x^2 - xy - 3y^2$ and then multiply that product by $(x + 2y)$ to obtain the same result.

Practice Problem 6   Multiply. $(3x - 2)(2x + 3)(3x + 2)$
(*Hint:* Rearrange the factors.)

5.5 Exercises

Verbal and Writing Skills

1. In the special case of $(a + b)(a - b)$, a binomial times a binomial is a _____.

2. Identify which of the following could be the answer to a problem using the formula for $(a + b)(a - b)$. Why?
(a) $9x^2 - 16$ **(b)** $4x^2 + 25$
(c) $9x^2 + 12x + 4$ **(d)** $x^4 - 1$

3. A student evaluated $(4x - 7)^2$ as $16x^2 + 49$. What is missing? State the correct answer.

4. The square of a binomial, $(a - b)^2$, always produces which of the following?
(a) binomial   **(b)** trinomial
(c) four-term polynomial

Use the formula $(a + b)(a - b) = a^2 - b^2$ to multiply.

5. $(y - 7)(y + 7)$

6. $(x + 5)(x - 5)$

7. $(x - 9)(x + 9)$

8. $(x + 6)(x - 6)$

9. $(8x + 3)(8x - 3)$

10. $(6x + 5)(6x - 5)$

11. $(2x - 7)(2x + 7)$

12. $(3x - 10)(3x + 10)$

13. $(5x - 3y)(5x + 3y)$

14. $(8a - 3b)(8a + 3b)$

15. $(0.6x + 3)(0.6x - 3)$

16. $(5x - 0.2)(5x + 0.2)$

Use the formula for a binomial squared to multiply.

17. $(3y + 1)^2$

18. $(4x - 1)^2$

19. $(5x - 4)^2$

20. $(6x + 5)^2$

21. $(9x + 5)^2$

22. $(5x - 7)^2$

23. $(3x - 7)^2$

24. $(2x + 3y)^2$

25. $\left(\dfrac{2}{3}x + \dfrac{1}{4}\right)^2$

26. $\left(\dfrac{3}{4}x + \dfrac{1}{2}\right)^2$

27. $(6w + 5z)^2$

28. $(5xy - 6z)^2$

Multiply. Use the special formula that applies.

29. $(7x + 3y)(7x - 3y)$

30. $(12a - 5b)(12a + 5b)$

31. $(7c^3 - 6d)^2$

32. $(4f^2 - 5g^2)^2$

Multiply.

33. $(x^2 + 3x - 2)(x - 3)$

34. $(x^2 + 3x - 1)(x + 3)$

35. $(4x + 1)(x^3 - 2x^2 + x - 1)$

36. $(3x - 1)(x^3 + x^2 - 4x - 2)$

37. $(x + 3)(x - 1)(3x - 8)$

38. $(x - 7)(x + 4)(2x - 5)$

39. $(3x + 5)(x - 2)(x - 4)$

40. $(2x - 7)(x + 1)(x - 2)$

41. $(x + 4)(2x - 7)(x - 4)$

42. $(x - 2)(5x + 3)(x - 2)$

43. $(a^2 - 3a + 2)(a^2 + 4a - 3)$

44. $(x^2 + 4x - 5)(x^2 - 3x + 4)$

To Think About

Find the volume.

 45.

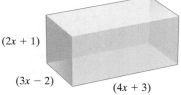

$(2x + 1)$

$(3x - 2)$

$(4x + 3)$

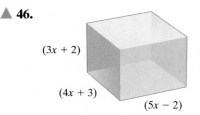 **46.**

$(3x + 2)$

$(4x + 3)$

$(5x - 2)$

Cumulative Review Problems

47. An executive invested $18,000 in two accounts, one a cash-on-reserve account that yielded 7% simple interest, and the other a long-term certificate paying 11% simple interest. At the end of one year, the two accounts had earned her a total of $1540. How much had she invested in each account?

48. The perimeter of a rectangular room measures 34 meters. The width is 2 meters more than half the length. Find the dimensions of the room.

49. Ammonia has a boiling point of $-33.4°$C. What is the boiling point of ammonia measured in degrees Fahrenheit? (Use $F = 1.8C + 32$.)

50. A man with 5.5 liters of blood has approximately 2.75×10^{10} red blood cells. Each red blood cell is a small disk that measures 7.5×10^{-6} meters in diameter. If all of the red blood cells were arranged in a long straight line, how long would that line be?

Putting Your Skills to Work

The Mathematics of DNA

The genetic material that organisms inherit from their parents consists of DNA. Each time a cell reproduces itself by dividing, its DNA is copied and passed on to the next generation. The DNA molecule consists of two strands twisted in the shape of a double helix. Each strand is composed of fundamental units called nucleotides. Each nucleotide on one strand is bonded (base paired) to a complementary nucleotide on the opposing strand. (*Source:* Dr. Russ Camp, Biology Department, Gordon College, Wenham, MA.)

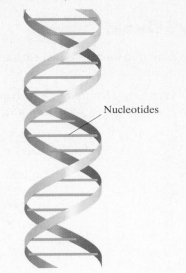

Nucleotides

Problems for Individual Investigation and Analysis

1. *E. coli*, the most popular bacterial cell used in molecular biology research, has a single chromosome made up of one double-stranded DNA molecule. The linear distance from one nucleotide pair to the next is 0.34 nanometer (one nanometer = 1×10^{-9} meter). The entire *E. coli* chromosome contains 4.5×10^6 nucleotide pairs (or base pairs). How long is the chromosome in nanometers?

2. How long is the chromosome in millimeters?

Problems for Cooperative Group Investigation

Together with other members of your class, see if you can determine the following.

3. If a human cell has 46 chromosomes containing a total of 8.0×10^9 nucleotide pairs, and if all the DNA in these chromosomes were added together, how long would the DNA be in meters?

4. How many times larger is the DNA in a human cell than the DNA in an *E. coli* cell?

Internet Connections

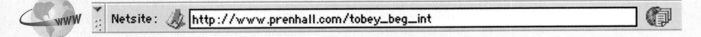

Netsite: http://www.prenhall.com/tobey_beg_int

Introduction to Primer on Molecular Genetics (Department of Energy) or a related site

This site provides some basic information about genes and DNA. Use the information to answer the following questions.

▲ 5. Estimate the volume of DNA in a human genome by assuming that, when the strands of DNA are end to end unwound and tied together, the resulting singular strand has the shape of a right circular cylinder.

6. Tell what percent of a human genome is made up of intron sequences and other noncoding regions.

5.6 Division of Polynomials

Student Learning Objectives

After studying this section, you will be able to:

1 Divide a polynomial by a monomial.

2 Divide a polynomial by a binomial.

SSM · PH TUTOR · CD & VIDEO · MATH PRO · WEB
CENTER

---

1 Dividing a Polynomial by a Monomial

To divide a polynomial by a monomial, divide each term of the numerator by the denominator; then write the sum of the results. We are using the property of fractions that states that

$$\frac{a + b}{c} = \frac{a}{c} + \frac{b}{c}.$$

Dividing a Polynomial by a Monomial

1. Divide each term of the polynomial by the monomial.
2. When dividing variables, use the property $\frac{x^a}{x^b} = x^{a-b}$.

 EXAMPLE 1 Divide. $\dfrac{8y^6 - 8y^4 + 24y^2}{8y^2}$

$$\frac{8y^6 - 8y^4 + 24y^2}{8y^2} = \frac{8y^6}{8y^2} - \frac{8y^4}{8y^2} + \frac{24y^2}{8y^2} = y^4 - y^2 + 3$$

Practice Problem 1 Divide. $\dfrac{15y^4 - 27y^3 - 21y^2}{3y^2}$

---

2 Dividing a Polynomial by a Binomial

Division of a polynomial by a binomial is similar to long division in arithmetic. Notice the similarity in the following division problems.

Division of a three-digit number by a two-digit number

$$
\begin{array}{r}
32 \\
21\overline{)672} \\
\underline{63} \\
42 \\
\underline{42} \\
0
\end{array}
$$

Division of a polynomial by a binomial

$$
\begin{array}{r}
3x + 2 \\
2x + 1\overline{)6x^2 + 7x + 2} \\
\underline{6x^2 + 3x} \\
4x + 2 \\
\underline{4x + 2} \\
0
\end{array}
$$

Dividing a Polynomial by a Binomial

1. Place the terms of the polynomial and binomial in descending order. Insert a 0 for any missing term.
2. Divide the first term of the polynomial by the first term of the binomial. The result is the first term of the answer.
3. Multiply the first term of the answer by the binomial and subtract the result from the first two terms of the polynomial. Bring down the next term to obtain a new polynomial.
4. Divide the new polynomial by the binomial using the process described in step 2.
5. Continue dividing, multiplying, and subtracting until the degree of the remainder is less than the degree of the binomial divisor.
6. Write the remainder as the numerator of a fraction that has the binomial divisor as its denominator.

EXAMPLE 2 Divide. $(x^3 + 5x^2 + 11x + 14) \div (x + 2)$

Step 1 The terms are arranged in descending order. No terms are missing.

Step 2 Divide the first term of the polynomial by the first term of the binomial. In this case, divide x^3 by x to get x^2.

$$x + 2 \overline{)x^3 + 5x^2 + 11x + 14}^{\displaystyle x^2}$$

Step 3 Multiply x^2 by $x + 2$ and subtract the result from the first two terms of the polynomial, $x^3 + 5x^2$ in this case.

$$
\begin{array}{r}
x^2 \\
x + 2 \overline{)x^3 + 5x^2 + 11x + 14} \\
\underline{x^3 + 2x^2} \\
3x^2 + 11x
\end{array}
$$

Bring down the next term.

Step 4 Continue to use the step 2 process. Divide $3x^2$ by x. Write the resulting $3x$ as the next term of the answer.

$$
\begin{array}{r}
x^2 + 3x \\
x + 2 \overline{)x^3 + 5x^2 + 11x + 14} \\
\underline{x^3 + 2x^2} \\
3x^2 + 11x
\end{array}
$$

Step 5 Continue multiplying, dividing, and subtracting until the degree of the remainder is less than the degree of the divisor. In this case, we stop when the remainder does not have an x.

$$
\begin{array}{r}
x^2 + 3x + 5 \\
x + 2 \overline{)x^3 + 5x^2 + 11x + 14} \\
\underline{x^3 + 2x^2} \\
3x^2 + 11x \\
\underline{3x^2 + 6x} \\
5x + 14 \\
\underline{5x + 10} \\
4
\end{array}
$$

$\longleftarrow$ The remainder is 4.

Step 6 The answer is $x^2 + 3x + 5 + \dfrac{4}{x + 2}$.

To check the answer, we multiply $(x + 2)(x^2 + 3x + 5)$ and add the remainder 4.

$$(x + 2)(x^2 + 3x + 5) + 4 = x^3 + 5x^2 + 11x + 10 + 4 = x^3 + 5x^2 + 11x + 14$$

This is the original polynomial. It checks.

Practice Problem 2 Divide. $(x^3 + 10x^2 + 31x + 35) \div (x + 4)$

Take great care with the subtraction step when negative numbers are involved.

EXAMPLE 3 Divide. $(5x^3 - 24x^2 + 9) \div (5x + 1)$

We must first insert $0x$ to represent the missing x-term. Then we divide $5x^3$ by $5x$.

$$
\begin{array}{r}
x^2 \\
5x + 1 \overline{)\, 5x^3 - 24x^2 + 0x + 9} \\
\underline{5x^3 + x^2} \\
- 25x^2
\end{array}
$$

Note that we are subtracting:
$-24x^2 - (+1x^2) = -24x^2 - 1x^2$
$ = -25x^2$

Next we divide $-25x^2$ by $5x$.

$$
\begin{array}{r}
x^2 - 5x \\
5x + 1 \overline{)\, 5x^3 - 24x^2 + 0x + 9} \\
\underline{5x^3 + x^2} \\
- 25x^2 + 0x \\
\underline{- 25x^2 - 5x} \\
5x
\end{array}
$$

Note that we are subtracting:
$0x - (-5x) = 0x + 5x = 5x$

Finally, we divide $5x$ by $5x$.

$$
\begin{array}{r}
x^2 - 5x + 1 \\
5x + 1 \overline{)\, 5x^3 - 24x^2 + 0x + 9} \\
\underline{5x^3 + x^2} \\
- 25x^2 + 0x \\
\underline{- 25x^2 - 5x} \\
5x + 9 \\
\underline{5x + 1} \\
8
\end{array}
$$

⟵——— The remainder is 8.

The answer is $x^2 - 5x + 1 + \dfrac{8}{5x + 1}$.

To check, multiply $(5x + 1)(x^2 - 5x + 1)$ and add the remainder 8.

$$(5x + 1)(x^2 - 5x + 1) + 8 = 5x^3 - 24x^2 + 1 + 8 = 5x^3 - 24x^2 + 9$$

This is the original polynomial. Our answer is correct.

Practice Problem 3 Divide. $(2x^3 - x^2 + 1) \div (x - 1)$

Now we will perform the division by writing a minimum of steps. See if you can follow each step.

EXAMPLE 4   Divide and check. $(12x^3 - 11x^2 + 8x - 4) \div (3x - 2)$

$$
\begin{array}{r}
4x^2 - x + 2 \\
3x - 2 \overline{)\, 12x^3 - 11x^2 + 8x - 4} \\
\underline{12x^3 - 8x^2} \\
-3x^2 + 8x \\
\underline{-3x^2 + 2x} \\
6x - 4 \\
\underline{6x - 4} \\
0
\end{array}
$$

Check. $(3x - 2)(4x^2 - x + 2) = 12x^3 - 3x^2 + 6x - 8x^2 + 2x - 4$

$ = 12x^3 - 11x^2 + 8x - 4$ Our answer is correct.

Practice Problem 4   Divide and check. $(20x^3 - 11x^2 - 11x + 6) \div (4x - 3)$

Developing Your Study Skills

Exam Time: Taking the Exam

Allow yourself plenty of time to get to your exam. You may even find it helpful to arrive a little early in order to collect your thoughts and ready yourself. This will help you feel more relaxed.

After you get your exam, you will find it helpful to do the following.

1. Take two or three moderately deep breaths. Inhale, then exhale slowly. You will feel your entire body begin to relax.

2. Write down on the back of the exam any formulas or ideas that you need to remember.

3. Look over the entire test quickly in order to pace yourself and use your time wisely. Notice how many points each question is worth. Spend more time on items of greater worth.

4. Read directions carefully and be sure to answer all questions clearly. Keep your work neat and easy to read.

5. Ask your instructor about anything that is not clear to you.

6. Answer the questions that are easiest for you first. Then come back to the more difficult ones.

7. Do not get bogged down on one question for too long because it may jeopardize your chances of finishing other problems. Leave the tough question and come back to it when you have time later.

8. Check your work. This will help you to catch minor errors.

9. Stay calm if others leave before you do. You are entitled to use the full amount of allotted time. You will do better on the exam if you take your time and work carefully.

Divide.

1. $\dfrac{25x^4 - 15x^2 + 20x}{5x}$

2. $\dfrac{18b^5 - 12b^3 + 6b^2}{3b^2}$

3. $\dfrac{8y^4 - 12y^3 - 4y^2}{4y^2}$

4. $\dfrac{10y^4 - 35y^3 + 5y^2}{5y^2}$

5. $\dfrac{49x^6 - 21x^4 + 56x^2}{7x^2}$

6. $\dfrac{36x^6 + 54x^4 - 6x^2}{6x^2}$

7. $(48x^7 - 54x^4 + 36x^3) \div 6x^3$

8. $(72x^8 - 56x^5 - 40x^3) \div 8x^3$

Divide. Check your answers for exercises 9–16 by multiplying.

9. $\dfrac{6x^2 + 13x + 5}{2x + 1}$

10. $\dfrac{12x^2 + 19x + 5}{3x + 1}$

11. $\dfrac{x^2 - 9x - 6}{x - 7}$

12. $\dfrac{x^2 - 8x - 4}{x - 6}$

13. $\dfrac{3x^3 - x^2 + 4x - 2}{x + 1}$

14. $\dfrac{2x^3 - 3x^2 - 3x + 6}{x - 1}$

15. $\dfrac{4x^3 + 4x^2 - 19x - 15}{2x + 5}$

16. $\dfrac{6x^3 + 11x^2 - 8x + 5}{2x + 5}$

17. $\dfrac{6x^3 + x^2 - 7x + 2}{3x - 1}$

18. $\dfrac{3x^3 + 8x^2 - 6x + 2}{3x - 1}$

19. $\dfrac{12y^3 - 12y^2 - 25y + 31}{2y - 3}$

20. $\dfrac{9y^3 - 30y^2 + 31y - 4}{3y - 5}$

21. $\left(y^3 - y^2 - 13y - 12\right) \div (y + 3)$

22. $\left(4y^3 - 17y^2 + 7y + 10\right) \div (4y - 5)$

23. $\left(y^4 - 9y^2 - 5\right) \div (y - 2)$

24. $\left(2y^4 + 3y^2 - 5\right) \div (y - 2)$

To Think About

25. Divide. $(8y^3 + 3y - 7) \div (4y - 1)$

(*Hint:* The answer contains fractions.)

26. Divide. $(6y^3 - 3y^2 + 4) \div (2y + 1)$

(*Hint:* The answer contains fractions.)

Cumulative Review Problems

27. Most environmental groups feel that with improved efficiency of water-using devices (washers, toilets, faucets, shower heads), the number of gallons of water used in the average United States home can be reduced to 77,000 per year. This would reflect a 30% decrease. How many gallons of water is the average home using now?

28. From 1900 to 2000, the population of New York City increased by 123% to a total of 7.6 million. (*Source:* U.S. Census Bureau.) What was the population of the city in 1900?

29. In a veterinary study of a group of cats over age 5 who experienced regular exercise and playtime, 184 of the cats, or 92%, had no health problems. How many cats were in the study?

30. Marlena's job requires doing research on the Internet for a marketing company. Today, she has downloaded a huge file. She has been reading all morning and has just completed two consecutive pages whose page numbers add to 1039. What are the page numbers?

Putting Your Skills to Work

The Enrollment Polynomial

The U.S. National Center for Education Statistics maintains records of past enrollments of students in school as well as projected enrollments for the future. The following bar graph below indicates the number of students in school for several selected years and includes some projected figures for the future.

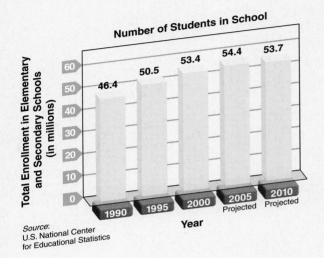

Number of Students in School

Source:
U.S. National Center
for Educational Statistics

Problems for Individual Investigation and Study

1. How many more students were enrolled in school in 1995 than in 1990?

2. How many fewer students are projected to be enrolled in school in 2010 than in 2005?

Problems for Group Investigation and Cooperative Learning

The approximate total number of students enrolled in the United States during a given year can be obtained by using the Enrollment Polynomial. It is written as

$$E = -0.00031x^3 - 0.024268x^2 + 0.975171x + 46.3761,$$

where x is the number of years since 1990 and E is the total number of students in millions. Use a scientific calculator, a graphing calculator, or a computer to evaluate the Enrollment Polynomial and answer the following questions. Round to the nearest tenth of a million.

3. How many students were enrolled in 1998?

4. How many students were enrolled in 1994?

5. What is the projected number of students who will be enrolled in 2006?

6. What is the projected number of students who will be enrolled in 2009?

Internet Connections

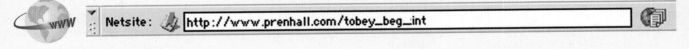

WWW Netsite: http://www.prenhall.com/tobey_beg_int

Site: The U.S. Department of Education

7. Based on the current projections, what is the number of classroom teachers that will be needed in 2006? Give the figures for the low, medium, and high estimates.

8. Comparing elementary teachers with secondary teachers, which group will have experienced the higher percentage growth in the period 1994–2006? The higher absolute growth? Consider the low, medium, and high estimates.

Math in the Media

Tallest Hotel in the World

Source: Guinness World Book®.com

According to the *Guinness World Book®*, the Burj Al Arab Hotel (Arabian Tower) is the world's tallest hotel. It is located south of Dubai, United Arab Emirates. The hotel stands 1052 feet tall and is shaped like a sail. This structure also features the world's fastest elevators which travel at 15 miles per hour, 22 feet per second.

EXERCISES

If an object is dropped from the top of the hotel, could you calculate the height of the object at specified times? Try the following questions.

1. Neglecting air resistance, the height of the object at a time t seconds after the object is dropped is given by the polynomial $-16t^2 + 1052$. Find the height of the object when $t = 1$ second and when $t = 5$ seconds.

2. Using the polynomial from exercise 1, how long before the object hits the ground? Use a scientific calculator or a graphing calculator to assist you. Find the answer to the nearest tenth of a second.

3. Use your results from exercise 1. Suppose an elevator is traveling at maximum speed from the height at $t = 1$ second to the height when $t = 5$ seconds. How much longer does it take the elevator to travel this distance than it took the object to fall the same distance? Round your answer to the nearest tenth of a second.

Chapter 5 Organizer

Topic	Procedure	Examples
Multiplying monomials, p. 264.	$x^a \cdot x^b = x^{a+b}$ **1.** Multiply the numerical coefficients. **2.** Add the exponents of a given base.	$3^{12} \cdot 3^{15} = 3^{27}$ $x^3 \cdot x^4 = x^7$ $(-3x^2)(6x^3) = -18x^5$ $(2ab)(4a^2b^3) = 8a^3b^4$
Dividing monomials, p. 267.	$\dfrac{x^a}{x^b} = \begin{cases} x^{a-b} & \text{Use if } a \text{ is greater than } b. \\ \dfrac{1}{x^{b-a}} & \text{Use if } b \text{ is greater than } a. \end{cases}$ **1.** Divide or reduce the fraction created by the quotient of the numerical coefficients. **2.** Subtract the exponents of a given base.	$\dfrac{16x^7}{8x^3} = 2x^4$ $\dfrac{5x^3}{25x^5} = \dfrac{1}{5x^2}$ $\dfrac{-12x^5y^7}{18x^3y^{10}} = -\dfrac{2x^2}{3y^3}$
Exponent of zero, p. 267.	$x^0 = 1 \qquad \text{if } x \neq 0$	$5^0 = 1 \qquad \dfrac{x^6}{x^6} = 1$ $w^0 = 1 \qquad 3x^0y = 3y$
Raising a power to a power, p. 268.	$(x^a)^b = x^{ab}$ $(xy)^a = x^a y^a$ $\left(\dfrac{x}{y}\right)^a = \dfrac{x^a}{y^a} \quad (y \neq 0)$ **1.** Raise the numerical coefficient to the power outside the parentheses. **2.** Multiply the exponent outside the parentheses times the exponent inside the parentheses.	$(x^9)^3 = x^{27}$ $(3x^2)^3 = 27x^6$ $\left(\dfrac{2x^2}{y^3}\right)^3 = \dfrac{8x^6}{y^9}$ $(-3x^4y^5)^4 = 81x^{16}y^{20}$ $(-5ab)^3 = -125a^3b^3$
Negative exponents, p. 273.	If $x \neq 0$ and $y \neq 0$, then $x^{-n} = \dfrac{1}{x^n}$ $\dfrac{1}{x^{-n}} = x^n$ $\dfrac{x^{-m}}{y^{-n}} = \dfrac{y^n}{x^m}$	Write with positive exponents. $3^{-4} = \dfrac{1}{3^4} = \dfrac{1}{81}$ $x^{-6} = \dfrac{1}{x^6}$ $\dfrac{1}{w^{-3}} = w^3$ $\dfrac{w^{-12}}{z^{-5}} = \dfrac{z^5}{w^{12}}$ $(2x^2)^{-3} = 2^{-3}x^{-6} = \dfrac{1}{2^3x^6} = \dfrac{1}{8x^6}$
Scientific notation, p. 274.	A positive number is written in scientific notation if it is in the form $a \times 10^n$, where $1 \leq a < 10$ and n is an integer.	$128 = 1.28 \times 10^2$ $2{,}568{,}000 = 2.568 \times 10^6$ $13{,}200{,}000{,}000 = 1.32 \times 10^{10}$ $0.16 = 1.6 \times 10^{-1}$ $0.00079 = 7.9 \times 10^{-4}$ $0.0000034 = 3.4 \times 10^{-6}$
Adding polynomials, p. 282.	To add two polynomials, we add their like terms.	$(-7x^3 + 2x^2 + 5) + (x^3 + 3x^2 + x)$ $= -6x^3 + 5x^2 + x + 5$

Topic	Procedure	Examples
Subtracting polynomials, p. 283.	To subtract polynomials, change all signs of the second polynomial and add the result to the first polynomial. $$(a) - (b) = (a) + (-b)$$	$(5x^2 - 6) - (-3x^2 + 2) = (5x^2 - 6) + (+3x^2 - 2)$ $\qquad\qquad\qquad\qquad = 8x^2 - 8$
Multiplying a monomial by a polynomial, p. 287.	Use the distributive property. $$a(b + c) = ab + ac$$ $$(b + c)a = ba + ca$$	Multiply. $-5x(2x^2 + 3x - 4) = -10x^3 - 15x^2 + 20x$ $(6x^3 - 5xy - 2y^2)(3xy) = 18x^4y - 15x^2y^2 - 6xy^3$
Multiplying two binomials, p. 287, 293.	1. The product of the sum and difference of the same two terms yields the difference of their squares. $$(a + b)(a - b) = a^2 - b^2$$ 2. The square of a binomial yields a trinomial: the square of the first term plus twice the product of the first and second terms, plus the square of the second term. $$(a + b)^2 = a^2 + 2ab + b^2$$ $$(a - b)^2 = a^2 - 2ab + b^2$$ 3. Use FOIL for other binomial multiplication. The middle terms can often be combined, giving a trinomial answer.	Multiply. $(3x + 7y)(3x - 7y) = 9x^2 - 49y^2$ $(3x + 7y)^2 = 9x^2 + 42xy + 49y^2$ $(3x - 7y)^2 = 9x^2 - 42xy + 49y^2$ $(3x - 5)(2x + 7) = 6x^2 + 21x - 10x - 35$ $\qquad\qquad\qquad\quad = 6x^2 + 11x - 35$
Multiplying two polynomials, p. 295.	To multiply two polynomials, multiply each term of one by each term of the other. This method is similar to the multiplication of many-digit numbers.	Vertical method: $$\begin{array}{r} 3x^2 - 7x + 4 \\ \times \qquad 3x - 1 \\ \hline -3x^2 + 7x - 4 \\ 9x^3 - 21x^2 + 12x \qquad \\ \hline 9x^3 - 24x^2 + 19x - 4 \end{array}$$ Horizontal method: $(5x + 2)(2x^2 - x + 3)$ $\quad = 10x^3 - 5x^2 + 15x + 4x^2 - 2x + 6$ $\quad = 10x^3 - x^2 + 13x + 6$
Multiplying three or more polynomials, p. 296.	1. Multiply any two polynomials. 2. Multiply the result by any remaining polynomials.	$(2x + 1)(x - 3)(x + 4) = (2x^2 - 5x - 3)(x + 4)$ $$\begin{array}{r} 2x^2 - 5x - 3 \\ x + 4 \\ \hline 8x^2 - 20x - 12 \\ 2x^3 - 5x^2 - 3x \qquad \\ \hline 2x^3 + 3x^2 - 23x - 12 \end{array}$$
Dividing a polynomial by a monomial, p. 300.	1. Divide each term of the polynomial by the monomial. 2. When dividing variables, use the property $$\frac{x^a}{x^b} = x^{a-b}.$$	Divide. $(15x^3 + 20x^2 - 30x) \div (5x)$ $\qquad = \dfrac{15x^3}{5x} + \dfrac{20x^2}{5x} + \dfrac{-30x}{5x}$ $\qquad = 3x^2 + 4x - 6$

Topic	Procedure	Examples
Dividing a polynomial by a binomial, p. 300.	1. Place the terms of the polynomial and binomial in descending order. Insert a 0 for any missing term. 2. Divide the first term of the polynomial by the first term of the binomial. 3. Multiply the partial answer by the binomial, and subtract the result from the first two terms of the polynomial. Bring down the next term to obtain a new polynomial. 4. Divide the new polynomial by the binomial using the process described in step 2. 5. Continue dividing, multiplying, and subtracting until the degree of the remainder is less than the degree of the binomial divisor. 6. Write the remainder as the numerator of a fraction that has the binomial divisor as its denominator.	Divide. $$(8x^3 - 13x + 2x^2 + 7) \div (4x - 1)$$ We rearrange the terms. $$\begin{array}{r} 2x^2 + x - 3 \\ 4x - 1\overline{)8x^3 + 2x^2 - 13x + 7} \\ \underline{8x^3 - 2x^2} \\ 4x^2 - 13x \\ \underline{4x^2 - x} \\ -12x + 7 \\ \underline{-12x + 3} \\ 4 \end{array}$$ The answer is $$2x^2 + x - 3 + \frac{4}{4x - 1}.$$

Chapter 5 Review Problems

5.1 *Simplify. Leave your answer in exponent form.*

1. $(-6a^2)(3a^5)$

2. $(5^{10})(5^{13})$

3. $(3xy^2)(2x^3y^4)$

4. $\dfrac{8^{20}}{8^3}$

5. $\dfrac{7^{15}}{7^{27}}$

6. $\dfrac{x^{12}}{x^{17}}$

7. $\dfrac{y^{30}}{y^{16}}$

8. $\dfrac{3x^8y^0}{9x^4}$

9. $\dfrac{-15xy^2}{25x^6y^6}$

10. $\dfrac{-12a^3b^6}{18a^2b^{12}}$

11. $(x^3)^8$

12. $(5xy^2)^3$

13. $(-3a^3b^2)^2$

14. $\dfrac{2x^4}{3y^2}$

15. $\left(\dfrac{5ab^2}{c^3}\right)^2$

16. $\left(\dfrac{x^0y^3}{4w^5z^2}\right)^3$

5.2 *Simplify. Write with positive exponents.*

17. x^{-3}

18. $x^{-5}y^{-11}$

19. $\dfrac{2x^{-6}}{y^{-3}}$

20. $2^{-1}x^5y^{-6}$

21. $\left(2x^3\right)^{-2}$

22. $\dfrac{3x^{-3}}{y^{-2}}$

23. $\dfrac{4x^{-5}y^{-6}}{w^{-2}z^8}$

24. $\dfrac{3^{-3}a^{-2}b^5}{c^{-3}d^{-4}}$

Write in scientific notation.

25. 156,340,200,000

26. 179,632

27. 0.0078

28. 0.00006173

Write in decimal notation.

29. 1.2×10^5

30. 8.367×10^{10}

31. 3×10^6

32. 2.5×10^{-1}

33. 5.708×10^{-8}

34. 6×10^{-9}

Perform the indicated calculation. Leave your answer in scientific notation.

35. $\dfrac{(28,000,000)(5,000,000,000)}{7000}$

36. $\left(3.12 \times 10^5\right)\left(2.0 \times 10^6\right)\left(1.5 \times 10^8\right)$

37. $\left(1.6 \times 10^{-3}\right)\left(3.0 \times 10^{-5}\right)\left(2.0 \times 10^{-2}\right)$

38. $\dfrac{(0.00078)(0.000005)(0.00004)}{0.002}$

39. If a space probe travels at 40,000 kilometers per hour for 1 year, how far will it travel? (Assume that 1 year = 365 days.)

40. An atomic clock is based on the fact that cesium emits 9,192,631,770 cycles of radiation in one second. How many of these cycles occur in one day? Round to three significant digits.

41. Today's fastest modern computers can perform one operation in 1×10^{-8} second. How many operations can such a computer perform in 1 minute?

5.3 *Combine.*

42. $\left(2x^2 - 3x + 5\right) + \left(-7x^2 - 8x - 23\right)$

43. $\left(1.2x^2 - 3.4x + 6\right) + \left(5.5x^2 - 7.6x - 3\right)$

44. $\left(x^3 + x^2 - 6x + 2\right) - \left(2x^3 - x^2 - 5x - 6\right)$

45. $\left(4x^3 - x^2 - x + 3\right) - \left(-3x^3 + 2x^2 + 5x - 1\right)$

46. $\left(9x^3y^3 + 3xy - 4\right) - \left(4x^3y^3 + 2x^2y^2 - 7xy\right)$

47. $\dfrac{1}{2}x^2 - \dfrac{3}{4}x + \dfrac{1}{5} - \left(\dfrac{1}{4}x^2 - \dfrac{1}{2}x + \dfrac{1}{10}\right)$

48. $\left(5x^2 + 3x\right) + \left(-6x^2 + 2\right) - \left(5x - 8\right)$

49. $\left(2x^2 - 7\right) - \left(3x^2 - 4\right) + \left(-5x^2 - 6x\right)$

5.4 *Multiply.*

50. $(3x + 1)(5x - 1)$ **51.** $(7x - 2)(4x - 3)$ **52.** $(2x + 3)(10x + 9)$

53. $5x(2x^2 - 6x + 3)$ **54.** $(3x^2y^2 - 5xy + 6)(-2xy)$ **55.** $(x^3 - 3x^2 + 5x - 2)(4x)$

56. $(5a + 7b)(a - 3b)$ **57.** $(2x^2 - 3)(4x^2 - 5y)$ **58.** $-3x^2y(5x^4y + 3x^2 - 2)$

5.5 *Multiply.*

59. $(3x - 2)^2$ **60.** $(5x + 3)(5x - 3)$ **61.** $(7x + 6y)(7x - 6y)$

62. $(5a - 2b)^2$ **63.** $(8x + 9y)^2$ **64.** $(x^2 + 7x + 3)(4x - 1)$

65. $(x - 6)(2x - 3)(x + 4)$

5.6 *Divide.*

66. $(12y^3 + 18y^2 + 24y) \div (6y)$ **67.** $(30x^5 + 35x^4 - 90x^3) \div (5x^2)$

68. $(16x^3y^2 - 24x^2y + 32xy^2) \div (4xy)$ **69.** $(106x^6 - 24x^5 + 38x^4 + 26x^3) \div (2x^3)$

70. $(16x^2 - 8x - 3) \div (4x - 3)$ **71.** $(15x^2 + 41x + 14) \div (5x + 2)$

72. $(6x^3 + x^2 + 6x + 5) \div (2x - 1)$ **73.** $(2x^3 - x^2 + 3x - 1) \div (x + 2)$

74. $(12x^2 + 11x + 2) \div (3x + 2)$ **75.** $(8x^2 - 6x + 6) \div (2x + 1)$

76. $(x^3 - x - 24) \div (x - 3)$ **77.** $(2x^3 - 3x + 1) \div (x - 2)$

Applications

Solve. Express your answer in scientific notation.

78. In 2000, the estimated population of China was 1.256×10^9 people while the estimated population of Brazil was 1.74×10^8 people. (*Source:* United Nations Statistical Bureau.) What was the total population in those two countries?

79. In 2000, the estimated population of India was 1.018×10^9 people while the estimated population of Bangladesh was 1.29×10^8 people. (*Source:* United Nations Statistical Bureau.) What was the total population in those two countries?

80. The mass of an electron is approximately 9.11×10^{-28} gram. Find the mass of 30,000 electrons.

81. The sun radiates energy into space at the rate of 3.9×10^{26} joules per second. How many joules are emitted in a day?

To Think About

Find a polynomial that describes the shaded area.

▲ **82.**

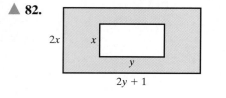

▲ **83.**

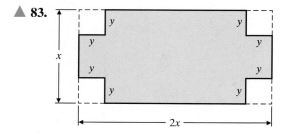

Simplify. Leave your answer in exponent form.

1. $(3^{10})(3^{24})$

2. $\dfrac{3^{35}}{3^7}$

3. $(8^4)^6$

In questions 4–8, simplify.

4. $(-3xy^4)(-4x^3y^6)$

5. $\dfrac{-35x^8y^{10}}{25x^5y^{10}}$

6. $(-5xy^6)^3$

7. $\left(\dfrac{7a^7b^2}{3c^0}\right)^2$

8. $\dfrac{4a^5b^6}{16a^{10}b^{12}}$

9. Evaluate. 4^{-3}

In questions 10 and 11, write with only positive exponents.

10. $6a^{-4}b^{-3}c^5$

11. $\dfrac{2x^{-3}y^{-4}}{w^{-6}z^8}$

12. Write in scientific notation. 0.0005482

13. Write in decimal notation. 5.82×10^8

14. Multiply. Leave your answer in scientific notation.
$(4.0 \times 10^{-3})(3.0 \times 10^{-8})(2.0 \times 10^4)$

Combine.

15. $(2x^2 - 3x - 6) + (-4x^2 + 8x + 6)$

16. $(5x^2 + 6xy - 7y^2) - (2x^2 + 3xy + 6y)$

Multiply.

17. $-7x^2(3x^3 - 4x^2 + 6x - 2)$

18. $(5x^2y^2 - 6xy + 2)(3x^2y)$

19. $(5a - 4b)(2a + 3b)$

20. $(3x + 2)(2x + 1)(x - 3)$

21. $(7x^2 + 2y^2)^2$

22. $(9x - 2y)(9x + 2y)$

23. $(3x - 2)(4x^3 - 2x^2 + 7x - 5)$

24. $(3x^2 - 5xy)(x^2 + 3xy)$

Divide.

25. $15x^6 - 5x^4 + 25x^3 \div 5x^3$

26. $(8x^3 - 22x^2 - 5x + 12) \div (4x + 3)$

27. $(2x^3 - 6x - 36) \div (x - 3)$

Solve. Express your answer in scientific notation. Round to the nearest hundredth.

28. The estimated population of the United States in 2000 was 2.749×10^8 people. The area of the United States is approximately 3.50×10^6 square miles. (*Source:* U.S. Census Bureau.) How many people per square mile were there in the United States in 2000?

29. A space probe is traveling from Earth to the planet Pluto at a speed of 2.49×10^4 miles per hour. How far would this space probe travel in one week?

1. _____

2. _____

3. _____

4. _____

5. _____

6. _____

7. _____

8. _____

9. _____

10. _____

11. _____

12. _____

13. _____

14. _____

15. _____

16. _____

17. _____

18. _____

19. _____

20. _____

21. _____

22. _____

23. _____

24. _____

25. _____

26. _____

27. _____

28. _____

29. _____

1. _____

2. _____

3. _____

4. _____

5. _____

6. _____

7. _____

8. _____

9. _____

10. _____

Approximately one-half of this test covers the content of Chapters 1 through 4. The remainder covers the content of Chapter 5.

1. Simplify. $7x(3x - 4) - 5x(2x - 3) - (3x)^2$

2. Evaluate $2x^2 - 3xy + y^2$ when $x = -2$ and $y = 3$.

In questions 3–5, solve.

3. $7x - 3(4 - 2x) = 14x - (3 - x)$

4. $\dfrac{x - 5}{3} = \dfrac{1}{4}x + 2$

5. $4 - 7x < 11$

6. Solve for f. $B = \frac{1}{2}a(c + 3f)$

7. A national walkout of 11,904 employees of the VBM Corp. occurred last month. This was 96% of the total number of employees. How many employees does VBM have?

8. Graph. $2y = 3x + 4$

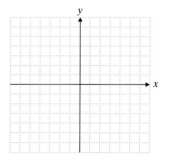

9. Find an equation for the line passing through $(2, 3)$ and $(-1, 4)$.

10. Graph. $y = x^2 + 2$

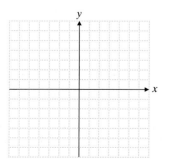

11. For $f(x) = -x^2 - 5x + 7$, find $f(-1)$.

12. Multiply. $(3x - 7)(5x - 4)$

13. Multiply. $(3x - 5)^2$

14. Multiply. $(3x + 2)(2x + 1)(x - 4)$

In questions 15–17, simplify.

15. $(-4x^4y^5)(5xy^3)$

16. $\dfrac{14x^8y^3}{-21x^5y^{12}}$

17. $(-3xy^4z^2)^3$

18. Write with only positive exponents. $\dfrac{9x^{-3}y^{-4}}{w^2z^{-8}}$

19. Write in scientific notation. $1{,}360{,}000{,}000{,}000{,}000$

20. Write in scientific notation. 0.00056

21. Calculate. Leave your answer in scientific notation. $\dfrac{(2.0 \times 10^{-12})(8.0 \times 10^{-20})}{4.0 \times 10^3}$

22. Subtract. $(x^3 - 3x^2 - 5x + 20) - (-4x^3 - 10x^2 + x - 30)$

Multiply.

23. $-6xy^2(6x^2 - 3xy + 8y^2)$

24. $(x^2 - 6x + 3)(2x^2 - 3x + 4)$

25. $(x^2 + 2x - 12) \div (x - 3)$

11. _____

12. _____

13. _____

14. _____

15. _____

16. _____

17. _____

18. _____

19. _____

20. _____

21. _____

22. _____

23. _____

24. _____

25. _____

Factoring

I f you buy new office equipment or a new vehicle and you use it for your business, you will need to keep track of how much these items depreciate in value each year for tax purposes. The IRS has three methods for calculating the depreciation. One of them is called the sum-of-the-year's-digits method. This common math calculation often gives people some difficulty. Do you think you could perform this calculation? Turn to the Putting Your Skills to Work problems on page 366 to find out.

1. _____

2. _____

3. _____

4. _____

5. _____

6. _____

7. _____

8. _____

9. _____

10. _____

11. _____

12. _____

13. _____

14. _____

15. _____

16. _____

17. _____

18. _____

19. _____

20. _____

21. _____

22. _____

23. _____

24. _____

If you are familiar with the topics in this chapter, take this test now. Check your answers with those in the back of the book. If an answer is wrong or you can't answer a question, study the appropriate section of the chapter.

If you are not familiar with the topics in this chapter, don't take this test now. Instead, study the examples, work the practice problems, and then take the test.

This test will help you to identify which concepts you have mastered and which you need to study further.

Section 6.1

Factor out the greatest common factor.

1. $2x^2 - 6xy + 12xy^2$

2. $3x(a - 2b) + 4y(a - 2b)$

3. $36ab^2 - 18ab$

Section 6.2

Factor.

4. $5a - 10b - 3ax + 6xb$

5. $3x^2 - 4y + 3xy - 4x$

6. $21x^2 - 14x - 9x + 6$

Section 6.3

Factor.

7. $x^2 - 22x - 48$

8. $x^2 - 8x + 15$

9. $x^2 + 9x + 8$

10. $2x^2 + 8x - 24$

11. $3x^2 - 6x - 189$

Section 6.4

Factor.

12. $15x^2 - 16x + 4$

13. $6y^2 + 5yz - 6z^2$

14. $12x^2 + 44x + 40$

Section 6.5

Factor completely.

15. $81x^4 - 16$

16. $49x^2 - 28xy + 4y^2$

17. $25x^2 + 80x + 64$

Section 6.6

Factor completely. If not possible, so state.

18. $6x^3 + 15x^2 - 9x$

19. $32x^2y^2 - 48xy^2 + 18y^2$

20. $25x^2 + 81$

Section 6.7

Solve for the roots of each quadratic equation.

21. $2x^2 + x - 3 = 0$

22. $x^2 + 30 = 11x$

23. $\dfrac{3x^2 - 7x}{2} = 3$

▲ **24.** The walking trail to the top of Mount Washington is marked by triangular signs. The area of each sign is 60 square centimeters. The base of each sign is 1 centimeter less than double the altitude. Find the altitude and the base of the triangular signs.

6.1 Introduction to Factoring

1 *Factoring Polynomials Whose Terms Contain a Common Factor*

Student Learning Objectives

After studying this section, you will be able to:

1   Factor polynomials whose terms contain a common factor.

SSM   PH TUTOR CENTER   CD & VIDEO   MATH PRO   WEB

Recall that when two or more numbers, variables, or algebraic expressions are multiplied, each is called a **factor**.

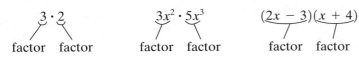

$$3 \cdot 2 \qquad 3x^2 \cdot 5x^3 \qquad (2x - 3)(x + 4)$$

$$\text{factor} \quad \text{factor} \qquad \text{factor} \quad \text{factor} \qquad \text{factor} \quad \text{factor}$$

When you are asked **to factor** a number or an algebraic expression, you are being asked, "What factors, when multiplied, will give that number or expression?"

For example, you can factor 6 as $3 \cdot 2$ since $3 \cdot 2 = 6$. You can factor $15x^5$ as $3x^2 \cdot 5x^3$ since $3x^2 \cdot 5x^3 = 15x^5$. Factoring is simply the reverse of multiplying. 6 and $15x^5$ are simple expressions to factor and can be factored in different ways.

The factors of the polynomial $2x^2 + x - 12$ are not so easy to recognize. In this chapter we will be learning techniques for finding the factors of a polynomial. We will begin with **common factors**.

EXAMPLE 1   Factor.   **(a)** $3x - 6y$ **(b)** $9x + 2xy$

Begin by looking for a common factor, a factor that both terms have in common. Then rewrite the expression as a product.

(a) $3x - 6y = 3(x - 2y)$ This is true because $3(x - 2y) = 3x - 6y$.
(b) $9x + 2xy = x(9 + 2y)$ This is true because $x(9 + 2y) = 9x + 2xy$.

Some people find it helpful to think of factoring as the distributive property in reverse. When we write $3x - 6y = 3(x - 2y)$, we are doing the reverse of distributing the 3.

Practice Problem 1   Factor.   **(a)** $21a - 7b$ **(b)** $p + prt$

When we factor, we begin by looking for the **greatest common factor**. For example, in the polynomial $48x - 16y$, a common factor is 2. We could factor $48x - 16y$ as $2(24x - 8y)$. However, this is not complete. To factor $48x - 16y$ completely, we look for the greatest common factor of 48 and of 16.

$$48x - 16y = 16(3x - y)$$

EXAMPLE 2   Factor $24xy + 12x^2 + 36x^3$. Be careful to remove the greatest common factor.

Find the greatest common factor of 24, 12, and 36. You may want to factor each number, or you may notice that 12 is a common factor. 12 is the greatest numerical common factor.

Notice also that x is a factor of each term. Thus, $12x$ is the greatest common factor.

$$24xy + 12x^2 + 36x^3 = 12x(2y + x + 3x^2)$$

Practice Problem 2   Factor $12a^2 + 16ab^2 - 12a^2b$. Be careful to remove the greatest common factor.

Common Factors of a Polynomial

1. You can determine the greatest common numerical factor by asking, "What is the largest integer that will divide into the coefficient of all the terms?"
2. You can determine the greatest common variable factor by asking, "What variables are common to all the terms, and what is the smallest exponent on each of those variables?"

EXAMPLE 3 Factor. **(a)** $12x^2 + 18y^2$ **(b)** $x^2y^2 + 3xy^2 + y^3$

(a) Note that the largest integer that is common to both terms is 6 (not 3 or 2).

$$12x^2 + 18y^2 = 6(2x^2 + 3y^2)$$

(b) Although y is common to all of the terms, we factor out y^2 since 2 is the largest exponent of y that is common to all terms. We do not factor out x since x is not common to all of the terms.

$$x^2y^2 + 3xy^2 + y^3 = y^2(x^2 + 3x + y)$$

Practice Problem 3 Factor. **(a)** $16a^3 - 24b^3$ **(b)** $r^3s^2 - 4r^4s + 7r^5$

Checking

You can check any factoring problem by multiplying the factors you obtain. The result should be the same as the original polynomial.

EXAMPLE 4 Factor. $8x^3y + 16x^2y^2 + 24x^3y^3$

We see that 8 is the largest integer that will divide evenly into the three numerical coefficients. We can factor an x^2 out of each term. We can also factor y out of each term.

$$8x^3y + 16x^2y^2 + 24x^3y^3 = 8x^2y(x + 2y + 3xy^2)$$

Check.

$$8x^2y(x + 2y + 3xy^2) = 8x^3y + 16x^2y^2 + 24x^3y^3 \checkmark$$

Practice Problem 4 Factor. $18a^3b^2c - 27ab^3c^2 - 45a^2b^2c^2$

EXAMPLE 5 Factor. $9a^3b^2 + 9a^2b^2$

We observe that both terms contain a common factor of 9. We can factor a^2 and b^2 out from each term.

$$9a^3b^2 + 9a^2b^2 = 9a^2b^2(a + 1)$$

🚫 **WARNING** Don't forget to include the 1 inside the parentheses in Example 5. The solution is wrong without it. You will see why if you try to check a result written without the 1.

Practice Problem 5 Factor and check. $30x^3y^2 - 24x^2y^2 + 6xy^2$

● **EXAMPLE 6** Factor. $3x(x - 4y) + 2(x - 4y)$

Be sure you understand what are terms and what are factors of the polynomial in this example. There are two terms. The expression $3x(x - 4y)$ is one term. The expression $2(x - 4y)$ is the second term. Each term is made up of two factors. Observe that the binomial $(x - 4y)$ is a common factor of the terms. A common factor may be any type of polynomial. Thus we can factor out the common factor $(x - 4y)$.

$$3x(x - 4y) + 2(x - 4y) = (x - 4y)(3x + 2)$$

Practice Problem 6 Factor. $3(a + b) + x(a + b)$ ●

● **EXAMPLE 7** Factor. $7x^2(2x - 3y) - (2x - 3y)$

The common factor of the terms is $(2x - 3y)$. What happens when we factor out $(2x - 3y)$? What are we left with in the second term?

Recall that $(2x - 3y) = 1(2x - 3y)$. Thus

$7x^2(2x - 3y) - (2x - 3y) = 7x^2(2x - 3y) - 1(2x - 3y)$ Rewrite the original expression.

$= (2x - 3y)(7x^2 - 1)$ Factor out $(2x - 3y)$.

Practice Problem 7 Factor. $8y(9y^2 - 2) - (9y^2 - 2)$ ●

▲ ● **EXAMPLE 8** A computer programmer is writing a program to find the area of 4 circles. She uses the formula $A = \pi r^2$. The radii of the circles are a, b, c, and d, respectively. She wants the final answer to be in factored form with the value of π occurring only once to minimize round-off error. Write the total area with a formula that has π occurring only once.

For each circle, $A = \pi r^2$, where $r = a, b, c,$ or d.
The total area is $\pi a^2 + \pi b^2 + \pi c^2 + \pi d^2$.
In factored form the total area $= \pi(a^2 + b^2 + c^2 + d^2)$.

▲ **Practice Problem 8** Use $A = \pi r^2$ to find the shaded area. The radius of the larger circle is b. The radius of the smaller circle is a. Write the total area formula in factored form so that π appears only once.

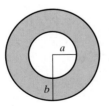

●

6.1 Exercises

Verbal and Writing Skills

In exercises 1 and 2, write a word or words to complete each sentence.

1. In the expression $3x^2 \cdot 5x^3$, $3x^2$ and $5x^3$ are called _____

2. In the expression $3x^2 + 5x^3$, $3x^2$ and $5x^3$ are called _____

3. We can factor $30a^4 + 15a^3 - 45a^2$ as $5a(6a^3 + 3a^2 - 9a)$. Is the factoring complete? Why or why not?

Remove the largest possible common factor. Check your answers for exercises 4–15 by multiplication.

4. $3a^2 + 3a$ **5.** $2c^2 + 2c$ **6.** $21abc - 14ab^2$ **7.** $18wz - 27w^2z$

8. $5x^3 + 25x^2 - 15x$ **9.** $8x^3 - 10x^2 - 14x$ **10.** $12ab - 28bc + 20ac$ **11.** $12xy - 18yz - 36xz$

12. $3xy^2 - 2ay + 5xy - 2y$ **13.** $2ab^3 + 3xb^2 - 5b^4 + 2b^2$ **14.** $60x^3 - 50x^2 + 25x$ **15.** $6x^9 - 8x^7 + 4x^5$

16. $2\pi rh + 2\pi r^2$ **17.** $9a^2b^2 - 36ab$ **18.** $14x^2y - 35xy - 63x$ **19.** $40a^2 - 16ab - 24a$

20. $54x^2 - 45xy + 18x$ **21.** $48xy - 24y^2 + 40y$

Hint: In exercises 22–35, refer to Examples 6 and 7.

22. $7a(x + 2y) - b(x + 2y)$ **23.** $6(3a + b) - z(3a + b)$ **24.** $3x(x - 4) - 2(x - 4)$

25. $5x(x - 7) + 3(x - 7)$ **26.** $6b(2a - 3c) - 5d(2a - 3c)$ **27.** $7x(3y + 5z) - 6t(3y + 5z)$

28. $3(x^2 + 1) + 2y(x^2 + 1) + w(x^2 + 1)$ **29.** $5a(bc - 1) + b(bc - 1) + c(bc - 1)$

30. $2a(xy - 3) - 4(xy - 3) - z(xy - 3)$ **31.** $3c(bc - 3a) - 2(bc - 3a) - 6b(bc - 3a)$

32. $4y(x + 2y) + (x + 2y)$ **33.** $3x^2(x - 2y) - (x - 2y)$

34. $(2a + 3) - 7x(2a + 3)$ **35.** $d(5x - 3) - (5x - 3)$

To Think About

▲ **36.** Find a formula for the area of four rectangles of width 2.786 inches. The lengths of the rectangles are a, b, c, and d inches. Write the formula in factored form.

37. Find a formula for the total cost of all purchases by four people. Each person went to the local wholesale warehouse and spent $29.95 per item. Harry bought a items, Richard bought b items, Lyle bought c items, and Selena bought d items. Write the formula in factored form.

Cumulative Review Problems

38. The sum of three consecutive odd integers is 40 more than the smallest of these integers. Find each of these three consecutive odd integers. (*Hint*: Consecutive odd integers are numbers included in the following pattern: $-3, -1, 1, 3, 5, 7, 9, \ldots, x, x + 2, \ldots$).

39. Tuition, room, and board at a college in Colorado cost Lisa $27,040. This was a 4% increase over the cost of these items last year. What did Lisa pay last year for tuition, room, and board?

Simplify.

40. $\left(-3x^0yz^3\right)^3$

41. $\left(\dfrac{2a^2b^{-1}}{3a^3b^{-3}}\right)^2$

In a recent poll, we asked average people what they would do if they won the lottery. Specifically, we asked what would be the first thing that they would do with the money.

28% said that they would buy a new car.

20% said that they would buy themselves a house.

18% said that they would travel.

13% said that they would pay off their debts.

12% said that they would buy a house for their parents.

7% said that they would give the money to a church or to a charity.

2% said that they would invest the money.

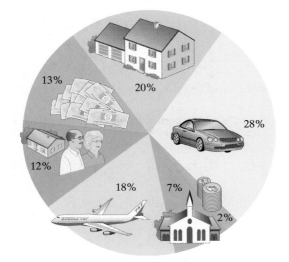

42. Lottery officials estimate that next year 750 people will win a state lottery with a prize of one million dollars or more. If this happens, how many people will first think of buying a car or traveling?

43. Of those who won a state lottery last year, 130 said they would buy themselves a house. Seventy-eight said they would buy a house for their parents. Use the information given to determine the total number of people who won a state lottery last year.

6.2 Factor by Grouping

Student Learning Objectives

After studying this section, you will be able to:

1 Factor expressions with four terms by grouping.

SSM
PH TUTOR CENTER
CD & VIDEO
MATH PRO
WEB

1 *Factoring Expressions with Four Terms by Grouping*

A common factor of a polynomial can be a number, a variable, or an algebraic expression. Sometimes the polynomial is written so that it is easy to recognize the common factor. This is especially true when the common factor is enclosed by parentheses.

EXAMPLE 1 Factor. $x(x - 3) + 2(x - 3)$

Observe each term:

$$\underbrace{x(x - 3)}_{\substack{\text{first} \\ \text{term}}} + \underbrace{2(x - 3)}_{\substack{\text{second} \\ \text{term}}}$$

The common factor of the first and second terms is the quantity $(x - 3)$, so we have

$$x(x - 3) + 2(x - 3) = (x - 3)(x + 2).$$

Practice Problem 1 Factor. $3y(2x - 7) - 8(2x - 7)$

Suppose the polynomial in Example 1, $x(x - 3) + 2(x - 3)$, were written in the form $x^2 - 3x + 2x - 6$. (Note that this form is obtained by multiplying the factors of the first and second terms.) How would we factor a four-term polynomial like this?

In such cases we remove a common factor from the first two terms and a different common factor from the second two terms. That is, we would factor x from $x^2 - 3x$ and 2 from $2x - 6$.

$$x^2 - 3x + 2x - 6 = x(x - 3) + 2(x - 3)$$

Because the resulting terms have a common factor (the binomial enclosed by the parentheses), we would then proceed as we did in Example 1. This procedure for factoring is often called **factoring by grouping**.

EXAMPLE 2 Factor. $2x^2 + 3x + 6x + 9$

$$\begin{array}{cc} 2x^2 + 3x & + & 6x + 9 \\ \text{Factor out a common} & & \text{Factor out a common} \\ \text{factor of } x \text{ from} & & \text{factor of 3 from} \\ \text{the first two terms.} & & \text{the second two terms.} \end{array}$$

$$\underbrace{x(2x + 3)}_{\big\uparrow} \qquad\qquad \underbrace{3(2x + 3)}_{\big\uparrow}$$

Note that the sets of parentheses in the two terms contain the same expression at this step.

The expression in parentheses is now a common factor of the terms. Now we finish the factoring.

$$2x^2 + 3x + 6x + 9 = x(2x + 3) + 3(2x + 3) = (2x + 3)(x + 3)$$

Practice Problem 2 Factor. $6x^2 - 15x + 4x - 10$

EXAMPLE 3   Factor. $4x + 8y + ax + 2ay$

Factor out a common
factor of 4 from
the first two terms.

$$\overline{4x + 8y} + \underbrace{ax + 2ay} = \overline{4(x + 2y)} + \underbrace{a(x + 2y)}$$

Factor out a common
factor of a from
the second two terms.

$4(x + 2y) + a(x + 2y) = (x + 2y)(4 + a)$ The common factor of the terms
is the expression in parentheses,
$x + 2y$.

Practice Problem 3   Factor by grouping. $ax + 2a + 4bx + 8b$

In practice, these problems are done in just two steps.
In some problems the terms are out of order. We have to rearrange the order
of the terms first so that the first two terms have a common factor.

EXAMPLE 4   Factor. $bx + 4y + 4b + xy$

$bx + 4y + 4b + xy = bx + 4b + xy + 4y$ Rearrange the terms so that the first
terms have a common factor.

$\qquad\qquad = b(x + 4) + y(x + 4)$ Factor out the common factor of b
from the first two terms.
Factor out the common factor of y
from the second two terms.

$\qquad\qquad = (x + 4)(b + y)$

Practice Problem 4   Factor. $6a^2 + 5bc + 10ab + 3ac$

Sometimes you will need to factor out a negative common factor from the second two terms to obtain two terms that contain the same parenthetical expression.

EXAMPLE 5   Factor. $2x^2 + 5x - 4x - 10$

$2x^2 + 5x - 4x - 10 = x(2x + 5) - 2(2x + 5)$ Factor out the common factor of
x from the first two terms and
the common factor of -2 from
the second two terms.

$\qquad\qquad\qquad = (2x + 5)(x - 2)$

Notice that if you factored out a common factor of $+2$ in the first step, the two
resulting terms would not contain the same parenthetical expression. If the expressions inside the two sets of parentheses are not exactly the same, you cannot express
the polynomial as a product of two factors!

Practice Problem 5   Factor. $6xy + 14x - 15y - 35$

EXAMPLE 6 Factor. $2ax - a - 2bx + b$

$$2ax - a - 2bx + b = a(2x - 1) - b(2x - 1)$$

Factor out the common factor of a from the first two terms. Factor out the common factor of $-b$ from the second two terms.

$$= (2x - 1)(a - b)$$

Since the two resulting terms contain the same parenthetical expression, we can complete the factoring.

Practice Problem 6 Factor. $3x - 10ay + 6y - 5ax$

⊘ **WARNING** Many students find that they make a factoring error in the first step of problems like Example 6. Multiplying the results of the first step (even if you do it in your head) will usually help you to detect any error you may have made.

EXAMPLE 7 Factor and check your answer. $8ad + 21bc - 6bd - 28ac$

We observe that the first two terms do not have a common factor.

$$8ad + 21bc - 6bd - 28ac = 8ad - 6bd - 28ac + 21bc$$

Rearrange the order using the commutative property of addition.

$$= 2d(4a - 3b) - 7c(4a - 3b)$$

Factor out the common factor of $2d$ from the first two terms and the common factor of $-7c$ from the last two terms.

$$= (4a - 3b)(2d - 7c)$$

Factor out the common factor of $(4a - 3b)$.

To check, we multiply the two binomials using the FOIL procedure.

$$(4a - 3b)(2d - 7c) = 8ad - 28ac - 6bd + 21bc$$
$$= 8ad + 21bc - 6bd - 28ac \quad ✓$$

Rearrange the order of the terms. This is the original problem. Thus it checks.

Practice Problem 7 Factor and check your answer. $10ad + 27bc - 6bd - 45ac$

Factor by grouping. Check your answers for exercises 1–12.

1. $ab - 3a + 4b - 12$

2. $xy - x + 4y - 4$

3. $2ax + 6bx - ay - 3by$

4. $4x + 8y - 3wx - 6wy$

5. $x^3 - 4x^2 + 3x - 12$

6. $x^3 - 6x^2 + 2x - 12$

7. $3ax + bx - 6a - 2b$

8. $4ax + bx - 28a - 7b$

9. $5a + 12bc + 10b + 6ac$

10. $2x + 15yz + 6y + 5xz$

11. $5a - 5b - 2ax + 2xb$

12. $xy - 4x - 3y + 12$

13. $y^2 - 2y - 3y + 6$

14. $12 + 3x - 4x - x^2$

15. $14 - 7y + 2y - y^2$

16. $xa + 2bx - a - 2b$

17. $6ax - y + 2ay - 3x$

18. $6tx - 3t - 2rx + r$

19. $2x^2 + 8x - 3x - 12$

20. $3y^2 - y + 9y - 3$

21. $28x^2 + 8xy^2 + 21xw + 6y^2w$

22. $8xw + 10x^2 + 35xy^2 + 28y^2w$

Verbal and Writing Skills

23. Although $6a^2 - 12bd - 8ad + 9ab = 6(a^2 - 2bd) - a(8d - 9b)$ is true, it is not the correct solution to the problem "Factor $6a^2 - 12bd - 8ad + 9ab$." Explain. Can this expression be factored?

Cumulative Review Problems

24. Solve for *y*. $4x - 2y = -1$

25. Solve and graph. $\frac{1}{3}(x - 6) \geq \frac{1}{6}x - 3$

26. In 1998, the Recording Industry Association of America reported a 6.8% increase from the previous year in U.S. unit sales of all musical recordings. This translates to an 11.9% increase in dollar value, or a 15.1 million increase. (*Source*: Bureau of Economic Analysis.) If the dollar value increase is equal to $15.1 million dollars, what was the total dollar value of musical recordings in the United States in 1997? Round to the nearest tenth of a million.

27. Using the data from exercise 26, if there was a 20% increase in the dollar value of sales from 1998 to 2000, what was the dollar value of sales for the year 2000?

329

Student Learning Objectives

After studying this section, you will be able to:

1 Factor polynomials of the form $x^2 + bx + c$.

2 Factor polynomials that have a common factor and a factor of the form $x^2 + bx + c$.

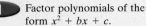

SSM   PH TUTOR   CD & VIDEO   MATH PRO   WEB
      CENTER

1 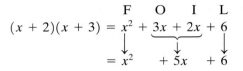 *Factoring Polynomials of the Form $x^2 + bx + c$*

Suppose that you wanted to factor $x^2 + 5x + 6$. After some trial and error you *might* obtain $(x + 2)(x + 3)$, or you might get discouraged and not get an answer. If you did get these factors, you could check this answer by the FOIL method.

$$(x + 2)(x + 3) = x^2 + 3x + 2x + 6$$
$$= x^2 + 5x + 6$$

But trial and error can be a long process. There is another way. Let's look at the preceding equation again.

$$\begin{array}{cccc} & \text{F} & \text{O} \quad \text{I} & \text{L} \\ (x + 2)(x + 3) = & x^2 & + 3x + 2x & + 6 \\ = & x^2 & + 5x & + 6 \end{array}$$

The first thing to notice is that the product of the first terms in the factors gives the first term of the polynomial. That is, $x \cdot x = x^2$.

The first term is the product of these terms.

$$x^2 + 5x + 6 \quad = \quad (x + 2)(x + 3)$$

The next thing to notice is that the sum of the products of the outer and inner terms in the factors produces the middle term of the polynomial. That is, $(x \cdot 3) + (2 \cdot x) = 3x + 2x = 5x$. Thus we see that the sum of the second terms in the factors, $2 + 3$, gives the coefficient of the middle term, 5.

Finally, note that the product of the last terms of the factors gives the last term of the polynomial. That is, $2 \cdot 3 = 6$.

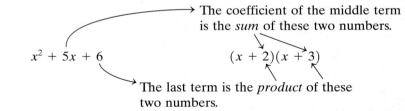

The coefficient of the middle term is the *sum* of these two numbers.

$$x^2 + 5x + 6 \qquad\qquad (x + 2)(x + 3)$$

The last term is the *product* of these two numbers.

Let's summarize our observations in general terms and then try a few examples.

Factoring Trinomials of the Form $x^2 + bx + c$

1. The answer will be of the form $(x + m)(x + n)$.

2. m and n are numbers such that:
  (a) When you multiply them, you get the last term, which is c.
  (b) When you add them, you get the coefficient of the middle term, which is b.

EXAMPLE 1   Factor. $x^2 + 7x + 12$

The answer is of the form $(x + m)(x + n)$. We want to find the two numbers, m and n, that you can multiply to get 12 and add to get 7. The numbers are 3 and 4.

$$x^2 + 7x + 12 = (x + 3)(x + 4)$$

Practice Problem 1   Factor. $x^2 + 8x + 12$

EXAMPLE 2   Factor. $x^2 + 12x + 20$

We want two numbers that have a product of 20 and a sum of 12. The numbers are 10 and 2.

$$x^2 + 12x + 20 = (x + \underline{10})(x + \underline{2})$$

Note: If you cannot think of the numbers in your head, write down the possible factors whose product is 20.

$$\overbrace{}^{\textit{Product}} \qquad \overbrace{}^{\textit{Sum}}$$

$$1 \cdot 20 = 20 \qquad 1 + 20 = 21$$
$$2 \cdot 10 = 20 \qquad 2 + 10 = 12 \leftarrow$$
$$4 \cdot 5 = 20 \qquad 4 + 5 = 9$$

Then select the pair whose sum is 12.   Select this pair.

Practice Problem 2   Factor. $x^2 + 17x + 30$

You may find that it is helpful to list all the factors whose product is 30 first.

So far we have factored only trinomials of the form $x^2 + bx + c$, where b and c are positive numbers. The same procedure applies if b is a negative number and c is positive. Because m and n have a positive product and a negative sum, they must both be negative.

EXAMPLE 3   Factor. $x^2 - 8x + 15$

We want two numbers that have a product of $+15$ and a sum of -8. They must be negative numbers since the sign of the middle term is negative and the sign of the last term is positive.

$$\overset{\text{the sum } - 5 + (-3)}{x^2 - 8x + 15 = (x - 5)(x - 3)} \qquad \begin{array}{l} \textit{Think}: (-5)(-3) = +15 \\ \textit{and } -5 + (-3) = -8. \end{array}$$

the product $(-5)(-3)$

Multiply using FOIL to check.

Practice Problem 3   Factor. $x^2 - 11x + 18$

EXAMPLE 4   Factor. $x^2 - 9x + 14$

We want two numbers whose product is 14 and whose sum is -9. The numbers are -7 and -2. So

$$x^2 - 9x + 14 = (x - 7)(x - 2) \text{ or } (x - 2)(x - 7).$$

Practice Problem 4   Factor. $y^2 - 11y + 24$

All the examples so far have had a positive last term. What happens when the last term is negative? If the last term is negative, one of the numbers m or n must be a positive and the other must be a negative. Why? The product of a positive number and a negative number is negative.

EXAMPLE 5 Factor. $x^2 - 3x - 10$

We want two numbers whose product is -10 and whose sum is -3. The two numbers are -5 and $+2$.

$$x^2 - 3x - 10 = (x - 5)(x + 2)$$

Practice Problem 5 Factor. $a^2 - 5a - 24$

What if we made a sign error and *incorrectly* factored the trinomial $x^2 - 3x - 10$ as $(x + 5)(x - 2)$? We could detect the error immediately since the sum of $+5$ and -2 is 3. We need a sum of -3!

EXAMPLE 6 Factor. $x^2 + 10x - 24$ Check your answer.

The two numbers whose product is -24 and whose sum is $+10$ are the numbers $+12$ and -2.

$$x^2 + 10x - 24 = (x + 12)(x - 2)$$

⊘ **WARNING** It is very easy to make a sign error in these problems. Make sure that you mentally multiply your answer back to obtain the original expression. Check each sign carefully.

Check: $(x + 12)(x - 2) = x^2 - 2x + 12x - 24 = x^2 + 10x - 24$ ✓

Practice Problem 6 Factor. $x^2 + 17x - 60$ Multiply your answer to check.

EXAMPLE 7 Factor. $x^2 - 16x - 36$

We want two numbers whose product is -36 and whose sum is -16.

List all the possible factors of 36 (without regard to sign). Find the pair that has a difference of 16. We are looking for a difference because the signs of the factors are different.

Factors of 36	*The Difference between the Factors*
36 and 1	35
18 and 2	16 ← This is the value we want.
12 and 3	9
9 and 4	5
6 and 6	0

Once we have picked the pair of numbers (18 and 2), it is easy to find the signs. For the coefficient of the middle term to be -16, we will have to add the numbers -18 and $+2$.

$$x^2 - 16x - 36 = (x - 18)(x + 2)$$

Practice Problem 7 Factor. $x^2 - 7x - 60$ You may find it helpful to list the pairs of numbers whose product is 60.

At this point you should work several problems to develop your factoring skills. This is one section where you really need to drill by doing many problems.

Feel a little confused about the signs? If you do, you may find these facts helpful.

Facts About Factoring Trinomials of the Form $x^2 + bx + c$

The *two numbers* m and n will have the *same sign* if the last term of the polynomial is *positive*.

1. They will both be *positive* if the *coefficient* of the *middle* term is *positive*.

2. They will both be *negative* if the *coefficient* of the *middle* term is *negative*.

The two numbers m and n will have *opposite signs* if the last term is *negative*.

1. The *larger* of the absolute values of the two numbers will be given a plus sign if the coefficient of the *middle term* is *positive*.

2. The larger of the absolute values of the two numbers will be given a negative sign if the coefficient of the *middle term* is *negative*.

$x^2 + bx + c = (x \quad m)(x \quad n)$

$x^2 + 5x + 6 = (x + 2)(x + 3)$

$x^2 - 5x + 6 = (x - 2)(x - 3)$

$x^2 + 6x - 7 = (x + 7)(x - 1)$

$x^2 - 6x - 7 = (x - 7)(x + 1)$

Do not memorize these facts; rather, try to understand the pattern.

Sometimes the exponent of the first term of the polynomial will be greater than 2. If the exponent is an even power, it is a square. For example, $x^4 = (x^2)(x^2)$. Likewise, $x^6 = (x^3)(x^3)$.

EXAMPLE 8 Factor. $y^4 - 2y^2 - 35$

Think: $y^4 = (y^2)(y^2)$ This will be the first term of each parentheses.

$(y^2 \quad)(y^2 \quad)$

$(y^2 + \quad)(y^2 - \quad)$ The last term of the polynomial is negative.

$(y^2 + 5)(y^2 - 7)$ Thus the signs of m and n will be different.
Now think of factors of 35 whose difference is 2.

Practice Problem 8 Factor. $a^4 + a^2 - 42$

---

2 Factoring Polynomials That Have a Common Factor and a Factor of the Form $x^2 + bx + c$

Some factoring problems require two steps. Often we must first factor out a common factor from each term of the polynomial. Once this is done, we may find that the other factor is a trinomial that can be factored using the methods previously discussed in this section.

EXAMPLE 9 Factor. $2x^2 + 36x + 160$

$2x^2 + 36x + 160 = 2(x^2 + 18x + 80)$ First factor out the common factor of 2 from each term of the polynomial.

$= 2(x + 8)(x + 10)$ Then factor the remaining polynomial.

The final answer is $2(x + 8)(x + 10)$. *Be sure to list all parts of the answer.*

Check: $2(x + 8)(x + 10) = 2(x^2 + 18x + 80) = 2x^2 + 36x + 160$ ✓

Thus we are sure that the answer is $2(x + 8)(x + 10)$.

Practice Problem 9   Factor. $3x^2 + 45x + 150$

EXAMPLE 10   Factor. $3x^2 + 9x - 162$

$3x^2 + 9x - 162 = 3(x^2 + 3x - 54)$ First factor out the common factor of 3 from each term of the polynomial.

$= 3(x - 6)(x + 9)$ Then factor the remaining polynomial.

The final answer is $3(x - 6)(x + 9)$.

Check: $3(x - 6)(x + 9) = 3(x^2 + 3x - 54) = 3x^2 + 9x - 162$ ✓

Thus we are sure that the answer is $3(x - 6)(x + 9)$.

Practice Problem 10   Factor. $4x^2 - 8x - 140$

It is quite easy to forget to look for a greatest common factor as the first step of factoring a trinomial. Therefore, it is a good idea to examine your final answer in any factoring problem and ask yourself, "Can I factor out a common factor from any binomial contained inside a set of parentheses?" Often you will be able to see a common factor at that point if you missed it in the first step of the problem.

▲ **EXAMPLE 11**   Find a polynomial in factored form for the shaded area in the figure.

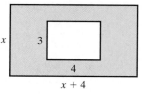

To obtain the shaded area, we find the area of the larger rectangle and subtract from it the area of the smaller rectangle. Thus we have the following:

$$\text{shaded area} = x(x + 4) - (4)(3)$$
$$= x^2 + 4x - 12$$

Now we factor this polynomial to obtain the shaded area $= (x + 6)(x - 2)$.

▲ **Practice Problem 11**   Find a polynomial in factored form for the shaded area in the figure.

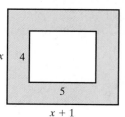

Verbal and Writing Skills

Fill in the blanks.

1. To factor $x^2 + 5x + 6$, find two numbers whose _____ is 6 and whose _____ is 5.

2. To factor $x^2 + 5x - 6$, find two numbers whose _____ is -6 and whose _____ is 5.

Factor.

3. $x^2 + 2x + 1$

4. $x^2 + 11x + 30$

5. $x^2 + 12x + 35$

6. $x^2 + 11x + 24$

7. $x^2 - 4x + 3$

8. $x^2 - 6x + 8$

9. $x^2 - 11x + 28$

10. $x^2 - 13x + 12$

11. $x^2 + x - 12$

12. $x^2 + 4x - 5$

13. $x^2 - 13x - 14$

14. $x^2 - 6x - 16$

15. $x^2 + 2x - 35$

16. $x^2 - 4x - 12$

17. $x^2 - 2x - 24$

18. $x^2 - 11x - 26$

Look over your answers to exercises 3–18 carefully. Be sure that you are clear on your sign rules. Exercises 19–42 contain a mixture of all the types of problems in this section. Make sure you can do them all. Check your answers by multiplication.

19. $x^2 + 5x - 14$

20. $x^2 - 2x - 15$

21. $x^2 - 10x + 24$

22. $x^2 - 13x + 42$

23. $x^2 + 13x + 30$

24. $x^2 - 3x - 28$

25. $y^2 - 4y - 5$

26. $y^2 - 8y + 7$

27. $a^2 + 6a - 16$

28. $a^2 - 13a + 30$

29. $x^2 - 12x + 32$

30. $x^2 - 6x - 27$

31. $x^2 + 4x - 21$

32. $x^2 - 9x + 18$

33. $x^2 + 13x + 40$

34. $x^2 + 15x + 50$

35. $x^2 - 21x - 46$

36. $x^2 + 12x - 45$

37. $x^2 + 9x - 36$

38. $x^2 - 13x + 36$

39. $x^2 - 2xy - 15y^2$

40. $x^2 - 2xy - 35y^2$

41. $x^2 - 16xy + 63y^2$

42. $x^2 + 19xy + 48y^2$

In exercises 43–54, first factor out the greatest common factor from each term. Then factor the remaining polynomial. Refer to Examples 9 and 10.

43. $2x^2 - 12x + 16$

44. $2x^2 - 14x + 24$

45. $3x^2 - 6x - 72$

46. $3x^2 - 12x - 63$

47. $4x^2 + 24x + 20$

48. $4x^2 + 28x + 40$

49. $7x^2 + 21x - 70$

50. $7x^2 + 7x - 84$

51. $6x^2 + 18x + 12$

52. $6x^2 + 24x + 18$

53. $3x^2 - 18x + 15$

54. $3x^2 - 33x + 54$

Find a polynomial in factored form for the shaded area.

▲ **55.** The circle has radius $2x$. The square has diagonals of $4x$.

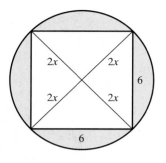

▲ **56.** Both figures are rectangles with dimensions as labeled.

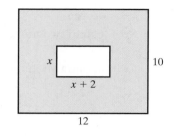

Cumulative Review Problems

57. In a triangle, the measure of the first angle is two degrees more than double the second angle. The third angle is two degrees less than triple the second angle. What is the measure of each angle?

58. Kerri works as a radio advertising sales rep. She earns a guaranteed minimum salary of $500 per month plus 3% commission on her sales. She wants to earn $4400 or more this month. At least how much must she generate in sales?

59. Write an equation of the line passing through $(2, 5)$ having a slope of $\frac{1}{2}$.

60. Write an equation of the line passing through $(-1, 3)$ and $(2, 0)$.

The Golden Sounds DJ Service charges $65 per hour plus $85 for each 30-minute period after midnight.

61. If the bill for Marcia's wedding reception was $515 and the reception started at 8 P.M., when did the reception end?

62. If the bill for Melissa's wedding reception was $535 and the reception ended at 2 A.M., when did the reception start?

The equation $T = 19 + 2M$ has been used by some meteorologists to predict the monthly average temperature for the small island of Menorca off the coast of Spain during the first 6 months of the year. The variable T represents the average monthly temperature measured in degrees Celsius. The variable M represents the number of months since January.

63. What is the average temperature of Menorca during the month of April?

64. During what month will the average temperature be $29°C$?

6.4 Factoring Trinomials of the Form $ax^2 + bx + c$

1 Using the Trial-and-Error Method

When the coefficient of the x^2-term in a trinomial of the form $ax^2 + bx + c$ is not 1, the trinomial is more difficult to factor. Several possibilities must be considered.

EXAMPLE 1   Factor. $2x^2 + 5x + 3$

In order for the coefficient of the x^2-term of the polynomial to be 2, the coefficients of the x-terms in the factors must be 2 and 1. Thus $2x^2 + 5x + 3 = (2x \quad)(x \quad)$.

In order for the last term of the polynomial to be 3, the constants in the factors must be 3 and 1.

Since all signs in the polynomial are positive, we know that each factor in parentheses will contain only positive signs. However, we still have two possibilities. They are as follows.

$$(2x + 3)(x + 1)$$
$$(2x + 1)(x + 3)$$

We check them by multiplying by the FOIL method.

$$(2x + 1)(x + 3) = 2x^2 + 7x + 3 \quad \text{Wrong middle term}$$
$$(2x + 3)(x + 1) = 2x^2 + 5x + 3 \quad \text{Correct middle term}$$

Thus the correct answer is

$$(2x + 3)(x + 1) \text{ or } (x + 1)(2x + 3).$$

Practice Problem 1   Factor. $2x^2 - 7x + 5$

Some problems have many more possibilities.

EXAMPLE 2   Factor. $4x^2 - 13x + 3$

The Different Factors of 4 Are:	**The Factors of 3 Are:**
2 and 2	1 and 3
1 and 4	

Let us list the possible factoring combinations and compute the middle term by the FOIL method. Note that the signs of the constants in both factors will be negative. Why?

Possible Factors	**Middle Term**	**Correct?**
$(2x - 3)(2x - 1)$	$-8x$	No
$(4x - 3)(x - 1)$	$-7x$	No
$(4x - 1)(x - 3)$	$-13x$	Yes

The correct answer is $(4x - 1)(x - 3)$ or $(x - 3)(4x - 1)$.
This method is called the **trial-and-error method**.

Practice Problem 2   Factor. $9x^2 - 64x + 7$

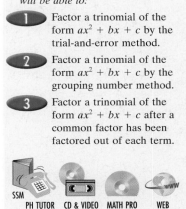

EXAMPLE 3 Factor. $3x^2 - 2x - 8$

Factors of 3	*Factors of 8*
3 and 1	8 and 1
	4 and 2

Let us list only one-half of the possibilities. We'll let the constant in the first factor of each product be positive.

Possible Factors	*Middle Term*	*Correct Factors?*
$(x + 8)(3x - 1)$	$+23x$	No
$(x + 1)(3x - 8)$	$-5x$	No
$(x + 4)(3x - 2)$	$+10x$	No
$(x + 2)(3x - 4)$	$+2x$	No (but only because the sign is wrong)

So we just *reverse* the signs of the constants in the factors.

	Middle Term	*Correct Factor?*
$(x - 2)(3x + 4)$	$-2x$	Yes

The correct answer is

$$(x - 2)(3x + 4) \text{ or } (3x + 4)(x - 2).$$

Practice Problem 3 Factor. $3x^2 - x - 14$

It takes a good deal of practice to readily factor problems of this type. The more problems you do, the more proficient you will become. The following method will help you factor more quickly.

2 *Using the Grouping Number Method*

One way to factor a trinomial of the form $ax^2 + bx + c$ is to write it with four terms and factor by grouping, as we did in Section 6.2. For example, the trinomial $2x^2 + 13x + 20$ can be written as $2x^2 + 5x + 8x + 20$. Using the methods of Section 6.2, we factor it as follows.

$$2x^2 + 5x + 8x + 20 = x(2x + 5) + 4(2x + 5)$$
$$= (2x + 5)(x + 4)$$

We can factor all factorable trinomials of the form $ax^2 + bx + c$ in this way. We will use the following procedure.

> **Grouping Number Method for Factoring Trinomials of the Form $ax^2 + bx + c$**
>
> 1. Obtain the grouping number ac.
> 2. Find the two numbers whose product is the grouping number and whose sum is b.
> 3. Use those numbers to write bx as the sum of two terms.
> 4. Factor by grouping.
> 5. Multiply to check.

Let's try the problem from Example 1.

EXAMPLE 4   Factor by grouping. $2x^2 + 5x + 3$

1. The grouping number is $(2)(3) = 6$.
2. The factors of 6 are $6 \cdot 1$ and $3 \cdot 2$. We choose the numbers 3 and 2 because their product is 6 and their sum is 5.
3. We write $5x$ as the sum $3x + 2x$.
4. Factor by grouping.

$$2x^2 + 5x + 3 = 2x^2 + 2x + 3x + 3$$
$$= 2x(x + 1) + 3(x + 1)$$
$$= (x + 1)(2x + 3)$$

5. Multiply to check.

$$(x + 1)(2x + 3) = 2x^2 + 3x + 2x + 3$$
$$= 2x^2 + 5x + 3 \quad \checkmark$$

Practice Problem 4   Factor by grouping. $2x^2 - 7x + 5$

EXAMPLE 5   Factor by grouping. $4x^2 - 13x + 3$

1. The grouping number is $(4)(3) = 12$.
2. The factors of 12 are $(12)(1)$ or $(4)(3)$ or $(6)(2)$. Note that the middle term of the polynomial is negative. Thus we choose the numbers -12 and -1 because their product is still 12 and their sum is -13.
3. We write $-13x$ as the sum $-12x + (-1x)$.
4. Factor by grouping.

$$4x^2 - 13x + 3 = 4x^2 - 12x - 1x + 3$$
$$= 4x(x - 3) - 1(x - 3) \quad \text{Remember to factor out a } -1 \text{ from the last two terms so that both sets of parentheses contain the same expression.}$$
$$= (x - 3)(4x - 1)$$

Practice Problem 5   Factor by grouping. $9x^2 - 64x + 7$

EXAMPLE 6   Factor by grouping. $3x^2 - 2x - 8$

1. The grouping number is $(3)(-8) = -24$.
2. We want two numbers whose product is -24 and whose sum is -2. They are -6 and 4.
3. We write $-2x$ as a sum $-6x + 4x$.
4. Factor by grouping.

$$3x^2 - 6x + 4x - 8 = 3x(x - 2) + 4(x - 2)$$
$$= (x - 2)(3x + 4)$$

Practice Problem 6   Factor by grouping. $3x^2 + 4x - 4$

To factor polynomials of the form $ax^2 + bx + c$, use the method, either trial-and-error or grouping, that works best for you.

 **3** *Using the Common Factor Method*

Some problems require first factoring out a common factor and then factoring the trinomial by one of the two methods of this section.

EXAMPLE 7 Factor. $9x^2 + 3x - 30$

$9x^2 + 3x - 30 = 3(3x^2 + 1x - 10)$ We first factor out the common factor of 3 from each term of the trinomial.

$= 3(3x - 5)(x + 2)$ We then factor the trinomial by the grouping method or by the trial-and-error method.

Practice Problem 7 Factor. $8x^2 - 8x - 6$

EXAMPLE 8 Factor. $32x^2 - 40x + 12$

$32x^2 - 40x + 12 = 4(8x^2 - 10x + 3)$ We first factor out the greatest common factor of 4 from each term of the trinomial.

$= 4(2x - 1)(4x - 3)$ We then factor the trinomial by the grouping method or by the trial-and-error method.

Practice Problem 8 Factor. $24x^2 - 38x + 10$

Factor. Check your answers for exercises 1–20 using the FOIL method.

1. $4x^2 + 13x + 3$

2. $3x^2 + 11x + 10$

3. $2x^2 - 5x + 2$

4. $3x^2 - 8x + 4$

5. $3x^2 - 4x - 7$

6. $5x^2 + 7x - 6$

7. $2x^2 - 5x - 3$

8. $2x^2 - x - 6$

9. $5x^2 + 3x - 2$

10. $6x^2 + x - 2$

11. $15x^2 - 34x + 15$

12. $10x^2 - 29x + 10$

13. $2x^2 + 3x - 20$

14. $6x^2 + 11x - 10$

15. $9x^2 + 9x + 2$

16. $4x^2 + 11x + 6$

17. $6x^2 - 5x - 6$

18. $3x^2 - 13x - 10$

19. $6x^2 - 19x + 10$

20. $10x^2 - 19x + 6$

21. $7x^2 - 5x - 18$

22. $9x^2 - 22x - 15$

23. $9y^2 - 13y + 4$

24. $5y^2 - 11y + 2$

25. $5a^2 - 13a - 6$

26. $3a^2 - 10a - 8$

27. $12x^2 - 20x + 3$

28. $9x^2 + 5x - 4$

29. $15x^2 + 4x - 4$

30. $8x^2 - 11x + 3$

31. $12x^2 + 28x + 15$

32. $24x^2 + 17x + 3$

33. $12x^2 - 16x - 3$

34. $12x^2 + x - 6$

35. $2x^4 + 15x^2 - 8$

36. $4x^4 - 11x^2 - 3$

37. $4x^2 + 8xy - 5y^2$

38. $3x^2 + 8xy + 4y^2$

39. $5x^2 + 16xy - 16y^2$

40. $12x^2 + 11xy - 5y^2$

Factor by first factoring out the greatest common factor. See Examples 7 and 8.

41. $10x^2 + 22x + 12$ **42.** $4x^2 + 34x + 42$ **43.** $12x^2 - 24x + 9$ **44.** $8x^2 - 26x + 6$

45. $10x^2 - 25x - 15$ **46.** $20x^2 - 25x - 30$ **47.** $6x^3 - 16x^2 - 6x$ **48.** $6x^3 + 9x^2 - 60x$

Factor.

49. $12x^2 + 16x - 35$ **50.** $20x^2 - 53x + 18$ **51.** $20x^2 - 27x + 9$ **52.** $12x^2 - 29x + 15$

Cumulative Review Problems

53. Solve. $7x - 3(4 - 2x) = 2(x - 3) - (5 - x)$

54. In the 2000–2001 school year, 33.8 million children were enrolled in kindergarten through grade 8 in the United States. This represented an increase of 80% in the number of enrolled children since the school year 1939–1940. (*Source*: U.S. National Center for Education Statistics.) How many children were enrolled in grades K–8 during 1939–1940? Round to the nearest tenth of a million.

55. In the 2002–2003 school year, it is projected that 18% more students will be enrolled in kindergarten through grade 8 in the United States than there were in the year 1992–1993. In the 1992–1993 school year, 31.1 million children were enrolled in kindergarten through grade 8. (*Source*: U.S. National Center for Education Statistics.) How many children will be enrolled in grades K–8 during 2002–2003? Round to the nearest tenth of a million.

6.5 Special Cases of Factoring

As we proceed in this section you will be able to reduce the time it takes you to factor a polynomial by quickly recognizing and factoring two special types of polynomials: the difference of two squares and perfect-square trinomials.

① Factoring the Difference of Two Squares

Recall the formula from Section 4.5:

$$(a + b)(a - b) = a^2 - b^2.$$

In reverse form we can use it for factoring.

> **Difference of Two Squares**
>
> $$a^2 - b^2 = (a + b)(a - b)$$

We can state it in words in this way: "The difference of two squares can be factored into the sum and difference of those values that were squared."

EXAMPLE 1 Factor. $9x^2 - 1$

We see that the problem is in the form of the difference of two squares. $9x^2$ is a square and 1 is a square. So using the formula we can write the following.

$$9x^2 - 1 = (3x + 1)(3x - 1) \quad \text{Because } 9x^2 = (3x)^2 \text{ and } 1 = (1)^2$$

Practice Problem 1 Factor. $1 - 64x^2$

EXAMPLE 2 Factor. $25x^2 - 16$

Again we use the formula for the difference of squares.

$$25x^2 - 16 = (5x + 4)(5x - 4) \quad \text{Because } 25x^2 = (5x)^2 \text{ and } 16 = (4)^2$$

Practice Problem 2 Factor. $36x^2 - 49$

Sometimes the polynomial contains two variables.

EXAMPLE 3 Factor. $4x^2 - 49y^2$

We see that

$$4x^2 - 49y^2 = (2x + 7y)(2x - 7y).$$

Practice Problem 3 Factor. $100x^2 - 81y^2$

Some problems may involve more than one step.

EXAMPLE 4 Factor. $81x^4 - 1$

We see that

$$81x^4 - 1 = (9x^2 + 1)(9x^2 - 1) \quad \text{Because } 81x^4 = (9x^2)^2 \text{ and } 1 = (1)^2$$

Is the factoring complete? We can factor $9x^2 - 1$.

$$81x^4 - 1 = (9x^2 + 1)(3x - 1)(3x + 1) \quad \text{Because } (9x^2 - 1) = (3x - 1)(3x + 1)$$

Practice Problem 4 Factor. $x^8 - 1$

2 Factoring Perfect-Square Trinomials

There is a formula that will help us to factor very quickly certain trinomials, called **perfect-square trinomials**. Recall from Section 5.5 the formulas for a binomial squared.

$$(a + b)^2 = a^2 + 2ab + b^2$$
$$(a - b)^2 = a^2 - 2ab + b^2$$

We can use these two equations in reverse form for factoring.

> **Perfect-Square Trinomials**
>
> $$a^2 + 2ab + b^2 = (a + b)^2$$
> $$a^2 - 2ab + b^2 = (a - b)^2$$

A perfect-square trinomial is a trinomial that is the result of squaring a binomial. How can we recognize a perfect-square trinomial?

1. The first and last terms are *perfect squares*.
2. The middle term is twice the product of the values whose squares are the first and last terms.

EXAMPLE 5 Factor. $x^2 + 6x + 9$

This is a perfect-square trinomial.

1. The first and last terms are perfect squares because $x^2 = (x)^2$ and $9 = (3)^2$.
2. The middle term, $6x$, is twice the product of x and 3.

Since $x^2 + 6x + 9$ is a perfect-square trinomial, we can use the formula

$$a^2 + 2ab + b^2 = (a + b)^2$$

with $a = x$ and $b = 3$. So we have

$$x^2 + 6x + 9 = (x + 3)^2.$$

Practice Problem 5 Factor. $16x^2 + 8x + 1$

EXAMPLE 6 Factor. $4x^2 - 20x + 25$

This is a perfect-square trinomial. Note that $20x = 2(2x \cdot 5)$. Also note the negative sign. Thus we have the following.

$$4x^2 - 20x + 25 = (2x - 5)^2 \quad \text{Since } a^2 - 2ab + b^2 = (a - b)^2$$

Practice Problem 6 Factor. $25x^2 - 30x + 9$

A polynomial may have more than one variable and its exponents may be higher than 2. The same principles apply.

EXAMPLE 7   Factor.   **(a)** $49x^2 + 42xy + 9y^2$ **(b)** $36x^4 - 12x^2 + 1$

(a) This is a perfect-square trinomial. Why?

$$49x^2 + 42xy + 9y^2 = (7x + 3y)^2$$ Because $49x^2 = (7x)^2$, $9y^2 = (3y)^2$, and $42xy = 2(7x \cdot 3y)$

(b) This is a perfect-square trinomial. Why?

$$36x^4 - 12x^2 + 1 = \left(6x^2 - 1\right)^2$$ Because $36x^4 = \left(6x^2\right)^2$, $1 = (1)^2$, and $12x^2 = 2\left(6x^2 \cdot 1\right)$

Practice Problem 7   Factor. **(a)** $25x^2 - 60xy + 36y^2$ **(b)** $64x^6 - 48x^3 + 9$

Some polynomials appear to be perfect-square trinomials but are not. They were factored in other ways in Section 6.4.

EXAMPLE 8   Factor. $49x^2 + 35x + 4$

This is *not* a perfect-square trinomial! Although the first and last terms are perfect squares since $(7x)^2 = 49x^2$ and $(2)^2 = 4$, the middle term, $35x$, is not double the product of 2 and $7x$! $35x \neq 28x$! So we must factor by trial and error or by grouping to obtain

$$49x^2 + 35x + 4 = (7x + 4)(7x + 1).$$

Practice Problem 8   Factor. $9x^2 - 15x + 4$

③ *Factoring Out a Common Factor and Then Using a Special-Case Formula*

For some polynomials, we will first factor out a common factor. Then we will find an opportunity to use the difference-of-two-squares formula or one of the perfect-square trinomial formulas.

EXAMPLE 9   Factor. $12x^2 - 48$

$$12x^2 - 48 = 12(x^2 - 4)$$ First we factor out the greatest common factor, 12.

$$= 12(x + 2)(x - 2)$$ Then we use the difference-of-two-squares formula, $a^2 - b^2 = (a + b)(a - b)$.

Practice Problem 9   Factor. $20x^2 - 45$

EXAMPLE 10   Factor. $24x^2 + 72x + 54$

$$24x^2 + 72x + 54 = 6\left(4x^2 + 12x + 9\right)$$ First we factor out the greatest common factor, 6.

$$= 6(2x + 3)^2$$ Then we use the perfect-square trinomial formula, $a^2 + 2ab + b^2 = (a + b)^2$.

Practice Problem 10   Factor. $75x^2 - 60x + 12$

Factor.

1. $81x^2 - 16$

2. $100x^2 - 49$

3. $16 - 9x^2$

4. $49 - 25x^2$

5. $9x^2 - 25$

6. $81x^2 - 1$

7. $4x^2 - 25$

8. $16x^2 - 25$

9. $36x^2 - 25$

10. $1 - 25x^2$

11. $1 - 49x^2$

12. $1 - 36x^2$

13. $16x^2 - 49y^2$

14. $25x^4 - 81y^4$

15. $25 - 121x^2$

16. $9x^2 - 49$

17. $81x^2 - 100y^2$

18. $25a^2 - 1$

19. $25a^2 - 49$

20. $9x^2 - 49y^2$

21. $9x^2 + 6x + 1$

22. $25x^2 + 10x + 1$

23. $y^2 - 6y + 9$

24. $y^2 - 8y + 16$

25. $9x^2 - 24x + 16$

26. $4x^2 + 20x + 25$

27. $49x^2 + 28x + 4$

28. $25x^2 + 30x + 9$

29. $x^2 + 14x + 49$

30. $x^2 + 8x + 16$

31. $25x^2 - 40x + 16$

32. $49x^2 - 42x + 9$

33. $81x^2 + 36xy + 4y^2$

34. $36x^2 + 60xy + 25y^2$

35. $25x^2 - 30xy + 9y^2$

36. $4x^2 - 28xy + 49y^2$

37. $16a^2 + 72ab + 81b^2$

38. $169a^2 + 26ab + b^2$

39. $9x^4 - 6x^2y + y^2$

40. $y^4 - 22y^2 + 121$

41. $49x^2 + 70x + 9$

42. $25x^2 - 50x + 16$

43. $16x^4 - 1$

44. $81x^4 - 1$

45. $x^{10} - 36y^{10}$

46. $x^4 - 49y^6$

47. $9x^{10} - 12x^5 + 4$

48. $36x^8 - 36x^4 + 9$

To Think About

49. In Example 4, first we factored $81x^4 - 1$ as $(9x^2 + 1)(9x^2 - 1)$, then we factored $9x^2 - 1$ as $(3x + 1)(3x - 1)$. Show why you cannot factor $9x^2 + 1$.

50. What two numbers could replace the b in $25x^2 + bx + 16$ so that the resulting trinomial would be a perfect square? (*Hint*: One number is negative.)

51. What value could you give to c so that $16y^2 - 56y + c$ would become a perfect-square trinomial? Is there only one answer or more than one?

52. Jerome says that he can find two values of b so that $100x^2 + bx - 9$ will be a perfect square. Kesha says there is only one that fits, and Larry says there are none. Who is correct and why?

Factor by first looking for a greatest common factor. See Examples 9 and 10.

53. $16x^2 - 36$

54. $27x^2 - 75$

55. $147x^2 - 3y^2$

56. $16y^2 - 100x^2$

57. $12x^2 - 36x + 27$

58. $125x^2 - 100x + 20$

59. $98x^2 + 84x + 18$

60. $128x^2 + 96x + 18$

Mixed Practice

Factor. Be sure to look for common factors first.

61. $x^2 - 9x + 14$

62. $x^2 - 9x - 36$

63. $2x^2 + 5x - 3$

64. $15x^2 - 11x + 2$

65. $16x^2 - 121$

66. $9x^2 - 100y^2$

67. $9x^2 + 42x + 49$

68. $9x^2 + 30x + 25$

69. $3x^2 + 6x - 45$

70. $4x^2 + 24x + 32$

71. $5x^2 - 80$

72. $13x^2 - 13$

73. $5x^2 + 20x + 20$

74. $8x^2 + 48x + 72$

75. $2x^2 - 32x + 126$

76. $2x^2 - 32x + 110$

Cumulative Review Problems

77. Divide. $(x^3 + x^2 - 2x - 11) \div (x - 2)$

78. Divide. $(6x^3 + 11x^2 - 11x - 20) \div (3x + 4)$

The green iguana can reach a length of 6 feet and weigh up to 18 pounds. Of the basic diet of the iguana, 40% should consist of greens such as lettuce, spinach, and parsley; 35% should consist of bulk vegetables such as broccoli, zucchini, and carrots; and 25% should consist of fruit.

79. If a certain iguana weighing 150 ounces has a daily diet equal to 2% of its body weight, compose a diet for it in ounces that will meet the iguana's one-day requirement for nutrition.

80. Ethan's iguana has a daily diet equal to 3% of its body weight. Ethan has prepared the following diet for the iguana: 1.44 ounces of greens, 1.26 ounces of bulk vegetables, and 0.9 ounces of fruit. How much does the iguana weigh?

The peak of Mount Washington is at an altitude of 6288 feet above sea level. The altitude A in feet of a car driving down the mountain road from the peak a distance M measured in miles is given by $A = 6288 - 700M$.

81. What is the altitude of a car that has driven from the mountain peak down a distance of 3.5 miles?

82. A car drives from the peak down the mountain road to a point where the altitude is 2788 feet above sea level. How many miles down the road has the car driven?

6.6 A Brief Review of Factoring

Student Learning Objectives

After studying this section, you will be able to:

1. Identify and factor any polynomial that can be factored.

2. Determine whether a polynomial is prime.

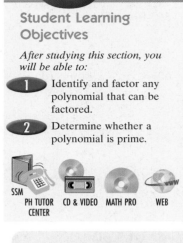

SSM PH TUTOR CD & VIDEO MATH PRO WEB
CENTER

1 Identifying and Factoring Polynomials

Often the various types of factoring problems are all mixed together. We need to be able to identify each type of polynomial quickly. The following table summarizes the information we have learned about factoring.

Many polynomials require more than one factoring method. When you are asked to factor a polynomial, it is expected that you will factor it completely. Usually, the first step is factoring out a common factor; then the next step will become apparent.

Carefully go through each example in the following Factoring Organizer. Be sure you understand each step that is involved.

Factoring Organizer

Number of Terms in the Polynomial	*Identifying Name and/or Formula*	*Example*
A. Any number of terms	**Common factor** The terms have a common factor consisting of a number, a variable, or both.	$2x^2 - 16x = 2x(x - 8)$ $3x^2 + 9y - 12 = 3(x^2 + 3y - 4)$ $4x^2y + 2xy^2 - wxy + xyz = xy(4x + 2y - w + z)$
B. Two terms	**Difference of two squares** First and last terms are perfect squares. $a^2 - b^2 = (a + b)(a - b)$	$16x^2 - 1 = (4x + 1)(4x - 1)$ $25y^2 - 9x^2 = (5y + 3x)(5y - 3x)$
C. Three terms	**Perfect-square trinomial** First and last terms are perfect squares. $a^2 + 2ab + b^2 = (a + b)^2$ $a^2 - 2ab + b^2 = (a - b)^2$	$25x^2 - 10x + 1 = (5x - 1)^2$ $16x^2 + 24x + 9 = (4x + 3)^2$
D. Three terms	**Trinomial of the form $x^2 + bx + c$** It starts with x^2. The constants of the two factors are numbers whose product is c and whose sum is b.	$x^2 - 7x + 12 = (x - 3)(x - 4)$ $x^2 + 11x - 26 = (x + 13)(x - 2)$ $x^2 - 8x - 20 = (x - 10)(x + 2)$
E. Three terms	**Trinomial of the form $ax^2 + bx + c$** It starts with ax^2, where a is any number but 1.	Use trial-and-error or the grouping number method to factor $12x^2 - 5x - 2$. 1. The grouping number is -24. 2. The two numbers whose product is -24 and whose sum is -5 are -8 and 3. 3. $12x^2 - 5x - 2 = 12x^2 + 3x - 8x - 2$ $= 3x(4x + 1) - 2(4x + 1)$ $= (4x + 1)(3x - 2)$
F. Four terms	**Factor by grouping** Rearrange the order if the first two terms do not have a common factor.	$wx - 6yz + 2wy - 3xz = wx + 2wy - 3xz - 6yz$ $= w(x + 2y) - 3z(x + 2y)$ $= (x + 2y)(w - 3z)$

EXAMPLE 1 Factor.

(a) $25x^3 - 10x^2 + x$

(b) $20x^2y^2 - 45y^2$

(c) $2ax + 4ay + 4x + 8y$

(d) $15x^2 - 3x^3 + 18x$

(a) $25x^3 - 10x^2 + x = x(25x^2 - 10x + 1)$ Factor out the common factor of x. The other factor is a perfect-square trinomial.

$$= x(5x - 1)^2$$

(b) $20x^2y^2 - 45y^2 = 5y^2(4x^2 - 9)$ Factor out the common factor of $5y^2$. The other factor is a difference of squares.

$$= 5y^2(2x - 3)(2x + 3)$$

(c) $2ax + 4ay + 4x + 8y = 2[ax + 2ay + 2x + 4y]$ Factor out the common factor of 2.

$$= 2[a(x + 2y) + 2(x + 2y)]$$ Factor the terms inside the bracket by the grouping method.

$$= 2[(x + 2y)(a + 2)]$$ Factor out the common factor of $(x + 2y)$.

(d) $15x^2 - 3x^3 + 18x = -3x^3 + 15x^2 + 18x$ Rearrange the terms in descending order of powers of x.

$$= -3x(x^2 - 5x - 6)$$ Factor out the common factor of $-3x$.

$$= -3x(x - 6)(x + 1)$$ Factor the trinomial.

Practice Problem 1   Factor. Be careful. These practice problems are mixed.

(a) $9x^4y^2 - 9y^2$

(b) $12x - 9 - 4x^2$

(c) $3x^2 - 36x + 108$

(d) $5x^3 - 15x^2y + 10x^2 - 30xy$

2 Determining Whether a Polynomial Is Prime

Not all polynomials can be factored using the methods in this chapter. If we cannot factor a polynomial by elementary methods, we will identify it as a **prime** polynomial. If, after you have mastered the factoring techniques in this chapter, you encounter a polynomial that you cannot factor with these methods, you should feel comfortable enough to say, "The polynomial cannot be factored with the methods in this chapter, so it is prime," rather than "I can't do it—I give up!"

EXAMPLE 2   Factor, if possible. $x^2 + 6x + 12$

The factors of 12 are

$$(1)(12) \text{ or } (2)(6) \text{ or } (3)(4).$$

None of these pairs add up to 6, the coefficient of the middle term. Thus the problem cannot be factored by the methods of this chapter. It is prime.

Practice Problem 2   Factor. $x^2 - 9x - 8$

EXAMPLE 3   Factor, if possible. $25x^2 + 4$

We have a formula to factor the difference of two squares. There is no way to factor the sum of two squares. That is, $a^2 + b^2$ cannot be factored. Thus

$$25x^2 + 4 \text{ is prime.}$$

Practice Problem 3   Factor, if possible. $25x^2 + 82x + 4$

6.6 Exercises

Review the six basic types of factoring in the Factoring Organizer on page 348. Each of the six types is included in exercises 1–12. Be sure you can find two of each type.

Factor. Check your answer by multiplying.

1. $6a^2 + 2ab - 3a$

2. $6x^2 - 3xy + 5x$

3. $36x^2 - 9y^2$

4. $100x^2 - 1$

5. $9x^2 - 12xy + 4y^2$

6. $16x^2 + 24xy + 9y^2$

7. $x^2 + 8x + 15$

8. $x^2 + 15x + 54$

9. $15x^2 + 7x - 2$

10. $6x^2 + 13x - 5$

11. $ax - 3ay - 6by + 2bx$

12. $ax - 20y - 5x + 4ay$

Factor, if possible. Be sure to factor completely. Always factor out the greatest common factor first, if one exists.

13. $3x^4 - 12$

14. $y^2 + 16y + 64$

15. $4x^2 - 12x + 9$

16. $108x^2 - 3$

17. $2x^2 - 11x + 12$

18. $2xy^2 - 50x$

19. $x^2 - 3xy - 70y^2$

20. $2x^3 - 7x^2 + 4x - 14$

21. $ax - 5a + 3x - 15$

22. $by + 7b - 6y - 42$

23. $45x - 5x^3$

24. $18y^2 + 3y - 6$

25. $5x^3y^3 - 10x^2y^3 + 5xy^3$

26. $12x^2 - 36x + 27$

27. $27xyz^2 - 12xy$

28. $12x^2 - 2x - 18x^3$

29. $3x^2 + 6x - 105$

30. $4x^2 - 28x - 72$

31. $5x^2 - 30x + 40$

32. $7x^2 + 3x - 2$

33. $7x^2 - 2x^4 + 4$

34. $2x^4 - 9x^2 - 5$

35. $6x^2 - 3x + 2$

36. $4x^3 + 8x^2 - 60x$

Remove the greatest common factor first. Then continue to factor.

37. $5x^2 + 10xy - 30y$

38. $7a^2 + 21b - 42$

39. $30x^3 + 3x^2y - 6xy^2$

40. $56x^2 - 14xy - 7y^2$

41. $8x^2 + 28x - 16$

42. $12x^2 - 30x + 12$

To Think About

43. A polynomial that cannot be factored by the methods of this chapter is called _____.

44. A binomial of the form $x^2 - d$ can be quickly factored or identified as prime. If it can be factored, what is true of the number d?

Cumulative Review Problems

45. When Dave Barry decided to leave the company and work as an independent contractor, he took a pay cut of 14%. He earned $24,080 this year. What did he earn in his previous job?

46. Nina and Mario bought their house in 1986. In 2000 they sold it for $210,000. This was a 32% increase in the price. How much did Nina and Mario pay for their house in 1986? (Round your answer to the nearest dollar.)

47. Solve the system.
$$\frac{1}{2}x - y = 7$$
$$-3x + 2y = -22$$

48. Solve the system.
$$x + 2y - z = 6$$
$$-2x - y + 4z = -12$$
$$y + \frac{1}{3}z = 1$$

49. Gary loves to read. In his living room he has hardcover books, softcover books, and magazines. He has 37 fewer hardcover books than softcover books. He has twice as many softcover books as magazines. If there are 198 total books and magazines in his bookcase, how many of each type did he have?

50. Gary took some of the items listed in exercise 47 from his bookcase. It now contains a total of 193 books and magazines. The statements comparing the numbers of hardcover and softcover books and the numbers of softcover books and magazines given in exercise 47 still apply. How many of each type does he have now?

Student Learning Objectives

After studying this section, you will be able to:

1 Solve a quadratic equation by factoring.

2 Use quadratic equations to solve applied problems.

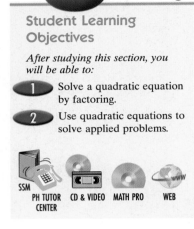

SSM
PH TUTOR   CD & VIDEO   MATH PRO   WEB
CENTER

1   Solving a Quadratic Equation by Factoring

In Chapter 2 we learned how to solve linear equations such as $3x + 5 = 0$ by finding the root (or value of x) that satisfied the equation. Now we turn to the question of how to solve equations like $3x^2 + 5x + 2 = 0$. Such equations are called **quadratic equations**. A quadratic equation is a polynomial equation in one variable that contains a variable term of degree 2 and no terms of higher degree.

> The *standard form* of a quadratic equation is $ax^2 + bx + c = 0$, where a, b, and c are real numbers and $a \neq 0$.

In this section, we will study quadratic equations in standard form, where a, b, and c are integers.

Many quadratic equations have two real number solutions (also called real **roots**). But how can we find them? The most direct approach is the factoring method. This method depends on a very powerful property.

> **Zero Factor Property**
>
> If $a \cdot b = 0$, then $a = 0$ or $b = 0$.

Notice the word "or" in the zero factor property. When we make a statement in mathematics using this word, we intend it to mean *one or the other or both*. Therefore, the zero factor property states that if the product $a \cdot b$ is zero, then a can equal zero, or b can equal zero, or *both a and b can equal zero*. We can use this principle to solve a quadratic equation. Before you start, make sure that the equation is in standard form.

> **1.** Factor, if possible, the quadratic expression that equals 0.
> **2.** Set each factor equal to 0.
> **3.** Solve the resulting equations to find each root.
> **4.** Check each root.

EXAMPLE 1   Solve the equation to find the two roots. $3x^2 + 5x + 2 = 0$

$3x^2 + 5x + 2 = 0$	The equation is in standard form.
$(3x + 2)(x + 1) = 0$	Factor the quadratic expression.
$3x + 2 = 0 \qquad x + 1 = 0$	Set each factor equal to 0.
$3x = -2 \qquad\quad x = -1$	Solve the equations to find the two roots.
$x = -\dfrac{2}{3}$	

The two roots (that is, solutions) are $-\frac{2}{3}$ and -1.

Check.   We can determine if the two numbers $-\frac{2}{3}$ and -1 are solutions to the equation. Substitute $-\frac{2}{3}$ for x in the *original equation*. If an identity results, $-\frac{2}{3}$ is a solution. Do the same for -1.

$$3x^2 + 5x + 2 = 0 \qquad\qquad\qquad 3x^2 + 5x + 2 = 0$$

$$3\left(-\frac{2}{3}\right)^2 + 5\left(-\frac{2}{3}\right) + 2 \overset{?}{=} 0 \qquad\qquad 3(-1)^2 + 5(-1) + 2 \overset{?}{=} 0$$

$$3\left(\frac{4}{9}\right) + 5\left(-\frac{2}{3}\right) + 2 \overset{?}{=} 0 \qquad\qquad 3(1) + 5(-1) + 2 \overset{?}{=} 0$$

$$\frac{4}{3} - \frac{10}{3} + 2 \overset{?}{=} 0 \qquad\qquad\qquad\qquad 3 - 5 + 2 \overset{?}{=} 0$$

$$\frac{4}{3} - \frac{10}{3} + \frac{6}{3} \overset{?}{=} 0 \qquad\qquad\qquad\qquad -2 + 2 \overset{?}{=} 0$$

$$0 = 0 \;\checkmark \qquad\qquad\qquad\qquad\qquad\quad 0 = 0 \;\checkmark$$

Thus $-\frac{2}{3}$ and -1 are both roots of the equation $3x^2 + 5x + 2 = 0$.

Practice Problem 1   Solve the equation by factoring to find the two roots and check.
$10x^2 - x - 2 = 0$

EXAMPLE 2   Solve the equation to find the two roots. $2x^2 + 13x - 7 = 0$

$2x^2 + 13x - 7 = 0$	The equation is in standard form.
$(2x - 1)(x + 7) = 0$	Factor.
$2x - 1 = 0 \qquad x + 7 = 0$	Set each factor equal to 0.
$2x = 1 \qquad\qquad x = -7$	Solve the equations to find the two roots.
$x = \dfrac{1}{2}$	

The two roots are $\frac{1}{2}$ and -7.

Check.   If $x = \frac{1}{2}$, then we have the following.

$$2\left(\frac{1}{2}\right)^2 + 13\left(\frac{1}{2}\right) - 7 = 2\left(\frac{1}{4}\right) + 13\left(\frac{1}{2}\right) - 7$$

$$= \frac{1}{2} + \frac{13}{2} - \frac{14}{2} = 0 \ \checkmark$$

If $x = -7$, then we have the following.

$$2(-7)^2 + 13(-7) - 7 = 2(49) + 13(-7) - 7$$

$$= 98 - 91 - 7 = 0 \ \checkmark$$

Thus $\frac{1}{2}$ and -7 are both roots of the equation $2x^2 + 13x - 7 = 0$.

Practice Problem 2   Solve the equation to find the two roots. $3x^2 - 5x + 2 = 0$

If the quadratic equation $ax^2 + bx + c = 0$ has no visible constant term, then $c = 0$. All such quadratic equations can be solved by factoring out a common factor and then using the zero factor property to obtain two solutions that are real numbers.

EXAMPLE 3   Solve the equation to find the two roots. $7x^2 - 3x = 0$

$7x^2 - 3x = 0$	The equation is in standard form. Here $c = 0$.
$x(7x - 3) = 0$	Factor out the common factor.
$x = 0 \qquad 7x - 3 = 0$	Set each factor equal to 0 by the zero factor property.
$7x = 3$	Solve the equations to find the two roots.
$x = \dfrac{3}{7}$	

The two roots are 0 and $\frac{3}{7}$.

Check.   Verify that 0 and $\frac{3}{7}$ are the roots of $7x^2 - 3x = 0$.

Practice Problem 3   Solve the equation to find the two roots. $7x^2 + 11x = 0$

If the quadratic equation is not in standard form, we use the same basic algebraic methods we studied in Sections 2.1–2.4 to place the terms on one side and zero on the other so that we can use the zero factor property.

EXAMPLE 4   Solve. $x^2 = 12 - x$

$$x^2 = 12 - x$$ The equation is not in standard form.

$$x^2 + x - 12 = 0$$ Add x and -12 to both sides of the equation so that the left side is equal to zero; we can now factor.

$$(x - 3)(x + 4) = 0$$ Factor.

$$x - 3 = 0 \qquad x + 4 = 0$$ Set each factor equal to 0 by the zero factor property.

$$x = 3 \qquad\qquad x = -4$$ Solve the equations for x.

Check.   If $x = 3$: $(3)^2 \overset{?}{=} 12 - 3$ If $x = -4$: $(-4)^2 \overset{?}{=} 12 - (-4)$

$$9 \overset{?}{=} 12 - 3 \qquad\qquad\qquad 16 \overset{?}{=} 12 + 4$$

$$9 = 9 \ \checkmark \qquad\qquad\qquad\qquad 16 = 16 \ \checkmark$$

Both roots check.

Practice Problem 4   Solve. $x^2 - 6x + 4 = -8 + x$

EXAMPLE 5   Solve. $\dfrac{x^2 - x}{2} = 6$

We must first clear the fractions from the equation.

$$2\left(\frac{x^2 - x}{2}\right) = 2(6)$$ Multiply each side by 2.

$$x^2 - x = 12$$ Simplify.

$$x^2 - x - 12 = 0$$ Place in standard form.

$$(x - 4)(x + 3) = 0$$ Factor.

$$x - 4 = 0 \qquad x + 3 = 0$$ Set each factor equal to zero.

$$x = 4 \qquad\qquad x = -3$$ Solve the equations for x.

The check is left to the student.

Practice Problem 5   Solve. $\dfrac{2x^2 - 7x}{3} = 5$

---

② Using Quadratic Equations to Solve Applied Problems

Certain types of word problems—for example, some geometry applications—lead to quadratic equations. We'll show how to solve such word problems in this section.

It is particularly important to check the apparent solutions to the quadratic equation with conditions stated in the word problem. Often a particular solution to the quadratic equation will be eliminated by the conditions of the word problem.

EXAMPLE 6   Carlos lives in Mexico City. He has a rectangular brick walkway in front of his house. The length of the walkway is 3 meters longer than twice the width. The area of the walkway is 44 square meters. Find the length and width of the rectangular walkway.

1. *Understand the problem.*
   Draw a picture.

   Let w = the width in meters

   Then $2w + 3$ = the length in meters

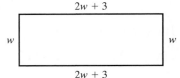

2. *Write an equation.*

$$\text{area} = (\text{width})(\text{length})$$

$$44 = w(2w + 3)$$

3. *Solve and state the answer.*

$$44 = w(2w + 3)$$

$$44 = 2w^2 + 3w$$ Remove parentheses.

$$0 = 2w^2 + 3w - 44$$ Subtract 44 from both sides.

$$0 = (2w + 11)(w - 4)$$ Factor.

$$2w + 11 = 0 \qquad w - 4 = 0$$ Set each factor equal to 0.

$$2w = -11 \qquad w = 4$$ Simplify and solve.

$$w = -5\frac{1}{2}$$ Although $-5\frac{1}{2}$ is a solution to the quadratic equation, it is not a valid solution to the word problem. It would not make sense to have a rectangle with a negative number as a width.

Since $w = 4$, the width of the walkway is 4 meters. The length is $2w + 3$, so we have $2(4) + 3 = 8 + 3 = 11$. Thus the length of the walkway is 11 meters.

4. *Check.*   Is the length 3 meters more than twice the width?

$$11 \overset{?}{=} 3 + 2(4) \qquad 11 = 3 + 8 \ ✓$$

Is the area of the rectangle 44 square meters?

$$4 \times 11 \overset{?}{=} 44 \qquad 44 = 44 \ ✓$$

▲ **Practice Problem 6**   The length of a rectangle is 2 meters longer than triple the width. The area of the rectangle is 85 square meters. Find the length and width of the rectangle.   ●

▲ ● **EXAMPLE 7**   The top of a local cable television tower has several small triangular reflectors. The area of each triangle is 49 square centimeters. The altitude of each triangle is 7 centimeters longer than the base. Find the altitude and the base of one of the triangles.

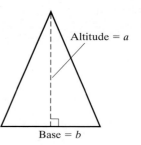

Let $b =$ the length of the base in centimeters

$b + 7 =$ the length of the altitude in centimeters

To find the area of a triangle, we use

$$\text{area} = \frac{1}{2}(\text{altitude})(\text{base}) = \frac{1}{2}ab = \frac{ab}{2}.$$

$$\frac{ab}{2} = 49$$ Write an equation.

$$\frac{(b + 7)(b)}{2} = 49$$ Substitute the expressions for altitude and base.

$$\frac{b^2 + 7b}{2} = 49$$ Simplify.

$$b^2 + 7b = 98$$ Multiply each side of the equation by 2.

$$b^2 + 7b - 98 = 0$$ Place the quadratic equation in standard form.

$$(b - 7)(b + 14) = 0$$ Factor.

$$b - 7 = 0 \qquad b + 14 = 0$$ Set each factor equal to zero.

$$b = 7 \qquad b = -14$$ Solve the equations for b.

We cannot have a base of -14 centimeters, so we reject the negative answer. The only possible solution is 7. So the base is 7 centimeters. The altitude is $b + 7 = 7 + 7 = 14$. The altitude is 14 centimeters. The triangular reflector has a base of 7 centimeters and an altitude of 14 centimeters.

Check. When you do the check, answer the following two questions.

1. Is the altitude 7 centimeters longer than the base?

2. Is the area of a triangle with a base of 7 centimeters and an altitude of 14 centimeters actually 49 square centimeters?

▲ **Practice Problem 7**   A triangle has an area of 35 square centimeters. The altitude of the triangle is 3 centimeters shorter than the base. Find the altitude and the base of the triangle.

Many problems in the sciences require the use of quadratic equations. You will study these in more detail if you take a course in physics or calculus in college. Often a quadratic equation is given as part of the problem.

When an object is thrown upward, its height (S) in meters is given, approximately, by the quadratic equation

$$S = -5t^2 + vt + h.$$

The letter h represents the initial height in meters. The letter v represents the initial velocity of the object thrown. The letter t represents the time in seconds starting from the time the object is thrown.

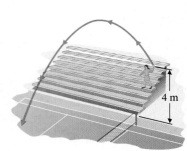

4 m

EXAMPLE 8   A tennis ball is thrown upward with an initial velocity of 8 meters/second. Suppose that the initial height above the ground is 4 meters. At what time t will the ball hit the ground?

In this case $S = 0$ since the ball will hit the ground. The initial upward velocity is $v = 8$ meters/second. The initial height is 4 meters, so $h = 4$.

$S = -5t^2 + vt + h$	Write an equation.
$0 = -5t^2 + 8t + 4$	Substitute all values into the equation.
$5t^2 - 8t - 4 = 0$	Isolate the terms on the left side. (Most students can factor more readily if the squared variable is positive.)
$(5t + 2)(t - 2) = 0$	Factor.
$5t + 2 = 0 \qquad t - 2 = 0$	Set each factor $= 0$.
$5t = -2 \qquad\quad t = 2$	Solve the equations for t.
$t = -\dfrac{2}{5}$	

We want a positive time t in seconds; thus we do not use $t = -\frac{2}{5}$. Therefore the ball will strike the ground 2 seconds after it is thrown.

Check.   Verify the solution.

Practice Problem 8   A Mexican cliff diver does a dive from a cliff 45 meters above the ocean. This constitutes free fall, so the initial velocity is $v = 0$, and if there is no upward spring, then $h = 45$ meters. How long will it be until he breaks the water's surface?

Exercises

Using the factoring method, solve for the roots of each quadratic equation. Be sure to place the equation in standard form before factoring. Check your answers.

1. $x^2 - 4x - 21 = 0$ **2.** $x^2 - x - 20 = 0$ **3.** $2x^2 - 5x - 3 = 0$ **4.** $3x^2 - 5x - 2 = 0$

5. $2x^2 - 7x + 6 = 0$ **6.** $2x^2 - 11x + 12 = 0$ **7.** $6x^2 - 13x = -6$ **8.** $10x^2 + 19x = 15$

9. $x^2 + 13x = 0$ **10.** $8x^2 - x = 0$ **11.** $8x^2 = 72$ **12.** $9x^2 = 81$

13. $5x^2 + 3x = 8x$ **14.** $6x^2 - 4x = 3x$

15. $(x - 5)(x + 2) = -4(x + 1)$ **16.** $(x - 5)(x + 4) = 2(x - 5)$

17. $4x^2 - 3x + 1 = -7x$ **18.** $9x^2 - 2x + 4 = 10x$ **19.** $\dfrac{x^2}{2} - 8 + x = -8$ **20.** $4 + \dfrac{x^2}{3} = 2x + 4$

21. $\dfrac{x^2 + 7x}{4} = -3$ **22.** $\dfrac{x^2 + 5x}{6} = 4$ **23.** $\dfrac{9x^2 - 12x}{3} = 15$ **24.** $\dfrac{4x^2 - 10x}{2} = 12$

To Think About

25. Why can an equation in standard form with $c = 0$ (that is, an equation of the form $ax^2 + bx = 0$) always be solved?

26. Martha solved $(x + 3)(x - 2) = 14$ as follows:

$$x + 3 = 14 \quad \text{or} \quad x - 2 = 14$$
$$x = 11 \quad \text{or} \quad x = 16$$

Josette said this had to be wrong because these values do not check. Explain what is wrong with Martha's method.

Applications

▲ **27.** The area of a rectangular garden is 140 square meters. The width is 3 meters longer than one-half of the length. Find the length and the width of the garden.

▲ **28.** The area of a triangular sign is 33 square meters. The base of the triangle is 1 meter less than double the altitude. Find the altitude and the base of the sign.

Suppose the number of teams competing in a sports league is x. In this league each team plays each other team twice. The total number G of games to be played is given by the equation $G = x^2 - x$. Use this information in solving exercises 29–32.

29. A women's basketball league has a total of 14 teams. How many games will be played during the season?

30. A men's baseball league has a total of 12 teams. How many games will be played during the season?

31. A city tennis league has a total of 210 games. How many teams are in the league?

32. A pee-wee football league has a total of 156 games. How many teams are in the league?

Use the following information for exercises 33 and 34. When an object is thrown upward, its height (S), in meters, is given (approximately) by the quadratic equation

$$S = -5t^2 + vt + h,$$

where v = the upward initial velocity in meters/second,

t = the time of flight in seconds, and

h = the height above level ground from which the object is thrown.

33. Johnny is standing on a platform 6 meters high and throws a ball straight up as high as he can at a velocity of 13 meters/second. At what time t will the ball hit the ground? How far from the ground is the ball after 2 seconds have elapsed from the time of the throw? (Assume that the ball is 6 meters from the ground when it leaves Johnny's hand.)

34. You are standing on the edge of a cliff near Acapulco, overlooking the ocean. The place where you stand is 180 meters from the ocean. You drop a pebble into the water. ("Dropping" the pebble implies that there is no initial velocity, so $v = 0$.) How many seconds will it take to hit the water? How far has the pebble dropped after 3 seconds?

35. Paying overtime to employees can be very expensive, but if profits are realized, it is worth putting a shift on overtime. In a high-tech helicopter company, the extra hourly cost in dollars for producing x additional helicopters is given by the cost equation $C = 2x^2 - 7x$. If the extra hourly cost is $15, how many additional helicopters are produced?

36. A boat generator on a Gloucester fishing boat is required to produce 64 watts of power. The amount of current I measured in amperes needed to produce the power for this generator is given by the equation $P = 40I - 4I^2$. What is the *minimum* number of amperes required to produce the necessary power?

The technology and communication office of a local company has set up a new telephone system so that each employee has a separate telephone and extension number. They are studying the possible number of telephone calls that can be made from people in the office to other people in the office. They have discovered that the total number of possible telephone calls T is described by the equation $T = 0.5(x^2 - x)$, where x is the number of people in the office. Use this information to answer exercises 37–40.

37. If 70 people are presently employed at the office, how many possible telephone calls can be made between these 70 people?

38. If the company hires 10 new employees next year, how many possible telephone calls can be made between the 80 people that will be employed next year?

39. One Saturday, only a small number of employees were working at the office. It has been determined that on that day, a total of 153 different phone calls could have been made from people working in the office to other people working in the office. How many people worked on that Saturday?

40. On the day after Thanksgiving, only a small number of employees were working at the office. It has been determined that on that day, a total of 105 different phone calls could have been made from people working in the office to other people working in the office. How many people worked on the day after Thanksgiving?

Cumulative Review Problems

Simplify.

41. $(2x^2y^3)(-6xy^4)$

42. $(3a^4b^5)(4a^6b^8)$

43. $\dfrac{21a^5b^{10}}{-14ab^{12}}$

44. $\dfrac{18x^3y^6}{54x^8y^{10}}$

Putting Your Skills to Work

Predicting Total Earnings by a Mathematical Series

When assessing the financial health and future of a company or person, it is often helpful to look at the total income of that company or person over a long period of time. If the income tends to increase or decrease at a fairly constant rate, then the sum of this income can be determined by a formula for a mathematical series.

Angela Sanchez has been employed by West Coast Cable Vision for 9 years. She started at an annual salary of $18,000. Her income has increased about $1200 per year for the last 9 years. She is now earning $27,600. To find her total earnings for 9 years we can use the formula in factored form

$$T = 0.5n(s + e),$$

where n = the number of years she has worked,

s = her starting salary, and

e = her salary at the end of the time period.

Thus we have the following.

$$T = 0.5(9)(18,000 + 27,600)$$
$$T = 0.5(9)(45,600)$$
$$T = \$205,200$$

Thus Angela has earned $205,200 in the last 9 years.

Use the preceding formula to determine the following.

Problems for Individual Study and Investigation

1. Monique Bussell has worked for the last 20 years at a salary that has increased about $1500 per year. Her starting salary was $24,000. Approximately how much has she earned in this 20-year period?

2. The Software Service Center has been in operation for 8 years. During the first year they made a profit of $150,000. Each year the profit has increased by $70,000. What is their total profit for 8 years?

Problems for Cooperative Group Investigation

3. James Kerr opened his own travel business. Last year he earned $40,000. This year he earned $56,000. If his business earnings continue to increase each year by the same amount, how many years will it be until he has earned a total income of $1 million?

4. A national chain of department stores made a profit of $780,000,000 ten years ago. Over the past ten years, the profit has been decreasing at the rate of $15,000,000 per year. If this rate of decrease continues, what will be the total earnings of the store for the next 15 years (starting with this present year)?

Internet Connections

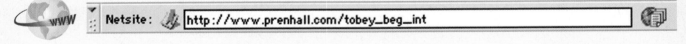

Site: DataMasters Salary Survey or a related site

This site gives salary information for executives and professionals in the computer industry.

5. Karl is an experienced software engineer whose salary has kept up with the regional median for the West Coast since 1990. Use the DataMasters site to determine Karl's salary for 1990 and for the current year. Then use the formula $T = 0.5n(s + e)$ to find his total salary for all years from 1990 to the current year, assuming that his income has increased by a constant amount each year.

Math in the Media

Media Literacy Tips

We receive information daily from a number of media sources—radio, newspapers, television, and on-line content providers. It's important to be able to use math skills to understand the data presented. But using your math skills alone is not enough. You also need to be able to interpret the message in context of the article.

The media can be very persuasive when promoting a viewpoint. There are many factors to consider when evaluating the article. Following are a few tips that may help you to filter out the "hype" and focus on the meaning.

1. Consider the source. Who is the author? Does he or she have a particular platform that may influence the message?

2. Look for the key points. Has anything been done to either attract attention to certain issues and facts or distract the reader from them?

3. Be alert for key omissions. Sometimes what is left unsaid is the most compelling statement.

4. Consider the supporting information. Does the author support his or her argument with facts, opinions, both? How is math used to support the argument?

EXERCISES

1. Locate an article or news feature that interests you. Read the article and write down your responses to each of the 4 tips above.

2. What response do you think the author is trying to elicit?

3. Suppose you read an article with the following information: "Data released yesterday by the FBI indicates that the number of murders committed in the United States has decreased by 33.7% in the last five years. The murder rate for 1998 was 6.3% per 100,000 inhabitants. This rate is the lowest the rate has been since 1979. What a wonderful improvement. Truly the United States is becoming a safe place to live." What concerns or objections would you raise concerning this article?

Chapter 6 Organizer

Topic	Procedure	Examples
A. Common factor, p. 322.	Factor out the largest common factor from each term.	$2x^2 - 2x = 2x(x - 1)$ $3a^2 + 3ab - 12a = 3a(a + b - 4)$ $8x^4y - 24x^3 = 8x^3(xy - 3)$
Special cases **B. Difference of squares,** p. 343. **C. Perfect-square trinomials,** p. 344.	If you recognize the special cases, you will be able to factor quickly. $$a^2 - b^2 = (a + b)(a - b)$$ $$a^2 + 2ab + b^2 = (a + b)^2$$ $$a^2 - 2ab + b^2 = (a - b)^2$$	$25x^2 - 36y^2 = (5x + 6y)(5x - 6y)$ $16x^4 - 1 = (4x^2 + 1)(2x + 1)(2x - 1)$ $25x^2 + 10x + 1 = (5x + 1)^2$ $49x^2 - 42xy + 9y^2 = (7x - 3y)^2$
D. Trinomials of the form $x^2 + bx + c$, p. 330.	Factor trinomials of the form $x^2 + bx + c$ by asking what two numbers have a product of c and a sum of b. If each term of the trinomial has a common factor, factor it out as the first step.	$x^2 - 18x + 77 = (x - 7)(x - 11)$ $x^2 + 7x - 18 = (x + 9)(x - 2)$ $5x^2 - 10x - 40 = 5(x^2 - 2x - 8)$ $\qquad = 5(x - 4)(x + 2)$
E. Trinomials of the form $ax^2 + bx + c$, where $a \neq 1$, p. 337, 338.	Factor trinomials of the form $ax^2 + bx + c$ by the grouping number method or by the trial-and-error method.	$6x^2 + 11x - 10$ Grouping number $= -60$ Two numbers whose product is -60 and whose sum is $+11$ are $+15$ and -4. $6x^2 + 15x - 4x - 10 = 3x(2x + 5) - 2(2x + 5)$ $\qquad = (2x + 5)(3x - 2)$
F. Four terms. **Factor by grouping,** p. 326.	Rearrange the terms if necessary so that the first two terms have a common factor. Then factor out the common factors. $$ax + ay - bx - by = a(x + y) - b(x + y)$$ $$= (a - b)(x + y)$$	$2ax^2 + 21 + 3a + 14x^2$ $= 2ax^2 + 14x^2 + 3a + 21$ $= 2x^2(a + 7) + 3(a + 7)$ $= (a + 7)(2x^2 + 3)$
Prime polynomials, p. 349.	A polynomial that is not factorable is called prime.	$x^2 + y^2$ is prime. $x^2 + 5x + 7$ is prime.
Multistep factoring, p. 348.	Many problems require two or three steps of factoring. Always try to factor out a common factor as the first step.	$3x^2 - 21x + 36 = 3(x^2 - 7x + 12)$ $\qquad = 3(x - 4)(x - 3)$ $2x^3 - x^2 - 6x = x(2x^2 - x - 6)$ $\qquad = x(2x + 3)(x - 2)$ $25x^3 - 49x = x(25x^2 - 49)$ $\qquad = x(5x + 7)(5x - 7)$ $8x^2 - 24x + 18 = 2(4x^2 - 12x + 9)$ $\qquad = 2(2x - 3)^2$
Solving quadratic equations by factoring, p. 352.	1. Write as $ax^2 + bx + c = 0$. 2. Factor. 3. Set each factor equal to 0. 4. Solve the resulting equations.	Solve: $3x^2 + 5x = 2$ $3x^2 + 5x - 2 = 0$ $(3x - 1)(x + 2) = 0$ $3x - 1 = 0$ or $x + 2 = 0$ $x = \dfrac{1}{3}$ or $x = -2$

Topic	Procedure	Examples
Using quadratic equations to solve applied problems, *p. 354.*	Some word problems, like those involving the product of two numbers, area, and formulas with a squared variable, can be solved using the factoring methods we have shown.	The length of a rectangle is 4 less than three times the width. Find the length and width if the area is 55 square inches. Let w = width. Then $3w - 4$ = length. $$55 = w(3w - 4)$$ $$55 = 3w^2 - 4w$$ $$0 = 3w^2 - 4w - 55$$ $$0 = (3w + 11)(w - 5)$$ $$w = -\tfrac{11}{3} \quad \text{or} \quad w = 5$$ $-\tfrac{11}{3}$ is not a valid solution. Thus width = 5 inches and length = 11 inches.

Chapter 6 Review Problems

6.1 *Factor out the greatest common factor.*

1. $15x^3y - 9x^2y^2$

2. $40x^2 - 32x$

3. $7x^2y - 14xy^2 - 21x^3y^3$

4. $50a^4b^5 - 25a^4b^4 + 75a^5b^5$

5. $27x^3 - 9x^2$

6. $2x - 4y + 6z + 12$

7. $2a(a + 3b) - 5(a + 3b)$

8. $15x^3y + 6xy^2 + 3xy$

6.2 *Factor by grouping.*

9. $3ax - 7a - 6x + 14$

10. $a^2 + 5ab - 4a - 20b$

11. $x^2y + 3y - 2x^2 - 6$

12. $4ad - 25c - 5d + 20ac$

13. $15x^2 - 3x + 10x - 2$

14. $30w^2 - 18w + 5wz - 3z$

6.3 *Factor completely. Be sure to factor out any common factors as your first step.*

15. $x^2 - 2x - 35$

16. $x^2 - 10x + 24$

17. $x^2 + 14x + 48$

18. $x^2 + 8xy + 15y^2$

19. $x^4 + 13x^2 + 42$

20. $x^2 - 14x - 51$

21. $5x^2 + 20x + 15$

22. $3x^2 + 39x + 36$

23. $2x^2 - 28x + 96$

24. $4x^2 - 44x + 120$

6.4 *Factor completely. Be sure to factor out any common factors as your first step.*

25. $4x^2 + 7x - 15$

26. $12x^2 + 11x - 5$

27. $15x^2 + 7x - 4$

28. $6x^2 - 13x + 6$

29. $2x^2 - x - 3$

30. $3x^2 + 2x - 8$

31. $20x^2 + 48x - 5$

32. $20x^2 + 21x - 5$

33. $4a^2 - 11a - 3$

34. $4a^2 - a - 3$

35. $6x^2 + 4x - 10$

36. $6x^2 - 4x - 10$

37. $4x^2 - 26x + 30$

38. $4x^2 - 20x - 144$

39. $12x^2 + 5x - 3$

40. $16x^2 + 14x - 30$

41. $6x^2 - 19xy + 10y^2$

42. $6x^2 - 32xy + 10y^2$

6.5 *Factor these special cases. Be sure to factor out any common factors.*

43. $49x^2 - y^2$

44. $16x^2 - 36y^2$

45. $9x^2 - 12x + 4$

46. $64x^2 - 1$

47. $25x^2 - 36$

48. $100x^2 - 9$

49. $1 - 49x^2$

50. $4 - 49x^2$

51. $36x^2 + 12x + 1$

52. $25x^2 - 20x + 4$

53. $16x^2 - 24xy + 9y^2$

54. $49x^2 - 28xy + 4y^2$

55. $2x^2 - 18$

56. $3x^2 - 75$

57. $8x^2 + 40x + 50$

58. $50x^2 - 120x + 72$

6.6 *If possible, factor each polynomial completely. If a polynomial cannot be factored, state that it is prime.*

59. $4x^2 - 9y^2$

60. $x^2 + 6x + 9$

61. $x^2 - 9x + 18$

62. $x^2 + 13x - 30$

63. $6x^2 + x - 7$

64. $10x^2 + x - 2$

65. $12x + 16$

66. $8x^2y^2 - 4xy$

67. $50x^3y^2 + 30x^2y^2 - 10x^2y^2$

68. $26a^3b - 13ab^3 + 52a^2b^4$

69. $x^3 - 16x^2 + 64x$

70. $2x^2 + 40x + 200$

71. $3x^2 - 18x + 27$

72. $25x^3 - 60x^2 + 36x$

73. $7x^2 - 9x - 10$

74. $4x^2 - 13x - 12$

75. $9x^3y - 4xy^3$

76. $3x^3a^3 - 11x^4a^2 - 20x^5a$

77. $12a^2 + 14ab - 10b^2$

78. $16a^2 - 40ab + 25b^2$

79. $7a - 7 - ab + b$

80. $3d - 4 - 3cd + 4c$

81. $2x - 1 + 2bx - b$

82. $5xb - 35x + 4by - 28y$

83. $2a^2x - 15ax + 7x$

84. $x^5 - 17x^3 + 16x$

85. $x^4 - 81y^{12}$

86. $6x^4 - x^2 - 15$

87. $28yz - 16xyz + x^2yz$

88. $12x^3 + 17x^2 + 6x$

89. $16w^2 - 2w - 5$

90. $12w^2 - 12w + 3$

91. $4y^3 + 10y^2 - 6y$

92. $10y^2 + 33y - 7$

93. $8y^{10} - 16y^8$

94. $49x^4 - 49$

95. $x^2 + 13x + 54$

96. $8x^2 - 19x - 6$

97. $8y^5 + 4y^3 - 60y$

98. $9xy^2 + 3xy - 42x$

99. $16x^4y^2 - 56x^2y + 49$

100. $128x^3y - 2xy$

101. $2ax + 5a - 10b - 4bx$

102. $2x^3 - 9 + x^2 - 18x$

6.7 *Solve the following equations by factoring.*

103. $x^2 - 3x - 18 = 0$

104. $x^2 + 6x - 27 = 0$

105. $5x^2 = 2x - 7x^2$

106. $8x^2 + 5x = 2x^2 - 6x$

107. $2x^2 + 9x - 5 = 0$

108. $x^2 + 11x + 24 = 0$

109. $x^2 + 14x + 45 = 0$

110. $5x^2 = 7x + 6$

111. $3x^2 + 6x = 2x^2 - 9$

112. $4x^2 + 9x - 9 = 0$

113. $5x^2 - 11x + 2 = 0$

Solve.

▲ **114.** The area of a triangle is 35 square centimeters. The base is 3 centimeters longer than the altitude of the triangle. Find the length of the base and the altitude.

▲ **115.** The area of a rectangle is 105 square feet. The length of the rectangle is 1 foot longer than double the width. Find the length and width of the rectangle.

116. The height in feet that a model rocket attains is given by $h = -16t^2 + 80t + 96$, where t is the time measured in seconds. How many seconds will it take until the rocket finally reaches the ground? (*Hint:* At ground level $h = 0$.)

117. An electronic technician is working with a 100-volt electric generator. The output power of the generator is given by the equation $p = -5x^2 + 100x$, where x is the amount of current measured in amperes and p is measured in watts. The technician wants to find the value for x when the power is 480 watts. Can you find the two answers?

The Sum-of-the-Digits Depreciation Method

All of us find that cars and trucks depreciate in value too quickly. The amount of time from when a car is new to when it ends up in the scrap heap is all too brief. However, in mathematics, we need to understand precisely how much decrease in value each year occurs for a given car or truck. The following method is one possible way to measure this.

Many people have a car or a truck that they use for business. When figuring out their annual income tax, it is important that they determine how much these vehicles have decreased in value. One such method used by the IRS is called the sum-of-the-digits depreciation method. Suppose that someone bought a new Ford Taurus for $20,500 and wanted to use it in a business for six years and then sell it for $5,500. The usable value of the vehicle for six years would be $20,500 − $5,500, which is $15,000.

Now how do we spread this usable value over six years?

The sum-of-the-digits depreciation method states that we can spread a usable value over n years by multiplying this value by a series of fractions. We form these fractions in the following way.

Find the sum of the integers from 1 to n. The result will become the denominators of the fractions. The numerators are the integers from 1 to n in descending order. Thus we have the fractions

$$\frac{n}{1 + 2 + \ldots + n}, \quad \frac{n-1}{1 + 2 + \ldots + n}, \quad \ldots,$$
$$\frac{1}{1 + 2 + \ldots + n}.$$

To find the depreciation for:

the first year, multiply the usable value by the first fraction.

the second year, multiply the usable value by the second fraction.

(Continue each year following the same pattern.)

the last year, multiply the usable value by the last fraction.

So, for the example given, we can spread the usable value of $15,000 over six years by first finding the sum of the integers from 1 to 6.

$$1 + 2 + 3 + 4 + 5 + 6 = 21$$

Next we multiply the $15,000 by the fractions $\frac{6}{21}$, $\frac{5}{21}$, $\frac{4}{21}$, $\frac{3}{21}$, $\frac{2}{21}$, and $\frac{1}{21}$.

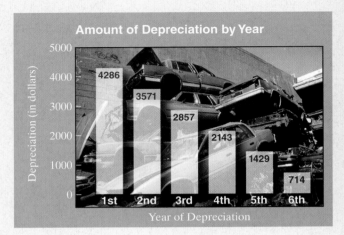

Amount of Depreciation by Year

We will round each value to the nearest dollar, a method suggested by the IRS.

First year's depreciation: $\frac{6}{21} \times \$15,000 = \$4286$

Second year's depreciation: $\frac{5}{21} \times \$15,000 = \$3571$

Third year's depreciation: $\frac{4}{21} \times \$15,000 = \$2857$

Fourth year's depreciation: $\frac{3}{21} \times \$15,000 = \$2143$

Fifth year's depreciation: $\frac{2}{21} \times \$15,000 = \$1429$

Sixth year's depreciation: $\frac{1}{21} \times \$15,000 = \$714$

The dollar amounts by which the car will depreciate each year are shown on the bar graph above.

Problems for Individual Investigation and Analysis

1. How many times greater is the first year's depreciation than the fifth year's depreciation? Round to the nearest tenth.

2. How many times greater is the second year's depreciation than the sixth year's depreciation? Round to the nearest tenth.

Problems for Group Investigation and Cooperative Study

3. If a new Honda Civic is purchased for Sal's Delivery Service for $16,750, used for five years, and then sold for $5,750, what will be the depreciation each year for these five years? Round to the nearest dollar.

4. If a new Ford F-250 truck is purchased for Camp Property Care for $26,250, used for seven years, and then sold for $12,250, what will be the depreciation each year for these seven years? Round to the nearest dollar.

Some of the students at North Shore Community College wrote an equation that described the approximate trade-in value of a three-year-old Ford Taurus. The equation was $V = 5700 - 30x$, where V represents the value of the car in dollars and x represents the number of miles in thousands in excess of 20,000 miles. For example, if the car had 50,000 miles on the odometer, it would have 30,000 miles in excess of the 20,000 figure. Measured in thousands, then, the value of x would be 30. To find the trade-in value of the car, substitute $x = 30$ and obtain

$$V = 5700 - 30(30) = 5700 - 900 = 4800.$$

A used Ford Taurus with 50,000 miles would have an approximate trade-in value of $4800.

5. To simplify calculations write the equation $V = 5700 - 30x$ in factored form. Use the factored form to find the trade-in value for a used three-year-old Ford Taurus with 80,000 miles on the odometer.

6. Use your result from problem 5 to find the number of miles that would be registered on the odometer of a used three-year-old Taurus that had a trade-in value of $4500.

Internet Connections

Site: Kelley Blue Book of Car and Truck Values

7. Use this site to determine the amount of depreciation for a one-year-old Honda Civic. (Subtract the list price from the wholesale value after one year.) What percent of the value of the Honda Civic is lost in one year?

8. Use this site to determine the amount of depreciation for a one-year-old Ford F-250 truck. (Subtract the list price from the wholesale value after one year.) What percent of the value of the truck is lost in one year?

Chapter 6 Test

1. _____

2. _____

3. _____

4. _____

5. _____

6. _____

7. _____

8. _____

9. _____

10. _____

11. _____

12. _____

13. _____

14. _____

15. _____

16. _____

17. _____

18. _____

19. _____

20. _____

21. _____

22. _____

23. _____

24. _____

If possible, factor each polynomial completely. If a polynomial cannot be factored, state that it is prime.

1. $x^2 + 12x - 28$

2. $25x^2 - 49y^2$

3. $10x^2 + 27x + 5$

4. $9a^2 - 30ab + 25b^2$

5. $7x - 9x^2 + 14xy$

6. $3x^2 - 4wy - 2wx + 6xy$

7. $6x^3 - 20x^2 + 16x$

8. $5a^2c - 11abc + 2b^2c$

9. $100x^4 - 16y^4$

10. $9x^2 - 15xy + 4y^2$

11. $7x^2 - 42x$

12. $36x^2 + 1$

13. $3x^2 + 5x + 1$

14. $60xy^2 - 20x^2y - 45y^3$

15. $81x^2 - 1$

16. $x^{16} - 1$

17. $2ax + 6a - 5x - 15$

18. $aw^2 - 8b + 2bw^2 - 4a$

19. $3x^2 - 3x - 90$

20. $2x^3 - x^2 - 15x$

Solve.

21. $x^2 + 14x + 45 = 0$

22. $14 + 3x(x + 2) = -7x$

23. $2x^2 + x - 10 = 0$

Solve using a quadratic equation.

▲ **24.** The park service is studying a rectangular piece of land that has an area of 91 square miles. The length of this piece of land is 1 mile shorter than double the width. Find the length and width of this rectangular piece of land.

Approximately one-half of this test covers the content of Chapters 1–5. The remainder covers the content of Chapter 6.

Simplify.

1. $-3.2 - 6.4 + 0.24 - 1.8 + 0.8$

2. $(-2x^3y^4)(-4xy^6)$

3. $(-3)^4$

4. $(9x - 4)(3x + 2)$

5. $(2x^2 - 6x + 1)(x - 3)$

Solve.

6. $3x - 4 \geq 6x + 5$

7. $3x - (7 - 5x) = 3(4x - 5)$

8. $\dfrac{1}{2}x - 3 = \dfrac{1}{4}(3x + 3)$

9. Solve for t: $s = \dfrac{1}{2}(2a + 3t)$.

10. Find the slope of the line passing through $(-2, 3)$ and $(0, 5)$.

11. For $h(x) = \dfrac{5}{x - 7}$, find $h(-3)$.

12. Solve the system. $\quad 2x + \dfrac{2}{3}y = -8$

$$\dfrac{1}{5}x + y = 2$$

Factor each polynomial completely. If a polynomial cannot be factored, state that it is prime.

13. $6x^2 - 5x + 1$

14. $6x^2 + 5x - 4$

15. $9x^2 + 3x - 2$

16. $121x^2 - 64y^2$

17. $4x + 120 - 80x^2$

18. $x^2 + 5x + 9$

19. $16x^3 + 40x^2 + 25x$

20. $81x^4 - 16b^4$

21. $2ax - 4bx + 3a - 6b$

22. $x^4 + 8x^2 + 15$

Solve.

23. $x^2 + 5x - 24 = 0$

24. $3x^2 - 11x + 10 = 0$

Solve using a quadratic equation.

▲ **25.** The park service is studying a triangular piece of land that contains 57 square miles. The altitude of this triangle is 7 miles longer than double the base. Find the altitude and the base of this triangular piece of land.

1.	
2.	
3.	
4.	
5.	
6.	
7.	
8.	
9.	
10.	
11.	
12.	
13.	
14.	
15.	
16.	
17.	
18.	
19.	
20.	
21.	
22.	
23.	
24.	
25.	

Rational Expressions and Equations

Do you realize that the United States might have a surplus of $1.3 trillion by 2010? Furthermore, if economic growth drops just a little to 2.1% per year, we might have instead a deficit of $286 billion by 2010? Do you think you could use mathematical equations to predict the future of the United States' budget surplus or deficit? Turn to the Putting Your Skills to Work problems on page 392 to find out.

Pretest Chapter 7

1. _____

2. _____

3. _____

4. _____

5. _____

6. _____

7. _____

8. _____

9. _____

10. _____

11. _____

12. _____

13. _____

14. _____

15. _____

16. _____

17. _____

18. _____

If you are familiar with the topics in this chapter, take this test now. Check your answers with those in the back of the book. If an answer is wrong or you can't answer a question, study the appropriate section of the chapter.

If you are not familiar with the topics in this chapter, don't take this test now. Instead, study the examples, work the practice problems, and then take the test.

This test will help you identify which concepts you have mastered and which you need to study further.

Simplify the algebraic expressions in questions 1–14.

Section 7.1

1. $\dfrac{6a - 4b}{2b - 3a}$

2. $\dfrac{x^2 - x - 6}{2x^2 + 7x + 6}$

3. $\dfrac{a^2b + 2ab^2}{2a^3 + 3a^2b - 2ab^2}$

Section 7.2

4. $\dfrac{4a^2 - b^2}{6a - 6b} \cdot \dfrac{3a - 3b}{6a + 3b}$

5. $\dfrac{x^2 - 6x + 9}{x^2 - x - 6} \div \dfrac{x^2 + 2x - 15}{x^2 + 2x}$

6. $\dfrac{x^2 + 5x + 6}{3x^2 + 8x + 4} \cdot \dfrac{6x^2 - 11x - 10}{2x^2 + x - 15}$

7. $\dfrac{xy + 3y}{x^2 - x} \div \dfrac{x + 3}{x}$

Section 7.3

8. $\dfrac{3y + 1}{y + 2} + \dfrac{5}{y + 2}$

9. $\dfrac{2y - 1}{2y^2 + y - 3} - \dfrac{2}{y - 1}$

10. $\dfrac{4x + 10}{x^2 - 25} - \dfrac{3}{x - 5}$

11. $\dfrac{x + 3}{x^2 - 6x + 9} + \dfrac{2x + 3}{3x^2 - 9x}$

Section 7.4

12. $\dfrac{\dfrac{2}{a} - \dfrac{3}{a^2}}{5 + \dfrac{1}{a}}$

13. $\dfrac{\dfrac{a}{a + 1} - \dfrac{2}{a}}{3a}$

14. $\dfrac{\dfrac{x - y}{x} - \dfrac{x + y}{y}}{\dfrac{x - y}{x} + \dfrac{x + y}{y}}$

Section 7.5

Solve for x. If the equation has no solution, so state.

15. $\dfrac{5}{2x} = 2 - \dfrac{2x}{x + 1}$

16. $\dfrac{24}{x^2 - 3x - 10} = \dfrac{2}{x - 5} + \dfrac{3}{x + 2}$

Section 7.6

17. Solve for x and round to the nearest tenth. $\dfrac{5}{x} = \dfrac{7}{13}$

18. When Marcia went to England, she found that $3.10 in U.S. money was needed at the current exchange rate to receive 2 British pounds. How much U.S. money was needed to receive 170 British pounds?

7.1 Simplifying Rational Expressions

Student Learning Objectives

After studying this section, you will be able to:

1. Simplify rational expressions by factoring.

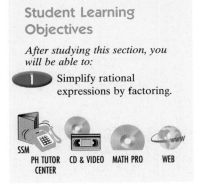

SSM PH TUTOR CENTER   CD & VIDEO   MATH PRO   WEB

Recall that a rational number is a number that can be written as one integer divided by another integer, such as $3 \div 4$ or $\frac{3}{4}$. We usually use the word *fraction* to mean $\frac{3}{4}$. We can extend this idea to algebraic expressions. A **rational expression** is a polynomial divided by another polynomial, such as

$$\frac{7}{x-3} \quad \text{or} \quad \frac{3x+2}{x+4}.$$

Recall that division by 0 is undefined. In the first expression above, x cannot equal 3 since $3 - 3 = 0$. Any other value of x is allowed. We say that the domain of $\frac{7}{x-3}$ is all real numbers except 3. The domain of an expression or function is the set of values that can replace the variable. Similarly, the domain of $\frac{3x+2}{x+4}$ is all real numbers except -4. The concept of domain will be revisited in later chapters. The following important restriction will apply throughout this chapter. We state it here to avoid having to mention it repeatedly throughout this chapter.

Restriction

The denominator of a rational expression cannot be zero. Any value of the variable that would make the denominator zero is not allowed.

We have discovered that fractions can be simplified (or reduced) in the following way.

$$\frac{15}{25} = \frac{3 \cdot \cancel{5}}{5 \cdot \cancel{5}} = \frac{3}{5}$$

This is sometimes referred to as the **basic rule of fractions** and can be stated as follows.

Basic Rule of Fractions

For any rational expression $\frac{a}{b}$ and any polynomial a, b, and c (where $b \neq 0$ and $c \neq 0$),

$$\frac{ac}{bc} = \frac{a}{b}.$$

We will examine several examples where a, b, and c are real numbers, as well as more involved examples where a, b, and c are polynomials. In either case we shall make extensive use of our factoring skills in this section.

One essential property is revealed by the basic rule of fractions: If the numerator and denominator of a given fraction are multiplied by the same nonzero quantity, an equivalent fraction is obtained. The rule can be used two ways. You can start with $\frac{ac}{bc}$ and end with the equivalent fraction $\frac{a}{b}$. Or, you can start with $\frac{a}{b}$ and end with the equivalent fraction $\frac{ac}{bc}$.

● EXAMPLE 1

(a) Write a fraction equivalent to $\frac{3}{5}$ with a denominator of 10.

(b) Reduce $\frac{21}{39}$.

(a) $\dfrac{3}{5} = \dfrac{3 \cdot \cancel{2}}{5 \cdot \cancel{2}} = \dfrac{6}{10}$ Use the rule $\frac{a}{b} = \frac{ac}{bc}$. Let $c = 2$ since $5 \cdot 2 = 10$.

(b) $\dfrac{21}{39} = \dfrac{7 \cdot \cancel{3}}{13 \cdot \cancel{3}} = \dfrac{7}{13}$ Use the rule $\frac{ac}{bc} = \frac{a}{b}$. Let $c = 3$ because 3 is the greatest common factor of 21 and 39.

Practice Problem 1

(a) Find a fraction equivalent to $\dfrac{7}{3}$ with a denominator of 18.

(b) Reduce $\dfrac{28}{63}$.

1 Simplifying Rational Expressions by Factoring

The process of reducing the fraction shown previously is sometimes called *dividing out* common factors. Remember, only factors of both the numerator and the denominator can be divided out. To apply the basic rule of fractions, it is usually necessary that the numerator and denominator of the fraction be completely factored. You will need to use your factoring skills from Chapter 6 to accomplish this step. When you apply this rule, you are **simplifying the fraction**.

EXAMPLE 2 Simplify. $\dfrac{4x + 12}{5x + 15}$

$\dfrac{4x + 12}{5x + 15} = \dfrac{4(x + 3)}{5(x + 3)}$ Factor 4 from the numerator.
   Factor 5 from the denominator.

$\qquad = \dfrac{4\cancel{(x + 3)}}{5\cancel{(x + 3)}}$ Apply the basic rule of fractions.

$\qquad = \dfrac{4}{5}$

Practice Problem 2 Simplify. $\dfrac{12x - 6}{14x - 7}$

EXAMPLE 3 Simplify. $\dfrac{x^2 + 9x + 14}{x^2 - 4}$

$\qquad = \dfrac{(x + 7)(x + 2)}{(x - 2)(x + 2)}$ Factor the numerator.
   Factor the denominator.

$\qquad = \dfrac{(x + 7)\cancel{(x + 2)}}{(x - 2)\cancel{(x + 2)}}$ Apply the basic rule of fractions.

$\qquad = \dfrac{x + 7}{x - 2}$

Practice Problem 3 Simplify. $\dfrac{4x - 6}{2x^2 - x - 3}$

Some problems may involve more than one step of factoring. Always remember to factor out any common factors as the first step, if it is possible to do so.

EXAMPLE 4 Simplify. $\dfrac{9x - x^3}{x^3 + x^2 - 6x}$

$= \dfrac{x\left(9 - x^2\right)}{x\left(x^2 + x - 6\right)}$ Factor out a common factor from the polynomials in the numerator and the denominator.

$= \dfrac{\cancel{x}\cancel{(3 + x)}(3 - x)}{\cancel{x}\cancel{(3 + x)}(x - 2)}$ Factor each polynomial and apply the basic rule of fractions. Note that $(3 + x)$ is equivalent to $(x + 3)$ since addition is commutative.

$= \dfrac{3 - x}{x - 2}$

Practice Problem 4 Simplify. $\dfrac{x^3 + 11x^2 + 30x}{3x^3 + 17x^2 - 6x}$

When you are simplifying, be on the lookout for the special situation where *a factor in the denominator is the opposite of a factor in the numerator.* In such a case you should factor a negative number from one of the factors so that it becomes equivalent to the other factor and it can be divided out. Look carefully at the following two examples.

EXAMPLE 5 Simplify. $\dfrac{5x - 15}{6 - 2x}$

Notice that the variable term in the numerator, $5x$, and the variable term in the denominator, $-2x$, *are opposite in sign.* Likewise, the numerical terms -15 and 6 *are opposite in sign.* Factor out a negative number from the denominator.

$\dfrac{5x - 15}{6 - 2x} = \dfrac{5\,(x - 3)}{-2\,(-3 + x)}$ Factor 5 from the numerator.
    Factor -2 from the denominator.

    Note that $(x - 3)$ and $(-3 + x)$ are equivalent since $(+x - 3) = (-3 + x)$.

$= \dfrac{5(x - 3)}{-2(-3 + x)}$ Apply the basic rule of fractions.

$= -\dfrac{5}{2}$

Note that $\dfrac{5}{-2}$ is not considered to be in simple form. We usually avoid leaving a negative number in the denominator. Therefore, to simplify, give the result as $-\dfrac{5}{2}$ or $\dfrac{-5}{2}$.

Practice Problem 5 Simplify. $\dfrac{2x - 5}{5 - 2x}$

EXAMPLE 6 Simplify. $\dfrac{2x^2 - 11x + 12}{16 - x^2}$

$= \dfrac{(x - 4)(2x - 3)}{(4 - x)(4 + x)}$ Factor the numerator and the denominator. Observe that $(x - 4)$ and $(4 - x)$ are opposites.

$= \dfrac{(x - 4)(2x - 3)}{-1\,(-4 + x)(4 + x)}$ Factor -1 out of $(+4 - x)$ to obtain $-1(-4 + x)$.

$= \dfrac{\cancel{(x - 4)}(2x - 3)}{-1\cancel{(-4 + x)}(4 + x)}$ Apply the basic rule of fractions since $(x - 4)$ and $(-4 + x)$ are equivalent.

$= \dfrac{2x - 3}{-1(4 + x)}$

$= -\dfrac{2x - 3}{4 + x}$

Practice Problem 6 Simplify. $\dfrac{4x^2 + 3x - 10}{25 - 16x^2}$

After doing Examples 5 and 6, you will notice a pattern. Whenever the factor in the numerator and the factor in the denominator are opposites, the value -1 results. We could actually make this a definition of that property.

> For all monomials A and B where $A \neq B$, it is true that
>
> $$\frac{A - B}{B - A} = -1.$$

You may use this definition in reducing fractions if it is helpful to you. Otherwise, you may use the factoring method discussed in Examples 5 and 6.

Some problems will involve two or more variables. In such cases, you will need to factor carefully and make sure that each set of parentheses contains the correct letters.

EXAMPLE 7 Simplify. $\dfrac{x^2 - 7xy + 12y^2}{2x^2 - 7xy - 4y^2}$

$$= \frac{(x - 4y)(x - 3y)}{(2x + y)(x - 4y)} \qquad \text{Factor the numerator.}$$
$$\qquad\qquad\qquad\qquad\quad \text{Factor the denominator.}$$

$$= \frac{\cancel{(x - 4y)}(x - 3y)}{(2x + y)\cancel{(x - 4y)}} \qquad \text{Apply the basic rule of fractions.}$$

$$= \frac{x - 3y}{2x + y}$$

Practice Problem 7 Simplify. $\dfrac{4x^2 - 9y^2}{4x^2 + 12xy + 9y^2}$

EXAMPLE 8 Simplify. $\dfrac{6a^2 + ab - 7b^2}{36a^2 - 49b^2}$

$$= \frac{(6a + 7b)(a - b)}{(6a + 7b)(6a - 7b)} \qquad \text{Factor the numerator.}$$
$$\qquad\qquad\qquad\qquad\qquad \text{Factor the denominator.}$$

$$= \frac{\cancel{(6a + 7b)}(a - b)}{\cancel{(6a + 7b)}(6a - 7b)} \qquad \text{Apply the basic rule of fractions.}$$

$$= \frac{a - b}{6a - 7b}$$

Practice Problem 8 Simplify. $\dfrac{12x^3 - 48x}{6x - 3x^2}$

7.1 Exercises

Simplify.

1. $\dfrac{3a - 9b}{a - 3b}$

2. $\dfrac{5x + 2y}{35x + 14y}$

3. $\dfrac{6x + 18}{x^2 + 3x}$

4. $\dfrac{8x - 12}{2x^3 - 3x^2}$

5. $\dfrac{2x - 8}{x^2 - 8x + 16}$

6. $\dfrac{4x^2 + 4x + 1}{1 - 4x^2}$

7. $\dfrac{xy(x + y^2)}{x^2 y^2}$

8. $\dfrac{6x^2}{2x(x - 3y)}$

9. $\dfrac{x^2 + x - 2}{x^2 - x}$

10. $\dfrac{x^2 + x - 12}{2x^2 - 3x - 9}$

11. $\dfrac{x^2 - 3x - 10}{3x^2 + 5x - 2}$

12. $\dfrac{4x^2 - 10x + 6}{2x^2 + x - 3}$

13. $\dfrac{x^3 - 8x^2 + 16x}{x^2 + 2x - 24}$

14. $\dfrac{x^2 - 10x - 24}{x^3 + 9x^2 + 14x}$

15. $\dfrac{3x^2 + 7x - 6}{x^2 + 7x + 12}$

16. $\dfrac{x^2 - 5x - 14}{2x^2 - x - 10}$

17. $\dfrac{3x^2 - 8x + 5}{4x^2 - 5x + 1}$

18. $\dfrac{3y^2 + 10y + 3}{3y^2 - 14y - 5}$

19. $\dfrac{5x^2 - 27x + 10}{5x^2 + 3x - 2}$

20. $\dfrac{2x^2 + 2x - 12}{x^2 + 3x - 4}$

21. $\dfrac{6 - 3x}{2x - 4}$

22. $\dfrac{5 - ay}{ax^2 y - 5x^2}$

23. $\dfrac{2x^2 - 7x - 15}{25 - x^2}$

24. $\dfrac{49 - x^2}{2x^2 - 9x - 35}$

25. $\dfrac{(4x + 5)^2}{8x^2 + 6x - 5}$

26. $\dfrac{6x^2 - 13x - 8}{(3x - 8)^2}$

27. $\dfrac{2x^2 + 9x - 18}{30 - x - x^2}$

28. $\dfrac{4y^2 - y - 3}{8 - 7y - y^2}$

29. $\dfrac{a^2 + 2ab - 3b^2}{2a^2 + 5ab - 3b^2}$

30. $\dfrac{a^2 + 3ab - 10b^2}{3a^2 - 7ab + 2b^2}$

31. $\dfrac{9x^2 - 4y^2}{9x^2 + 12xy + 4y^2}$

32. $\dfrac{16x^2 - 24xy + 9y^2}{16x^2 - 9y^2}$

33. $\dfrac{6x^4 - 9x^3 - 6x^2}{12x^3 + 42x^2 + 18x}$

34. $\dfrac{xa - yb - ya + xb}{xa - ya + 2bx - 2by}$

Cumulative Review Problems

Multiply.

35. Multiply. $(3x - 7)^2$

36. $(2x + 3)(x - 4)(x - 2)$

37. Simplify. $\left(-2ab^0c^3\right)^3$

38. Simplify. $\dfrac{20xy^2z^5}{24x^2yz^8}$

The number of people on our planet who speak the Chinese Mandarin language is 221,000,000 more than twice the number of people who speak Spanish. The number of people on Earth who speak English is 10,000,000 less than the number of those who speak Spanish. There are 322,000,000 people who speak English on our planet.

39. How many people speak Chinese Mandarin?

40. How many people speak Spanish?

7.2 Multiplication and Division of Rational Expressions

1 Multiplying Rational Expressions

To multiply two rational expressions, we multiply the numerators and we multiply the denominators. As before, the denominators cannot equal zero.

> For any two rational expressions $\frac{a}{b}$ and $\frac{c}{d}$ where $b \neq 0$ and $d \neq 0$,
>
> $$\frac{a}{b} \cdot \frac{c}{d} = \frac{ac}{bd}.$$

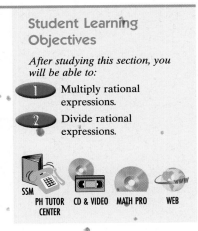
Simplifying or reducing fractions *prior to multiplying them* usually makes the computations easier to do. Leaving the reducing step until the end makes the simplifying process longer and increases the chance for error. This long approach should be avoided.

As an example, let's do the same problem two ways to see which one is easier. Let's simplify the following problem by multiplying first and then reducing the result.

$$\frac{5}{7} \times \frac{49}{125}$$

$$\frac{5}{7} \times \frac{49}{125} = \frac{245}{875} \qquad \text{Multiply the numerators and multiply the denominators.}$$

$$= \frac{7}{25} \qquad \text{Reduce the fraction. (\textit{Note:} It takes a bit of trial and error to discover how to reduce it.)}$$

Compare this with the following method, where we reduce the fractions prior to multiplying them.

$$\frac{5}{7} \times \frac{49}{125}$$

$$\frac{5}{7} \times \frac{7 \cdot 7}{5 \cdot 5 \cdot 5} \qquad \textbf{Step 1.} \text{ It is easier to factor first. We factor the numerator and denominator of the second fraction.}$$

$$\frac{5 \cdot 7 \cdot 7}{7 \cdot 5 \cdot 5 \cdot 5} \qquad \textbf{Step 2.} \text{ We express the product as one fraction (by the definition of multiplication of fractions).}$$

$$\frac{\cancel{5} \cdot \cancel{7} \cdot 7}{\cancel{7} \cdot \cancel{5} \cdot 5 \cdot 5} = \frac{7}{25} \qquad \textbf{Step 3.} \text{ Then we apply the basic rule of fractions to divide the common factors of 5 and 7 that appear in the numerator and in the denominator.}$$

A similar approach can be used with the multiplication of rational expressions. We first factor the numerator and denominator of each fraction. Then we divide out any factor that is common to a numerator and a denominator. Finally, we multiply the remaining numerators and the remaining denominators.

EXAMPLE 1 Multiply. $\dfrac{x^2 - x - 12}{x^2 - 16} \cdot \dfrac{2x^2 + 7x - 4}{x^2 - 4x - 21}$

$$\frac{(x - 4)(x + 3)}{(x - 4)(x + 4)} \cdot \frac{(x + 4)(2x - 1)}{(x + 3)(x - 7)} \qquad \text{Factoring is always the first step.}$$

$$= \frac{\cancel{(x - 4)}\cancel{(x + 3)}}{\cancel{(x - 4)}\cancel{(x + 4)}} \cdot \frac{\cancel{(x + 4)}(2x - 1)}{\cancel{(x + 3)}(x - 7)} \qquad \text{Apply the basic rule of fractions. (Three pairs of factors divide out.)}$$

$$= \frac{2x - 1}{x - 7} \qquad \text{The final answer.}$$

379

Practice Problem 1 Multiply. $\dfrac{10x}{x^2 - 7x + 10} \cdot \dfrac{x^2 + 3x - 10}{25x}$

In some cases, a given numerator can be factored more than once. You should always check for a *common factor* as your first step.

EXAMPLE 2 Multiply. $\dfrac{x^4 - 16}{x^3 + 4x} \cdot \dfrac{2x^2 - 8x}{4x^2 + 2x - 12}$

$= \dfrac{(x^2 + 4)(x^2 - 4)}{x(x^2 + 4)} \cdot \dfrac{2x(x - 4)}{2(2x^2 + x - 6)}$ Factor each numerator and denominator. Factoring out the common factor first is very important.

$= \dfrac{(x^2 + 4)(x + 2)(x - 2)}{x(x^2 + 4)} \cdot \dfrac{2x(x - 4)}{2(x + 2)(2x - 3)}$ Factor again where possible.

$= \dfrac{\cancel{(x^2 + 4)}\cancel{(x + 2)}(x - 2)}{\cancel{x}\cancel{(x^2 + 4)}} \cdot \dfrac{\cancel{2}\cancel{x}(x - 4)}{\cancel{2}\cancel{(x + 2)}(2x - 3)}$ Divide out factors that appear in both the numerator and the denominator. (There are four such pairs of factors.)

$= \dfrac{(x - 2)(x - 4)}{(2x - 3)}$ or $\dfrac{x^2 - 6x + 8}{2x - 3}$ Write the answer as one fraction. (Usually, if there is more than one factor in a numerator, the answer is left in factored form.)

Practice Problem 2 Multiply. $\dfrac{2y^2 - 6y - 8}{y^2 - y - 2} \cdot \dfrac{y^2 - 5y + 6}{2y^2 - 4y - 6}$

2   *Dividing Rational Expressions*

For any two fractions $\frac{a}{b}$ and $\frac{c}{d}$, the operation of division can be performed by inverting the second fraction and multiplying it by the first fraction. When we invert a fraction, we are finding its *reciprocal*. Two numbers are **reciprocals** of each other if their product is 1. The reciprocal of $\frac{3}{5}$ is $\frac{5}{3}$. The reciprocal of 7 is $\frac{1}{7}$. The reciprocal of $\frac{a}{b}$ is $\frac{b}{a}$. Sometimes people state the rule for dividing fractions this way: "To divide two fractions, keep the first fraction unchanged and multiply by the reciprocal of the second fraction."

The definition for division of fractions is

$$\frac{a}{b} \div \frac{c}{d} = \frac{a}{b} \cdot \frac{d}{c}.$$

This property holds whether a, b, c, and d are polynomials or numerical values. (It is assumed, of course, that no denominator is zero.)

In the first step for dividing two rational expressions, invert the second fraction and rewrite the quotient as a product. Then follow the procedure for multiplying rational expressions.

EXAMPLE 3 Divide. $\dfrac{6x + 12y}{2x - 6y} \div \dfrac{9x^2 - 36y^2}{4x^2 - 36y^2}$

$= \dfrac{6x + 12y}{2x - 6y} \cdot \dfrac{4x^2 - 36y^2}{9x^2 - 36y^2}$ Invert the second fraction and write the problem as the product of two fractions.

$$= \frac{6(x + 2y)}{2(x - 3y)} \cdot \frac{4(x^2 - 9y^2)}{9(x^2 - 4y^2)}$$ Factor each numerator and denominator.

$$= \frac{(3)(2)(x + 2y)}{2(x - 3y)} \cdot \frac{(2)(2)(x + 3y)(x - 3y)}{(3)(3)(x + 2y)(x - 2y)}$$ Factor again where possible.

$$= \frac{\cancel{(3)}\,\cancel{(2)}\,\cancel{(x + 2y)}}{2\cancel{(x - 3y)}} \cdot \frac{(2)(2)(x + 3y)\cancel{(x - 3y)}}{\cancel{(3)}\,(3)\,\cancel{(x + 2y)}(x - 2y)}$$ Divide out factors that appear in both numerator and denominator.

$$= \frac{(2)(2)(x + 3y)}{3(x - 2y)}$$ Write the result as one fraction.

$$= \frac{4(x + 3y)}{3(x - 2y)}$$ Simplify. (Usually, answers are left in this form.)

Although it is correct to write this answer as $\dfrac{4x + 12y}{3x - 6y}$, it is customary to leave the answer in factored form to ensure that the final answer is simplified.

Practice Problem 3   Divide. $\dfrac{x^2 + 5x + 6}{x^2 + 8x} \div \dfrac{2x^2 + 5x + 2}{2x^2 + x}$

A polynomial that is not in fraction form can be written as a fraction if you give it a denominator of 1.

EXAMPLE 4   Divide. $\dfrac{15 - 3x}{x + 6} \div (x^2 - 9x + 20)$

Note that $x^2 - 9x + 20$ can be written as $\dfrac{x^2 - 9x + 20}{1}$.

$$= \frac{15 - 3x}{x + 6} \cdot \frac{1}{x^2 - 9x + 20}$$ Invert and multiply.

$$= \frac{-3(-5 + x)}{x + 6} \cdot \frac{1}{(x - 5)(x - 4)}$$ Factor where possible. Note that we had to factor -3 from the first numerator so that it would contain a factor in common with the second denominator.

$$= \frac{-3\cancel{(-5 + x)}}{x + 6} \cdot \frac{1}{\cancel{(x - 5)}(x - 4)}$$ Divide out the common factor. $(-5 + x)$ is equivalent to $(x - 5)$.

$$= \frac{-3}{(x + 6)(x - 4)}$$ The final answer. Note that the answer can be written in several equivalent forms.

or $-\dfrac{3}{(x + 6)(x - 4)}$ or $\dfrac{3}{(x + 6)(4 - x)}$

Practice Problem 4   Divide. $\dfrac{x + 3}{x - 3} \div (9 - x^2)$

A word of caution:

It is logical to assume that the problems included in Section 7.2 have at least one common factor that can be divided out. Therefore, if after factoring, you do not observe any common factors, you should be somewhat suspicious. In such cases, it would be wise to double check your factoring steps to see if an error has been made.

7.2 Exercises

Verbal and Writing Skills

1. Before multiplying rational expressions, we should always first try to

   _____

   _____ .

2. Division of two rational expressions is done by keeping the first fraction unchanged and then

   _____ .

Perform the operation indicated.

3. $\dfrac{2x - 10}{x - 4} \cdot \dfrac{x^2 + 5x + 4}{x^2 - 4x - 5}$

4. $\dfrac{7x + 7}{x + 4} \cdot \dfrac{x^2 - x - 20}{7x^2 - 42x - 49}$

5. $\dfrac{x^2 + 2x}{6x} \cdot \dfrac{3x^2}{x^2 - 4}$

6. $\dfrac{3x + 12}{8x^3} \cdot \dfrac{16x^2}{9x + 36}$

7. $\dfrac{x^2 + 3x - 10}{x^2 + x - 20} \cdot \dfrac{x^2 - 3x - 4}{x^2 + 4x + 3}$

8. $\dfrac{x^2 - x - 20}{x^2 - 3x - 10} \cdot \dfrac{x^2 + 7x + 10}{x^2 + 4x - 5}$

9. $(6x - 5) \div \dfrac{36x^2 - 25}{6x^2 + 17x + 10}$

10. $\dfrac{4x^2 - 9}{4x^2 + 12x + 9} \div (6x - 9)$

11. $\dfrac{3x^2 + 12xy + 12y^2}{x^2 + 4xy + 3y^2} \div \dfrac{4x + 8y}{x + y}$

12. $\dfrac{5x^2 + 10xy + 5y^2}{x^2 + 5xy + 6y^2} \div \dfrac{3x + 3y}{x + 2y}$

13. $\dfrac{x^2 + 5x - 14}{x - 5} \div \dfrac{x^2 + 12x + 35}{15 - 3x}$

14. $\dfrac{3x^2 + 13x + 4}{16 - x^2} \div \dfrac{3x^2 - 5x - 2}{3x - 12}$

15. $\dfrac{(x + 5)^2}{3x^2 - 7x + 2} \cdot \dfrac{x^2 - 4x + 4}{x + 5}$

16. $\dfrac{3x^2 - 10x - 8}{(4x + 5)^2} \cdot \dfrac{4x + 5}{(x - 4)^2}$

17. $\dfrac{y^2 + 4y - 12}{y^2 + 2y - 24} \cdot \dfrac{y^2 - 16}{y^2 + 2y - 8}$

18. $\dfrac{5y^2 + 17y + 6}{10y^2 + 9y + 2} \cdot \dfrac{4y^2 - 1}{2y^2 + 5y - 3}$

To Think About

19. Consider the problem $\dfrac{x+5}{x-2} \div \dfrac{x+7}{x-6}$. Explain why 2, −7, and 6 are not allowable replacements for the variable x.

20. Consider the problem $\dfrac{x-8}{x+5} \div \dfrac{x-9}{x+4}$. Explain why −5, 9, and −4 are not allowable replacements for the variable x.

Cumulative Review Problems

21. Solve. $5x^2 - 7x + 11 = 5x^2 - x + 2$

22. Multiply. $(7x^2 - x - 1)(x - 3)$

▲ **23.** The Golden Gate Bridge has a total length (including approaches) of 8981 feet and a road width of 90 feet. The width of the sidewalk is 10.5 feet. (The sidewalk spans the entire length of the bridge.) Assume it would cost $\$x$ per square foot to resurface the road or the sidewalk. Write an expression for how much more it would cost to resurface the road than the sidewalk.

24. Denise Abrahamson purchased 4 gallons of milk. The price of 4 gallons of milk is the same as the price of 2 gallons of milk plus some other groceries that cost $4.76. What is the price of a gallon of milk?

After studying this section, you will be able to:

1. Add and subtract rational expressions with the same denominator.

2. Determine the LCD for two or more rational expressions with different denominators.

3. Add and subtract rational expressions with different denominators.

SSM
PH TUTOR CENTER | CD & VIDEO | MATH PRO | WEB

1 **Adding and Subtracting Rational Expressions with the Same Denominator**

If rational expressions have the same denominator, they can be combined in a fashion similar to that used for fractions in arithmetic. The numerators are added or subtracted and the denominator remains the same.

Adding Rational Expressions

For any rational expressions $\frac{a}{b}$ and $\frac{c}{b}$,

$$\frac{a}{b} + \frac{c}{b} = \frac{a+c}{b} \qquad \text{where } b \neq 0.$$

EXAMPLE 1 Add. $\dfrac{5a}{a+2b} + \dfrac{6a}{a+2b}$

$$\frac{5a}{a+2b} + \frac{6a}{a+2b} = \frac{5a+6a}{a+2b} = \frac{11a}{a+2b}$$
Note that the denominators are the same. Add the numerators.

Practice Problem 1 Add. $\dfrac{2s+t}{2s-t} + \dfrac{s-t}{2s-t}$

Subtracting Rational Expressions

For any rational expressions $\frac{a}{b}$ and $\frac{c}{b}$,

$$\frac{a}{b} - \frac{c}{b} = \frac{a-c}{b} \qquad \text{where } b \neq 0.$$

EXAMPLE 2 Subtract. $\dfrac{3x}{(x+y)(x-2y)} - \dfrac{8x}{(x+y)(x-2y)}$

$$\frac{3x}{(x+y)(x-2y)} - \frac{8x}{(x+y)(x-2y)} = \frac{3x-8x}{(x+y)(x-2y)} \quad \text{Write as one fraction.}$$

$$= \frac{-5x}{(x+y)(x-2y)} \quad \text{Simplify.}$$

Practice Problem 2 Subtract. $\dfrac{b}{(a-2b)(a+b)} - \dfrac{2b}{(a-2b)(a+b)}$

2 **Determining the LCD for Two or More Rational Expressions with Different Denominators**

How do we add or subtract rational expressions when the denominators are not the same? First we must find the **least common denominator (LCD)**, as we would to add or subtract fractions.

How to Find the LCD of Two or More Rational Expressions

1. Factor each denominator completely.
2. The LCD is a product containing each *different factor*.
3. If a factor occurs more than once in any one denominator, the LCD will contain that factor repeated the greatest number of times that it occurs in any one denominator.

EXAMPLE 3 Find the LCD. $\dfrac{5}{2x - 4}, \dfrac{6}{3x - 6}$

Factor each denominator.

$$2x - 4 = 2(x - 2) \qquad 3x - 6 = 3(x - 2)$$

The different factors are 2, 3, and $(x - 2)$. Since no factor appears more than once in any one denominator, the LCD is the product of these three factors.

$$\text{LCD} = (2)(3)(x - 2) = 6(x - 2)$$

Practice Problem 3 Find the LCD. $\dfrac{7}{6x + 21}, \dfrac{13}{10x + 35}$

EXAMPLE 4 Find the LCD. **(a)** $\dfrac{5}{12ab^2c}, \dfrac{13}{18a^3bc^4}$

(b) $\dfrac{8}{x^2 - 5x + 4}, \dfrac{12}{x^2 + 2x - 3}$

If a factor occurs more than once in any one denominator, the LCD will contain that factor repeated the greatest number of times that it occurs in any one denominator.

(a) $12ab^2c = 2 \cdot 2 \cdot 3 \cdot \quad a \cdot \qquad b \cdot b \cdot c$

$18a^3bc^4 = \Big| \quad 2 \cdot 3 \cdot 3 \cdot a \cdot a \cdot a \cdot b \cdot \Big| \quad c \cdot c \cdot c \cdot c$

$\qquad\qquad 2 \cdot 2 \cdot 3 \cdot 3 \cdot a \cdot a \cdot a \cdot b \cdot b \cdot c \cdot c \cdot c \cdot c$

$\text{LCD} = 2^2 \cdot 3^2 \cdot a^3 \cdot b^2 \cdot c^4 = 36a^3b^2c^4$

(b) $x^2 - 5x + 4 = (x - 4)(x - 1)$

$x^2 + 2x - 3 = \quad\Big| \quad (x - 1)(x + 3)$

$\text{LCD} = (x - 4)(x - 1)(x + 3)$

Practice Problem 4 Find the LCD. **(a)** $\dfrac{3}{50xy^2z}, \dfrac{19}{40x^3yz}$

(b) $\dfrac{2}{x^2 + 5x + 6}, \dfrac{6}{3x^2 + 5x - 2}$

3 ➤ *Adding and Subtracting Rational Expressions with Different Denominators*

If two rational expressions have different denominators, we first change them to equivalent rational expressions with the least common denominator. Then we add or subtract the numerators and keep the common denominator.

EXAMPLE 5 Add. $\dfrac{5}{xy} + \dfrac{2}{y}$

The denominators are different. We must find the LCD. The two factors are x and y. We observe that the LCD is xy.

$$\frac{5}{xy} + \frac{2}{y} = \frac{5}{xy} + \frac{2}{y} \cdot \frac{x}{x} \qquad \text{Multiply the second fraction by } \frac{x}{x}.$$

$$= \frac{5}{xy} + \frac{2x}{xy} \qquad \text{Now each fraction has a common denominator of } xy.$$

$$= \frac{5 + 2x}{xy} \qquad \text{Write the sum as one fraction.}$$

Practice Problem 5 Add. $\dfrac{7}{a} + \dfrac{3}{abc}$

EXAMPLE 6 Add. $\dfrac{3x}{(x + y)(x - y)} + \dfrac{5}{x + y}$

The factors of the denominators are $(x + y)$ and $(x - y)$. We observe that the LCD $= (x + y)(x - y)$.

$$\frac{3x}{(x + y)(x - y)} + \frac{5}{(x + y)} \cdot \frac{x - y}{x - y} \qquad \text{Multiply the second fraction by } \frac{x - y}{x - y}.$$

$$= \frac{3x}{(x + y)(x - y)} + \frac{5x - 5y}{(x + y)(x - y)} \qquad \begin{array}{l}\text{Now each fraction has a common} \\ \text{denominator of } (x + y)(x - y).\end{array}$$

$$= \frac{3x + 5x - 5y}{(x + y)(x - y)} \qquad \begin{array}{l}\text{Write the sum of the numerators over one} \\ \text{common denominator.}\end{array}$$

$$= \frac{8x - 5y}{(x + y)(x - y)} \qquad \text{Collect like terms.}$$

Practice Problem 6 Add. $\dfrac{2a - b}{(a + 2b)(a - 2b)} + \dfrac{2}{(a + 2b)}$

It is important to remember that the LCD is the smallest algebraic expression into which each denominator can be divided. For rational expressions the LCD must contain *each factor* that appears in any denominator. If the factor is repeated, the LCD must contain that factor the greatest number of times that it appears in any one denominator.

In many cases, the denominators in an addition or subtraction problem are not in factored form. You must factor each denominator to determine the LCD. Collect like terms in the numerator; then determine whether that final numerator can be factored. If so, you may be able to simplify the fraction.

⬤ **EXAMPLE 7** Add. $\dfrac{5}{x^2 - y^2} + \dfrac{3x}{x^3 + x^2y}$

$\dfrac{5}{x^2 - y^2} + \dfrac{3x}{x^3 + x^2y}$

Factor the two denominators. Observe that the LCD is $x^2(x + y)(x - y)$.

$= \dfrac{5}{(x + y)(x - y)} + \dfrac{3x}{x^2(x + y)}$

$= \dfrac{5}{(x + y)(x - y)} \cdot \dfrac{x^2}{x^2} + \dfrac{3x}{x^2(x + y)} \cdot \dfrac{x - y}{x - y}$

Multiply each fraction by the appropriate value to obtain a common denominator of $x^2(x + y)(x - y)$.

$= \dfrac{5x^2}{x^2(x + y)(x - y)} + \dfrac{3x^2 - 3xy}{x^2(x + y)(x - y)}$

$= \dfrac{5x^2 + 3x^2 - 3xy}{x^2(x + y)(x - y)}$

Write the sum of the numerators over one common denominator.

$= \dfrac{8x^2 - 3xy}{x^2(x + y)(x - y)}$

Collect like terms.

$= \dfrac{x(8x - 3y)}{x^2(x + y)(x - y)}$

Divide out the common factor x in the numerator and denominator to simplify.

$= \dfrac{8x - 3y}{x(x + y)(x - y)}$

Practice Problem 7 Add. $\dfrac{7a}{a^2 + 2ab + b^2} + \dfrac{4}{a^2 + ab}$ ⬤

   It is very easy to make a sign mistake when subtracting two fractions. You will find it helpful to place parentheses around the numerator of the second fraction so that you will not forget to subtract the entire numerator.

⬤ **EXAMPLE 8** Subtract. $\dfrac{3x + 4}{x - 2} - \dfrac{x - 3}{2x - 4}$

Factor the second denominator.

$= \dfrac{3x + 4}{x - 2} - \dfrac{x - 3}{2(x - 2)}$

Observe that the LCD is $2(x - 2)$.

$= \dfrac{2}{2} \cdot \dfrac{(3x + 4)}{x - 2} - \dfrac{x - 3}{2(x - 2)}$

Multiply the first fraction by $\frac{2}{2}$ so that the resulting fraction will have the common denominator.

$= \dfrac{2(3x + 4) - (x - 3)}{2(x - 2)}$

Write the indicated subtraction as one fraction. Note the parentheses around $x - 3$.

$= \dfrac{6x + 8 - x + 3}{2(x - 2)}$

Remove the parentheses in the numerator.

$= \dfrac{5x + 11}{2(x - 2)}$

Collect like terms.

Practice Problem 8 Subtract. $\dfrac{x + 7}{3x - 9} - \dfrac{x - 6}{x - 3}$

To avoid making errors when subtracting two fractions, some students find it helpful to change subtraction to addition of the opposite of the second fraction.

EXAMPLE 9 Subtract. $\dfrac{8x}{x^2 - 16} - \dfrac{4}{x - 4}$

$\dfrac{8x}{(x + 4)(x - 4)} + \dfrac{-4}{x - 4}$
Factor the first denominator. Use the property that $\dfrac{a}{b} - \dfrac{c}{b} = \dfrac{a}{b} + \dfrac{-c}{b}$.

$= \dfrac{8x}{(x + 4)(x - 4)} + \dfrac{-4}{x - 4} \cdot \dfrac{x + 4}{x + 4}$
Multiply the second fraction by $\dfrac{x + 4}{x + 4}$.

$= \dfrac{8x + (-4)(x + 4)}{(x + 4)(x - 4)}$
Write the sum of the numerators over one common denominator.

$= \dfrac{8x - 4x - 16}{(x + 4)(x - 4)}$
Remove parentheses.

$= \dfrac{4x - 16}{(x + 4)(x - 4)}$
Collect like terms. Note that the numerator can be factored.

$= \dfrac{4\cancel{(x - 4)}}{(x + 4)\cancel{(x - 4)}}$
Since $(x - 4)$ is a *factor* of the numerator *and* the denominator, we may divide out the common factor.

$= \dfrac{4}{x + 4}$

Practice Problem 9 Subtract and simplify. $\dfrac{x - 2}{x^2 - 4} - \dfrac{x + 1}{2x^2 + 4x}$

Verbal and Writing Skills

1. Suppose two rational expressions have denominators of $(x + 3)(x + 5)$ and $(x + 3)^2$. Explain how you would determine the LCD.

2. Suppose two rational expressions have denominators of $(x - 4)^2(x + 7)$ and $(x - 4)^3$. Explain how you would determine the LCD.

Perform the operation indicated. Be sure to simplify.

3. $\dfrac{x}{x + 5} + \dfrac{2x + 1}{5 + x}$

4. $\dfrac{8}{7 + 2x} + \dfrac{x + 3}{2x + 7}$

5. $\dfrac{x}{x - 4} - \dfrac{x + 1}{x - 4}$

6. $\dfrac{x + 3}{x + 5} - \dfrac{x - 8}{x + 5}$

7. $\dfrac{8x + 3}{5x + 7} - \dfrac{6x + 10}{5x + 7}$

Find the LCD. Do not combine the fractions.

8. $\dfrac{13}{3ab}, \dfrac{7}{a^2b^2}$

9. $\dfrac{12}{5a^2}, \dfrac{9}{a^3}$

10. $\dfrac{5}{18x^2y^5}, \dfrac{7}{30x^3y^3}$

11. $\dfrac{11}{16x^2y^3}, \dfrac{17}{56xy^4}$

12. $\dfrac{8}{x + 3}, \dfrac{15}{x^2 - 9}$

13. $\dfrac{13}{x^2 - 16}, \dfrac{7}{x - 4}$

14. $\dfrac{6}{2x^2 + x - 3}, \dfrac{5}{4x^2 + 12x + 9}$

15. $\dfrac{3}{16x^2 - 8x + 1}, \dfrac{6}{4x^2 + 7x - 2}$

Add or subtract.

16. $\dfrac{3}{x + 7} + \dfrac{8}{x^2 - 49}$

17. $\dfrac{5}{x^2 - 2x + 1} + \dfrac{3}{x - 1}$

18. $\dfrac{3y}{y + 2} + \dfrac{y}{y - 2}$

19. $\dfrac{2}{y - 1} + \dfrac{2}{y + 1}$

20. $\dfrac{4}{a + 3} + \dfrac{2}{3a}$

21. $\dfrac{5}{2ab} + \dfrac{1}{2a + b}$

22. $\dfrac{2}{3xy} + \dfrac{1}{6yz}$

23. $\dfrac{x - 3}{4x} + \dfrac{6}{x^2}$

24. $\dfrac{2}{x^2 + 5x + 6} + \dfrac{3}{x^2 + 7x + 10}$

25. $\dfrac{3}{x^2 - x - 12} + \dfrac{2}{x^2 + x - 20}$

26. $\dfrac{3x - 8}{x^2 - 5x + 6} + \dfrac{x + 2}{x^2 - 6x + 8}$

27. $\dfrac{3x + 5}{x^2 + 4x + 3} + \dfrac{-x + 5}{x^2 + 2x - 3}$

28. $\dfrac{a + b}{3} - \dfrac{a - 2b}{4}$

29. $\dfrac{a + 1}{2} - \dfrac{a - 1}{3}$

30. $\dfrac{8}{2x - 3} - \dfrac{6}{x + 2}$

31. $\dfrac{6}{3x - 4} - \dfrac{5}{4x - 3}$

32. $\dfrac{x}{x^2 + 2x - 3} - \dfrac{x}{x^2 - 5x + 4}$

33. $\dfrac{1}{x^2 - 2x} - \dfrac{5}{x^2 - 4x + 4}$

34. $\dfrac{3y}{8y^2 + 2y - 1} - \dfrac{5y}{2y^2 - 9y - 5}$

35. $\dfrac{2x}{x^2 + 5x + 6} - \dfrac{x + 1}{x^2 + 2x - 3}$

36. $\dfrac{6a}{a - 7} - \dfrac{3a}{7 - a}$

37. $\dfrac{7y}{y - 3} - \dfrac{2y}{3 - y}$

38. $\dfrac{4y}{y^2 + 4y + 3} + \dfrac{2}{y + 1}$

39. $\dfrac{y - 23}{y^2 - y - 20} + \dfrac{2}{y - 5}$

40. $\dfrac{2x}{x - 3} + \dfrac{3x}{x + 2} + \dfrac{7}{x^2 - x - 6}$

41. $\dfrac{x}{x - 4} + \dfrac{5}{x + 4} + \dfrac{10}{x^2 - 16}$

42. $\dfrac{3x}{x^2 + 3x - 10} + \dfrac{5}{4 - 2x}$

43. $\dfrac{2y}{3y^2 - 8y - 3} + \dfrac{1}{6y - 2y^2}$

44. $\dfrac{6}{7x + 14} - \dfrac{2}{5x + 10}$

45. $\dfrac{8}{15x + 10} - \dfrac{3}{6x + 4}$

Cumulative Review Problems

46. Solve. $\dfrac{1}{3}(x - 2) + \dfrac{1}{2}(x + 3) = \dfrac{1}{4}(3x + 1)$

47. Solve for y. $5ax = 2(ay - 3bc)$

48. Find the slope of the line passing through $(2, -5)$ and $(-1, -2)$.

49. Find the slope and y-intercept. $x + 2y = -6$

50. A subway token costs $1.50. A monthly unlimited ride subway pass costs $50. How many days per month would you have to use the subway to go to work (assume one subway token to get to work and one subway token to get back home) in order for it to be cheaper to buy a monthly subway pass?

51. In Finland the unemployment rate went from 3.5% in 1990 to 14.6% in 1999. (*Source:* United Nations Statistics Division.) If the working age population of Finland was 5,100,000 during this period, how many more people were unemployed in 1999 than in 1990?

52. The government of Finland is hoping to lower the number of people unemployed in Finland to 400,000 or less by the year 2003. (*Source:* United Nations Statistics Division.) Assuming the population of Finland remains constant during the entire time period from 1999 to 2003, what will be the unemployment rate in Finland in 2003 if this goal is met? Refer to exercise 51.

Putting Your Skills to Work

Estimating the Surplus or Deficit

In March 1999, the Congressional Budget Office issued the following surprising projection: assuming a moderate level of economic growth, we can expect a surplus of $489 billion in the U.S. budget in 2010. Then the office further amazed people with the statement that the surplus could grow to $1.3 trillion if the U.S. experiences stronger economic growth. Finally, the office projected a deficit of $286 billion by 2010 if the United States experiences slow economic growth. A three-dimensional bar graph visualizing these three possibilities is shown for 2000, 2005, and 2010.

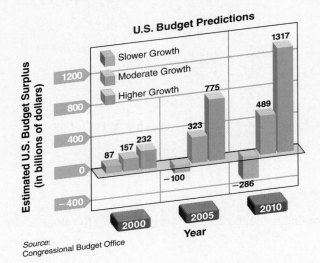

The three possibilities are defined more precisely as follows: The higher surplus figures would result from strong economic growth (3.2% per year) and limited congressional spending (that is, holding to agreed-upon budgeted spending limits). The moderate or lower surplus figures would result from moderate economic growth (2.65% per year) and increased congressional spending (whose annual rate of increase would equal that of inflation). The deficit figures would result from slow economic growth (2.1% per year) and increased congressional spending (whose annual rate of increase would equal that of inflation).

Each of these projections can be represented by an equation that predicts the budget surplus (or deficit) in billions of dollars in terms of the number of years x since 1999.

For strong economic growth, the equation is

$$y = 124 + 108.45x.$$

For moderate economic growth, the equation is

$$y = 124 + 33.18x.$$

For slow economic growth, the equation is

$$y = 124 - 37.27x.$$

Problems for Individual Study and Investigation

1. Use one of the preceding equations to find the budget surplus for 2006 if there is strong economic growth. Round to the nearest billion.

2. Use one of the preceding equations to find the budget surplus for 2008 if there is moderate economic growth. Round to the nearest billion.

Problems for Cooperative Group Investigation and Analysis

3. Use one of the preceding equations to find, for 2004, the difference between the budget surplus that would result from moderate economic growth and the budget deficit that would result from slow economic growth. Round to the nearest billion.

4. Use one of the preceding equations to find, for 2007, the difference between the budget surplus that would result from strong economic growth and the budget deficit that would result from slow economic growth. Round to the nearest billion.

One of the major factors in predicting a government deficit or surplus is Social Security. With larger numbers of older people retiring and drawing Social Security funds, accurately predicting the income to the Social Security trust fund in the coming years becomes an extremely critical governmental function. Under one model studied by the Social Security Administration, the net income to the Social Security trust fund is described by the equation

$$I = \frac{5.6x + 10.8}{x + 3},$$

where I is measured in hundreds of billions of dollars and x is the number of years since 2000. For example, in 2000 x would be 0. Thus

$$I = \frac{5.6(0) + 10.8}{0 + 3} = \frac{10.8}{3} = 3.6.$$

Thus in 2000 the projected income was $3.6 hundred billion or $360,000,000,000. (*Source:* U.S. Social Security Administration)

5. Use the preceding equation to find how much larger the income to the Social Security trust fund will be in 2005 than in 2001.

6. Use the preceding equation to find how much larger the income to the Social Security trust fund will be in 2012 than in 2009.

Internet Connections

Netsite: http://www.prenhall.com/tobey_beg_int

Site: U.S. Congressional Budget Office

7. Use the site to find projections about the size of the national debt. The national debt is "Debt Held by the Public." According to current projections, how much will the national debt decrease from 2005 to 2010?

8. Use the above site to find data about the revenues and spending for social insurance (Social Security and Medicare together). How much less are the projected revenues associated with social insurance in 2008 than the projected costs?

Student Learning Objectives

After studying this section, you will be able to:

1. Simplify complex rational expressions by adding or subtracting in the numerator and denominator.

2. Simplify complex rational expressions by multiplying by the LCD of all the denominators.

SSM PH TUTOR   CD & VIDEO MATH PRO   WEB
CENTER

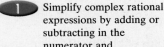 **Simplifying Complex Rational Expressions by Adding or Subtracting in the Numerator and Denominator**

A **complex rational expression** (also called a **complex fraction**) has a fraction in the numerator or in the denominator, or both. Look at the following complex rational expressions.

$$\frac{3 + \dfrac{2}{x}}{\dfrac{x}{7} + 2} \qquad \frac{\dfrac{x}{y} + 1}{2} \qquad \frac{\dfrac{a + b}{3}}{\dfrac{x - 2y}{4}}$$

The bar in a complex rational expression is both a grouping symbol and a symbol for division.

$$\frac{\dfrac{a + b}{3}}{\dfrac{x - 2y}{4}} \qquad \text{is equivalent to} \qquad \left(\frac{a + b}{3}\right) \div \left(\frac{x - 2y}{4}\right).$$

We need a procedure for simplifying complex rational expressions.

> **Procedure to Simplify a Complex Rational Expression: Adding and Subtracting**
>
> 1. Add or subtract so that you have a single fraction in the numerator and in the denominator.
> 2. Divide the fraction in the numerator by the fraction in the denominator. This is done by inverting the fraction in the denominator and multiplying it by the numerator.

● **EXAMPLE 1** Simplify. $\dfrac{\dfrac{1}{x}}{\dfrac{2}{y^2} + \dfrac{1}{y}}$

Step 1 Add the two fractions in the denominator.

$$\frac{\dfrac{1}{x}}{\dfrac{2}{y^2} + \dfrac{1}{y} \cdot \dfrac{y}{y}} = \frac{\dfrac{1}{x}}{\dfrac{2 + y}{y^2}}$$

Step 2 Divide the fraction in the numerator by the fraction in the denominator.

$$\frac{1}{x} \div \frac{2 + y}{y^2} = \frac{1}{x} \cdot \frac{y^2}{2 + y} = \frac{y^2}{x(2 + y)}$$

Practice Problem 1 Simplify. $\dfrac{\dfrac{1}{a} + \dfrac{1}{b}}{\dfrac{2}{ab^2}}$

A complex rational expression may contain two or more fractions in the numerator and the denominator.

EXAMPLE 2 Simplify. $\dfrac{\dfrac{1}{x} + \dfrac{1}{y}}{\dfrac{3}{a} - \dfrac{2}{b}}$

We observe that the LCD of the fractions in the numerator is xy. The LCD of the fractions in the denominator is ab.

$= \dfrac{\dfrac{1}{x} \cdot \dfrac{y}{y} + \dfrac{1}{y} \cdot \dfrac{x}{x}}{\dfrac{3}{a} \cdot \dfrac{b}{b} - \dfrac{2}{b} \cdot \dfrac{a}{a}}$ Multiply each fraction by the appropriate value to obtain common denominators.

$= \dfrac{\dfrac{y + x}{xy}}{\dfrac{3b - 2a}{ab}}$ Add the two fractions in the numerator.

Subtract the two fractions in the denominator.

$= \dfrac{y + x}{xy} \cdot \dfrac{ab}{3b - 2a}$ Invert the fraction in the denominator and multiply it by the numerator.

$= \dfrac{ab(y + x)}{xy(3b - 2a)}$ Write the answer as one fraction.

Practice Problem 2 Simplify. $\dfrac{\dfrac{1}{a} + \dfrac{1}{b}}{\dfrac{1}{a} - \dfrac{1}{b}}$

For some complex rational expressions, factoring may be necessary to determine the LCD and to combine fractions.

EXAMPLE 3 Simplify. $\dfrac{\dfrac{1}{x^2 - 1} + \dfrac{2}{x + 1}}{x}$

We need to factor $x^2 - 1$.

$= \dfrac{\dfrac{1}{(x + 1)(x - 1)} + \dfrac{2}{(x + 1)} \cdot \dfrac{x - 1}{x - 1}}{x}$ The LCD for the fractions in the numerator is $(x + 1)(x - 1)$.

$= \dfrac{\dfrac{1 + 2x - 2}{(x + 1)(x - 1)}}{x}$ Add the two fractions in the numerator.

$= \dfrac{2x - 1}{(x + 1)(x - 1)} \cdot \dfrac{1}{x}$ Simplify the numerator. Invert the fraction in the denominator and multiply.

$= \dfrac{2x - 1}{x(x + 1)(x - 1)}$ Write the answer as one fraction.

Practice Problem 3 Simplify. $\dfrac{\dfrac{x}{x^2 + 4x + 3} + \dfrac{2}{x + 1}}{x + 1}$

When simplifying complex rational expressions, always check to see if the final fraction can be reduced or simplified.

EXAMPLE 4 Simplify. $\dfrac{\dfrac{3}{a+b} - \dfrac{3}{a-b}}{\dfrac{5}{a^2-b^2}}$

The LCD of the two fractions in the numerator is $(a+b)(a-b)$.

$$\dfrac{\dfrac{3}{a+b} \cdot \dfrac{a-b}{a-b} - \dfrac{3}{a-b} \cdot \dfrac{a+b}{a+b}}{\dfrac{5}{a^2-b^2}}$$

$$= \dfrac{\dfrac{3a-3b}{(a+b)(a-b)} - \dfrac{3a+3b}{(a+b)(a-b)}}{\dfrac{5}{a^2-b^2}}$$

Study carefully how we combine the two fractions in the numerator. Do you see how we obtain $-6b$?

$$= \dfrac{\dfrac{-6b}{(a+b)(a-b)}}{\dfrac{5}{(a+b)(a-b)}}$$ Factor $a^2 - b^2$ as $(a+b)(a-b)$.

$$= \dfrac{-6b}{(a+b)(a-b)} \cdot \dfrac{(a+b)(a-b)}{5}$$ Since $(a+b)(a-b)$ are factors in both numerator and denominator, they may be divided out.

$$= \dfrac{-6b}{5} \quad \text{or} \quad -\dfrac{6b}{5}$$

Practice Problem 4 Simplify. $\dfrac{\dfrac{6}{x^2-y^2}}{\dfrac{1}{x-y} + \dfrac{3}{x+y}}$

2 Simplifying Complex Rational Expressions by Multiplying by the LCD of All the Denominators

There is another way to simplify complex rational expressions: Multiply the numerator and denominator of the complex fraction by the least common denominator of all the denominators appearing in the complex fraction.

Procedure to Simplify a Complex Rational Expression: Multiplying by the LCD

1. Determine the LCD of all individual denominators occurring in the numerator and denominator of the complex rational expression.
2. Multiply both the numerator and the denominator of the complex rational expression by the LCD.
3. Simplify, if possible.

● **EXAMPLE 5**   Simplify by multiplying by the LCD. $\dfrac{\dfrac{5}{ab^2} - \dfrac{2}{ab}}{3 - \dfrac{5}{2a^2b}}$

The LCD of all the denominators in the complex rational expression is $2a^2b^2$.

$$\dfrac{2a^2b^2\left(\dfrac{5}{ab^2} - \dfrac{2}{ab}\right)}{2a^2b^2\left(3 - \dfrac{5}{2a^2b}\right)}$$

$$= \dfrac{2a^2b^2\left(\dfrac{5}{ab^2}\right) - 2a^2b^2\left(\dfrac{2}{ab}\right)}{2a^2b^2(3) - 2a^2b^2\left(\dfrac{5}{2a^2b}\right)} \qquad \text{Multiply each term by } 2a^2b^2.$$

$$= \dfrac{10a - 4ab}{6a^2b^2 - 5b} \qquad\qquad\qquad \text{Simplify.}$$

Practice Problem 5   Simplify by multiplying by the LCD. $\dfrac{\dfrac{2}{3x^2} - \dfrac{3}{y}}{\dfrac{5}{xy} - 4}$ ●

So that you can compare the two methods, we will redo Example 4 by multiplying by the LCD.

● **EXAMPLE 6**   Simplify by multiplying by the LCD. $\dfrac{\dfrac{3}{a+b} - \dfrac{3}{a-b}}{\dfrac{5}{a^2-b^2}}$

The LCD of all individual fractions contained in the complex fraction is $(a+b)(a-b)$.

$$\dfrac{(a+b)(a-b)\left(\dfrac{3}{a+b}\right) - (a+b)(a-b)\left(\dfrac{3}{a-b}\right)}{(a+b)(a-b)\left(\dfrac{5}{(a+b)(a-b)}\right)} \qquad \begin{array}{l}\text{Multiply each term}\\ \text{by the LCD.}\end{array}$$

$$= \dfrac{3(a-b) - 3(a+b)}{5} \qquad\qquad \text{Simplify.}$$

$$= \dfrac{3a - 3b - 3a - 3b}{5} \qquad\qquad \text{Remove parentheses.}$$

$$= -\dfrac{6b}{5} \qquad\qquad\qquad\qquad \text{Simplify.}$$

Practice Problem 6   Simplify by multiplying by the LCD. $\dfrac{\dfrac{6}{x^2-y^2}}{\dfrac{7}{x-y} + \dfrac{3}{x+y}}$ ●

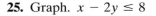

7.4 Exercises

Simplify.

1. $\dfrac{\dfrac{5}{xy}}{\dfrac{7}{y}}$

2. $\dfrac{\dfrac{4a}{b}}{\dfrac{6}{ab}}$

3. $\dfrac{\dfrac{1}{x} + \dfrac{1}{y}}{\dfrac{1}{xy}}$

4. $\dfrac{\dfrac{1}{x} + 1}{\dfrac{1}{x}}$

5. $\dfrac{\dfrac{1}{x} + y}{\dfrac{1}{y} + x}$

6. $\dfrac{\dfrac{1}{x} + \dfrac{1}{y}}{x + y}$

7. $\dfrac{1 - \dfrac{9}{x^2}}{\dfrac{3}{x} + 1}$

8. $\dfrac{\dfrac{4}{x} + 1}{1 - \dfrac{16}{x^2}}$

9. $\dfrac{\dfrac{2}{x + 1} - 2}{3}$

10. $\dfrac{1 - \dfrac{3}{xy}}{x + 2y}$

11. $\dfrac{a + \dfrac{3}{a}}{\dfrac{a^2 + 2}{3a}}$

12. $\dfrac{a + \dfrac{1}{a}}{\dfrac{3}{a} - a}$

13. $\dfrac{y - \dfrac{4}{y}}{1 + \dfrac{3}{2y + 1}}$

14. $\dfrac{a + 1 - \dfrac{12}{a - 2}}{\dfrac{-2}{a - 2} + a - 1}$

15. $\dfrac{\dfrac{x}{6} - \dfrac{1}{3}}{\dfrac{2}{3x} + \dfrac{5}{6}}$

16. $\dfrac{\dfrac{7}{5x} - \dfrac{1}{x}}{\dfrac{3}{5} + \dfrac{2}{x}}$

17. $\dfrac{\dfrac{1}{x^2 - 9} + \dfrac{2}{x + 3}}{\dfrac{3}{x - 3}}$

18. $\dfrac{\dfrac{5}{x + 4}}{\dfrac{1}{x - 4} - \dfrac{2}{x^2 - 16}}$

19. $\dfrac{\dfrac{2}{y - 1} + 2}{\dfrac{2}{y + 1} - 2}$

20. $\dfrac{\dfrac{y}{y + 1} + 1}{\dfrac{2y + 1}{y - 1}}$

To Think About

21. Consider the complex fraction $\dfrac{\dfrac{4}{x + 3}}{\dfrac{5}{x} - 1}$. What values are not allowable replacements for the variable x?

22. Consider the complex fraction $\dfrac{\dfrac{5}{x - 2}}{\dfrac{6}{x} + 1}$. What values are not allowable replacements for the variable x?

Cumulative Review Problems

23. Solve for w. $P = 2(l + w)$

24. Solve and graph. $7 + x < 11 + 5x$

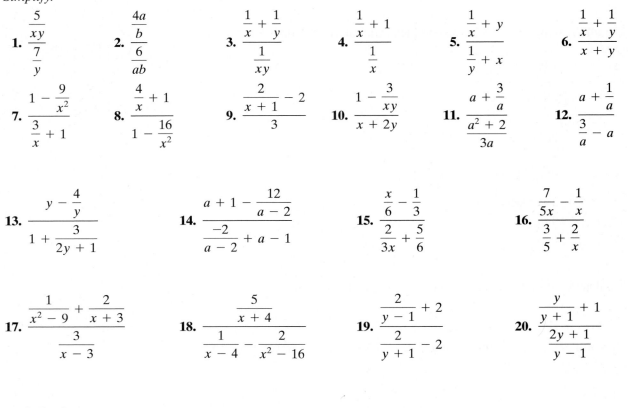

25. Graph. $x - 2y \le 8$

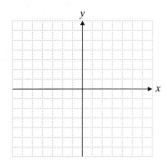

26. Solve the system. $2x + y = -\dfrac{11}{2}$

$-x + 2y = 4$

7.5 Equations Involving Rational Expressions

1 Solving Equations Involving Rational Expressions That Have Solutions

Student Learning Objectives

After studying this section, you will be able to:

1 Solve equations involving rational expressions that have solutions.

2 Determine whether an equation involving rational expressions has no solution.

In Section 2.4 we developed procedures to solve linear equations containing fractions whose denominators were numerical values. In this section we use a similar approach to solve equations containing fractions whose denominators are polynomials. It would be wise for you to review Section 2.4 briefly *before you begin this section.* It will be especially helpful to carefully study Examples 1 and 2.

SSM   PH TUTOR   CD & VIDEO   MATH PRO   WEB
CENTER

> **To Solve an Equation Containing Rational Expressions**
>
> 1. Determine the LCD of all the denominators.
> 2. Multiply each term of the equation by the LCD.
> 3. Solve the resulting equation.
> 4. Check your solution. Exclude from your solution any value that would make the LCD equal to zero. If such a value is obtained, there is *no solution.*

EXAMPLE 1 Solve for x and check your solution. $\dfrac{5}{x} + \dfrac{2}{3} = 2 - \dfrac{2}{x} - \dfrac{1}{6}$

$6x\left(\dfrac{5}{x}\right) + 6x\left(\dfrac{2}{3}\right) = 6x(2) - 6x\left(\dfrac{2}{x}\right) - 6x\left(\dfrac{1}{6}\right)$ Observe that the LCD is $6x$. Multiply each term by $6x$.

$\qquad 30 + 4x = 12x - 12 - x$ Simplify. Do you see how each term is obtained?

$\qquad 30 + 4x = 11x - 12$ Collect like terms.

$\qquad\qquad 30 = 7x - 12$ Subtract $4x$ from both sides.

$\qquad\qquad 42 = 7x$ Add 12 to both sides.

$\qquad\qquad 6 = x$ Divide both sides by 7.

Check. $\dfrac{5}{6} + \dfrac{2}{3} \overset{?}{=} 2 - \dfrac{2}{6} - \dfrac{1}{6}$ Replace each x by 6.

$\qquad\quad \dfrac{5}{6} + \dfrac{4}{6} \overset{?}{=} \dfrac{12}{6} - \dfrac{2}{6} - \dfrac{1}{6}$

$\qquad\qquad\quad \dfrac{9}{6} = \dfrac{9}{6}$ ✓   It checks.

Practice Problem 1 Solve for x and check your solution. $\dfrac{2}{x} + \dfrac{4}{x} = 3 - \dfrac{1}{x} - \dfrac{17}{8}$

EXAMPLE 2 Solve and check. $\dfrac{6}{x+3} = \dfrac{3}{x}$

Observe that the LCD $= x(x+3)$.

$x(x+3)\left(\dfrac{6}{x+3}\right) = x(x+3)\left(\dfrac{3}{x}\right)$ Multiply both sides by $x(x+3)$.

$\qquad\qquad 6x = 3(x+3)$ Simplify. Do you see how this is done?

$\qquad\qquad 6x = 3x + 9$ Remove parentheses.

$\qquad\qquad 3x = 9$ Subtract $3x$ from both sides.

$\qquad\qquad\; x = 3$ Divide both sides by 3.

Check. $\dfrac{6}{3+3} \stackrel{?}{=} \dfrac{3}{3}$ Replace each x by 3.

$\dfrac{6}{6} = \dfrac{3}{3}$ ✓   It checks.

Practice Problem 2   Solve and check. $\dfrac{4}{2x+1} = \dfrac{6}{2x-1}$

It is sometimes necessary to factor denominators before the correct LCD can be determined.

EXAMPLE 3   Solve and check. $\dfrac{3}{x+5} - 1 = \dfrac{4-x}{2x+10}$

$\dfrac{3}{x+5} - 1 = \dfrac{4-x}{2(x+5)}$ Factor $2x+10$. We determine that the LCD is $2(x+5)$.

$2(x+5)\left(\dfrac{3}{x+5}\right) - 2(x+5)(1) = 2(x+5)\left[\dfrac{4-x}{2(x+5)}\right]$ Multiply each term by the LCD.

$2(3) - 2(x+5) = 4 - x$ Simplify.

$6 - 2x - 10 = 4 - x$ Remove parentheses.

$-2x - 4 = 4 - x$ Collect like terms.

$-4 = 4 + x$ Add $2x$ to both sides.

$-8 = x$ Subtract 4 from both sides.

Check. $\dfrac{3}{-8+5} - 1 \stackrel{?}{=} \dfrac{4-(-8)}{2(-8)+10}$ Replace each x in the original equation by -8.

$\dfrac{3}{-3} - 1 \stackrel{?}{=} \dfrac{4+8}{-16+10}$

$-1 - 1 \stackrel{?}{=} \dfrac{12}{-6}$

$-2 = -2$ ✓   It checks. The solution is -8.

Practice Problem 3   Solve and check. $\dfrac{x-1}{x^2-4} = \dfrac{2}{x+2} + \dfrac{4}{x-2}$

  2 *Determining Whether an Equation Involving Rational Expressions Has No Solution*

Equations containing rational expressions sometimes appear to have solutions when in fact they do not. By this we mean that the "solutions" we get by using completely correct methods are, in actuality, not solutions.

In the case where a value makes a denominator in the equation equal to zero, we say it is not a solution to the equation. Such a value is called an **extraneous solution**. An extraneous solution is an apparent solution that does *not* satisfy the original equation. If all of the apparent solutions of an equation are extraneous solutions, we say that the equation has **no solution**. It is important that you check all apparent solutions in the original equation.

EXAMPLE 4   Solve and check. $\dfrac{y}{y-2} - 4 = \dfrac{2}{y-2}$

Observe that the LCD is $y - 2$.

$$(y-2)\left(\frac{y}{y-2}\right) - (y-2)(4) = (y-2)\left(\frac{2}{y-2}\right)$$ Multiply each term by $(y-2)$.

$$y - 4(y-2) = 2$$ Simplify. Do you see how this is done?

$$y - 4y + 8 = 2$$ Remove parentheses.

$$-3y + 8 = 2$$ Collect like terms.

$$-3y = -6$$ Subtract 8 from both sides.

$$\frac{-3y}{-3} = \frac{-6}{-3}$$ Divide both sides by -3.

$$y = 2$$ 2 is only an apparent solution.

This equation has no solution.
Why? We can see immediately that $y = 2$ is not a solution for the original equation. When we substitute 2 for y in a denominator, the denominator is equal to zero and the expression is undefined.

Check. $\dfrac{y}{y-2} - 4 = \dfrac{2}{y-2}$ Suppose that you try to check the apparent solution by substituting 2 for y.

$$\frac{2}{2-2} - 4 \overset{?}{=} \frac{2}{2-2}$$

$$\frac{2}{0} - 4 = \frac{2}{0}$$ This does not check since you do not obtain a real number when you divide by zero.

These expressions are not defined.

There is no such number as $2 \div 0$. We see that 2 does *not* check. This equation has **no solution.**

Practice Problem 4   Solve and check. $\dfrac{2x}{x+1} = \dfrac{-2}{x+1} + 1$

Solve and check.

1. $\dfrac{x+1}{2x} = \dfrac{2}{3}$

2. $\dfrac{6}{3x-5} = \dfrac{3}{2x}$

3. $\dfrac{1}{x+6} = \dfrac{4}{x}$

4. $\dfrac{x-2}{4x} = \dfrac{1}{6}$

5. $\dfrac{5}{3x-4} = \dfrac{3}{x-3}$

6. $\dfrac{5}{x-1} = \dfrac{3}{x+1}$

7. $\dfrac{2}{x} + \dfrac{x}{x+1} = 1$

8. $\dfrac{5}{2} = 3 + \dfrac{2x+7}{x+6}$

9. $\dfrac{x+1}{x} = 1 + \dfrac{x-2}{2x}$

10. $\dfrac{7x-4}{5x} = \dfrac{9}{5} - \dfrac{4}{x}$

11. $\dfrac{x+1}{x} = \dfrac{1}{2} - \dfrac{4}{3x}$

12. $\dfrac{2x-1}{3x} + \dfrac{1}{9} = \dfrac{x+2}{x} + \dfrac{1}{9x}$

13. $\dfrac{2}{x-6} - 5 = \dfrac{2(x-5)}{x-6}$

14. $7 - \dfrac{x}{x+5} = \dfrac{5}{5+x}$

15. $\dfrac{2}{x+1} - \dfrac{1}{x-1} = \dfrac{2x}{x^2-1}$

16. $\dfrac{8x}{4x^2-1} = \dfrac{3}{2x+1} + \dfrac{3}{2x-1}$

17. $\dfrac{y+1}{y^2+2y-3} = \dfrac{1}{y+3} - \dfrac{1}{y-1}$

18. $\dfrac{2y}{y+1} + \dfrac{1}{3y-2} = 2$

19. $\dfrac{79-x}{x} = 5 + \dfrac{7}{x}$

20. $\dfrac{6}{x-3} = \dfrac{-5}{x-2} + \dfrac{-5}{x^2-5x+6}$

21. $\dfrac{x+11}{x^2-5x+4} + \dfrac{3}{x-1} = \dfrac{5}{x-4}$

22. $\dfrac{52-x}{x} = 9 + \dfrac{2}{x}$

23. $\dfrac{2x}{x+4} - \dfrac{8}{x-4} = \dfrac{2x^2+32}{x^2-16}$

24. $\dfrac{4x}{x+3} - \dfrac{12}{x-3} = \dfrac{4x^2+36}{x^2-9}$

25. $\dfrac{6}{x-5} + \dfrac{3x+1}{x^2-2x-15} = \dfrac{5}{x+3}$

26. $\dfrac{4}{x^2-1} + \dfrac{7}{x+1} = \dfrac{5}{x-1}$

To Think About

In each of the following equations, what values are not allowable replacements for the variable x? Do not solve the equation.

27. $\dfrac{3x}{x-2} - \dfrac{4x}{x-4} = \dfrac{3}{x^2 - 6x + 8}$

28. $\dfrac{2x}{x+2} - \dfrac{x}{x+3} = \dfrac{-3}{x^2 + 5x + 6}$

Cumulative Review Problems

▲ **29.** The perimeter of a rectangular sign is 54 meters. The length is 1 meter less than triple the width. Find the dimensions of the sign.

30. Determine the domain and range of the relation. Is the relation a function?
$\{(7, 3), (2, 2), (-2, 0), (2, -2), (7, -3)\}$

31. Graph. $x = y^2 - 2$

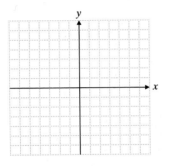

32. Factor. $6x^2 - x - 12$

33. Factor. $4y^2 - 121z^2$

34. Solve. $x^2 - 4x - 21 = 0$

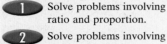

SSM
PH TUTOR
CENTER
CD & VIDEO
MATH PRO
WEB

1 *Solving Problems Involving Ratio and Proportion*

A **ratio** is a comparison of two quantities. You may be familiar with ratios that compare miles to hours or miles to gallons. A ratio is often written as a quotient in the form of a fraction. For example, the ratio of 7 to 9 can be written as $\frac{7}{9}$.

A **proportion** is an equation that states that two ratios are equal. For example,

$$\frac{7}{9} = \frac{21}{27}, \quad \frac{2}{3} = \frac{10}{15}, \quad \text{and} \quad \frac{a}{b} = \frac{c}{d} \quad \text{are proportions.}$$

Let's take a closer look at the last proportion. We can see that the LCD of the fractional equation is bd.

$$(bd)\frac{a}{b} = (bd)\frac{c}{d} \qquad \text{Multiply each side by the LCD.}$$
$$da = bc$$
$$ad = bc \qquad \text{Since multiplication is commutative, } da = ad.$$

Thus we have proved the following.

The Proportion Equation

If

$$\frac{a}{b} = \frac{c}{d}, \quad \text{then} \quad ad = bc$$

for all real numbers a, b, c, and d, where $b \neq 0$ and $d \neq 0$.

This is sometimes called **cross multiplying**. It can be applied only if you have *one* fraction and nothing else on each side of the equation.

EXAMPLE 1 Michael took 5 hours to drive 245 miles on the turnpike. At the same rate, how many hours will it take him to drive a distance of 392 miles?

1. **Understand the problem.**
   Let x = the number of hours it will take to drive 392 miles. If 5 hours are needed to drive 245 miles, then x hours are needed to drive 392 miles.

2. **Write an equation.**
   We can write this as a proportion. Compare time to distance in each ratio.

$$\begin{array}{ccc} \text{Time} & \longrightarrow & \dfrac{5 \text{ hours}}{245 \text{ miles}} = \dfrac{x \text{ hours}}{392 \text{ miles}} & \longleftarrow & \text{Time} \\ \text{Distance} & \longrightarrow & & \longleftarrow & \text{Distance} \end{array}$$

3. **Solve and state the answer.**

$$5(392) = 245x \qquad \text{Cross multiply.}$$
$$\frac{1960}{245} = x \qquad \text{Divide both sides by 245.}$$
$$8 = x$$

It will take Michael 8 hours to drive 392 miles.

4. **Check.** Is $\frac{5}{245} = \frac{8}{392}$? Do the computation and see.

Practice Problem 1 It took Brenda 8 hours to drive 420 miles. At the same rate, how long would it take her to drive 315 miles?

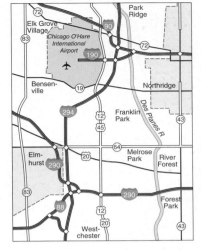

 EXAMPLE 2 If $\frac{3}{4}$ inch on a map represents an actual distance of 20 miles, how long is the distance represented by $4\frac{1}{8}$ inches on the same map?

Let x = the distance represented by $4\frac{1}{8}$ inches.

Initial measurement on map $\longrightarrow \dfrac{3}{4}$ $4\frac{1}{8}$ $\longleftarrow$ Second measurement on the map

Initial distance $\longrightarrow \dfrac{3}{4} = \dfrac{4\frac{1}{8}}{x}$ $\longleftarrow$ Second distance

$$\left(\frac{3}{4}\right)(x) = (20)\left(4\frac{1}{8}\right) \qquad \text{Cross multiply.}$$

$$\left(\frac{3}{4}\right)(x) = (\overset{5}{\cancel{20}})\left(\frac{33}{\underset{2}{\cancel{8}}}\right) \qquad \text{Write } 4\frac{1}{8} \text{ as } \frac{33}{8} \text{ and simplify.}$$

$$\frac{3x}{4} = \frac{165}{2} \qquad \text{Multiply fractions.}$$

$$4\left(\frac{3x}{4}\right) = \overset{2}{\cancel{4}}\left(\frac{165}{\cancel{2}}\right) \qquad \text{Multiply each side by 4.}$$

$$3x = 330 \qquad \text{Simplify.}$$

$$x = 110 \qquad \text{Divide both sides by 3.}$$

$4\frac{1}{8}$ inches on the map represents an actual distance of 110 miles.

Practice Problem 2 If $\frac{5}{8}$ inch on a map represents an actual distance of 30 miles, how long is the distance represented by $2\frac{1}{2}$ inches on the same map?

2 *Solving Problems Involving Similar Triangles*

Similar triangles are triangles that have the same shape but may be different sizes. For example, if you draw a triangle on a sheet of paper, place the paper in a photocopy machine, and make a copy that is reduced by 25%, you would create a triangle that is similar to the original triangle. The two triangles will have the same shape. The corresponding sides of the triangles will be proportional. The corresponding angles of the triangles will also be equal.

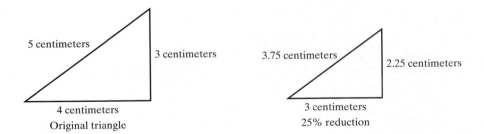

You can use the proportion equation to show that the corresponding sides of the preceding triangles are proportional. In fact, you can use the proportion equation to find an unknown length of a side of one of the two similar triangles.

▲ **EXAMPLE 3** A ramp is 32 meters long and rises up 15 meters. A ramp at the same angle is 9 meters long. How high is the second ramp? To answer this question, we find the length of side x in the following two similar triangles.

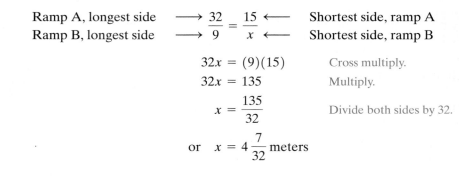

Ramp A, longest side $\longrightarrow \dfrac{32}{9} = \dfrac{15}{x} \longleftarrow$ Shortest side, ramp A

Ramp B, longest side $\longrightarrow$ $\longleftarrow$ Shortest side, ramp B

$$32x = (9)(15) \qquad \text{Cross multiply.}$$

$$32x = 135 \qquad \text{Multiply.}$$

$$x = \frac{135}{32} \qquad \text{Divide both sides by 32.}$$

or $\quad x = 4\dfrac{7}{32}$ meters

▲ **Practice Problem 3** Triangle C is similar to triangle D. Find the length of side x. Leave your answer as a fraction.

13 centimeters   16 centimeters x 18 centimeters

Triangle C   Triangle D

We can also use similar triangles for indirect measurement—for instance, to find the measure of an object that is too tall to measure using standard measuring devices. When the sun shines on two vertical objects at the same time, the shadows and the objects form similar triangles.

▲ ⬤ **EXAMPLE 4** A woman who is 5 feet tall casts a shadow that is 8 feet long. At the same time of day, a building casts a shadow that is 72 feet long. How tall is the building?

1. *Understand the problem.*

First we draw a sketch. We do not know the height of the building, so we call it x.

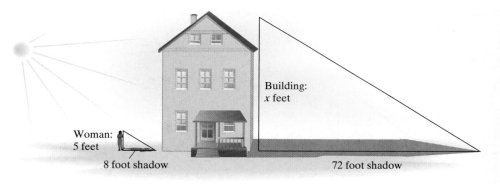

Building: x feet

Woman: 5 feet

8 foot shadow   72 foot shadow

2. *Write an equation and solve.*

Height of woman $\longrightarrow \dfrac{5}{8} = \dfrac{x}{72} \longleftarrow$ Height of building

Length of woman's shadow $\longrightarrow$ $\longleftarrow$ Length of building's shadow

$$(5)(72) = 8x \quad \text{Cross multiply.}$$

$$360 = 8x$$

$$45 = x$$

The height of the building is 45 feet.

▲ **Practice Problem 4** A man who is 6 feet tall casts a shadow that is 7 feet long. At the same time of day, a large flagpole casts a shadow that is 38.5 feet long. How tall is the flagpole?

In problems such as Example 4, we are assuming that the building and the person are standing exactly perpendicular to the ground. In other words, each triangle is assumed to be a right triangle. In other similar triangle problems, if the triangles are not right triangles you must be careful that the angles between the objects and the ground are the same.

3 Solving Distance Problems Involving Rational Expressions

Some distance problems are solved using equations with rational expressions. We will need the formula Distance = Rate × Time, $D = RT$, which we can write in the form $T = \dfrac{D}{R}$.

EXAMPLE 5 Plane A flies at a speed that is 50 kilometers per hour faster than plane B. Plane A flies 500 kilometers in the amount of time that plane B flies 400 kilometers. Find the speed of each plane.

1. **Understand the problem.**
   Let s = the speed of plane B in kilometers per hour.
   Let $s + 50$ = the speed of plane A in kilometers per hour.
   Make a simple table for D, R, and T.

	D	R	$T = \dfrac{D}{R}$
Plane A	500	$s + 50$	?
Plane B	400	s	?

   Since $T = \dfrac{D}{R}$, for each plane we divide the expression for D by the expression for R and write it in the table in the column for time.

	D	R	$T = \dfrac{D}{R}$
Plane A	500	$s + 50$	$\dfrac{500}{s + 50}$
Plane B	400	s	$\dfrac{400}{s}$

2. **Write an equation and solve.**
   Each plane flies for the same amount of time. That is, the time for plane A equals the time for plane B.

$$\frac{500}{s + 50} = \frac{400}{s}$$

   You can solve this equation using the methods in Section 7.5 or you may cross multiply. Here we will cross multiply.

$$500s = (s + 50)(400) \qquad \text{Cross multiply.}$$
$$500s = 400s + 20{,}000 \qquad \text{Remove parentheses.}$$
$$100s = 20{,}000 \qquad \text{Subtract } 400s \text{ from each side.}$$
$$s = 200 \qquad \text{Divide each side by 100.}$$

   Plane B travels 200 kilometers per hour. Since

$$s + 50 = 200 + 50 = 250,$$

   plane A travels 250 kilometers per hour.

Practice Problem 5 Two European freight trains traveled toward Paris for the same amount of time. Train A traveled 180 kilometers while train B traveled 150 kilometers. Train A traveled 10 kilometers per hour faster than train B. What was the speed of each train?

4 *Solving Work Problems*

Some applied problems involve the length of time needed to do a job. These problems are often referred to as work problems.

EXAMPLE 6 Reynaldo can sort a huge stack of mail on an old sorting machine in 9 hours. His brother Carlos can sort the same amount of mail using a newer sorting machine in 8 hours. How long would it take them to do the job working together? Express your answer in hours and minutes. Round to the nearest minute.

1. *Understand the problem.*
   Let's do a little reasoning.
   If Reynaldo can do the job in 9 hours, then in *1 hour* he could do $\frac{1}{9}$ of the job.
   If Carlos can do the job in 8 hours, then in *1 hour* he could do $\frac{1}{8}$ of the job.
   Let $x =$ the number of hours it takes Reynaldo and Carlos to do the job together. In *1 hour* together they could do $\frac{1}{x}$ of the job.

2. *Write an equation and solve.*
   The amount of work Reynaldo can do in 1 hour plus the amount of work Carlos can do in 1 hour must be equal to the amount of work they could do together in 1 hour.

Amount of work done by Reynaldo		Amount of work done by Carlos		Amount of work done together
$\frac{1}{9}$	$+$	$\frac{1}{8}$	$=$	$\frac{1}{x}$

Let us solve for x. We observe that the LCD is $72x$.

$$72x\left(\frac{1}{9}\right) + 72x\left(\frac{1}{8}\right) = 72x\left(\frac{1}{x}\right) \qquad \text{Multiply each term by the LCD.}$$

$$8x + 9x = 72 \qquad \text{Simplify.}$$

$$17x = 72 \qquad \text{Collect like terms.}$$

$$x = \frac{72}{17} \qquad \text{Divide each side by 17.}$$

$$x = 4\frac{4}{17}$$

To change $\frac{4}{17}$ of an hour to minutes we multiply.
$$\frac{4}{17} \; \text{hour} \times \frac{60 \text{ minutes}}{1 \text{ hour}} = \frac{240}{17} \text{ minutes, which is approximately 14.118 minutes.}$$
To the nearest minute this is 14 minutes. Thus doing the job together will take 4 hours and 14 minutes.

Practice Problem 6 John Tobey and Dave Wells obtained night custodian jobs at a local factory while going to college part-time. Using the buffer machine, John can buff all the floors in the building in 6 hours. Dave takes a little longer and can do all the floors in the building in 7 hours. Their supervisor bought another buffer machine. How long will it take John and Dave to do all the floors in the building working together, each with his own machine? Express your answer in hours and minutes. Round to the nearest minute.

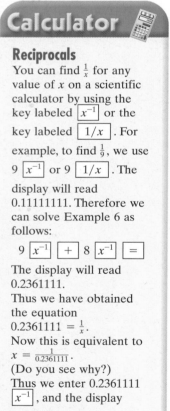

Calculator

Reciprocals
You can find $\frac{1}{x}$ for any value of x on a scientific calculator by using the key labeled $\boxed{x^{-1}}$ or the key labeled $\boxed{1/x}$. For example, to find $\frac{1}{9}$, we use $9 \boxed{x^{-1}}$ or $9 \boxed{1/x}$. The display will read 0.11111111. Therefore we can solve Example 6 as follows:

$$9 \boxed{x^{-1}} \boxed{+} 8 \boxed{x^{-1}} \boxed{=}$$

The display will read 0.2361111.
Thus we have obtained the equation
$0.2361111 = \frac{1}{x}$.
Now this is equivalent to $x = \frac{1}{0.2361111}$.
(Do you see why?)
Thus we enter 0.2361111 $\boxed{x^{-1}}$, and the display reads 4.235294118.
If we round to the nearest hundredth, we have $x \approx 4.24$ hours, which is approximately equal to our answer of $4\frac{4}{17}$ hours.

Solve.

1. $\dfrac{4}{9} = \dfrac{8}{x}$

2. $\dfrac{5}{12} = \dfrac{x}{8}$

3. $\dfrac{x}{17} = \dfrac{12}{5}$

4. $\dfrac{16}{x} = \dfrac{3}{4}$

5. $\dfrac{5}{3} = \dfrac{x}{8}$

6. $\dfrac{150}{70} = \dfrac{9}{x}$

7. $\dfrac{7}{x} = \dfrac{40}{130}$

8. $\dfrac{x}{18} = \dfrac{13}{2}$

Applications

Use a proportion to answer exercises 9–16.

9. The scale on the AAA map of Colorado is approximately $\frac{3}{4}$ inch to 15 miles. If the distance from Denver to Pueblo measures 5.5 inches on the map, how far apart are the two cities?

10. Nella Coastes' recipe for shoofly pie contains $\frac{3}{4}$ cup of unsulfured molasses. This recipe makes a small pie that serves 8 people. If she makes the larger pie that serves 12 people and the ratio of molasses to people remains the same, how much molasses will she need for the larger pie?

11. James LeBlanc spent a summer semester in England. When he arrived in England, he converted $800 to pounds. The posted exchange rate the day he arrived that summer was that 1 British pound was worth $1.53 in American currency.
(a) How many pounds did James receive for his $800?
(b) At the end of the summer he had 140 pounds left and he changed it back into dollars. The exchange was the same that day as it was the day he arrived. How many dollars did he receive?

12. Christine Maney spent a summer studying in Paris as an exchange student. The exchange rate the day she arrived that summer was 5.56 francs per American dollar. When she arrived, she exchanged $900.
(a) How many francs was her $900 worth?
(b) At the end of the summer she had 350 francs left and exchanged them for American dollars. The exchange rate was the same that day as it was the day she arrived. How many dollars did she receive?

13. Alfonse and Melinda are taking a drive in Mexico. They know that a speed of 100 kilometers per hour is approximately equal to 62 miles per hour. They are now driving on a Mexican road that has a speed limit of 90 kilometers per hour. How many miles per hour is the speed limit? Round to the nearest mile per hour.

14. Dick and Anne took a trip to France. Their suitcases were weighed at the airport and the weight recorded was 39 kilograms. If 50 kilograms is equivalent to 110 pounds, how many pounds did their suitcases weigh? Round to the nearest pound.

15. On a map the distance between two mountains is $3\frac{1}{2}$ inches. The actual distance between the mountains is 136 miles. Russ is camped at a location that on the map is $\frac{3}{4}$ inch from the base of the mountain. How many miles is he from the base of the mountain? Round to the nearest mile.

16. Maria is adding a porch to her house that is 18 feet long. On the drawing done by the carpenter the length is shown as 11 inches. The drawing shows that the width of the porch is 8 inches. How many feet wide will the porch be? Round to the nearest foot.

Triangles A and B are similar. Use them to answer exercises 17 and 18. Leave your answers as fractions.

Triangle A   Triangle B

▲ **17.** If $x = 20$ in., $y = 29$ in., and $m = 13$ in., find the length of side n.

▲ **18.** If $p = 14$ in., $m = 17$ in., and $z = 23$ in., find the length of side x.

Just as we have discussed similar triangles, other geometric shapes can be similar. Similar geometric shapes will have sides that are proportional. Quadrilaterals abcd and ghjk are similar. Use them to answer exercises 19–22. Leave your answers as fractions.

▲ **19.** If $a = 7$ in., $g = 9$ in., and $k = 12$ in., find the length of side d.

▲ **20.** If $b = 8$ ft, $c = 7$ ft, and $j = 11$ ft, find the length of side h.

▲ **21.** If $b = 20$ m, $h = 24$ m, and $d = 32$ m, find the length of side k.

▲ **22.** If $a = 16$ cm, $d = 19$ cm, and $k = 23$ cm, find the length of side g.

Use a proportion to solve.

▲ **23.** A rectangle whose width-to-length ratio is approximately 5 to 8 is called a **golden rectangle** and is said to be pleasing to the eye. Using this ratio, what should the length of a rectangular picture be if its width is to be 30 inches?

▲ **24.** A 5-foot-tall woman casts a shadow of 4 feet. At the same time, a tree casts a shadow of 31 feet. How tall is the tree? Round to the nearest foot.

▲ **25.** A kite is held out on a line that is almost perfectly straight. When the kite is held out on 7 meters of line, it is 5 meters off the ground. How high would the kite be if it is held out on 120 meters of line at the same angle from the ground? Round to the nearest meter.

▲ **26.** A wire line helps to secure a radio transmission tower. The wire measures 23 meters from the tower to the ground anchor pin. The wire is secured 14 meters up on the tower. If a second wire is secured 130 meters up on the tower and is extended from the tower at the same angle as the first wire, how long would the second wire need to be to reach an anchor pin on the ground? Round to the nearest meter.

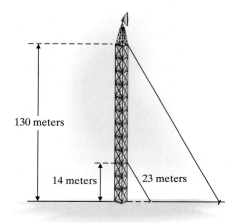

130 meters

14 meters   23 meters

27. Ben Hale is driving his new Toyota Camry on Interstate 90 at 45 miles per hour. He accelerates at the rate of 3 miles per hour every 2 seconds. How fast will he be traveling after accelerating for 11 seconds?

28. Tim Newitt is driving a U-Haul truck to Chicago. He is driving at 55 miles per hour and has to hit the brakes because of heavy traffic. His truck slows at the rate of 2 miles per hour for every 3 seconds. How fast will he be traveling 10 seconds after he hits the brakes?

29. A Montreal commuter airliner travels 40 kilometers per hour faster than the television news helicopter over the city. The commuter airliner travels 1250 kilometers during the same time that the television news helicopter travels only 1050 kilometers. How fast does the commuter airliner fly? How fast does the television news helicopter fly?

30. Melissa drove to Dallas while Marcia drove to Houston in the same amount of time. Melissa drove 360 kilometers while Marcia drove 280 kilometers. Melissa traveled 20 kilometers per hour faster than Marcia on her trip. What was the average speed in kilometers per hour for each woman?

31. A famous brand of men's cologne is priced at $75 for a 3.4-ounce size. A 1.7-ounce bottle is priced at $42.
  (a) How much does the cologne in the larger bottle cost per ounce? Round to the nearest cent.

  (b) How much does the cologne in the smaller bottle cost per ounce? Round to the nearest cent.

  (c) How much would you save if you purchased 10.2 ounces of the cologne in the larger bottles rather than the smaller bottles? Find the exact answer; do not round off.

32. Won Ling is a Chinese tea importer in Boston's Chinatown. He charges $12.25 for four sample packs of his famous green tea. The packs are in the following sizes: 25 grams, 40 grams, 50 grams, and 60 grams.
  (a) How much is Won Ling charging per gram for his green tea?
  (b) How much would you pay for a 60-gram pack if he were willing to sell that one by itself?
  (c) How much would you pay for an 800-gram package of green tea if it cost the same amount per gram?

33. It takes a person using a large rotary mower 4 hours to mow all the lawns at the town park. It takes a person using a small rotary mower 5 hours to mow these same lawns. How long should it take two people using these mowers to mow these lawns together? Round to the nearest minute.

34. It takes a secretary with a typewriter 7 hours to address envelopes for a business mailing list. It takes 3 hours for the computer to print out the addresses and the secretary to put the computer labels on the envelopes. The company boss is in a big rush and wants one secretary to do part of the job with the computer and another secretary to do the rest of the job with a typewriter. How long should it take the two secretaries working together? Round to the nearest minute.

35. When all the leaves have fallen at Fred and Suzie's house in Concord, New Hampshire, Suzie can rake the entire yard in 6 hours. When Fred does it alone, it takes him 8 hours. How long would it take them to rake the yard together? Round to the nearest minute.

36. Professor Matthews can type his course syllabus in 12 hours. The departmental secretary can type it in 7 hours. How long would it take the two people to type it together? Round to the nearest minute.

Cumulative Review Problems

37. Write in scientific notation. 0.0000006316

38. Write in decimal notation. 5.82×10^8

39. Write with positive exponents. $\dfrac{x^{-3}y^{-2}}{z^4w^{-8}}$

40. Evaluate. $\left(\dfrac{2}{3}\right)^{-3}$

Putting Your Skills to Work

Mathematical Measurement of Planet Orbit Time

The closer a planet is to the sun, the fewer the number of days it takes to complete one orbit around the sun. The Earth takes approximately 365 days to complete one orbit. Only two of the nine planets have a shorter orbit time. The orbit times of the four planets closest to the sun are shown in the bar graph below. Use this bar graph to answer the following questions. Round all answers to the nearest whole number.

Number of Days for a Planet to Orbit the Sun

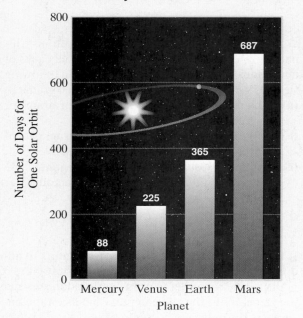

Problems for Individual Investigation

1. How many more days does it take Mars than Venus to complete one rotation around the sun?

2. In 5 years, the Earth will rotate around the sun five times. In that time period, how many orbits will Mercury make around the sun?

Problems for Group Investigation and Cooperative Learning

Together with some other members of your class, see if you can answer the following.

The **synodic period** of a planet is the number of days it takes the planet to gain one orbit on a planet farther from the Sun. The synodic period (S) can be found by the formula

$$\frac{1}{S} = \frac{1}{a} - \frac{1}{b}$$

where a = the number of days it takes the planet closer to the sun to complete one orbit and
b = the number of days it takes the planet farther from the sun to complete one orbit.

3. What is the synodic period for Mercury to gain one orbit on Venus?

4. What is the synodic period for the Earth to gain one orbit on Jupiter if the number of days it takes Jupiter to orbit the sun is approximately 4333 days?

Internet Connections

Site: The Nine Planets or a related site

This site provides a wealth of information about our solar system. Note that many of the numbers here are given in an abbreviated form of scientific notation; for example, 1.35e23 means 1.35×10^{23}.

5. Tell how many days it takes each of the planets Saturn, Uranus, Neptune, and Pluto to complete one rotation about the sun. This information is listed as "O-Period" in the Solar System Data appendix of the Web site.

6. How many more days does it take Neptune than Saturn to complete one orbit about the sun?

7. Use the formula to find the synodic period for Uranus to gain one orbit on Pluto.

Math Behind the Scenes

At close to 200 million dollars, *Titanic* is one of the most expensive movies ever made. For the movie, a replica of the original ship was constructed. *Entertainment Weekly Online* reported the replica to be 770 feet long—close to 90% to scale of the real ship.

Details of the original ship were followed closely to create the replica. A special studio was built for filming. The ship "sailed" in an enormous tank that held 17 million gallons of water.

Although movies frequently use models when filming, replicas tend to be on a much smaller scale than that used for the *Titanic*. In the questions that follow, you can use your own calculations to compare model and actual scales.

EXERCISES

1. The original Titanic was 880 feet long and 92 feet high. Use the exact length of the original Titanic and the exact length of the model to determine the percentage to scale of the real ship. Using this information, determine the height of the replica.

2. Does your calculation in Exercise 1 verify that the replica is 90% to scale of the actual ship? How much does your answer differ from the original estimate of 90%?

3. In the movie *The Hunt for Red October*, the story centers on the search for a Russian ballistic missile submarine that is reported to be 610 feet long and has a beam (width) of 46 feet. In filming some of the underwater sequences, a model of the hull of a submarine was used that was reported to be 14 feet wide. If that report is true what percent to scale is the model to the actual submarine in the story? If the entire model was built to that scale, what would be the expected length of the model submarine used in the filming?

Chapter 7 Organizer

Topic	Procedure	Examples
Simplifying rational expressions, p. 373.	1. Factor the numerator and denominator. 2. Divide out any factor common to both the numerator and denominator.	$\dfrac{36x^2 - 16y^2}{18x^2 + 24xy + 8y^2} = \dfrac{4(3x - 2y)(3x + 2y)}{2(3x + 2y)(3x + 2y)}$ $= \dfrac{2(3x - 2y)}{3x + 2y}$
Multiplying rational expressions, p. 379.	1. Factor all numerators and denominators and rewrite the product as one fraction. 2. Simplify the resulting rational expression as described above.	$\dfrac{x^2 - y^2}{x^2 + 2xy + y^2} \cdot \dfrac{x^2 + 4xy + 3y^2}{x^2 - 4xy + 3y^2}$ $= \dfrac{(x + y)(x - y)(x + y)(x + 3y)}{(x + y)(x + y)(x - y)(x - 3y)}$ $= \dfrac{x + 3y}{x - 3y}$
Dividing rational expressions, p. 380.	1. Invert the second fraction and rewrite the problem as a product. 2. Multiply the rational expressions.	$\dfrac{14x^2 + 17x - 6}{x^2 - 25} \div \dfrac{4x^2 - 8x - 21}{x^2 + 10x + 25}$ $= \dfrac{(2x + 3)(7x - 2)}{(x + 5)(x - 5)} \cdot \dfrac{(x + 5)(x + 5)}{(2x - 7)(2x + 3)}$ $= \dfrac{(7x - 2)(x + 5)}{(x - 5)(2x - 7)}$
Adding rational expressions, p. 384.	1. If the denominators differ, factor them and determine the least common denominator (LCD). 2. Change each fraction by multiplication into an equivalent one with the LCD. 3. Add numerators; put the answer over the LCD. 4. Simplify as needed.	$\dfrac{x - 1}{x^2 - 4} + \dfrac{x - 1}{3x + 6} = \dfrac{x - 1}{(x + 2)(x - 2)} + \dfrac{x - 1}{3(x + 2)}$ $\text{LCD} = 3(x + 2)(x - 2)$ $\dfrac{x - 1}{(x + 2)(x - 2)} = \dfrac{?}{3(x + 2)(x - 2)}$ Need to multiply by $\dfrac{3}{3}$. $\dfrac{(x - 1)}{3(x + 2)} = \dfrac{?}{3(x + 2)(x - 2)}$ Need to multiply by $\dfrac{x - 2}{x - 2}$. $\dfrac{x - 1}{(x + 2)(x - 2)} + \dfrac{x - 1}{3(x + 2)}$ $= \dfrac{(x - 1) \cdot 3}{3(x + 2)(x - 2)} + \dfrac{(x - 1)(x - 2)}{3(x + 2)(x - 2)}$ $= \dfrac{3x - 3 + x^2 - 3x + 2}{3(x + 2)(x - 2)}$ $= \dfrac{x^2 - 1}{3(x + 2)(x - 2)}$
Subtracting rational expressions, p. 388.	Move a subtraction sign to the numerator of the second fraction. Add. Simplify if possible. $\dfrac{a}{b} - \dfrac{c}{b} = \dfrac{a}{b} + \dfrac{-c}{b}$	$\dfrac{5x}{x - 2} - \dfrac{3x + 4}{x - 2} = \dfrac{5x}{x - 2} + \dfrac{-(3x + 4)}{x - 2}$ $= \dfrac{5x - 3x - 4}{x - 2}$ $= \dfrac{2x - 4}{x - 2}$ $= \dfrac{2(x - 2)}{x - 2} = 2$

Topic	Procedure	Examples
Simplifying complex rational expressions, p. 394.	1. Add the two fractions in the numerator. 2. Add the two fractions in the denominator. 3. Divide the fraction in the numerator by the fraction in the denominator. This is done by inverting the fraction in the denominator and multiplying by the numerator. 4. Simplify.	$$\dfrac{\dfrac{x}{x^2-4}+\dfrac{1}{x+2}}{\dfrac{3}{x+2}-\dfrac{4}{x-2}}$$ $$=\dfrac{\dfrac{x}{(x+2)(x-2)}+\dfrac{1(x-2)}{(x+2)(x-2)}}{\dfrac{3(x-2)}{(x+2)(x-2)}+\dfrac{-4(x+2)}{(x+2)(x-2)}}$$ $$=\dfrac{\dfrac{x+x-2}{(x+2)(x-2)}}{\dfrac{3x-6-4x-8}{(x+2)(x-2)}}$$ $$=\dfrac{2x-2}{(x+2)(x-2)}\div\dfrac{-x-14}{(x+2)(x-2)}$$ $$=\dfrac{2(x-1)}{(x+2)(x-2)}\cdot\dfrac{(x+2)(x-2)}{-x-14}$$ $$=\dfrac{2x-2}{-x-14}\ \text{or}\ \dfrac{-2x+2}{x+14}$$
Solving equations involving rational expressions, p. 399.	1. Determine the LCD of all denominators. 2. Note what values will make the LCD equal to 0. These are excluded from your solutions. 3. Multiply each side by the LCD, distributing as needed. 4. Solve the resulting polynomial equation. 5. Check. Be sure to exclude those values found in step 2.	$$\dfrac{3}{x-2}=\dfrac{4}{x+2}$$ LCD $=(x-2)(x+2)$. Since LCD $\neq 0$, then $x\neq 2,-2$. $$(x-2)(x+2)\dfrac{3}{x-2}=\dfrac{4}{x+2}(x-2)(x+2)$$ $$3(x+2)=4(x-2)$$ $$3x+6=4x-8$$ $$-x=-14$$ $$x=14$$ (Since $x\neq 2,-2$, this solution should check unless an error has been made.) $Check:$ $$\dfrac{3}{14-2}\overset{?}{=}\dfrac{4}{14+2}$$ $$\dfrac{3}{12}\overset{?}{=}\dfrac{4}{16}$$ $$\dfrac{1}{4}=\dfrac{1}{4}\ \checkmark$$
Solving applied problems with proportions, p. 404.	1. Organize the data. 2. Write a proportion equating the respective parts. Let x represent the value that is not known. 3. Solve the proportion.	Renee can make five cherry pies with 3 cups of flour. How many cups of flour does she need to make eight cherry pies? $$\dfrac{5\text{ cherry pies}}{3\text{ cups flour}}=\dfrac{8\text{ cherry pies}}{x\text{ cups flour}}$$ $$\dfrac{5}{3}=\dfrac{8}{x}$$ $$5x=24$$ $$x=\dfrac{24}{5}$$ $$x=4\dfrac{4}{5}$$ $4\dfrac{4}{5}$ cups of flour are needed for eight cherry pies.

Chapter 7 Review Problems

7.1 *Simplify.*

1. $\dfrac{4x - 4y}{5y - 5x}$

2. $\dfrac{bx}{bx - by}$

3. $\dfrac{2x^2 + 5x - 3}{2x^2 - 9x + 4}$

4. $\dfrac{3x^2 + 7x + 2}{3x^2 + 13x + 4}$

5. $\dfrac{x^2 - 9}{x^2 - 10x + 21}$

6. $\dfrac{2x^2 + 18x + 40}{3x + 15}$

7. $\dfrac{4x^2 + 4x - 3}{4x^2 - 2x}$

8. $\dfrac{x^3 + 3x^2}{x^3 - 2x^2 - 15x}$

9. $\dfrac{2x^2 - 2xy - 24y^2}{2x^2 + 5xy - 3y^2}$

10. $\dfrac{4 - y^2}{3y^2 + 5y - 2}$

11. $\dfrac{5x^3 - 10x^2}{25x^4 + 5x^3 - 30x^2}$

12. $\dfrac{16x^2 - 4y^2}{4x - 2y}$

7.2 *Multiply or divide.*

13. $\dfrac{3x^2 - 13x - 10}{3x^2 + 2x} \cdot \dfrac{x^2 - 25x}{x^2 - 25}$

14. $\dfrac{2y^2 - 18}{3y^2 + 3y} \div \dfrac{y^2 + 6y + 9}{y^2 + 4y + 3}$

15. $\dfrac{2y^2 + 3y - 2}{2y^2 + y - 1} \div \dfrac{2y^2 + y - 1}{2y^2 - 3y - 2}$

16. $\dfrac{6y^2 + 13y - 5}{9y^2 + 3y} \div \dfrac{4y^2 + 20y + 25}{12y^2}$

17. $\dfrac{3xy^2 + 12y^2}{2x^2 - 11x + 5} \div \dfrac{2xy + 8y}{8x^2 + 2x - 3}$

18. $\dfrac{11}{x - 2} \cdot \dfrac{2x^2 - 8}{44}$

19. $\dfrac{2x^2 + 10x + 2}{8x - 8} \cdot \dfrac{3x - 3}{4x^2 + 20x + 4}$

20. $\dfrac{x^2 - 5xy - 24y^2}{2x^2 - 2xy - 24y^2} \cdot \dfrac{4x^2 + 4xy - 24y^2}{x^2 - 10xy + 16y^2}$

7.3 *Add or subtract.*

21. $\dfrac{7}{x + 1} + \dfrac{4}{2x}$

22. $5 + \dfrac{1}{x} + \dfrac{1}{x + 1}$

23. $\dfrac{2}{x^2 - 9} + \dfrac{x}{x + 3}$

24. $\dfrac{7}{x + 2} + \dfrac{3}{x - 4}$

25. $\dfrac{x}{y} + \dfrac{3}{2y} + \dfrac{1}{y + 2}$

26. $\dfrac{4}{a} + \dfrac{2}{b} + \dfrac{3}{a + b}$

27. $\dfrac{3x + 1}{3x} - \dfrac{1}{x}$

28. $\dfrac{x + 4}{x + 2} - \dfrac{1}{2x}$

29. $\dfrac{1}{x^2 + 7x + 10} - \dfrac{x}{x + 5}$

30. $\dfrac{27}{x^2 - 81} + \dfrac{3}{2(x + 9)}$

7.4 *Simplify.*

31. $\dfrac{\dfrac{3}{2y} - \dfrac{1}{y}}{\dfrac{4}{y} + \dfrac{3}{2y}}$

32. $\dfrac{\dfrac{2}{x} + \dfrac{1}{2x}}{x + \dfrac{x}{2}}$

33. $\dfrac{w - \dfrac{4}{w}}{1 + \dfrac{2}{w}}$

34. $\dfrac{1 - \dfrac{w}{w - 1}}{1 + \dfrac{w}{1 - w}}$

35. $\dfrac{1 + \dfrac{1}{y^2 - 1}}{\dfrac{1}{y + 1} - \dfrac{1}{y - 1}}$

36. $\dfrac{\dfrac{1}{y} + \dfrac{1}{x + y}}{1 + \dfrac{2}{x + y}}$

37. $\dfrac{\dfrac{1}{a + b} - \dfrac{1}{a}}{b}$

38. $\dfrac{\dfrac{2}{a + b} - \dfrac{3}{b}}{\dfrac{1}{a + b}}$

39. $\left(\dfrac{1}{x + 2y} - \dfrac{1}{x - y} \right) \div \dfrac{2x - 4y}{x^2 - 3xy + 2y^2}$

40. $\dfrac{x + 5y}{x - 6y} \div \left(\dfrac{1}{5y} - \dfrac{1}{x + 5y} \right)$

7.5 *Solve for the variable. If there is no solution, say so.*

41. $\dfrac{8}{a - 3} = \dfrac{12}{a + 3}$

42. $\dfrac{8a - 1}{6a + 8} = \dfrac{3}{4}$

43. $\dfrac{2x - 1}{x} - \dfrac{1}{2} = -2$

44. $\dfrac{5 - x}{x} - \dfrac{7}{x} = -\dfrac{3}{4}$

45. $\dfrac{5}{2} - \dfrac{2y + 7}{y + 6} = 3$

46. $\dfrac{5}{4} - \dfrac{1}{2x} = \dfrac{1}{x} + 2$

47. $\dfrac{7}{8x} - \dfrac{3}{4} = \dfrac{1}{4x} + \dfrac{1}{2}$

48. $\dfrac{1}{3x} + 2 = \dfrac{5}{6x} - \dfrac{1}{2}$

49. $\dfrac{3}{y - 3} = \dfrac{3}{2} + \dfrac{y}{y - 3}$

50. $\dfrac{x - 8}{x - 2} = \dfrac{2x}{x + 2} - 2$

51. $\dfrac{3y - 1}{3y} - \dfrac{6}{5y} = \dfrac{1}{y} - \dfrac{4}{15}$

52. $\dfrac{9}{2} - \dfrac{7y - 4}{y + 2} = -\dfrac{1}{4}$

53. $\dfrac{y + 18}{y^2 - 16} = \dfrac{y}{y + 4} - \dfrac{y}{y - 4}$

54. $\dfrac{4}{x^2 - 1} = \dfrac{2}{x - 1} + \dfrac{2}{x + 1}$

55. $\dfrac{9y - 3}{y^2 + 2y} - \dfrac{5}{y + 2} = \dfrac{3}{y}$

56. $\dfrac{2}{3 - 3y} + \dfrac{2}{2y - 1} = \dfrac{4}{3y - 3}$

7.6 *Solve. Round to the nearest tenth.*

57. $\dfrac{8}{5} = \dfrac{2}{x}$

58. $\dfrac{x}{4} = \dfrac{12}{17}$

59. $\dfrac{33}{10} = \dfrac{x}{8}$

60. $\dfrac{5}{x} = \dfrac{22}{9}$

61. $\dfrac{13.5}{0.6} = \dfrac{360}{x}$

62. $\dfrac{2\frac{1}{2}}{3\frac{1}{4}} = \dfrac{7}{x}$

Use a proportion to answer each question.

▲ **63.** A 5-gallon can of paint will cover 240 square feet. How many gallons of paint will be needed to cover 400 square feet? Round to the nearest tenth of a gallon.

64. Aunt Lexie uses 3 pounds of sugar to make 100 cookies. How many cookies can she make with 5 pounds of sugar? Round to the nearest whole cookie.

65. Ron found that his car used 7 gallons of gas to travel 200 miles. He plans to drive 1300 miles from his home to Denver, Colorado. How many gallons of gas will he use if his car continues to consume gas at the same rate? Round to the nearest gallon.

66. The distance on a map between two cities is 4 inches. The actual distance between these cities is 122 miles. Two lakes are 3 inches apart on the same map. How many miles apart are these two lakes?

67. A train travels 180 miles in the same time that a car travels 120 miles. The speed of the train is 20 miles per hour faster than the speed of the car. Find the speed of the train and the speed of the car.

68. A professional painter can paint the interior of the Jacksons' house in 5 hours. John Jackson can do the same job in 8 hours. How long would it take these two people working together on the painting job? Round to the nearest minute.

▲ **69.** A flagpole that is 8 feet tall casts a shadow of 3 feet. At the same time of day, a tall office building in the city casts a shadow of 450 feet. How tall is the office building?

▲ **70.** Mary takes a walk across a canyon in New Mexico. She stands 5.75 feet tall and her shadow is 3 feet long. At the same time, the shadow from the peak of the canyon wall casts a shadow that is 95 feet long. How tall is the peak of the canyon? Round to the nearest foot.

71. Fred is an experienced painter. He can paint the sides of an average house in 5 hours. His new assistant is still being trained. It takes the assistant 10 hours to paint the sides of an average house. How long would it take Fred and his assistant to paint the sides of an average house if they worked together?

72. Sally runs the family farm in Boone, Iowa. She can plow the fields of the farm in 20 hours. Her daughter Brenda can plow the fields of the farm in 30 hours. If they have two identical tractors, how long would it take Brenda and Sally to plow the fields of the farm if they worked together?

Perform the indicated operation. Simplify.

1. $\dfrac{2ac + 2ad}{3a^2c + 3a^2d}$

2. $\dfrac{8x^2 - 2x^2y^2}{y^2 + 4y + 4}$

3. $\dfrac{x^2 + 2x}{2x - 1} \cdot \dfrac{10x^2 - 5x}{12x^3 + 24x^2}$

4. $\dfrac{x + 2y}{12y^2} \cdot \dfrac{4y}{x^2 + xy - 2y^2}$

5. $\dfrac{2a^2 - 3a - 2}{4a^2 + a - 14} \div \dfrac{2a^2 + 5a + 2}{16a^2 - 49}$

6. $\dfrac{3}{x^2 + x - 6} + \dfrac{1}{x^2 + 3x - 10}$

7. $\dfrac{x - y}{xy} - \dfrac{a - y}{ay}$

8. $\dfrac{3x}{x^2 - 3x - 18} - \dfrac{x - 4}{x - 6}$

9. $\dfrac{\dfrac{x}{3y} - \dfrac{1}{2}}{\dfrac{4}{3y} - \dfrac{2}{x}}$

10. $\dfrac{\dfrac{2}{x + 3} + \dfrac{1}{x}}{3x + 9}$

11. $\dfrac{2x^2 + 3xy - 9y^2}{4x^2 + 13xy + 3y^2}$

12. $\dfrac{1}{x + 4} - \dfrac{2}{x^2 + 6x + 8}$

In questions 13–18, solve for x. Check your answers. If there is no solution, say so.

13. $\dfrac{15}{x} + \dfrac{9x - 7}{x + 2} = 9$

14. $\dfrac{x - 3}{x - 2} = \dfrac{2x^2 - 15}{x^2 + x - 6} - \dfrac{x + 1}{x + 3}$

15. $3 - \dfrac{7}{x + 3} = \dfrac{x - 4}{x + 3}$

16. $\dfrac{3}{3x - 5} = \dfrac{7}{5x + 4}$

17. $\dfrac{9}{x} = \dfrac{13}{5}$

18. $\dfrac{9.3}{2.5} = \dfrac{x}{10}$

19. Katie is typing a term paper for her English class. She typed 3 pages of the paper in 55 minutes. If she continues at the same rate, how long will it take her to type the entire 21-page paper? Express your answer in hours and minutes.

20. In northern Michigan the Gunderson family heats their home with firewood. They used $100 worth of wood in 25 days. Mr. Gunderson estimates that he needs to burn wood at that rate for about 92 days during the winter. If that is so, how much will the 92-day supply of wood cost?

▲ **21.** A hiking club is trying to construct a rope bridge across a canyon. A 6-foot construction pole held upright casts a 7-foot shadow. At the same time of day, a tree at the edge of the canyon casts a shadow that exactly covers the distance that is needed for the rope bridge. The tree is exactly 87 feet tall. How long should the rope bridge be? Round to the nearest foot.

1. _____

2. _____

3. _____

4. _____

5. _____

6. _____

7. _____

8. _____

9. _____

10. _____

11. _____

12. _____

13. _____

14. _____

15. _____

16. _____

17. _____

18. _____

19. _____

20. _____

21. _____

1. _____

2. _____

3. _____

4. _____

5. _____

6. _____

7. _____

8. _____

9. _____

10. _____

11. _____

Approximately one-half of this test covers the content of Chapters 1–6.
The remainder covers the content of Chapter 7.

1. Solve. $5(x - 3) - 2(4 - 2x) = 7(x - 1) - (x - 2)$

2. Solve for h. $A = \pi r^2 h$

3. Solve and graph on a number line. $4(2 - x) < 3$

4. Solve for x. $\dfrac{1}{4} x + \dfrac{3}{4} < \dfrac{2}{3} x - \dfrac{4}{3}$

5. Graph. $y = 2x - 1$

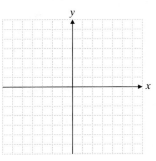

6. Find an equation of the line that passes through $(-3, 0)$ and is parallel to $y = x + 5$.

7. Solve the system.
$$x - y - z = 8$$
$$2y - z = 1$$
$$3x + y + z = -4$$

8. Simplify. $(-3x^2 y)(4x^3 y^6)$

9. Multiply. $(2x - 7)^2$

10. Factor. $3ax + 3bx - 2ay - 2by$

11. Simplify. $\dfrac{4x^2 - 25}{2x^2 + 9x - 35}$

Perform the indicated operations.

12. $\dfrac{x^2 - 4}{x^2 - 25} \cdot \dfrac{3x^2 - 14x - 5}{3x^2 + 6x}$

13. $\dfrac{2x^2 - 9x + 9}{8x - 12} \div \dfrac{x^2 - 3x}{2x}$

14. $\dfrac{5}{2x + 4} + \dfrac{3}{x - 3}$

15. $\dfrac{2x + 1}{x^2 + x - 12} - \dfrac{x - 1}{x^2 - x - 6}$

Solve for x.

16. $\dfrac{3x - 2}{3x + 2} = 2$

17. $\dfrac{x - 3}{x} = \dfrac{x + 2}{x + 3}$

In questions 18 and 19, simplify.

18. $\dfrac{\dfrac{1}{x - 3} + \dfrac{5}{x^2 - 9}}{\dfrac{6x}{x + 3}}$

19. $\dfrac{\dfrac{3}{a} + \dfrac{2}{b}}{\dfrac{5}{a^2} - \dfrac{2}{b^2}}$

20. Solve for x. $\dfrac{5}{7} = \dfrac{14}{x}$

21. Jane is looking at a road map. The distance between two cities is 130 miles. This distance is represented by $2\frac{1}{2}$ inches on the map. She sees that the distance she has to drive today is on a totally straight interstate highway 4 inches long on the map. How many miles does she have to drive?

22. Roberto is working as a telemarketing salesperson for a large corporation. In the last 22 telephone calls he has made, he was able to make a sale five times. His goal is to make 110 sales this month. If his rate for making a sale continues as in the past, how many phone calls must he make?

12. _____

13. _____

14. _____

15. _____

16. _____

17. _____

18. _____

19. _____

20. _____

21. _____

22. _____

Rational Exponents and Radicals

Y ou have probably seen a picture of a sailor standing in the crow's nest of a tall sailing ship and looking out over the horizon. Do you have any idea how far that sailor is able to see? Do you know that the distance can be estimated with a mathematical equation that sailors have used for over 200 years? Turn to the Putting Your Skills to Work exercises on page 479 to find out how this is done.

If you are familiar with the topics in this chapter, take this test now. Check your answers with those in the back of the book. If an answer is wrong or you can't do an exercise, study the appropriate section of the chapter.

If you are not familiar with the topics in this chapter, don't take this test now. Instead, study the examples, work the practice exercises, and then take the test.

This test will help you identify those concepts that you have mastered and those that need more study.

Section 8.1

1. Multiply and simplify your answer. $(-3x^{1/4}y^{1/2})(-2x^{-1/2}y^{1/3})$

Simplify.

2. $(-4x^{-1/4}y^{1/3})^3$

3. $\dfrac{-18x^{-2}y^2}{-3x^{-5}y^{1/3}}$

4. $\left(\dfrac{27x^2y^{-5}}{x^{-4}y^4}\right)^{2/3}$

Section 8.2

Evaluate.

5. $27^{-4/3}$

6. $\sqrt{169} + \sqrt[3]{-64}$

7. $\sqrt[3]{27a^{12}b^6c^{15}}$

Section 8.3

8. Simplify $\sqrt[4]{32x^8y^{15}}$.

9. Combine like terms where possible. $3\sqrt{48y^3} - 2\sqrt[3]{16} + 3\sqrt[3]{54} - 5y\sqrt{12y}$

Section 8.4

10. Multiply and simplify $(3\sqrt{3} - 5\sqrt{6})(\sqrt{12} - 3\sqrt{6})$.

11. Rationalize the denominator and simplify your answer. $\dfrac{6}{\sqrt[3]{9x}}$

12. Rationalize the denominator and simplify your answer. $\dfrac{\sqrt{2} + \sqrt{3}}{\sqrt{2} - \sqrt{3}}$

Section 8.5

Solve and check your solution(s).

13. $\sqrt{3x + 4} + 2 = x$

14. $\sqrt{2x + 3} - \sqrt{x - 2} = 2$

Section 8.6

Perform the operations indicated.

15. $(3 - 2i) - (-1 + 3i)$

16. $i^{15} + \sqrt{-25}$

17. $(3 + 5i)^2$

18. $\dfrac{3 + 2i}{2 + 3i}$

Section 8.7

19. If y varies directly with x^2, and $y = 18$ when $x = 3$, find the value of y when $x = 5$.

20. If y varies inversely with x, and $y = 12$ when $x = 6$, find the value of y when $x = 10$.

1. _____

2. _____

3. _____

4. _____

5. _____

6. _____

7. _____

8. _____

9. _____

10. _____

11. _____

12. _____

13. _____

14. _____

15. _____

16. _____

17. _____

18. _____

19. _____

20. _____

8.1 Rational Exponents

1 Simplifying Expressions with Rational Exponents

Student Learning Objectives

After studying this section, you will be able to apply the laws of exponents to:

1 Simplify expressions with rational exponents.

2 Add expressions with rational exponents.

3 Factor expressions with rational exponents.

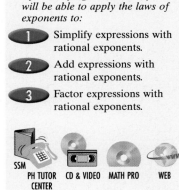

SSM   PH TUTOR   CD & VIDEO   MATH PRO   WEB
CENTER

Before studying this section, you may need to review Sections 4.1 and 4.2. For convenience, we list the rules of exponents that we learned there.

$$x^m x^n = x^{m+n} \qquad\qquad x^0 = 1$$

$$\frac{x^m}{x^n} = x^{m-n} \qquad\qquad (x^m)^n = x^{mn}$$

$$x^{-n} = \frac{1}{x^n} \qquad\qquad (xy)^n = x^n y^n$$

$$\frac{x^{-n}}{y^{-m}} = \frac{y^m}{x^n} \qquad\qquad \left(\frac{x}{y}\right)^n = \frac{x^n}{y^n}$$

To ensure that you understand these rules, study Example 1 carefully and work Practice Problem 1.

EXAMPLE 1 Simplify $\left(\dfrac{5xy^{-3}}{2x^{-4}y}\right)^{-2}$.

$$\left(\frac{5xy^{-3}}{2x^{-4}y}\right)^{-2} = \frac{(5xy^{-3})^{-2}}{(2x^{-4}y)^{-2}} \qquad \left(\frac{x}{y}\right)^n = \frac{x^n}{y^n}.$$

$$= \frac{5^{-2}x^{-2}(y^{-3})^{-2}}{2^{-2}(x^{-4})^{-2}y^{-2}} \qquad (xy)^n = x^n y^n.$$

$$= \frac{5^{-2}x^{-2}y^6}{2^{-2}x^8 y^{-2}} \qquad (x^m)^n = x^{mn}.$$

$$= \frac{5^{-2}}{2^{-2}} \cdot \frac{x^{-2}}{x^8} \cdot \frac{y^6}{y^{-2}}$$

$$= \frac{2^2}{5^2} \cdot x^{-2-8} \cdot y^{6+2} \qquad \frac{x^{-n}}{y^{-m}} = \frac{y^m}{x^n}; \frac{x^m}{x^n} = x^{m-n}.$$

$$= \frac{4}{25} x^{-10}y^8$$

The answer can also be written as $\dfrac{4y^8}{25x^{10}}$. Explain why.

Practice Problem 1 Simplify $\left(\dfrac{3x^{-2}y^4}{2x^{-5}y^2}\right)^{-3}$.

Sidelight Deciding when to use the rule $\dfrac{x^{-n}}{y^{-m}} = \dfrac{y^m}{x^n}$ is entirely up to you. In Example 1, we could have begun by writing

$$\left(\frac{5xy^{-3}}{2x^{-4}y}\right)^{-2} = \left(\frac{5x \cdot x^4}{2y \cdot y^3}\right)^{-2} = \left(\frac{5x^5}{2y^4}\right)^{-2}.$$

Complete the steps to simplify this expression.

Likewise, in the fourth step in Example 1, we could have written

$$\frac{5^{-2}x^{-2}y^6}{2^{-2}x^8 y^{-2}} = \frac{2^2 y^6 y^2}{5^2 x^8 x^2}.$$

Complete the steps to simplify this expression. Are the two answers the same as the answer in Example 1? Why or why not?

We generally begin to simplify a rational expression with exponents by raising a power to a power because sometimes negative powers become positive. The order in which you use the rules of exponents is up to you. Work carefully. Keep track of your exponents and where you are as you simplify the rational expression.

These rules for exponents can also be extended to include rational exponents—that is, exponents that are fractions. As you recall, rational numbers are of the form $\frac{a}{b}$, where a and b are integers and b does not equal zero. We will write fractional exponents using diagonal lines. Thus, we will write $\frac{5}{6}$ as 5/6 and $\frac{a}{b}$ as a/b throughout this chapter when writing fractional exponents. For now we restrict the base to *positive* real numbers. Later we will talk about negative bases.

EXAMPLE 2   Simplify.

(a) $\left(x^{2/3}\right)^4$ **(b)** $\dfrac{x^{5/6}}{x^{1/6}}$ **(c)** $x^{2/3} \cdot x^{-1/3}$ **(d)** $5^{3/7} \cdot 5^{2/7}$

We will not write out every step or every rule of exponents that we use. You should be able to follow the solutions.

(a) $\left(x^{2/3}\right)^4 = x^{(2/3)(4/1)} = x^{8/3}$ **(b)** $\dfrac{x^{5/6}}{x^{1/6}} = x^{5/6-1/6} = x^{4/6} = x^{2/3}$

(c) $x^{2/3} \cdot x^{-1/3} = x^{2/3-1/3} = x^{1/3}$ **(d)** $5^{3/7} \cdot 5^{2/7} = 5^{3/7+2/7} = 5^{5/7}$

Practice Problem 2   Simplify.

(a) $\left(x^4\right)^{3/8}$ **(b)** $\dfrac{x^{3/7}}{x^{2/7}}$ **(c)** $x^{-7/5} \cdot x^{4/5}$

Sometimes fractional exponents will not have the same denominator. Remember that you need to change the fractions to equivalent fractions with the same denominator when the rule of exponents requires you to add or to subtract them.

EXAMPLE 3   Simplify. Express your answers with positive exponents only.

(a) $\left(2x^{1/2}\right)\left(3x^{1/3}\right)$ **(b)** $\dfrac{18x^{1/4}y^{-1/3}}{-6x^{-1/2}y^{1/6}}$

(a) $\left(2x^{1/2}\right)\left(3x^{1/3}\right) = 6x^{1/2+1/3} = 6x^{3/6+2/6} = 6x^{5/6}$

(b) $\dfrac{18x^{1/4}y^{-1/3}}{-6x^{-1/2}y^{1/6}} = -3x^{1/4-(-1/2)}y^{-1/3-1/6}$

$$= -3x^{1/4+2/4}y^{-2/6-1/6}$$
$$= -3x^{3/4}y^{-3/6}$$
$$= -3x^{3/4}y^{-1/2}$$
$$= \dfrac{-3x^{3/4}}{y^{1/2}}$$

Practice Problem 3   Simplify. Express your answers with positive exponents only.

(a) $\left(-3x^{1/4}\right)\left(2x^{1/2}\right)$ **(b)** $\dfrac{13x^{1/12}y^{-1/4}}{26x^{-1/3}y^{1/2}}$

EXAMPLE 4 Multiply and simplify $-2x^{5/6}(3x^{1/2} - 4x^{-1/3})$.

We will need to be very careful when we add the exponents for x as we use the distributive property. Study each step of the following example. Be sure you understand each operation.

$$-2x^{5/6}(3x^{1/2} - 4x^{-1/3}) = -6x^{5/6+1/2} + 8x^{5/6-1/3}$$

$$= -6x^{5/6+3/6} + 8x^{5/6-2/6}$$

$$= -6x^{8/6} + 8x^{3/6}$$

$$= -6x^{4/3} + 8x^{1/2}$$

Practice Problem 4 Multiply and simplify $-3x^{1/2}(2x^{1/4} + 3x^{-1/2})$.

Sometimes we can use the rules of exponents to simplify numerical values raised to rational powers.

EXAMPLE 5 Evaluate: **(a)** $(25)^{3/2}$ **(b)** $(27)^{2/3}$

(a) $(25)^{3/2} = (5^2)^{3/2} = 5^{2/1 \cdot 3/2} = 5^3 = 125$

(b) $(27)^{2/3} = (3^3)^{2/3} = 3^{3/1 \cdot 2/3} = 3^2 = 9$

Practice Problem 5 Evaluate: **(a)** $(4)^{5/2}$ **(b)** $(27)^{4/3}$

2 *Adding Expressions with Rational Exponents*

Adding expressions with rational exponents may require several steps. Sometimes this involves removing negative exponents. For example, to add $2x^{-1/2} + x^{1/2}$, we begin by writing $2x^{-1/2}$ as $\dfrac{2}{x^{1/2}}$. This is a rational expression. Recall that to add rational expressions we need to have a common denominator. Take time to look at the steps needed to write $2x^{-1/2} + x^{1/2}$ as one term.

EXAMPLE 6 Write as one fraction with positive exponents. $2x^{-1/2} + x^{1/2}$

$$2x^{-1/2} + x^{1/2} = \frac{2}{x^{1/2}} + \frac{x^{1/2} \cdot x^{1/2}}{x^{1/2}} = \frac{2}{x^{1/2}} + \frac{x^1}{x^{1/2}} = \frac{2 + x}{x^{1/2}}$$

Practice Problem 6 Write as one fraction with only positive exponents. $3x^{1/3} + x^{-1/3}$

3 *Factoring Expressions with Rational Exponents*

To factor expressions, we need to be able to recognize common factors. If the terms of the expression contain exponents, we look for the same exponential factor in each term. For example, in the expression $6x^5 + 4x^3 - 8x^2$, the common factor of each term is $2x^2$. Thus, we can factor out the common factor $2x^2$ from each term. The expression then becomes $2x^2(3x^3 + 2x - 4)$.

We do exactly the same thing when we factor expressions with rational exponents. The key is to identify the exponent of the common factor. In the expression $6x^{3/4} + 4x^{1/2} - 8x^{1/4}$, the common factor is $2x^{1/4}$. Thus, we factor the expression $6x^{3/4} + 4x^{1/2} - 8x^{1/4}$ as $2x^{1/4}(3x^{1/2} + 2x^{1/4} - 4)$. We do not always need to factor out the greatest common factor. In the following examples we simply factor out a common factor.

EXAMPLE 7 Factor out the common factor of $2x$ from $2x^{3/2} + 4x^{5/2}$.

We rewrite the exponent of each term so that we can see that each term contains the factor $2x$ or $2x^{2/2}$.

$$2x^{3/2} + 4x^{5/2} = 2x^{2/2+1/2} + 4x^{2/2+3/2}$$

$$= 2(x^{2/2})(x^{1/2}) + 4(x^{2/2})(x^{3/2})$$

$$= 2x(x^{1/2} + 2x^{3/2})$$

Practice Problem 7 Factor out the common factor of $4y$ from $4y^{3/2} - 8y^{5/2}$.

For convenience we list here the properties of exponents that we have discussed in this section, as well as the property $x^0 = 1$.

When x and y are **positive real numbers** and a and b are **rational numbers**:

$$x^a x^b = x^{a+b} \qquad \frac{x^a}{x^b} = x^{a-b} \qquad x^0 = 1$$

$$x^{-a} = \frac{1}{x^a} \qquad \frac{x^{-a}}{y^{-b}} = \frac{y^b}{x^a}$$

$$(x^a)^b = x^{ab} \qquad (xy)^a = x^a y^a \qquad \left(\frac{x}{y}\right)^a = \frac{x^a}{y^a}$$

Simplify.

1. $\left(\dfrac{4x^2y^{-3}}{x}\right)^2$

2. $\left(\dfrac{3xy^{-2}}{x^3}\right)^2$

3. $\left(\dfrac{2a^{-1}b^3}{-3b^2}\right)^3$

4. $\left(\dfrac{-a^{-2}b}{5b^2}\right)^2$

5. $\left(x^{3/4}\right)^2$

6. $\left(x^{4/3}\right)^6$

7. $\left(y^{12}\right)^{2/3}$

8. $\left(y^2\right)^{5/2}$

9. $\dfrac{x^{7/12}}{x^{1/12}}$

10. $\dfrac{x^{7/8}}{x^{3/8}}$

11. $\dfrac{x^3}{x^{1/2}}$

12. $\dfrac{x^2}{x^{1/3}}$

13. $x^{1/7} \cdot x^{3/7}$

14. $x^{3/5} \cdot x^{1/5}$

15. $y^{3/5} \cdot y^{-1/10}$

16. $y^{7/10} \cdot y^{-1/5}$

Write each expression with positive exponents.

17. $x^{-3/4}$

18. $x^{-5/6}$

19. $a^{-5/6}b^{1/3}$

20. $2a^{-1/6}b^{3/4}$

21. $6^{-1/2}$

22. $4^{-1/3}$

23. $2a^{-1/4}$

24. $3^{-2/5} \cdot 2^{1/3}$

Mixed Practice

Simplify and express your answers with positive exponents. Evaluate or simplify the numerical expressions.

25. $\left(x^{1/2}y^{1/3}\right)\left(x^{1/3}y^{2/3}\right)$

26. $\left(x^{-1/3}y^{2/3}\right)\left(x^{1/3}y^{1/4}\right)$

27. $\left(7x^{1/3}y^{1/4}\right)\left(-2x^{1/4}y^{-1/6}\right)$

28. $\left(8x^{-1/5}y^{1/3}\right)\left(-3x^{-1/4}y^{1/6}\right)$

29. $6^2 \cdot 6^{-2/3}$

30. $11^{1/2} \cdot 11^3$

31. $\dfrac{2x^{1/5}}{x^{-1/2}}$

32. $\dfrac{3y^{2/3}}{y^{-1/4}}$

33. $\dfrac{-20x^2y^{-1/5}}{5x^{-1/2}y}$

34. $\dfrac{12x^{-2/3}y}{-6xy^{-3/4}}$

35. $\left(\dfrac{8a^2b^6}{a^{-1}b^3}\right)^{1/3}$

36. $\left(\dfrac{16a^5b^{-2}}{a^{-1}b^{-6}}\right)^{1/2}$

37. $\left(-3x^{2/5}y^{3/2}z^{1/3}\right)^2$ **38.** $\left(5x^{-1/2}y^{1/3}z^{4/5}\right)^3$ **39.** $x^{2/3}\left(x^{4/3} - x^{1/5}\right)$

40. $y^{-2/3}\left(y^{2/3} + y^{3/2}\right)$ **41.** $m^{7/8}\left(m^{-1/2} + 2m\right)$ **42.** $\dfrac{\left(x^{-1/6}x\right)^{3/2}}{x^2}$

43. $\dfrac{\left(x^2 \cdot x^{-3/2}\right)^{1/2}}{x^{1/2}}$ **44.** $(25)^{1/2}$ **45.** $(27)^{2/3}$

46. $(16)^{3/4}$ **47.** $(4)^{3/2}$ **48.** $9^{3/2} + 4^{1/2}$

49. $(81)^{3/4} + (25)^{1/2}$

Write each expression as one fraction with positive exponents.

50. $3y^{1/2} + y^{-1/2}$ **51.** $2y^{1/3} + y^{-2/3}$ **52.** $x^{-1/3} + 6^{4/3}$ **53.** $5^{-1/4} + x^{-1/2}$

Factor out the common factor of 2a.

54. $10a^{5/4} - 4a^{8/5}$ **55.** $6a^{4/3} - 8a^{3/2}$

To Think About

56. What is the value of a if $x^a \cdot x^{1/4} = x^{-1/8}$?   **57.** What is the value of b if $x^b \div x^{1/3} = x^{-1/12}$?

Applications

The radius needed to create a sphere with a given volume V can be approximated by the equation $r = 0.62(V)^{1/3}$. Find the radius of the spheres with the following volumes.

▲ **58.** 27 cubic meters   ▲ **59.** 64 cubic meters

The radius required for a cone to have a volume V and a height h is given by the equation

$$r = \left(\frac{3V}{\pi h}\right)^{1/2}.$$

Find the necessary radius to have a cone with the properties below. Use $\pi \approx 3.14$.

▲ **60.** $V = 314$ cubic feet and $h = 12$ feet.

▲ **61.** $V = 3140$ cubic feet and $h = 30$ feet.

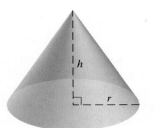

Cumulative Review Problems

Solve for x.

62. $-4(x + 1) = \frac{1}{3}(3 - 2x)$

Solve for b.

63. $A = \frac{h}{2}(a + b)$

Giving a young patient the wrong amount of medication can have serious and even fatal consequences. A formula used by doctors, nurses, and pharmacists to verify the correct dosage of a prescription drug for a child is

$$y = \frac{ax}{a + 12},$$

where y = the child dosage, x = the adult dosage, and a = the age of the child in years.

64. If the adult dosage of a medication is 400 milligrams, how much should a 7-year-old child receive? Round your answer to the nearest milligram.

65. The adult dosage of a medication is 250 milligrams, and a certain child was assigned the correct dosage level of 75 milligrams. How old was the child? Round your answer to the nearest year.

8.2 Radical Expressions and Functions

1 Evaluating Radical Expressions and Functions

How long is the side of a square whose area is 9? Recall the formula for the area of a square.

$$\text{area of a square} = s^2$$

Our question then becomes, what number times itself is 9?

$$s^2 = 9$$
$$s = 3 \quad \text{(because } (3)(3) = 9)$$
$$s = -3 \quad \text{(because } (-3)(-3) = 9)$$

We say that 3 is a **square root** of 9 because $(3)(3) = 9$. We can also say that -3 is a square root of 9 because $(-3)(-3) = 9$. Note that 9 is a **perfect square**. A square root of a perfect square is an integer.

The symbol $\sqrt{}$ is called a **radical sign** and is used for the **principal square root** of a number, which is the nonnegative square root. It is also used to denote positive higher-order roots. A negative square root is written $-\sqrt{}$. Thus, we have the following:

$$\sqrt{9} = 3 \qquad -\sqrt{9} = -3$$
$$\sqrt{64} = 8 \quad \text{(because } 8 \cdot 8 = 64)$$
$$\sqrt{121} = 11 \quad \text{(because } 11 \cdot 11 = 121)$$

Because $\sqrt{9} = \sqrt{3 \cdot 3} = \sqrt{3^2} = 3$, we can say the following:

> ### Definition of Square Root
>
> If x is a nonnegative real number, then $\sqrt{x}$ is the *nonnegative* (or principal) *square root* of x; in other words, $\left(\sqrt{x}\right)^2 = x$.

Note that x must be *nonnegative*. Why? Suppose we want to find $\sqrt{-36}$. We must find a number that when multiplied by itself gives -36. Is there one? No, because

$$6 \cdot 6 = 36 \quad \text{and}$$
$$(-6)(-6) = 36.$$

So there is no real number that we can square to get -36.

We call $\sqrt[n]{x}$ a **radical expression**. The $\sqrt{}$ symbol is the radical sign, the x is the **radicand**, and the n is the **index** of the radical. When no number for n appears in the radical expression, it is understood that 2 is the index, which means that we are looking for the square root. For example, in the radical expression $\sqrt{25}$, with no number given for the index n we take the index to be 2. Thus, $\sqrt{25}$ is the principal square root of 25.

We can extend the notion of square root to **higher-order roots**, such as cube roots, fourth roots, and so on. A **cube root** of a number is a value that when cubed is equal to the original number. The index n of the radical is 3, and the radical used is $\sqrt[3]{}$. Similarly, a **fourth root** of a number is a value that when raised to the fourth power is equal to the original number. The index n of the radical is 4, and the radical used is $\sqrt[4]{}$. Thus, we have the following:

$$\sqrt[3]{27} = 3 \qquad \text{because } 3 \cdot 3 \cdot 3 = 3^3 = 27.$$
$$\sqrt[3]{8} = 2 \qquad \text{because } 2 \cdot 2 \cdot 2 = 2^3 = 8.$$
$$\sqrt[4]{81} = 3 \qquad \text{because } 3 \cdot 3 \cdot 3 \cdot 3 = 3^4 = 81.$$
$$\sqrt[5]{32} = 2 \qquad \text{because } 2 \cdot 2 \cdot 2 \cdot 2 \cdot 2 = 2^5 = 32.$$
$$\sqrt[3]{-64} = -4 \qquad \text{because } (-4)(-4)(-4) = (-4)^3 = -64.$$

You should be able to see a pattern here.

$$\sqrt[3]{27} = \sqrt[3]{3^3} = 3 \qquad\qquad \sqrt[6]{729} = \sqrt[6]{3^6} = 3$$
$$\sqrt[4]{81} = \sqrt[4]{3^4} = 3 \qquad\qquad \sqrt[3]{-64} = \sqrt[3]{(-4)^3} = -4$$
$$\sqrt[5]{32} = \sqrt[5]{2^5} = 2$$

In these cases, we see that $\sqrt[n]{x^n} = x$. We now give the following definition.

Definition of Higher-Order Roots

1. If x is a *nonnegative* real number, then $\sqrt[n]{x}$ is a nonnegative nth root and has the property that

$$\left(\sqrt[n]{x}\right)^n = x.$$

2. If x is a *negative* real number, then
   (a) $\left(\sqrt[n]{x}\right)^n = x$ when n is an *odd integer*.
   (b) $\left(\sqrt[n]{x}\right)^n$ is *not* a real number when n is an *even integer*.

EXAMPLE 1 If possible, find the root of each negative number. If there is no real number root, say so.

(a) $\sqrt[3]{-216}$ (b) $\sqrt[5]{-32}$ (c) $\sqrt[4]{-16}$ (d) $\sqrt[6]{-64}$

(a) $\sqrt[3]{-216} = \sqrt[3]{(-6)^3} = -6$ (b) $\sqrt[5]{-32} = \sqrt[5]{(-2)^5} = -2$

(c) $\sqrt[4]{-16}$ is not a real number because n is even and x is negative.

(d) $\sqrt[6]{-64}$ is not a real number because n is even and x is negative.

Practice Problem 1 If possible, find the roots. If there is no real number root, say so.

(a) $\sqrt[3]{216}$ (b) $\sqrt[5]{32}$ (c) $\sqrt[3]{-8}$ (d) $\sqrt[4]{-81}$

Because the symbol $\sqrt{x}$ represents exactly one real number for all real numbers x that are nonnegative, we can use it to define the **square root function** $f(x) = \sqrt{x}$.

Recall that the domain of a function is the set of values that can replace the variable. The function $f(x) = \sqrt{x}$ has a domain of all real numbers x that are greater than or equal to zero.

EXAMPLE 2 Find the indicated function values of the function $f(x) = \sqrt{2x + 4}$.

Round your answers to the nearest tenth when necessary.

(a) $f(-2)$ (b) $f(6)$ (c) $f(3)$

(a) $f(-2) = \sqrt{2(-2) + 4} = \sqrt{-4 + 4} = \sqrt{0} = 0$ The square root of zero is zero.

(b) $f(6) = \sqrt{2(6) + 4} = \sqrt{12 + 4} = \sqrt{16} = 4$

(c) $f(3) = \sqrt{2(3) + 4} = \sqrt{6 + 4} = \sqrt{10} \approx 3.2$ We use a calculator or a square root table to approximate $\sqrt{10}$.

Practice Problem 2 Find the indicated values of the function $f(x) = \sqrt{4x - 3}$.

Round your answers to the nearest tenth when necessary.

(a) $f(3)$ (b) $f(4)$ (c) $f(7)$

EXAMPLE 3 Find the domain of the function $f(x) = \sqrt{3x - 6}$.

We know that the expression $3x - 6$ must be nonnegative. That is, $3x - 6 \geq 0$.

$$3x - 6 \geq 0$$
$$3x \geq 6$$
$$x \geq 2$$

Thus, the domain is all real numbers x where $x \geq 2$.

Practice Problem 3 Find the domain of the function $f(x) = \sqrt{0.5x + 2}$.

⬤ **EXAMPLE 4**   Graph the function $f(x) = \sqrt{x + 2}$. Use the values $f(-2), f(-1)$, $f(0), f(1), f(2)$, and $f(7)$. Round your answers to the nearest tenth when necessary. We show the table of values here.

x	$f(x)$
-2	0
-1	1
0	1.4
1	1.7
2	2
7	3

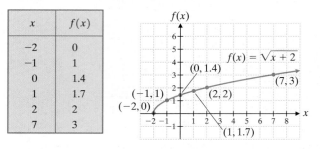

Practice Problem 4   Graph the function $f(x) = \sqrt{3x - 9}$. Use the values $f(3), f(4)$, $f(5), f(6)$, and $f(15)$.   ⬤

2   ***Changing Radical Expressions to Expressions with Rational Exponents***

Now we want to extend our definition of roots to rational exponents. By the laws of exponents we know that

$$x^{1/2} \cdot x^{1/2} = x^{1/2+1/2} = x^1 = x.$$

Since $x^{1/2}x^{1/2} = x$, $x^{1/2}$ must be a square root of x. That is, $x^{1/2} = \sqrt{x}$. Is this true? By the definition of square root, $\left(\sqrt{x}\right)^2 = x$. Does $\left(x^{1/2}\right)^2 = x$? Using the law of exponents we have

$$\left(x^{1/2}\right)^2 = x^{(1/2)(2)} = x^1 = x.$$

We conclude that

$$x^{1/2} = \sqrt{x}.$$

In the same way we can write the following:

$$x^{1/3} \cdot x^{1/3} \cdot x^{1/3} = x \qquad x^{1/3} = \sqrt[3]{x}$$
$$x^{1/4} \cdot x^{1/4} \cdot x^{1/4} \cdot x^{1/4} = x \qquad x^{1/4} = \sqrt[4]{x}$$
$$\vdots \qquad\qquad \vdots$$
$$\underbrace{x^{1/n} \cdot x^{1/n} \cdot \cdots \cdot x^{1/n}}_{n \text{ factors}} = x \qquad x^{1/n} = \sqrt[n]{x}$$

Therefore, we are ready to define fractional exponents in general.

> **Definition**
>
> If n is a positive integer and x is a nonnegative real number, then
>
> $$x^{1/n} = \sqrt[n]{x}.$$

⬤ **EXAMPLE 5**   Change to rational exponents and simplify. Assume that all variables are nonnegative real numbers.

(a) $\sqrt[4]{x^4}$ **(b)** $\sqrt[5]{(32)^5}$

(a) $\sqrt[4]{x^4} = \left(x^4\right)^{1/4} = x^{4/4} = x^1 = x$ **(b)** $\sqrt[5]{(32)^5} = \left(32^5\right)^{1/5} = 32^{5/5} = 32^1 = 32$

Practice Problem 5   Change to rational exponents and simplify. Assume that all variables are nonnegative real numbers.

(a) $\sqrt[3]{x^3}$ **(b)** $\sqrt[4]{y^4}$ ⬤

EXAMPLE 6 Replace all radicals with rational exponents.

(a) $\sqrt[3]{x^2}$ **(b)** $\left(\sqrt[5]{w}\right)^7$

(a) $\sqrt[3]{x^2} = \left(x^2\right)^{1/3} = x^{2/3}$ **(b)** $\left(\sqrt[5]{w}\right)^7 = \left(w^{1/5}\right)^7 = w^{7/5}$

Practice Problem 6 Replace all radicals with rational exponents.

(a) $\sqrt[4]{x^3}$ **(b)** $\sqrt[9]{(xy)^7}$

EXAMPLE 7 Evaluate or simplify. Assume that all variables are positive.

(a) $\sqrt[5]{32x^{10}}$ **(b)** $\sqrt[3]{125x^9}$ **(c)** $\left(16x^4\right)^{3/4}$

(a) $\sqrt[5]{32x^{10}} = \left(2^5 x^{10}\right)^{1/5} = 2x^2$ **(b)** $\sqrt[3]{125x^9} = \left[(5)^3 x^9\right]^{1/3} = 5x^3$

(c) $\left(16x^4\right)^{3/4} = \left(2^4 x^4\right)^{3/4} = 2^3 x^3 = 8x^3$

Practice Problem 7 Evaluate or simplify. Assume that all variables are nonnegative.

(a) $\sqrt[4]{81x^{12}}$ **(b)** $\sqrt[3]{27x^6}$ **(c)** $\left(32x^5\right)^{3/5}$

3 *Changing Expressions with Rational Exponents to Radical Expressions*

Sometimes we need to change an expression with rational exponents to a radical expression. This is especially helpful because the value of the radical form of an expression is sometimes more recognizable. For example, because of our experience with radicals, we know that $\sqrt{25} = 5$. It is not as easy to see that $25^{1/2} = 5$. Therefore, we simplify expressions with rational exponents by first rewriting them as radical expressions. Recall that

$$x^{1/n} = \sqrt[n]{x}.$$

Again, using the laws of exponents, we know that

$$x^{m/n} = \left(x^m\right)^{1/n} = \left(x^{1/n}\right)^m,$$

when x is nonnegative. We can make the following general definition.

Definition

For positive integers m and n and any real number x for which $x^{1/n}$ is defined,

$$x^{m/n} = \left(\sqrt[n]{x}\right)^m = \sqrt[n]{x^m}.$$

If it is also true that $x \neq 0$, then

$$x^{-m/n} = \frac{1}{x^{m/n}} = \frac{1}{\left(\sqrt[n]{x}\right)^m} = \frac{1}{\sqrt[n]{x^m}}.$$

EXAMPLE 8 Change to radical form.

(a) $(xy)^{5/7}$ **(b)** $w^{-2/3}$ **(c)** $3x^{3/4}$ **(d)** $(3x)^{3/4}$

(a) $(xy)^{5/7} = \sqrt[7]{(xy)^5} = \sqrt[7]{x^5 y^5}$ **(b)** $w^{-2/3} = \dfrac{1}{w^{2/3}} = \dfrac{1}{\sqrt[3]{w^2}}$

    or $(xy)^{5/7} = \left(\sqrt[7]{xy}\right)^5$ or $w^{-2/3} = \dfrac{1}{w^{2/3}} = \dfrac{1}{\left(\sqrt[3]{w}\right)^2}$

(c) $3x^{3/4} = 3\sqrt[4]{x^3}$ **(d)** $(3x)^{3/4} = \sqrt[4]{(3x)^3} = \sqrt[4]{27x^3}$

    or $3x^{3/4} = 3\left(\sqrt[4]{x}\right)^3$ or $(3x)^{3/4} = \left(\sqrt[4]{3x}\right)^3$

Practice Problem 8 Change to radical form.

(a) $x^{3/4}$ **(b)** $y^{-1/3}$ **(c)** $(2x)^{4/5}$ **(d)** $2x^{4/5}$

EXAMPLE 9   Change to radical form and evaluate.

(a) $125^{2/3}$ **(b)** $(-16)^{5/2}$ **(c)** $144^{-1/2}$

(a) $125^{2/3} = \left(\sqrt[3]{125}\right)^2 = (5)^2 = 25$

(b) $(-16)^{5/2} = \left(\sqrt{-16}\right)^5$; however, $\sqrt{-16}$ is not a real number. Thus, $(-16)^{5/2}$ is not a real number.

(c) $144^{-1/2} = \dfrac{1}{144^{1/2}} = \dfrac{1}{\sqrt{144}} = \dfrac{1}{12}$

Practice Problem 9   Change to radical form and evaluate.

(a) $8^{2/3}$ **(b)** $(-8)^{4/3}$ **(c)** $100^{-3/2}$

---

4   *Evaluating Higher-Order Radicals Containing a Variable Radicand That Represents Any Real Number (Including a Negative Real Number)*

We now give a definition of higher-order radicals that works for all radicals, no matter what their signs are.

> **Definition**
>
> For all real numbers x (including negative real numbers),
>
> $$\sqrt[n]{x^n} = |x| \quad \text{when } n \text{ is an } even \text{ positive integer, and}$$
> $$\sqrt[n]{x^n} = x \quad \text{when } n \text{ is an } odd \text{ positive integer.}$$

EXAMPLE 10   Evaluate; x may be any real number.

(a) $\sqrt[3]{(-2)^3}$ **(b)** $\sqrt[4]{(-2)^4}$ **(c)** $\sqrt[5]{x^5}$ **(d)** $\sqrt[6]{x^6}$

(a) $\sqrt[3]{(-2)^3} = -2$ because the index is odd.
(b) $\sqrt[4]{(-2)^4} = |-2| = 2$ because the index is even.
(c) $\sqrt[5]{x^5} = x$ because the index is odd.
(d) $\sqrt[6]{x^6} = |x|$ because the index is even.

Practice Problem 10   Evaluate; y and w may be any real numbers.

(a) $\sqrt[5]{(-3)^5}$ **(b)** $\sqrt[4]{(-5)^4}$ **(c)** $\sqrt[4]{w^4}$ **(d)** $\sqrt[7]{y^7}$

EXAMPLE 11   Simplify. Assume that x and y may be any real numbers.

(a) $\sqrt{49x^2}$ **(b)** $\sqrt[4]{81y^{16}}$ **(c)** $\sqrt[3]{27x^6y^9}$

(a) We observe that the index is an even positive number. We will need the absolute value. $\sqrt{49x^2} = 7|x|$

(b) Again, we need the absolute value. $\sqrt[4]{81y^{16}} = 3\left|y^4\right|$
Since we know that $3y^4$ is positive (anything to the fourth power will be positive), we can write $3\left|y^4\right|$ without the absolute value symbol. Thus, $\sqrt[4]{81y^{16}} = 3y^4$.

(c) The index is an odd integer. The absolute value is never needed in such a case.
$$\sqrt[3]{27x^6y^9} = \sqrt[3]{(3)^3(x^2)^3(y^3)^3} = 3x^2y^3$$

Practice Problem 11   Simplify. Assume that x and y may be any real numbers.

(a) $\sqrt{36x^2}$ **(b)** $\sqrt[3]{125x^3y^6}$ **(c)** $\sqrt[4]{16y^8}$

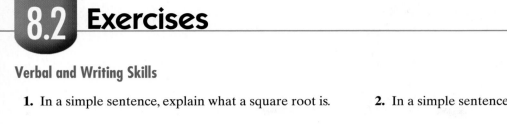

8.2 Exercises

Verbal and Writing Skills

1. In a simple sentence, explain what a square root is.

2. In a simple sentence, explain what a cube root is.

3. Give an example to show why the cube root of a negative number is a negative number.

4. Give an example to show why it is not possible to find a real number that is the square root of a negative number.

Evaluate if possible.

5. $\sqrt{64}$

6. $\sqrt{100}$

7. $\sqrt{25} + \sqrt{49}$

8. $\sqrt{16} + \sqrt{81}$

9. $-\sqrt{\dfrac{1}{9}}$

10. $-\sqrt{\dfrac{4}{25}}$

11. $\sqrt{36} - \sqrt{25}$

12. $\sqrt{49} + \sqrt{100}$

13. $\sqrt{0.04}$

14. $\sqrt{0.16}$

For the given function, find the indicated function values. Find the domain of each function. Round your answers to the nearest tenth when necessary.

15. $f(x) = \sqrt{10x + 5}$; $f(0), f(1), f(2), f(3)$

16. $f(x) = \sqrt{3x + 21}$; $f(0), f(1), f(5), f(-4)$

17. $f(x) = \sqrt{0.5x - 3}$; $f(6), f(8), f(14), f(16)$

18. $f(x) = \sqrt{1.5x - 4}$; $f(4), f(6), f(8), f(14)$

Graph each of the following functions. Plot at least four points for each function.

19. $f(x) = \sqrt{x - 3}$

20. $f(x) = \sqrt{x - 1}$

21. $f(x) = \sqrt{2x + 4}$

22. $f(x) = \sqrt{3x + 9}$

437

Evaluate if possible.

23. $\sqrt[3]{216}$ **24.** $\sqrt[3]{27}$ **25.** $\sqrt[3]{64}$ **26.** $\sqrt[3]{-125}$ **27.** $\sqrt[3]{-8}$

28. $\sqrt[4]{625}$ **29.** $\sqrt[4]{81}$ **30.** $-\sqrt[6]{64}$ **31.** $\sqrt[5]{(8)^5}$ **32.** $\sqrt[6]{(9)^6}$

33. $\sqrt[8]{(5)^8}$ **34.** $\sqrt[7]{(11)^7}$ **35.** $\sqrt[3]{-\dfrac{1}{64}}$ **36.** $\sqrt[3]{-\dfrac{8}{27}}$ **37.** $\sqrt[3]{\dfrac{27}{125}}$

For Exercises 38–83, assume that variables represent positive real numbers.
Replace all radicals with rational exponents.

38. $\sqrt[3]{y}$ **39.** $\sqrt{a}$ **40.** $\sqrt[5]{2x}$ **41.** $\sqrt[4]{3y}$ **42.** $\sqrt[7]{(a+b)^3}$

43. $\sqrt[9]{(a-b)^5}$ **44.** $\sqrt{\sqrt[3]{x}}$ **45.** $\sqrt[5]{\sqrt{y}}$ **46.** $\left(\sqrt[6]{3x}\right)^5$ **47.** $\left(\sqrt[5]{2x}\right)^3$

Simplify.

48. $\sqrt[6]{(12)^6}$ **49.** $\sqrt[5]{(-11)^5}$ **50.** $\sqrt[3]{x^3y^6}$ **51.** $\sqrt[4]{a^8b^4}$ **52.** $\sqrt{36x^8y^4}$

53. $\sqrt{49x^2y^8}$ **54.** $\sqrt[4]{16a^8b^4}$ **55.** $\sqrt[4]{81a^{12}b^{20}}$ **56.** $\sqrt[3]{-125x^{30}}$ **57.** $\sqrt[3]{8x^3y^9}$

Change to radical form.

58. $y^{4/7}$ **59.** $x^{5/6}$ **60.** $7^{-2/3}$ **61.** $5^{-3/5}$ **62.** $(2a+b)^{5/7}$

63. $(x+3y)^{4/7}$ **64.** $(-x)^{3/5}$ **65.** $(-y)^{5/7}$ **66.** $(2xy)^{3/5}$ **67.** $(3ab)^{2/7}$

Mixed Practice

Evaluate or simplify.

68. $4^{3/2}$ **69.** $27^{2/3}$ **70.** $(-64)^{1/3}$ **71.** $\left(\dfrac{4}{25}\right)^{1/2}$

72. $\left(\dfrac{16}{81}\right)^{3/4}$ **73.** $(-125)^{2/3}$ **74.** $(25x^4)^{-1/2}$ **75.** $(36y^8)^{-1/2}$

76. $\sqrt{121x^4}$ **77.** $\sqrt{49x^8}$ **78.** $\sqrt{144a^6b^{24}}$ **79.** $\sqrt{25a^{14}b^{18}}$

80. $\sqrt{36x^6y^8z^{10}}$ **81.** $\sqrt{100x^{10}y^{12}z^2}$ **82.** $\sqrt[3]{216a^3b^9c^{12}}$ **83.** $\sqrt[3]{-125a^6b^{15}c^{21}}$

To Think About

Simplify. Assume that the variables represent any positive or negative real number.

84. $\sqrt{25x^2}$ **85.** $\sqrt{100x^2}$ **86.** $\sqrt[3]{-8x^6}$ **87.** $\sqrt[3]{-27x^9}$ **88.** $\sqrt[4]{x^8y^{16}}$

89. $\sqrt[4]{x^{16}y^{40}}$ **90.** $\sqrt[4]{a^{12}b^4}$ **91.** $\sqrt[4]{a^4b^{20}}$ **92.** $\sqrt{4x^8y^4}$ **93.** $\sqrt{49a^{12}b^4}$

Applications

A company finds that the daily cost of producing appliances at one of its factories is represented by the equation $C = 120\sqrt[3]{n} + 375$, *where n is the number of parts produced in a day and C is the cost in dollars.*

94. Find the cost if 343 parts are produced per day.   **95.** Find the cost if 216 parts are produced per day.

Cumulative Review Problems

96. In the year 2000 the world produced 4.027×10^{17} Btu of energy. Use the pie graph to determine how much energy was produced that year in the Middle East.

97. The total world production of energy is predicted to increase by 52% of its year 2000 level by the year 2020. If North America increases its production of energy by only 24%, what percent of the world's energy in 2020 will be produced in North America?

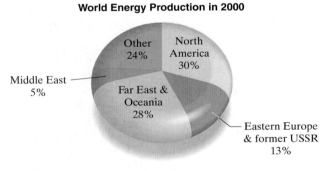

World Energy Production in 2000

Other 24%
North America 30%
Middle East 5%
Far East & Oceania 28%
Eastern Europe & former USSR 13%

Source: U.S. Energy Information Administration

98 Factor. $2x^2 + 12xy + 18y^2$ **99.** Factor. $2x^2 - x - 21$

100. Divide. $\dfrac{x^2 + 6x + 9}{2x^2y - 18y} \div \dfrac{6xy + 18y}{3x^2y - 27y}$ **101.** Add. $\dfrac{6}{2x - 4} + \dfrac{1}{12x + 4}$

Student Learning Objectives

After studying this section, you will be able to:

1 Simplify a radical by using the product rule.

2 Add and subtract like radical terms.

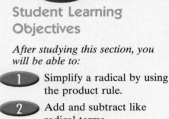

SSM
PH TUTOR CENTER | CD & VIDEO | MATH PRO | WEB

1 *Simplifying a Radical by Using the Product Rule*

When we simplify a radical, we want to get an equivalent expression with the smallest possible quantity in the radicand. We can use the product rule for radicals to simplify radicals.

> **Product Rule for Radicals**
>
> For all nonnegative real numbers a and b and positive integers n,
> $$\sqrt[n]{a}\,\sqrt[n]{b} = \sqrt[n]{ab}.$$

You should be able to derive the product rule from your knowledge of the laws of exponents. We have
$$\sqrt[n]{a}\,\sqrt[n]{b} = a^{1/n}b^{1/n} = (ab)^{1/n} = \sqrt[n]{ab}.$$

Throughout the remainder of this chapter, assume that all variables in any radicand represent nonnegative numbers, unless a specific statement is made to the contrary.

EXAMPLE 1 Simplify $\sqrt{32}$.

Solution 1: $\sqrt{32} = \sqrt{16 \cdot 2} = \sqrt{16}\,\sqrt{2} = 4\sqrt{2}$

Solution 2: $\sqrt{32} = \sqrt{4 \cdot 8} = \sqrt{4}\,\sqrt{8} = 2\sqrt{8} = 2\sqrt{4 \cdot 2} = 2\sqrt{4}\,\sqrt{2} = 4\sqrt{2}$

Although we obtained the same answer both times, the first solution is much shorter. You should try to use the largest factor that is a perfect square when you use the product rule.

Practice Problem 1 Simplify $\sqrt{20}$.

EXAMPLE 2 Simplify $\sqrt{48}$.
$$\sqrt{48} = \sqrt{16}\,\sqrt{3} = 4\sqrt{3}$$

Practice Problem 2 Simplify $\sqrt{27}$.

EXAMPLE 3 Simplify. **(a)** $\sqrt[3]{16}$ **(b)** $\sqrt[3]{-81}$

(a) $\sqrt[3]{16} = \sqrt[3]{8}\,\sqrt[3]{2} = 2\sqrt[3]{2}$ **(b)** $\sqrt[3]{-81} = \sqrt[3]{-27}\,\sqrt[3]{3} = -3\sqrt[3]{3}$

Practice Problem 3 Simplify. **(a)** $\sqrt[3]{24}$ **(b)** $\sqrt[3]{-108}$

EXAMPLE 4 Simplify $\sqrt[4]{48}$.
$$\sqrt[4]{48} = \sqrt[4]{16}\,\sqrt[4]{3} = 2\sqrt[4]{3}$$

Practice Problem 4 Simplify $\sqrt[4]{64}$.

EXAMPLE 5 Simplify.

(a) $\sqrt{27x^3y^4}$ **(b)** $\sqrt[3]{16x^4y^3z^6}$

(a) $\sqrt{27x^3y^4} = \sqrt{9 \cdot 3 \cdot x^2 \cdot x \cdot y^4} = \sqrt{9x^2y^4}\,\sqrt{3x}$ Factor out the perfect squares.

$$= 3xy^2\sqrt{3x}$$

(b) $\sqrt[3]{16x^4y^3z^6} =$

$\sqrt[3]{8 \cdot 2 \cdot x^3 \cdot x \cdot y^3 \cdot z^6} = \sqrt[3]{8x^3y^3z^6}\,\sqrt[3]{2x}$ Factor out the perfect cubes.

$$= 2xyz^2\sqrt[3]{2x}$$ Why is z^6 a perfect cube?

Practice Problem 5 Simplify.

(a) $\sqrt{45x^6y^7}$ **(b)** $\sqrt{27a^7b^8c^9}$

2   *Adding and Subtracting Like Radical Terms*

Only like radicals can be added or subtracted. Two radicals are **like radicals** if they have the same radicand and the same index. $2\sqrt{5}$ and $3\sqrt{5}$ are like radicals. $2\sqrt{5}$ and $2\sqrt{3}$ are not like radicals; $2\sqrt{5}$ and $2\sqrt[3]{5}$ are not like radicals. When we combine radicals, we combine like terms by using the distributive property.

EXAMPLE 6   Combine $2\sqrt{5} + 3\sqrt{5} - 4\sqrt{5}$.

$$2\sqrt{5} + 3\sqrt{5} - 4\sqrt{5} = (2 + 3 - 4)\sqrt{5} = 1\sqrt{5} = \sqrt{5}$$

Practice Problem 6   Combine $19\sqrt{xy} + 5\sqrt{xy} - 10\sqrt{xy}$.

Sometimes when you simplify radicands, you may find you have like radicals.

EXAMPLE 7   Combine $5\sqrt{3} - \sqrt{27} + 2\sqrt{48}$.

$$\begin{aligned} 5\sqrt{3} - \sqrt{27} + 2\sqrt{48} &= 5\sqrt{3} - \sqrt{9}\sqrt{3} + 2\sqrt{16}\sqrt{3} \\ &= 5\sqrt{3} - 3\sqrt{3} + 2(4)\sqrt{3} \\ &= 5\sqrt{3} - 3\sqrt{3} + 8\sqrt{3} \\ &= 10\sqrt{3} \end{aligned}$$

Practice Problem 7   Combine $4\sqrt{2} - 5\sqrt{50} - 3\sqrt{98}$.

EXAMPLE 8   Combine $6\sqrt{x} + 4\sqrt{12x} - \sqrt{75x} + 3\sqrt{x}$.

$$\begin{aligned} 6\sqrt{x} + 4\sqrt{12x} - \sqrt{75x} + 3\sqrt{x} &= 6\sqrt{x} + 4\sqrt{4}\sqrt{3x} - \sqrt{25}\sqrt{3x} + 3\sqrt{x} \\ &= 6\sqrt{x} + 8\sqrt{3x} - 5\sqrt{3x} + 3\sqrt{x} \\ &= 6\sqrt{x} + 3\sqrt{x} + 8\sqrt{3x} - 5\sqrt{3x} \\ &= 9\sqrt{x} + 3\sqrt{3x} \end{aligned}$$

Practice Problem 8   Combine $4\sqrt{2x} + \sqrt{18x} - 2\sqrt{125x} - 6\sqrt{20x}$.

EXAMPLE 9   Combine $2\sqrt[3]{81x^3y^4} + 3xy\sqrt[3]{24y}$.

$$\begin{aligned} 2\sqrt[3]{81x^3y^4} + 3xy\sqrt[3]{24y} &= 2\sqrt[3]{27x^3y^3}\sqrt[3]{3y} + 3xy\sqrt[3]{8}\sqrt[3]{3y} \\ &= 2(3xy)\sqrt[3]{3y} + 3xy(2)\sqrt[3]{3y} \\ &= 6xy\sqrt[3]{3y} + 6xy\sqrt[3]{3y} \\ &= 12xy\sqrt[3]{3y} \end{aligned}$$

Practice Problem 9   Combine $3x\sqrt[3]{54x^4} - 3\sqrt[3]{16x^7}$.

Simplify. Assume that all variables are nonnegative real numbers.

1. $\sqrt{8}$ **2.** $\sqrt{12}$ **3.** $\sqrt{18}$ **4.** $\sqrt{75}$

5. $\sqrt{120}$ **6.** $\sqrt{80}$ **7.** $\sqrt{44}$ **8.** $\sqrt{90}$

9. $\sqrt{9x^3}$ **10.** $\sqrt{16x^5}$ **11.** $\sqrt{60a^4b^5}$

12. $\sqrt{45a^3b^8}$ **13.** $\sqrt{98x^5y^6z}$ **14.** $\sqrt{24xy^8z^3}$

15. $\sqrt[3]{8}$ **16.** $\sqrt[3]{27}$ **17.** $\sqrt[3]{108}$

18. $\sqrt[3]{128}$ **19.** $\sqrt[3]{56y}$ **20.** $\sqrt[3]{54x}$

21. $\sqrt[3]{8a^3b^8}$ **22.** $\sqrt[3]{125a^6b^2}$ **23.** $\sqrt[3]{24x^6y^{11}}$

24. $\sqrt[3]{40x^7y^{26}}$ **25.** $\sqrt[4]{81kp^{23}}$ **26.** $\sqrt[4]{16k^{12}p^{18}}$

27. $\sqrt[5]{-32x^5y^6}$ **28.** $\sqrt[5]{-243x^4y^{10}}$

To Think About

29. $\sqrt[4]{1792} = a\sqrt[4]{7}$. What is the value of a?   **30.** $\sqrt[3]{3072} = b\sqrt[3]{6}$. What is the value of b?

Combine. Assume that all variables represent nonnegative real numbers.

31. $\sqrt{49} + \sqrt{100}$ **32.** $\sqrt{25} + \sqrt{81}$ **33.** $\sqrt{3} + 7\sqrt{3} - 2\sqrt{3}$

34. $\sqrt{11} - 5\sqrt{11} + 3\sqrt{11}$ **35.** $3\sqrt{18} - \sqrt{2}$ **36.** $\sqrt{40} - \sqrt{10}$

37. $-2\sqrt{50} + \sqrt{32} - 3\sqrt{8}$

38. $-\sqrt{12} + 2\sqrt{48} - \sqrt{75}$

39. $-5\sqrt{45} + 6\sqrt{20} + 3\sqrt{5}$

40. $-7\sqrt{10} + 4\sqrt{40} - 8\sqrt{90}$

41. $\sqrt{44} - 3\sqrt{63x} + 4\sqrt{28x}$

42. $\sqrt{75x} + 2\sqrt{108x} - 6\sqrt{3x}$

43. $\sqrt{200x^3} - x\sqrt{32x}$

44. $\sqrt{75a^3} + a\sqrt{12a}$

45. $\sqrt[3]{16} + 3\sqrt[3]{54}$

46. $\sqrt[3]{128} - 4\sqrt[3]{16}$

47. $-2\sqrt[3]{125x^3y^4} + 3y^2\sqrt[3]{8x^3}$

48. $2x\sqrt[3]{40xy} - 3\sqrt[3]{5x^4y}$

To Think About

49. Use a calculator to show that
$\sqrt{48} + \sqrt{27} + \sqrt{75} = 12\sqrt{3}$.

50. Use a calculator to show that
$\sqrt{98} + \sqrt{50} + \sqrt{128} = 20\sqrt{2}$.

Applications

We can approximate the amount of current in amps (amperes) I drawn by an appliance in the home using the formula

$$I = \sqrt{\frac{P}{R}},$$

where P is the power measured in watts and R is the resistance measured in ohms. In Exercises 51 and 52 round your answers to three decimal places.

51. What is the current I if $P = 500$ watts and $R = 10$ ohms?

52. What is the current I if $P = 480$ watts and $R = 8$ ohms?

*The **period** of a pendulum is the amount of time it takes the pendulum to make one complete swing back and forth. If the length of the pendulum L is measured in feet, then its period T measured in seconds is given by the formula*

$$T = 2\pi\sqrt{\frac{L}{32}}.$$

Use $\pi \approx 3.14$ for exercises 53–54.

53. Find the period of a pendulum if its length is 8 feet.

54. A person suspended on a rope swinging back and forth acts like a human pendulum. What is the period of a person swinging on a rope that is 128 feet long?

Cumulative Review Problems

Graph.

55. $3y - 2x = 9$

56. $3y - 2x \le 9$

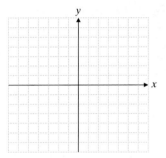

Simplify.

57. $\left(\dfrac{2x^2y}{3xy^3}\right)^{-2}$

58. $(-3a^2b^0x^4)^3$

8.4 Multiplication and Division of Radicals

Student Learning Objectives

After studying this section, you will be able to:

1 Multiply radical expressions.

2 Divide radical expressions.

3 Simplify radical expressions by rationalizing the denominator.

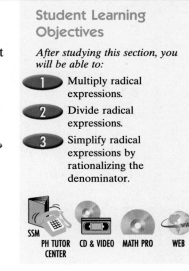

SSM PH TUTOR CD & VIDEO MATH PRO WEB
CENTER

1 Multiplying Radical Expressions

We use the product rule for radicals to multiply radical expressions. Recall that $\sqrt[n]{a}\,\sqrt[n]{b} = \sqrt[n]{ab}$.

EXAMPLE 1 Multiply $(3\sqrt{2})(5\sqrt{11x})$.

$$(3\sqrt{2})(5\sqrt{11x}) = (3)(5)\sqrt{2 \cdot 11x} = 15\sqrt{22x}$$

Practice Problem 1 Multiply $(-4\sqrt{2})(-3\sqrt{13x})$.

EXAMPLE 2 Multiply $\sqrt{6x}(\sqrt{3} + \sqrt{2x} + \sqrt{5})$.

$$\begin{aligned}
\sqrt{6x}(\sqrt{3} + \sqrt{2x} + \sqrt{5}) &= (\sqrt{6x})(\sqrt{3}) + (\sqrt{6x})(\sqrt{2x}) + (\sqrt{6x})(\sqrt{5}) \\
&= \sqrt{18x} + \sqrt{12x^2} + \sqrt{30x} \\
&= \sqrt{9}\sqrt{2x} + \sqrt{4x^2}\sqrt{3} + \sqrt{30x} \\
&= 3\sqrt{2x} + 2x\sqrt{3} + \sqrt{30x}
\end{aligned}$$

Practice Problem 2 Multiply $\sqrt{2x}(\sqrt{5} + 2\sqrt{3x} + \sqrt{8})$.

To multiply two binomials containing radicals, we can use the distributive property. Most students find that the FOIL method is helpful in remembering how to find the four products.

EXAMPLE 3 Multiply $(\sqrt{2} + 3\sqrt{5})(2\sqrt{2} - \sqrt{5})$.

By FOIL:

$$\begin{aligned}
(\sqrt{2} + 3\sqrt{5})(2\sqrt{2} - \sqrt{5}) &= 2\sqrt{4} - \sqrt{10} + 6\sqrt{10} - 3\sqrt{25} \\
&= 4 + 5\sqrt{10} - 15 \\
&= -11 + 5\sqrt{10}
\end{aligned}$$

By the distributive property:

$$\begin{aligned}
(\sqrt{2} + 3\sqrt{5})(2\sqrt{2} - \sqrt{5}) &= (\sqrt{2} + 3\sqrt{5})(2\sqrt{2}) - (\sqrt{2} + 3\sqrt{5})\sqrt{5} \\
&= (\sqrt{2})(2\sqrt{2}) + (3\sqrt{5})(2\sqrt{2}) - (\sqrt{2})(\sqrt{5}) - (3\sqrt{5})(\sqrt{5}) \\
&= 2\sqrt{4} + 6\sqrt{10} - \sqrt{10} - 3\sqrt{25} \\
&= 4 + 5\sqrt{10} - 15 \\
&= -11 + 5\sqrt{10}
\end{aligned}$$

Practice Problem 3 Multiply $(\sqrt{7} + 4\sqrt{2})(2\sqrt{7} - 3\sqrt{2})$.

EXAMPLE 4 Multiply $(7 - 3\sqrt{2})(4 - \sqrt{3})$.

$$(7 - 3\sqrt{2})(4 - \sqrt{3}) = 28 - 7\sqrt{3} - 12\sqrt{2} + 3\sqrt{6}$$

Practice Problem 4 Multiply $(2 - 5\sqrt{5})(3 - 2\sqrt{2})$.

EXAMPLE 5 Multiply $(\sqrt{7} + \sqrt{3x})^2$.

Method 1: We can use the FOIL method or the distributive property.

$$\begin{aligned}
(\sqrt{7} + \sqrt{3x})(\sqrt{7} + \sqrt{3x}) &= \sqrt{49} + \sqrt{21x} + \sqrt{21x} + \sqrt{9x^2} \\
&= 7 + \sqrt{21x} + \sqrt{21x} + 3x \\
&= 7 + 2\sqrt{21x} + 3x
\end{aligned}$$

445

Method 2: We could also use the Chapter 5 formula.

$$(a + b)^2 = a^2 + 2ab + b^2,$$

where $a = \sqrt{7}$ and $b = \sqrt{3x}$. Then

$$(\sqrt{7} + \sqrt{3x})^2 = (\sqrt{7})^2 + 2\sqrt{7}\sqrt{3x} + (\sqrt{3x})^2$$
$$= 7 + 2\sqrt{21x} + 3x.$$

Practice Problem 5   Multiply $(\sqrt{5x} + \sqrt{10})^2$. Use the approach that seems easiest to you.

EXAMPLE 6   Multiply.

(a) $\sqrt[3]{3x}\left(\sqrt[3]{x^2} + 3\sqrt[3]{4y}\right)$ **(b)** $(\sqrt[3]{2y} + \sqrt[3]{4})(2\sqrt[3]{4y^2} - 3\sqrt[3]{2})$

(a) $\sqrt[3]{3x}\left(\sqrt[3]{x^2} + 3\sqrt[3]{4y}\right) = (\sqrt[3]{3x})(\sqrt[3]{x^2}) + 3(\sqrt[3]{3x})(\sqrt[3]{4y})$
$$= \sqrt[3]{3x^3} + 3\sqrt[3]{12xy}$$
$$= x\sqrt[3]{3} + 3\sqrt[3]{12xy}$$

(b) $(\sqrt[3]{2y} + \sqrt[3]{4})(2\sqrt[3]{4y^2} - 3\sqrt[3]{2}) = 2\sqrt[3]{8y^3} - 3\sqrt[3]{4y} + 2\sqrt[3]{16y^2} - 3\sqrt[3]{8}$
$$= 2(2y) - 3\sqrt[3]{4y} + 2\sqrt[3]{8}\sqrt[3]{2y^2} - 3(2)$$
$$= 4y - 3\sqrt[3]{4y} + 4\sqrt[3]{2y^2} - 6$$

Practice Problem 6   Multiply.

(a) $\sqrt[3]{2x}\left(\sqrt[3]{4x^2} + 3\sqrt[3]{y}\right)$ **(b)** $(\sqrt[3]{7} + \sqrt[3]{x^2})(2\sqrt[3]{49} - \sqrt[3]{x})$

2   *Dividing Radical Expressions*

We can use the laws of exponents to develop a rule for dividing two radicals.

$$\sqrt[n]{\frac{a}{b}} = \left(\frac{a}{b}\right)^{1/n} = \frac{a^{1/n}}{b^{1/n}} = \frac{\sqrt[n]{a}}{\sqrt[n]{b}}$$

This quotient rule is very useful. We now state it more formally.

> **Quotient Rule for Radicals**
>
> For all nonnegative real numbers a, all positive real numbers b, and positive integers n,
>
> $$\frac{\sqrt[n]{a}}{\sqrt[n]{b}} = \sqrt[n]{\frac{a}{b}}.$$

Sometimes it will be best to change $\sqrt[n]{\frac{a}{b}}$ to $\frac{\sqrt[n]{a}}{\sqrt[n]{b}}$, whereas at other times it will be best to change $\frac{\sqrt[n]{a}}{\sqrt[n]{b}}$ to $\sqrt[n]{\frac{a}{b}}$. To use the quotient rule for radicals, you need to have good number sense. You should know your squares up to 15^2 and your cubes up to 5^3.

EXAMPLE 7   Divide.

(a) $\dfrac{\sqrt{48}}{\sqrt{3}}$ **(b)** $\sqrt[3]{\dfrac{125}{8}}$ **(c)** $\dfrac{\sqrt{28x^5y^3}}{\sqrt{7x}}$

(a) $\dfrac{\sqrt{48}}{\sqrt{3}} = \sqrt{\dfrac{48}{3}} = \sqrt{16} = 4$

(b) $\sqrt[3]{\dfrac{125}{8}} = \dfrac{\sqrt[3]{125}}{\sqrt[3]{8}} = \dfrac{5}{2}$

(c) $\dfrac{\sqrt{28x^5y^3}}{\sqrt{7x}} = \sqrt{\dfrac{28x^5y^3}{7x}} = \sqrt{4x^4y^3} = 2x^2y\sqrt{y}$

Practice Problem 7 Divide.

(a) $\dfrac{\sqrt{75}}{\sqrt{3}}$ **(b)** $\sqrt[3]{\dfrac{27}{64}}$ **(c)** $\dfrac{\sqrt{54a^3b^7}}{\sqrt{6b^5}}$

3 Simplifying Radical Expressions by Rationalizing the Denominator

Recall that to simplify a radical we want to get the smallest possible quantity in the radicand. Whenever possible, we find the square root of a perfect square. Thus, to simplify $\sqrt{\tfrac{7}{16}}$ we have

$$\sqrt{\dfrac{7}{16}} = \dfrac{\sqrt{7}}{\sqrt{16}} = \dfrac{\sqrt{7}}{4}.$$

Notice that the denominator does not contain a square root. The expression $\dfrac{\sqrt{7}}{4}$ is in simplest form.

Let's look at $\sqrt{\tfrac{16}{7}}$. We have

$$\sqrt{\dfrac{16}{7}} = \dfrac{\sqrt{16}}{\sqrt{7}} = \dfrac{4}{\sqrt{7}}.$$

Notice that the denominator contains a square root. If an expression contains a square root in the denominator, it is not considered to be simplified. How can we rewrite $\dfrac{4}{\sqrt{7}}$ as an equivalent expression that does not contain the $\sqrt{7}$ in the denominator? Since $\sqrt{7}\,\sqrt{7} = 7$, we can multiply the numerator and the denominator by the radical in the denominator.

$$\dfrac{4}{\sqrt{7}} \cdot \dfrac{\sqrt{7}}{\sqrt{7}} = \dfrac{4\sqrt{7}}{\sqrt{49}} = \dfrac{4\sqrt{7}}{7}$$

This expression is considered to be in simplest form. We call this process rationalizing the denominator.

   Rationalizing the denominator is the process of transforming a fraction with one or more radicals in the denominator into an equivalent fraction without a radical in the denominator.

EXAMPLE 8 Simplify by rationalizing the denominator $\dfrac{3}{\sqrt{2}}$.

$$\begin{aligned}
\dfrac{3}{\sqrt{2}} &= \dfrac{3}{\sqrt{2}} \cdot \dfrac{\sqrt{2}}{\sqrt{2}} && \text{Since } \dfrac{\sqrt{2}}{\sqrt{2}} = 1. \\[2mm]
&= \dfrac{3\sqrt{2}}{\sqrt{4}} && \text{Product rule for radicals.} \\[2mm]
&= \dfrac{3\sqrt{2}}{2}
\end{aligned}$$

Practice Problem 8 Simplify by rationalizing the denominator $\dfrac{7}{\sqrt{3}}$.

We can rationalize the denominator either before or after we simplify the denominator.

EXAMPLE 9 Simplify $\dfrac{3}{\sqrt{12x}}$.

Method 1: First we simplify the radical in the denominator, and then we multiply in order to rationalize the denominator.

$$\frac{3}{\sqrt{12x}} = \frac{3}{\sqrt{4}\sqrt{3x}} = \frac{3}{2\sqrt{3x}} \cdot \frac{\sqrt{3x}}{\sqrt{3x}} = \frac{3\sqrt{3x}}{2(3x)} = \frac{\sqrt{3x}}{2x}$$

Method 2: We can multiply numerator and denominator by a value that will make the denominator a perfect square (i.e., rationalize the denominator).

$$\frac{3}{\sqrt{12x}} = \frac{3}{\sqrt{12x}} \cdot \frac{\sqrt{3x}}{\sqrt{3x}}$$

$$= \frac{3\sqrt{3x}}{\sqrt{36x^2}} \qquad \text{Since } \sqrt{12x}\,\sqrt{3x} = \sqrt{36x^2}.$$

$$= \frac{3\sqrt{3x}}{6x} = \frac{\sqrt{3x}}{2x}$$

Practice Problem 9 Simplify $\dfrac{8}{\sqrt{20x}}$.

If the radicand has a fraction, it is not considered to be simplified. We can use the quotient rule for radicals and then rationalize the denominator to simplify the radical. We have already rationalized denominators when they contain square roots. Now we will rationalize denominators when they contain radical expressions that are cube roots or higher-order roots.

EXAMPLE 10 Simplify $\sqrt[3]{\dfrac{2}{3x^2}}$.

Method 1: $\sqrt[3]{\dfrac{2}{3x^2}} = \dfrac{\sqrt[3]{2}}{\sqrt[3]{3x^2}}$ Quotient rule for radicals.

$$= \frac{\sqrt[3]{2}}{\sqrt[3]{3x^2}} \cdot \frac{\sqrt[3]{9x}}{\sqrt[3]{9x}} \qquad \begin{array}{l}\text{Multiply the numerator and denominator} \\ \text{by an appropriate value so that the new} \\ \text{denominator will be a perfect cube.}\end{array}$$

$$= \frac{\sqrt[3]{18x}}{\sqrt[3]{27x^3}} \qquad \begin{array}{l}\text{Observe that we can evaluate the cube root} \\ \text{in the denominator.}\end{array}$$

$$= \frac{\sqrt[3]{18x}}{3x}$$

Method 2: $\sqrt[3]{\dfrac{2}{3x^2}} = \sqrt[3]{\dfrac{2}{3x^2} \cdot \dfrac{9x}{9x}}$

$$= \sqrt[3]{\frac{18x}{27x^3}}$$

$$= \frac{\sqrt[3]{18x}}{\sqrt[3]{27x^3}}$$

$$= \frac{\sqrt[3]{18x}}{3x}$$

Practice Problem 10 Simplify $\sqrt[3]{\dfrac{6}{5x}}$.

If the denominator of a radical expression contains a sum or difference with radicals, we multiply the numerator and denominator by the *conjugate* of the denominator. For example, the conjugate of $x + \sqrt{y}$ is $x - \sqrt{y}$; similarly, the conjugate of $x - \sqrt{y}$ is $x + \sqrt{y}$. What is the conjugate of $3 + \sqrt{2}$? It is $3 - \sqrt{2}$. How about $\sqrt{11} + \sqrt{xyz}$? It is $\sqrt{11} - \sqrt{xyz}$.

Conjugates

The expressions $a + b$ and $a - b$, where a and b represent any algebraic term, are called **conjugates**. Each expression is the conjugate of the other expression.

Multiplying by conjugates is simply an application of the formula

$$(a + b)(a - b) = a^2 - b^2.$$

For example,

$$\left(\sqrt{x} + \sqrt{y}\right)\left(\sqrt{x} - \sqrt{y}\right) = \left(\sqrt{x}\right)^2 - \left(\sqrt{y}\right)^2 = x - y.$$

EXAMPLE 11   Simplify $\dfrac{5}{3 + \sqrt{2}}$.

$$\frac{5}{3 + \sqrt{2}} = \frac{5}{3 + \sqrt{2}} \cdot \frac{3 - \sqrt{2}}{3 - \sqrt{2}} \qquad \text{Multiply the numerator and denominator by the conjugate of } 3 + \sqrt{2}.$$

$$= \frac{15 - 5\sqrt{2}}{(3)^2 - \left(\sqrt{2}\right)^2}$$

$$= \frac{15 - 5\sqrt{2}}{9 - 2} = \frac{15 - 5\sqrt{2}}{7}$$

Practice Problem 11   Simplify $\dfrac{4}{2 + \sqrt{5}}$.

EXAMPLE 12   Simplify $\dfrac{\sqrt{7} + \sqrt{3}}{\sqrt{7} - \sqrt{3}}$.

The conjugate of $\sqrt{7} - \sqrt{3}$ is $\sqrt{7} + \sqrt{3}$.

$$\frac{\sqrt{7} + \sqrt{3}}{\sqrt{7} - \sqrt{3}} \cdot \frac{\sqrt{7} + \sqrt{3}}{\sqrt{7} + \sqrt{3}} = \frac{\sqrt{49} + 2\sqrt{21} + \sqrt{9}}{\left(\sqrt{7}\right)^2 - \left(\sqrt{3}\right)^2}$$

$$= \frac{7 + 2\sqrt{21} + 3}{7 - 3}$$

$$= \frac{10 + 2\sqrt{21}}{4}$$

$$= \frac{2(5 + \sqrt{21})}{2 \cdot 2}$$

$$= \frac{\cancel{2}(5 + \sqrt{21})}{\cancel{2} \cdot 2}$$

$$= \frac{5 + \sqrt{21}}{2}$$

Practice Problem 12   Simplify $\dfrac{\sqrt{11} + \sqrt{2}}{\sqrt{11} - \sqrt{2}}$.

Multiply and simplify. Assume that all variables represent nonnegative numbers.

1. $(2\sqrt{6})(-3\sqrt{2})$

2. $(-4\sqrt{5})(2\sqrt{10})$

3. $(-3\sqrt{y})(\sqrt{5x})$

4. $(\sqrt{2x})(-7\sqrt{3y})$

5. $7\sqrt{x}(2\sqrt{3} - 5\sqrt{x})$

6. $3\sqrt{y}(4\sqrt{6} + 11\sqrt{y})$

7. $(3 - \sqrt{2})(8 + \sqrt{2})$

8. $(\sqrt{5} + 4)(\sqrt{5} - 1)$

9. $(2\sqrt{3} + \sqrt{2})(2\sqrt{3} - 4\sqrt{2})$

10. $(3\sqrt{3} + \sqrt{5})(\sqrt{3} - 2\sqrt{5})$

11. $(\sqrt{7} + 4\sqrt{5x})(2\sqrt{7} + 3\sqrt{5x})$

12. $(\sqrt{6} + 3\sqrt{3y})(5\sqrt{6} + 2\sqrt{3y})$

13. $(\sqrt{3} + 2\sqrt{2})(\sqrt{5} + \sqrt{3})$

14. $(3\sqrt{5} + \sqrt{3})(\sqrt{2} + 2\sqrt{5})$

15. $(\sqrt{x} - 2\sqrt{3x})(\sqrt{x} + 2\sqrt{3x})$

16. $(2\sqrt{x} + \sqrt{5x})(2\sqrt{x} - \sqrt{5x})$

17. $(\sqrt{5} - 2\sqrt{6})^2$

18. $(\sqrt{3} + 4\sqrt{7})^2$

19. $(\sqrt{3x + 4} + 3)^2$

20. $(\sqrt{2x + 1} - 2)^2$

21. $(6 - 5\sqrt{a})^2$

22. $(3\sqrt{b} + 4)^2$

23. $(\sqrt[3]{x^2})(3\sqrt[3]{4x} - 4\sqrt[3]{x^5})$

24. $(2\sqrt[3]{x})(\sqrt[3]{4x^2} - \sqrt[3]{14x})$

Divide and simplify. Assume that all variables represent positive numbers.

25. $\sqrt{\dfrac{49}{25}}$

26. $\sqrt{\dfrac{16}{36}}$

27. $\sqrt{\dfrac{12x}{49y^6}}$

28. $\sqrt{\dfrac{27a^4}{64x^2}}$

29. $\sqrt[3]{\dfrac{8x^5y^6}{27}}$

30. $\sqrt[3]{\dfrac{125a^3b^4}{64}}$

31. $\dfrac{\sqrt[3]{24x^3y^5}}{\sqrt[3]{3y^2}}$

32. $\dfrac{\sqrt[3]{250a^4b^6}}{\sqrt[3]{2a}}$

Simplify by rationalizing the denominator.

33. $\dfrac{3}{\sqrt{2}}$

34. $\dfrac{5}{\sqrt{7}}$

35. $\sqrt{\dfrac{4}{3}}$

36. $\sqrt{\dfrac{25}{2}}$

37. $\dfrac{1}{\sqrt{5y}}$

38. $\dfrac{1}{\sqrt{3x}}$

39. $\dfrac{x}{\sqrt{5} - \sqrt{2}}$

40. $\dfrac{y}{\sqrt{7} + \sqrt{3}}$

41. $\dfrac{\sqrt{3}}{\sqrt{5} - 2}$

42. $\dfrac{\sqrt{7}}{\sqrt{7} - 1}$

43. $\dfrac{\sqrt{x}}{\sqrt{3x} + \sqrt{2}}$

44. $\dfrac{\sqrt{x}}{\sqrt{5} + \sqrt{2x}}$

45. $\dfrac{\sqrt{5} + \sqrt{3}}{\sqrt{5} - \sqrt{3}}$

46. $\dfrac{\sqrt{11} - \sqrt{5}}{\sqrt{11} + \sqrt{5}}$

47. $\dfrac{\sqrt{3x} - 2\sqrt{y}}{\sqrt{3x} + \sqrt{y}}$

48. $\dfrac{\sqrt{x} + \sqrt{y}}{\sqrt{x} - 2\sqrt{y}}$

49. $\dfrac{x\sqrt{5} + 1}{\sqrt{5} + 2}$

50. $\dfrac{y\sqrt{2} - 1}{2\sqrt{2} + 1}$

51. $\dfrac{5}{\sqrt{2} - \sqrt{3}}$

52. $\dfrac{ab}{\sqrt{6} - \sqrt{7}}$

53. $\dfrac{\sqrt[3]{x^2}}{\sqrt[3]{7x^2}}$

54. $\dfrac{\sqrt[3]{6y^4}}{\sqrt[3]{4x^5}}$

To Think About

55. A student rationalized the denominator of $\dfrac{\sqrt{6}}{2\sqrt{3} - \sqrt{2}}$ and obtained $\dfrac{\sqrt{3} + 3\sqrt{2}}{5}$. Find a decimal approximation of each expression. Are the decimals equal? Did the student do the work correctly?

56. A student rationalized the denominator of $\dfrac{\sqrt{5}}{\sqrt{5} + \sqrt{3}}$ and obtained $\dfrac{5 - \sqrt{15}}{2}$. Find a decimal approximation of each expression. Are the decimals equal? Did the student do the work correctly?

In calculus, students are sometimes required to rationalize the numerator of an expression. In this case the numerator will not have a radical in the answer. Rationalize the numerator in each of the following:

57. $\dfrac{\sqrt{3} + 2\sqrt{7}}{8}$

58. $\dfrac{\sqrt{5} - 4\sqrt{3}}{6}$

Applications

The cost of fertilizing a lawn is $0.18 per square foot. Find the cost to fertilize each of the following triangular lawns. Round your answers to the nearest cent.

▲ **59.** The base of the triangle is $\sqrt{21}$ feet, and the altitude is $\sqrt{50}$ feet.

▲ **60.** The base of the triangle is $\sqrt{17}$ feet, and the altitude is $\sqrt{40}$ feet.

▲ **61.** A medical doctor has designed a pacemaker that has a rectangular control panel. This rectangle has a width of $\sqrt{x} + 3$ millimeters and a length of $\sqrt{x} + 5$ millimeters. Find the area in square millimeters of this rectangle.

▲ **62.** An FBI agent has designed a secret listening device that has a rectangular base. The rectangle has a width of $\sqrt{x} + 7$ centimeters and a length of $\sqrt{x} + 11$ centimeters. Find the area in square centimeters of this rectangle.

Cumulative Review Problems

Solve for x and y.

63. $2x + 3y = 13$
$5x - 2y = 4$

Solve for x, y, and z.

64. $3x - y - z = 5$
$2x + 3y - z = -16$
$x + 2y + 2z = -3$

65. A cup of strong coffee contains about 200 milligrams of caffeine. A cup of strong tea contains 80 milligrams of caffeine. Juanita used to drink 1 cup of each every day. However, she resolved on January 1 to reduce her intake of caffeine to less than 18 milligrams per day. On January 2 she cut her consumption of both coffee and tea in half. Three days later she again cut her consumption in half. If she continues this pattern of reduction, on what day will she reach her goal?

66. Juanita's husband, Carlos, has several cups of coffee and tea each day. On January 1, he had 11 cups in total and consumed a total of 1480 milligrams of caffeine. Using the information in Exercise 65, find out how many cups of coffee and how many cups of tea he consumed. If he cut his consumption of coffee and tea in half on January 2 and continues to cut his consumption of coffee and tea in half every 4 days after that, when will he reach his goal of fewer than 24 milligrams per day?

Student Learning Objectives

After studying this section, you will be able to:

1 Solve a radical equation that requires squaring each side once.

2 Solve a radical equation that requires squaring each side twice.

SSM PH TUTOR CD & VIDEO MATH PRO WEB
CENTER

---

1 Solving a Radical Equation by Squaring Each Side Once

A **radical equation** is an equation with a variable in one or more of the radicals. $3\sqrt{x} = 8$ and $\sqrt{3x - 1} = 5$ are radical equations. We solve radical equations by raising each side of the equation to the appropriate power. In other words, we square both sides if the radicals are square roots, cube both sides if the radicals are cube roots, and so on. Once we have done this, solving for the unknown becomes routine.

Sometimes after we square each side, we obtain a quadratic equation. In this case we collect all terms on one side and use the zero factor method that we developed in Section 6.7. After solving the equation, *always* check your answers to see whether extraneous solutions have been introduced.

We will now generalize this rule because it is very useful in higher-level mathematics courses.

> **Raising Each Side of an Equation to a Power**
>
> If $y = x$, then $y^n = x^n$, for all natural numbers n.

EXAMPLE 1 Solve $\sqrt{2x + 9} = x + 3$.

$$(\sqrt{2x + 9})^2 = (x + 3)^2 \quad \text{Square each side.}$$
$$2x + 9 = x^2 + 6x + 9 \quad \text{Simplify.}$$
$$0 = x^2 + 4x \quad \text{Collect all terms on one side.}$$
$$0 = x(x + 4) \quad \text{Factor.}$$
$$x = 0 \quad \text{or} \quad x + 4 = 0$$
$$x = 0 \qquad\qquad x = -4 \quad \text{Solve for } x.$$

Check:

For $x = 0$: $\sqrt{2(0) + 9} \stackrel{?}{=} 0 + 3$ For $x = -4$: $\sqrt{2(-4) + 9} \stackrel{?}{=} -4 + 3$

$$\sqrt{9} \stackrel{?}{=} 3 \qquad\qquad\qquad \sqrt{1} \stackrel{?}{=} -1$$
$$3 = 3 \checkmark \qquad\qquad\qquad 1 \neq -1$$

Therefore, 0 is the only solution to this equation.

Practice Problem 1 Solve and check your solution(s). $\sqrt{3x - 8} = x - 2$

As you begin to solve more complicated radical equations, it is important to make sure that one radical expression is alone on one side of the equation. This is often referred to as **isolating the radical term**.

EXAMPLE 2 Solve $\sqrt{10x + 5} - 1 = 2x$.

$$\sqrt{10x + 5} = 2x + 1 \quad \text{Isolate the radical term.}$$
$$(\sqrt{10x + 5})^2 = (2x + 1)^2 \quad \text{Square each side.}$$
$$10x + 5 = 4x^2 + 4x + 1 \quad \text{Simplify.}$$
$$0 = 4x^2 - 6x - 4 \quad \text{Collect all terms on one side.}$$
$$0 = 2(2x^2 - 3x - 2) \quad \text{Factor out the common factor.}$$
$$0 = 2(2x + 1)(x - 2) \quad \text{Factor completely.}$$
$$2x + 1 = 0 \quad \text{or} \quad x - 2 = 0 \quad \text{Solve for } x.$$
$$2x = -1 \qquad\qquad x = 2$$
$$x = -\frac{1}{2}$$

Graphing Calculator

Solving Radical Equations

On a graphing calculator Example 1 can be solved in two ways. Let $y_1 = \sqrt{2x + 9}$ and let $y_2 = x + 3$. Use your graphing calculator to determine where y_1 intersects y_2. What value of x do you obtain? Now let $y = \sqrt{2x + 9} - x - 3$ and find the value of x when $y = 0$. What value of x do you obtain? Which method seems more efficient?

Use the method above that you found most efficient to solve the following equations, and round your answers to the nearest tenth.

$$\sqrt{x + 9.5} = x - 2.3$$
$$\sqrt{6x + 1.3} = 2x - 1.5$$

Check:

$$x = -\frac{1}{2}: \quad \sqrt{10\left(-\frac{1}{2}\right) + 5} - 1 \overset{?}{=} 2\left(-\frac{1}{2}\right) \qquad x = 2: \quad \sqrt{10(2) + 5} - 1 \overset{?}{=} 2(2)$$

$$\sqrt{-5 + 5} - 1 \overset{?}{=} -1 \qquad\qquad\qquad \sqrt{25} - 1 \overset{?}{=} 4$$

$$\sqrt{0} - 1 \overset{?}{=} -1 \qquad\qquad\qquad 5 - 1 \overset{?}{=} 4$$

$$-1 = -1 \;\checkmark \qquad\qquad\qquad 4 = 4 \;\checkmark$$

Both answers check, so $-\dfrac{1}{2}$ and 2 are roots of the equation.

Practice Problem 2 Solve and check your solution(s). $\sqrt{x + 4} = x + 4$ ⬤

2 *Solving a Radical Equation by Squaring Each Side Twice*

In some exercises, we must square each side twice in order to remove all the radicals. It is important to isolate at least one radical before squaring each side.

⬤ **EXAMPLE 3** Solve $\sqrt{5x + 1} - \sqrt{3x} = 1$.

$$\sqrt{5x + 1} = 1 + \sqrt{3x} \qquad\qquad \text{Isolate one of the radicals.}$$

$$\left(\sqrt{5x + 1}\right)^2 = \left(1 + \sqrt{3x}\right)^2 \qquad\qquad \text{Square each side.}$$

$$5x + 1 = \left(1 + \sqrt{3x}\right)\left(1 + \sqrt{3x}\right)$$

$$5x + 1 = 1 + 2\sqrt{3x} + 3x$$

$$2x = 2\sqrt{3x} \qquad\qquad \text{Isolate the remaining radical.}$$

$$x = \sqrt{3x} \qquad\qquad \text{Divide each side by 2.}$$

$$(x)^2 = \left(\sqrt{3x}\right)^2 \qquad\qquad \text{Square each side.}$$

$$x^2 = 3x$$

$$x^2 - 3x = 0 \qquad\qquad \text{Collect all terms on one side.}$$

$$x(x - 3) = 0$$

$$x = 0 \quad \text{or} \quad x - 3 = 0 \qquad\qquad \text{Solve for } x.$$

$$x = 3$$

Check:

$$x = 0: \quad \sqrt{5(0) + 1} - \sqrt{3(0)} \overset{?}{=} 1 \qquad x = 3: \quad \sqrt{5(3) + 1} - \sqrt{3(3)} \overset{?}{=} 1$$

$$\sqrt{1} - \sqrt{0} \overset{?}{=} 1 \qquad\qquad\qquad \sqrt{16} - \sqrt{9} \overset{?}{=} 1$$

$$1 = 1 \;\checkmark \qquad\qquad\qquad 1 = 1 \;\checkmark$$

Both answers check. The solutions are 0 and 3.

Practice Problem 3 Solve and check your solution(s). $\sqrt{2x + 5} - 2\sqrt{2x} = 1$ ⬤

We will now formalize the procedure for solving radical equations.

Procedure for Solving Radical Equations

1. Perform algebraic operations to obtain one radical by itself on one side of the equation.
2. If the equation contains square roots, square each side of the equation. Otherwise, raise each side to the appropriate power for third- and higher-order roots.
3. Simplify, if possible.
4. If the equation still contains a radical, repeat steps 1 to 3.
5. Collect all terms on one side of the equation.
6. Solve the resulting equation.
7. Check all apparent solutions. Solutions to radical equations must be verified.

● **EXAMPLE 4** Solve $\sqrt{2y + 5} - \sqrt{y - 1} = \sqrt{y + 2}$.

$$\left(\sqrt{2y + 5} - \sqrt{y - 1}\right)^2 = \left(\sqrt{y + 2}\right)^2$$

$$\left(\sqrt{2y + 5} - \sqrt{y - 1}\right)\left(\sqrt{2y + 5} - \sqrt{y - 1}\right) = y + 2$$

$$2y + 5 - 2\sqrt{(y - 1)(2y + 5)} + y - 1 = y + 2$$

$$-2\sqrt{(y - 1)(2y + 5)} = -2y - 2$$

$$\sqrt{(y - 1)(2y + 5)} = y + 1 \qquad \text{Divide each side by } -2.$$

$$\left(\sqrt{2y^2 + 3y - 5}\right)^2 = (y + 1)^2 \qquad \text{Square each side.}$$

$$2y^2 + 3y - 5 = y^2 + 2y + 1$$

$$y^2 + y - 6 = 0 \qquad \text{Collect all terms on one side.}$$

$$(y + 3)(y - 2) = 0$$

$$y = -3 \quad \text{or} \quad y = 2$$

Check: Verify that 2 is a valid solution but -3 is not a valid solution.

Practice Problem 4 Solve and check your solution(s).

$$\sqrt{y - 1} + \sqrt{y - 4} = \sqrt{4y - 11}$$

8.5 Exercises

Verbal and Writing Skills

1. Before squaring each side of a radical equation, what step should be taken first?

2. Why do we have to check the solutions when we solve radical equations?

Solve each radical equation. Check your solution(s).

3. $\sqrt{8x + 1} = 5$

4. $\sqrt{5x - 4} = 6$

5. $2x = \sqrt{x + 3}$

6. $3x = \sqrt{9x - 2}$

7. $y - \sqrt{y - 3} = 5$

8. $\sqrt{2y - 4} + 2 = y$

9. $\sqrt{y + 1} - 1 = y$

10. $5 + \sqrt{2y + 5} = y$

11. $x - 2\sqrt{x - 3} = 3$

12. $2\sqrt{4x + 1} + 5 = x + 9$

13. $\sqrt{3x^2 - x} = x$

14. $\sqrt{5x^2 - 3x} = 2x$

15. $\sqrt[3]{2x + 3} = 2$

16. $\sqrt[3]{3x - 6} = 3$

17. $\sqrt[3]{4x - 1} = 3$

18. $\sqrt[3]{3 - 5x} = 2$

19. $\sqrt{x + 4} = 1 + \sqrt{x - 3}$

20. $\sqrt{5x + 1} = 1 + \sqrt{3x}$

21. $\sqrt{x + 6} = 1 + \sqrt{x + 2}$

22. $\sqrt{3x + 1} - \sqrt{x - 4} = 3$

23. $\sqrt{6x + 6} = 1 + \sqrt{4x + 5}$

24. $\sqrt{8x + 17} = \sqrt{2x + 8} + 3$

25. $\sqrt{2x + 9} - \sqrt{x + 1} = 2$ **26.** $\sqrt{4x + 6} = \sqrt{x + 1} - \sqrt{x + 5}$ **27.** $\sqrt{3x + 4} + \sqrt{x + 5} = \sqrt{7 - 2x}$

28. $\sqrt{2x + 6} = \sqrt{7 - 2x} + 1$ **29.** $2\sqrt{x} - \sqrt{x - 5} = \sqrt{2x - 2}$ **30.** $\sqrt{3 - 2\sqrt{x}} = \sqrt{x}$

Optional Graphing Calculator Problems

Solve for x. Round your answer to four decimal places.

31. $x = \sqrt{5.326x - 1.983}$

32. $\sqrt[3]{5.62x + 9.93} = 1.47$

Applications

33. When a car traveling on wet pavement at a speed V in miles per hour stops suddenly, it will produce skid marks of length S feet according to the formula $V = 2\sqrt{3S}$.
  (a) Solve the equation for S.
  (b) Use your result from **(a)** to find the length of the skid mark S if the car is traveling at 18 miles per hour.

34. The volume V of a steel container inside a flight data recorder is defined by the equation

$$x = \sqrt{\frac{V}{5}},$$

where x is the sum of the length and the width of the container in inches and the height of the container is 5 inches.
  (a) Solve the equation for V.
  (b) Use the result from **(a)** to find the volume of the container whose length and width total 3.5 inches.

Recently an experiment was conducted relating the speed a car is traveling and the stopping distance. In this experiment, a car is traveling on dry pavement at a constant rate of speed. From the instant that a driver recognizes the need to stop, the number of feet it takes for him to stop the car is recorded. For example, for a driver traveling at 50 miles per hour, a stopping distance of 190 feet is required. In general, the stopping distance x in feet is related to the speed of the car y in miles per hour by the equation

$$0.11y + 1.25 = \sqrt{3.7625 + 0.22x}.$$

Source: National Highway Traffic Safety Administration.

35. Solve this equation for x.

36. Use your answer from Exercise 35 to find what the stopping distance x would have been for a car traveling at $y = 60$ miles per hour.

To Think About

37. The solution to the equation

$$\sqrt{x^2 - 4x + c} = x - 1$$

is $x = 4$. What is the value of c?

38. The solution to the equation

$$\sqrt{x + b} - \sqrt{x} = -2$$

is $x = 16$. What is the value of b?

Cumulative Review Problems

Simplify.

39. $\left(4^3 x^6\right)^{2/3}$

40. $\left(2^{-3} x^{-6}\right)^{1/3}$

41. $\sqrt[3]{-216 x^6 y^9}$

42. $\sqrt[4]{64 x^{12} y^{16}}$

▲ **43.** The area of the top of a solid rectangular coffee table measures $(4x^2 + 2x + 9)$ square centimeters. The height of the coffee table measures $(2x + 3)$ centimeters. Find the volume of the solid coffee table.

▲ **44.** A rectangular display case has $(2r^2 + 5r + 3)$ boxes of cereal on each shelf. The display case has $(2r + 4)$ shelves. Find the number of cereal boxes in the display case.

45. The Mississippi Magic paddleboat can travel 12 miles per hour in still water. After traveling for 3 hours downstream with the current, it takes 5 hours to get upstream with the current and return to its original starting point. What is the speed of the current?

46. Louise Elton rides the ski lift for 1.75 miles to the top of Mount Gray. Once she is there, she immediately skis directly down the mountain. The ski trail winding down the mountain is 2.5 miles long. If she skis five times as fast as the lift runs and the round trip takes 45 minutes, find the rate at which she skis.

Complex Numbers

Student Learning
Objectives

*After studying this section, you
will be able to:*

1. Simplify expressions
involving complex
numbers.

2. Add and subtract complex
numbers.

3. Multiply complex
numbers.

4. Evaluate complex numbers
of the form i^n.

5. Divide two complex
numbers.

SSM
PH TUTOR   CD & VIDEO   MATH PRO   WEB
CENTER

1 *Simplifying Expressions Involving Complex Numbers*

Until now we have not been able to solve an equation such as $x^2 = -4$ because there is no *real* number that satisfies this equation. However, this equation *does* have a nonreal solution. This solution is an *imaginary number*.

We define a new number:

$$i = \sqrt{-1} \text{ or } i^2 = -1.$$

Now let us use this procedure

$$\sqrt{-a} = \sqrt{-1}\sqrt{a}$$

and see if it is valid.

Then $\sqrt{-4} = \sqrt{4(-1)} = \sqrt{4}\sqrt{-1} = \sqrt{4} \cdot i = 2i.$

Thus, one solution to the equation $x^2 = -4$ is $2i$. Let's check it.

$$x^2 = -4$$
$$(2i)^2 \overset{?}{=} -4$$
$$4i^2 \overset{?}{=} -4$$
$$4(-1) \overset{?}{=} -4$$
$$-4 = -4 \checkmark$$

The value $-2i$ is also a solution. You should verify this.

Now we formalize our definitions and give some examples of imaginary numbers.

Definition of Imaginary Number

The **imaginary number *i*** is defined as follows:

$$i = \sqrt{-1} \quad \text{and} \quad i^2 = -1.$$

The set of imaginary numbers consists of numbers of the form bi, where b is a real number and $b \neq 0$.

Definition

For all positive real numbers a,

$$\sqrt{-a} = \sqrt{-1}\sqrt{a} = i\sqrt{a}.$$

EXAMPLE 1 Simplify. **(a)** $\sqrt{-36}$ **(b)** $\sqrt{-17}$

(a) $\sqrt{-36} = \sqrt{-1}\sqrt{36} = (i)(6) = 6i$

(b) $\sqrt{-17} = \sqrt{-1}\sqrt{17} = i\sqrt{17}$

Practice Problem 1 Simplify. **(a)** $\sqrt{-49}$ **(b)** $\sqrt{-31}$

To avoid confusing $\sqrt{17}i$ with $\sqrt{17i}$, we write the *i* before the radical. That is, we write $i\sqrt{17}$. The *i* is not part of the radicand in the first expression.

⬭ **EXAMPLE 2** Simplify $\sqrt{-45}$.

$$\sqrt{-45} = \sqrt{-1}\sqrt{45} = i\sqrt{45} = i\sqrt{9}\sqrt{5} = 3i\sqrt{5}$$

Practice Problem 2 Simplify $\sqrt{-98}$. ⬭

The rule $\sqrt{a}\sqrt{b} = \sqrt{ab}$ requires that $a \geq 0$ and $b \geq 0$. Therefore, we cannot use our product rule when the radicands are negative unless we first use the definition of $\sqrt{-1}$. Recall that

$$\sqrt{-1} \cdot \sqrt{-1} = i \cdot i = i^2 = -1.$$

⬭ **EXAMPLE 3** Multiply $\sqrt{-16} \cdot \sqrt{-25}$.

First we must use the definition $\sqrt{-1} = i$. Thus, we have the following:

$$\begin{aligned}(\sqrt{-16})(\sqrt{-25}) &= (i\sqrt{16})(i\sqrt{25}) \\ &= i^2(4)(5) \\ &= -1(20) \qquad i^2 = -1. \\ &= -20\end{aligned}$$

Practice Problem 3 Multiply $\sqrt{-8} \cdot \sqrt{-2}$. ⬭

Now we formally define a complex number.

Definition

A number that can be written in the form $a + bi$, where a and b are real numbers, is a **complex number**. We say that a is the **real part** and bi is the **imaginary part**.

A real number is any rational or irrational number. Rational numbers are numbers that can be written as fractions. Irrational numbers cannot be expressed as fractions, such as $\sqrt{2}$ and π.

Under the above definition, every real number is also a complex number. For example, the real number 5 can be written as $5 + 0i$. Therefore, 5 is a complex number. In a similar fashion, the imaginary number $2i$ can be written as $0 + 2i$. So $2i$ is a complex number. Thus, the set of complex numbers includes the set of real numbers and the set of imaginary numbers.

Definition

Two complex numbers $a + bi$ and $c + di$ are equal if and only if $a = c$ and $b = d$.

This definition means that two complex numbers are equal if and only if their real parts are equal *and* their imaginary parts are equal.

⬭ **EXAMPLE 4** Find the real numbers x and y if $x + 3i\sqrt{7} = -2 + yi$.

By our definition, the real parts must be equal, so x must be -2; the imaginary parts must also be equal, so y must be $3\sqrt{7}$.

Practice Problem 4 Find the real numbers x and y if $-7 + 2yi\sqrt{3} = x + 6i\sqrt{3}$. ⬭

Graphing Calculator

Complex Numbers

Some graphing calculators, such as the TI-83, have a complex number mode. If your graphing calculator has this capability, you will be able to use it to do complex number operations. First you must use the Mode command to transfer selection from "Real" to "Complex" or "a + bi." To verify your status, try to find $\sqrt{-7}$ on your graphing calculator. If you obtain an approximate answer of "2.645751311 *i*," then your calculator is operating in the complex number mode. If you obtain "ERROR: NONREAL ANSWER," then your calculator is not operating in the complex number mode.

2 Adding and Subtracting Complex Numbers

Adding and Subtracting Complex Numbers

For all real numbers a, b, c, and d,

$$(a + bi) + (c + di) = (a + c) + (b + d)i \quad \text{and}$$
$$(a + bi) - (c + di) = (a - c) + (b - d)i.$$

In other words, to combine complex numbers we add (or subtract) the real parts, and we add (or subtract) the imaginary parts.

EXAMPLE 5 Subtract $(6 - 2i) - (3 - 5i)$.

$$(6 - 2i) - (3 - 5i) = (6 - 2i) + (-3 + 5i) = (6 - 3) + (-2 + 5)i = 3 + 3i$$

Practice Problem 5 Subtract $(3 - 4i) - (-2 - 18i)$.

3 Multiplying Complex Numbers

As we might expect, the procedure for multiplying complex numbers is similar to the procedure for multiplying polynomials. We will see that the complex numbers obey the associative, commutative, and distributive properties.

EXAMPLE 6 Multiply $(7 - 6i)(2 + 3i)$.

Use FOIL.

$$\begin{aligned}
(7 - 6i)(2 + 3i) &= (7)(2) + (7)(3i) + (-6i)(2) + (-6i)(3i) \\
&= 14 + 21i - 12i - 18i^2 \\
&= 14 + 21i - 12i - 18(-1) \\
&= 14 + 21i - 12i + 18 \\
&= 32 + 9i
\end{aligned}$$

Practice Problem 6 Multiply $(4 - 2i)(3 - 7i)$.

EXAMPLE 7 Multiply $3i(4 - 5i)$.

Use the distributive property.

$$\begin{aligned}
3i(4 - 5i) &= (3)(4)i + (3)(-5)i^2 \\
&= 12i - 15i^2 \\
&= 12i - 15(-1) \\
&= 15 + 12i
\end{aligned}$$

Practice Problem 7 Multiply $-2i(5 + 6i)$.

It is important to rewrite any square roots with negative radicands using i notation for complex numbers before attempting to do any multiplication.

● **EXAMPLE 8** Multiply and simplify your answers. $\sqrt{-25} \cdot \sqrt{-36}$

Before any other steps are taken, we must first rewrite the expression using i notation.

$$\sqrt{-25} \cdot \sqrt{-36} = (i\sqrt{25})(i\sqrt{36})$$

Now we can finish the calculations using the properties of complex numbers.

$$\begin{aligned}(i\sqrt{25})(i\sqrt{36}) &= (5i)(6i) \\ &= 30i^2 \\ &= 30(-1) \\ &= -30\end{aligned}$$

Practice Problem 8 Multiply. Then simplify your answer. $\sqrt{-50} \cdot \sqrt{-4}$ ●

● **4** *Evaluating Complex Numbers of the Form i^n*

How would you evaluate i^n, where n is any positive integer? What if n is a negative integer? We look for a pattern. We have defined

$$i^2 = -1.$$

We could write

$$i^3 = i^2 \cdot i = (-1)i = -i.$$

We also have the following:

$$i^4 = i^2 \cdot i^2 = (-1)(-1) = +1$$
$$i^5 = i^4 \cdot i = (+1)i = +i$$

We notice that $i^5 = i$. Let's look at i^6.

$$i^6 = i^4 \cdot i^2 = (+1)(-1) = -1$$

We begin to see a pattern that starts with i and repeats itself for i^5. Will $i^7 = -i$? Why or why not?

Values			
of i^n	$i = i$	$i^5 = i$	$i^9 = i$
	$i^2 = -1$	$i^6 = -1$	$i^{10} = -1$
	$i^3 = -i$	$i^7 = -i$	$i^{11} = -i$
	$i^4 = +1$	$i^8 = +1$	$i^{12} = +1$

We can use this pattern to evaluate powers of i.

● **EXAMPLE 9** Evaluate. **(a)** i^{36} **(b)** i^{27}

(a) $i^{36} = (i^4)^9 = (1)^9 = 1$

(b) $i^{27} = (i^{24+3}) = (i^{24})(i^3) = (i^4)^6(i^3) = (1)^6(-i) = -i$

This suggests a quick method for evaluating powers of i. Divide the exponent by 4. i^4 raised to any power will be 1. Then use the first column of the values of i^n chart above to evaluate the remainder.

Practice Problem 9 Evaluate. **(a)** i^{42} **(b)** i^{53} ●

5 *Dividing Two Complex Numbers*

The complex numbers $a + bi$ and $a - bi$ are called **conjugates**. The product of two complex conjugates is always a real number.

$$(a + bi)(a - bi) = a^2 - abi + abi - b^2i^2$$
$$= a^2 - b^2(-1)$$
$$= a^2 + b^2$$

When dividing two complex numbers, we want to remove any expression involving i from the denominator. So we multiply the numerator and denominator by the conjugate of the denominator. This is just what we did when we rationalized the denominator in a radical expression.

EXAMPLE 10 Divide $\dfrac{7 + i}{3 - 2i}$.

$$\frac{(7 + i)}{(3 - 2i)} \cdot \frac{(3 + 2i)}{(3 + 2i)} = \frac{21 + 14i + 3i + 2i^2}{9 - 4i^2} = \frac{21 + 17i + 2(-1)}{9 - 4(-1)} = \frac{21 + 17i - 2}{9 + 4}$$
$$= \frac{19 + 17i}{13} \quad \text{or} \quad \frac{19}{13} + \frac{17}{13}i$$

Practice Problem 10 Divide $\dfrac{4 + 2i}{3 + 4i}$.

EXAMPLE 11 Divide $\dfrac{3 - 2i}{4i}$.

The conjugate of $0 + 4i$ is $0 - 4i$ or simply $-4i$.

$$= \frac{(3 - 2i)}{(4i)} \cdot \frac{(-4i)}{(-4i)} = \frac{-12i + 8i^2}{-16i^2} = \frac{-12i + 8(-1)}{-16(-1)}$$
$$= \frac{-8 - 12i}{16} = \frac{\cancel{4}(-2 - 3i)}{\cancel{4} \cdot 4}$$
$$= \frac{-2 - 3i}{4} \quad \text{or} \quad -\frac{1}{2} - \frac{3}{4}i$$

Practice Problem 11 Divide $\dfrac{5 - 6i}{-2i}$.

Verbal and Writing Skills

1. Does $x^2 = -9$ have a real number solution? Why or why not?

2. Describe a complex number and give an example(s).

3. Are the complex numbers $2 + 3i$ and $3 + 2i$ equal? Why or why not?

4. Describe in your own words how to add or subtract complex numbers.

Simplify. Express in terms of i.

5. $\sqrt{-36}$

6. $\sqrt{-81}$

7. $\sqrt{-50}$

8. $\sqrt{-48}$

9. $\sqrt{-\dfrac{1}{4}}$

10. $\sqrt{-\dfrac{1}{9}}$

11. $-\sqrt{-81}$

12. $-\sqrt{-36}$

13. $2 + \sqrt{-3}$

14. $5 + \sqrt{-7}$

15. $-3 + \sqrt{-24}$

16. $-6 - \sqrt{-32}$

Find the real numbers x and y.

17. $x - 3i = 5 + yi$

18. $x - 6i = 7 + yi$

19. $1.3 - 2.5yi = x - 5i$

20. $3.4 - 0.8i = 2x - yi$

21. $23 + yi = 17 - x + 3i$

22. $2 + x - 11i = 19 + yi$

Perform the addition or subtraction.

23. $\left(-\dfrac{3}{2} + \dfrac{1}{2}i \right) + \left(\dfrac{5}{2} - \dfrac{3}{2}i \right)$

24. $\left(\dfrac{3}{4} - \dfrac{3}{4}i \right) + \left(\dfrac{9}{4} + \dfrac{5}{4}i \right)$

25. $(2.8 - 0.7i) - (1.6 - 2.8i)$

26. $(5.4 + 4.1i) - (4.8 + 2.6i)$

Multiply and simplify your answers. Place in i notation before doing any other operations.

27. $(2 + 3i)(2 - i)$

28. $(4 - 6i)(2 + i)$

29. $5i - 2(-4 + i)$

30. $12i - 6(3 + i)$

31. $2i(5i - 6)$

32. $4i(7 - 2i)$

33. $\left(\dfrac{1}{2} + i\right)^2$

34. $\left(\dfrac{1}{3} - i\right)^2$

35. $(i\sqrt{3})(i\sqrt{7})$

36. $(i\sqrt{2})(i\sqrt{6})$

37. $(\sqrt{-3})(\sqrt{-2})$

38. $(\sqrt{-5})(\sqrt{-3})$

39. $(\sqrt{-36})(\sqrt{-4})$

40. $(\sqrt{-25})(\sqrt{-9})$

41. $(3 + \sqrt{-2})(4 + \sqrt{-5})$

42. $(2 + \sqrt{-3})(6 + \sqrt{-2})$

Evaluate.

43. i^{17}

44. i^{21}

45. i^{24}

46. i^{16}

47. i^{46}

48. i^{83}

49. $i^{30} + i^{28}$

50. $i^{26} + i^{24}$

Divide.

51. $\dfrac{2 + i}{3 - i}$

52. $\dfrac{4 + 2i}{2 - i}$

53. $\dfrac{3i}{4 + 2i}$

54. $\dfrac{-2i}{3 + 5i}$

55. $\dfrac{5 - 2i}{6i}$

56. $\dfrac{7 + 10i}{3i}$

57. $\dfrac{7}{5 - 6i}$

58. $\dfrac{3}{4 + 2i}$

59. $\dfrac{5 - 2i}{3 + 2i}$

60. $\dfrac{6 + 3i}{6 - 3i}$

61. $\dfrac{2 - 3i}{2 + i}$

62. $\dfrac{4 - 3i}{5 + 2i}$

Optional Graphing Calculator Problems

Perform each operation to obtain approximate answers.

63. $(29.3 + 56.2i)^2$

64. $\dfrac{196 - 34.8i}{24.9 + 56.4i}$

Applications

The impedance Z in an alternating current circuit (like the one used in your home and in your classroom) is given by the formula Z = V/I, where V is the voltage and I is the current.

65. Find the value of Z if $V = 3 + 2i$ and $I = 3i$.

66. Find the value of Z if $V = 4 + 2i$ and $I = -3i$.

Cumulative Review Problems

67. A grape juice factory produces juice in three different types of containers. $x + 3$ hours per week are spent on producing juice in glass bottles. $2x - 5$ hours per week are spent on producing juice in cans. $4x + 2$ hours per week are spent on producing juice in plastic bottles. If the factory operates 105 hours per week, how much time is spent producing juice in each type of container?

68. One summer day, Sunday's Ice Cream Parlor sold 120 bowls of ice cream. A small bowl costs $1.50 and a large bowl costs $2.50. A profit of $243.00 was made. How many bowls of each size were sold?

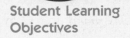

8.7 Variation

Student Learning Objectives

After studying this section, you will be able to:

1 Solve problems requiring the use of direct variation.

2 Solve problems requiring the use of inverse variation.

3 Solve problems requiring the use of joint or combined variation.

SSM
PH TUTOR CENTER • CD & VIDEO • MATH PRO • WEB

1 Solving Problems Using Direct Variation

Many times in daily life we observe how a change in one quantity produces a change in another. If we order one large pepperoni pizza, we pay $8.95. If we order two large pepperoni pizzas, we pay $17.90. For three large pepperoni pizzas, it is $26.85. The change in the number of pizzas we order results in a corresponding change in the price we pay.

Notice that the price we pay for each pizza stays the same. That is, each pizza costs $8.95. The number of pizzas changes, and the corresponding price of the order changes. From our experience with functions and with equations, we see that the cost of the order is $y = \$8.95x$, where the price y depends on the number of pizzas x. We see that the variable y is a constant multiple of x. The two variables are said to *vary directly*. That is, y varies directly with x. We write a general equation that represents this idea as follows:

When we solve problems using direct variation, we usually are not given the value of the constant of variation k. This is something that we must find. Usually all we are given is a point of reference. That is, we are given the value of y for a specific value of x. Using this information, we can find k.

EXAMPLE 1
The time of a pendulum's swing varies directly with the square root of its length. If the pendulum is 1 foot long when the time is 0.2 second, find the time if the length is 4 feet.

Let t = the time and L = the length.
We then have the equation

$$t = k\sqrt{L}.$$

We can evaluate k by substituting $L = 1$ and $t = 0.2$ into the equation.

$$t = k\sqrt{L}$$
$$0.2 = k(\sqrt{1})$$
$$0.2 = k \qquad \text{Because } \sqrt{1} = 1.$$

Now we know the value of k and can write the equation more completely.

$$t = 0.2\sqrt{L}$$

When $L = 4$, we have the following:

$$t = 0.2\sqrt{4}$$
$$t = (0.2)(2)$$
$$t = 0.4 \text{ second}$$

Practice Problem 1 The maximum speed of a racing car varies directly with the square root of the horsepower of the engine. If the maximum speed of a car with 256 horsepower is 128 miles per hour, what is the maximum speed of a car with 225 horsepower?

2 Solving Problems Using Inverse Variation

In some cases when one variable increases, another variable decreases. For example, as the amount of money you earn each year increases, the percentage of your income that you get to keep after taxes decreases. If one variable is a constant multiple of the reciprocal of the other, the two variables are said to *vary inversely*.

468

EXAMPLE 2   If y varies inversely with x, and $y = 12$ when $x = 5$, find the value of y when $x = 14$.

If y varies inversely with x, we can write the equation $y = \dfrac{k}{x}$. We can find the value of k by substituting the values $y = 12$ and $x = 5$.

$$12 = \frac{k}{5}$$
$$60 = k$$

We can now write the equation

$$y = \frac{60}{x}.$$

To find the value of y when $x = 14$, we substitute 14 for x in the equation.

$$y = \frac{60}{14}$$
$$y = \frac{30}{7}$$

Practice Problem 2   If y varies inversely with x, and $y = 45$ when $x = 16$, find the value of y when $x = 36$.

EXAMPLE 3   The amount of light from a light source varies inversely with the square of the distance to the light source. If an object receives 6.25 lumens when the light source is 8 meters away, how much light will the object receive if the light source is 4 meters away?

Let $L = $ the amount of light and $d = $ the distance to the light source.
Since the amount of light varies inversely with the *square of the distance* to the light source, we have

$$L = \frac{k}{d^2}.$$

Substituting the known values of $L = 6.25$ and $d = 8$, we can find the value of k.

$$6.25 = \frac{k}{8^2}$$
$$6.25 = \frac{k}{64}$$
$$400 = k$$

We are now able to write a more specific equation,

$$L = \frac{400}{d^2} .$$

We will use this to find L when $d = 4$ meters.

$$L = \frac{400}{4^2}$$

$$L = \frac{400}{16}$$

$$L = 25 \text{ lumens}$$

Check: Does this answer seem reasonable? Would we expect to have more light if we move closer to the light source? ✓

Practice Problem 3 If the amount of power in an electrical circuit is held constant, the resistance in the circuit varies inversely with the square of the amount of current. If the amount of current is 0.01 ampere, the resistance is 800 ohms. What is the resistance if the amount of current is 0.04 ampere?

3 ⬤ Solving Problems Using Joint or Combined Variation

Sometimes a quantity depends on the variation of two or more variables. This is called joint or **combined variation**.

⬤ **EXAMPLE 4** y varies directly with x and z and inversely with d^2. When $x = 7$, $z = 3$, and $d = 4$, the value of y is 20. Find the value of y when $x = 5$, $z = 6$, and $d = 2$.

We can write the equation

$$y = \frac{kxz}{d^2} .$$

To find the value of k, we substitute into the equation $y = 20$, $x = 7$, $z = 3$, and $d = 4$.

$$20 = \frac{k(7)(3)}{4^2}$$

$$20 = \frac{21k}{16}$$

$$320 = 21k$$

$$\frac{320}{21} = k$$

Now we substitute $\frac{320}{21}$ for k into our original equation.

$$y = \frac{\frac{320}{21} xz}{d^2} \quad \text{or} \quad y = \frac{320xz}{21d^2}$$

We use this equation to find y for the known values of x, z, and d. We want to find y when $x = 5$, $z = 6$, and $d = 2$.

$$y = \frac{320(5)(6)}{21(2)^2} = \frac{9600}{84}$$

$$y = \frac{800}{7}$$

Practice Problem 4 y varies directly with z and w^2 and inversely with x. $y = 20$ when $z = 3$, $w = 5$, and $x = 4$. Find y when $z = 4$, $w = 6$, and $x = 2$.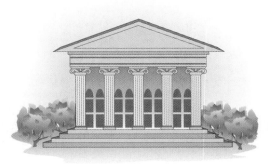

Many applied problems involve joint variation. For example, a cylindrical cement column has a safe load capacity that varies directly with the diameter raised to the fourth power and inversely with the square of its length.

Therefore, if d = diameter and l = length, the equation would be of the form

$$y = \frac{kd^4}{l^2}.$$

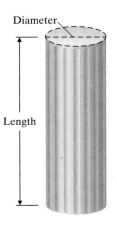

Diameter

Length

8.7 Exercises

Verbal and Writing Skills

1. Give an example in everyday life of direct variation and write an equation as a mathematical model.

2. The general equation $y = kx$ means that y varies _____ with x. k is called the _____ of variation.

3. If y varies inversely with x, we write the equation _____ .

4. Write a mathematical model for the following situation: The strength of a rectangular beam varies directly with its width and the square of its depth.

Round all answers to the nearest tenth.

5. If y varies directly with x and $y = 15$ when $x = 40$, find y when $x = 64$.

6. If y varies directly with x and $y = 42$ when $x = 35$, find y when $x = 100$.

7. A marine biology submarine was searching the waters for blue whales at 50 feet below the surface, where it experienced a pressure of 21 pounds per square inch (psi). If the pressure of water on a submerged object varies directly with its distance beneath the surface, how much pressure would the submarine have experienced if it had to dive to 170 feet?

8. A car's stopping distance varies directly with the square of its speed. A car that is traveling 30 miles per hour can stop in 40 feet. What distance will it take to stop if it is traveling 60 miles per hour?

9. When an object is dropped, the distance it falls in feet varies directly with the square of the duration of the fall in seconds. An apple that falls from a tree falls 1 foot in $\frac{1}{4}$ second. How far will it fall in 1 second? How far will it fall in 2 seconds?

10. A veterinarian specializing in marine biology is faced with operating on a dolphin. A cube-shaped aquarium has been built, which will allow the doctor to operate while keeping most of the dolphin's body submerged. The time it takes to fill a tank with water varies directly with the cube of each length of the side of the container. If a cube 2 meters on each side can be filled in 7 minutes, how long will it take to fill the dolphin tank, which is 3.5 meters on each side?

11. If y varies inversely with the square of x, and $y = 10$ when $x = 2$, find y when $x = 0.5$.

12. If y varies inversely with the square root of x, and $y = 1.8$ when $x = 0.04$, find y when $x = 0.3$.

13. Engineers have decided that part of the structure that houses certain elements of a satellite will use 9 inches of special insulation. The heat lost through a certain type of insulation varies inversely as the thickness of the insulation. If the heat lost through 6 inches of insulation is 2000 Btu per hour, how much heat will be lost with the 9 inches on the satellite?

14. The weight of an object on the Earth's surface varies inversely with the square of its distance from the center of the Earth. An object weighs 1000 pounds on the Earth's surface. This is approximately 4000 miles from the center of the Earth. How much would an object weigh 4500 miles from the center of the Earth?

15. Police officers can detect speeding by using variation. The speed of a car varies inversely with the time it takes to cover a certain fixed distance. Between two points on a highway, a car travels 45 miles per hour in 6 seconds. What is the speed of a car that travels the same distance in 9 seconds?

16. If the voltage in an electric circuit is kept at the same level, the current varies inversely with the resistance. The current measures 40 amperes when the resistance is 270 ohms. Find the current when the resistance is 100 ohms.

17. The weight that can be safely supported by a 2- by 6-inch support beam varies inversely with its length. A builder finds that a support beam that is 8 feet long will support 900 pounds. Find the weight that can be safely supported by a beam that is 18 feet long.

18. The speed that is required to maintain a satellite in a circular orbit around the Earth varies directly with the square root of the distance of the satellite from the center of the Earth. We will assume that the radius of the Earth is approximately 4000 miles. A satellite that is 100 miles above the surface of the Earth is orbiting at approximately 18,000 miles per hour. What speed would be necessary for the satellite to orbit 500 miles above the surface of the Earth? Round to nearest mile per hour.

19. The field intensity of a magnetic field varies directly with the force acting on it and inversely with the strength of the pole. If the intensity of the magnetic field is 4 oersteds when the force is 700 dynes and the strength of the pole is 200, find the intensity of the field if the force is 500 dynes and the strength of the pole is 250.

20. The attraction F of two masses m_1 and m_2 varies directly with the product of m_1 and m_2 and inversely with the square of the distance between the two bodies. If a force of 10 pounds attracts two bodies weighing 80 tons and 100 tons that are 100 miles apart, how great would the force be if the two bodies weighed 8 tons and 15 tons and were 20 miles apart?

21. Atmospheric drag tends to slow down moving objects. Atmospheric drag varies jointly with an object's surface area A and velocity v. If a Dodge Intrepid, traveling at a speed of 45 mph with a surface area of 37.8 square feet, experiences a drag of 222 newtons, how fast must a Dodge Caravan, with a surface area of 55 square feet, travel in order to experience a drag force of 450 newtons?

22. The force on a blade of a wind generator varies jointly with the product of the blade's area and the square of the wind velocity. The force of the wind is 20 pounds when the area is 3 square feet and the velocity is 30 feet per second. Find the force when the area is increased to 5 square feet and the velocity is reduced to 25 feet per second.

Cumulative Review Problems

Solve each of the following equations or word problems.

23. $3x^2 - 8x + 4 = 0$

24. $4x^2 = -28x + 32$

25. In Champaign, Illinois, the sales tax is 6.25%. Donny bought an amplifier for his stereo that cost $488.75 after tax. What was the original price of the amplifier?

26. It takes 7.5 gallons of white paint to properly paint lines on three tennis courts. How much paint is needed to paint twenty-two tennis courts?

27. Craig Emanuel has a photography studio in Los Angeles. He wants to frame his collection of 110 antique photographs in special gold leaf and silver frames. The price of each gold leaf frame is $140, and the price of each silver frame is $95. If Craig has $13,375 for the frames, how many of each frame can he buy to decorate his studio?

▲ **28.** A triangular New Year's noisemaking toy has a perimeter of 50 centimeters. The first side is $\frac{4}{5}$ as long as the second side. The third side is 2 centimeters shorter than the first side. Find the length of each side.

The Pythagorean Winning Percentage Formula

The Pythagorean winning percentage formula, developed by Bill James, is used to predict the record of a team based on its runs scored and runs allowed. The formula is:

$$\frac{(\text{Runs Scored})^2}{(\text{Runs Scored})^2 + (\text{Runs Allowed})^2}$$

In each part of the formula, an exponent of 1.83 is sometimes used instead of 2 because it provides a slightly more precise calculation.

The Pythagorean winning percentage formula was put to just such a use in a recent article, *So Who Would Win?* by Devin Clancy in USATODAY.com.

To get some hands on experience with this formula try the following exercises.

EXERCISES

1. The Cincinnati Reds scored 865 runs and allowed 711 runs in the 1999 season. Using the Pythagorean winning percentage formula (with the exponent of 2), calculate the team's predicted winning percent. Round to 3 decimal places.

   exponent of 2 in the winning percentage formula. What is the predicted winning percent? Did the accuracy of the prediction improve?

   2000 season. Using the Pythagorean winning percentage formula, with an exponent of 2, calculate the team's predicted winning percentage. Repeat the formula with an exponent of 1.83. Which calculation is closer to the exact value?

2. The Cincinnati Reds ended the 1999 season with a 0.589 win percentage. Calculate exercise 1 again and use 1.83 rather than an

3. The New York Yankees ended the 2000 season with a 0.540 win percentage. The team scored 871 runs and allowed 814 runs in the

Chapter 8 Organizer

Topic	Procedure	Examples
Multiplication of variables with rational exponents, p. 426.	$x^m x^n = x^{m+n}$	$(3x^{1/5})(-2x^{3/5}) = -6x^{4/5}$
Division of variables with rational exponents, p. 426.	$\dfrac{x^m}{x^n} = x^{m-n}, \qquad n \neq 0, x \neq 0$	$\dfrac{-16x^{3/20}}{24x^{5/20}} = -\dfrac{2x^{-1/10}}{3}$
Removing negative exponents, p. 425.	$x^{-n} = \dfrac{1}{x^n}, \qquad m$ and $n \neq 0,\ x$ and $y \neq 0$ $\dfrac{x^{-n}}{y^{-m}} = \dfrac{y^m}{x^n}$	Write with positive exponents. $3x^{-4} = \dfrac{3}{x^4}$ $\dfrac{2x^{-6}}{5y^{-8}} = \dfrac{2y^8}{5x^6}$ $4^{-2} = \dfrac{1}{4^2} = \dfrac{1}{16}$
Zero exponent, p. 428.	$x^0 = 1 \quad (\text{if } x \neq 0)$	$(3x^{1/2})^0 = 1$
Raising a variable with an exponent to a power, p. 425.	$(x^m)^n = x^{mn}$ $(xy)^n = x^n y^n$ $\left(\dfrac{x}{y}\right)^n = \dfrac{x^n}{y^n}, \quad y \neq 0$	$\left(x^{-1/2}\right)^{-2/3} = x^{1/3}$ $(3x^{-2}y^{-1/2})^{2/3} = 3^{2/3}x^{-4/3}y^{-1/3}$ $\left(\dfrac{4x^{-2}}{3^{-1}y^{-1/2}}\right)^{1/4} = \dfrac{4^{1/4}x^{-1/2}}{3^{-1/4}y^{-1/8}}$
Multiplication of expressions with rational exponents, p. 426.	Add exponents whenever expressions with the same base are multiplied.	$x^{2/3}(x^{1/3} - x^{1/4}) = x^{3/3} - x^{2/3+1/4} = x - x^{11/12}$
Higher-order roots, p. 432.	If x is a nonnegative real number, $\sqrt[n]{x}$ is a nonnegative nth root and has the property that $\left(\sqrt[n]{x}\right)^n = x.$ If x is a negative real number, $\left(\sqrt[n]{x}\right)^n = x$ when n is an odd integer. If x is a negative real number, $\left(\sqrt[n]{x}\right)^n$ is not a real number when n is an even integer. $\sqrt[3]{27} = 3$ because $3^3 = 27$.	$\sqrt[5]{-32} = -2$ because $(-2)^5 = -32$. $\sqrt[4]{-16}$ is *not* a real number.
Rational exponents and radicals, p. 435.	For positive integers m and n and any real number x for which $x^{1/n}$ is defined, $x^{m/n} = \left(\sqrt[n]{x}\right)^m = \sqrt[n]{x^m}.$ If it is also true that $x \neq 0$, then $x^{1/n} = \sqrt[n]{x}.$	Write as a radical: $x^{3/7} = \sqrt[7]{x^3},\ 3^{1/5} = \sqrt[5]{3}$ Write as an expression with a fractional exponent: $\sqrt[4]{w^3} = w^{3/4}$ Evaluate. $25^{3/2} = \left(\sqrt{25}\right)^3 = (5)^3 = 125$
Higher-order roots and absolute value, p. 436.	$\sqrt[n]{x^n} = |x|$ when n is an even positive integer. $\sqrt[n]{x^n} = x$ when n is an odd positive integer.	$\sqrt[6]{x^6} = |x|$ $\sqrt[5]{x^5} = x$
Evaluation of higher-order roots, p. 436.	Use exponent notation.	$\sqrt[5]{-32x^{15}} = \sqrt[5]{(-2)^5 x^{15}}$ $= \left[(-2)^5 x^{15}\right]^{1/5} = (-2)^1 x^3 = -2x^3$

Topic	Procedure	Examples
Simplification of radicals with the product rule, p. 440.	For nonnegative real numbers a and b and positive integers n, $$\sqrt[n]{a}\,\sqrt[n]{b} = \sqrt[n]{ab}.$$	Simplify when $x \geq 0$, $y \geq 0$. $$\sqrt{75x^3} = \sqrt{25x^2}\,\sqrt{3x}$$ $$= 5x\sqrt{3x}$$ $$\sqrt[3]{16x^5y^6} = \sqrt[3]{8x^3y^6}\,\sqrt[3]{2x^2}$$ $$= 2xy^2\sqrt[3]{2x^2}$$
Combining radicals, p. 441.	Simplify radicals and combine them if they have the same index and the same radicand.	Combine. $$2\sqrt{50} - 3\sqrt{98} = 2\sqrt{25}\,\sqrt{2} - 3\sqrt{49}\,\sqrt{2}$$ $$= 2(5)\sqrt{2} - 3(7)\sqrt{2}$$ $$= 10\sqrt{2} - 21\sqrt{2} = -11\sqrt{2}$$
Multiplying radicals, p. 445.	1. Multiply coefficients outside the radical and then multiply the radicands. 2. Simplify your answer.	$$(2\sqrt{3})(4\sqrt{5}) = 8\sqrt{15}$$ $$2\sqrt{6}(\sqrt{2} - 3\sqrt{12}) = 2\sqrt{12} - 6\sqrt{72}$$ $$= 2\sqrt{4}\,\sqrt{3} - 6\sqrt{36}\,\sqrt{2}$$ $$= 4\sqrt{3} - 36\sqrt{2}$$ $$(\sqrt{2} + \sqrt{3})(2\sqrt{2} - \sqrt{3}) \quad \text{By the FOIL method.}$$ $$= 2\sqrt{4} - \sqrt{6} + 2\sqrt{6} - \sqrt{9}$$ $$= 4 + \sqrt{6} - 3$$ $$= 1 + \sqrt{6}$$
Simplifying quotients of radicals with the quotient rule, p. 446.	For nonnegative real numbers a, positive real numbers b, and positive integers n, $$\sqrt[n]{\frac{a}{b}} = \frac{\sqrt[n]{a}}{\sqrt[n]{b}}.$$	$$\sqrt[3]{\frac{5}{27}} = \frac{\sqrt[3]{5}}{\sqrt[3]{27}} = \frac{\sqrt[3]{5}}{3}$$
Rationalizing denominators, p. 447.	Multiply numerator and denominator by a value that eliminates the radical in the denominator.	$$\frac{2}{\sqrt{7}} = \frac{2}{\sqrt{7}} \cdot \frac{\sqrt{7}}{\sqrt{7}} = \frac{2\sqrt{7}}{7}$$ $$\frac{3}{\sqrt{5} + \sqrt{2}} = \frac{3}{\sqrt{5} + \sqrt{2}} \cdot \frac{\sqrt{5} - \sqrt{2}}{\sqrt{5} - \sqrt{2}} = \frac{3\sqrt{5} - 3\sqrt{2}}{(\sqrt{5})^2 - (\sqrt{2})^2}$$ $$= \frac{3\sqrt{5} - 3\sqrt{2}}{5 - 2}$$ $$= \frac{3\sqrt{5} - 3\sqrt{2}}{3}$$ $$= \sqrt{5} - \sqrt{2}$$
Solving radical equations, p. 454.	1. Perform algebraic operations to obtain one radical by itself on one side of the equation. 2. If the equation contains square roots, square each side of the equation. Otherwise, raise each side to the appropriate power for third- and higher-order roots. 3. Simplify, if possible. 4. If the equation still contains a radical, repeat steps 1 to 3. 5. Collect all terms on one side of the equation. 6. Solve the resulting equation. 7. Check all apparent solutions. Solutions to radical equations must be verified.	Solve. $$x = \sqrt{2x + 9} - 3$$ $$x + 3 = \sqrt{2x + 9}$$ $$(x + 3)^2 = (\sqrt{2x + 9})^2$$ $$x^2 + 6x + 9 = 2x + 9$$ $$x^2 + 6x - 2x + 9 - 9 = 0$$ $$x^2 + 4x = 0$$ $$x(x + 4) = 0$$ $$x = 0 \quad \text{or} \quad x = -4$$ *Check:* $$x = 0: \quad 0 \overset{?}{=} \sqrt{2(0) + 9} - 3$$ $$0 \overset{?}{=} \sqrt{9} - 3$$ $$0 = 3 - 3 \quad \checkmark$$ $$x = -4: \quad -4 \overset{?}{=} \sqrt{2(-4) + 9} - 3$$ $$-4 \overset{?}{=} \sqrt{1} - 3$$ $$-4 \neq -2$$ The only solution is 0.

Topic	Procedure	Examples
Simplifying imaginary numbers, p. 460.	Use $i = \sqrt{-1}$ and $i^2 = -1$ and $\sqrt{-a} = \sqrt{a}\,\sqrt{-1}$.	$\sqrt{-16} = \sqrt{-1}\,\sqrt{16} = 4i$ $\sqrt{-18} = \sqrt{-1}\,\sqrt{18} = i\sqrt{9}\,\sqrt{2} = 3i\sqrt{2}$
Adding and subtracting complex numbers, p. 462.	Combine real parts and imaginary parts separately.	$(5 + 6i) + (2 - 4i) = 7 + 2i$ $(-8 + 3i) - (4 - 2i) = -8 + 3i - 4 + 2i$ $= -12 + 5i$
Multiplying complex numbers, p. 462.	Use the FOIL method and $i^2 = -1$.	$(5 - 6i)(2 - 4i) = 10 - 20i - 12i + 24i^2$ $= 10 - 32i + 24(-1)$ $= 10 - 32i - 24$ $= -14 - 32i$
Dividing complex numbers, p. 464.	Multiply the numerator and denominator by the conjugate of the denominator.	$\dfrac{5 + 2i}{4 - i} = \dfrac{5 + 2i}{4 - i} \cdot \dfrac{4 + i}{4 + i} = \dfrac{20 + 5i + 8i + 2i^2}{16 - i^2}$ $= \dfrac{20 + 13i + 2(-1)}{16 - (-1)}$ $= \dfrac{20 + 13i - 2}{16 + 1}$ $= \dfrac{18 + 13i}{17}$ or $\dfrac{18}{17} + \dfrac{13}{17}i$
Raising i to a power, p. 463.	$i^1 = i$ $i^2 = -1$ $i^3 = -i$ $i^4 = 1$	Evaluate. $i^{27} = i^{24} \cdot i^3$ $= (i^4)^6 \cdot i^3$ $= (1)^6(-i)$ $= -i$
Direct variation, p. 468.	If y varies directly with x, there is a constant of variation k such that $y = kx$. After k is determined, other values of y or x can easily be computed.	y varies directly with x. When $x = 2$, $y = 7$. $y = kx$ $7 = k(2)$ Substitute. $k = \dfrac{7}{2}$ Solve. $y = \dfrac{7}{2}x$ What is y when $x = 18$? $y = \dfrac{7}{2}x = \dfrac{7}{2} \cdot 18 = 63$
Inverse variation, p. 469.	If y varies inversely with x, the constant k is such that $$y = \frac{k}{x}.$$	y varies inversely with x. When x is 5, y is 12. What is y when x is 30? $y = \dfrac{k}{x}$ $12 = \dfrac{k}{5}$ Substitute. $k = 60$ Solve. $y = \dfrac{60}{x}$ Substitute. When $x = 30$, $y = \dfrac{60}{30} = 2$.

The Sailor's Observation Formula

For centuries sailors have used a formula to determine how many miles they can see in the distance D given the number of feet they are above sea level H. The formula is

$$D = \sqrt{\frac{3H}{2}}.$$

Use this formula to answer the following questions. Round your answers to the nearest tenth.

Problems for Individual Investigation and Analysis

1. Find the distance in miles that a sailor can see if he is 10 feet above sea level, 50 feet above sea level, and 100 feet above sea level.

2. Find the distance a sailor can see if he is 150 feet and 250 feet above sea level. Use these answers as well as the answers obtained in Question 1 to graph the equation.

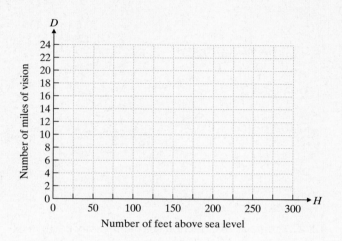

Problems for Group Investigation and Cooperative Learning

3. Solve the equation for H. Use the result to find the height H that a sailor must be above sea level in order to see a distance of 20 miles.

4. On the USS *Constitution*, sailors were normally stationed 150 feet above sea level in order to serve as lookouts. However, it was possible to climb higher in order to obtain a better view. How much was the viewing distance increased if a sailor climbed from 150 feet to 210 feet above sea level?

Internet Connections

Site: USS Constitution site

This site provides you with history and data regarding the oldest ship in the U.S. Navy.

5. Determine the height of the tallest mast on the USS *Constitution*. If a sailor stood on the top of the tallest mast with his head level with the top of the mast, how far would he be able to see into the distance?

6. How much further would a sailor on top of the main mast be able to see than one on top of the mizzenmast?

Chapter 8 Review Problems

In all exercises assume that the variables represent positive real numbers unless otherwise stated. Simplify using only positive exponents in your answers.

1. $(3xy^{1/2})(5x^2y^{-3})$

2. $\dfrac{3x^{2/3}}{6x^{1/6}}$

3. $(25a^3b^4)^{1/2}$

4. $5^{1/4} \cdot 5^{1/2}$

5. $(2a^{1/3}b^{1/4})(-3a^{1/2}b^{1/2})$

6. $\dfrac{6x^{2/3}y^{1/10}}{12x^{1/6}y^{-1/5}}$

7. $(2x^{-1/5}y^{1/10}z^{4/5})^{-5}$

8. $\left(\dfrac{49a^3b^6}{a^{-7}b^4}\right)^{1/2}$

9. $\dfrac{(x^{3/4}y^{2/5})^{1/2}}{x^{-1/8}}$

10. $\left(\dfrac{27x^{5n}}{x^{2n-3}}\right)^{1/3}$

11. $(5^{6/5})^{10/7}$

12. Combine as one fraction containing only positive exponents. $2x^{1/3} + x^{-2/3}$

13. Factor out a common factor of $3x$ from $6x^{3/2} - 9x^{1/2}$.

In Exercises 14–23, assume that all variables are positive real numbers.

14. Write in exponential form $\sqrt{\sqrt[5]{2x}}$.

15. Write in radical form $(2x + 3y)^{4/9}$.

16. Evaluate $\sqrt[3]{125} + \sqrt[4]{81}$.

17. Explain the difference between $-\sqrt[6]{64}$ and $\sqrt[6]{-64}$.

Evaluate or simplify each expression.

18. $27^{-4/3}$

19. $\left(\dfrac{4}{9}\right)^{3/2}$

20. $\sqrt{99x^3y^6z^{10}}$

21. $\sqrt[3]{-56a^8b^{10}c^{12}}$

22. $\sqrt{144x^{10}y^{12}z^0}$

23. $\sqrt[3]{125a^9b^6c^{300}}$

*In Exercises 24–29, assume that x and y can be any **positive** or **negative** real number.*

Simplify the following:

24. $\sqrt[3]{y^3}$

25. $\sqrt{y^2}$

26. $\sqrt[4]{x^4y^4}$

27. $\sqrt[5]{x^{10}}$

28. $\sqrt[3]{x^{21}}$

29. $\sqrt{x^8}$

480

Combine where possible.

30. $\sqrt{50} + 2\sqrt{32} - \sqrt{8}$

31. $\sqrt{28} - 4\sqrt{7} + 5\sqrt{63}$

32. $\sqrt[3]{8} + 3\sqrt[3]{16} - 4\sqrt[3]{54}$

33. $2\sqrt{32x} - 5x\sqrt{2} + \sqrt{18x} + 2\sqrt{8x^2}$

Multiply and simplify.

34. $(5\sqrt{12})(3\sqrt{6})$

35. $3\sqrt{x}(2\sqrt{8x} - 3\sqrt{48})$

36. $(5\sqrt{2} + \sqrt{3})(\sqrt{2} - 2\sqrt{3})$

37. $(5\sqrt{6} - 2\sqrt{2})(\sqrt{6} - \sqrt{2})$

38. $(2\sqrt{5} - 3\sqrt{6})^2$

39. $(\sqrt[3]{2x} + \sqrt[3]{6})(\sqrt[3]{4x^2} - \sqrt[3]{y})$

40. Let $f(x) = \sqrt{5x + 20}$.
   (a) Find $f(16)$.
   (b) What is the domain of $f(x)$?

41. Let $f(x) = \sqrt{36 - 4x}$.
   (a) Find $f(5)$.
   (b) What is the domain of $f(x)$?

42. Let $f(x) = \sqrt{\dfrac{3}{4}x - \dfrac{1}{2}}$.

   (a) Find $f(1)$.

   (b) What is the domain of $f(x)$?

Rationalize the denominator and simplify the expression.

43. $\sqrt{\dfrac{3x^2}{y}}$

44. $\dfrac{2}{\sqrt{3y}}$

45. $\dfrac{3\sqrt{7x}}{\sqrt{21x}}$

46. $\dfrac{2}{\sqrt{6} - \sqrt{5}}$

47. $\dfrac{\sqrt{x}}{3\sqrt{x} + \sqrt{y}}$

48. $\dfrac{\sqrt{5}}{\sqrt{7} - 3}$

49. $\dfrac{2\sqrt{3} + \sqrt{6}}{\sqrt{3} + 2\sqrt{6}}$

50. $\dfrac{5\sqrt{2} - \sqrt{3}}{\sqrt{6} - \sqrt{3}}$

51. $\dfrac{3\sqrt{x} + \sqrt{y}}{\sqrt{x} - \sqrt{y}}$

52. $\dfrac{2xy}{\sqrt[3]{16xy^5}}$

53. Simplify $\sqrt{-16} + \sqrt{-45}$.

54. Find x and y. $2x - 3i + 5 = yi - 2 + \sqrt{6}$

Simplify by performing the operation indicated.

55. $(-12 - 6i) + (3 - 5i)$ **56.** $(2 - i) - (12 - 3i)$ **57.** $(7 + 3i)(2 - 5i)$ **58.** $(8 - 4i)^2$

59. $2i(3 + 4i)$ **60.** $3 - 4(2 + i)$ **61.** Evaluate i^{34}.   **62.** Evaluate i^{65}.

Divide.

63. $\dfrac{7 - 2i}{3 + 4i}$ **64.** $\dfrac{5 - 2i}{1 - 3i}$ **65.** $\dfrac{4 - 3i}{5i}$ **66.** $\dfrac{12}{3 - 5i}$ **67.** $\dfrac{10 - 4i}{2 + 5i}$

Solve and check your solution(s).

68. $2\sqrt{6x + 1} = 10$

69. $\sqrt[3]{3x - 1} = \sqrt[3]{5x + 1}$

70. $\sqrt{2x + 1} = 2x - 5$

71. $1 + \sqrt{3x + 1} = x$

72. $\sqrt{3x + 1} - \sqrt{2x - 1} = 1$

73. $\sqrt{7x + 2} = \sqrt{x + 3} + \sqrt{2x - 1}$

Round all answers to the nearest tenth.

74. If y varies directly with x, and $y = 16$ when $x = 5$, find the value of y when $x = 3$.

75. If y varies directly with x, and $y = 5$ when $x = 20$, find the value of y when $x = 50$.

76. A car's stopping distance varies directly with the square of its speed. A car traveling on wet pavement can stop in 50 feet when traveling at 30 miles per hour. What distance will it take the car to stop if it is traveling at 55 miles per hour?

77. The time it takes a falling object to drop a given distance varies directly with the square root of the distance traveled. A steel ball takes 2 seconds to drop a distance of 64 feet. How many seconds will it take to drop a distance of 196 feet?

78. If y varies inversely with x, and $y = 8$ when $x = 3$, find the value of y when $x = 48$.

79. The volume of a gas varies inversely with the pressure of the gas on its container. If a pressure of 24 pounds per square inch corresponds to a volume of 70 cubic inches, what pressure corresponds to a volume of 100 cubic inches?

80. Suppose that y varies directly with x and inversely with the square of z. When $x = 8$ and $z = 4$, then $y = 1$. Find y when $x = 6$ and $z = 3$.

81. The capacity of a cylinder varies directly with the height and the square of the radius. A cylinder with a radius of 3 centimeters and a height of 5 centimeters has a capacity of 50 cubic centimeters. What is the capacity of a cylinder with a height of 9 centimeters and a radius of 4 centimeters?

Simplify.

1. $(2x^{1/2}y^{1/3})(-3x^{1/3}y^{1/6})$

2. $\dfrac{7x^3}{4x^{3/4}}$

3. $(8x^{1/3})^{3/2}$

4. $6^{1/5} \cdot 6^{3/5}$

Evaluate.

5. $8^{-2/3}$

6. $16^{5/4}$

Simplify. Assume that all variables are nonnegative.

7. $\sqrt{75a^4b^9}$

8. $\sqrt{64x^6y^5}$

9. $\sqrt[3]{250x^4y^6}$

Combine like terms where possible.

10. $3\sqrt{48} - \sqrt[3]{54x^5} + 2\sqrt{27} + 2x\sqrt[3]{16x^2}$

11. $\sqrt{32} - 3\sqrt{8} + 2\sqrt{72}$

Multiply and simplify.

12. $2\sqrt{3}(3\sqrt{6} - 5\sqrt{2})$

13. $(5\sqrt{3} - \sqrt{6})(2\sqrt{3} + 3\sqrt{6})$

Rationalize the denominator.

14. $\dfrac{8}{\sqrt{20x}}$

15. $\sqrt{\dfrac{xy}{3}}$

16. $\dfrac{5 + 2\sqrt{3}}{4 - \sqrt{3}}$

Solve and check your solution(s).

17. $\sqrt{3x - 2} = x$

18. $5 + \sqrt{x + 15} = x$

19. $5 - \sqrt{x - 2} = \sqrt{x + 3}$

1. _____

2. _____

3. _____

4. _____

5. _____

6. _____

7. _____

8. _____

9. _____

10. _____

11. _____

12. _____

13. _____

14. _____

15. _____

16. _____

17. _____

18. _____

19. _____

20. _____

21. _____

22. _____

23. _____

24. _____

25. _____

26. _____

27. _____

28. _____

Simplify by using the properties of complex numbers.

20. $(8 + 2i) - 3(2 - 4i)$

21. $i^{18} + \sqrt{-16}$

22. $(3 - 2i)(4 + 3i)$

23. $\dfrac{2 + 5i}{1 - 3i}$

24. $(6 + 3i)^2$

25. i^{43}

26. If y varies inversely with x, and $y = 9$ when $x = 2$, find the value of y when $x = 6$.

27. Suppose y varies directly with x and inversely with the square of z. When $x = 8$ and $z = 4$, then $y = 3$. Find y when $x = 5$ and $z = 6$.

28. A car's stopping distance varies directly with the square of its speed. A car traveling on pavement can stop in 30 feet when traveling at 30 miles per hour. What distance will it take the car to stop if it is traveling at 50 miles per hour?

Approximately one-half of this test covers the content of Chapters 1–7. The remainder covers the content of Chapter 8.

1. Simplify. $\dfrac{7}{8} \cdot \dfrac{1}{2} + \dfrac{3}{8} \div \dfrac{1}{8}$

2. Remove parentheses and collect like terms. $2a(3a^3 - 4) - 3a^2(a - 5)$

3. Simplify. $7(12 - 14)^3 - 7 + 3 \div (-3)$

4. Solve for x. $y = -\dfrac{3}{4}x + 2$

5. Graph $3x - 5y = 15$.

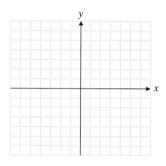

6. Factor completely $16x^2 + 24x - 16$.

7. Solve for x, y, and z.

$$x + 4y - z = 10$$
$$3x + 2y + z = 4$$
$$2x - 3y + 2z = -7$$

8. Combine $\dfrac{7x}{x^2 - 2x - 15} - \dfrac{2}{x - 5}$.

9. The length of a rectangle is 3 meters longer than twice its width. The perimeter of the rectangle is 48 meters. Find the dimensions of the rectangle.

10. Solve for b. $56x + 2 = 8b + 4x$

1. _____

2. _____

3. _____

4. _____

5. _____

6. _____

7. _____

8. _____

9. _____

10. _____

11. _____

12. _____

13. _____

14. _____

15. _____

16. _____

17. _____

18. _____

19. _____

20. _____

21. _____

22. _____

23. _____

24. _____

Simplify.

11. $\dfrac{2x^{-3}y^{-4}}{4x^{-5/2}y^{7/2}}$

12. $\left(3x^{-1/2}y^2\right)^{-1/3}$

13. Evaluate $64^{-1/3}$.

14. Simplify $\sqrt[3]{40x^5y^9}$.

15. Combine like terms. $\sqrt{80x} + 2\sqrt{45x} - 3\sqrt{20x}$

16. Multiply and simplify. $\left(2\sqrt{3} - 5\sqrt{2}\right)\left(\sqrt{3} + 4\sqrt{2}\right)$

17. Rationalize the denominator. $\dfrac{\sqrt{3} + 2}{2\sqrt{3} - 5}$

18. Simplify $i^{21} + \sqrt{-16} + \sqrt{-49}$.

19. Simplify $(3 - 4i)^2$.

20. Simplify $\dfrac{1 + 4i}{1 + 3i}$.

Solve for x and check your solutions.

21. $x - 3 = \sqrt{3x + 1}$

22. $1 + \sqrt{x + 1} = \sqrt{x + 2}$

23. If y varies directly with the square of x, and $y = 12$ when $x = 2$, find the value of y if $x = 5$.

24. The amount of light provided by a lightbulb varies inversely with the square of the distance from the lightbulb. A lightbulb provides 120 lumens at a distance of 10 feet from the light. How many lumens are provided if the distance from the light is 15 feet?

Quadratic Equations, Inequalities, and Absolute Value

T he Hoover Dam stands as one of the largest
construction projects ever completed in the United
States in the last century. This immense structure holds
back 35,200,000 cubic meters of water and can provide up
to 2074 megawatts of electric power if needed. How many
cubic feet of concrete was used to build this dam? How
thick is the dam at the point where the dam is 400 feet
high? Do you think you could use your mathematical skills
to solve these problems? Please turn to the Putting Your
Skills to Work exercises on page 536 to find out.

1. _____

2. _____

3. _____

4. _____

5. _____

6. _____

7. _____

8. _____

9. _____

10. _____

If you are familiar with the topics in this chapter, take this test now. Check your answers with those in the back of the book. If an answer is wrong or you can't do an exercise, study the appropriate section of the chapter.

    If you are not familiar with the topics in this chapter, don't take this test now. Instead, study the examples, work the practice exercises, and then take the test.

    This test will help you to identify those concepts that you have mastered and those that need more study.

Section 9.1

1. Solve by the square root property: $2x^2 + 3 = 39$

2. Solve by completing the square: $2x^2 - 4x - 3 = 0$

Section 9.2

Solve by the quadratic formula.

3. $8x^2 - 2x - 7 = 0$

4. $(x - 1)(x + 5) = 2$

5. Solve for the *nonreal* complex roots of the equation: $4x^2 = -12x - 17$

Sections 9.1 and 9.2

Solve by any method.

6. $5x^2 + 4x - 12 = 0$

7. $7x^2 + 9x = 14x^2 - 3x$

8. $\dfrac{18}{x} + \dfrac{12}{x + 1} = 9$

Section 9.3

Solve for any real roots and check your answers.

9. $x^6 - 7x^3 - 8 = 0$

10. $w^{4/3} - 6w^{2/3} + 8 = 0$

Section 9.4

11. Solve for x. Assume that w is a positive constant.

$$3x^2 + 2wx + 8w = 0$$

▲ **12.** The area of a rectangle is 52 square meters. The length of the rectangle is 1 meter longer than three times its width. Find the length and width of the rectangle.

Section 9.5

13. Find the vertex and the intercepts of the quadratic function $f(x) = 3x^2 + 6x - 9$.

14. Draw a graph of the quadratic function $g(x) = -x^2 + 6x - 5$. Label the vertex and intercepts.

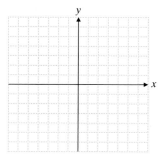

Section 9.6

Solve each quadratic inequality and graph your solutions.

15. $x^2 - x - 6 > 0$

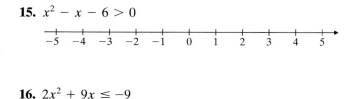

16. $2x^2 + 9x \leq -9$

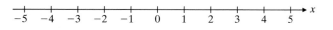

17. Use a square root table or a calculator to approximate to the nearest tenth the solution for the quadratic inequality.

$$4x^2 - 3x - 5 < 0$$

11. _____

12. _____

13. _____

14. _____

15. _____

16. _____

17. _____

18.

19.

20.

21.

22.

23.

24.

25.

Section 9.7

Solve for x.

18. $|3x - 2| = 7$

19. $|8 - x| - 3 = 1$

20. $\left| \dfrac{2x + 3}{4} \right| = 2$

21. $|5x - 8| = |3x + 2|$

Section 9.8

Solve for x.

22. $|3x + 2| < 8$

23. $\left| \dfrac{2}{3}x - \dfrac{1}{2} \right| \le 3$

24. $|2 - 5x - 4| > 13$

25. $|2x - 7| \le 11$

9.1 Quadratic Equations

1 ⟩ Solving Quadratic Equations by the Square Root Property

Recall from Section 6.7 that an equation written in the form $ax^2 + bx + c = 0$, where a, b, and c are real numbers and $a \neq 0$, is called a **quadratic equation**. Recall also that we call this the **standard form** of a quadratic equation. We have previously solved quadratic equations using the zero factor property. This has allowed us to factor the left side of an equation such as $x^2 - 7x + 12 = 0$ and obtain $(x - 3)(x - 4) = 0$ and then solve to find that $x = 3$ and $x = 4$. In this chapter we develop new methods of solving quadratic equations.

The first method is often called the **square root property**.

The Square Root Property

If $x^2 = a$, then $x = \pm\sqrt{a}$ for all real numbers a.

The notation $\pm\sqrt{a}$ is a shorthand way of writing "$+\sqrt{a}$ or $-\sqrt{a}$." The symbol $\pm$ is read "plus or minus." We can justify this property by using the zero factor property. If we write $x^2 = a$ in the form $x^2 - a = 0$, we can factor it to obtain $(x + \sqrt{a})(x - \sqrt{a}) = 0$ and thus, $x = -\sqrt{a}$ or $x = +\sqrt{a}$. This can be written more compactly as $x = \pm\sqrt{a}$.

EXAMPLE 1 Solve and check: $x^2 - 36 = 0$

If we add 36 to each side, we have $x^2 = 36$.

$$x = \pm\sqrt{36}$$
$$= \pm 6$$

Thus, the two roots are 6 and −6.

Check: $6^2 = 36$ and $(-6)^2 = 36$

Practice Problem 1 Solve and check: $x^2 - 121 = 0$

EXAMPLE 2 Solve $x^2 = 48$.

$$x = \pm\sqrt{48} = \pm\sqrt{16 \cdot 3}$$
$$x = \pm 4\sqrt{3}$$

The roots are $4\sqrt{3}$ and $-4\sqrt{3}$.

Practice Problem 2 Solve $x^2 = 18$.

EXAMPLE 3 Solve and check: $3x^2 + 2 = 77$

$$3x^2 + 2 = 77$$
$$3x^2 = 75$$
$$x^2 = 25$$
$$x = \pm\sqrt{25}$$
$$x = \pm 5$$

The roots are 5 and −5.

Check:

$$3(5)^2 + 2 \overset{?}{=} 77 \qquad\qquad 3(-5)^2 + 2 \overset{?}{=} 77$$
$$3(25) + 2 \overset{?}{=} 77 \qquad\qquad 3(25) + 2 \overset{?}{=} 77$$
$$75 + 2 \overset{?}{=} 77 \qquad\qquad 75 + 2 \overset{?}{=} 77$$
$$77 = 77 \ \checkmark \qquad\qquad 77 = 77 \ \checkmark$$

Student Learning Objectives

After studying this section, you will be able to:

1 ⟩ Solve quadratic equations by the square root property.

2 ⟩ Solve quadratic equations by completing the square.

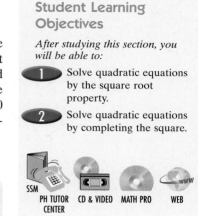

SSM PH TUTOR CD & VIDEO MATH PRO WEB
CENTER

Practice Problem 3   Solve $5x^2 + 1 = 46$.

Sometimes we obtain roots that are complex numbers.

EXAMPLE 4   Solve and check: $4x^2 = -16$

$$x^2 = -4$$
$$x = \pm\sqrt{-4}$$
$$x = \pm 2i \qquad \text{Simplify using } \sqrt{-1} = i.$$

The roots are $2i$ and $-2i$.

$$\begin{array}{ll}
\textit{Check:} & 4(2i)^2 \overset{?}{=} -16 \\
& 4(4i^2) \overset{?}{=} -16 \\
& 4(-4) \overset{?}{=} -16 \\
& -16 = -16 \;\checkmark
\end{array}
\qquad
\begin{array}{l}
4(-2i)^2 \overset{?}{=} -16 \\
4(4i^2) \overset{?}{=} -16 \\
4(-4) \overset{?}{=} -16 \\
-16 = -16 \;\checkmark
\end{array}$$

Practice Problem 4   Solve and check: $3x^2 = -27$

EXAMPLE 5   Solve $(4x - 1)^2 = 5$.

$$4x - 1 = \pm\sqrt{5}$$
$$4x = 1 \pm \sqrt{5}$$
$$x = \frac{1 \pm \sqrt{5}}{4}$$

The roots are $\dfrac{1 + \sqrt{5}}{4}$ and $\dfrac{1 - \sqrt{5}}{4}$.

Practice Problem 5   Solve $(2x + 3)^2 = 7$.

2   *Solving Quadratic Equations by Completing the Square*

Often, a quadratic equation cannot be factored (or it may be difficult to factor). So we use another method of solving the equation, called **completing the square**. When we complete the square, we are changing the polynomial to a perfect-square trinomial. The form of the equation then becomes $(x + d)^2 = e$.

We already know that

$$(x + d)^2 = x^2 + 2dx + d^2.$$

Notice three things about the quadratic equation on the right-hand side.

   1. The coefficient of the quadratic term (x^2) is 1.
   2. The coefficient of the linear (x) term is $2d$.
   3. The constant term (d^2) is the square of *half* the coefficient of the linear term.

For example, in the trinomial $x^2 + 6x + 9$, the coefficient of the linear term is 6 and the constant term is $\left(\dfrac{6}{2}\right)^2 = (3)^2 = 9$.

For the trinomial $x^2 - 10x + 25$, the coefficient of the linear term is -10 and the constant term is $\left(\dfrac{-10}{2}\right)^2 = (-5)^2 = 25$.

What number n makes the trinomial $x^2 + 12x + n$ a perfect square?

$$n = \left(\frac{12}{2}\right)^2 = 6^2 = 36$$

Hence, the trinomial $x^2 + 12x + 36$ is a perfect-square trinomial and can be written as $(x + 6)^2$.

Now let's solve some equations.

EXAMPLE 6 Solve and check: $x^2 + 6x + 1 = 0$

Step 1 First we rewrite the equation in the form $ax^2 + bx = c$ by adding -1 to each side of the equation. Thus, we obtain

$$x^2 + 6x = -1.$$

Step 2 We want to complete the square of $x^2 + 6x$. That is, we want to add a constant term to $x^2 + 6x$ so that we get a perfect-square trinomial. We do this by taking half the coefficient of x and squaring it.

$$\left(\frac{6}{2}\right)^2 = 3^2 = 9$$

Adding 9 to $x^2 + 6x$ gives the perfect-square trinomial $x^2 + 6x + 9$, which we factor to $(x + 3)^2$. But we cannot just add 9 to the left side of our equation unless we also add 9 to the right side. (Why?) We now have

$$x^2 + 6x + 9 = -1 + 9.$$

Step 3 Now we factor.

$$(x + 3)^2 = 8$$

Step 4 We now use the square root property.

$$(x + 3) = \pm\sqrt{8}$$
$$x + 3 = \pm 2\sqrt{2}$$

Step 5 Next we solve for x by adding -3 to each side of the equation.

$$x = -3 \pm 2\sqrt{2}$$

The roots are $-3 + 2\sqrt{2}$ and $-3 - 2\sqrt{2}$.

Step 6 We *must* check our solution in the *original* equation (not the perfect-square trinomial we constructed).

$$x^2 + 6x + 1 = 0 \qquad\qquad\qquad x^2 + 6x + 1 = 0$$
$$(-3 + 2\sqrt{2})^2 + 6(-3 + 2\sqrt{2}) + 1 \overset{?}{=} 0 \qquad (-3 - 2\sqrt{2})^2 + 6(-3 - 2\sqrt{2}) + 1 \overset{?}{=} 0$$
$$9 - 12\sqrt{2} + 8 - 18 + 12\sqrt{2} + 1 \overset{?}{=} 0 \qquad 9 + 12\sqrt{2} + 8 - 18 - 12\sqrt{2} + 1 \overset{?}{=} 0$$
$$18 - 18 - 12\sqrt{2} + 12\sqrt{2} \overset{?}{=} 0 \qquad\qquad 18 - 18 + 12\sqrt{2} - 12\sqrt{2} \overset{?}{=} 0$$
$$0 = 0 \ \checkmark \qquad\qquad\qquad\qquad\qquad 0 = 0 \ \checkmark$$

Practice Problem 6 Solve by completing the square: $x^2 + 8x + 3 = 0$

Let us summarize for future reference the six steps we have performed to solve a quadratic equation by completing the square.

Completing the Square

1. Put the equation in the form $ax^2 + bx = -c$.
2. If $a \neq 1$, divide each term of the equation by a.
3. Square half of the numerical coefficient of the linear term. Add the result to both sides of the equation.
4. Factor the left side; then take the square root of both sides of the equation.
5. Solve each resulting equation for x.
6. Check the solutions in the original equation.

EXAMPLE 7 Solve: $3x^2 - 8x + 1 = 0$

$$3x^2 - 8x = -1 \qquad \text{Add } -1 \text{ to each side.}$$

$$\frac{3x^2}{3} - \frac{8x}{3} = -\frac{1}{3} \qquad \text{Divide each term by 3. (Remember that the coefficient of the quadratic term must be 1.)}$$

$$x^2 - \frac{8}{3}x + \frac{16}{9} = -\frac{1}{3} + \frac{16}{9}$$

$$\left(x - \frac{4}{3}\right)^2 = \frac{13}{9}$$

$$x - \frac{4}{3} = \pm\sqrt{\frac{13}{9}}$$

$$x - \frac{4}{3} = \pm\frac{\sqrt{13}}{3}$$

$$x = \frac{4}{3} \pm \frac{\sqrt{13}}{3}$$

$$x = \frac{4 \pm \sqrt{13}}{3}$$

Check: For $x = \dfrac{4 + \sqrt{13}}{3}$,

$$3\left(\frac{4 + \sqrt{13}}{3}\right)^2 - 8\left(\frac{4 + \sqrt{13}}{3}\right) + 1 \overset{?}{=} 0$$

$$\frac{16 + 8\sqrt{13} + 13}{3} - \frac{32 + 8\sqrt{13}}{3} + 1 \overset{?}{=} 0$$

$$\frac{16 + 8\sqrt{13} + 13 - 32 - 8\sqrt{13}}{3} + 1 \overset{?}{=} 0$$

$$\frac{29 - 32}{3} + 1 \overset{?}{=} 0$$

$$-\frac{3}{3} + 1 \overset{?}{=} 0$$

$$-1 + 1 = 0 \quad \checkmark$$

See whether you can check the solution $\dfrac{4 - \sqrt{13}}{3}$.

Practice Problem 7 Solve by completing the square: $2x^2 + 4x + 1 = 0$

Solve the equations by using the square root property. Express any complex numbers using i notation.

1. $x^2 = 100$

2. $x^2 = 49$

3. $x^2 + 81 = 0$

4. $x^2 + 144 = 0$

5. $3x^2 - 45 = 0$

6. $4x^2 - 68 = 0$

7. $5x^2 - 10 = 0$

8. $2x^2 - 14 = 0$

9. $x^2 = -81$

10. $x^2 = -64$

11. $6x^2 + 4 = 4x^2$

12. $(x - 3)^2 = 12$

13. $(x + 2)^2 = 18$

14. $(2x + 1)^2 = 7$

15. $(3x + 2)^2 = 5$

16. $(4x - 3)^2 = 36$

17. $(5x - 2)^2 = 25$

18. $\left(\dfrac{x}{2} + 5\right)^2 = 8$

19. $\left(\dfrac{x}{3} - 1\right)^2 = 45$

Solve the equations by completing the square. Simplify your answers. Express any complex numbers using i notation.

20. $x^2 + 10x + 5 = 0$

21. $x^2 + 6x + 2 = 0$

22. $x^2 - 8x = 17$

23. $x^2 - 12x = 4$

24. $\dfrac{x^2}{2} + \dfrac{5}{2}x = 2$

25. $\dfrac{x^2}{3} - \dfrac{x}{3} = 3$

26. $2y^2 + 10y = -11$

27. $7x^2 + 4x - 5 = 0$

28. $3x^2 + 10x - 2 = 0$

29. $5x^2 + 4x - 3 = 0$

30. $2y^2 - y = 6$

31. $2y^2 - y = 15$

32. $x^2 + 1 = x$ **33.** $2x^2 + 2 = 3x$ **34.** $3x^2 + 8x + 3 = 2$

35. Check the solution $x = -1 + \sqrt{6}$ in the equation $x^2 + 2x - 5 = 0$.

36. Check the solution $x = 2 + \sqrt{3}$ in the equation $x^2 - 4x + 1 = 0$.

Applications

The sides of the box shown are labeled with the dimensions in feet.

▲ **37.** What is the value of x if the volume of the box is 648 cubic feet?

▲ **38.** What is the value of x if the volume of the box is 1800 cubic feet?

The time a basketball player spends in the air when shooting a basket is called "the hang time." The vertical leap L measured in feet is related to the hang time t measured in seconds by the equation $L = 4t^2$.

39. During his career as a Boston Celtics player, Larry Bird often displayed a leap of 3.1 feet. Find the hang time for that leap.

40. Shaquille O'Neal of the Los Angeles Lakers has often shown a vertical leap of 3.3 feet. Find the hang time for that leap.

The formula $D = 16t^2$ is used to approximate the distance in feet that an object falls in t seconds.

41. A parachutist jumps from an airplane, falls 3600 feet, and then opens her parachute. For how many seconds was the parachutist falling before she opened the parachute?

42. How long would it take an object to fall to the ground from a helicopter hovering at 1936 feet above the ground?

Cumulative Review Problems

Evaluate the expressions for the given values.

43. $\sqrt{b^2 - 4ac}$; $b = 4, a = 3, c = -4$

44. $\sqrt{b^2 - 4ac}$; $b = -5, a = 2, c = -3$

45. $5x^2 - 6x + 8$; $x = -2$

46. $2x^2 + 3x - 5$; $x = -3$

1 Solving a Quadratic Equation by Using the Quadratic Formula

Student Learning Objectives

After studying this section, you will be able to:

1 Solve a quadratic equation by using the quadratic formula.

2 Use the discriminant to determine the nature of the roots of a quadratic equation.

3 Write a quadratic equation given the solutions of the equation.

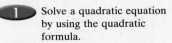

SSM PH TUTOR CENTER   CD & VIDEO   MATH PRO   WEB

The last method we'll study for solving quadratic equations is the **quadratic formula**. This method works for *any* quadratic equation.

The quadratic formula is developed from completing the square. We begin with the **standard form** of the quadratic equation.

$$ax^2 + bx + c = 0$$

To complete the square, we want the equation to be in the form $x^2 + dx = e$. Thus, we divide by a.

$$\frac{ax^2}{a} + \frac{b}{a}x + \frac{c}{a} = 0$$

$$x^2 + \frac{b}{a}x = -\frac{c}{a}$$

Now we complete the square by adding $\left(\dfrac{b}{2a}\right)^2$ to each side.

$$x^2 + \frac{b}{a}x + \left(\frac{b}{2a}\right)^2 = -\frac{c}{a} + \left(\frac{b}{2a}\right)^2$$

We factor the left side and write the right side as one fraction.

$$\left(x + \frac{b}{2a}\right)^2 = \frac{b^2 - 4ac}{4a^2}$$

Now we use the square root property.

$$x + \frac{b}{2a} = \pm\sqrt{\frac{b^2 - 4ac}{4a^2}}$$

We solve for x and simplify.

$$x = -\frac{b}{2a} \pm \sqrt{\frac{b^2 - 4ac}{4a^2}}$$

$$= \frac{-b \pm \sqrt{b^2 - 4ac}}{2a}$$

This is the quadratic formula.

Quadratic Formula

For all equations $ax^2 + bx + c = 0$,

$$x = \frac{-b \pm \sqrt{b^2 - 4ac}}{2a}, \qquad \text{where } a \neq 0.$$

EXAMPLE 1 Solve by using the quadratic formula: $x^2 + 8x = -3$

The standard form is $x^2 + 8x + 3 = 0$. We substitute $a = 1$, $b = 8$, and $c = 3$.

$$x = \frac{-b \pm \sqrt{b^2 - 4ac}}{2a}$$

$$= \frac{-8 \pm \sqrt{8^2 - 4(1)(3)}}{2(1)}$$

$$= \frac{-8 \pm \sqrt{64 - 12}}{2} = \frac{-8 \pm \sqrt{52}}{2} = \frac{-8 \pm \sqrt{4}\sqrt{13}}{2}$$

497

$$= \frac{-8 \pm 2\sqrt{13}}{2} = \frac{\cancel{2}(-4 \pm \sqrt{13})}{\cancel{2}}$$
$$= -4 \pm \sqrt{13}$$

Practice Problem 1 Solve by using the quadratic formula: $x^2 + 5x = -1 + 2x$

EXAMPLE 2 Solve by using the quadratic formula: $3x^2 - x - 2 = 0$

Here $a = 3, b = -1,$ and $c = -2$.

$$x = \frac{-b \pm \sqrt{b^2 - 4ac}}{2a}$$

$$= \frac{-(-1) \pm \sqrt{(-1)^2 - 4(3)(-2)}}{2(3)}$$

$$= \frac{1 \pm \sqrt{1 + 24}}{6} = \frac{1 \pm \sqrt{25}}{6}$$

$$x = \frac{1 + 5}{6} = \frac{6}{6} \quad \text{or} \quad x = \frac{1 - 5}{6} = -\frac{4}{6}$$

$$x = 1 \qquad\qquad x = -\frac{2}{3}$$

Practice Problem 2 Solve by using the quadratic formula: $2x^2 + 7x + 6 = 0$

EXAMPLE 3 Solve by using the quadratic formula: $2x^2 - 48 = 0$

This equation is equivalent to $2x^2 - 0x - 48 = 0$. Therefore, we know that $a = 2, b = 0,$ and $c = -48$.

$$x = \frac{-b \pm \sqrt{b^2 - 4ac}}{2a}$$

$$= \frac{-0 \pm \sqrt{(0)^2 - 4(2)(-48)}}{2(2)}$$

$$= \frac{\pm \sqrt{384}}{4}$$

$$= \frac{\pm \sqrt{64}\sqrt{6}}{4} = \frac{\pm 8\sqrt{6}}{4}$$

$$= \pm 2\sqrt{6}$$

Practice Problem 3 Solve by using the quadratic formula: $2x^2 - 26 = 0$

EXAMPLE 4 A small company that manufactures canoes makes a daily profit p according to the equation $p = -100x^2 + 3400x - 26,196$, where p is measured in dollars and x is the number of canoes made per day. Find the number of canoes that must be made each day to produce a zero profit for the company. Round your answer to the nearest whole number.

Since $p = 0$, we are solving the equation $0 = -100x^2 + 3400x - 26,196$.

In this case we have $a = -100, b = 3400,$ and $c = -26,196$.

Now we substitute these into the quadratic formula.

$$x = \frac{-b \pm \sqrt{b^2 - 4ac}}{2a}$$

$$= \frac{-3400 \pm \sqrt{(3400)^2 - 4(-100)(-26,196)}}{2(-100)}$$

We will use a calculator to assist us with computation in this problem.

$$= \frac{-3400 \pm \sqrt{11,560,000 - 10,478,400}}{-200}$$

$$= \frac{-3400 \pm \sqrt{1,081,600}}{-200}$$

$$= \frac{-3400 \pm 1040}{-200}$$

We now obtain two answers.

$$x = \frac{-3400 + 1040}{-200} = \frac{-2360}{-200} = 11.8 \approx 12$$

$$x = \frac{-3400 - 1040}{-200} = \frac{-4440}{-200} = 22.2 \approx 22$$

A zero profit is obtained when approximately twelve canoes are produced or when approximately twenty-two canoes are produced. Actually a slight profit of $204 is made when these numbers of canoes are produced. The discrepancy is due to the round-off error that occurs when we approximate. By methods that we will learn later in this chapter, the maximum profit is produced when seventeen canoes are made at the factory. We will investigate exercises of this kind later.

Practice Problem 4   A company that manufactures modems makes a daily profit p according to the equation $p = -100x^2 + 4800x - 52,559$, where p is measured in dollars and x is the number of modems made per day. Find the number of modems that must be made each day to produce a zero profit for the company. Round your answer to the nearest whole number.

When a quadratic equation contains fractions, eliminate them by multiplying each term by the LCD. Then rewrite the equation in standard form before using the quadratic formula.

EXAMPLE 5   Solve by using the quadratic formula: $\dfrac{2x}{x + 2} = 1 - \dfrac{3}{x + 4}$

The LCD is $(x + 2)(x + 4)$.

$$\frac{2}{x + 2} = 1 - \frac{3}{x + 4}$$

$$\frac{2x}{x+2}(x+2)(x + 4) = 1(x + 2)(x + 4) - \frac{3}{x+4}(x + 2)(x+4)$$

$$2x(x + 4) = (x + 2)(x + 4) - 3(x + 2)$$

$$2x^2 + 8x = x^2 + 6x + 8 - 3x - 6 \quad \text{Now we have an equation that is quadratic.}$$

$$2x^2 + 8x = x^2 + 3x + 2$$

$$x^2 + 5x - 2 = 0$$

Now the equation is in standard form, and we can use the quadratic formula with $a = 1, b = 5$, and $c = -2$.

$$x = \frac{-5 \pm \sqrt{5^2 - 4(1)(-2)}}{2(1)} = \frac{-5 \pm \sqrt{25 + 8}}{2}$$

$$x = \frac{-5 \pm \sqrt{33}}{2}$$

Practice Problem 5   Solve by using the quadratic formula: $\dfrac{1}{x} + \dfrac{1}{x - 1} = \dfrac{5}{6}$

Some quadratic equations will have solutions that are not real numbers. You should use i notation to simplify the solutions of nonreal complex numbers.

EXAMPLE 6   Solve and simplify your answer: $8x^2 - 4x + 1 = 0$
$a = 8, b = -4$, and $c = 1$.

$$x = \frac{-(-4) \pm \sqrt{(-4)^2 - 4(8)(1)}}{2(8)}$$

$$= \frac{4 \pm \sqrt{16 - 32}}{16} = \frac{4 \pm \sqrt{-16}}{16}$$

$$= \frac{4 \pm 4i}{16} = \frac{4(1 \pm i)}{16} = \frac{1 \pm i}{4}$$

Practice Problem 6   Solve by using the quadratic formula: $2x^2 - 4x + 5 = 0$

You may have noticed that complex roots come in pairs. In other words, if $a + bi$ is a solution of a quadratic equation, its conjugate $a - bi$ is also a solution.

2 Using the Discriminant to Determine the Nature of the Roots of a Quadratic Equation

So far we have used the quadratic formula to solve quadratic equations that had two real roots. Sometimes the roots were rational, and sometimes they were irrational. We have also solved equations like Example 6 with nonreal complex numbers. Such solutions occur when the expression $b^2 - 4ac$, the radicand in the quadratic formula, is negative.

$$x = \frac{-b \pm \sqrt{b^2 - 4ac}}{2a}$$

The expression $b^2 - 4ac$ is called the **discriminant**. Depending on the value of the discriminant and whether the discriminant is positive, zero, or negative, the roots of the quadratic equation will be rational, irrational, or complex. We summarize the types of solutions in the following table.

If the discriminant $b^2 - 4ac$ is:	Then the quadratic equation $ax^2 + bx + c = 0$, where a, b, and c are integers, will have:
A positive number that is also a perfect square	Two different rational solutions (Such an equation can always be factored.)
A positive number that is not a perfect square	Two different irrational solutions
Zero	One rational solution
Negative	Two complex solutions containing i (They will be complex conjugates.)

EXAMPLE 7   What type of solutions does the equation $2x^2 - 9x - 35 = 0$ have? Do not solve the equation.

$a = 2, b = -9$, and $c = -35$. Thus,

$$b^2 - 4ac = (-9)^2 - 4(2)(-35) = 361.$$

Since the discriminant is positive, the equation has two real roots.
Since $(19)^2 = 361$, 361 is a perfect square. Thus, the equation has two different rational solutions. This type of quadratic equation can always be factored.

Practice Problem 7   Use the discriminant to find what type of solutions the equation $9x^2 + 12x + 4 = 0$ has. Do not solve the equation.

EXAMPLE 8 Use the discriminant to determine the type of solutions each of the following equations has.

(a) $3x^2 - 4x + 2 = 0$ **(b)** $5x^2 - 3x - 5 = 0$

(a) Here $a = 3, b = -4$, and $c = 2$. Thus,

$$b^2 - 4ac = (-4)^2 - 4(3)(2)$$
$$= 16 - 24 = -8.$$

Since the discriminant is negative, the equation will have two complex solutions containing i.

(b) Here $a = 5, b = -3$, and $c = -5$. Thus,

$$b^2 - 4ac = (-3)^2 - 4(5)(-5)$$
$$= 9 + 100 = 109.$$

Since this positive number is not a perfect square, the equation will have two different irrational solutions.

Practice Problem 8 Use the discriminant to determine the type of solutions each of the following equations has.

(a) $x^2 - 4x + 13 = 0$ **(b)** $9x^2 + 6x + 7 = 0$

3 Writing a Quadratic Equation Given the Solutions of the Equation

By using the zero factor property in reverse, we can find a quadratic equation that contains two given solutions. To illustrate, if 3 and 7 are the two solutions, then we could write the equation $(x - 3)(x - 7) = 0$, and therefore, a quadratic equation that has these two solutions is $x^2 - 10x + 21 = 0$. This answer is not unique. Any constant multiple of $x^2 - 10x + 21 = 0$ would also have roots of 3 and 7. Thus, $2x^2 - 20x + 42 = 0$ also has roots of 3 and 7.

EXAMPLE 9 Find a quadratic equation whose roots are 5 and -2.

$$x = 5 \qquad\qquad x = -2$$
$$x - 5 = 0 \qquad\qquad x + 2 = 0$$
$$(x - 5)(x + 2) = 0$$
$$x^2 - 3x - 10 = 0$$

Practice Problem 9 Find a quadratic equation whose roots are -10 and -6.

EXAMPLE 10 Find a quadratic equation whose solutions are $3i$ and $-3i$.
First we write the two equations.

$$x - 3i = 0 \quad \text{and} \quad x + 3i = 0$$
$$(x - 3i)(x + 3i) = 0$$
$$x^2 + 3ix - 3ix - 9i^2 = 0$$
$$x^2 - 9(-1) = 0 \quad \text{Use } i^2 = -1.$$
$$x^2 + 9 = 0$$

Practice Problem 10 Find a quadratic equation whose solutions are $2i\sqrt{3}$ and $-2i\sqrt{3}$.

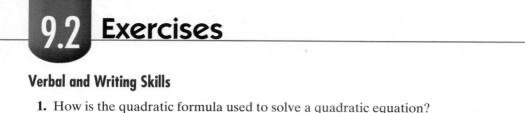

9.2 Exercises

Verbal and Writing Skills

1. How is the quadratic formula used to solve a quadratic equation?

2. The discriminant in the quadratic formula is the expression _____.

3. If the discriminant in the quadratic formula is zero, then the quadratic equation will have _____ solution(s).

4. If the discriminant in the quadratic formula is a perfect square, then the quadratic equation will have _____ solution(s).

Solve by the quadratic formula. Simplify your answers. Use i notation for nonreal complex numbers.

5. $x^2 - x - 3 = 0$ **6.** $x^2 + 5x + 2 = 0$ **7.** $2x^2 + x - 4 = 0$ **8.** $5x^2 - x - 1 = 0$

9. $x^2 = \dfrac{2}{3}x$ **10.** $\dfrac{4}{5}x^2 = x$ **11.** $6x^2 - x - 1 = 0$ **12.** $4x^2 + 11x - 3 = 0$

13. $4x^2 + 3x - 2 = 0$ **14.** $6x^2 - 2x - 1 = 0$ **15.** $3x^2 + 1 = 8$

16. $5x^2 - 1 = 5$ **17.** $2x(x + 3) - 3 = 4x - 2$ **18.** $5 + 3x(x - 2) = 4$

19. $3x^2 + 5x + 1 = 5x + 4$ **20.** $2x^2 - 7x - 3 = 9 - 7x$ **21.** $(x - 2)(x + 1) = \dfrac{2x + 3}{2}$

22. $3x(x + 1) = \dfrac{7x + 1}{3}$

23. $\dfrac{1}{x + 2} + \dfrac{1}{x} = \dfrac{1}{3}$

24. $y + \dfrac{3}{y + 3} = \dfrac{5}{2}$

25. $\dfrac{1}{y} - y = \dfrac{5}{3}$

26. $\dfrac{1}{15} + \dfrac{3}{y} = \dfrac{4}{y + 1}$

27. $\dfrac{1}{4} + \dfrac{6}{y + 2} = \dfrac{6}{y}$

28. $x^2 - 4x + 8 = 0$

29. $x^2 - 2x + 4 = 0$

30. $2x^2 + 15 = 0$

31. $5x^2 = -3$

32. $3x^2 - 8x + 7 = 0$

Use the discriminant to find what type of solutions (two rational, two irrational, one rational, or two nonreal complex) each of the following equations has. Do not solve the equation.

33. $3x^2 + 4x = 2$

34. $4x^2 - 20x + 25 = 0$

35. $2x^2 + 10x + 8 = 0$

36. $2x^2 - 7x - 4 = 0$

37. $9x^2 + 4 = 12x$

38. $5x^2 - 8x - 2 = 0$

Write a quadratic equation having the given solutions.

39. $13, -2$

40. $5, -11$

41. $-5, -12$

42. $-6, -10$

43. $4i, -4i$

44. $6i, -6i$

45. $3, -\dfrac{5}{2}$

46. $-2, \dfrac{5}{6}$

Solve for x by using the quadratic formula. Approximate your answers to four decimal places.

47. $3x^2 + 5x - 9 = 0$

48. $1.2x^2 - 12.3x - 4.2 = 0$

49. $20.6x^2 - 73.4x + 41.8 = 0$

50. $0.162x^2 + 0.094x - 0.485 = 0$

Applications

51. A company that manufactures sport parachutes makes a daily profit p according to the equation $p = -100x^2 + 4200x - 39{,}476$, where p is measured in dollars and x is the number of parachutes made per day. Find the number of parachutes that must be made each day to produce a zero profit for the company. Round your answer to the nearest whole number.

52. A company that manufactures mountain bikes makes a daily profit p according to the equation $p = -100x^2 + 4800x - 54{,}351$, where p is measured in dollars and x is the number of mountain bikes made per day. Find the number of mountain bikes that must be made each day to produce a zero profit for the company. Round your answer to the nearest whole number.

To Think About

53. The company described in Exercise 51 earns a maximum profit when $x = 21$. What profit does the company make per day if it produces twenty-one parachutes? Speculate how you could have predicted that the maximum profit occurs when $x = 21$ based on the answers you obtained in the Exercise 51.

54. The company described in Exercise 52 earns a maximum profit when $x = 24$. What profit does the company make per day if it produces twenty-four mountain bikes? Speculate how you could have predicted that the maximum profit occurs when $x = 24$ based on the answers you obtained in Exercise 52.

Cumulative Review Problems

Simplify.

55. $9x^2 - 6x + 3 - 4x - 12x^2 + 8$

56. $3y(2 - y) + \dfrac{1}{5}(10y^2 - 15y)$

57. Music Galaxy sells compact discs, cassettes, and everything else you could possibly want from a music supply superstore. The management plans to expand its compact disc section. Presently, it takes 50 feet of an inner security fence to enclose the rectangular section. The expansion plans call for tripling the width and doubling the length. The new CD section will need 118 feet of inner security fencing. What is the length and width of the current compact disc section?

58. Last year, Cecile, a professional mountain bike racer, purchased three new padded riding suits to protect her from injury and compress her muscles while riding. In addition, she purchased two pairs of racing goggles. The cost for these items was $343. This year, suits cost $10 more and goggles cost $5 more than last year. This year she purchased two new suits and three pairs of goggles for $312. How much did each suit cost last year? How much did each pair of goggles cost last year?

9.3 Equations That Can Be Transformed into Quadratic Form

1 Solve Equations of Degree Greater than 2

Student Learning Objectives

After studying this section, you will be able to:

1. Solve equations of degree greater than 2 that can be transformed into quadratic form.

2. Solve equations with fractional exponents that can be transformed into quadratic form.

SSM PH TUTOR CD & VIDEO MATH PRO WEB
CENTER

Some higher-order equations can be solved by writing them in the form of a quadratic equation. An equation is **quadratic in form** if we can substitute a linear term for the variable raised to the lowest power and get an equation of the form $ay^2 + by + c = 0$.

EXAMPLE 1 Solve $x^4 - 13x^2 + 36 = 0$.

Let $y = x^2$. Then $y^2 = x^4$. Thus, we obtain a new equation and solve it as follows:

$y^2 - 13y + 36 = 0$	Replace x^2 by y and x^4 by y^2.
$(y - 4)(y - 9) = 0$	Factor.
$y - 4 = 0$ or $y - 9 = 0$	Solve for y.
$y = 4 \qquad y = 9$	These are *not* the roots to the original equation. We must replace y by x^2.
$x^2 = 4 \qquad x^2 = 9$	
$x = \pm\sqrt{4} \qquad x = \pm\sqrt{9}$	
$x = \pm 2 \qquad x = \pm 3$	

Thus, there are *four* solutions to the original equation: $x = +2$, $x = -2$, $x = +3$, and $x = -3$. Check these values to verify that they are solutions.

Practice Problem 1 Solve $x^4 - 5x^2 - 36 = 0$.

EXAMPLE 2 Solve for all real roots: $2x^6 - x^3 - 6 = 0$

Let $y = x^3$. Then $y^2 = x^6$. Thus, we have the following:

$2y^2 - y - 6 = 0$	Replace x^3 by y and x^6 by y^2.
$(2y + 3)(y - 2) = 0$	Factor.
$2y + 3 = 0$ or $y - 2 = 0$	Solve for y.
$y = -\dfrac{3}{2} \qquad y = 2$	
$x^3 = -\dfrac{3}{2}$ or $x^3 = 2$	Replace y by x^3.
$x = \sqrt[3]{-\dfrac{3}{2}} \qquad x = \sqrt[3]{2}$	Take the cube root of each side of the equation.
$x = \dfrac{\sqrt[3]{-12}}{2}$	Simplify $\sqrt[3]{-\dfrac{3}{2}}$ by rationalizing the denominator.

Check these solutions.

Practice Problem 2 Solve for all real roots: $x^6 - 5x^3 + 4 = 0$

2 Solving Equations with Fractional Exponents

EXAMPLE 3 Solve and check your solutions: $x^{2/3} - 3x^{1/3} + 2 = 0$

Let $y = x^{1/3}$. Then $y^2 = x^{2/3}$.

$$y^2 - 3y + 2 = 0 \qquad \text{Replace } x^{1/3} \text{ by } y \text{ and } x^{2/3} \text{ by } y^2.$$
$$(y - 2)(y - 1) = 0 \qquad \text{Factor.}$$
$$y - 2 = 0 \quad \text{or} \quad y - 1 = 0$$
$$y = 2 \qquad\qquad y = 1 \qquad \text{Solve for } y.$$

$$x^{1/3} = 2 \quad \text{or} \quad x^{1/3} = 1 \qquad \text{Replace } y \text{ by } x^{1/3}.$$
$$\left(x^{1/3}\right)^3 = (2)^3 \qquad \left(x^{1/3}\right)^3 = (1)^3 \qquad \text{Cube each side of the equation.}$$
$$x = 8 \qquad\qquad x = 1$$

Check:

$x = 8$: $\quad (8)^{2/3} - 3(8)^{1/3} + 2 \overset{?}{=} 0 \qquad\qquad x = 1$: $\quad (1)^{2/3} - 3(1)^{1/3} + 2 \overset{?}{=} 0$

$$\left(\sqrt[3]{8}\right)^2 - 3\left(\sqrt[3]{8}\right) + 2 \overset{?}{=} 0 \qquad\qquad \left(\sqrt[3]{1}\right)^2 - 3\left(\sqrt[3]{1}\right) + 2 \overset{?}{=} 0$$

$$(2)^2 - 3(2) + 2 \overset{?}{=} 0 \qquad\qquad\qquad 1 - 3 + 2 \overset{?}{=} 0$$

$$4 - 6 + 2 \overset{?}{=} 0 \qquad\qquad\qquad\qquad 0 = 0 \quad \checkmark$$

$$0 = 0 \quad \checkmark$$

The exercises that appear in this section are somewhat difficult to solve. Part of the difficulty lies in the fact that the equations have different numbers of solutions. A fourth-degree equation like the one in Example 1 has four different solutions. Whereas a sixth-degree equation such as the one in Example 2 has only two solutions, some sixth-degree equations will have as many as six solutions. Although the equation that we examined in Example 3 has only two solutions, other equations with fractional exponents may have one solution or even no solution at all. It is good to take some time to carefully examine your work to determine that you have obtained the correct number of solutions.

A graphing program on a computer such as TI Interactive, Derive, or Maple can be very helpful in determining or verifying the solutions to these types of problems. Of course a graphing calculator can be most helpful, particularly in verifying the value of a solution and the number of solutions.

Optional Graphing Calculation Exploration: If you have a graphing calculator, verify the solutions for Example 3 by graphing the equation

$$y = x^{2/3} - 3x^{1/3} + 2.$$

Determine from your graph whether the curve does in fact cross the *x*-axis (that is, $y = 0$ when $x = 1$ and $x = 8$). You will have to carefully select the window so that you can see the behavior of the curve clearly. For this equation a useful window is $[-1, 12, -1, 2]$. Remember that with most graphing calculators, you will need to surround the exponents with parentheses.

Practice Problem 3 Solve and check your solutions: $3x^{4/3} - 5x^{2/3} + 2 = 0$

EXAMPLE 4 Solve and check your solutions: $2x^{1/2} = 5x^{1/4} + 12$

$2x^{1/2} - 5x^{1/4} - 12 = 0$	Place in standard form.
$2y^2 - 5y - 12 = 0$	Replace $x^{1/4}$ by y and $x^{1/2}$ by y^2.
$(2y + 3)(y - 4) = 0$	Factor.

$$2y = -3 \qquad \text{or} \qquad y = 4$$

$$y = -\frac{3}{2} \qquad\qquad\qquad \text{Solve for } y.$$

$$x^{1/4} = -\frac{3}{2} \qquad \text{or} \qquad x^{1/4} = 4 \qquad \text{Replace } y \text{ by } x^{1/4}.$$

$$\left(x^{1/4}\right)^4 = \left(-\frac{3}{2}\right)^4 \qquad \left(x^{1/4}\right)^4 = (4)^4 \qquad \text{Solve for } x.$$

$$x = \frac{81}{16} \qquad\qquad\qquad x = 256$$

Check:

$$x = \frac{81}{16}: \quad 2\left(\frac{81}{16}\right)^{1/2} - 5\left(\frac{81}{16}\right)^{1/4} - 12 \overset{?}{=} 0$$

$$2\left(\frac{9}{4}\right) - 5\left(\frac{3}{2}\right) - 12 \overset{?}{=} 0$$

$$\frac{9}{2} - \frac{15}{2} - 12 \overset{?}{=} 0$$

$$-15 \neq 0$$

$$x = 256: \quad 2(256)^{1/2} - 5(256)^{1/4} - 12 \overset{?}{=} 0$$

$$2(16) - 5(4) - 12 \overset{?}{=} 0$$

$$32 - 20 - 12 \overset{?}{=} 0$$

$$0 = 0 \;\checkmark$$

$\frac{81}{16}$ is extraneous and not a valid solution. The only valid solution is 256.

Practice Problem 4 Solve and check your solutions: $3x^{1/2} = 8x^{1/4} - 4$

Although we have covered just four basic examples here, this substitution technique can be extended to other types of equations. In each case we substitute y for an appropriate expression in order to obtain a quadratic equation. The following table lists some substitutions that would be appropriate.

If You Want to Solve:	Then You Would Use the Substitution:
$x^4 - 13x^2 + 36 = 0$	$y = x^2$
$2x^6 - x^3 - 6 = 0$	$y = x^3$
$x^{2/3} - 3x^{1/3} + 2 = 0$	$y = x^{1/3}$
$6(x - 1)^{-2} + (x - 1)^{-1} - 2 = 0$	$y = (x - 1)^{-1}$
$(2x^2 + x)^2 + 4(2x^2 + x) + 3 = 0$	$y = 2x^2 + x$
$\left(\dfrac{1}{x - 1}\right)^2 + \dfrac{1}{x - 1} - 6 = 0$	$y = \dfrac{1}{x - 1}$
$2x - 5x^{1/2} + 2 = 0$	$y = x^{1/2}$

A collection of exercises like these is provided in the exercise set.

Solve. Express any nonreal complex numbers with i notation.

1. $x^4 - 9x^2 + 20 = 0$

2. $x^4 - 11x^2 + 18 = 0$

3. $x^4 + x^2 - 12 = 0$

4. $x^4 - 2x^2 - 8 = 0$

5. $3x^4 = 10x^2 + 8$

6. $5x^4 = 4x^2 + 1$

In Exercises 7–10, find all valid real roots for each equation.

7. $x^6 - 7x^3 - 8 = 0$

8. $x^6 - 3x^3 - 4 = 0$

9. $x^6 - 5x^3 - 14 = 0$

10. $x^6 + 2x^3 - 15 = 0$

Solve for real roots.

11. $x^8 = 3x^4 - 2$

12. $x^8 = 7x^4 - 12$

13. $3x^8 + 13x^4 = 10$

14. $3x^8 - 10x^4 = 8$

15. $x^{2/3} + 2x^{1/3} - 8 = 0$ **16.** $x^{2/3} + x^{1/3} - 12 = 0$ **17.** $2x^{2/3} - 7x^{1/3} - 4 = 0$ **18.** $12x^{2/3} + 5x^{1/3} - 2 = 0$

19. $3x^{1/2} - 14x^{1/4} - 5 = 0$ **20.** $2x^{1/2} - 5x^{1/4} - 3 = 0$ **21.** $2x^{1/2} - x^{1/4} - 6 = 0$

22. $2x^{1/2} - x^{1/4} - 1 = 0$ **23.** $x^{2/5} + x^{1/5} - 2 = 0$ **24.** $2x^{2/5} + 7x^{1/5} + 3 = 0$

In each exercise make an appropriate substitution in order to obtain a quadratic equation. Find all complex values for x.

25. $\left(x^2 + x\right)^2 - 5\left(x^2 + x\right) = -6$ **26.** $\left(x^2 - 2x\right)^2 + 2\left(x^2 - 2x\right) = 3$ **27.** $x - 5x^{1/2} + 6 = 0$

28. $x - 5x^{1/2} - 36 = 0$ **29.** $10x^{-2} + 7x^{-1} + 1 = 0$ **30.** $20x^{-2} + 9x^{-1} + 1 = 0$

To Think About

Solve. Find all valid real roots for each equation.

31. $15 - \dfrac{2x}{x - 1} = \dfrac{x^2}{x^2 - 2x + 1}$

32. $4 - \dfrac{x^3 + 1}{x^3 + 6} = \dfrac{x^3 - 3}{x^3 + 2}$

Cumulative Review Problems

Simplify.

33. $\sqrt{8x} + 3\sqrt{2x} - 4\sqrt{50x}$

34. $\sqrt{27x} + 6\sqrt{3x} - 2\sqrt{48x}$

Multiply and simplify.

35. $3\sqrt{2}(\sqrt{5} - 2\sqrt{6})$

36. $(\sqrt{2} + \sqrt{6})(3\sqrt{2} - 2\sqrt{5})$

9.4 Formulas and Applications

Student Learning Objectives

After studying this section, you will be able to:

1 Solve a quadratic equation containing several variables.

2 Solve problems requiring the use of the Pythagorean theorem.

3 Solve applied problems requiring the use of a quadratic equation.

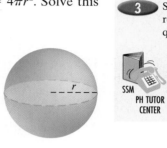

SSM PH TUTOR CD & VIDEO MATH PRO WEB
CENTER

1 Solving a Quadratic Equation Containing Several Variables

In mathematics, physics, and engineering, we must often solve an equation for a variable in terms of other variables. You recall we solved linear equations in several variables in Section 2.5. We now examine several cases where the variable that we are solving for is squared. If the variable we are solving for is squared, and there is no other term containing that variable, then the equation can be solved using the square root property.

EXAMPLE 1 The surface area of a sphere is given by $A = 4\pi r^2$. Solve this equation for r. (You do not need to rationalize the denominator.)

$$A = 4\pi r^2$$

$$\frac{A}{4\pi} = r^2$$

$$\pm\sqrt{\frac{A}{4\pi}} = r \qquad \text{Use the square root property.}$$

$$\pm\frac{1}{2}\sqrt{\frac{A}{\pi}} = r \qquad \text{Simplify.}$$

Since the radius of a sphere must be a positive value, we use only the principal root.

$$r = \frac{1}{2}\sqrt{\frac{A}{\pi}}$$

Practice Problem 1 The volume of a cylindrical cone is $V = \frac{1}{3}\pi r^2 h$. Solve this equation for r. (You do not need to rationalize the denominator.)

Some quadratic equations containing many variables can be solved for one variable by factoring.

EXAMPLE 2 Solve for y: $y^2 - 2yz - 15z^2 = 0$

$$(y + 3z)(y - 5z) = 0 \qquad \text{Factor.}$$

$$y + 3z = 0 \qquad y - 5z = 0 \qquad \text{Set each factor equal to 0.}$$

$$y = -3z \qquad\quad y = 5z \qquad \text{Solve for } y.$$

Practice Problem 2 Solve for y: $2y^2 + 9wy + 7w^2 = 0$

Sometimes the quadratic formula is required in order to solve the equation.

EXAMPLE 3 Solve for x: $2x^2 + 3wx - 4z = 0$

We use the quadratic formula where the variable is considered to be x and the letters w and z are considered constants. Thus, $a = 2$, $b = 3w$, and $c = -4z$.

$$x = \frac{-b \pm \sqrt{b^2 - 4ac}}{2a}$$

$$= \frac{-3w \pm \sqrt{(3w)^2 - 4(2)(-4z)}}{2(2)} = \frac{-3w \pm \sqrt{9w^2 + 32z}}{4}$$

Note that this answer cannot be simplified any further.

Practice Problem 3 Solve for y: $3y^2 + 2fy - 7g = 0$

511

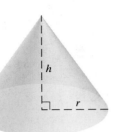

▲ ⬭ **EXAMPLE 4**   The formula for the curved surface area S of a right circular cone of altitude h and with base of radius r is $S = \pi r \sqrt{r^2 + h^2}$.

Solve for r^2.

$$S = \pi r \sqrt{r^2 + h^2}$$

$$\frac{S}{\pi r} = \sqrt{r^2 + h^2} \qquad \text{Isolate the radical.}$$

$$\frac{S^2}{\pi^2 r^2} = r^2 + h^2 \qquad \text{Square both sides.}$$

$$\frac{S^2}{\pi^2} = r^4 + h^2 r^2 \qquad \text{Multiply each term by } r^2.$$

$$0 = r^4 + h^2 r^2 - \frac{S^2}{\pi^2} \qquad \text{Subtract } S^2/\pi^2.$$

This equation is quadratic in form. If we let $y = r^2$, then we have

$$0 = y^2 + h^2 y - \frac{S^2}{\pi^2}.$$

By the quadratic formula we have the following:

$$y = \frac{-h^2 \pm \sqrt{(h^2)^2 - 4(1)\left(-\dfrac{S^2}{\pi^2}\right)}}{2}$$

$$= \frac{-h^2 \pm \sqrt{\dfrac{\pi^2 h^4}{\pi^2} + \dfrac{4S^2}{\pi^2}}}{2}$$

$$= \frac{-h^2 \pm \dfrac{1}{\pi} \sqrt{\pi^2 h^4 + 4S^2}}{2}$$

$$= \frac{-\pi h^2 \pm \sqrt{\pi^2 h^4 + 4S^2}}{2\pi}$$

Since $y = r^2$, we have

$$r^2 = \frac{-\pi h^2 \pm \sqrt{\pi^2 h^4 + 4S^2}}{2\pi}.$$

Practice Problem 4   The formula for the number of diagonals d in a polygon of n sides is $d = \dfrac{n^2 - 3n}{2}$. Solve for n.

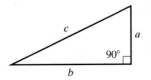 **2**   *Solving Problems Requiring the Use of the Pythagorean Theorem*

A very useful formula is the Pythagorean theorem for right triangles.

Pythagorean Theorem

If c is the length of the longest side of a right triangle and a and b are the lengths of the other two sides, then $a^2 + b^2 = c^2$.

The longest side of a right triangle is called the **hypotenuse**. The other two sides are called the **legs** of the triangle.

▲ ◉ EXAMPLE 5

(a) Solve the Pythagorean theorem $a^2 + b^2 = c^2$ for a.

(b) Find the value of a if $c = 13$ and $b = 5$.

(a) $a^2 = c^2 - b^2$ Subtract b^2 from each side.

$a = \pm\sqrt{c^2 - b^2}$ Use the square root property.

Since a, b, and c must be positive numbers because they represent lengths, we use only the positive root, $a = \sqrt{c^2 - b^2}$.

(b) $a = \sqrt{c^2 - b^2}$

$\qquad = \sqrt{(13)^2 - (5)^2} = \sqrt{169 - 25} = \sqrt{144} = 12$

Thus, $a = 12$.

Practice Problem 5

(a) Solve the Pythagorean theorem for b.

(b) Find the value of b if $c = 26$ and $a = 24$.　　　　　　◉

▲ ◉ EXAMPLE 6 The perimeter of a triangular piece of land is 12 miles. One leg of the triangle is 1 mile longer than the other leg. Find the length of each boundary of the land if the triangle is a right triangle.

1. *Understand the problem.*
   Draw a picture of the piece of land and label the sides of the triangle.

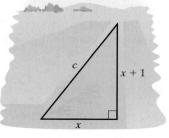

2. *Write an equation.*
   We can use the Pythagorean theorem. First, we want only one variable in our equation. (Right now, both c and x are not known.)
   We are given that the perimeter is 12 miles, so

   $$x + (x + 1) + c = 12.$$

   Thus,

   $$c = -2x + 11.$$

   By the Pythagorean theorem,

   $$x^2 + (x + 1)^2 = (-2x + 11)^2.$$

3. *Solve the equation and state the answer.*

   $$x^2 + (x + 1)^2 = (-2x + 11)^2$$
   $$x^2 + x^2 + 2x + 1 = 4x^2 - 44x + 121$$
   $$0 = 2x^2 - 46x + 120$$
   $$0 = x^2 - 23x + 60$$

By the quadratic formula, we have the following:

$$x = \frac{-(-23) \pm \sqrt{(-23)^2 - 4(1)(60)}}{2(1)}$$

$$x = \frac{23 \pm \sqrt{289}}{2}$$

$$x = \frac{23 \pm 17}{2}$$

$$x = \frac{40}{2} = 20 \quad \text{or} \quad x = \frac{6}{2} = 3$$

The answer $x = 20$ cannot be right because the perimeter (the sum of *all* the sides) is only 12. The only answer that makes sense is $x = 3$. Thus, the sides of the triangle are $x = 3$, $x + 1 = 3 + 1 = 4$, and $-2x + 11 = -2(3) + 11 = 5$. The longest boundary of this triangular piece of land is 5 miles. The other two boundaries are 4 miles and 3 miles.

Notice that we could have factored the quadratic equation instead of using the quadratic formula. $x^2 - 23x + 60 = 0$ can be written as $(x - 20)(x - 3) = 0$.

4. **Check.**
Is the perimeter 12 miles?

$$5 + 4 + 3 = 12 \quad \checkmark$$

Is one leg 1 mile longer than the other?

$$4 = 3 + 1 \quad \checkmark$$

Practice Problem 6   The perimeter of a triangular piece of land is 30 miles. One leg of the triangle is 7 miles shorter than the other leg. Find the length of each boundary of the land if the triangle is a right triangle.

3 Solving Applied Problems Requiring the Use of a Quadratic Equation

Many types of area problems can be solved with quadratic equations as shown in the next two examples.

▲ **EXAMPLE 7**   The radius of an old circular pipe under a roadbed is 10 inches. Designers want to replace it with a smaller pipe and have decided they can use one with a cross-sectional area that is 36π square inches smaller. What should the radius of the new pipe be?

First we need the formula for the area of a circle,

$$A = \pi r^2,$$

where A is the area and r is the radius. The area of the cross section of the old pipe is as follows:

$$A_{\text{old}} = \pi(10)^2$$
$$= 100\pi$$

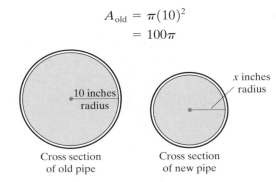

10 inches radius

x inches radius

Cross section of old pipe

Cross section of new pipe

Let x = the radius of the new pipe.

$$(\text{area of old pipe}) - (\text{area of new pipe}) = 36\pi$$
$$100\pi \quad - \quad \pi x^2 \quad = 36\pi$$

$64\pi = \pi x^2$ Add πx^2 to each side and subtract 36π from each side.

$\dfrac{64\pi}{\pi} = \dfrac{\pi x^2}{\pi}$ Divide each side by π.

$64 = x^2$

$\pm 8 = x$ Use the square root property.

Since the radius must be positive, we select $x = 8$. The radius of the new pipe is 8 inches. Check to verify this solution.

Practice Problem 7   Redo Example 7 when the radius of the pipe under the roadbed is 6 inches and the designers want to replace it with a pipe that has a cross-sectional area that is 45π square inches larger. What should the radius of the new pipe be?

▲ ⬭ **EXAMPLE 8**   A triangular sign marks the edge of the rocks in Rockport Harbor. The sign has an area of 35 square meters. Find the base and altitude of this triangular sign if the base is 3 meters shorter than the altitude.

The area of a triangle is given by

$$A = \frac{1}{2}ab.$$

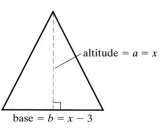

altitude = $a = x$

base = $b = x - 3$

Let x = the length in meters of the altitude. Then $x - 3$ = the length in meters of the base.

$35 = \dfrac{1}{2}x(x - 3)$ Replace A (area) by 35, a (altitude) by x, and b (base) by $x - 3$.

$70 = x(x - 3)$ Multiply each side by 2.

$70 = x^2 - 3x$ Use the distributive property.

$0 = x^2 - 3x - 70$ Subtract 70 from each side.

$0 = (x - 10)(x + 7)$

$x = 10 \quad \text{or} \quad x = -7$

The length of a side of a triangle must be a positive number, so we disregard -7. Thus,

$$\text{altitude} = x = 10 \text{ meters and}$$
$$\text{base} = x - 3 = 7 \text{ meters.}$$

The check is left to the student.

Practice Problem 8   The length of a rectangle is 3 feet shorter than twice the width. The area of the rectangle is 54 square feet. Find the dimensions of the rectangle.

We will now examine a few word problems that require the use of the formula distance = (rate)(time) or $d = rt$.

⬭ **EXAMPLE 9**   When Barbara was training for a bicycle race, she rode a total of 135 miles on Monday and Tuesday. On Monday she rode for 75 miles in the rain. On Tuesday she rode 5 miles per hour faster because the weather was better. Her total cycling time for the 2 days was 8 hours. Find her speed for each day.

It would be helpful to organize a few basic facts.

We can find each distance. If Barbara rode 75 miles on Monday and a total of 135 miles during the 2 days, then she rode $135 - 75 = 60$ miles on Tuesday.

Let x = the cycling rate in miles per hour on Monday. Since Barbara rode 5 miles per hour faster on Tuesday, $x + 5$ = the cycling rate in miles per hour on Tuesday.

Since distance divided by rate is equal to time $\left(\dfrac{d}{r} = t\right)$, we can determine that the time Barbara cycled on Monday was $\dfrac{75}{x}$ and the time she cycled on Tuesday was $\dfrac{60}{x + 5}$.

We put these facts into a table.

Day	Distance	Rate	Time
Monday	75	x	$\dfrac{75}{x}$
Tuesday	60	$x + 5$	$\dfrac{60}{x + 5}$
Totals	135	(not used)	8

Since the total cycling time was 8 hours, we have the following:

$$\text{time cycling Monday} \ + \ \text{time cycling Tuesday} = 8 \text{ hours}$$
$$\frac{75}{x} \quad + \quad \frac{60}{x + 5} = 8$$

The LCD of this equation is $x(x + 5)$. Multiply each term by the LCD.

$$x(x + 5)\left(\frac{75}{x}\right) + x(x + 5)\left(\frac{60}{x + 5}\right) = x(x + 5)(8)$$
$$75(x + 5) + 60x = 8x(x + 5)$$
$$75x + 375 + 60x = 8x^2 + 40x$$
$$0 = 8x^2 - 95x - 375$$
$$0 = (x - 15)(8x + 25)$$
$$x - 15 = 0 \quad \text{or} \quad 8x + 25 = 0$$
$$x = 15 \qquad\qquad x = \frac{-25}{8}$$

We disregard the negative answer. The cyclist did not have a negative rate of speed—unless she was pedaling backward! Thus, $x = 15$. So Barbara's rate of speed on Monday was 15 mph, and her rate of speed on Tuesday was $x + 5 = 15 + 5 = 20$ mph.

Practice Problem 9 Carlos traveled in his car at a constant speed on a secondary road for 150 miles. Then he traveled 10 mph faster on a better road for 240 miles. If Carlos drove for 7 hours, find the car's speed for each part of the trip.

Solve for the variable specified. Assume that all other variables are nonzero.

1. $S = 16t^2$; for t

2. $E = mc^2$; for c

3. $A = \pi\left(\dfrac{d}{2}\right)^2$; for d

4. $V = \dfrac{1}{3}\pi r^2 h$; for r

5. $3H = \dfrac{1}{2}ax^2$; for x

6. $5B = \dfrac{2}{3}hx^2$; for x

7. $4(y^2 + w) - 5 = 7R$; for y

8. $9x^2 - 2 = 3B$; for x

9. $Q = \dfrac{3mwM^2}{2c}$; for M

10. $H = \dfrac{5abT^2}{7k}$; for T

11. $V = \pi(r^2 + R^2)h$; for r

12. $H = b(a^2 + w^2)$; for w

13. $x^2 + 3bx - 10b^2 = 0$; for x

14. $y^2 - 4yw - 45w^2 = 0$; for y

15. $P = EI - RI^2$; for I

16. $A = P(1 + r)^2$; for r

17. $10w^2 - 3qw - 4 = 0$; for w

18. $7w^2 + 5qw - 1 = 0$; for w

19. $S = 2\pi rh + \pi r^2$; for r

20. $B = 3abx^2 - 5x$; for x

21. $(a + 1)x^2 + 5x + 2w = 0$; for x

22. $(b - 2)x^2 - 3x + 5y = 0$; for x

In Exercises 23–32 use the Pythagorean theorem to find the missing side(s).

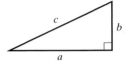

▲ **23.** $b = 7, a = \sqrt{3}$; find c

▲ **24.** $a = 2\sqrt{3}, b = 3$; find c

▲ **25.** $c = \sqrt{34}, b = \sqrt{19}$; find a

▲ **26.** $c = \sqrt{21}, a = \sqrt{5}$; find b

▲ **27.** $c = 12, b = 2a$; find b and a

▲ **28.** $c = 15, a = 2b$; find b and a

▲ **29.** A racing sailboat is traveling along a triangular course. The three straight line distances form a right triangle. One leg of the triangle represents a distance of 12 miles. The other leg of the triangle is 6 miles shorter than the hypotenuse. What is the length of the hypotenuse of this triangle? What is the length of the other leg?

▲ **30.** Tony Pitkin has a cornfield in South Dakota that is shaped like a right triangle. The hypotenuse of this triangle is 10 miles long. One leg of the triangular field is 2 miles shorter than the other leg. Find the length of the other two sides of the field.

31. An airplane flew from London a distance of 11 kilometers due north. The plane banked to the right and flew the second leg of the journey. Finally, the plane then banked to the right again and flew back to the starting point. The entire course was shaped like a right triangle with the 11-kilometer distance serving as the hypotenuse. The final leg of this journey was 3 miles longer than the second leg. How many miles were the second and third legs of the journey? Round your answers to the nearest hundredth of a mile.

32. A Norwegian freighter travels 12 miles due east toward the United States. Seeking to avoid an underwater obstruction, the vessel turns to the left abruptly and travels for a certain distance. Having discovered engine trouble, the captain turns abruptly again and heads for home port. The entire course is shaped like a right triangle, with the 12-mile distance serving as the hypotenuse and with the final leg of the trip being 4 miles longer than the second leg of the trip. What are the lengths of the second and third legs of the trip? Round your answers to the nearest hundredth of a mile.

33. The area of a rectangular wall of a barn is 126 square feet. Its length is 4 feet longer than twice its width. Find the length and width of the wall of the barn.

34. The area of a rectangular tennis court is 140 square meters. Its length is 6 meters shorter than twice its width. Find the length and width of the tennis court.

35. The area of a triangular flag is 72 square centimeters. Its altitude is 2 centimeters longer than twice its base. Find the lengths of the altitude and the base.

36. A children's playground is triangular in shape. Its altitude is 2 yards shorter than its base. The area of the playground is 60 square yards. Find the base and altitude of the playground.

37. Roberto drove at a constant speed in a rainstorm for 225 miles. He took a break, and the rain stopped. He then drove 150 miles at a speed that was 5 miles per hour faster than his previous speed. If he drove for 8 hours, find the car's speed for each part of the trip.

38. Benita traveled at a constant speed on an old road for 160 miles. She then traveled 5 miles per hour faster on a newer road for 90 miles. If she drove for 6 hours, find the car's speed for each part of the trip.

39. Bob drove from home to work at 50 mph. After work the traffic was heavier, and he drove home at 45 mph. His driving time to and from work was 1 hour and 16 minutes. How far does he live from his job?

40. A driver drove his heavily loaded truck from the company warehouse to a delivery point at 35 mph. He unloaded the truck and drove back to the warehouse at 45 mph. The total trip took 5 hours and 20 minutes. How far is the delivery point from the warehouse?

The number of inmates N (measured in thousands) in federal and state prisons in the United States can be approximated by the equation $N = 1.11x^2 + 33.39x + 304.09$, where x is the number of years since 1980. For example, when $x = 1$, $N = 338.59$. This tells us that in 1981 there were approximately 338,590 inmates in federal and state prisons. Use this equation to answer the following questions. (Source: U.S. Bureau of Justice Statistics.)

41. How many inmates does the equation predict there will be in the year 2003?

43. In what year is the number of inmates expected to be 1,744,800?

42. How many inmates does the equation predict there will be in the year 2006?

44. In what year is the number of inmates expected to be 1,832,600?

To Think About

45. Solve for w: $w = \dfrac{12b^2}{\dfrac{5}{2}w + \dfrac{7}{2}b + \dfrac{21}{2}}$

46. The formula $A = P(1 + r)^2$ gives the amount A in dollars that will be obtained in 2 years if P dollars are invested at an annual compound interest rate of r. If you invest $P = \$1400$ and it grows to $\$1514.24$ in 2 years, what is the annual interest rate r?

Cumulative Review Problems

Rationalize the denominators.

47. $\dfrac{4}{\sqrt{3x}}$

48. $\dfrac{5\sqrt{6}}{2\sqrt{5}}$

49. $\dfrac{3}{\sqrt{x} + \sqrt{y}}$

50. $\dfrac{2\sqrt{3}}{\sqrt{3} - \sqrt{6}}$

51. $\dfrac{3ab}{\sqrt[3]{8ab^2}}$

9.5 Quadratic Functions

1 Finding the Vertex and the Intercepts of a Quadratic Function

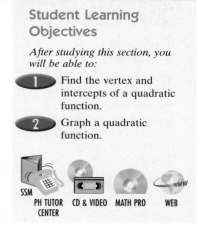

Student Learning Objectives

After studying this section, you will be able to:

1 Find the vertex and intercepts of a quadratic function.

2 Graph a quadratic function.

SSM PH TUTOR CD & VIDEO MATH PRO WEB
       CENTER

In Section 3.6 we graphed functions such as $p(x) = x^2$ and $g(x) = (x + 2)^2$. We will now study quadratic functions in more detail.

Definition of a Quadratic Function

A **quadratic function** is a function of the form

$$f(x) = ax^2 + bx + c, \text{ where } a, b, \text{ and } c \text{ are real numbers and } a \neq 0.$$

Graphs of quadratic functions written in this form will be parabolas opening upward if $a > 0$ or downward if $a < 0$. The **vertex** of a parabola is the lowest point on a parabola opening upward or the highest point on a parabola opening downward. The vertex will occur at $x = \dfrac{-b}{2a}$. To find the y-value, or $f(x)$, when $x = \dfrac{-b}{2a}$, we find $f\left(\dfrac{-b}{2a}\right)$. Therefore, we can say that a quadratic function has its vertex at $\left(\dfrac{-b}{2a}, f\left(\dfrac{-b}{2a}\right)\right)$.

It is helpful to know the x-intercept and the y-intercept when graphing a quadratic function.

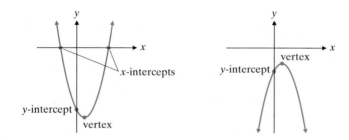

A quadratic function will always have exactly one y-intercept. However, it may have zero, one, or two x-intercepts. Why?

When Graphing Quadratic Functions of the Form $f(x) = ax^2 + bx + c, a \neq 0$

1. The coordinates of the vertex are $\left(\dfrac{-b}{2a}, f\left(\dfrac{-b}{2a}\right)\right)$.

2. The y-intercept is at $f(0)$.

3. The x-intercepts (if they exist) occur where $f(x) = 0$. They can always be found with the quadratic formula and can sometimes be found by factoring.

Since we may replace y by $f(x)$, the graph is equivalent to the graph of $y = ax^2 + bx + c$.

Graphing Calculator

**Finding the
x-intercepts
and the vertex**

You can use a graphing
calculator to find the
x-intercepts and vertex of
a quadratic function. To
find the intercepts of the
quadratic function

$$f(x) = x^2 - 4x + 3,$$

graph $y = x^2 - 4x + 3$ on
a graphing calculator
using an appropriate
window.
Display:

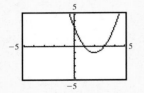

Next you can use the
Trace and Zoom features
or zero command of your
calculator to find the
x-intercepts.
You can also use the
Trace and Zoom features
to determine the vertex.
Some calculators have
a feature that will
calculate the maximum or
minimum point on the
graph. Use the feature
that calculates the
minimum point to find
the vertex of
$f(x) = x^2 - 4x + 3$.
Display:

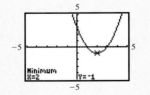

Thus, the vertex is at
$(2, -1)$.

● **EXAMPLE 1**   Find the coordinates of the vertex and the intercepts of the quadratic function $f(x) = x^2 - 8x + 15$.

For this function $a = 1, b = -8$, and $c = 15$.

Step 1   The vertex occurs at $x = \dfrac{-b}{2a}$. Thus,

$$x = \frac{-(-8)}{2(1)} = \frac{8}{2} = 4.$$

The vertex has an x-coordinate of 4. To find the y-coordinate, we evaluate $f(4)$.

$$f(4) = 4^2 - 8(4) + 15 = 16 - 32 + 15 = -1$$

Thus, the vertex is $(4, -1)$.

Step 2   The y-intercept is at $f(0)$. We evaluate $f(0)$ to find the y-coordinate when x is 0.

$$f(0) = 0^2 - 8(0) + 15 = 15$$

The y-intercept is $(0, 15)$.

Step 3   If there are x-intercepts, they will occur when $f(x) = 0$—that is, when $x^2 - 8x + 15 = 0$. We solve for x.

$$(x - 5)(x - 3) = 0$$

$$x - 5 = 0 \qquad x - 3 = 0$$

$$x = 5 \qquad\quad x = 3$$

Thus, we conclude that the x-intercepts are $(5, 0)$ and $(3, 0)$. We list these four important points of the function in table form.

Name	x	$f(x)$
Vertex	4	−1
y-intercept	0	15
x-intercept	5	0
x-intercept	3	0

Practice Problem 1   Find the coordinates of the vertex and the intercepts of the quadratic function $f(x) = x^2 - 6x + 5$.   ●

2 *Graphing a Quadratic Function*

It is helpful to find the vertex and the intercepts of a quadratic function before graphing it.

EXAMPLE 2 Find the vertex and the intercepts, and then graph the function $f(x) = x^2 + 2x - 4$.

Here $a = 1, b = 2$, and $c = -4$. Since $a > 0$, the parabola opens *upward*.

Step 1 We find the vertex.

$$x = \frac{-b}{2a} = \frac{-2}{2(1)} = \frac{-2}{2} = -1$$

$$f(-1) = (-1)^2 + 2(-1) - 4 = 1 + (-2) - 4 = -5$$

The vertex is $(-1, -5)$.

Step 2 We find the y-intercept. The y-intercept is at $f(0)$.

$$f(0) = (0)^2 + 2(0) - 4 = -4$$

The y-intercept is $(0, -4)$.

Step 3 We find the x-intercepts. The x-intercepts occur when $f(x) = 0$.

Thus, we solve $x^2 + 2x - 4 = 0$ for x. We cannot factor this equation, so we use the quadratic formula.

$$x = \frac{-b \pm \sqrt{b^2 - 4ac}}{2a} = \frac{-2 \pm \sqrt{2^2 - 4(1)(-4)}}{2(1)} = \frac{-2 \pm \sqrt{20}}{2} = -1 \pm \sqrt{5}$$

To aid our graphing, we will approximate the value of x to the nearest tenth by using a square root table or a scientific calculator.

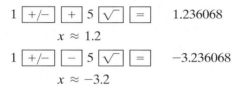

$$x \approx 1.2$$

$$x \approx -3.2$$

The x-intercepts are approximately $(-3.2, 0)$ and $(1.2, 0)$.

We have found that the vertex is $(-1, -5)$; the y-intercept is $(0, -4)$; and the x-intercepts are approximately $(-3.2, 0)$ and $(1.2, 0)$. We connect these points by a smooth curve to graph the parabola.

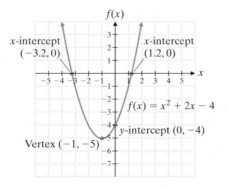

PRACTICE PROBLEM 2

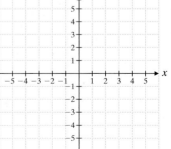

Practice Problem 2 Find the vertex and the intercepts, and then graph the function $g(x) = x^2 - 2x - 2$.

EXAMPLE 3 Find the vertex and the intercepts, and then graph the function $f(x) = -2x^2 + 4x - 3$.

Here $a = -2$, $b = 4$, and $c = -3$. Since $a < 0$, the parabola opens *downward*. The vertex occurs at $x = \dfrac{-b}{2a}$.

$$x = \frac{-4}{2(-2)} = \frac{-4}{-4} = 1$$

$$f(1) = -2(1)^2 + 4(1) - 3 = -2 + 4 - 3 = -1$$

The vertex is $(1, -1)$.

The y-intercept is at $f(0)$.

$$f(0) = -2(0)^2 + 4(0) - 3 = -3$$

The y-intercept is $(0, -3)$.

If there are any x-intercepts, they will occur when $f(x) = 0$. We use the quadratic formula to solve $-2x^2 + 4x - 3 = 0$ for x.

$$x = \frac{-4 \pm \sqrt{4^2 - 4(-2)(-3)}}{2(-2)} = \frac{-4 \pm \sqrt{-8}}{-4}$$

Because $\sqrt{-8}$ yields an imaginary number, there are no real roots. Thus, there are no x-intercepts for the graph of the function. That is, the graph does not intersect the x-axis.

We know that the parabola opens *downward*. Thus, the vertex is a maximum value at $(1, -1)$. Since this graph has no x-intercepts, we will look for three additional points to help us in drawing the graph. We try $f(2)$, $f(3)$, and $f(-1)$.

$$f(2) = -2(2)^2 + 4(2) - 3 = -8 + 8 - 3 = -3$$
$$f(3) = -2(3)^2 + 4(3) - 3 = -18 + 12 - 3 = -9$$
$$f(-1) = -2(-1)^2 + 4(-1) - 3 = -2 - 4 - 3 = -9$$

We plot the vertex, the y-intercept, and the points $(2, -3)$, $(3, -9)$, and $(-1, -9)$.

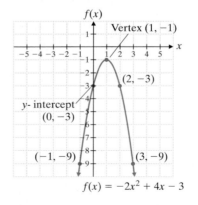

PRACTICE PROBLEM 3

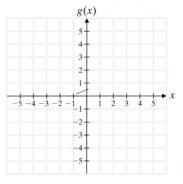

Practice Problem 3 Find the vertex and the intercepts, and then graph the function $g(x) = -2x^2 - 8x - 6$.

Find the coordinates of the vertex and the intercepts of each of the following quadratic functions. When necessary, approximate the x-intercepts to the nearest tenth.

1. $f(x) = x^2 - 2x - 8$

2. $f(x) = x^2 - 4x - 5$

3. $g(x) = -x^2 - 4x + 12$

4. $g(x) = x^2 + 10x - 24$

5. $p(x) = 3x^2 + 12x + 3$

6. $p(x) = 2x^2 + 4x + 1$

7. $r(x) = -3x^2 - 2x - 6$

8. $s(x) = -2x^2 + 6x + 5$

9. $f(x) = 2x^2 + 2x - 4$

10. $f(x) = 5x^2 + 2x - 3$

In each of the following exercises, find the vertex, the y-intercept, and the x-intercepts (if any exist), and then graph the function.

11. $f(x) = x^2 - 6x + 8$

12. $f(x) = x^2 + 6x + 8$

13. $g(x) = x^2 + 2x - 8$

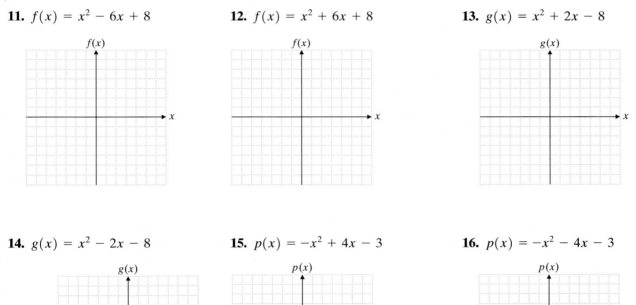

14. $g(x) = x^2 - 2x - 8$

15. $p(x) = -x^2 + 4x - 3$

16. $p(x) = -x^2 - 4x - 3$

17. $r(x) = x^2 + 4x + 6$

$r(x)$

x

18. $r(x) = -x^2 + 4x - 5$

$r(x)$

x

19. $f(x) = x^2 - 6x + 5$

$f(x)$

x

20. $f(x) = x^2 - 4x + 4$

$f(x)$

x

21. $g(x) = -x^2 + 6x - 9$

$g(x)$

x

22. $g(x) = 2x^2 - 2x + 1$

$g(x)$

x

Applications

Some sports such as scuba diving are expensive. People with larger incomes are more likely to participate in such a sport. The number of people N (measured in thousands) who engage in scuba diving can be described by the function $N(x) = 0.18x^2 - 3.18x + 102.25$, *where* x *is the mean income (measured in thousands) and* $x \geq 20$. *Use this information to answer problems 23–27. (Source: U.S. Census Bureau.)*

23. Find $N(20), N(40), N(60), N(80),$ and $N(100)$.

24. Use the results of Exercise 23 to graph the function from $x = 20$ to $x = 100$. You may use the graph grid provided on page 463.

25. Find $N(70)$ from your graph. Explain what $N(70)$ means.

26. Find $N(70)$ from the equation for $N(x)$. Compare your answers for Exercises 25 and 26.

27. Use your graph to determine for what value of x $N(x)$ is equal to 390. Explain what this means.

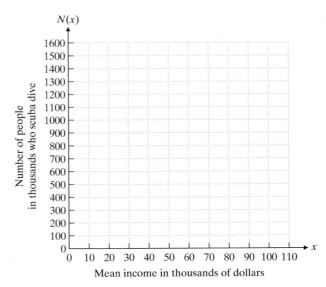

The daily profit P in dollars of the Pine Tree Table Company is described by the function $P(x) = -6x^2 + 312x - 3672$, where x is the number of tables that are manufactured in 1 day. Use this information to answer Exercises 28–32.

28. Find $P(16), P(20), P(24), P(30)$, and $P(35)$.

29. Use the results of Exercise 28 to graph the function from $x = 16$ to $x = 35$.

30. The maximum profit of the company occurs at the vertex of the parabola. How many tables should be made per day in order to obtain the maximum profit for the company? What is the maximum profit?

31. How many tables per day should be made in order to obtain a daily profit of $360? Why are there two answers to this question?

32. How many tables are made per day if the company has a daily profit of zero dollars?

33. Susan throws a softball upward into the air at a speed of 32 feet per second from a 40-foot platform. The distance d upward that the ball travels is given by the function $d(t) = -16t^2 + 32t + 40$, where t is the time in seconds. What is the maximum height of the softball? How many seconds does it take to reach the ground after first being thrown upward? (Round your answer to the nearest tenth.)

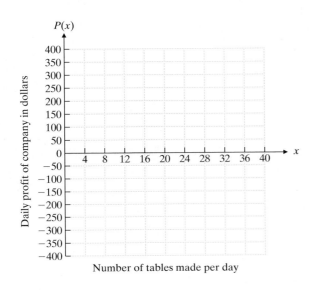

34. Henry is standing on a platform overlooking a baseball stadium. It is 160 feet above the playing field. When he throws a baseball upward at 64 feet per second, the distance d from the baseball to the ground is given by the function $d(t) = -16t^2 + 64t + 160$, where t is the time in seconds. What is the maximum height of the baseball if he throws it upward? How many seconds does it take until the ball finally hits the ground? (Round your answer to the nearest tenth.)

Optional Graphing Calculator Problems

Find the vertex and intercepts for each of the following. If your answers are not exact, round them to the nearest tenth.

35. $y = x^2 - 4.4x + 7.59$

36. $y = x^2 + 7.8x + 13.8$

37. Graph $y = 2.3x^2 - 5.4x - 1.6$. Find the x-intercepts to the nearest tenth.

38. Graph $y = -4.6x^2 + 7.2x - 2.3$. Find the x-intercepts to the nearest tenth.

Cumulative Review Problems

Solve each system.

39. $9x + 5y = 6$
$2x - 5y = -17$

40. $x + y = 16$
$95x + 143y = 1760$

41. $3x - y + 2z = 12$
$2x - 3y + z = 5$
$x + 3y + 8z = 22$

42. $7x + 3y - z = -2$
$x + 5y + 3z = 2$
$x + 2y + z = 1$

9.6 Quadratic Inequalities in One Variable

Student Learning Objectives

After studying this section, you will be able to:

1. Solve a factorable quadratic inequality in one variable.

2. Solve a nonfactorable quadratic inequality in one variable.

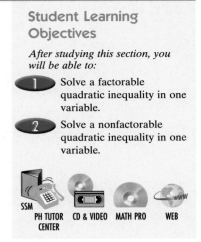

SSM PH TUTOR CD & VIDEO MATH PRO WEB
CENTER

1 ▸ Solving a Factorable Quadratic Inequality in One Variable

We will now solve quadratic inequalities such as $x^2 - 2x - 3 > 0$ and $2x^2 + x - 15 < 0$. A **quadratic inequality** has the form $ax^2 + bx + c < 0$ (or replace $<$ by $>$, $\leq$, or $\geq$), where a, b, and c are real numbers $a \neq 0$. We use our knowledge of solving quadratic equations to solve quadratic inequalities.

Let's solve the inequality $x^2 - 2x - 3 > 0$. We want to find the two points where the expression on the left side is equal to zero. We call these the **critical points**. To do this, we replace the inequality symbol by an equal sign and solve the resulting equation.

$$x^2 - 2x - 3 = 0$$
$$(x + 1)(x - 3) = 0 \qquad \text{Factor.}$$
$$x + 1 = 0 \quad \text{or} \quad x - 3 = 0 \qquad \text{Zero factor property}$$
$$x = -1 \qquad\qquad x = 3$$

These two solutions form critical points that divide the number line into three segments.

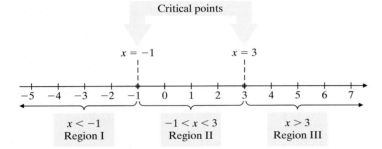

We will show as an exercise that all values of x in a given segment produce results that are greater than zero, or all values of x in a given segment produce results that are less than zero.

To solve the quadratic inequality, we pick an arbitrary test point in each region and then substitute it into the inequality to determine whether it satisfies the inequality. If one point in a region satisfies the inequality, then *all* points in the region satisfy the inequality. We will test three values of x in the expression $x^2 - 2x - 3$.

$\boxed{x < -1, \text{region I:}}$ A sample point is $x = -2$.

$$(-2)^2 - 2(-2) - 3 = 4 + 4 - 3 = 5 > 0$$

$\boxed{-1 < x < 3, \text{region II:}}$ A sample point is $x = 0$.

$$(0)^2 - 2(0) - 3 = 0 + 0 - 3 = -3 < 0$$

$\boxed{x > 3, \text{region III:}}$ A sample point is $x = 4$.

$$(4)^2 - 2(4) - 3 = 16 - 8 - 3 = 5 > 0$$

Thus, we see that $x^2 - 2x - 3 > 0$ when $x < -1$ or $x > 3$. No points in region II satisfy the inequality. The graph of the solution is shown next.

We summarize our method.

Solving a Quadratic Inequality

1. Replace the inequality symbol by an equal sign. Solve the resulting equation to find the critical points.
2. Use the critical points to separate the number line into three distinct regions.
3. Evaluate the quadratic expression at a test point in each region.
4. Determine which regions satisfy the original conditions of the quadratic inequality.

EXAMPLE 1 Solve and graph $x^2 - 10x + 24 > 0$.

1. We replace the inequality symbol by an equal sign and solve the resulting equation.

$$x^2 - 10x + 24 = 0$$
$$(x - 4)(x - 6) = 0$$
$$x - 4 = 0 \quad \text{or} \quad x - 6 = 0$$
$$x = 4 \qquad\qquad x = 6$$

2. We use the critical points to separate the number line into distinct regions.

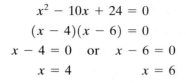

3. We evaluate the quadratic expression at a test point in each of the regions.

$$x^2 - 10x + 24$$

$\boxed{x < 4, region\ I:}$ We pick the sample point $x = 1$.

$$(1)^2 - 10(1) + 24 = 1 - 10 + 24 = 15 > 0$$

$\boxed{4 < x < 6, region\ II:}$ We pick the sample point $x = 5$.

$$(5)^2 - 10(5) + 24 = 25 - 50 + 24 = -1 < 0$$

$\boxed{x > 6, region\ III:}$ We pick the sample point $x = 7$.

$$(7)^2 - 10(7) + 24 = 49 - 70 + 24 = 3 > 0$$

4. We determine which regions satisfy the original conditions of the quadratic inequality.

$$x^2 - 10x + 24 > 0 \text{ when } x < 4 \text{ or when } x > 6.$$

The graph of the solution is shown next.

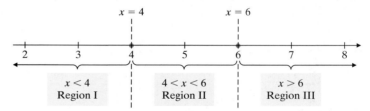

PRACTICE PROBLEM 1

Practice Problem 1 Solve and graph $x^2 - 2x - 8 < 0$.

● **EXAMPLE 2** Solve and graph $2x^2 + x - 6 \le 0$.

We replace the inequality symbol by an equal sign and solve the resulting equation.

$$2x^2 + x - 6 = 0$$

$$(2x - 3)(x + 2) = 0$$

$$2x - 3 = 0 \qquad \text{or} \qquad x + 2 = 0$$

$$2x = 3 \qquad\qquad\qquad x = -2$$

$$x = \frac{3}{2} = 1.5$$

We use the critical points to separate the number line into distinct regions. The critical points are $x = -2$ and $x = 1.5$. Now we arbitrarily pick a test point in each region.

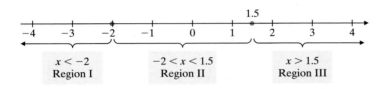

$$2x^2 + x - 6$$

Region I: We pick $x = -3$.

$$2(-3)^2 + (-3) - 6 = 18 - 3 - 6 = 9 > 0$$

Region II: We pick $x = 0$.

$$2(0)^2 + (0) - 6 = 0 + 0 - 6 = -6 < 0$$

Region III: We pick $x = 2$.

$$2(2)^2 + (2) - 6 = 8 + 2 - 6 = 4 > 0$$

Since our inequality is $\le$ and not just $<$, we need to include the critical points. Thus, $2x^2 + x - 6 \le 0$ when $-2 \le x \le 1.5$. The graph of our solution is shown next.

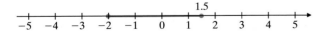

Practice Problem 2 Solve and graph $3x^2 - x - 2 \ge 0$.

Graphing Calculator

Solving Quadratic Inequalities

To solve Example 2 on a graphing calculator, graph $y = 2x^2 + x - 6$, zoom in on the two x-intercepts, and use the Trace feature to find the value of x where y changes from a negative value to a positive value or zero. You can use the Zero command if your calculator has it. Can you verify from your graph that the solution to Example 2 is $-2 \le x \le 1.5$? Use your graphing calculator and this method to solve the exercises below. Round your answers to the nearest hundredth.

1. $3x^2 - 3x - 60 \ge 0$

2. $2.1x^2 + 4.3x - 29.7 > 0$

3. $15.3x^2 - 20.4x + 6.8 \ge 0$

4. $16.8x^2 > -16.8x - 35.7$

If you are using a graphing calculator and you think there is no solution to a quadratic inequality, how can you *verify* this from the graph? Why is this possible?

PRACTICE PROBLEM 2

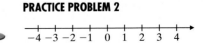

2 Solving a Nonfactorable Quadratic Inequality in One Variable

If the quadratic expression in a quadratic inequality cannot be factored, then we will use the quadratic formula to obtain the critical points.

EXAMPLE 3 Solve and graph $x^2 + 4x > 6$. Round your answer to the nearest tenth.

First we write $x^2 + 4x - 6 > 0$. Because we cannot factor $x^2 + 4x - 6$, we use the quadratic formula to find the critical points.

$$x = \frac{-4 \pm \sqrt{4^2 - 4(1)(-6)}}{2(1)} = \frac{-4 \pm \sqrt{16 + 24}}{2}$$

$$= \frac{-4 \pm \sqrt{40}}{2} = \frac{-4 \pm 2\sqrt{10}}{2} = -2 \pm \sqrt{10}$$

Using a calculator or our table of square roots, we find the following:

$$-2 + \sqrt{10} \approx -2 + 3.162 \approx 1.162 \text{ or about } 1.2$$
$$-2 - \sqrt{10} \approx -2 - 3.162 \approx -5.162 \text{ or about } -5.2$$

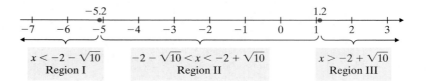

We will see where $x^2 + 4x - 6 > 0$.

Region I: $x = -6$

$$(-6)^2 + 4(-6) - 6 = 36 - 24 - 6 = 6 > 0$$

Region II: $x = 0$

$$(0)^2 + 4(0) - 6 = 0 + 0 - 6 = -6 < 0$$

Region III: $x = 2$

$$(2)^2 + 4(2) - 6 = 4 + 8 - 6 = 6 > 0$$

Thus, $x^2 + 4x > 6$ when $x^2 + 4x - 6 > 0$, and this occurs when $x < -2 - \sqrt{10}$ or $x > -2 + \sqrt{10}$. Rounding to the nearest tenth, our answer is

$$x < -5.2 \quad \text{or} \quad x > 1.2.$$

PRACTICE PROBLEM 3

Practice Problem 3 Solve and graph $x^2 + 2x < 7$. Round your answer to the nearest tenth.

Verbal and Writing Skills

1. When solving a quadratic inequality, why is it necessary to find the critical points?

2. What is the difference between solving an exercise like $ax^2 + bx + c > 0$ and an exercise like $ax^2 + bx + c \geq 0$?

Solve and graph.

3. $x^2 + x - 12 < 0$

4. $x^2 - x - 6 > 0$

5. $2x^2 + x - 3 < 0$

6. $6x^2 - 5x + 1 < 0$

7. $x^2 \geq 4$

8. $x^2 - 9 \leq 0$

Solve.

9. $5x^2 \leq 4x + 1$

10. $7x^2 \leq 5x + 2$

11. $20 - x - x^2 > 0$

12. $28 - 3x - x^2 > 0$

13. $6x^2 - 5x > 6$

14. $3x^2 + 17x > -10$

15. $-2x + 30 \geq x(x + 5)$
Hint: Put variables on the right and zero on the left in your first step.

16. $55 - x^2 \geq 6x$
Hint: Put variables on the right and zero on the left in your first step.

17. $x^2 - 4x \leq -4$

18. $x^2 - 6x \leq -9$

Solve each of the following quadratic inequalities if possible. Round your answers to the nearest tenth.

19. $x^2 - 2x > 4$ **20.** $x^2 + 6x > 8$ **21.** $x^2 - 6x < -7$

22. $x^2 < 2x + 1$ **23.** $2x^2 \geq x^2 - 4$ **24.** $4x^2 \geq 3x^2 - 9$

Applications

In Exercises 25 and 26, a projectile is fired vertically with an initial velocity of 640 feet per second. The distance s in feet above the ground after t seconds is given by the equation $s = -16t^2 + 640t$.

25. For what range of time (measured in seconds) will the height s be greater than 6000 feet?

26. For what range of time (measured in seconds) will the height s be less than 4800 feet?

*In Exercises 27 and 28, the profit of a manufacturing company is determined by the number of units x manufactured each day according to the given equation. **(a)** Find when the profit is greater than zero. **(b)** Find the daily profit when 50 units are manufactured. **(c)** Find the daily profit when 60 units are manufactured.*

27. Profit $= -20(x^2 - 220x + 2400)$ **28.** Profit $= -25(x^2 - 280x + 4000)$

Cumulative Review Problems

29. The university's synchronized swimming team will not let Mona participate unless she passes biology with a C (70 or better) average. There are six tests in the semester, and she failed the first one (with a score of 0). She decided to find a tutor. Since then, she received an 81, 92, and 80 on the next three tests. What must her minimum scores be on the last two tests to pass the course with a minimum grade of 70 and participate in synchronized swimming?

30. In a huge bowl at a college party, there are 360 ounces of mixed potato chips, peanuts, pretzels, and popcorn. There are 70 more ounces of peanuts than potato chips. There are twice as many ounces of pretzels as ounces of popcorn. There are 10 more ounces of popcorn than potato chips. How many ounces of each ingredient are in the snack mix?

Graph.

31. $\dfrac{1}{2}y + \dfrac{1}{3}x = -1$

32. $f(x) = (x - 5)^2$

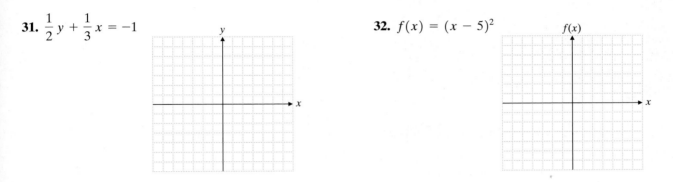

Factor completely.

33. $16y^2 - 1$

34. $-3x^2 - 18xy - 27y^2$

Putting Your Skills to Work

The Mathematics of the Hoover Dam

One of the most impressive construction projects completed in the United States in the twentieth century is the Hoover Dam. It weighs 6.6 million tons and contains 3.25 million cubic yards of concrete. The builders used so much concrete for this construction project that with it they could have built a two-lane highway from New York City to San Francisco.

During the construction of the dam, the water had to be diverted away from the project. This required the building of four diversion tunnels. The average length of the tunnels was 4000 feet.

Problems for Individual Investigation and Analysis

▲ **1.** Find the volume of rock that had to be removed for the four diversion tunnels if each tunnel was 56 feet in diameter. (Use $\pi \approx 3.14$.)

▲ **2.** Each diversion tunnel was lined with 3 feet of concrete. Thus, the outer shell of the tunnels was 56 feet in diameter while the inside shaft of the tunnels was 50 feet in diameter. Find the volume of concrete that was used to build the four diversion tunnels. (Use $\pi \approx 3.14$.)

Problems for Group Investigation and Cooperative Learning

The Hoover Dam is 726.4 feet tall. It is 45 feet thick at the top and 660 feet thick at the bottom. The force of water on the bottom edge of the dam is 22.5 tons per square foot.

3. If the cross section of the center of the dam were superimposed on a rectangular coordinate system, its outside edges would be approximately defined by the equations $y = 0.00205x^2$, $y = 726$, $x = 660$, and $y = 0$, where x and y are measured in feet. Graph these four equations to show a cross-sectional view of the main wall of the center of the Hoover Dam.

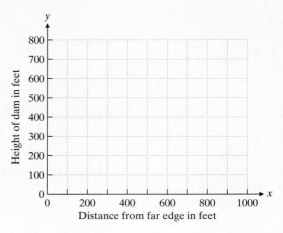

4. Use your graph to estimate the thickness of the Hoover Dam at the point where the dam is 400 feet high. Find the exact answer using the equations in Exercise 3. Compare your answers.

Internet Connections

WWW   Netsite: http://www.prenhall.com/tobey_beg_int

Site: Hoover Dam Visitor Site

5. Of all the excavation that took place in the construction of the Hoover Dam, what percent was to dig the earth and rockfill needed for the cofferdams, the temporary dams that held back the water while the construction took place?

6. With the same number of cubic yards of concrete in the Hoover Dam, power plant, and other structures on the site, how many monuments could be built with a cross-section of 100 square feet and a height of 30 feet?

9.7 Absolute Value Equations

① Solving Absolute Value Equations of the Form $|ax + b| = c$

From Section 1.1, you know that the absolute value of a number x is the distance between 0 and x on the number line. Let's look at a simple absolute value equation, $|x| = 4$, and draw a picture.

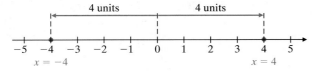

Thus, the equation $|x| = 4$ has two solutions, $x = 4$ and $x = -4$. Let's look at another example.

$$\text{If} \quad |x| = \frac{2}{3},$$

$$\text{then} \quad x = \frac{2}{3} \quad \text{or} \quad x = -\frac{2}{3},$$

$$\text{because} \quad \left|\frac{2}{3}\right| = \frac{2}{3} \quad \text{and} \quad \left|-\frac{2}{3}\right| = \frac{2}{3}.$$

We can solve these relatively simple absolute value equations by recalling the definition of absolute value.

$$|x| = \begin{cases} x, & \text{if } x \geq 0 \\ -x, & \text{if } x < 0 \end{cases}$$

Now let's take a look at a more complicated absolute value equation: $|ax + b| = c$.

> The solutions of an equation of the form $|ax + b| = c$, where $a \neq 0$ and c is a positive number, are those values that satisfy
>
> $$ax + b = c \quad \text{or} \quad ax + b = -c.$$

EXAMPLE 1 Solve $|2x + 5| = 11$ and check your solutions.

Using the rule established in the box, we have the following:

$$
\begin{array}{ll}
2x + 5 = 11 \quad \text{or} & 2x + 5 = -11 \\
2x = 6 & 2x = -16 \\
x = 3 & x = -8
\end{array}
$$

The two solutions are 3 and -8.

Check: if $x = 3$ if $x = -8$

$$
\begin{array}{ll}
|2x + 5| = 11 & |2x + 5| = 11 \\
|2(3) + 5| \overset{?}{=} 11 & |2(-8) + 5| \overset{?}{=} 11 \\
|6 + 5| \overset{?}{=} 11 & |-16 + 5| \overset{?}{=} 11 \\
|11| \overset{?}{=} 11 & |-11| \overset{?}{=} 11 \\
11 = 11 \; \checkmark & 11 = 11 \; \checkmark
\end{array}
$$

Practice Problem 1 Solve $|3x - 4| = 23$ and check your solutions.

Student Learning Objectives

After studying this section, you will be able to:

① Solve absolute value equations of the form $|ax + b| = c$.

② Solve absolute value equations of the form $|ax + b| + c = d$.

③ Solve absolute value equations of the form $|ax + b| = |cx + d|$.

SSM   PH TUTOR   CD & VIDEO   MATH PRO   WEB
CENTER

EXAMPLE 2 Solve $\left|\dfrac{1}{2}x - 1\right| = 5$ and check your solutions.

The solutions of the given absolute value equation must satisfy

$$\frac{1}{2}x - 1 = 5 \quad \text{or} \quad \frac{1}{2}x - 1 = -5.$$

If we multiply each term of both equations by 2, we obtain the following:

$$x - 2 = 10 \quad \text{or} \quad x - 2 = -10$$
$$x = 12 \qquad\qquad x = -8$$

Check: if $x = 12$ if $x = -8$

$$\left|\frac{1}{2}(12) - 1\right| \overset{?}{=} 5 \qquad\qquad \left|\frac{1}{2}(-8) - 1\right| \overset{?}{=} 5$$
$$|6 - 1| \overset{?}{=} 5 \qquad\qquad |-4 - 1| \overset{?}{=} 5$$
$$|5| \overset{?}{=} 5 \qquad\qquad |-5| \overset{?}{=} 5$$
$$5 = 5 \;\checkmark \qquad\qquad 5 = 5 \;\checkmark$$

Practice Problem 2   Solve and check your solutions.

$$\left|\frac{2}{3}x + 4\right| = 2$$

2   *Solving Absolute Value Equations of the Form* $|ax + b| + c = d$

Notice that in each of the previous examples the absolute value expression is on one side of the equation and a positive real number is on the other side of the equation. What happens when we encounter an equation of the form $|ax + b| + c = d$?

EXAMPLE 3   Solve $|3x - 1| + 2 = 5$ and check your solutions.

First we will rewrite the equation so that the absolute value expression is alone on one side of the equation.

$$|3x - 1| + 2 - 2 = 5 - 2$$
$$|3x - 1| = 3$$

Now we solve $|3x - 1| = 3$.

$$3x - 1 = 3 \quad \text{or} \quad 3x - 1 = -3$$
$$3x = 4 \qquad\qquad 3x = -2$$
$$x = \frac{4}{3} \qquad\qquad x = -\frac{2}{3}$$

Check: if $x = \dfrac{4}{3}$ if $x = -\dfrac{2}{3}$

$$\left|3\left(\frac{4}{3}\right) - 1\right| + 2 \overset{?}{=} 5 \qquad\qquad \left|3\left(-\frac{2}{3}\right) - 1\right| + 2 \overset{?}{=} 5$$
$$|4 - 1| + 2 \overset{?}{=} 5 \qquad\qquad |-2 - 1| + 2 \overset{?}{=} 5$$
$$|3| + 2 \overset{?}{=} 5 \qquad\qquad |-3| + 2 \overset{?}{=} 5$$
$$3 + 2 \overset{?}{=} 5 \qquad\qquad 3 + 2 \overset{?}{=} 5$$
$$5 = 5 \;\checkmark \qquad\qquad 5 = 5 \;\checkmark$$

Practice Problem 3   Solve $|2x + 1| + 3 = 8$ and check your solutions.

3   *Solving Absolute Value Equations of the Form*
$|ax + b| = |cx + d|$

Let us now consider the possibilities for a and b if $|a| = |b|$.

Suppose $a = 5$; then $b = 5$ or -5.
If $a = -5$, then $b = 5$ or -5.

To generalize, if $|a| = |b|$, then $a = b$ or $a = -b$.

This rule allows for all the possibilities described in the box. We now apply this property to solve more complex equations.

EXAMPLE 4   Solve and check: $|3x - 4| = |x + 6|$

The solutions of the given equation must satisfy

$$3x - 4 = x + 6 \quad \text{or} \quad 3x - 4 = -(x + 6).$$

Now we solve each equation in the normal fashion.

$$3x - 4 = x + 6 \quad \text{or} \quad 3x - 4 = -x - 6$$

$$3x - x = 4 + 6 \qquad 3x + x = 4 - 6$$

$$2x = 10 \qquad\qquad 4x = -2$$

$$x = 5 \qquad\qquad x = -\frac{1}{2}$$

We will check each solution by substituting it into the *original equation*.

Check: if $x = 5$ if $x = -\dfrac{1}{2}$

$$|3(5) - 4| \stackrel{?}{=} |5 + 6| \qquad\qquad \left|3\left(-\frac{1}{2}\right) - 4\right| \stackrel{?}{=} \left|-\frac{1}{2} + 6\right|$$

$$|15 - 4| \stackrel{?}{=} |11| \qquad\qquad\qquad \left|-\frac{3}{2} - 4\right| \stackrel{?}{=} \left|-\frac{1}{2} + 6\right|$$

$$|11| \stackrel{?}{=} |11| \qquad\qquad\qquad \left|-\frac{3}{2} - \frac{8}{2}\right| \stackrel{?}{=} \left|-\frac{1}{2} + \frac{12}{2}\right|$$

$$11 = 11 \;\checkmark \qquad\qquad\qquad\qquad \left|-\frac{11}{2}\right| \stackrel{?}{=} \left|\frac{11}{2}\right|$$

$$\frac{11}{2} = \frac{11}{2} \;\checkmark$$

Practice Problem 4   Solve and check: $|x - 6| = |5x + 8|$

To Think About   Explain how you would solve an absolute value equation of the form $|ax + b| = 0$. Give an example. Does $|3x + 2| = -4$ have a solution? Why or why not?

Solve each absolute value equation. Check your solutions for Exercises 1–18.

1. $|x| = 30$

2. $|x| = 14$

3. $|2x - 5| = 13$

4. $|2x + 1| = 15$

5. $|5 - 4x| = 11$

6. $|2 - 3x| = 13$

7. $\left|\dfrac{1}{2}x - 3\right| = 2$

8. $\left|\dfrac{1}{4}x + 5\right| = 3$

9. $|2.3 - 0.3x| = 1$

10. $|0.7 - 0.3x| = 2$

11. $|x + 2| - 1 = 7$

12. $|x + 3| - 4 = 8$

13. $\left|\dfrac{x - 1}{3}\right| - 2 = 1$

14. $\left|\dfrac{3 - 2x}{2}\right| + 5 = 7$

15. $\left|1 - \dfrac{3}{4}x\right| + 4 = 7$

16. $\left|4 - \dfrac{5}{2}x\right| + 3 = 15$

17. $|x + 6| = |2x - 3|$

18. $|x - 4| = |2x + 5|$

19. $\left|\dfrac{x - 1}{2}\right| = |2x + 3|$

20. $\left|\dfrac{2x + 7}{3}\right| = |x + 2|$

21. $|1.5x - 2| = |x - 0.5|$

22. $|2.2x + 2| = |1 - 2.8x|$

23. $|3 - x| = \left|\dfrac{x}{2} + 3\right|$

Solve for x. Round to the nearest hundredth.

24. $|1.62x + 3.14| = 2.19$

25. $|-0.74x - 8.26| = 5.36$

Mixed Practice

Solve each equation, if possible. Check your solutions.

26. $|3(x + 4)| + 2 = 14$

27. $|4(x - 2)| + 1 = 19$

28. $\left|\dfrac{5x}{3} - 1\right| = 0$

29. $\left|\dfrac{3}{4}x + 9\right| = 0$

30. $\left|\dfrac{4}{3}x - \dfrac{1}{8}\right| = -5$

31. $\left|\dfrac{3}{4}x - \dfrac{2}{3}\right| = -8$

32. $\left|\dfrac{2x - 1}{3}\right| = \dfrac{5}{6}$

33. $\left|\dfrac{3x + 1}{2}\right| = \dfrac{3}{4}$

34. $|1.5x - 2.5| = |x + 3|$

35. $|6x - 0.3| = |5.6x + 11.9|$

36. $\dfrac{|x + 2|}{-3} = -5$

37. $\dfrac{|x - 3|}{-4} = -2$

Cumulative Review Problems

Solve for x.

38. $\dfrac{1}{2}(3x + 1) - \dfrac{1}{6}(7x + 3) = \dfrac{1}{3}(3 - x)$

Simplify. Do not leave negative exponents in your answer.

39. $\left(\dfrac{2x^{-2}y}{z^{-1}}\right)^3$

40. A scientist bought three new Bunsen burners and twenty-five new beakers last month for $975. This month she bought three Bunsen burners and twenty new beakers for $825. How much did each beaker cost? How much did each Bunsen burner cost?

41. The new mechanic at Speedy Lube can perform 30 oil changes on cars coming through the lube center in 4 hours. His assistant can perform the same number of oil changes in 6 hours. How long would it take them to do the work together?

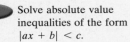
1 *Solving Absolute Value Inequalities of the Form* $|ax + b| < c$

We begin by looking at $|x| < 3$. What does this mean? The inequality $|x| < 3$ means that x is less than 3 units from 0 on the number line. We draw a picture.

This picture shows all possible values of x such that $|x| < 3$. We see that this occurs when $-3 < x < 3$. We conclude that $|x| < 3$ and $-3 < x < 3$ are equivalent statements.

> **Definition**
>
> If a is a positive real number and $|x| < a$, then $-a < x < a$.

The solutions to the problems in this section will be compound inequalities. A **compound inequality** consists of two inequalities connected by the word *and* or the word *or*. Consider the inequality above, $-3 < x < 3$. Another way to write this is $-3 < x$ *and* $x < 3$. Since $-3 < x < 3$ is shorter to write, it is the preferred way to indicate all values x between -3 and 3. Both statements, $-3 < x < 3$ and $-3 < x$ *and* $x < 3$ are compound inequalities. ($-3 < x < 3$ can be read 'x is greater than -3 *and* less than 3.')

EXAMPLE 1   Solve $|x| \leq 4.5$.

The inequality $|x| \leq 4.5$ means that x is less than or equal to 4.5 units from 0 on the number line. We draw a picture.

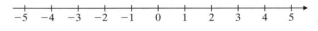

Thus, the solution is the compound inequality $-4.5 \leq x \leq 4.5$.

Practice Problem 1   Solve $|x| < 2$.

This same technique can be used to solve more complicated inequalities.

EXAMPLE 2   Solve and graph the solution: $|x + 5| \leq 10$

We want to find the values of x that make $-10 \leq x + 5 \leq 10$ a true statement. We need to solve the compound inequality.

To solve this inequality, we add -5 to each part.

$$-10 - 5 \leq x + 5 - 5 \leq 10 - 5$$
$$-15 \leq x \leq 5$$

Thus, the solution is $-15 \leq x \leq 5$. We graph this solution.

Practice Problem 2   Solve $|x - 6| < 15$. Graph your solution. (*Hint*: Choose a convenient scale.)

EXAMPLE 3 Solve and graph the solution: $\left| x - \dfrac{2}{3} \right| \le \dfrac{5}{2}$

$$-\dfrac{5}{2} \le x - \dfrac{2}{3} \le \dfrac{5}{2}$$ If $|x| < a$, then $-a < x < a$.

$$6\left(-\dfrac{5}{2}\right) \le 6(x) - 6\left(\dfrac{2}{3}\right) \le 6\left(\dfrac{5}{2}\right)$$ Multiply each part of the inequality by 6.

$$-15 \le 6x - 4 \le 15$$ Simplify.

$$-15 + 4 \le 6x - 4 + 4 \le 15 + 4$$ Add 4 to each part.

$$-11 \le 6x \le 19$$ Simplify.

$$-\dfrac{11}{6} \le \dfrac{6x}{6} \le \dfrac{19}{6}$$ Divide each part by 6.

$$-1\dfrac{5}{6} \le x \le 3\dfrac{1}{6}$$ Change to mixed numbers to facilitate graphing.

Practice Problem 3 Solve and graph the solution: $\left| x + \dfrac{3}{4} \right| \le \dfrac{7}{6}$

EXAMPLE 4 Solve and graph the solution: $|2(x - 1) + 4| < 8$

First we simplify the expression within the absolute value symbol.

$$|2x - 2 + 4| < 8$$

$$|2x + 2| < 8$$

$$-8 < 2x + 2 < 8$$ If $|x| < a$, then $-a < x < a$.

$$-8 - 2 < 2x + 2 - 2 < 8 - 2$$ Add -2 to each part.

$$-10 < 2x < 6$$ Simplify.

$$\dfrac{-10}{2} < \dfrac{2x}{2} < \dfrac{6}{2}$$ Divide each part by 2.

$$-5 < x < 3$$

Practice Problem 4 Solve and graph the solution: $|2 + 3(x - 1)| < 20$

2 *Solving Absolute Value Inequalities of the Form* $|ax + b| > c$

Now consider $|x| > 3$. What does this mean? This inequality $|x| > 3$ means that x is greater than 3 units from 0 on the number line. We draw a picture.

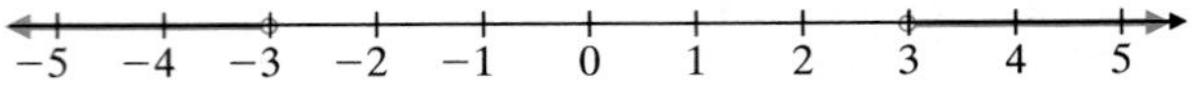

This picture shows all possible values of x such that $|x| > 3$. This occurs when $x < -3$ or when $x > 3$. (Note that a solution can be either in the region to the left of -3 on the number line or in the region to the right of 3 on the number line.) We conclude that $|x| > 3$ and $x < -3$ or $x > 3$ are equivalent statements. They are both compound inequalities using the word *or*.

Definition

If a is a positive real number and $|x| > a$, then $x < -a$ or $x > a$.

EXAMPLE 5 Solve $|x| \geq 5\frac{1}{4}$.

The inequality $|x| \geq 5\frac{1}{4}$ means that x is more than $5\frac{1}{4}$ units from 0 on the number line. We draw a picture.

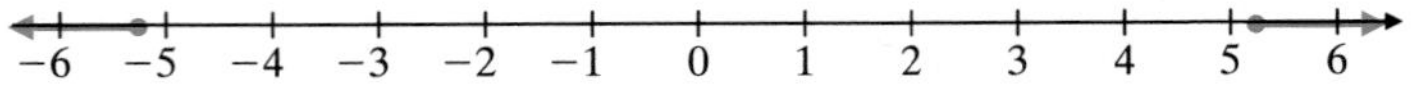

Thus, the solution is $x \leq -5\frac{1}{4}$ or $x \geq 5\frac{1}{4}$.

Practice Problem 5 Solve and graph the solution: $|x| > 2.5$

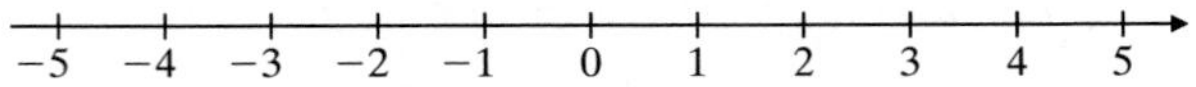

This same technique can be used to solve more complicated inequalities.

EXAMPLE 6 Solve and graph your solution: $|x - 4| > 5$

We want to find the values of x that make $x - 4 < -5$ or $x - 4 > 5$ a true statement. We need to solve the compound inequality.
We will solve each inequality separately.

$$
\begin{array}{ccc}
x - 4 < -5 & or & x - 4 > 5 \\
x - 4 + 4 < -5 + 4 & & x - 4 + 4 > 5 + 4 \\
x < -1 & & x > 9
\end{array}
$$

Thus, the solution is $x < -1$ *or* $x > 9$. We graph the solution on the number line.

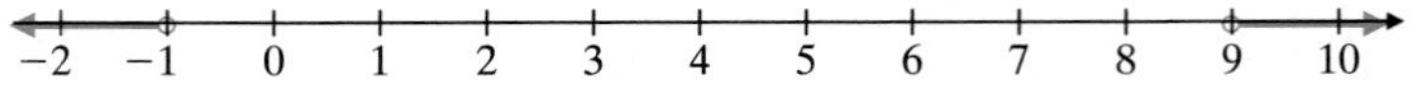

Practice Problem 6 Solve and graph the solution: $|x + 6| > 2$

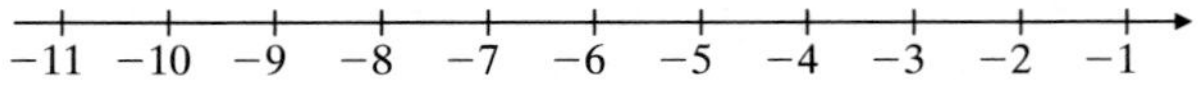

EXAMPLE 7   Solve and graph the solution: $|-3x + 6| > 18$

By definition, we have the following compound inequality.

$$-3x + 6 > 18 \qquad \qquad or \qquad \qquad -3x + 6 < -18$$

$$-3x > 12 \qquad \qquad \qquad \qquad -3x < -24$$

$$\frac{-3x}{-3} < \frac{12}{-3} \quad \longleftarrow \text{Division by a negative} \longrightarrow \quad \frac{-3x}{-3} > \frac{-24}{-3}$$
$$\qquad \qquad \qquad \text{number reverses the}$$
$$\qquad \qquad \qquad \text{inequality sign.}$$

$$x < -4 \qquad \qquad \qquad \qquad \qquad x > 8$$

The solution is $x < -4$ *or* $x > 8$.

Practice Problem 7   Solve and graph $|-5x - 2| > 13$.

EXAMPLE 8   Solve and graph the solution: $\left| 3 - \dfrac{2}{3} x \right| \geq 5$

By definition, we have the following compound inequality.

$$3 - \frac{2}{3} x \geq 5 \qquad or \qquad 3 - \frac{2}{3} x \leq -5$$

$$3(3) - 3\left(\frac{2}{3} x \right) \geq 3(5) \qquad 3(3) - 3\left(\frac{2}{3} x \right) \leq 3(-5)$$

$$9 - 2x \geq 15 \qquad \qquad 9 - 2x \leq -15$$

$$-2x \geq 6 \qquad \qquad -2x \leq -24$$

$$\frac{-2x}{-2} \leq \frac{6}{-2} \qquad \qquad \frac{-2x}{-2} \geq \frac{-24}{-2}$$

$$x \leq -3 \qquad \qquad x \geq 12$$

The solution is $x \leq -3$ *or* $x \geq 12$.

Practice Problem 8   Solve and graph $\left| 4 - \dfrac{3}{4} x \right| \geq 5$.

EXAMPLE 9   When a new car transmission is built, the diameter d of the transmission must not differ from the specified standard s by more than 0.37 millimeter. The engineers express this requirement as $|d - s| \leq 0.37$. If the standard s is 216.82 millimeters for a particular car, find the limits of d.

$$|d - s| \leq 0.37$$
$$|d - 216.82| \leq 0.37 \qquad \text{Substitute the known value of } s.$$
$$-0.37 \leq d - 216.82 \leq 0.37 \qquad \text{If } |x| \leq a, \text{ then}$$
$$\qquad\qquad\qquad\qquad\qquad\qquad\qquad -a \leq x \leq a.$$
$$-0.37 + 216.82 \leq d - 216.82 + 216.82 \leq 0.37 + 216.82$$
$$216.45 \leq d \leq 217.19$$

Thus, the diameter of the transmission must be at least 216.45 millimeters, but not greater than 217.19 millimeters.

Practice Problem 9   The diameter d of a transmission must not differ from the specified standard s by more than 0.37 millimeter. Solve to find the allowed limits of d for a truck transmission for which the standard s is 276.53 millimeters.

Summary of Absolute Value Equations and Inequalities

It may be helpful to review the key concepts of absolute value equations and inequalities that we have covered in Sections 9.7 and 9.8. For real numbers a, b, and c, where $a \neq 0$ and $c > 0$, we have the following:

Absolute value form of the equation or inequality	Equivalent form without the absolute value	Type of solution obtained	Graphed form of the solution on a number line
$|ax + b| = c$	$ax + b = c$ or $ax + b = -c$	Two distinct numbers: m and n	
$|ax + b| < c$	$-c < ax + b < c$	The set of numbers between the two numbers m and n: $m < x < n$	
$|ax + b| > c$	$ax + b < -c$ or $ax + b > c$	The set of numbers less than m or the set of numbers greater than n: $x < m$ or $x > n$	

9.8 Exercises

Solve and graph the solutions.

1. $|x| \le 8$

2. $|x| < 6$

3. $|x| > 5$

4. $|x| \ge 7$

5. $|x + 4.5| < 5$

6. $|x + 6| < 3.5$

Solve for x.

7. $|x - 3| \le 5$

8. $|x - 7| \le 10$

9. $|2x - 5| \le 7$

10. $|3x + 2| \le 12$

11. $|0.2x - 0.7| \le 0.3$

12. $|0.3x - 0.5| \le 0.1$

13. $\left| x - \dfrac{3}{2} \right| < \dfrac{1}{2}$

14. $\left| x - \dfrac{1}{2} \right| < \dfrac{5}{2}$

15. $\left| \dfrac{1}{4}x + 2 \right| < 6$

16. $\left| \dfrac{1}{5}x + 1 \right| < 5$

17. $\left| \dfrac{3}{4}(x - 1) \right| < 6$

18. $\left| \dfrac{4}{5}(x - 1) \right| < 8$

19. $\left| \dfrac{3x - 2}{4} \right| < 3$

20. $\left| \dfrac{5x - 3}{2} \right| < 4$

21. $|x + 2| > 5$

22. $|x + 4| > 7$

23. $|x - 1| \ge 2$

24. $|x - 2| \ge 3$

25. $|3x - 8| \ge 7$

26. $|5x - 2| \ge 13$

27. $\left| 3 - \dfrac{3}{4}x \right| > 9$

28. $\left| 3 - \dfrac{2}{3}x \right| > 5$

29. $\left| \dfrac{1}{5}x - \dfrac{1}{10} \right| > 2$

30. $\left| \dfrac{1}{4}x - \dfrac{3}{8} \right| > 1$

31. $\left| \dfrac{1}{3}(x - 2) \right| < 5$

32. $\left| \dfrac{2}{5}(x - 2) \right| \le 4$

Applications

In a certain company, the measured thickness m of a helicopter blade must not differ from the standard s by more than 0.12 millimeter. The manufacturing engineer expresses this as $|m - s| \le 0.12$.

33. Find the limits of m if the standard s is 18.65 millimeters.

34. Find the limits of m if the standard s is 17.48 millimeters.

A small computer microchip has dimension requirements. The manufacturing engineer has written the specification that the new length n of the chip can differ from the previous length p by only 0.05 centimeter or less. The equation is $|n - p| \leq 0.05$.

35. Find the limits of the new length if the previous length was 9.68 centimeters.

36. Find the limits of the new length if the previous length was 7.84 centimeters.

To Think About

37. A student tried to solve the inequality $|4x - 8| > 12$. Instead of writing $4x - 8 > 12$ or $4x - 8 < -12$ as he should have done, he wrote $12 < 4x - 8 < -12$. What was wrong with his approach?

38. A student tried to write the solution to the compound inequality $6 < 4 - 3x < 19$. He used the following steps.

Step 1 $6 - 4 < 4 - 4 - 3x < 19 - 4$

Step 2 $2 < -3x < 15$

Step 3 $\dfrac{2}{-3} < \dfrac{-3x}{-3} < \dfrac{15}{-3}$

Step 4 $-\dfrac{2}{3} < x < -5$

What error did he make?

Cumulative Review Problems

Perform the correct order of operations to simplify.

39. $(6 - 4)^3 \div (-4) + 2^2$

40. $12 \div (-2)(3) - (-5) + 2$

41. Add. $\dfrac{-2}{x^2 - 25} + \dfrac{3}{2x + 10}$

42. Solve. $\dfrac{1}{x + 2} - \dfrac{1}{x} = -\dfrac{2}{x}$

Math in the Media

Stopping Distances

A USA TODAY.com Snapshot, Quick Stops, reported that increased speeds on U.S. highways were accompanied by increased stopping distances.

Stopping distances are an important and universal element in highway safety. For this reason, this data is widely collected by many different highway safety agencies throughout the world.

To get some hands-on experience examining the relationship between speed and stopping distance, try answering the questions that follow which are based on data from the Traffic Board of Western Australia.

EXERCISES

While driving, you suddenly see the brake lights of the car in front of you come on. If you assume the car ahead is going to stop and you then apply the brakes, how far will your car travel before it comes to a complete stop? In other words, what is the stopping distance from the moment your brain receives the signal to stop until the car is no longer moving?

The table below contains data for stopping distance in this situation. Note that the total stopping distance is the sum of the reaction distance (the distance traveled from the time you realize that you must brake unitl your foot hits the brake pedal) and the braking distance (distance traveled after the brake pedal is pressed).

V Speed miles/hour	x Speed feet/second	R (reaction time: 0.7 sec) Reaction Distance feet	B Braking Distance feet	y Total Stopping Distance feet
55	81	57	219	276
65	95	67	301	368
75	110	77	403	480

Source: Based on data from the Traffic Board of Western Australia. *converted from Metric to US units.

1. Notice the pattern between the second and third columns and find an equation that gives the reaction distance R as a function of the speed x. R is rounded to the nearest foot.

2. The model for braking distance B as a function of the speed x is $B = 0.032512x^2 + 0.134975x - 5.24631$. Use this equation and the equation from your answer to question 1 to find an equation for total stopping distance y as a function of the speed x. Write the equation for total stopping distance in simplest form by combining like terms.

Chapter 9 Organizer

Topic	Procedure	Examples
Solving a quadratic equation by using the square root property, p. 491.	If $x^2 = a$, then $x = \pm\sqrt{a}$.	Solve. $$2x^2 - 50 = 0$$ $$2x^2 = 50$$ $$x^2 = 25$$ $$x = \pm\sqrt{25}$$ $$x = \pm 5$$
Solving a quadratic equation by completing the square, p. 492.	1. Rewrite the equation in the form $ax^2 + bx = c$. If $a \neq 1$, divide each term of the equation by a. 2. Square half of the numerical coefficient of the linear term. Add the result to both sides of the equation. 3. Factor the left side. 4. Take the square root of both sides of the equation. 5. Solve the resulting equation for x. 6. Check the solutions in the original equation.	Solve. $$2x^2 - 4x - 1 = 0$$ $$2x^2 - 4x = 1$$ $$\frac{2x^2}{2} - \frac{4x}{2} = \frac{1}{2}$$ $$x^2 - 2x + \underline{\hspace{1cm}} = \frac{1}{2} + \underline{\hspace{1cm}}$$ $$x^2 - 2x + 1 = \frac{1}{2} + 1$$ $$(x - 1)^2 = \frac{3}{2}$$ $$x - 1 = \pm\sqrt{\frac{3}{2}}$$ $$x - 1 = \frac{\pm\sqrt{6}}{2}$$ $$x = 1 \pm \frac{1}{2}\sqrt{6}$$
Placing a quadratic equation in standard form, p. 499.	A quadratic equation in standard form is an equation of the form $ax^2 + bx + c = 0$, where a, b, and c are real numbers and $a \neq 0$. It is often necessary to remove parentheses and clear away fractions by multiplying each term of the equation by the LCD to obtain the standard form.	Rewrite in quadratic form: $$\frac{2}{x - 3} + \frac{x}{x + 3} = \frac{5}{x^2 - 9}$$ $$(x + 3)(x - 3)\left[\frac{2}{x - 3}\right]$$ $$+ (x + 3)(x - 3)\left[\frac{x}{x + 3}\right]$$ $$= (x + 3)(x - 3)\left[\frac{5}{(x + 3)(x - 3)}\right]$$ $$2(x + 3) + x(x - 3) = 5$$ $$2x + 6 + x^2 - 3x = 5$$ $$x^2 - x + 1 = 0$$
Solve a quadratic equation by using the quadratic formula, p. 497.	If $ax^2 + bx + c = 0$, where $a \neq 0$, $$x = \frac{-b \pm \sqrt{b^2 - 4ac}}{2a}.$$ 1. Rewrite the equation in standard form. 2. Determine the values of a, b, and c. 3. Substitute the values of a, b, and c into the formula. 4. Simplify the result to obtain the values of x. 5. Any imaginary solutions to the quadratic equation should be simplified by using the definition $\sqrt{-a} = i\sqrt{a}$, where $a > 0$.	Solve. $$2x^2 = 3x - 2$$ $$2x^2 - 3x + 2 = 0$$ $$a = 2, b = -3, c = 2$$ $$x = \frac{-(-3) \pm \sqrt{(-3)^2 - 4(2)(2)}}{2(2)}$$ $$= \frac{3 \pm \sqrt{9 - 16}}{4}$$ $$= \frac{3 \pm \sqrt{-7}}{4}$$ $$= \frac{3 \pm i\sqrt{7}}{4}$$

Topic	Procedure	Examples
Equations that can be transformed into quadratic form, p. 505.	1. Find the variable with the smallest exponent. Let this quantity be replaced by y. 2. Continue to make substitutions for the remaining variable terms based on the first substitution. (You should be able to replace the variable with the largest exponent by y^2.) 3. Solve the resulting equation for y. 4. Reverse the substitution used in step 1. 5. Solve the resulting equation for x. 6. Check your solution in the *original* equation.	Solve: $x^{2/3} - x^{1/3} - 2 = 0$ Let $y = x^{1/3}$. Then $y^2 = x^{2/3}$. $$y^2 - y - 2 = 0$$ $$(y - 2)(y + 1) = 0$$ $y = 2 \quad$ or $\quad y = -1$ $x^{1/3} = 2 \quad$ or $\quad x^{1/3} = -1$ $\left(x^{1/3}\right)^3 = 2^3 \qquad \left(x^{1/3}\right)^3 = (-1)^3$ $x = 8 \qquad\qquad x = -1$
Checking solutions for equations in quadratic form (continued), p. 506.		*Check.* $x = 8$: $\quad (8)^{2/3} - (8)^{1/3} - 2 \overset{?}{=} 0$ $\qquad\qquad 2^2 - 2 - 2 \overset{?}{=} 0$ $\qquad\qquad\qquad 4 - 4 = 0 \;\checkmark$ $x = -1$: $\quad (-1)^{2/3} - (-1)^{1/3} - 2 \overset{?}{=} 0$ $\qquad\qquad (-1)^2 - (-1) - 2 \overset{?}{=} 0$ $\qquad\qquad\qquad 1 + 1 - 2 = 0 \;\checkmark$ Both 8 and -1 are solutions.
Solving quadratic equations containing two or more variables, p. 511.	Treat the letter to be solved for as a variable, but treat all other letters as constants. Solve the equation by factoring, by using the square root property, or by using the quadratic formula.	Solve for x. **(a)** $6x^2 - 11xw + 4w^2 = 0$ **(b)** $4x^2 + 5b = 2w^2$ **(c)** $2x^2 + 3xz - 10z = 0$ **(a)** By factoring: $$(3x - 4w)(2x - w) = 0$$ $3x - 4w = 0 \quad$ or $\quad 2x - w = 0$ $x = \dfrac{4w}{3} \qquad\qquad x = \dfrac{w}{2}$ **(b)** Using the square root property: $4x^2 = 2w^2 - 5b$ $x^2 = \dfrac{2w^2 - 5b}{4}$ $x = \pm\sqrt{\dfrac{2w^2 - 5b}{4}} = \pm\dfrac{1}{2}\sqrt{2w^2 - 5b}$ **(c)** By the quadratic formula, with $a = 2$, $b = 3z$, $c = -10z$: $x = \dfrac{-3z \pm \sqrt{9z^2 + 80z}}{4}$
The Pythagorean theorem, p. 512.	In any right triangle, if c is the length of the hypotenuse and a and b are the lengths of the two legs, then $$c^2 = a^2 + b^2.$$	Find a if $c = 7$ and $b = 5$. $49 = a^2 + 25$ $49 - 25 = a^2$ $24 = a^2$ $\sqrt{24} = a$ $2\sqrt{6} = a$

551

Topic	Procedure	Examples
Graphing quadratic functions, p. 523.	Graph quadratic functions of the form $f(x) = ax^2 + bx + c$ with $a \neq 0$ as follows: **1.** Find the vertex at $\left(\dfrac{-b}{2a}, f\left(\dfrac{-b}{2a}\right)\right)$. **2.** Find the y-intercept, which occurs at $f(0)$. **3.** Find the x-intercepts if they exist. Solve $f(x) = 0$ for x.	Graph $f(x) = x^2 + 6x + 8$. Vertex: $$x = \frac{-6}{2} = -3$$ $$f(-3) = (-3)^2 + 6(-3) + 8 = -1$$ The vertex is $(-3, -1)$. Intercepts: $f(0) = (0)^2 + 6(0) + 8 = 8$ The y-intercept is $(0, 8)$. $$x^2 + 6x + 8 = 0$$ $$(x + 2)(x + 4) = 0$$ $$x = -2, x = -4$$ The x-intercepts are $(-2, 0)$ and $(-4, 0)$. *(graph of $f(x) = x^2 + 6x + 8$ showing $(0,8)$, $(-2,0)$, $(-4,0)$, vertex $(-3,-1)$)*
Solving quadratic inequalities in one variable, p. 529.	**1.** Replace the inequality symbol by an equal sign. Solve the resulting equation to find the critical points. **2.** Use the critical points to separate the number line into distinct regions. **3.** Evaluate the quadratic expression at a test point in each region. **4.** Determine which regions satisfy the original conditions of the quadratic inequality.	Solve and graph: $3x^2 + 5x - 2 > 0$ **1.** $3x^2 + 5x - 2 = 0$ $$(3x - 1)(x + 2) = 0$$ $$3x - 1 = 0 \qquad x + 2 = 0$$ $$x = \frac{1}{3} \qquad\qquad x = -2$$ Critical points are -2 and $\dfrac{1}{3}$. **2.** *(number line showing Region I: $x < -2$, Region II: $-2 < x < \frac{1}{3}$, Region III: $x > \frac{1}{3}$)* **3.** $3x^2 + 5x - 2$ *Region I:* Pick $x = -3$. $3(-3)^2 + 5(-3) - 2 = 27 - 15 - 2 = 10 > 0$ *Region II:* Pick $x = 0$. $3(0)^2 + 5(0) - 2 = 0 + 0 - 2 = -2 < 0$ *Region III:* Pick $x = 3$. $3(3)^2 + 5(3) - 2 = 27 + 15 - 2 = 40 > 0$ **4.** We know that the expression is greater than zero (that is, $3x^2 + 5x - 2 > 0$) when $$x < -2 \text{ or } x > \frac{1}{3}.$$ *(number line graph)*

Topic	Procedure	Examples
Absolute value equations, p. 537.	To solve an equation that involves an absolute value, we rewrite the absolute value equation as two separate equations without the absolute value. We solve each equation. If $\lvert ax + b\rvert = c$, then $ax + b = c$ or $ax + b = -c$.	Solve for x: $\lvert 4x - 1\rvert = 17$ $$4x - 1 = 17 \qquad or \quad 4x - 1 = -17$$ $$4x = 17 + 1 \qquad\qquad 4x = -17 + 1$$ $$4x = 18 \qquad\qquad\qquad 4x = -16$$ $$x = \frac{18}{4} \qquad\qquad\qquad x = \frac{-16}{4}$$ $$x = \frac{9}{2} \qquad\qquad\qquad x = -4$$
Solving absolute value inequalities involving $<$ or $\leq$, p. 542.	Let a be a positive real number. If $\lvert x\rvert < a$, then $-a < x < a$. If $\lvert x\rvert \leq a$, then $-a \leq x \leq a$.	Solve and graph: $\lvert 3x - 2\rvert < 19$ $$-19 < 3x - 2 < 19$$ $$-19 + 2 < 3x - 2 + 2 < 19 + 2$$ $$-17 < 3x < 21$$ $$-\frac{17}{3} < \frac{3x}{3} < \frac{21}{3}$$ $$-5\frac{2}{3} < x < 7$$
Solving absolute value inequalities involving $>$ or $\geq$, p. 544.	Let a be a positive real number. If $\lvert x\rvert > a$, then $x > a$ or $x < -a$. If $\lvert x\rvert \geq a$, then $x \geq a$ or $x \leq -a$.	Solve and graph: $\left\lvert \frac{1}{3}(x - 2)\right\rvert \geq 2$ $$\frac{1}{3}(x - 2) \geq 2 \qquad or \quad \frac{1}{3}(x - 2) \leq -2$$ $$\frac{1}{3}x - \frac{2}{3} \geq 2 \qquad\qquad \frac{1}{3}x - \frac{2}{3} \leq -2$$ $$x - 2 \geq 6 \qquad\qquad\qquad x - 2 \leq -6$$ $$x \geq 6 + 2 \qquad\qquad\qquad x \leq -6 + 2$$ $$x \geq 8 \qquad or \qquad\qquad x \leq -4$$

Chapter 9 Review Problems

Solve each of the following exercises by the specified method. Simplify all answers.

Solve by the square root property.

1. $5x^2 = 100$

2. $(x + 8)^2 = 81$

Solve by completing the square.

3. $x^2 + 8x + 13 = 0$

4. $4x^2 - 8x + 1 = 0$

Solve by the quadratic formula.

5. $3x^2 - 10x + 6 = 0$

6. $x^2 - 6x - 4 = 0$

Solve by any appropriate method and simplify your answers. Express any nonreal complex solutions using i notation.

7. $4x^2 - 12x + 9 = 0$

8. $3x^2 - 8x + 6 = 0$

9. $6x^2 - 23x = 4x$

10. $12x^2 - 29x + 15 = 0$

11. $x^2 - 3x - 23 = 5$

12. $3x^2 + 7x + 13 = 13$

13. $3x^2 - 2x = 15x - 10$

14. $6x^2 + 12x - 24 = 0$

15. $4x^2 - 3x + 2 = 0$

16. $3x^2 + 5x + 1 = 0$

17. $3x(3x + 2) - 2 = 3x$

18. $10x(x - 2) + 10 = 2x$

19. $\dfrac{5}{6}x^2 - x + \dfrac{1}{3} = 0$

20. $\dfrac{4}{5}x^2 + x + \dfrac{1}{5} = 0$

21. $y + \dfrac{5}{3y} + \dfrac{17}{6} = 0$

22. $\dfrac{19}{y} - \dfrac{15}{y^2} + 10 = 0$

23. $\dfrac{15}{y^2} - \dfrac{2}{y} = 1$

24. $y - 18 + \dfrac{81}{y} = 0$

25. $(3y + 2)(y - 1) = 7(-y + 1)$

26. $y(y + 1) + (y + 2)^2 = 4$

27. $\dfrac{2x}{x + 3} + \dfrac{3x - 1}{x + 1} = 3$

28. $\dfrac{4x + 1}{2x + 5} + \dfrac{3x}{x + 4} = 2$

Determine the nature of each of the following quadratic equations. Do not solve the equation. Find the discriminant in each case and determine whether the equation has (a) one rational solution, (b) two rational solutions, (c) two irrational solutions, or (d) two nonreal complex solutions.

29. $2x^2 + 5x - 3 = 0$

30. $3x^2 - 7x - 12 = 0$

31. $4x^2 - 6x + 5 = 0$

32. $25x^2 - 20x + 4 = 0$

Write a quadratic equation having the given numbers as solutions.

33. $5, -5$

34. $3i, -3i$

35. $4\sqrt{2}, -4\sqrt{2}$

36. $-3/4, -1/2$

Solve for any valid real roots.

37. $x^4 - 6x^2 + 8 = 0$

38. $2x^6 - 5x^3 - 3 = 0$

39. $x^{2/3} + 9x^{1/3} = -8$

40. $3x^{1/2} - 11x^{1/4} = 4$

41. $(2x - 5)^2 + 4(2x - 5) + 3 = 0$

42. $1 + 4x^{-8} = 5x^{-4}$

Solve for the variable specified. Assume that all radical expressions obtained have a positive radicand.

43. $A = \dfrac{2B^2C}{3H}$; for B

44. $2H = 3g(a^2 + b^2)$; for b

45. $20d^2 - xd - x^2 = 0$; for d

46. $yx^2 - 3x - 7 = 0$; for x

47. $3y^2 - 4ay + 2a = 0$; for y

48. $PV = 5x^2 + 3y^2 + 2x$; for x

Use the Pythagorean theorem to find the missing side. Assume that c is the length of the hypotenuse of a right triangle and that a and b are the lengths of the legs. Leave your answers as a radical in simplified form.

▲ **49.** $c = 16, b = 4$; find a

▲ **50.** $a = 3\sqrt{2}, b = 2$; find c

▲ **51.** A plane is 6 miles away from an observer and exactly 5 miles above the ground. The plane is directly above a car. How far is the car from the observer? Round your answer to the nearest tenth of a mile.

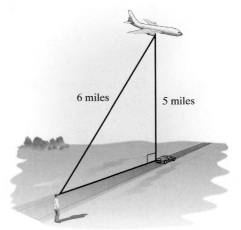

6 miles   5 miles

▲ **52.** The area of a rectangle is 203 square meters. Its length is 1 meter longer than four times its width. Find the length and width of the rectangle.

▲ **53.** The area of a triangle is 70 square centimeters. Its altitude is 6 meters longer than twice the length of the base. Find the dimensions of the altitude and base.

54. Jessica drove at a constant speed for 200 miles. Then it started to rain. So for the next 90 miles she traveled 5 miles per hour slower. The entire trip took 6 hours of driving time. Find her speed for each part of the trip.

55. John rode in a motorboat for 60 miles at constant cruising speed to get to his fishing grounds. Then for 5 miles he trolled to catch fish. His trolling speed was 15 miles per hour slower than his cruising speed. The trip took 4 hours. Find his speed for each part of the trip.

▲ **56.** Mr. and Mrs. Gomez are building a rectangular garden that is 10 feet by 6 feet. Around the outside of the garden, they will build a brick walkway. They have 100 square feet of brick. How wide should they make the brick walkway? Round your answer to the nearest tenth of a foot.

▲ **57.** The local YMCA is building a rectangular swimming pool that is 40 feet by 30 feet. The builders want to make a walkway around the pool with a nonslip cement surface. They have enough material to make 296 square feet of nonslip cement surface. How wide should the walkway be?

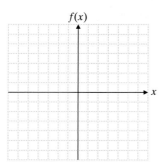

Find the vertex and the intercepts of the following quadratic functions.

58. $g(x) = -x^2 + 6x - 11$

59. $f(x) = x^2 + 10x + 25$

In each of the following exercises, find the vertex, the y-intercept, and the x-intercepts (if any exist) and then graph the function.

60. $f(x) = x^2 + 4x + 3$

61. $f(x) = x^2 + 6x + 5$

62. $f(x) = -x^2 + 6x - 5$

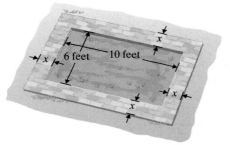

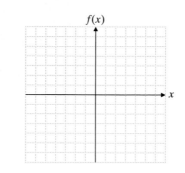

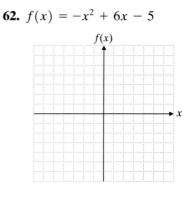

63. A model rocket is launched upward from a platform 40 feet above the ground. The height of the rocket h is given at any time t in seconds by the function $h(t) = -16t^2 + 400t + 40$. Find the maximum height of the rocket. How long will it take the rocket to go through its complete flight and then hit the ground? (Assume that the rocket does *not* have a parachute.) Round your answer to the nearest tenth.

64. A salesman for an electronics store finds that in 1 month he can sell $(1200 - x)$ compact disc players that each sell for x dollars. Write a function for the revenue. What is the price x that will result in the maximum revenue for the store?

Solve and graph your solutions.

65. $x^2 + 7x - 18 < 0$

+‒+‒+‒+‒+‒+‒+‒+‒+‒+‒+‒+‒+‒→

66. $x^2 + 4x - 21 < 0$

+‒+‒+‒+‒+‒+‒+‒+‒+‒+‒+‒+‒+‒→

67. $x^2 - 9x + 20 > 0$

+‒+‒+‒+‒+‒+‒+‒+‒+‒+‒+‒+‒→

68. $x^2 - 11x + 28 > 0$

+‒+‒+‒+‒+‒+‒+‒+‒+‒+‒+‒+‒→

Solve each of the following if possible. Approximate, if necessary, any irrational solutions to the nearest tenth.

69. $2x^2 - 5x - 3 \leq 0$ **70.** $3x^2 - 5x - 2 \leq 0$ **71.** $16x^2 - 25 > 0$ **72.** $9x^2 - 4 > 0$

73. $x^2 - 9x > 4 - 7x$ **74.** $4x^2 - 8x \leq 12 + 5x^2$ **75.** $x^2 + 13x > 16 + 7x$

76. $3x^2 - 12x > -11$ **77.** $-2x^2 + 7x + 12 \leq -3x^2 + x$ **78.** $4x^2 + 12x + 9 < 0$

Solve for x.

79. $|x + 1| = 8$

80. $|4x - 5| = 7$

81. $|3x + 2| = 20$

82. $|3 - x| = |5 - 2x|$

83. $\left| \dfrac{1}{4}x - 3 \right| = 8$

84. $|2x - 8| + 7 = 12$

Solve for x.

85. $|x + 7| < 15$

86. $|x + 9| < 18$

87. $\left| \dfrac{1}{2} x + 2 \right| < \dfrac{7}{4}$

88. $|2x - 1| \geq 9$

89. $|3x - 1| \geq 2$

90. $|3(x - 1)| \geq 5$

To Think About

91. $(x + 4)(x - 2)(3 - x) > 0$

92. $(x + 1)(x + 4)(2 - x) < 0$

Solve the quadratic equations and simplify your answers. Use i notation for any imaginary numbers.

1. $8x^2 + 9x = 0$

2. $8x^2 + 10x = 3$

3. $\dfrac{3x}{2} - \dfrac{8}{3} = \dfrac{2}{3x}$

4. $x(x - 3) - 30 = 5(x - 2)$

5. $7x^2 - 4 = 52$

6. $\dfrac{2x}{2x + 1} - \dfrac{6}{4x^2 - 1} = \dfrac{x + 1}{2x - 1}$

7. $2x^2 - 6x + 5 = 0$

8. $2x(x - 3) = -3$

Solve for any valid real roots.

9. $x^4 - 9x^2 + 14 = 0$

10. $3x^{-2} - 11x^{-1} - 20 = 0$

11. $x^{2/3} - 2x^{1/3} - 12 = 0$

1.

2.

3.

4.

5.

6.

7.

8.

9.

10.

11.

Solve for the variable specified.

12. $B = \dfrac{xyw}{z^2}$; for z **13.** $5y^2 + 2by + 6w = 0$; for y

▲ **14.** The area of a rectangle is 80 square miles. Its length is 1 mile longer than three times its width. Find its length and width.

▲ **15.** Find the hypotenuse of a right triangle if the lengths of its legs are 6 and $2\sqrt{3}$.

16. Shirley and Bill paddled a canoe at a constant speed for 6 miles. They rested, had lunch, and then paddled 1 mile per hour faster for an additional 3 miles. The travel time for the entire trip was 4 hours. How fast did they paddle during each part of the trip?

17. Find the vertex and the intercepts for $f(x) = -x^2 - 6x - 5$. Then graph the function.

Solve.

18. $2x^2 + 3x \geq 27$ **19.** $-3x^2 + 10x + 8 \geq 0$

20. Use a calculator or square root table to approximate to the nearest tenth a solution to $x^2 + 3x - 7 > 0$.

Solve for x.

21. $|5x - 2| = 37$ **22.** $\left|\dfrac{1}{2}x + 3\right| - 2 = 4$

Solve each absolute value inequality.

23. $|7x - 3| \leq 18$ **24.** $|3x + 1| > 7$

12. _____

13. _____

14. _____

15. _____

16. _____

17. _____

18. _____

19. _____

20. _____

21. _____

22. _____

23. _____

24. _____

Approximately one-half of this test is based on the content of Chapters 1–8. The remainder is based on the content of Chapter 9.

1. Simplify: $\left(-3x^{-2}y^3\right)^4$

2. Collect like terms.
$$\frac{1}{2}a^3 - 2a^2 + 3a - \frac{1}{4}a^3 - 6a + a^2$$

3. Solve for y: $a(2y + b) = 3ay - 4$

4. Graph: $6x - 3y = -12$

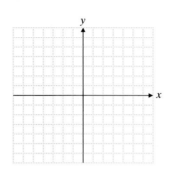

5. Write the equation of the line parallel to $2y + x = 8$ and passing through $(6, -1)$.

▲ **6.** Find the volume of a sphere of radius 2 inches.

7. Factor: $125x^3 - 27y^3$

8. Simplify: $\sqrt{72x^3y^6}$

9. Multiply: $(3 + \sqrt{2})(\sqrt{6} + \sqrt{3})$

10. Rationalize the denominator. $\dfrac{3}{\sqrt{11}}$

Solve and simplify your answers. Use i notation for imaginary numbers.

11. $3x^2 + 12x = 26x$

12. $12x^2 = 11x - 2$

13. $44 = 3(2x - 3)^2 + 8$

14. $3 - \dfrac{4}{x} + \dfrac{5}{x^2} = 0$

1. _____

2. _____

3. _____

4. _____

5. _____

6. _____

7. _____

8. _____

9. _____

10. _____

11. _____

12. _____

13. _____

14. _____

Solve and check.

15. $\sqrt{x - 12} = \sqrt{x} - 2$

16. $x^{2/3} + 9x^{1/3} + 18 = 0$

Solve for y.

17. $2y^2 + 5wy - 7z = 0$

18. $3y^2 + 16z^2 = 5w$

▲ **19.** The hypotenuse of a right triangle is $\sqrt{31}$. One leg of the triangle is 4. Find the length of the other leg.

▲ **20.** A triangle has an area of 45 square meters. The altitude is 3 meters longer than three times the length of the base. Find each dimension.

Exercises 21 and 22 refer to the quadratic function $f(x) = -x^2 + 8x - 12$.

21. Find the vertex and the intercepts of the function.

22. Graph the function.

Solve each of the following quadratic inequalities.

23. $6x^2 - x \leq 2$

24. $x^2 > -2x + 15$

Solve for x.

25. $|3x + 1| = 16$

Solve each absolute value inequality.

26. $\left| \dfrac{1}{2} x + 2 \right| \leq 8$

27. $|3x - 4| > 11$

The Conic Sections

Suppose a major earthquake takes place. Shock waves radiate from the epicenter, which is the center of where an earthquake occurs. These shock waves travel out across the Earth as concentric circles. How could these patterns of circles be used to locate the exact epicenter of the earthquake? Could you use your knowledge of mathematics to determine these locations? Turn to the Putting Your Skills to Work exercises on page 608 to find out.

1. _____

If you are familiar with the topics in this chapter, take this test now. Check your answers with those in the back of the book. If an answer is wrong or you can't do an exercise, study the appropriate section of the chapter.

If you are not familiar with the topics in this chapter, don't take this test now. Instead, study the examples, work the practice exercises, and then take the test.

This test will help you identify those concepts that you have mastered and those that need more study.

Section 10.1

2. _____

1. Write the standard form of the equation of a circle with center at $(8, -2)$ and a radius of $\sqrt{7}$.

2. Find the distance between $(-6, -2)$ and $(-3, 4)$.

3. Rewrite the equation $x^2 + y^2 - 2x - 4y + 1 = 0$ in standard form. Find the circle's center and radius and sketch its graph.

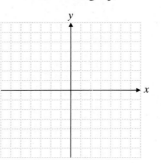

3. _____

Section 10.2

4. _____

Graph each parabola. Write the equation in standard form.

4. $x = (y + 1)^2 + 2$ **5.** $x^2 = y - 4x - 1$

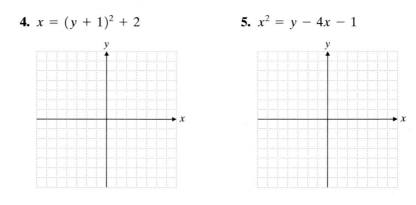

5. _____

Section 10.3

Graph each ellipse. Write the equations in standard form.

6. $4x^2 + y^2 - 36 = 0$

7. $\dfrac{(x + 3)^2}{25} + \dfrac{(y - 1)^2}{16} = 1$

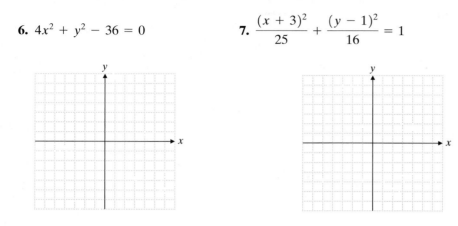

Section 10.4

Graph each hyperbola. Write the equations in standard form.

8. $25y^2 - 9x^2 = 225$

9. $\dfrac{(x - 2)^2}{4} - \dfrac{(y + 1)^2}{9} = 1$

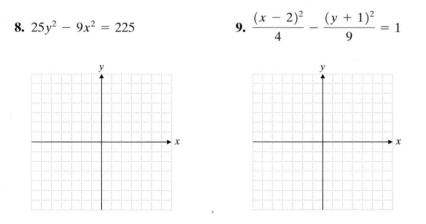

Section 10.5

Solve each nonlinear system.

10. $x^2 + y^2 = 25$
$3x + 4y = 0$

11. $y = x^2 + 1$
$4y^2 = 4 - x^2$

6. _____

7. _____

8. _____

9. _____

10. _____

11. _____

10.1 The Distance Formula and the Circle

Student Learning Objectives

After studying this section, you will be able to:

1 Find the distance between two points.

2 Find the center and radius of a circle and graph the circle if the equation is in standard form.

3 Write the equation of a circle in standard form given its center and radius.

4 Rewrite the equation of a circle in standard form.

SSM
PH TUTOR | CD & VIDEO | MATH PRO | WEB
CENTER

In this chapter we'll talk about the equations and graphs of four special geometric figures—the circle, the parabola, the ellipse, and the hyperbola. These shapes are called **conic sections** because they can be formed by slicing a cone with a plane. The equation of any conic section is of degree 2.

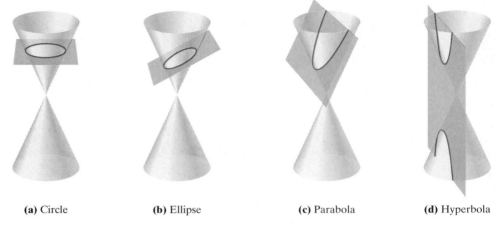

(a) Circle   **(b)** Ellipse   **(c)** Parabola   **(d)** Hyperbola

Conic sections are an important and interesting subject. They are studied along with many other things in a branch of mathematics called *analytic geometry*. Conic sections can be found in applications of physics and engineering. Satellite transmission dishes have parabolic shapes; the orbits of planets are ellipses, and the orbits of comets are hyperbolas; the path of a ball, rocket, or bullet is a parabola (if we neglect air resistance).

1 Finding the Distance Between Two Points

Before we investigate the conic sections, we need to know how to find the distance between two points in the xy-plane. (The rectangular coordinate system formed by the x- and y-axes is often called the xy-plane.) We will derive a *distance formula* and use it to find the equations for the conic sections.

To find the distance between two points on the real number line, we simply find the absolute value of the difference of the values of the points. For example, the distance from -3 to 5 on the x-axis is

$$|5 - (-3)| = |5 + 3| = 8.$$

Remember that absolute value is another name for distance. We could have written

$$|-3 - (5)| = |-8| = 8.$$

Similarly, the distance from -3 to 5 on the y-axis is

$$|5 - (-3)| = 8.$$

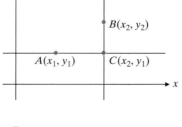

We use this simple fact to find the distance between two points in the xy-plane. Let $A(x_1, y_1)$ and $B(x_2, y_2)$ be points on a graph. First we draw a horizontal line through A, and then we draw a vertical line through B. (We could have drawn a horizontal line through B and a vertical line through A.) The lines intersect at point $C(x_2, y_1)$. Why are the coordinates x_2, y_1? The distance from A to C is $|x_2 - x_1|$ and from B to C $|y_2 - y_1|$.

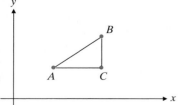

Now, if we draw a line from A to B, we have a right triangle ABC. We can use the Pythagorean theorem to find the length (distance) of the line from A to B. By the Pythagorean theorem,

$$(AB)^2 = (AC)^2 + (BC)^2.$$

566

Let's rename the distance AB as d. Then

$$d^2 = \left(|x_2 - x_1|\right)^2 + \left(|y_2 - y_1|\right)^2$$

and

$$d = \sqrt{(x_2 - x_1)^2 + (y_2 - y_1)^2}.$$

This is the **distance formula**.

Distance Formula

The distance between two points (x_1, y_1) and (x_2, y_2) is

$$d = \sqrt{(x_2 - x_1)^2 + (y_2 - y_1)^2}.$$

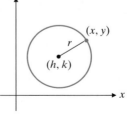 **EXAMPLE 1**   Find the distance between $(3, -4)$ and $(-2, -5)$.

To use the formula, we arbitrarily let $(x_1, y_1) = (3, -4)$ and $(x_2, y_2) = (-2, -5)$.

$$d = \sqrt{(x_2 - x_1)^2 + (y_2 - y_1)^2}$$

$$= \sqrt{[-2 - 3]^2 + [-5 - (-4)]^2}$$

$$= \sqrt{(-5)^2 + (-5 + 4)^2}$$

$$= \sqrt{(-5)^2 + (-1)^2}$$

$$= \sqrt{25 + 1} = \sqrt{26}$$

Practice Problem 1   Find the distance between $(-6, -2)$ and $(3, 1)$.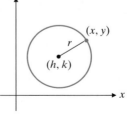

The choice of which point is (x_1, y_1) and which point is (x_2, y_2) is up to you. We would obtain exactly the same answer in Example 1 if $(x_1, y_1) = (-2, -5)$ and $(x_2, y_2) = (3, -4)$. Try it for yourself and see whether you obtain the same result.

2   Finding the Center and Radius of a Circle and Graphing the Circle

A **circle** is defined as the set of all points in a plane that are at a fixed distance from a point in that plane. The fixed distance is called the **radius**, and the point is called the **center** of the circle.

We can use the distance formula to find the equation of a circle. Let a circle of radius r have its center at (h, k). For any point (x, y) on the circle, the distance formula tells us that

$$\sqrt{(x - h)^2 + (y - k)^2} = r.$$

Squaring each side gives

$$(x - h)^2 + (y - k)^2 = r^2.$$

This is the equation of a circle with center at (h, k) and radius r.

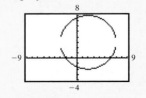

> **Standard Form of the Equation of a Circle**
>
> The standard form of the equation of a circle with center at (h, k) and radius r is
>
> $$(x - h)^2 + (y - k)^2 = r^2.$$

 EXAMPLE 2   Find the center and radius of the circle $(x - 2)^2 + (y - 3)^2 = 25$. Then sketch its graph.

From the equation of a circle,

$$(x - h)^2 + (y - k)^2 = r^2,$$

we see that $(h, k) = (2, 3)$. Thus, the center of the circle is at $(2, 3)$. Since $r^2 = 25$, the radius of the circle is $r = 5$.

The graph of this circle is shown on the right.

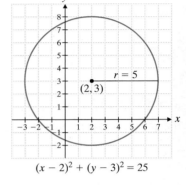

$(x - 2)^2 + (y - 3)^2 = 25$

PRACTICE PROBLEM 2

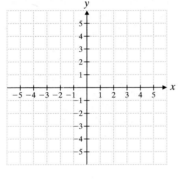

Practice Problem 2   Find the center and radius of the circle

$$(x + 1)^2 + (y + 2)^2 = 9.$$

Then sketch its graph.

3   Writing the Equation of a Circle in Standard Form Given the Center and Radius

We can write the standard form of the equation of a specific circle if we are given the center and the radius. We use the definition of the standard form of the equation of a circle to write the equation we want.

 EXAMPLE 3   Write the equation of the circle with center $(-1, 3)$ and radius $\sqrt{5}$. Put your answer in standard form.

We are given that $(h, k) = (-1, 3)$ and $r = \sqrt{5}$. Thus,

$$(x - h)^2 + (y - k)^2 = r^2$$

becomes the following:

$$[x - (-1)]^2 + [y - 3]^2 = (\sqrt{5})^2$$
$$(x + 1)^2 + (y - 3)^2 = 5$$

Be careful of the signs. It is easy to make a sign error in these steps.

Practice Problem 3   Write the equation of the circle with center $(-5, 0)$ and radius $\sqrt{3}$. Put your answer in standard form.

4 *Rewriting the Equation of a Circle in Standard Form*

The standard form of the equation of a circle helps us sketch the graph of the circle. Sometimes the equation of a circle is not given in standard form, and we need to rewrite the equation.

EXAMPLE 4   Write the equation of the circle $x^2 + 2x + y^2 + 6y + 6 = 0$ in standard form. Find the radius and center of the circle and sketch its graph.

The standard form of the equation of a circle is

$$(x - h)^2 + (y - k)^2 = r^2.$$

If we multiply out the terms in the equation, we get

$$(x^2 - 2hx + h^2) + (y^2 - 2ky + k^2) = r^2.$$

Comparing this with the equation we were given,

$$(x^2 + 2x) + (y^2 + 6y) = -6,$$

suggests that we can complete the square to put the equation in standard form.

$$x^2 + 2x + \underline{\hspace{1cm}} + y^2 + 6y + \underline{\hspace{1cm}} = -6$$
$$x^2 + 2x + 1 \quad\quad + y^2 + 6y + 9 \quad\quad = -6 + 1 + 9$$
$$x^2 + 2x + 1 + y^2 + 6y + 9 = 4$$
$$(x + 1)^2 + (y + 3)^2 = 4$$

Thus, the center is at $(-1, -3)$, and the radius is 2. The sketch of the circle is shown.

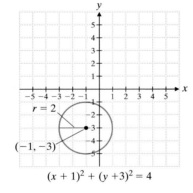

$$(x + 1)^2 + (y + 3)^2 = 4$$

Practice Problem 4   Write the equation of the circle $x^2 + 4x + y^2 + 2y - 20 = 0$ in standard form. Find its radius and center and sketch its graph.

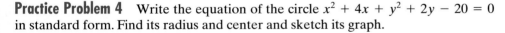

Graphing Calculator

Exploration

One way to graph the equation in Example 4 is to write it as a quadratic equation in y and then employ the quadratic formula.

$$y^2 + 6y + (x^2 + 2x + 6) = 0$$
$$ay^2 + by + c = 0$$

$a = 1, b = 6,$ and
$c = x^2 + 2x + 6$

$$y = \frac{-6 \pm \sqrt{36 - 4(1)(x^2 + 2x + 6)}}{2(1)}$$

$$y = \frac{-6 \pm \sqrt{12 - 8x - 4x^2}}{2}$$

Thus, we have the two halves of the circle.

$$y_1 = \frac{-6 + \sqrt{12 - 8x - 4x^2}}{2}$$

$$y_2 = \frac{-6 - \sqrt{12 - 8x - 4x^2}}{2}$$

We can graph these on one coordinate system to obtain the graph. Use the quadratic formula to obtain the graph of the following circle.

$$x^2 + y^2 + 6x - 4y - 12 = 0$$

From your graph, estimate the value of the radius and the coordinates of the center. Some graphing calculators have a feature for getting a background grid for your graph. If you have this feature, use it to find the coordinates of the center.

10.1 Exercises

Verbal and Writing Skills

1. Explain how you would find the distance from -2 to 4 on the y-axis.

2. Explain how you would find the distance between $(3, -1)$ and $(-4, 0)$ in the xy-plane.

3. $(x - 1)^2 + (y + 2)^2 = 9$ is the equation of a circle. Explain how to determine the center and the radius of the circle.

4. $x^2 - 6x + y^2 - 2y = 6$ is the equation of a circle. Explain how you would rewrite the equation in standard form.

Find the distance between each pair of points. Simplify your answers.

5. $(1, 6)$ and $(2, 4)$

6. $(4, 6)$ and $(7, 5)$

7. $\left(\frac{1}{2}, \frac{5}{2}\right)$ and $\left(\frac{3}{4}, \frac{3}{2}\right)$

8. $\left(\frac{2}{3}, \frac{7}{4}\right)$ and $\left(\frac{5}{6}, \frac{3}{4}\right)$

9. $(3, 9)$ and $(-2, -3)$

10. $(8, 4)$ and $(-4, -1)$

11. $(0, -3)$ and $(4, 1)$

12. $(-5, -6)$ and $(2, 0)$

13. $\left(\frac{1}{3}, \frac{3}{5}\right)$ and $\left(\frac{7}{3}, \frac{1}{5}\right)$

14. $\left(-\frac{1}{4}, \frac{1}{7}\right)$ and $\left(\frac{3}{4}, \frac{6}{7}\right)$

15. $(1.3, 2.6)$ and $(-5.7, 1.6)$

16. $(8.2, 3.5)$ and $(6.2, -0.5)$

Find the value of the unknown coordinate so that the distance between the points is as given.

17. $(7, 2)$ and $(1, y)$; distance is 10

18. $(3, y)$ and $(3, -5)$; distance is 9

19. $(1.5, 2)$ and $(0, y)$; distance is 2.5

20. $\left(1, \dfrac{15}{2}\right)$ and $\left(x, -\dfrac{1}{2}\right)$; distance is 10

21. $(7, 3)$ and $(x, 6)$; distance is $\sqrt{10}$

22. $(4, 5)$ and $(2, y)$; distance is $\sqrt{5}$

Applications

Use the following information to solve Exercises 23 and 24. An airport is located at point O. A short-range radar tower is located at point R. The maximum range at which the radar can detect a plane is 4 miles from point R.

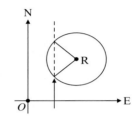

23. Assume that *R* is 6 miles east of *O* and 6 miles north of *O*. In other words, *R* is located at the point $(6, 6)$. An airplane is flying parallel to and 4 miles east of the north axis. (In other words, the plane is flying along the path $x = 4$.) What is the *greatest distance* north of the airport at which the plane can still be detected by the radar tower at *R*? Round your answer to the nearest tenth of a mile.

24. Assume that *R* is 5 miles east of *O* and 7 miles north of *O*. In other words, *R* is located at the point $(5, 7)$. An airplane is flying parallel to and 2 miles east of the north axis. (In other words, the plane is flying along the path $x = 2$.) What is the *shortest distance* north of the airport at which the plane can be detected by the radar tower at *R*? Round your answer to the nearest tenth of a mile.

Write in standard form the equation of the circle with the given center and radius.

25. center $(-1, -7)$; $r = \sqrt{5}$

26. center $(-3, -5)$; $r = \sqrt{2}$

27. center $(-3.5, 0)$; $r = 6$

28. center $\left(0, \dfrac{3}{2}\right)$; $r = \dfrac{1}{2}$

29. center $\left(\dfrac{7}{4}, 0\right)$; $r = \dfrac{1}{3}$

30. center $(0, -4.5)$; $r = 4$

Give the center and radius of each circle. Then sketch its graph.

31. $x^2 + y^2 = 25$

32. $x^2 + y^2 = 9$

33. $(x - 3)^2 + (y - 2)^2 = 4$

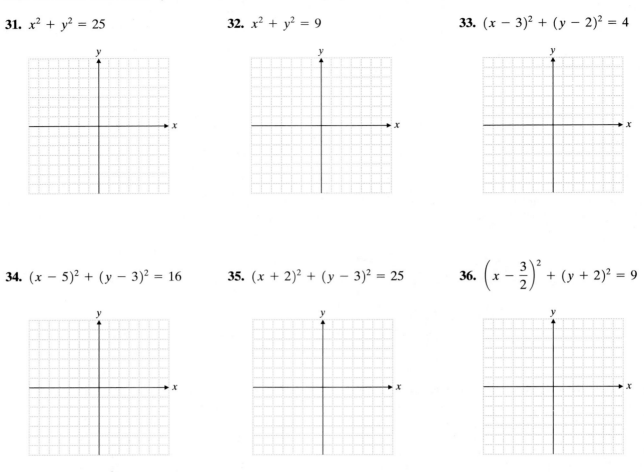

34. $(x - 5)^2 + (y - 3)^2 = 16$

35. $(x + 2)^2 + (y - 3)^2 = 25$

36. $\left(x - \dfrac{3}{2}\right)^2 + (y + 2)^2 = 9$

Rewrite each equation in standard form, using the approach of Example 4. Find the center and radius of each circle.

37. $x^2 + y^2 + 6x - 4y - 3 = 0$

38. $x^2 + y^2 + 8x - 6y - 24 = 0$

39. $x^2 + y^2 - 12x + 2y - 12 = 0$

40. $x^2 + y^2 + 4x - 4y + 7 = 0$

41. $x^2 + y^2 + 3x - 2 = 0$

42. $x^2 + y^2 - 5x - 1 = 0$

43. A Ferris wheel has a radius r of 25.1 feet. The height from the ground to the center of the tower t is 29.7 feet. The distance d from the origin to the base of the tower is 42.7 feet. Find the standard form of the equation of the circle represented by the Ferris wheel.

44. A Ferris wheel has a radius r of 25.3 feet. The height from the ground to the center of the tower t is 31.8 feet. The distance d from the origin to the base of the tower is 44.8 feet. Find the standard form of the equation of the circle represented by the Ferris wheel.

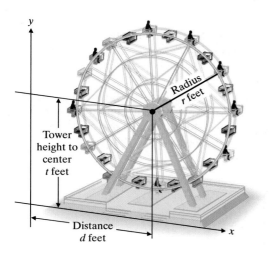

Optional Graphing Calculator Problems

Graph each circle with your graphing calculator.

45. $(x - 5.32)^2 + (y + 6.54)^2 = 47.28$

46. $x^2 + 9.56x + y^2 - 7.12y + 8.9995 = 0$

Cumulative Review Problems

Solve the following quadratic equations by factoring.

47. $9 + \dfrac{3}{x} = \dfrac{2}{x^2}$

48. $3x^2 - 5x + 2 = 0$

Solve the following quadratic equations by using the quadratic formula.

49. $5x^2 - 6x - 7 = 0$

50. $4x^2 + 2x = 1$

51. The 1980 eruptions of Mount Saint Helens blew down or scorched 230 square miles of forest. A deposit of rock and sediments soon filled up a 20-square-mile area to an average depth of 150 feet. How many cubic feet of rock and sediments settled in this region?

52. Within a 15-mile radius north of Mount Saint Helens, the blast of its 1980 eruption traveled up to 670 miles per hour. If an observer 15 miles north of the volcano saw the blast and attempted to run for cover, how many seconds did he have to run before the blast reached his original location?

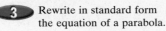

10.2 The Parabola

Student Learning Objectives

After studying this section, you will be able to:

 Graph vertical parabolas.

② Graph horizontal parabolas.

③ Rewrite in standard form the equation of a parabola.

SSM
PH TUTOR CENTER   **CD & VIDEO**   **MATH PRO**   **WEB**

If we pass a plane through a cone so that the plane is parallel to—but not touching—a side of the cone, we form a **parabola**. A **parabola** is defined as the set of points that are the same distance from some fixed line (called the **directrix**) and some fixed point (called the **focus**) that is *not* on the line.

The shape of a parabola is a common one. For example, the cables that are used to support the weight of a bridge are in the shape of parabolas.

The simplest form for the equation is one variable = (another variable)^2. That is, $y = x^2$ or $x = y^2$. We will make a table of values for each equation, plot the points, and draw a graph. For the first equation we choose values for x and find y. For the second equation we choose values for y and find x.

$y = x^2$

x	y
-2	4
-1	1
0	0
1	1
2	4

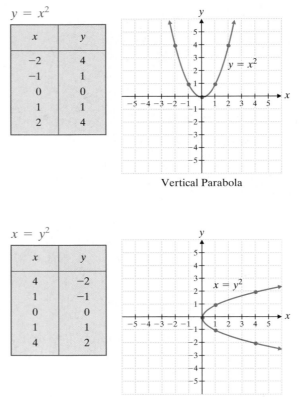

Vertical Parabola

$x = y^2$

x	y
4	-2
1	-1
0	0
1	1
4	2

Horizontal Parabola

Notice that the graph of $y = x^2$ is symmetric about the y-axis. That is, if you folded the graph along the y-axis, the two parts of the curve would coincide. For this parabola, the y-axis is the **axis of symmetry**.

What is the axis of symmetry for the parabola $x = y^2$? Every parabola has an axis of symmetry. This axis can be *any* line; it depends on the location and orientation of the parabola in the rectangular coordinate system. The point at which the parabola crosses the axis of symmetry is the **vertex**. What are the coordinates of the vertex for $y = x^2$? For $x = y^2$?

1   *Graphing Vertical Parabolas*

EXAMPLE 1   Graph $y = (x - 2)^2$. Identify the vertex and the axis of symmetry.

We make a table of values. We begin with $x = 2$ in the middle of the table of values because $(2 - 2)^2 = 0$. That is, when $x = 2$, $y = 0$. We then fill in the x- and y-values above and below $x = 2$. We plot the points and draw the graph.

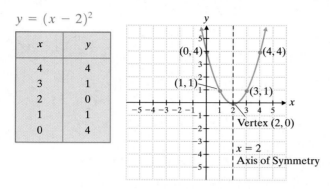

$y = (x - 2)^2$

x	y
4	4
3	1
2	0
1	1
0	4

PRACTICE PROBLEM 1

The vertex is $(2, 0)$, and the axis of symmetry is the line $x = 2$.

Practice Problem 1   Graph $y = -(x + 3)^2$. Identify the vertex and the axis of symmetry.

EXAMPLE 2   Graph $y = (x - 2)^2 + 3$. Find the vertex, the axis of symmetry, and the y-intercept.

This graph looks just like the graph of $y = x^2$, except that it is shifted 2 units to the right and 3 units up. The vertex is $(2, 3)$. The axis of symmetry is $x = 2$. We can find the y-intercept by letting $x = 0$ in the equation. We get

$$y = (0 - 2)^2 + 3 = 4 + 3 = 7.$$

Thus, the y-intercept is $(0, 7)$.

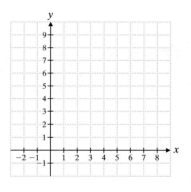

$y = (x - 2)^2 + 3$

PRACTICE PROBLEM 2

Practice Problem 2   Graph the parabola $y = (x - 6)^2 + 4$.

The examples we have studied illustrate the following properties of the standard form of the equation of a vertical parabola.

Standard Form of the Equation of a Vertical Parabola

1. The graph of $y = a(x - h)^2 + k$, where $a \neq 0$, is a vertical parabola.
2. The parabola opens upward ⌣ if $a > 0$ and downward ⌢ if $a < 0$.
3. The vertex of the parabola is (h, k).
4. The axis of symmetry is the line $x = h$.
5. The y-intercept is the point where the parabola crosses the y-axis (i.e., where $x = 0$).

We can use these properties as steps to graph a parabola. If we want greater accuracy, we should also plot a few other points.

● EXAMPLE 3   Graph $y = -\dfrac{1}{2}(x + 3)^2 - 1$.

Step 1   The equation has the form $y = a(x - h)^2 + k$, where $a = -\frac{1}{2}$, $h = -3$, and $k = -1$, so it is a vertical parabola.

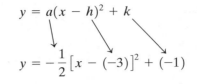

Step 2 $a < 0$; so the parabola opens downward.

Step 3   We have $h = -3$ and $k = -1$.
Therefore, the vertex of the parabola is $(-3, -1)$.

Step 4   The axis of symmetry is the line $x = -3$.
We plot a few points on either side of the axis of symmetry. We try $x = -1$ because $(-1 + 3)^2$ is 4 and $-\frac{1}{2}(4)$ is an integer. We avoid fractions. When $x = -1$, $y = -\frac{1}{2}(-1 + 3)^2 - 1 = -3$. Thus, the point is $(-1, -3)$. The image of this point on the other side of the axis of symmetry is $(-5, -3)$. We now try $x = 1$. When $x = 1$, $y = -\frac{1}{2}(1 + 3)^2 - 1 = -9$. Thus, the point is $(1, -9)$. The image of this point on the other side of the axis of symmetry is $(-7, -9)$.

PRACTICE PROBLEM 3

Step 5   When $x = 0$, we have the following:

$$y = -\frac{1}{2}(0 + 3)^2 - 1$$

$$= -\frac{1}{2}(9) - 1$$

$$= -4.5 - 1 = -5.5$$

Thus, the y-intercept is $(0, -5.5)$.
The graph is shown on the right.

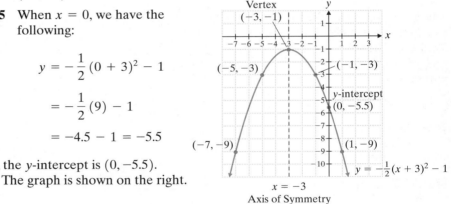

Practice Problem 3   Graph $y = \dfrac{1}{4}(x - 2)^2 + 3$.   ●

② Graphing Horizontal Parabolas

Recall that the equation $x = y^2$, in which the squared term is the y-variable, describes a horizontal parabola. Horizontal parabolas open to the left or right. They are symmetric about the x-axis or about a line parallel to the x-axis. We now look at examples of horizontal parabolas.

● EXAMPLE 4   Graph $x = -2y^2$.

Notice that the y-term is squared. This means that the parabola is horizontal. We make a table of values, plot points, and draw the graph. To make the table of values, we choose values for y and find x. We begin with $y = 0$.

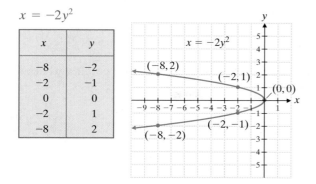

$x = -2y^2$

x	y
-8	-2
-2	-1
0	0
-2	1
-8	2

PRACTICE PROBLEM 4

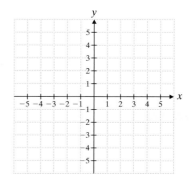

The parabola $x = -2y^2$ has its vertex at $(0, 0)$. The axis of symmetry is the x-axis.

Practice Problem 4   Graph the parabola $x = -2y^2 + 4$.

To Think About   Compare the graphs in Example 4 and Practice Problem 4 to the graph of $x = y^2$. How are they different? How are they the same? What does the coefficient -2 in the equation $x = -2y^2$ do to the graph of the equation $x = y^2$? What does the constant 4 in $x = -2y^2 + 4$ do to the graph of $x = -2y^2$?

Now we can make the same type of observations for horizontal parabolas as we did for vertical ones.

> ### Standard Form of the Equation of a Horizontal Parabola
>
> **1.** The graph of $x = a(y - k)^2 + h$, where $a \neq 0$, is a horizontal parabola.
>
> **2.** The parabola opens to the right $\big($ if $a > 0$ and opens to the left $\big)$ if $a < 0$.
>
> **3.** The vertex of the parabola is (h, k).
> **4.** The axis of symmetry is the line $y = k$.
> **5.** The x-intercept is the point where the parabola crosses the x-axis (i.e., where $y = 0$).

EXAMPLE 5   Graph $x = (y - 3)^2 - 5$. Find the vertex, the axis of symmetry, and the x-intercept.

Step 1   The equation has the form $x = a(y - k)^2 + h$, where $a = 1$, $k = 3$, and $h = -5$, so it is a horizontal parabola.

$$x = a(y - k)^2 + h$$
$$x = 1(y - 3)^2 + (-5)$$

Step 2 $a > 0$; so the parabola opens to the right.

Step 3   We have $k = 3$ and $h = -5$. Therefore, the vertex is $(-5, 3)$.

Step 4   The line $y = 3$ is the axis of symmetry.

We look for a few points on either side of the axis of symmetry. We will try y-values close to the vertex $(-5, 3)$. We try $y = 4$ and $y = 2$. When $y = 4$, $x = (4 - 3)^2 - 5 = -4$.   When $y = 2$, $x = (2 - 3)^2 - 5 = -4$. Thus, the points are $(-4, 4)$ and $(-4, 2)$. (Remember to list the x-value first in a coordinate pair.) We try $y = 5$ and $y = 1$. When $y = 5$, $x = (5 - 3)^2 - 5 = -1$. When $y = 1$, $x = (1 - 3)^2 - 5 = -1$. Thus, the points are $(-1, 5)$ and $(-1, 1)$. You may prefer to find one point, graph it, and find its image on the other side of the axis of symmetry, as was done in Example 3. In Example 5, we decided to look for both pairs of points using the equation.

Step 5   When $y = 0$,

$$x = (0 - 3)^2 - 5 = 9 - 5 = 4.$$

Thus, the x-intercept is $(4, 0)$. We plot the points and draw the graph.

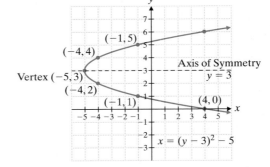

PRACTICE PROBLEM 5

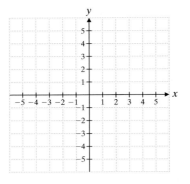

Notice that the graph also crosses the y-axis. You can find the y-intercepts by setting x equal to 0 and solving the resulting quadratic equation. Try it.

Practice Problem 5   Graph the parabola $x = -(y + 1)^2 - 3$. Find the vertex, the axis of symmetry, and the x-intercept.

3   Rewriting in Standard Form the Equation of a Parabola

So far, all the equations we have graphed have been in standard form. This rarely happens in the real world. How do you suppose we put the quadratic equation $y = ax^2 + bx + c$ in the standard form $y = a(x - h)^2 + k$? We do so by completing the square.

 EXAMPLE 6   Place the equation $x = y^2 + 4y + 1$ in standard form. Then graph it.

Since the y-term is squared, we have a horizontal parabola. So the standard form is

$$x = a(y - k)^2 + h.$$

Now we have the following:

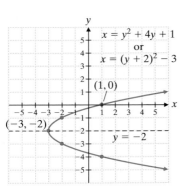

$x = y^2 + 4y + \underline{\hspace{1cm}} - \underline{\hspace{1cm}} + 1$ Whatever we add to the right side we must also subtract from the right side.

$= y^2 + 4y + \left(\dfrac{4}{2}\right)^2 - \left(\dfrac{4}{2}\right)^2 + 1$ Complete the square.

$= \left(y^2 + 4y + 4\right) - 3$ Simplify.

$= (y + 2)^2 - 3$ Standard form.

We see that $a = 1$, $k = -2$, and $h = -3$. Since a is positive, the parabola opens to the right. The vertex is $(-3, -2)$. The axis of symmetry is $y = -2$. If we let $y = 0$, we find that the x-intercept is $(1, 0)$. The graph is in the margin on the left.

Practice Problem 6 Place the equation $x = y^2 - 6y + 13$ in standard form and graph it.

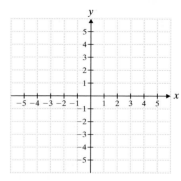

EXAMPLE 7 Place the equation $y = 2x^2 - 4x - 1$ in standard form. Then graph it.

This time the x-term is squared, so we have a vertical parabola. The standard form is

$$y = a(x - h)^2 + k.$$

We need to complete the square.

$$y = 2(x^2 - 2x + \underline{\quad\quad}) - \underline{\quad\quad} - 1$$
$$= 2[x^2 - 2x + (1)^2] - 2(1)^2 - 1$$
$$= 2(x - 1)^2 - 3$$

The parabola opens upward ($a > 0$), the vertex is $(1, -3)$, the axis of symmetry is $x = 1$, and the y-intercept is $(0, -1)$.

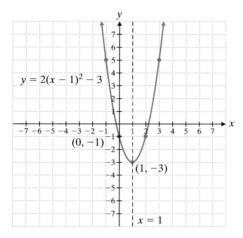

$y = 2(x - 1)^2 - 3$

$(0, -1)$

$(1, -3)$

$x = 1$

Graphing Calculator

Graphing Parabolas
Graphing horizontal parabolas such as the one in Example 6 on a graphing calculator requires dividing the curve into two halves. In this case the halves would be

$$y_1 = -2 + \sqrt{x + 3}$$

and

$$y_2 = -2 - \sqrt{x + 3}.$$

Vertical parabolas can be graphed immediately on a graphing calculator. Why is this? How can you tell whether it is necessary to divide a curve into two halves? Graph the equations below on a graphing calculator. Use the quadratic formula when needed.

1. $y^2 + 8x - 4y = 28$

2. $4x^2 - 4x + 32y = 47$

Practice Problem 7 Place $y = 2x^2 + 8x + 9$ in standard form and graph it.

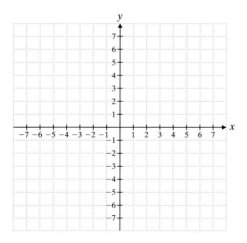

Verbal and Writing Skills

1. The graph of $y = x^2$ is symmetric about the _____. The graph of $x = y^2$ is symmetric about the _____.

2. Explain how to determine the axis of symmetry of the parabola $x = \frac{1}{2}(y + 5)^2 - 1$.

3. Explain how to determine the vertex of the parabola $y = 2(x - 3)^2 + 4$.

4. How does the coefficient -6 affect the graph of the parabola $y = -6x^2$?

Graph each parabola and label the vertex. Find the y-intercept. Place the x- and y-axes on the grid at a convenient place for your graphs.

5. $y = -4x^2$

6. $y = -3x^2$

7. $y = x^2 - 6$

8. $y = x^2 + 2$

9. $y = -2x^2 + 4$

10. $y = -3x^2 + 1$

11. $y = (x - 3)^2 - 2$

12. $y = (x - 2)^2 - 4$

13. $y = 2(x - 1)^2 + \dfrac{3}{2}$

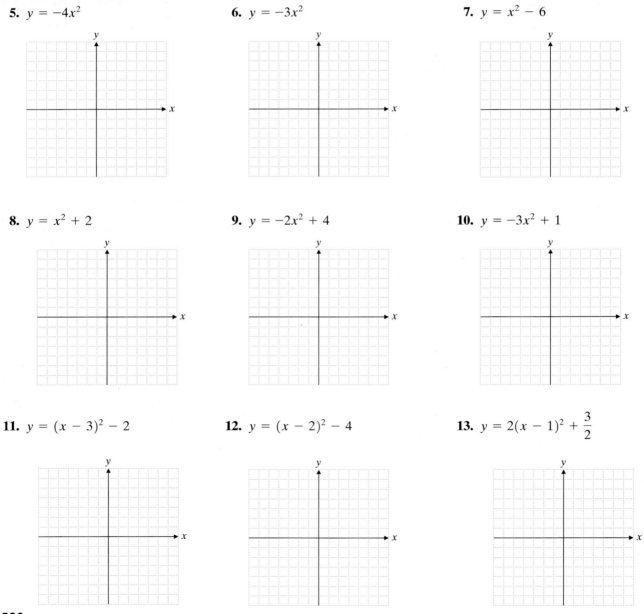

14. $y = 2(x - 2)^2 + \dfrac{5}{2}$

15. $y = -4\left(x + \dfrac{3}{2}\right)^2 + 5$

16. $y = -2\left(x + \dfrac{1}{2}\right)^2 - 1$

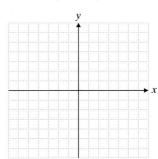

Graph each parabola and label the vertex. Find the x-intercept.

17. $x = \dfrac{1}{4}y^2 - 2$

18. $x = \dfrac{1}{3}y^2 + 1$

19. $x = (y - 2)^2 + 3$

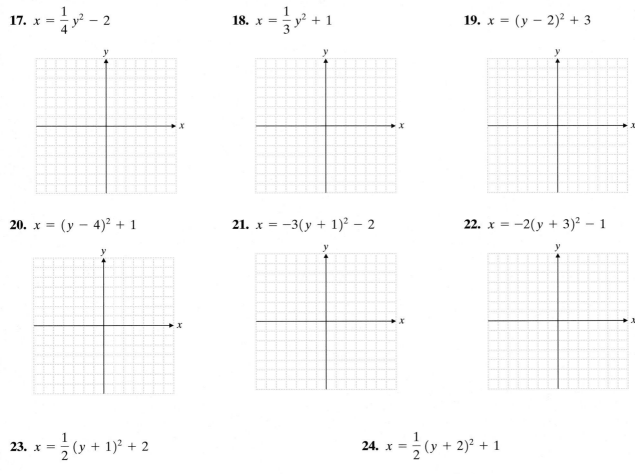

20. $x = (y - 4)^2 + 1$

21. $x = -3(y + 1)^2 - 2$

22. $x = -2(y + 3)^2 - 1$

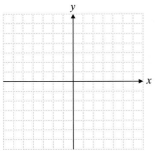

23. $x = \dfrac{1}{2}(y + 1)^2 + 2$

24. $x = \dfrac{1}{2}(y + 2)^2 + 1$

25. $x = -(y - 2)^2 + \dfrac{1}{2}$

26. $x = -(y - 3)^2 + \dfrac{9}{2}$

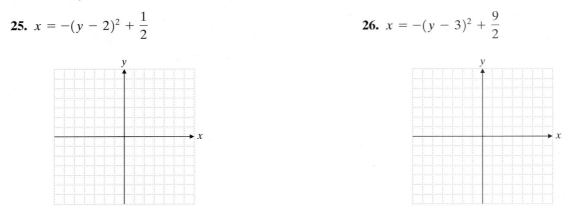

*Rewrite in standard form each equation. Determine **(a)** whether the parabola is horizontal or vertical, **(b)** the direction it opens, and **(c)** the vertex.*

27. $y = x^2 + 12x + 25$

28. $y = x^2 - 4x - 1$

29. $y = -2x^2 + 4x - 3$

30. $y = -2x^2 + 4x + 5$

31. $x = y^2 + 8y + 9$

32. $x = y^2 + 10y + 23$

Applications

33. Find an equation of the form $y = ax^2$ that describes the outline of a satellite dish such that the bottom of the dish passes through $(0, 0)$, the diameter of the dish is 32 inches, and the depth of the dish is 8 inches.

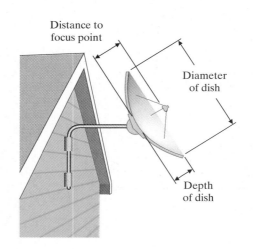

34. Find an equation of the form $y = ax^2$ that describes the outline of a satellite dish such that the bottom of the dish passes through $(0, 0)$, the diameter of the dish is 24 inches, and the depth of the dish is 4 inches.

35. If the outline of a satellite dish is described by the equation $y = ax^2$, then the distance p from the center of the dish to the focus point of the dish is given by the equation $a = \dfrac{1}{4p}$. Find the distance p for the dish in Exercise 33.

36. If the outline of a satellite dish is described by the equation $y = ax^2$, then the distance p from the center of the dish to the focus point of the dish is given by the equation $a = \dfrac{1}{4p}$. Find the distance p for the dish in Exercise 34.

Optional Graphing Calculator Problems

Find the vertex and y-intercept of each parabola. Find the two x-intercepts.

37. $y = 2x^2 + 6.48x - 0.1312$

38. $y = -3x^2 + 33.66x - 73.5063$

Applications

By writing a quadratic equation in the form $y = a(x - h)^2 + k$, we can find the maximum or minimum value of the equation and the value of x at which it occurs. Remember that the equation $y = a(x - h)^2 + k$ is a vertical parabola. For $a > 0$, the parabola opens upward. Thus, the y-coordinate of the vertex is the smallest (or minimum) value of x. Similarly, when $a < 0$, the parabola opens downward, so the y-coordinate of the vertex is the maximum value of the equation. Since the vertex occurs at (h, k), the maximum value of the equation occurs when $x = h$. Then

$$y = -a(x - h)^2 + k = a(0) + k = k.$$

For example, suppose the weekly profit of a manufacturing company in dollars is $P = -2(x - 45)^2 + 2300$ for x units manufactured. By looking at the equation, we see that the maximum profit per week is \$2300 and is attained when 45 units are manufactured. Use this approach for Exercises 39–42.

39. A company's monthly profit equation is
$$P = -2x^2 + 200x + 47,000,$$
where x is the number of items manufactured. Find the maximum monthly profit and the number of items that must be produced each month to attain maximum profit.

40. A company's monthly profit equation is
$$P = -3x^2 + 240x + 31,200,$$
where x is the number of items manufactured. Find the maximum monthly profit and the number of items that must be produced each month to attain maximum profit.

41. A research pharmacologist has determined that sensitivity S to a drug depends on the dosage d in milligrams, according to the equation $S = 650d - 2d^2$. What is the maximum sensitivity that will occur? What dosage will produce that maximum sensitivity?

42. The effective yield from a grove of orange trees is described by the equation $E = x(900 - x)$, where x is the number of orange trees per acre. What is the maximum effective yield? How many orange trees per acre should be planted to achieve the maximum yield?

Cumulative Review Problems

Simplify.

43. $\sqrt{50x^3}$

44. $\sqrt[3]{40x^3y^4}$

Add.

45. $\sqrt{98x} + x\sqrt{8} - 3\sqrt{50x}$

46. $\sqrt[3]{16x^4} + 4x\sqrt[3]{2} - 8x\sqrt[3]{54}$

47. Matthew drives from work to his home at 40 mph. One morning, an accident on the road delayed him for 15 minutes. The driving time including the delay was 56 minutes. How far does Matthew live from his job?

48. A driver delivering eggs drove from the farm to a supermarket warehouse at 30 mph. He unloaded the eggs and drove back to the farm at 50 mph. The total trip took 2 hours and 15 minutes. How far is the farm from the supermarket warehouse?

49. Sir George Tipkin of Sussex has a collection of eight large English rose bushes, each having approximately 1050 buds. In normal years this type of bush produces blooms from 73% of its buds. During years of drought this figure drops to 44%. During years of heavy rainfall the figure rises to 88%. How many blooms can Sir George expect on these bushes if there is heavy rainfall this year?

50. Last year Sir George had only six of the type of bushes described in Exercise 49. It was a drought year, and he counted 2900 blooms. Using the bloom rates given in Exercise 49, determine approximately how many buds appeared on each of these six bushes. (Round your answer to the nearest whole number.)

10.3 The Ellipse

Suppose a plane cuts a cone at an angle so that the plane intersects all sides of the cone. If the plane is not perpendicular to the axis of the cone, the conic section that is formed is called an ellipse.

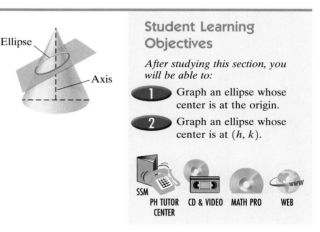

Ellipse

Axis

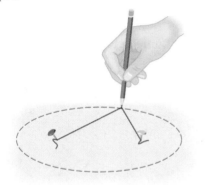

We define an **ellipse** as the set of points in a plane such that for each point in the set, the *sum* of its distances to two fixed points is constant. The fixed points are called **foci** (plural of *focus*).

We can use this definition to draw an ellipse using a piece of string tied at each end to a thumbtack. Place a pencil as shown in the drawing and draw the curve, keeping the pencil pushed tightly against the string. The two thumbtacks are the foci of the ellipse that results.

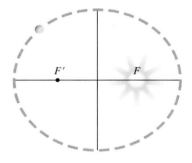

F' F

Examples of the ellipse can be found in the real world. The orbit of the Earth (and each of the other planets) is approximately an ellipse with the Sun at one focus.

An elliptical surface has a special reflecting property. When sound, light, or some other object originating at one focus reaches the ellipse, it is reflected in such a way that it passes through the other focus. This property can be found in the United States Capitol in a famous room known as the Statuary Hall. If a person whispers at the focus of one end of this elliptically shaped room, a person at the other focus can easily hear him or her.

1 Graphing an Ellipse Whose Center Is at the Origin

The equation of an ellipse is similar to the equation of a circle. The standard form of the equation of an ellipse centered at the origin is given next.

Standard Form of the Equation of an Ellipse

An ellipse with center at the origin has the equation

$$\frac{x^2}{a^2} + \frac{y^2}{b^2} = 1, \qquad \text{where } a \text{ and } b > 0.$$

The **vertices** of this ellipse are at $(a, 0)$, $(-a, 0)$, $(0, b)$, and $(0, -b)$.

To plot the ellipse, we need the x- and y-intercepts.

$$\frac{x^2}{a^2} + \frac{y^2}{b^2} = 1$$

$$\text{If } x = 0, \text{ then } \frac{y^2}{b^2} = 1. \qquad\qquad \text{If } y = 0, \text{ then } \frac{x^2}{a^2} = 1.$$

$$y^2 = b^2 \qquad\qquad\qquad x^2 = a^2$$

$$\pm\sqrt{y^2} = \pm\sqrt{b^2} \qquad\qquad \pm\sqrt{x^2} = \pm\sqrt{a^2}$$

$$\pm y = \pm b \text{ or } y = \pm b \qquad\qquad \pm x = \pm a \text{ or } x = \pm a$$

So the x-intercepts are $(a, 0)$ and $(-a, 0)$, and the y-intercepts are $(0, b)$ and $(0, -b)$ for an ellipse of the form $\dfrac{x^2}{a^2} + \dfrac{y^2}{b^2} = 1$.

A circle is a special case of an ellipse. If $a = b$, we get the following:

$$\frac{x^2}{a^2} + \frac{y^2}{a^2} = 1$$

$$x^2 + y^2 = a^2$$

This is the equation of a circle of radius a.

EXAMPLE 1 Graph $x^2 + 3y^2 = 12$. Label the intercepts.

Before we can graph this ellipse, we need to rewrite the equation in standard form.

$$\frac{x^2}{12} + \frac{3y^2}{12} = \frac{12}{12} \qquad \text{Divide each side by 12.}$$

$$\frac{x^2}{12} + \frac{y^2}{4} = 1 \qquad \text{Simplify.}$$

Thus, we have the following:

$$a^2 = 12 \qquad \text{so} \qquad a = 2\sqrt{3}$$

$$b^2 = 4 \qquad \text{so} \qquad b = 2$$

The x-intercepts are $\left(-2\sqrt{3}, 0\right)$ and $\left(2\sqrt{3}, 0\right)$, and the y-intercepts are $(0, 2)$ and $(0, -2)$. We plot these points and draw the ellipse.

PRACTICE PROBLEM 1

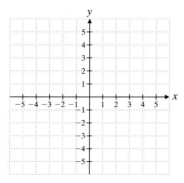

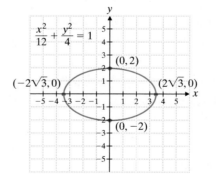

Practice Problem 1 Graph $4x^2 + y^2 = 16$. Label the intercepts.

2 *Graphing an Ellipse Whose Center Is at (h, k)*

If the center of the ellipse is not at the origin but at some point whose coordinates are (h, k), then the standard form of the equation is changed.

An ellipse with center at (h, k) has the equation

$$\frac{(x - h)^2}{a^2} + \frac{(y - k)^2}{b^2} = 1,$$

where a and $b > 0$.

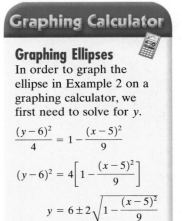

Note that a and b are *not* the x-intercepts now. Why is this? Look at the sketch. You'll see that a is the horizontal distance from the center of the ellipse to a point on the ellipse. Similarly, b is the vertical distance. Hence, when the center of the ellipse is not at the origin, the ellipse may not even cross either axis.

EXAMPLE 2   Graph $\dfrac{(x - 5)^2}{9} + \dfrac{(y - 6)^2}{4} = 1.$

The center of the ellipse is $(5, 6)$, $a = 3$, and $b = 2$. Therefore, we begin at $(5, 6)$. We plot points 3 units to the left, 3 units to the right, 2 units up, and 2 units down from $(5, 6)$.

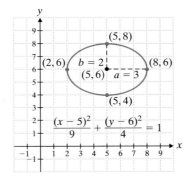

Practice Problem 2   Graph $\dfrac{(x - 2)^2}{16} + \dfrac{(y + 3)^2}{9} = 1.$

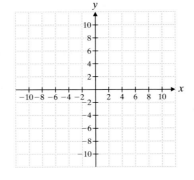

Graphing Calculator

Graphing Ellipses
In order to graph the ellipse in Example 2 on a graphing calculator, we first need to solve for y.

$$\frac{(y - 6)^2}{4} = 1 - \frac{(x - 5)^2}{9}$$

$$(y - 6)^2 = 4\left[1 - \frac{(x - 5)^2}{9}\right]$$

$$y = 6 \pm 2\sqrt{1 - \frac{(x - 5)^2}{9}}$$

Is it necessary to break up the curve into two halves in order to graph the ellipse? Why or why not?

Use the above concepts to graph

$$\frac{(x - 2)^2}{9} + \frac{(y - 1)^2}{4} = 1.$$

Using the Trace feature, determine from your graph the coordinates of the two x-intercepts and the two y-intercepts. Express your answers to the nearest hundredth.

10.3 Exercises

Verbal and Writing Skills

1. Explain how to determine the center of the ellipse $\dfrac{(x+2)^2}{4} + \dfrac{(y-3)^2}{9} = 1$.

2. Explain how to determine the x- and y-intercepts of the ellipse $\dfrac{x^2}{9} + \dfrac{y^2}{16} = 1$.

Graph each ellipse. Label the intercepts. You may need to use a scale other than 1 square = 1 unit.

3. $\dfrac{x^2}{36} + \dfrac{y^2}{4} = 1$

4. $\dfrac{x^2}{49} + \dfrac{y^2}{25} = 1$

5. $\dfrac{x^2}{81} + \dfrac{y^2}{100} = 1$

6. $\dfrac{x^2}{121} + \dfrac{y^2}{144} = 1$

7. $4x^2 + y^2 - 36 = 0$

8. $x^2 + 25y^2 - 25 = 0$

9. $x^2 + 9y^2 = 81$

10. $4x^2 + 25y^2 = 100$

11. $x^2 + 12y^2 = 36$

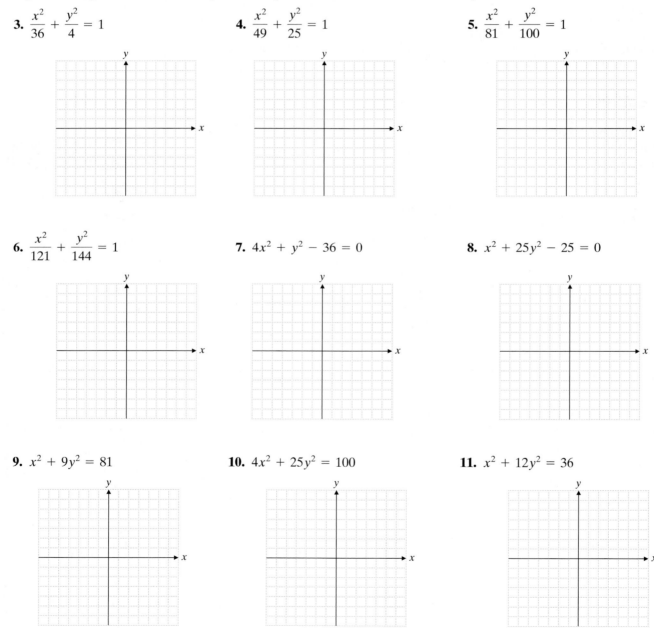

12. $8x^2 + y^2 = 16$

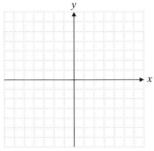

13. $\dfrac{x^2}{\frac{9}{4}} + \dfrac{y^2}{\frac{25}{4}} = 1$

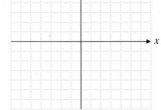

14. $\dfrac{x^2}{\frac{81}{4}} + \dfrac{y^2}{\frac{25}{16}} = 1$

15. $121x^2 + 64y^2 = 7744$

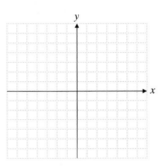

16. Write in standard form the equation of an ellipse with center at the origin, an x-intercept at $(8, 0)$, and a y-intercept at $(0, 10)$.

17. Write in standard form the equation of an ellipse with center at the origin, an x-intercept at $(9, 0)$, and a y-intercept at $(0, -2)$.

18. Write in standard form the equation of an ellipse with center at the origin, an x-intercept at $(2\sqrt{2}, 0)$, and a y-intercept at $(0, 8)$.

19. Write in standard form the equation of an ellipse with center at the origin, an x-intercept at $(9, 0)$, and a y-intercept at $(0, 3\sqrt{2})$.

Applications

20. The orbit of Venus is an ellipse with the Sun as a focus. If we say that the center of the ellipse is at the origin, an approximate equation for the orbit is

$$\frac{x^2}{5013} + \frac{y^2}{4970} = 1,$$

where x and y are measured in millions of miles. Find the largest possible distance across the ellipse. Round your answer to the nearest million miles.

21. The window shown in the sketch is in the shape of half of an ellipse. Find the equation for the ellipse if the center of the ellipse is at point $A = (0, 0)$.

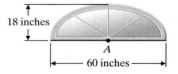

Graph each ellipse. Label the center. Place the x- and y-axes on the grid at a convenient location for each graph. You may need to use a scale other than 1 square = 1 unit.

22. $\dfrac{(x - 7)^2}{4} + \dfrac{(y - 6)^2}{9} = 1$

23. $\dfrac{(x - 5)^2}{9} + \dfrac{(y - 2)^2}{1} = 1$

24. $\dfrac{x^2}{25} + \dfrac{(y - 4)^2}{16} = 1$

25. $\dfrac{(x + 2)^2}{49} + \dfrac{y^2}{25} = 1$

26. $\dfrac{(x + 5)^2}{16} + \dfrac{(y + 2)^2}{36} = 1$

27. $\dfrac{(x + 1)^2}{36} + \dfrac{(y + 4)^2}{16} = 1$

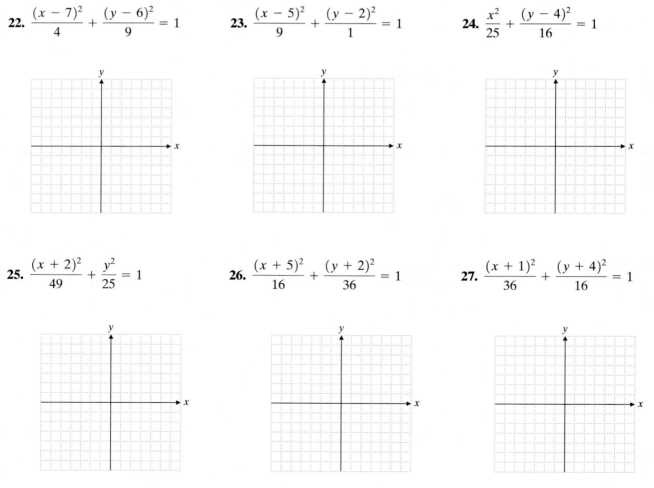

28. Write in standard form the equation of an ellipse whose vertices are $(-3,-2)$, $(5,-2)$, $(1, 1)$, and $(1,-5)$.

29. Write in standard form the equation of an ellipse whose vertices are $(2,3)$, $(6,3)$, $(4,7)$, and $(4,-1)$.

30. Bob's backyard is a rectangle 40 meters by 60 meters. He drove two posts into the ground and fastened a rope to each post, passing the rope through the metal ring on his dog's collar. When the dog pulls on the rope while running, its path is an ellipse. (See the figure.) If the dog can just reach all four sides of the rectangle, find the equation of the elliptical path.

31. For what value of a does the ellipse

$$\frac{(x+5)^2}{4} + \frac{(y+a)^2}{9} = 1$$

pass through the point $(-4,4)$?

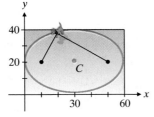

Optional Graphing Calculator Problems

Find the four intercepts, accurate to four decimal places, for each ellipse.

32. $\dfrac{x^2}{12} + \dfrac{y^2}{19} = 1$

33. $\dfrac{(x-3.6)^2}{14.98} + \dfrac{(y-5.3)^2}{28.98} = 1$

To Think About

The area enclosed by the ellipse $\dfrac{x^2}{a^2} + \dfrac{y^2}{b^2} = 1$ *is given by the equation* $A = \pi ab$. *Use the value* $\pi \approx 3.1416$ *to find an approximate value for each of the following answers.*

34. An oval mirror has an outer boundary in the shape of an ellipse. The width of the mirror is 20 inches, and the length of the mirror is 45 inches. Find the area of the mirror. Round your answer to the nearest tenth.

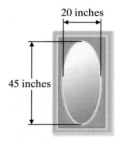

35. In Australia a type of football is played on Aussie Rules fields. These fields are in the shape of an ellipse. Suppose the distance from A to B for the field shown is 185 meters and the distance from C to D is 154 meters. Find the area of the playing field. Round your answer to the nearest tenth.

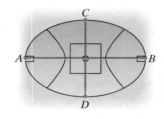

Cumulative Review Problems

Multiply and simplify.

36. $(2\sqrt{3} + 4\sqrt{2})(5\sqrt{6} - \sqrt{2})$

Rationalize the denominator.

37. $\dfrac{5}{\sqrt{2x} - \sqrt{y}}$

38. Write an equation for the line passing through $(-3, 5)$ with a slope of $\frac{1}{3}$.

39. Write an equation for the line passing through $(0, -2)$ and $(-2, 8)$.

40. Multiply. $(2 - 3i)^2$

41. Divide. $\dfrac{5 + 3i}{2i}$

10.4 The Hyperbola

By cutting two branches of a cone by a plane as shown in the sketch, we obtain the two branches of a hyperbola. A comet moving with more than enough kinetic energy to escape the Sun's gravitational pull will travel in a hyperbolic path. Similarly, a rocket traveling with more than enough velocity to escape the Earth's gravitational field will follow a hyperbolic path.

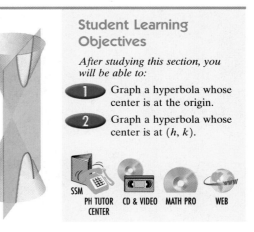

We define a **hyperbola** as the set of points in a plane such that for each point in the set, the absolute value of the *difference* of its distances to two fixed points (called **foci**) is constant.

1 Graphing a Hyperbola Whose Center Is at the Origin

Notice the similarity of the definition of a hyperbola to the definition of an ellipse. If we replace the word *difference* by *sum*, we have the definition of an ellipse. Hence, we should expect that the equation of a hyperbola will be that of an ellipse with the plus sign replaced by a minus sign. And it is. If the hyperbola has its center at the origin, its equation is

$$\frac{x^2}{a^2} - \frac{y^2}{b^2} = 1 \quad \text{or} \quad \frac{y^2}{b^2} - \frac{x^2}{a^2} = 1.$$

The hyperbola has two branches. If the center of the hyperbola is at the origin and the two branches have two x-intercepts but no y-intercepts, the hyperbola is a horizontal hyperbola, and its **axis** is the x-axis. If the center of the hyperbola is at the origin and the two branches have two y-intercepts but no x-intercepts, the hyperbola is a vertical hyperbola, and its axis is the y-axis.

The points where the hyperbola intersects its axis are called the **vertices** of the hyperbola.

For hyperbolas centered at the origin, the vertices are also the intercepts.

Standard Form of the Equation of a Hyperbola with Center at the Origin

Let a and b be any positive real numbers. A hyperbola with center at the origin and vertices $(-a, 0)$ and $(a, 0)$ has the equation

$$\frac{x^2}{a^2} - \frac{y^2}{b^2} = 1.$$

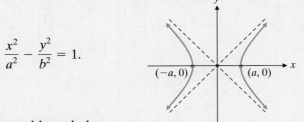

This is called a *horizontal hyperbola.*

A hyperbola with center at the origin and vertices $(0, b)$ and $(0, -b)$ has the equation

$$\frac{y^2}{b^2} - \frac{x^2}{a^2} = 1.$$

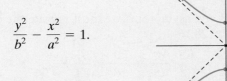

This is called a *vertical hyperbola.*

593

Notice that the two equations are slightly different. Be aware of this difference so that when you look at an equation you will be able to tell whether the hyperbola is horizontal or vertical.

Notice also the diagonal lines that we've drawn on the graphs of the hyperbolas. These lines are called **asymptotes**. The two branches of the hyperbola come increasingly closer to the asymptotes as the value of $|x|$ gets very large. By drawing the asymptotes and plotting the vertices, we can easily graph a hyperbola.

> ## Asymptotes of Hyperbolas
>
> The asymptotes of the hyperbolas $\dfrac{x^2}{a^2} - \dfrac{y^2}{b^2} = 1$ and $\dfrac{y^2}{b^2} - \dfrac{x^2}{a^2} = 1$ are
>
> $$y = \frac{b}{a}x \text{ and } y = -\frac{b}{a}x.$$

Note that $\dfrac{b}{a}$ and $-\dfrac{b}{a}$ are the slopes of the asymptotes.

An easy way to find the asymptotes is to draw extended diagonal lines through the rectangle whose center is at the origin and whose corners are at (a, b), $(a, -b)$, $(-a, b)$, and $(-a, -b)$. (This rectangle is sometimes called the **fundamental rectangle**.) We draw the fundamental rectangle and the asymptotes with a dashed line because they are not part of the curve.

EXAMPLE 1 Graph $\dfrac{x^2}{25} - \dfrac{y^2}{16} = 1$.

The equation has the form $\dfrac{x^2}{a^2} - \dfrac{y^2}{b^2} = 1$, so it is a horizontal hyperbola. $a^2 = 25$, so $a = 5$; $b^2 = 16$, so $b = 4$. Since the hyperbola is horizontal, it has vertices at $(a, 0)$ and $(-a, 0)$, or $(5, 0)$ and $(-5, 0)$.

To draw the asymptotes, we construct a fundamental rectangle with corners at $(5, 4)$, $(5, -4)$, $(-5, 4)$, and $(-5, -4)$. We draw extended diagonal lines through the rectangle as the asymptotes. We construct each branch of the curve so that it passes through a vertex and gets closer to the asymptotes as it moves away from the origin.

PRACTICE PROBLEM 1

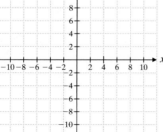

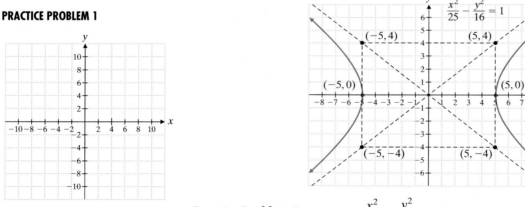

Practice Problem 1 Graph $\dfrac{x^2}{16} - \dfrac{y^2}{25} = 1$.

EXAMPLE 2 Graph $4y^2 - 7x^2 = 28$.

To find the vertices and asymptotes, we must rewrite the equation in standard form. Divide each term by 28.

$$\frac{4y^2}{28} - \frac{7x^2}{28} = \frac{28}{28}$$

$$\frac{y^2}{7} - \frac{x^2}{4} = 1$$

Thus, we have the standard form of a vertical hyperbola with center at the origin. Here $b^2 = 7$, so $b = \sqrt{7}$; $a^2 = 4$, so $a = 2$. The hyperbola has vertices at $(0, \sqrt{7})$ and $(0, -\sqrt{7})$. The fundamental rectangle has corners at $(2, \sqrt{7})$, $(2, -\sqrt{7})$, $(-2, \sqrt{7})$, and $(-2, -\sqrt{7})$. To aid us in graphing, we measure the distance $\sqrt{7}$ as approximately 2.6.

PRACTICE PROBLEM 2

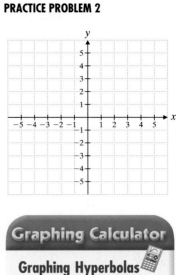

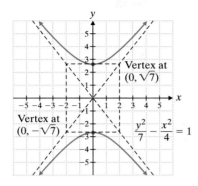

Practice Problem 2   Graph $y^2 - 4x^2 = 4$.

Graphing Calculator

Graphing Hyperbolas
Graph on your graphing calculator the hyperbola in Example 2 using

$$y_1 = \frac{\sqrt{28 + 7x^2}}{2}$$

and

$$y_2 = -\frac{\sqrt{28 + 7x^2}}{2}.$$

Do you see how we obtained y_1 and y_2?

② Graphing a Hyperbola Whose Center Is at (h, k)

If a hyperbola does not have its center at the origin but is shifted h units to the right or left and k units up or down, its equation is one of the following:

Standard Form of the Equation of a Hyperbola with Center at (h, k)

Let a and b be any positive real numbers. A horizontal hyperbola with center at (h, k) and vertices $(h - a, k)$ and $(h + a, k)$ has the equation

$$\frac{(x - h)^2}{a^2} - \frac{(y - k)^2}{b^2} = 1.$$

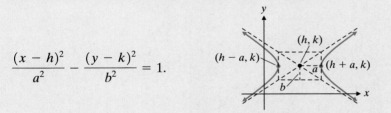

Horizontal Hyperbola

A vertical hyperbola with center at (h, k) and vertices $(h, k + b)$ and $(h, k - b)$ has the equation

$$\frac{(y - k)^2}{b^2} - \frac{(x - h)^2}{a^2} = 1.$$

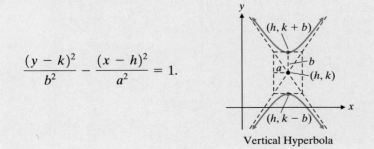

Vertical Hyperbola

EXAMPLE 3   Graph $\dfrac{(x-4)^2}{9} - \dfrac{(y-5)^2}{4} = 1.$

The center is at $(4, 5)$, and the hyperbola is horizontal. We have $a = 3$ and $b = 2$, so the vertices are $(4 \pm 3, 5)$, or $(7, 5)$ and $(1, 5)$. We can sketch the hyperbola more readily if we can draw a fundamental rectangle. Using $(4, 5)$ as the center, we construct a rectangle $2a$ units wide and $2b$ units high. We then draw and extend the diagonals of the rectangle. The extended diagonals are the asymptotes for the branches of the hyperbola.

In this example, since $a = 3$ and $b = 2$, we draw a rectangle $2a = 6$ units wide and $2b = 4$ units high with a center at $(4, 5)$. We draw extended diagonals through the rectangle. From the vertex at $(7, 5)$, we draw a branch of the hyperbola opening to the right. From the vertex at $(1, 5)$, we draw a branch of the hyperbola opening to the left. The graph of the hyperbola is shown.

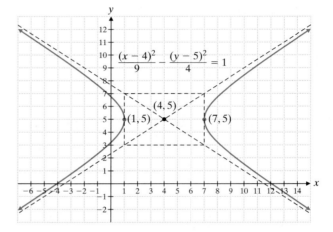

Practice Problem 3   Graph $\dfrac{(y+2)^2}{9} - \dfrac{(x-3)^2}{16} = 1.$

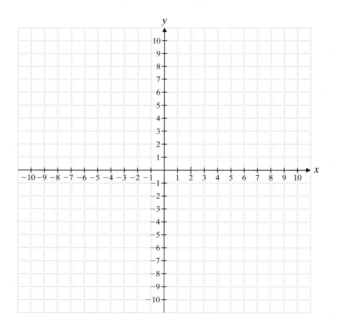

Verbal and Writing Skills

1. What is the standard form of the equation of a horizontal hyperbola centered at the origin?

2. What are the vertices of the hyperbola $\dfrac{y^2}{9} - \dfrac{x^2}{4} = 1$? Is this a horizontal hyperbola or a vertical hyperbola? Why?

3. Explain in your own words how you would draw the graph of the hyperbola $\dfrac{x^2}{16} - \dfrac{y^2}{4} = 1$.

4. Explain how you determine the center of the hyperbola $\dfrac{(x-2)^2}{4} - \dfrac{(y+3)^2}{25} = 1$?

Find the vertices and graph each hyperbola. If the equation is not in standard form, write it as such.

5. $\dfrac{x^2}{4} - \dfrac{y^2}{25} = 1$

6. $\dfrac{x^2}{9} - \dfrac{y^2}{36} = 1$

7. $\dfrac{y^2}{36} - \dfrac{x^2}{49} = 1$

8. $\dfrac{y^2}{64} - \dfrac{x^2}{25} = 1$

9. $4x^2 - y^2 = 64$

10. $49x^2 - 16y^2 = 196$

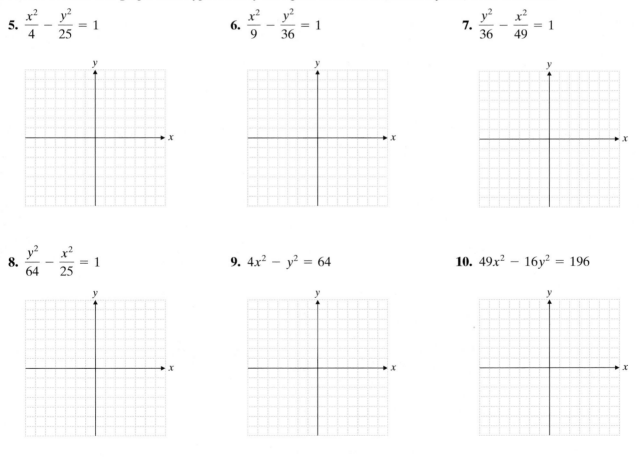

11. $8x^2 - y^2 = 16$

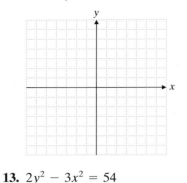

12. $12x^2 - y^2 = 36$

13. $2y^2 - 3x^2 = 54$

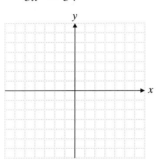

14. $8x^2 - 3y^2 = 24$

Find the equation of the hyperbola with center at the origin and with the following vertices and asymptotes.

15. Vertices at $(3, 0)$ and $(-3, 0)$; asymptotes $y = \dfrac{4}{3}x$, $y = -\dfrac{4}{3}x$

16. Vertices at $(2, 0)$ and $(-2, 0)$; asymptotes $y = \dfrac{3}{2}x$, $y = -\dfrac{3}{2}x$

17. Vertices at $(0, 7)$ and $(0, -7)$; asymptotes $y = \dfrac{7}{3}x$, $y = -\dfrac{7}{3}x$

18. Vertices at $(0, 6)$ and $(0, -6)$; asymptotes $y = \dfrac{6}{5}x$, $y = -\dfrac{6}{5}x$

Applications

19. Some comets have an orbit that is hyperbolic in shape with the Sun at the focus of the hyperbola. A comet is heading toward the Earth but then veers off as shown in the graph. It comes within 120 million miles of the Earth. As it travels into the distance, it moves closer and closer to the line $y = 3x$ with the Earth at the origin. Find the equation that describes the path of the comet.

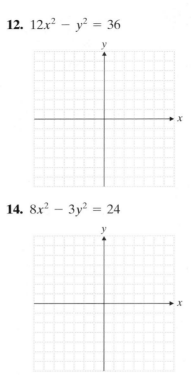

Scale on x axis: each square is 30 million miles.
Scale on y axis: each square is 90 million miles.

20. A rocket following the hyperbolic path shown in the graph turns rapidly at $(4, 0)$ and then moves closer and closer to the line $y = \dfrac{2}{3} x$ as the rocket gets farther from the tracking station at the origin. Find the equation that describes the path of the rocket if the center of the hyperbola is at $(0, 0)$.

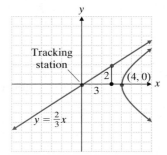

Find the center and then graph each hyperbola. Draw the axes on the grid at a convenient location. You may want to use a scale other than 1 square = 1 unit.

21. $\dfrac{(x - 6)^2}{25} - \dfrac{(y - 4)^2}{49} = 1$

22. $\dfrac{(x - 7)^2}{16} - \dfrac{(y - 5)^2}{25} = 1$

23. $\dfrac{(y + 2)^2}{36} - \dfrac{(x + 1)^2}{81} = 1$

24. $\dfrac{(y + 1)^2}{49} - \dfrac{(x + 3)^2}{81} = 1$

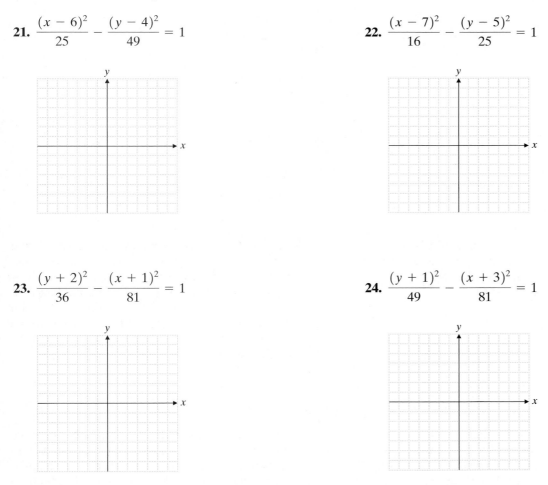

Find the center and the two vertices for each of the following hyperbolas.

25. $\dfrac{(x + 6)^2}{7} - \dfrac{y^2}{3} = 1$

26. $\dfrac{x^2}{5} - \dfrac{(y + 3)^2}{10} = 1$

27. A hyperbola's center is not at the origin. Its vertices are $(5, 0)$ and $(5, 14)$; one asymptote is $y = \frac{7}{5}x$. Find the equation of the hyperbola.

28. A hyperbola's center is not at the origin. Its vertices are $(4, -14)$ and $(4, 0)$. One asymptote is $y = -\frac{7}{4}x$. Find the equation of the hyperbola.

Optional Graphing Calculator Problems

29. For the hyperbola $8x^2 - y^2 = 16$, if $x = 3.5$, what are the two values of y?

30. For the hyperbola $x^2 - 12y^2 = 36$, if $x = 8.2$, what are the two values of y?

Cumulative Review Problems

Factor completely.

31. $12x^2 + x - 6$

32. $8x^2 - 24xy + 18y^2$

Combine.

33. $\dfrac{3}{x^2 - 5x + 6} + \dfrac{2}{x^2 - 4}$

34. $\dfrac{2x}{5x^2 + 9x - 2} - \dfrac{3}{5x - 1}$

▲ **35.** A house 49-feet tall casts a shadow of 14-feet long. At the same time a nearby flagpole casts a 9-foot shadow. How tall is the flagpole?

▲ **36.** A security fence encloses a rectangular area on one side of Fenway Park in Boston. Three sides of fencing are used, since the fourth side of the area is formed by a building. The enclosed area measures 1250 square feet. Exactly 100 feet of fencing is used to fence in three sides of this rectangle. What are the possible dimensions that could have been used to construct this area?

37. In 2000, approximately 2.1 billion pencils were produced in the United States by domestic manufacturers. If you add to this the number of pencils that were imported, then each American used ten pencils during the year. If there were 274 million people in the United States in 2000, how many pencils were imported to the United States in that year? (*Source: U.S. Department of Commerce.*)

38. It is estimated that the number of pencils produced in the United States will increase by 5% from 2000 to 2005. During that time the number of imported pencils will increase by 750 million. If these figures hold true, what percent of all the pencils used in the United States in 2005 will have been imported? (*Source: U.S. Department of Commerce.*)

10.5 Nonlinear Systems of Equations

1 Solving a Nonlinear System by the Substitution Method

Student Learning Objectives

After studying this section, you will be able to:

1 Solve a nonlinear system by the substitution method.

2 Solve a nonlinear system by the addition method.

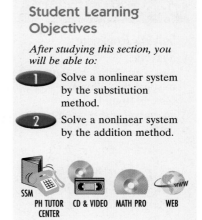

SSM   PH TUTOR CENTER   CD & VIDEO   MATH PRO   WEB

Any equation that is of second degree or higher is a **nonlinear equation**. In other words, the equation is not a straight line (which is what the word *nonlinear* means) and can't be written in the form $y = mx + b$. A **nonlinear system of equations** includes at least one nonlinear equation.

The most frequently used method for solving a nonlinear system is the method of substitution. This method works especially well when one equation of the system is linear. A sketch can often be used to verify the solution(s).

EXAMPLE 1 Solve the following nonlinear system and verify your answer with a sketch.

$$x + y - 1 = 0 \qquad (1)$$
$$y - 1 = x^2 + 2x \qquad (2)$$

We'll use the substitution method.

$$y = -x + 1 \qquad (3) \qquad \text{Solve for } y \text{ in equation (1).}$$
$$(-x + 1) - 1 = x^2 + 2x \qquad \text{Substitute (3) into equation (2).}$$
$$-x + 1 - 1 = x^2 + 2x$$
$$0 = x^2 + 3x \qquad \text{Solve the resulting quadratic equation.}$$
$$0 = x(x + 3)$$
$$x = 0 \quad \text{or} \quad x = -3$$

Now substitute the values for x in the equation $y = -x + 1$.

For $x = -3$: $\quad y = -(-3) + 1 = +3 + 1 = 4$

For $x = 0$: $\quad y = -(0) + 1 = +1 = 1$

Thus, the solutions of the system are $(-3, 4)$ and $(0, 1)$.

To sketch the system, we see that equation (2) describes a parabola. We can rewrite it in the form

$$y = x^2 + 2x + 1 = (x + 1)^2.$$

This is a parabola opening upward with its vertex at $(-1, 0)$. Equation (1) can be written as $y = -x + 1$, which is a straight line with slope $= -1$ and y-intercept $(0, 1)$.

A sketch shows the two graphs intersecting at $(0, 1)$ and $(-3, 4)$. Thus, the solutions are verified.

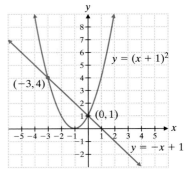

Practice Problem 1 Solve the system.

$$\frac{x^2}{4} - \frac{y^2}{4} = 1$$
$$x + y + 1 = 0$$

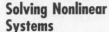

Graphing Calculator

Solving Nonlinear Systems

Use a graphing calculator to solve the following system. Round your answers to the nearest tenth.

$$30x^2 + 256y^2 = 7680$$

$$3x + y - 40 = 0$$

First we will need to obtain the equations

$$y_1 = \frac{\sqrt{7680 - 30x^2}}{16},$$

$$y_2 = -\frac{\sqrt{7680 - 30x^2}}{16},$$

and

$$y_3 = 40 - 3x$$

and graph them to approximate the solutions. Be sure that your window includes enough of the graphs to find the points of intersection.

◗ **EXAMPLE 2** Solve the following nonlinear system and verify your answer with a sketch.

$$y - 2x = 0 \qquad (1)$$

$$\frac{x^2}{4} + \frac{y^2}{9} = 1 \qquad (2)$$

$$y = 2x \qquad (3) \qquad \text{Solve equation (1) for } y.$$

$$\frac{x^2}{4} + \frac{(2x)^2}{9} = 1 \qquad \text{Substitute (3) into equation (2).}$$

$$\frac{x^2}{4} + \frac{4x^2}{9} = 1 \qquad \text{Simplify.}$$

$$36\left(\frac{x^2}{4}\right) + 36\left(\frac{4x^2}{9}\right) = 36(1) \qquad \text{Clear the fractions.}$$

$$9x^2 + 16x^2 = 36$$

$$25x^2 = 36$$

$$x^2 = \frac{36}{25}$$

$$x = \pm\sqrt{\frac{36}{25}}$$

$$x = \pm\frac{6}{5} = \pm 1.2$$

For $x = +1.2$: $\qquad y = 2(1.2) = 2.4$.

For $x = -1.2$: $\qquad y = 2(-1.2) = -2.4$.

Thus, the solutions are $(1.2, 2.4)$ and $(-1.2, -2.4)$.

We recognize $\frac{x^2}{4} + \frac{y^2}{9} = 1$ as an ellipse with center at the origin and vertices $(0, 3)$, $(0, -3)$, $(2, 0)$, and $(-2, 0)$. When we rewrite $y - 2x = 0$ as $y = 2x$, we recognize it as a straight line with slope 2 passing through the origin. The sketch shows that the points of intersection at $(1.2, 2.4)$ and $(-1.2, -2.4)$ seem reasonable.

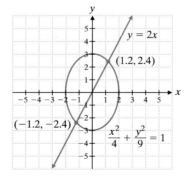

PRACTICE PROBLEM 2

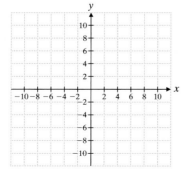

Practice Problem 2 Solve the system. Verify your answer with a sketch.

$$2x - 9 = y$$

$$xy = -4$$

2 *Solving a Nonlinear System by the Addition Method*

Sometimes a system may be solved more readily by adding the equations together. It should be noted that some systems have no solution.

EXAMPLE 3   Solve.

$$4x^2 + y^2 = 1 \quad (1)$$
$$x^2 + 4y^2 = 1 \quad (2)$$

Although we could use the substitution method, it is easier to use the addition method because neither equation is linear.

$$
\begin{aligned}
-16x^2 - 4y^2 &= -4 \\
x^2 + 4y^2 &= 1 \\
\hline
-15x^2 &= -3
\end{aligned}
$$

Multiply equation (1) by -4 and add to equation (2).

$$x^2 = \frac{-3}{-15}$$

$$x^2 = \frac{1}{5}$$

$$x = \pm\sqrt{\frac{1}{5}}$$

If $x = +\sqrt{\dfrac{1}{5}}$, then $x^2 = \dfrac{1}{5}$. Substituting this value into equation (2) gives

$$\frac{1}{5} + 4y^2 = 1$$

$$4y^2 = \frac{4}{5}$$

$$y^2 = \frac{1}{5}$$

$$y = \pm\sqrt{\frac{1}{5}}$$

Similarly, if $x = -\sqrt{\dfrac{1}{5}}$, then $y = \pm\sqrt{\dfrac{1}{5}}$. It is important to determine exactly how many solutions a nonlinear system of equations actually has. In this case, we have four solutions. When x is negative, there are two values for y. When x is positive, there are two values for y. If we rationalize each expression, the four solutions are $\left(\dfrac{\sqrt{5}}{5}, \dfrac{\sqrt{5}}{5}\right)$, $\left(\dfrac{\sqrt{5}}{5}, -\dfrac{\sqrt{5}}{5}\right)$, $\left(-\dfrac{\sqrt{5}}{5}, \dfrac{\sqrt{5}}{5}\right)$, and $\left(-\dfrac{\sqrt{5}}{5}, -\dfrac{\sqrt{5}}{5}\right)$.

Practice Problem 3   Solve the system.

$$
\begin{aligned}
x^2 + y^2 &= 12 \\
3x^2 - 4y^2 &= 8
\end{aligned}
$$

Solve each of the following systems by the substitution method. Graph each equation to verify that the answer seems reasonable.

1. $y^2 = 4x$
   $y = x + 1$

2. $y^2 = 2x$
   $y = -2x + 2$

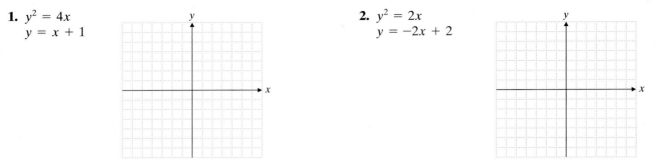

3. $y - 4x = 0$
   $4x^2 + y^2 = 20$

4. $x + 2y = 0$
   $x^2 + 4y^2 = 32$

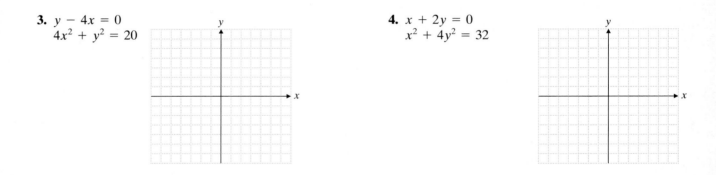

Solve each of the following systems by the substitution method.

5. $\dfrac{x^2}{1} - \dfrac{y^2}{3} = 1$
   $x + y = 1$

6. $y = (x + 3)^2 - 3$
   $2x - y + 2 = 0$

7. $x^2 + y^2 - 25 = 0$
   $3y = x + 5$

8. $x^2 + y^2 - 9 = 0$
   $2y = 3 - x$

9. $x^2 + 2y^2 = 4$

$y = -x + 2$

10. $2x^2 + 3y^2 = 27$

$y = x + 3$

11. $\dfrac{x^2}{4} - \dfrac{y^2}{4} = 1$

$x + y - 4 = 0$

12. $\dfrac{x^2}{3} - \dfrac{y^2}{12} = 1$

$y = -x$

Solve each of the following systems by the addition method. Graph each equation to verify that the answer seems reasonable.

13. $2x^2 - 5y^2 = -2$

$3x^2 + 2y^2 = 35$

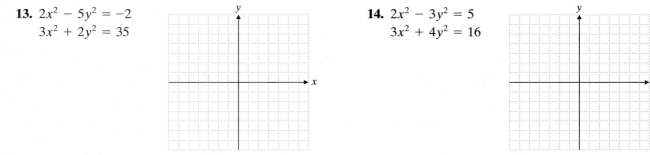

14. $2x^2 - 3y^2 = 5$

$3x^2 + 4y^2 = 16$

Solve each of the following systems by the addition method.

15. $2x^2 + 5y^2 = 42$

$3x^2 + 4y^2 = 35$

16. $x^2 + 2y^2 = 12$

$2x^2 + 3y^2 = 21$

17. $x^2 + 2y^2 = 8$

$x^2 - y^2 = 1$

18. $x^2 + 4y^2 = 13$

$x^2 - 3y^2 = -8$

Mixed Practice

Solve each of the following systems by any appropriate method. If there is no real number solution, so state.

19. $x^2 + y^2 = 7$

$\dfrac{x^2}{3} - \dfrac{y^2}{9} = 1$

20. $x^2 + 2y^2 = 4$

$x^2 + y^2 = 4$

21. $xy = 3$

$3y = 3x + 6$

22. $xy = 5$

$2y = 2x + 8$

23. $xy = 8$
$y = x + 2$

24. $xy = 1$
$3x - y + 2 = 0$

25. $x + y = 5$
$x^2 + y^2 = 4$

26. $x^2 + y^2 = 0$
$x - y = 6$

Applications

▲ **27.** The area of a rectangle is 540 square meters. The diagonal of the rectangle is 39 meters long. Find the dimensions of the rectangle.
Hint: Let x and y represent the length and width and write a system of two nonlinear equations.

28. In an experiment with a laser beam, the path of a particle orbiting a central object is described by the equation $\dfrac{x^2}{49} + \dfrac{y^2}{36} = 1$, where x and y are measured in centimeters from the center of the object. The laser beam follows the path $y = 2x - 6$. Find the coordinates at which the laser will illuminate the particle (that is, when the particle will pass through the beam).

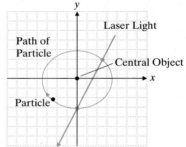

29. The outline of the Earth can be considered a circle with a radius of approximately 4000 miles. Thus, if we say that the center of the Earth is located at $(0, 0)$, then an equation for this circle is $x^2 + y^2 = 16,000,000$. Suppose that an incoming meteor is approaching the Earth in a hyperbolic path. The equation describing the meteor's path is $25,000,000x^2 - 9,000,000y^2 = 2.25 \times 10^{14}$. Will the meteor strike the Earth? Why or why not? If so, locate the point (x, y) where the meteor will strike the Earth. Assume that x and y are both positive. Round your answer to three significant digits.

30. Suppose that a second incoming meteor is approaching the Earth in a hyperbolic path. The equation for this second meteor's path is $16,000,000x^2 - 25,000,000y^2 = 4.0 \times 10^{14}$. Will the second meteor strike the Earth? Why or why not? If so, locate the point (x, y) where the meteor will strike the Earth. Assume that x and y are both positive. Round your answer to three significant digits.

Cumulative Review Problems

Simplify.

31. $\dfrac{6x^4 - 24x^3 - 30x^2}{3x^3 - 21x^2 + 30x}$

Divide.

32. $(3x^3 - 8x^2 - 33x - 10) \div (3x + 1)$

33. Highway patrol officers can trap speeders by various methods. Between two certain points on a back country road in a small town in Georgia, a speeding Audi travels 55 miles per hour for 5 seconds. What is the legal speed limit if at this speed, a car driving between those two points requires 11 seconds?

34. Ricardo is the staff accountant for a CD-ROM factory. He has determined that his monthly profit factor is given by the expression $11.5n - 290,000$. Here n represents the number of CD-ROMs manufactured each month. The profit this month was $1,187,750. How many CD-ROMs were produced this month?

Putting Your Skills to Work

Locating the Center of an Earthquake Mathematically

When an earthquake occurs, it is most important for scientists to locate its center. This location is known as the epicenter. Using a seismograph, scientists can calculate the distance from a recording station to the epicenter of the earthquake. If two recording stations at different locations measure an earthquake, scientists can use this data to narrow the possible locations of the epicenter of two places.

Now let us suppose that there is a scientific recording station for earthquake activity at location A and another at location B. Scientists read their instruments and determine that the epicenter of an earthquake is a miles from station A and b miles from station B. The two possible locations for the epicenter are at L and M on the diagram.

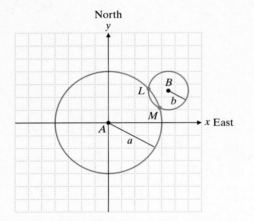

Problems for Individual Investigation and Analysis

1. Scientists determine that the epicenter of a strong earthquake is 40 miles away from station A. Station B is located 50 miles east and 30 miles north of station A. The epicenter is located 20 miles from station B. If station A is located at $(0, 0)$, determine the locations of L and M to the nearest tenth of a mile. Express your answers in terms of how many miles north and how many miles east of station A the possible locations of the epicenter are.

2. A second earthquake is recorded. This time a new recording station B is selected to make measurements. Station B is located 60 miles east and 20 miles north of station A, which is located at $(0, 0)$. The epicenter is located 50 miles from station A and 30 miles from station B. Determine the locations of L and M to the nearest tenth of a mile. Express your answers in terms of how many miles north and how many miles east of station A the possible locations of the epicenter are.

Problems for Group Investigation and Cooperative Learning

3. Station A (located at $(0, 0)$) determines that the epicenter of a strong earthquake is 30 miles away. Station B, located 40 miles east and 30 miles north of station A, determines that the epicenter is located 25 miles away. Station C, 30 miles east and 20 miles south of station A, determines that the epicenter is 27.55 miles away. Determine the location of the epicenter to the nearest tenth of a mile. Express your answer in terms of how many miles north, south, east, and/or west from station A the epicenter is.

4. During the night a stronger earthquake is recorded. Stations A and B from Exercise 3 record the quake. Station C is not operational. A new station D, which is 10 miles east and 15 miles north of station A, is used. This time the epicenter of the quake is recorded as 50 miles from station A, 40 miles from station B, and 45.92 miles from station D. Determine the location of the epicenter to the nearest tenth of a mile. Express your answer in terms of how many miles north, south, east, and/or west from station A the epicenter is.

Internet Connections

Netsite: http://www.prenhall.com/tobey_beg_int

Site: Earthquake Information of USGS

At this site you will find the exact locations of the epicenters of recent earthquakes, measured not only in terms of longitude and latitude but also in terms of depth beneath the Earth's surface. More than 4000 earthquake-measuring stations across the world are now available to record these events.

5. Find the number of major earthquakes (magnitude 7.0 or greater) that have occurred each year for the years 1960 to 1985. Construct a graph where x is the year and y is the number of major earthquakes recorded around the world for the years 1970 to 1985. Is the graph linear or nonlinear? Why? Did the number of major earthquakes around the world increase or decrease during this period?

6. Find the number of major earthquakes (magnitude 7.0 or greater) that have occurred each year for the years 1985 to 2000. Construct a graph where x is the year and y is the number of major earthquakes recorded around the world for the years 1985 to 2000. Is the graph linear or nonlinear? Why? Did the number of major earthquakes around the world increase or decrease during this period?

Math in the Media

Flexible Seating

In the late edition of *The New York Times* September 1999 article "Currents: Stadium; A Dome Where the Seats Move," it was reported that the Saitama Dome, north of Tokyo, will have a section of 10,000 seats, plus accompanying plumbing, lighting, concessions, and ceiling sections, that move on trucklike devices called "bogeys."

The dome will be able to seat 30,000 as a stadium, or 20,000 as an arena, or seat as few as 5,000.

Apply your knowledge from the chapter by taking the role of a design team member and answering the questions that follow.

EXERCISES

Suppose you are the member of the design team for a new stadium with moveable seating. You are doodling on graph paper. You draw a circle with center at (0, 0) and radius R. You want to "stretch" it into an ellipse with a center at (0, 0), with a distance R from the center to the y-intercepts, and with a distance R + M from the center to the x-intercepts. You plot the circle and the ellipse on the same axes and label the appropriate segments as R and M. You know that the area of the circle is $A_{circle} = [\pi]R^2$ and the area of the ellipse is $A_{ellipse} = [\pi]R(R + M)$.

1. If you wanted the ellipse to have 3 times the area of the circle, what would the distance M have to be, expressed in terms of R?

2. If you wanted the ellipse to have F times the area of the circle, what would the distance M be, expressed in terms of R?

3. If you drew a square around the circle (touching the circle at four points) and a rectangle around the ellipse (touching the ellipse at four points), how much more area would the rectangle contain than the square, expressed in terms of M and R?

Chapter 10 Organizer

Topic	Procedure	Examples
Distance between two points, p. 567.	The distance d between points (x_1, y_1) and (x_2, y_2) is $$d = \sqrt{(x_2 - x_1)^2 + (y_2 - y_1)^2}.$$	Find the distance between $(-6, -3)$ and $(5, -2)$. $$\begin{aligned} d &= \sqrt{[5 - (-6)]^2 + [-2 - (-3)]^2} \\ &= \sqrt{(5 + 6)^2 + (-2 + 3)^2} \\ &= \sqrt{121 + 1} \\ &= \sqrt{122} \end{aligned}$$
Standard form of the equation of a circle, p. 568.	The standard form of the equation of a circle with center at (h, k) and radius r is $$(x - h)^2 + (y - k)^2 = r^2.$$	Graph $(x - 3)^2 + (y + 4)^2 = 16$. Center at $(h, k) = (3, -4)$. Radius $= 4$.
Standard form of the equation of a vertical parabola, p. 575.	The equation of a vertical parabola with its vertex at (h, k) can be written in the form $y = a(x - h)^2 + k$. It opens upward if $a > 0$ and downward if $a < 0$.	Graph $y = \dfrac{1}{2}(x - 3)^2 + 5$. $a = \dfrac{1}{2}$, so parabola opens upward. Vertex at $(h, k) = (3, 5)$. If $x = 0$, $y = 9.5$.
Standard form of the equation of a horizontal parabola, p. 577.	The equation of a horizontal parabola with its vertex at (h, k) can be written in the form $x = a(y - k)^2 + h$. It opens to the right if $a > 0$ and to the left if $a < 0$.	Graph $x = \dfrac{1}{3}(y + 2)^2 - 4$. $a = \dfrac{1}{3}$, so parabola opens to the right. Vertex at $(h, k) = (-4, -2)$. If $x = 0$, $y = -2 - 2\sqrt{3} \approx -5.5$ and $y = -2 + 2\sqrt{3} \approx 1.5$.

Topic	Procedure	Examples
Standard form of the equation of an ellipse with center at (0, 0), p. 585.	An ellipse with center at the origin has the equation $$\frac{x^2}{a^2} + \frac{y^2}{b^2} = 1,$$ where $a > 0$ and $b > 0$.	Graph $\frac{x^2}{16} + \frac{y^2}{4} = 1$. $a^2 = 16$, $a = 4$; $b^2 = 4$, $b = 2$
Standard form of an ellipse with center at (h, k), p. 587.	An ellipse with center at (h, k) has the equation $$\frac{(x - h)^2}{a^2} + \frac{(y - k)^2}{b^2} = 1,$$ where $a > 0$ and $b > 0$.	Graph $\frac{(x + 2)^2}{9} + \frac{(y + 4)^2}{25} = 1$. $(h, k) = (-2, -4)$; $a = 3$, $b = 5$
Standard form of a horizontal hyperbola with center at (0, 0), p. 593.	Let a and b be positive real numbers. A horizontal hyperbola with center at the origin and vertices $(a, 0)$ and $(-a, 0)$ has the equation $$\frac{x^2}{a^2} - \frac{y^2}{b^2} = 1$$ and asymptotes $$y = \pm \frac{b}{a}x.$$	Graph $\frac{x^2}{25} - \frac{y^2}{9} = 1$. $a = 5$, $b = 3$
Standard form of a vertical hyperbola with center at (0, 0), p. 593.	Let a and b be positive real numbers. A vertical hyperbola with center at the origin and vertices $(0, b)$ and $(0, -b)$ has the equation $$\frac{y^2}{b^2} - \frac{x^2}{a^2} = 1$$ and asymptotes $$y = \pm \frac{b}{a}x.$$	Graph $\frac{y^2}{9} - \frac{x^2}{4} = 1$. $b = 3$, $a = 2$

Topic	Procedure	Examples
Standard form of a horizontal hyperbola with center at (h, k), p. 595.	Let a and b be positive real numbers. A horizontal hyperbola with center at (h, k) and vertices $(h - a, k)$ and $(h + a, k)$ has the equation $$\frac{(x - h)^2}{a^2} - \frac{(y - k)^2}{b^2} = 1.$$	Graph $\dfrac{(x - 2)^2}{4} - \dfrac{(y - 3)^2}{25} = 1.$ Center at $(2, 3)$; $a = 2$, $b = 5$
Standard form of a vertical hyperbola with center at (h, k), p. 595.	Let a and b be positive real numbers. A vertical hyperbola with center at (h, k) and vertices $(h, k + b)$ and $(h, k - b)$ has the equation $$\frac{(y - k)^2}{b^2} - \frac{(x - h)^2}{a^2} = 1.$$	Graph $\dfrac{(y - 5)^2}{9} - \dfrac{(x - 4)^2}{4} = 1.$ Center at $(4, 5)$; $b = 3$, $a = 2$
Nonlinear systems of equations, p. 601.	We can solve a nonlinear system by the substitution method or the addition method. In the addition method, we multiply one or more equations by a numerical value and then add them together so that one variable is eliminated. In the substitution method we solve one equation for one variable and substitute that expression into the other equation.	Solve by substitution. $$2x^2 + y^2 = 18$$ $$xy = 4$$ Solving the second equation for y, we have $y = \dfrac{4}{x}$. $$2x^2 + \left(\frac{4}{x}\right)^2 = 18$$ $$2x^2 + \frac{16}{x^2} = 18$$ $$2x^4 + 16 = 18x^2$$ $$2x^4 - 18x^2 + 16 = 0$$ $$x^4 - 9x^2 + 8 = 0$$ $$(x^2 - 1)(x^2 - 8) = 0$$ $x^2 - 1 = 0 \qquad x^2 - 8 = 0$ $x^2 = 1 \qquad\quad x^2 = 8$ $x = \pm 1 \qquad\quad x = \pm 2\sqrt{2}$ Since $xy = 4$, if $x = 1$, then $y = 4$. if $x = -1$, then $y = -4$. if $x = 2\sqrt{2}$, then $y = \sqrt{2}$. if $x = -2\sqrt{2}$, then $y = -\sqrt{2}$. The solutions are $(1, 4)$, $(-1, -4)$, $(2\sqrt{2}, \sqrt{2})$, and $(-2\sqrt{2}, -\sqrt{2})$.

Chapter 10 Review Problems

In Exercises 1 and 2, find the distance between the points.

1. $(10.5, -6)$ and $(7.5, -4)$

2. $(-7, 3)$ and $(-2, -1)$

3. Write in standard form the equation of a circle with center at $(-6, 3)$ and radius $\sqrt{15}$.

4. Write in standard form the equation of a circle with center at $(0, -7)$ and radius 5.

Rewrite each equation in standard form. Find the center and the radius of each circle.

5. $x^2 + y^2 - 6x - 8y + 3 = 0$

6. $x^2 + y^2 - 10x + 12y + 52 = 0$

Graph each parabola. Label its vertex and plot at least one intercept.

7. $x = \dfrac{1}{3} y^2$

8. $x = \dfrac{1}{2} (y - 2)^2 + 4$

9. $y = -2(x + 1)^2$

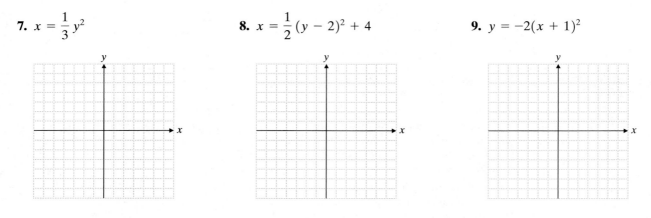

Rewrite each equation in standard form. Find the vertex and determine in which direction the parabola opens.

10. $x + 8y = y^2 + 10$

11. $x^2 + 6x = y - 4$

Graph each ellipse. Label its center and four other points.

12. $\dfrac{x^2}{\frac{1}{4}} + \dfrac{y^2}{1} = 1$

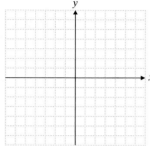

13. $16x^2 + y^2 - 32 = 0$

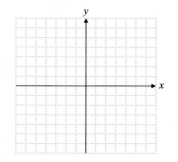

Determine the vertices and the center of each ellipse.

14. $\dfrac{(x + 5)^2}{4} + \dfrac{(y + 3)^2}{25} = 1$

15. $\dfrac{(x + 1)^2}{9} + \dfrac{(y - 2)^2}{16} = 1$

Find the center and vertices of each hyperbola and graph it.

16. $x^2 - 4y^2 - 16 = 0$

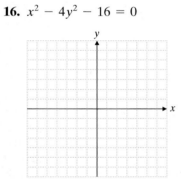

17. $9y^2 - 25x^2 = 225$

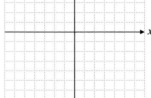

Determine the vertices and the center of each hyperbola.

18. $\dfrac{(x - 2)^2}{4} - \dfrac{(y + 3)^2}{25} = 1$

19. $9(y - 2)^2 - (x + 5)^2 - 9 = 0$

Solve each nonlinear system. If there is no real number solution, so state.

20. $x^2 + y = 9$
$\quad y - x = 3$

21. $y^2 + x^2 = 3$
$\quad x - 2y = 1$

22. $2x^2 + y^2 = 17$
$\quad x^2 + 2y^2 = 22$

23. $\quad xy = -2$
$\quad x^2 + y^2 = 5$

24. $3x^2 - 4y^2 = 12$
$\quad 7x^2 - y^2 = 8$

25. $y = x^2 + 1$
$\quad x^2 + y^2 - 8y + 7 = 0$

26. $2x^2 + y^2 = 18$
$\quad xy = 4$

27. $\quad y^2 - 2x^2 = 2$
$\quad 2y^2 - 3x^2 = 5$

28. $y^2 = \dfrac{1}{2}x$
$\quad y = x - 1$

29. $y^2 = 2x$
$\quad y = \dfrac{1}{2}x + 1$

Applications

30. The side view of a satellite dish on Jason and Wendy's house is shaped like a parabola. The signals that come from the satellite hit the surface of the dish and are then reflected to the point where the signal receiver is located. This point is the focus of the parabolic dish. The dish is 10 feet across at its opening and 4 feet deep at its center. How far from the center of the dish should the signal receiver be placed? Round your answer to the nearest hundredth.

31. The side view of an airport searchlight is shaped like a parabola. The center of the light source of the searchlight is located 2 feet from the base along the axis of symmetry, and the opening is 5 feet across. How deep should the searchlight be? Round your answer to the nearest hundredth.

1. Find the distance between $(-6, -8)$ and $(-2, 5)$.

Rewrite the equation in standard form. Find the center or vertex, plot at least one other point, identify the conic, and sketch the curve.

2. $y^2 - 6y - x + 13 = 0.$ **3.** $x^2 + y^2 + 6x - 4y + 9 = 0$

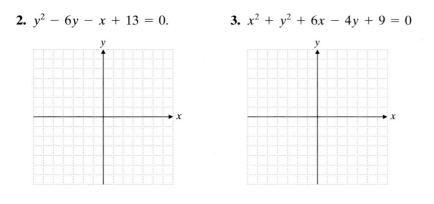

Identify and graph each conic section. Label the center and/or vertex as appropriate.

4. $\dfrac{x^2}{25} + \dfrac{y^2}{1} = 1$ **5.** $\dfrac{x^2}{10} - \dfrac{y^2}{9} = 1$

6. $y = -2(x + 3)^2 + 4$ **7.** $\dfrac{(x + 2)^2}{16} + \dfrac{(y - 5)^2}{4} = 1$

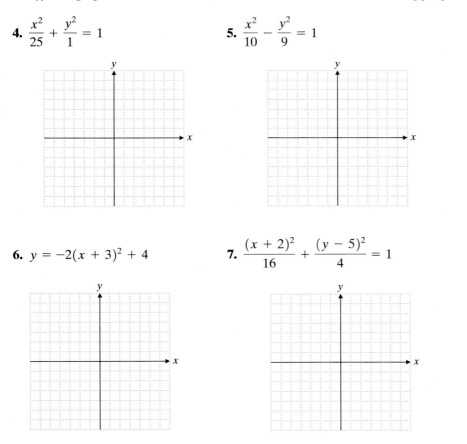

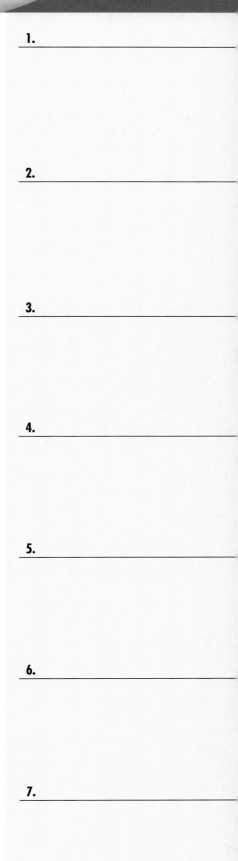

1. _____

2. _____

3. _____

4. _____

5. _____

6. _____

7. _____

8. $7y^2 - 7x^2 = 28$

Find the standard form of the equation of each of the following:

9. Circle of radius $\sqrt{8}$ with its center at $(3, -5)$

10. Ellipse with its center at $(-4, -2)$ and vertices at $(-4, 1)$, $(-4, -5)$, $(-5, -2)$, and $(-3, -2)$

11. Parabola with its vertex at $(-7, 3)$ and that opens to the right. This parabola crosses the x-axis at $(2, 0)$. It is of the form $x = (y - k)^2 + h$.

12. Hyperbola with its center at $(6, 7)$ and that is vertical. This hyperbola has vertices of $(6, 14)$ and $(6, 0)$. The value of a for this hyperbola is 3.

Solve each nonlinear system.

13. $-2x + y = 5$
   $x^2 + y^2 - 25 = 0$

14. $x^2 + y^2 = 9$
   $y = x - 3$

15. $4x^2 + y^2 - 4 = 0$
   $9x^2 - 4y^2 - 9 = 0$

16. $2x^2 + y^2 = 9$
   $xy = -3$

Approximately one-half of this test covers the content of Chapters 1–9. The remainder covers the content of Chapter 10.

1. Simplify: $2\{x - 3[x - 2(x + 1)]\}$

2. Evaluate: $3(4 - 6)^3 + \sqrt{25}$

3. Solve for p: $A = 3bt + prt$

4. Find the slope of the line passing through $\left(-\frac{1}{2}, 2\right)$ and $\left(-\frac{1}{2}, 4\right)$.

5. Factor: $25x^2 - 40xy + 16y^2$

6. Add: $\dfrac{3}{x - 4} + \dfrac{6}{x^2 - 16}$

7. Solve for x.

$$\frac{3}{2x + 3} = \frac{1}{2x - 3} + \frac{2}{4x^2 - 9}$$

8. Solve for (x, y, z).

$$3x - 2y - 9z = 9$$
$$x - y + z = 8$$
$$2x + 3y - z = -2$$

9. Multiply and simplify. $\left(\sqrt{2} + \sqrt{3}\right)\left(2\sqrt{6} - \sqrt{3}\right)$

10. Simplify: $\sqrt{8x} + 3x\sqrt{50} - 4x\sqrt{32}$

Solve the following inequalities.

11. $2x + (4x - 1) > 6 - x$

12. $\dfrac{6(x - 4)}{5} \geq \dfrac{3(x + 2)}{4}$

13. Find the distance between $(6, -1)$ and $(-3, -4)$.

1. _____

2. _____

3. _____

4. _____

5. _____

6. _____

7. _____

8. _____

9. _____

10. _____

11. _____

12. _____

13. _____

14. _____

15. _____

16. _____

17. _____

18. _____

Identify and graph each equation.

14. $y = -\dfrac{1}{2}(x + 2)^2 - 3$

15. $25x^2 + 25y^2 = 125$

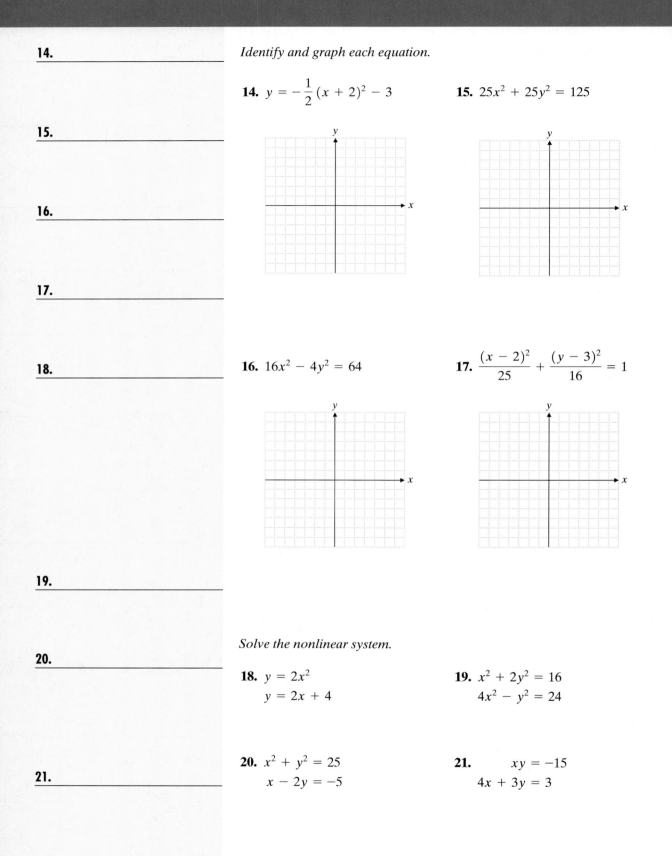

16. $16x^2 - 4y^2 = 64$

17. $\dfrac{(x - 2)^2}{25} + \dfrac{(y - 3)^2}{16} = 1$

19. _____

20. _____

21. _____

Solve the nonlinear system.

18. $y = 2x^2$
$y = 2x + 4$

19. $x^2 + 2y^2 = 16$
$4x^2 - y^2 = 24$

20. $x^2 + y^2 = 25$
$x - 2y = -5$

21. $xy = -15$
$4x + 3y = 3$

An Advanced Look at Functions

Polynomial functions are sometimes used to model or predict events or values in science and in business. For example, the number of cubic feet of lumber produced by a sawmill can be modeled with a third-degree polynomial equation. How accurate are such models? Could you use your knowledge of polynomials and functions to make accurate predictions based on such models? Turn to the Putting Your Skills to Work exercises on page 660 to find out.

1. (a) _____

(b) _____

(c) _____

(d) _____

2. (a) _____

(b) _____

(c) _____

3. (a) _____

(b) _____

4. _____

5. _____

6. _____

7. _____

If you are familiar with the topics in this chapter, take this test now. Check your answers with those in the back of the book. If an answer is wrong or you can't do an exercise, study the appropriate section of the chapter.

If you are not familiar with the topics in this chapter, don't take this test now. Instead, study the examples, work the practice exercises, and then take the test.

This test will help you identify those concepts that you have mastered and those that need more study.

Section 11.1

1. For the function $f(x) = 2x - 6$, find the following:
   (a) $f(-3)$
   (b) $f(a)$
   (c) $f(2a)$
   (d) $f(a + 2)$

2. For $f(x) = 5x^2 + 2x - 3$, find the following:
   (a) $f(-2)$
   (b) $f(a)$
   (c) $f(a + 1)$

3. For $f(x) = \dfrac{3x}{x + 2}$, find the following:

   (a) $f(a) + f(a - 2)$. Express your answer as one fraction.
   (b) $f(3a) - f(3)$. Express your answer as one fraction.

Section 11.2

Which of these graphs represent functions?

4.

5.

Graph the given functions on one coordinate plane.

6. $f(x) = |x|$ and $s(x) = |x - 3|$.

7. $f(x) = x^2$ and $h(x) = (x + 2)^2 + 3$.

Section 11.3

8. If $f(x) = \dfrac{2}{x + 6}$ and $g(x) = -3x + 1$, find the following:

   (a) $(fg)(x)$ **(b)** $(fg)(-4)$ **(c)** $f[g(x)]$

9. If $f(x) = 3x - 4$ and $g(x) = -2x^3 - 6x + 3$, find the following:

   (a) $(f + g)(x)$ **(b)** $(f + g)(2)$ **(c)** $f[g(x)]$

10. If $f(x) = 6x^2 - 5x - 4$ and $g(x) = 3x - 4$, find the following:

   (a) $\left(\dfrac{f}{g}\right)(x)$ **(b)** $\left(\dfrac{f}{g}\right)(-1)$ **(c)** $(f \circ g)(x)$ **(d)** $(g \circ f)(x)$

Section 11.4

Which graphs represent one-to-one functions?

11.

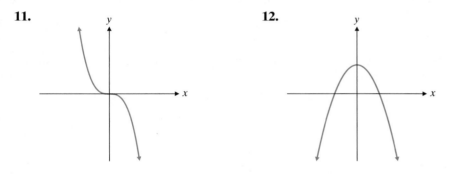

12.

13. Is function A a one-to-one function?

$$A = \{(-5, -3), (5, 3), (2, -1), (-2, 1)\}$$

14. Determine the inverse of the function $F = \{(7, 1), (6, 3), (2, -1), (-1, 5)\}$.

15. Find the inverse of $g(x) = 3 - 5x$ and graph g and its inverse on the same set of axes.

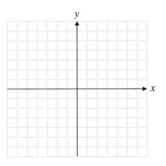

---

8. (a) _____

(b) _____

(c) _____

9. (a) _____

(b) _____

(c) _____

10. (a) _____

(b) _____

(c) _____

(d) _____

11. _____

12. _____

13. _____

14. _____

15. _____

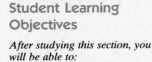

11.1 Function Notation

1 *Using Function Notation to Evaluate Expressions*

In Section 3.6, we studied the basic concepts of functions and function notation. We are now ready to examine functions more closely. Function notation is useful in solving a number of interesting exercises. Suppose you wanted to skydive from an airplane. Your instructor tells you that you must wait 20 seconds before you pull the cord to open the parachute. How far will you fall in that time?

The approximate distance that an object in free fall travels when there is no initial downward velocity is given by the distance function $d(t) = 16t^2$, where time t is measured in seconds and distance $d(t)$ is measured in feet. (Neglect air resistance.)

How far will a person in free fall travel in 20 seconds if he or she leaves the airplane with no downward velocity? We want to find the distance $d(20)$ where $t = 20$ seconds. To find the distance, we substitute 20 for t in the function $d(t) = 16t^2$.

$$d(20) = 16(20)^2 = 16(400) = 6400$$

Thus, a person in free fall will travel approximately 6400 feet in 20 seconds.

Now suppose that person waits longer than he or she should to pull the parachute cord. How will this affect the distance he or she falls? Suppose the person falls for e seconds beyond the 20-second mark before pulling the parachute cord.

$$\begin{aligned} d(20 + e) &= 16(20 + e)^2 \\ &= 16(20 + e)(20 + e) \\ &= 16(400 + 40e + e^2) \\ &= 6400 + 640e + 16e^2 \end{aligned}$$

Thus, if the person waited 5 seconds too long before pulling the cord, he or she would fall the following distance.

$$\begin{aligned} d(20 + 5) &= 6400 + 640(5) + 16(5)^2 \\ &= 6400 + 3200 + 16(25) \\ &= 6400 + 3200 + 400 \\ &= 10{,}000 \text{ feet} \end{aligned}$$

This is 3600 feet farther than the person would have fallen in 20 seconds. Obviously, a delay of 5 seconds could have life or death consequences.

We now revisit a topic that we first discussed in Section 3.6, evaluating a function for particular values of the variable.

EXAMPLE 1 If $g(x) = 5 - 3x$, find the following:

(a) $g(a)$ **(b)** $g(a + 3)$ **(c)** $g(a) + g(3)$

(a) $g(a) = 5 - 3a$

(b) $g(a + 3) = 5 - 3(a + 3) = 5 - 3a - 9 = -4 - 3a$

(c) This exercise requires us to find each addend separately. Then we add them together.
$g(a) = 5 - 3a$
$g(3) = 5 - 3(3) = 5 - 9 = -4$
Thus, $g(a) + g(3) = (5 - 3a) + (-4)$
$= 5 - 3a - 4$
$= 1 - 3a$
Notice that $g(a + 3) \neq g(a) + g(3)$.

Practice Problem 1 If $g(x) = \dfrac{1}{2}x - 3$, find the following:

(a) $g(a)$ **(b)** $g(a + 4)$ **(c)** $g(a) + g(4)$

To Think About Is $g(a + 4) = g(a) + g(4)$? Why or why not?

● **EXAMPLE 2**   If $p(x) = 2x^2 - 3x + 5$, find the following:

(a) $p(-2)$ **(b)** $p(a)$ **(c)** $p(3a)$ **(d)** $p(a - 2)$

(a) $p(-2) = 2(-2)^2 - 3(-2) + 5$
$= 2(4) - 3(-2) + 5$
$= 8 + 6 + 5$
$= 19$

(b) $p(a) = 2(a)^2 - 3(a) + 5 = 2a^2 - 3a + 5$

(c) $p(3a) = 2(3a)^2 - 3(3a) + 5$
$= 2(9a^2) - 3(3a) + 5$
$= 18a^2 - 9a + 5$

(d) $p(a - 2) = 2(a - 2)^2 - 3(a - 2) + 5$
$= 2(a - 2)(a - 2) - 3(a - 2) + 5$
$= 2(a^2 - 4a + 4) - 3(a - 2) + 5$
$= 2a^2 - 8a + 8 - 3a + 6 + 5$
$= 2a^2 - 11a + 19$

Practice Problem 2   If $p(x) = -3x^2 + 2x + 4$, find the following:

(a) $p(-3)$ **(b)** $p(a)$ **(c)** $p(2a)$ **(d)** $p(a - 3)$ ●

● **EXAMPLE 3**   If $r(x) = \dfrac{4}{x + 2}$, find   **(a)** $r(a + 3)$,   **(b)** $r(a)$, and
(c) $r(a + 3) - r(a)$. Express the last result as one fraction.

(a) $r(a + 3) = \dfrac{4}{a + 3 + 2} = \dfrac{4}{a + 5}$ **(b)** $r(a) = \dfrac{4}{a + 2}$

(c) $r(a + 3) - r(a) = \dfrac{4}{a + 5} - \dfrac{4}{a + 2}$

To express this as one fraction, we note that the LCD $= (a + 5)(a + 2)$.

$$r(a + 3) - r(a) = \frac{4(a + 2)}{(a + 5)(a + 2)} - \frac{4(a + 5)}{(a + 2)(a + 5)} = \frac{4a + 8}{(a + 5)(a + 2)} - \frac{4a + 20}{(a + 5)(a + 2)}$$

$$= \frac{4a - 4a + 8 - 20}{(a + 5)(a + 2)} = \frac{-12}{(a + 5)(a + 2)}$$

Practice Problem 3   If $r(x) = \dfrac{-3}{x + 1}$, find   **(a)** $r(a + 2)$,   **(b)** $r(a)$, and
(c) $r(a + 2) - r(a)$. Express the last result as one fraction.   ●

● **EXAMPLE 4**   Suppose that $f(x) = 3x - 7$. Find $\dfrac{f(x + h) - f(x)}{h}$.

First
$$f(x + h) = 3(x + h) - 7 = 3x + 3h - 7$$

and
$$f(x) = 3x - 7.$$

So
$$f(x + h) - f(x) = (3x + 3h - 7) - (3x - 7)$$
$$= 3x + 3h - 7 - 3x + 7$$
$$= 3h.$$

Therefore, $\dfrac{f(x + h) - f(x)}{h} = \dfrac{3h}{h} = 3.$

Practice Problem 4 Suppose that $g(x) = 2 - 5x$. Find $\dfrac{g(x + h) - g(x)}{h}$.

2 Using Function Notation to Solve Application Exercises

▲ **EXAMPLE 5** The surface area of a sphere is given by $S = 4\pi r^2$ where r is the radius. If we use $\pi = 3.14$ as an approximation, this becomes $S = 4(3.14)r^2$, or $S = 12.56r^2$.

(a) Write the surface area of a sphere as a function of radius r.

(b) Find the surface area of a sphere with a radius of 3 centimeters.

(c) Suppose that an error is made and the radius is calculated to be $(3 + e)$ centimeters. Find an expression for the surface area as a function of the error e.

(d) Evaluate the surface area for $r = (3 + e)$ centimeters when $e = 0.2$. Round your answer to the nearest hundredth of a centimeter. What is the difference in the surface area due to the error in measurement?

(a) $S(r) = 12.56r^2$

(b) $S(3) = 12.56(3)^2 = (12.56)(9) = 113.04$ square centimeters

(c) $S(e) = 12.56(3 + e)^2$
$\qquad = 12.56(3 + e)(3 + e)$
$\qquad = 12.56(9 + 6e + e^2)$
$\qquad = 113.04 + 75.36e + 12.56e^2$

(d) If an error in measure is made so that the radius is calculated to be $r = (3 + e)$ centimeters, where $e = 0.2$, we can use the function generated in part **(c)**.

$$S = 113.04 + 75.36e + 12.56e^2$$
$$S = 113.04 + 75.36(0.2) + 12.56(0.2)^2$$
$$S = 113.04 + 15.072 + 0.5024$$
$$S = 128.6144$$

Rounding, we have $S = 128.61$ square centimeters.

Thus, if the radius of 3 centimeters was incorrectly calculated as 3.2 centimeters, the surface area would be approximately $128.61 - 113.04 = 15.57$ square centimeters too large.

▲ **Practice Problem 5** The surface area of a cylinder of height 8 meters and radius r is given by $S = 16\pi r + 2\pi r^2$.

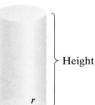

Height

(a) Write the surface area of a cylinder of height 8 meters (using $\pi = 3.14$) and radius r as a function of r.

(b) Find the surface area if the radius is 2 meters.

(c) Suppose that an error is made and the radius is calculated to be $(2 + e)$ meters. Find an expression for the surface area as a function of the error e.

(d) Evaluate the surface area for $r = (2 + e)$ meters when $e = 0.3$. Round your answer to the nearest hundredth of a meter. What is the difference in the surface area due to the error in measurement?

For the function $f(x) = 3x - 5$, find the following.

1. $f\left(-\dfrac{2}{3}\right)$

2. $f(1.5)$

3. $f(a - 4)$

4. $f(b + 3)$

For the function $g(x) = \frac{1}{2}x - 3$, find the following.

5. $g(4) + g(a)$

6. $g(6) + g(b)$

7. $g(2a)$

8. $g(8b)$

9. $g(2a - 4)$

10. $g(3a + 1)$

11. $g(a^2) - g\left(\dfrac{2}{5}\right)$

12. $g(b^2) - g\left(\dfrac{4}{3}\right)$

If $p(x) = 3x^2 + 4x - 2$, find the following.

13. $p(-2)$

14. $p(-3)$

15. $p\left(\dfrac{1}{2}\right)$

16. $p(2.5)$

17. $p(a + 1)$

18. $p(b - 1)$

19. $p\left(-\dfrac{a^2}{2}\right)$

20. $p\left(-\dfrac{b^2}{3}\right)$

If $h(x) = \sqrt{x + 5}$, find the following.

21. $h(-1)$

22. $h(-4)$

23. $h(3)$

24. $h(23)$

25. $h(a^2 - 1)$

26. $h(a^2 + 4)$

27. $h(3a)$

28. $h(5a)$

29. $h(4a - 1)$

30. $h(4a + 3)$

31. $h(b^2 + b)$

32. $h(b^2 + b - 5)$

If $r(x) = \dfrac{7}{x - 3}$, find the following and write your answers as one fraction.

33. $r(7)$

34. $r(-4)$

35. $r(1.5)$

36. $r(0.5)$

37. $r(a^2)$

38. $r(3b^2)$

39. $r(a + 2)$

40. $r(a - 3)$

41. $r\left(\dfrac{1}{2}\right) + r(8)$

42. $r\left(\dfrac{5}{3}\right) + r(7)$

Find $\dfrac{f(x + h) - f(x)}{h}$ *for the following functions.*

43. $f(x) = 5 - 2x$

44. $f(x) = 2x - 3$

45. $f(x) = 2x^2$

46. $f(x) = x^2 - x$

Applications

47. A turbine wind generator produces P kilowatts of power for wind speed w (measured in miles per hour) according to the equation $P = 2.5w^2$.
  (a) Write the number of kilowatts P as a function of w.
  (b) Find the power in kilowatts when the wind speed is $w = 20$ miles per hour.
  (c) Suppose that an error is made and the speed of the wind is calculated to be $(20 + e)$ miles per hour. Find an expression for the power as a function of error e.
  (d) Evaluate the power for $w = (20 + e)$ miles per hour when $e = 2$.

▲ 48. The area of a circle is $A = \pi r^2$.
  (a) Write the area of a circle as the function of the radius r. Use $\pi = 3.14$.
  (b) Find the area of a circle with a radius of 4.0 feet.
  (c) Suppose that an error is made and the radius is calculated to be $(4 + e)$ feet. Find an expression for the area as a function of error e.
  (d) Evaluate the area for $r = (4 + e)$ feet when $e = 0.4$. Round your answer to the nearest hundredth.

Because of the elimination of lead in automobile gasoline and increased use of emission controls in automobiles and industrial operations, the amount of lead in the air in the United States has shown a marked decrease. The percent of lead in the air $p(x)$ expressed in terms of 1984 levels is given in the line graph. The variable x indicates the number of years since 1984. The function value $p(x)$ indicates the amount of lead that remains in the air in selected regions of the United States, expressed as a percent of the amount of lead in the air in 1984.

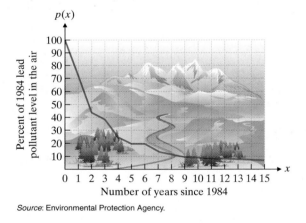

Source: Environmental Protection Agency.

49. If a new function were defined as $p(x) - 13$, what would happen to the function values associated with x? Find $p(3) - 13$.

50. If a new function were defined as $p(x + 2)$, what would happen to the function values associated with x? Find $p(x + 2)$ when $x = 4$.

If $f(x) = 3x^2 - 4.6x + 1.23$, find each of the following functions to the nearest thousandth.

51. $f(3.56a)$

52. $f(0.026a)$

53. $f(a - 0.152)$

54. $f(a + 2.23)$

▲ **55.** A rope 20 feet long is cut into two unequal pieces. Each piece is used to form a square. Write a function $A(x)$ that expresses the total area enclosed by the two squares. Assume that the shorter piece of rope is x feet long. Evaluate $A(2)$, $A(5)$, and $A(8)$.

▲ **56.** Assume that the smaller piece of rope in Exercise 55 is used to form a circle and the longer piece is used to form a square. Write a function $A(x)$ that expresses the total area enclosed by the circle and the square. Evaluate $A(3)$ and $A(9)$. Use $\pi = 3.14$. Round your answers to the nearest hundredth.

Cumulative Review Problems

Solve for x.

57. $\dfrac{7}{6} + \dfrac{5}{x} = \dfrac{3}{2x}$

58. $\dfrac{1}{6} - \dfrac{2}{3x + 6} = \dfrac{1}{2x + 4}$

▲ **59.** The diameter of Mercury is 3031 miles while that of Earth is 7927 miles. How many times greater is the volume of Earth compared to the volume of Mercury?

▲ **60.** The radius of Uranus is 14,584 miles. The radius of Jupiter is 43,348 miles. How many times greater is the volume of Jupiter compared to the volume of Uranus?

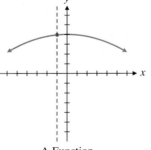
1 *Determining Using the Vertical Line Test Whether a Given Graph Represents a Function*

Not every graph we observe is that of a function. Recall from Section 3.6 that, by definition, a function must have no ordered pairs that have the same first coordinates and different second coordinates. A graph that includes the points $(4, 2)$ and $(4, -2)$, for example, would not be the graph of a function. Thus, the graph of $x = y^2$ would not be the graph of a function.

If any vertical line crosses a graph of a relation in more than one place, the relation is not a function. If no such line exists, the relation is a function.

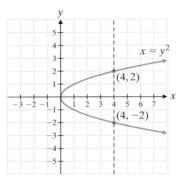

We restate the vertical line test.

Vertical Line Test

If any vertical line intersects the graph of a relation more than once, the relation is not a function. If no such line exists, the relation is a function.

In the following sketches, we observe that the dashed vertical line crosses the curve of a function no more than once. The dashed vertical line crosses the curve of a relation that is *not* a function more than once.

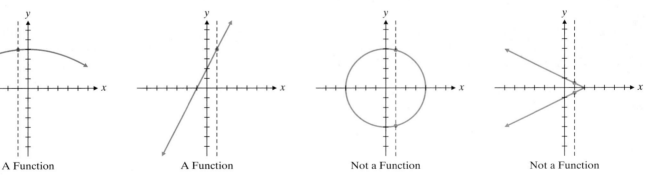

A Function   A Function   Not a Function   Not a Function

 EXAMPLE 1   Determine whether each of the following is the graph of a function.

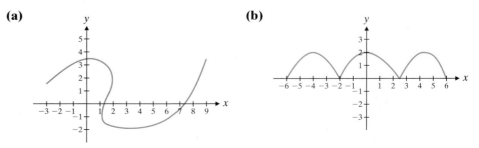

(a)

(b)

(a) By the vertical line test, this relation is not a function.

(b) By the vertical line test, this relation is a function.

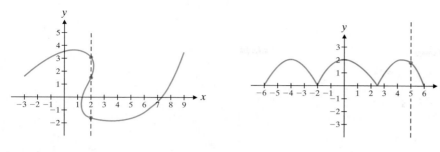

Practice Problem 1   Does this graph represent a function? Why or why not?

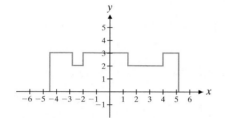

 Graphing a Function of the Form $f(x + h) + k$ by Means of Horizontal and Vertical Shifts from the Graph of $f(x)$

The graphs of some functions are simple vertical shifts of the graphs of similar functions.

EXAMPLE 2   Graph the functions on one coordinate plane.

$$f(x) = x^2 \quad \text{and} \quad h(x) = x^2 + 2$$

First we make a table of values for $f(x)$ and for $h(x)$.

x	$f(x) = x^2$
-2	4
-1	1
0	0
1	1
2	4

x	$h(x) = x^2 + 2$
-2	6
-1	3
0	2
1	3
2	6

Now we graph each function on the same coordinate plane.

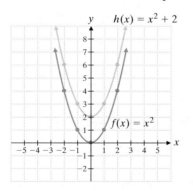

Notice that the graph of $h(x)$ is the graph of $f(x)$ moved 2 units upward.

To Think About   What would the graph of $j(x) = x^2 - 3$ look like? Verify by making a table of values and drawing a graph of the function.

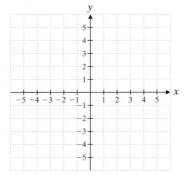

Practice Problem 2 Graph the functions on one coordinate plane.

$$f(x) = x^2 \quad \text{and} \quad h(x) = x^2 - 5$$

We have the following general summary.

Vertical Shifts

Suppose that k is a positive number.
1. To obtain the graph of $f(x) + k$, shift the graph of $f(x)$ up k units.
2. To obtain the graph of $f(x) - k$, shift the graph of $f(x)$ down k units.

Now we turn to the topic of horizontal shifts.

EXAMPLE 3 Graph the functions on one coordinate plane.

$$f(x) = |x| \quad \text{and} \quad p(x) = |x - 3|$$

First we make a table of values for $f(x)$ and $p(x)$.

x	$f(x) = |x|$	x	$p(x) = |x - 3|$
−2	2	−2	5
−1	1	−1	4
0	0	0	3
1	1	1	2
2	2	2	1
3	3	3	0
4	4	4	1

Now we graph each function on the same coordinate plane.

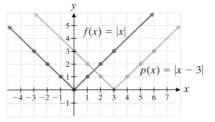

Notice that the graph of $p(x)$ is the graph of $f(x)$ shifted 3 units to the right.

Graphing Calculator

Exploration

Most graphing calculators have an absolute value function (abs). Use this function to graph $f(x)$ and $h(x)$ on one coordinate plane.

$$f(x) = |0.5x|$$

$$h(x) = |0.5x + 3.5| - 1.75$$

Describe how we shift the graph of $f(x)$ to obtain that of $h(x)$. Use your calculator to find the approximate coordinates of the point where $f(x)$ and $h(x)$ intersect. Find the coordinates to the nearest hundredth.

Practice Problem 3 Graph the functions on one coordinate plane.

$$f(x) = |x| \quad \text{and} \quad p(x) = |x + 2|$$

To Think About What would the graph of $h(x) = (x - 3)^2$ look like? What would the graph of $j(x) = (x + 2)^2$ look like? Verify by making tables of values and drawing the graphs.

Now we can write the following general summary.

Horizontal Shifts

Suppose that h is a positive number.
1. To obtain the graph of $f(x - h)$, shift the graph of $f(x)$ to the right h units.
2. To obtain the graph of $f(x + h)$, shift the graph of $f(x)$ to the left h units.

Some graphs will involve both horizontal and vertical shifts.

EXAMPLE 4   Graph the functions on one coordinate plane.

$$f(x) = x^3 \quad \text{and} \quad h(x) = (x - 3)^3 - 2$$

First we make a table of values for $f(x)$ and graph the function.

x	$f(x)$
-2	-8
-1	-1
0	0
1	1
2	8

Next we recognize that $h(x)$ will have a similar shape, but the curve will be shifted 3 units to the *right* and 2 units *downward*. We draw the graph of $h(x)$ using these shifts.

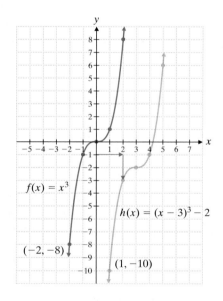

To Think About   The point $(-2, -8)$ has been shifted 3 units to the right and 2 units down to the point $\left(-2 + 3, -8 + (-2)\right)$ or $(1, -10)$. The point $(-1, -1)$ is a point on $f(x)$. Use the same reasoning to find the image of $(-1, -1)$ on the graph of $h(x)$. Verify by checking the graphs.

Practice Problem 4   Graph the functions on one coordinate plane.

$$f(x) = x^3 \quad \text{and} \quad h(x) = (x + 4)^3 + 3$$

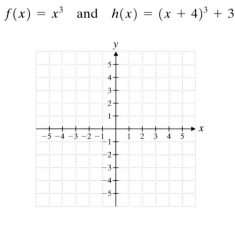

All the functions that we have sketched in this section so far have had a domain of all real numbers. Some functions have a restricted domain.

EXAMPLE 5   Graph the functions on one coordinate plane. State the domain of each function.

$$f(x) = \frac{4}{x} \quad \text{and} \quad g(x) = \frac{4}{x + 3} + 1$$

First we make a table of values for $f(x)$. The domain of $f(x)$ is all real numbers, where $x \neq 0$. Note that $f(x)$ is not defined when $x = 0$ since we cannot divide by 0.

x	$f(x)$
-4	-1
-2	-2
-1	-4
$-\frac{1}{2}$	-8
$\frac{1}{2}$	8
1	4
2	2
4	1

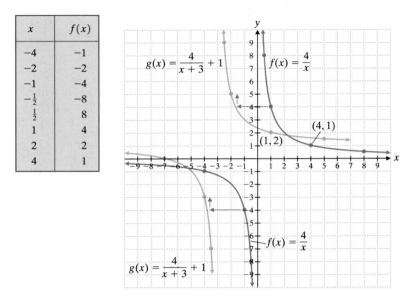

We draw $f(x)$ and note key points. From the equation, we see that the graph of $g(x)$ is 3 units to the left of and 1 unit above $f(x)$. We can find the image of each of the key points on $f(x)$ as a guide in graphing $g(x)$. For example, the image of $(4, 1)$ is $(4 - 3, 1 + 1)$ or $(1, 2)$.

Each point on $f(x)$ is shifted

$$\Leftarrow \text{ 3 units left and}$$
$$\Uparrow \text{ 1 unit up}$$

to form the graph of $g(x)$.

What is the domain of $g(x)$? Why? $g(x)$ contains the denominator $x + 3$. But $x + 3 \neq 0$. Therefore, $x \neq -3$. The domain of $g(x)$ is all real numbers, where $x \neq -3$.

Practice Problem 5   Graph the functions on one coordinate plane.

$$f(x) = \frac{2}{x} \quad \text{and} \quad g(x) = \frac{2}{x + 1} - 2$$

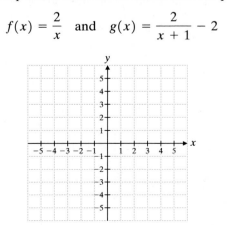

11.2 Exercises

Verbal and Writing Skills

1. Does $f(x + 2) = f(x) + f(2)$? Why or why not? Give an example.

2. Explain what the vertical line test is and why it works.

3. To obtain the graph of $f(x) + k$, shift the graph of $f(x)$ _____ k units.

4. To obtain the graph of $f(x - h)$, shift the graph of $f(x)$ _____ h units.

Determine whether or not each graph represents a function.

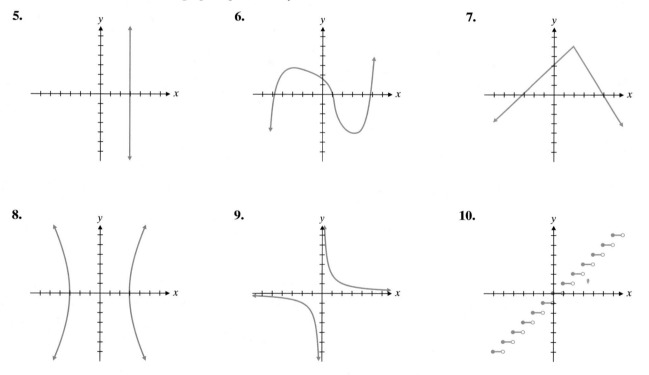

5. **6.** **7.**

8. **9.** **10.**

Hint: The open circle means that the function value does not exist at that point.

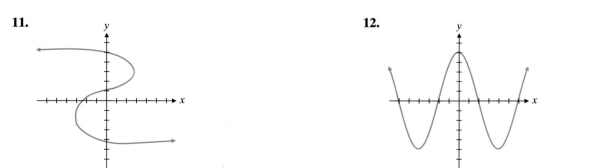

11. **12.**

13.

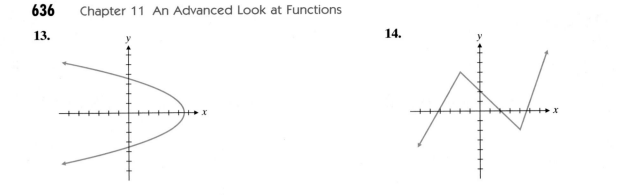

14.

For each of Exercises 15–30, graph the two functions on one coordinate plane.

15. $f(x) = x^2$
   $h(x) = x^2 + 4$

16. $f(x) = x^2$
   $h(x) = x^2 - 3$

17. $f(x) = x^2$
   $p(x) = (x + 1)^2$

18. $f(x) = x^2$
   $p(x) = (x - 2)^2$

19. $f(x) = x^2$
   $g(x) = (x - 2)^2 + 1$

20. $f(x) = x^2$
   $g(x) = (x + 1)^2 - 2$

21. $f(x) = |x|$
   $r(x) = |x| - 1$

22. $f(x) = |x|$
   $r(x) = |x| + 3$

23. $f(x) = |x|$
   $s(x) = |x + 4|$

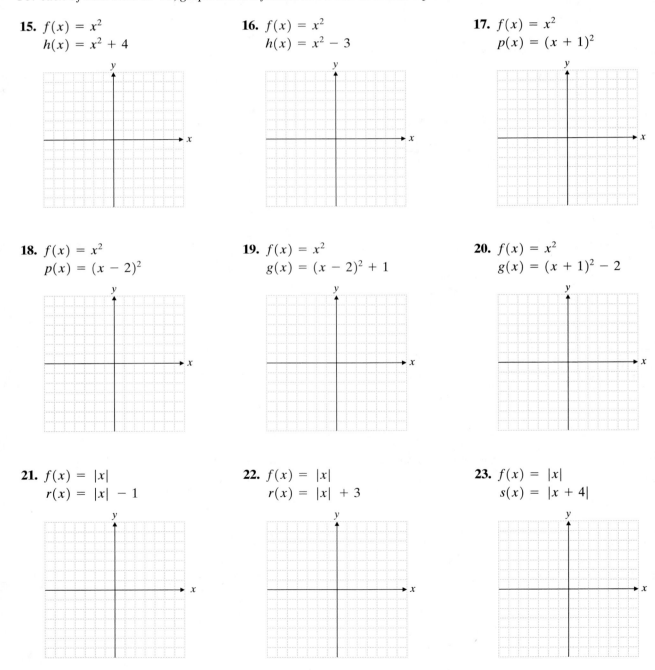

24. $f(x) = |x|$
$s(x) = |x - 2|$

25. $f(x) = |x|$
$t(x) = |x - 3| - 4$

26. $f(x) = |x|$
$t(x) = |x + 1| + 2$

27. $f(x) = x^3$
$j(x) = (x - 3)^3 + 3$

28. $f(x) = x^3$
$j(x) = (x + 3)^3 + 1$

29. $f(x) = \dfrac{2}{x}$
$g(x) = \dfrac{2}{x} + 3$

30. $f(x) = \dfrac{3}{x}$
$g(x) = \dfrac{3}{x} - 2$

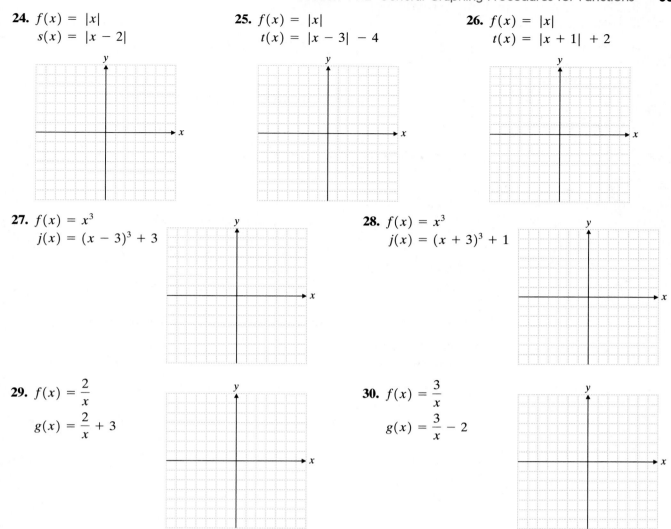

Optional Graphing Calculator Problems

31. Using your graphing calculator, graph $f(x) = x^4$ and $f(x) = (x - 3.2)^4 - 2.6$.

32. Using your graphing calculator, graph $f(x) = x^7$ and $f(x) = (x + 1.3)^7 + 3.3$.

Cumulative Review Problems

Simplify each expression. Assume that all variables are positive.

33. $\sqrt{12} + 3\sqrt{50} - 4\sqrt{27}$

34. $\left(\sqrt{3x} - \sqrt{y}\right)^2$

35. Rationalize the denominator. $\dfrac{2\sqrt{3} + \sqrt{5}}{\sqrt{3} - 2\sqrt{5}}$

36. Roy is on a diet and can have a maximum of 615 calories for lunch. A tuna fish sandwich on whole wheat bread has 315 calories, and 12 fluid ounces of a fruit juice soft drink has 120 calories. How many french fries can he eat if there are 10 calories in one french fry?

37. If y varies directly with x^2 and $y = 36$ when $x = 3$, find y when $x = 5$.

Student Learning Objectives

After studying this section, you will be able to:

1 Find the sum, difference, product, and quotient of two functions.

2 Find the composition of two functions.

SSM

PH TUTOR CENTER   CD & VIDEO   MATH PRO   WEB

1 Finding the Sum, Difference, Product, and Quotient of Functions

When two functions are given, new functions can be formed by combining them, as defined in the box:

> **If f represents one function and g represents a second function, we can define new functions as follows:**
>
> Sum of Functions $\quad (f + g)(x) = f(x) + g(x)$
>
> Difference of Functions $\quad (f - g)(x) = f(x) - g(x)$
>
> Product of Functions $\quad (fg)(x) = f(x) \cdot g(x)$
>
> Quotient of Functions $\quad \left(\dfrac{f}{g}\right)(x) = \dfrac{f(x)}{g(x)}, g(x) \neq 0$

EXAMPLE 1 Suppose that $f(x) = 3x^2 - 3x + 5$ and $g(x) = 5x - 2$.

(a) Find $(f + g)(x)$.

(b) Evaluate $(f + g)(x)$ when $x = 3$.

(a) $(f + g)(x) = f(x) + g(x)$
$$= (3x^2 - 3x + 5) + (5x - 2)$$
$$= 3x^2 - 3x + 5 + 5x - 2$$
$$= 3x^2 + 2x + 3$$

(b) To evaluate $(f + g)(x)$ when $x = 3$, we write $(f + g)(3)$ and use the formula obtained in **(a)**.

$$(f + g)(x) = 3x^2 + 2x + 3$$
$$(f + g)(3) = 3(3)^2 + 2(3) + 3$$
$$= 3(9) + 2(3) + 3$$
$$= 27 + 6 + 3 = 36$$

Practice Problem 1 Given $f(x) = 4x + 5$ and $g(x) = 2x^2 + 7x - 8$, find the following:

(a) $(f + g)(x)$ **(b)** $(f + g)(4)$

EXAMPLE 2 Given $f(x) = x^2 - 5x + 6$ and $g(x) = 2x - 1$, find the following:

(a) $(fg)(x)$ **(b)** $(fg)(-4)$

(a) $(fg)(x) = f(x) \cdot g(x)$
$$= (x^2 - 5x + 6)(2x - 1)$$
$$= 2x^3 - 10x^2 + 12x - x^2 + 5x - 6$$
$$= 2x^3 - 11x^2 + 17x - 6$$

(b) To evaluate $(fg)(x)$ when $x = -4$, we write $(fg)(-4)$ and use the formula obtained in **(a)**.

$$(fg)(x) = 2x^3 - 11x^2 + 17x - 6$$
$$(fg)(-4) = 2(-4)^3 - 11(-4)^2 + 17(-4) - 6$$
$$= 2(-64) - 11(16) + 17(-4) - 6$$
$$= -128 - 176 - 68 - 6$$
$$= -378$$

Practice Problem 2   Given $f(x) = 3x + 2$ and $g(x) = x^2 - 3x - 4$, find the following:

(a) $(fg)(x)$ **(b)** $(fg)(2)$

   When finding the quotient of a function, we must be careful to avoid division by zero. Thus, we always specify any values of x that must be eliminated from the domain.

 EXAMPLE 3   Given $f(x) = 3x + 1$, $g(x) = 2x - 1$, and $h(x) = 9x^2 + 6x + 1$, find the following:

(a) $\left(\dfrac{f}{g}\right)(x)$ **(b)** $\left(\dfrac{f}{h}\right)(x)$ **(c)** $\left(\dfrac{f}{h}\right)(-2)$

(a) $\left(\dfrac{f}{g}\right)(x) = \dfrac{3x + 1}{2x - 1}$

   The denominator of the quotient can never be zero. Since $2x - 1 \neq 0$, we know that $x \neq \frac{1}{2}$.

(b) $\left(\dfrac{f}{h}\right)(x) = \dfrac{3x + 1}{9x^2 + 6x + 1} = \dfrac{3x + 1}{(3x + 1)(3x + 1)} = \dfrac{1}{3x + 1}$

   Since $3x + 1 \neq 0$, we know that $x \neq -\frac{1}{3}$.

(c) To find $\left(\dfrac{f}{h}\right)(-2)$, we must evaluate $\left(\dfrac{f}{h}\right)(x)$ when $x = -2$.

$$\left(\frac{f}{h}\right)(x) = \frac{1}{3x + 1}$$

$$\left(\frac{f}{h}\right)(-2) = \frac{1}{(3)(-2) + 1} = \frac{1}{-6 + 1} = -\frac{1}{5}$$

Practice Problem 3   Given $p(x) = 5x^2 + 6x + 1$, $h(x) = 3x - 2$, and $g(x) = 5x + 1$, find the following:

(a) $\left(\dfrac{g}{h}\right)(x)$ **(b)** $\left(\dfrac{g}{p}\right)(x)$ **(c)** $\left(\dfrac{g}{h}\right)(3)$

---

2 ▸ **Finding the Composition of Two Functions**

Suppose that the music section of a department store finds that the number of sales of compact discs (CDs) on a given day is generally equal to 25% of the number of people who visit the store on that day. Thus, if x = the number of people who visit the store, then the sales S can be modeled by the equation $S(x) = 0.25x$.

   Suppose that the average CD in the store sells for $15. Then if S = the number of CD sales on a given day, the income for that day can be modeled by the equation $P(S) = 15S$. Suppose that eighty people came into the store.

$$S(x) = 0.25x$$
$$S(80) = 0.25(80) = 20$$

Thus, twenty CDs would be sold.

   If twenty CDs were sold and the average price of a CD is $15, then we would have the following:

$$P(S) = 15S$$
$$P(20) = 15(20) = 300$$

That is, the income from the sales of CDs would be $300.

Let us analyze the functions we have described and record a few values of x, $S(x)$, and $P(S)$.

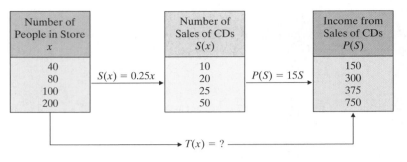

Is there a function $T(x)$ that describes the income from CD sales as a function of x, the number of people who visit the store?

The number of sales is

$$S(x) = 0.25x.$$

Thus, $0.25x$ is the number of sales.

If we replace S in $P(S) = 15S$ by $S(x)$, we have

$$P[S(x)] = P(0.25x) = 15(0.25x) = 3.75x.$$

Thus, the formula $T(x)$ that describes the income in terms of the number of visitors is

$$T(x) = 3.75x.$$

Is this correct? Let us check by finding $T(200)$. From our table the result should be 750.

$$\text{If} \qquad T(x) = 3.75x,$$
$$\text{then} \qquad T(200) = 3.75(200) = 750.$$

Thus, we have found a function T that is the composition of the functions P and S: $T(x) = P[S(x)]$.

We now state a definition of the composition of one function with another.

The **composition** of the functions f and g, denoted $f \circ g$, is defined as follows: $(f \circ g)(x) = f[g(x)]$. The domain of $f \circ g$ is the set of all x values in the domain of g such that $g(x)$ is in the domain of f.

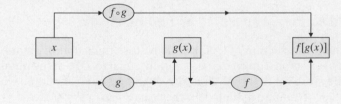

EXAMPLE 4 Given $f(x) = 3x - 2$ and $g(x) = 2x + 5$, find $f[g(x)]$.

$$
\begin{aligned}
f[g(x)] &= f(2x + 5) & &\text{Substitute } g(x) = 2x + 5. \\
&= 3(2x + 5) - 2 & &\text{Apply the formula for } f(x). \\
&= 6x + 15 - 2 & &\text{Remove parentheses.} \\
&= 6x + 13 & &\text{Simplify.}
\end{aligned}
$$

Practice Problem 4 Given $f(x) = 2x - 1$ and $g(x) = 3x - 4$, find $f[g(x)]$.

In most situations $f[g(x)]$ and $g[f(x)]$ are not the same.

EXAMPLE 5 Given $f(x) = \sqrt{x - 4}$ and $g(x) = 3x + 1$, find the following:

(a) $f[g(x)]$ **(b)** $g[f(x)]$

(a) $f[g(x)] = f[3x + 1]$ Substitute $g(x) = 3x + 1$.

$\qquad = \sqrt{(3x + 1) - 4}$ Apply the formula for $f(x)$.

$\qquad = \sqrt{3x + 1 - 4}$ Remove parentheses.

$\qquad = \sqrt{3x - 3}$ Simplify.

(b) $g[f(x)] = g[\sqrt{x - 4}]$ Substitute $f(x) = \sqrt{x - 4}$.

$\qquad = 3(\sqrt{x - 4}) + 1$ Apply the formula for $g(x)$.

$\qquad = 3\sqrt{x - 4} + 1$ Remove parentheses.

We note that $g[f(x)] \neq f[g(x)]$.

Practice Problem 5 Given $f(x) = 2x^2 - 3x + 1$ and $g(x) = x + 2$, find the following:

(a) $f[g(x)]$ **(b)** $g[f(x)]$

EXAMPLE 6 Given $f(x) = 2x$ and $g(x) = \dfrac{1}{3x - 4}$, $x \neq \dfrac{4}{3}$, find the following:

(a) $(f \circ g)(x)$ **(b)** $(f \circ g)(2)$

(a) $(f \circ g)(x) = f[g(x)] = f\left[\dfrac{1}{3x - 4}\right]$ Substitute $g(x) = \dfrac{1}{3x - 4}$.

$\qquad = 2\left(\dfrac{1}{3x - 4}\right)$ Apply the formula for $f(x)$.

$\qquad = \dfrac{2}{3x - 4}$ Simplify.

(b) $(f \circ g)(2) = \dfrac{2}{3(2) - 4} = \dfrac{2}{6 - 4} = \dfrac{2}{2} = 1$

Practice Problem 6 Given $f(x) = 3x + 1$ and $g(x) = \dfrac{2}{x - 3}$, find the following:

(a) $(g \circ f)(x)$ **(b)** $(g \circ f)(-3)$

Graphing Calculator

Composition of Functions

You can formulate the composition of functions on most graphing calculators by using the y-variable function (Y-VARS). To do Example 6 on most graphing calculators, you would use the following equations.

$$y_1 = \dfrac{1}{3x - 4}$$

$$y_2 = 2(y_1)$$

To find the function value, you can use the TableSet command to let $x = 2$. Then enter Table and you will see displayed $y_1 = 0.5$, which represents $g(2) = 0.5$, and $y_2 = 1$, which represents $f[g(2)] = 1$.

For the following functions, find **(a)** $(f + g)(x)$, **(b)** $(f - g)(x)$, **(c)** $(f + g)(2)$, *and* **(d)** $(f - g)(-1)$.

1. $f(x) = -2x + 3, g(x) = 2 + 4x$

2. $f(x) = 3x + 4, g(x) = 1 - 2x$

3. $f(x) = 2x^2 - 4x + 5, g(x) = 2x - 1$

4. $f(x) = 2 - x, g(x) = x^2 + 3x - 1$

5. $f(x) = x^3 - \dfrac{1}{2}x^2 + x, g(x) = x^2 - \dfrac{x}{4} - 5$

6. $f(x) = 2.4x^2 + x - 3.5, g(x) = 1.1x^3 - 2.2x$

7. $f(x) = 3\sqrt{3 - x}, g(x) = -5\sqrt{3 - x}$

8. $f(x) = -5\sqrt{x + 6}, g(x) = 8\sqrt{x + 6}$

For the following functions, find **(a)** $(fg)(x)$ *and* **(b)** $(fg)(-3)$.

9. $f(x) = x^2 - 3x + 2, g(x) = 1 - x$

10. $f(x) = 2x - 3, g(x) = -2x^2 - 3x + 1$

11. $f(x) = \dfrac{2}{x^2}, g(x) = x^2 - x$

12. $f(x) = \dfrac{6x}{x - 1}, g(x) = \dfrac{x}{2}$

13. $f(x) = \sqrt{-2x + 1}, g(x) = -3x$

14. $f(x) = 4x, g(x) = \sqrt{3x + 10}$

For the following functions, find **(a)** $\left(\dfrac{f}{g}\right)(x)$ *and* **(b)** $\left(\dfrac{f}{g}\right)(2)$.

15. $f(x) = 3x, g(x) = 4x - 1$

16. $f(x) = x - 6, g(x) = 3x$

17. $f(x) = x^2 - 1, g(x) = x - 1$

18. $f(x) = x, g(x) = x^2 - 5x$

19. $f(x) = x^2 + 10x + 25, g(x) = x + 5$

20. $f(x) = 4x^2 + 4x + 1, g(x) = 2x + 1$

21. $f(x) = 4x - 1, g(x) = 4x^2 + 7x - 2$

22. $f(x) = 3x + 2, g(x) = 3x^2 - x - 2$

Let $f(x) = 3x + 2$, $g(x) = x^2 - 2x$, and $h(x) = \dfrac{x-2}{3}$. Find the following:

23. $(f - g)(x)$ **24.** $(g - f)(x)$ **25.** $(fg)(x)$ **26.** $\left(\dfrac{f}{h}\right)(-1)$

27. $(fg)(-1)$ **28.** $(gh)(3)$ **29.** $\left(\dfrac{f}{h}\right)(x)$ **30.** $\left(\dfrac{g}{f}\right)(x)$

Find $f[g(x)]$ for each of the following:

31. $f(x) = 2 - 3x$, $g(x) = 2x + 5$ **32.** $f(x) = 3x + 2$, $g(x) = 4x - 1$

33. $f(x) = 3x^2$, $g(x) = x - 4$ **34.** $f(x) = x^2 + 5$, $g(x) = x - 3$

35. $f(x) = 4 - 3x$, $g(x) = 2x^2 - 1$ **36.** $f(x) = 1 - 2x$, $g(x) = 3x^2 + x - 1$

37. $f(x) = \dfrac{3}{x+1}$, $g(x) = 2x - 1$ **38.** $f(x) = \dfrac{4}{x-3}$, $g(x) = 4x + 1$

39. $f(x) = |x + 3|$, $g(x) = 2x - 1$ **40.** $f(x) = \left|\dfrac{1}{2}x - 5\right|$, $g(x) = 4x + 6$

Let $f(x) = x^2 + 2$, $g(x) = 3x + 5$, $h(x) = \dfrac{1}{x}$, and $p(x) = \sqrt{x - 1}$. Find each of the following:

41. $f[g(x)]$ **42.** $g[h(x)]$ **43.** $g[f(x)]$ **44.** $h[g(x)]$

45. $g[f(3)]$ **46.** $h\left[g\left(-\dfrac{1}{2}\right)\right]$ **47.** $(p \circ f)(x)$ **48.** $(f \circ h)(x)$

49. $(g \circ h)(\sqrt{2})$ **50.** $(f \circ p)(x)$ **51.** $(p \circ f)(-5)$ **52.** $(f \circ p)(7)$

Applications

53. Consider the Celsius function $C(F) = \dfrac{5F - 160}{9}$,
which converts degrees Fahrenheit to degrees Celsius. A different temperature scale, called the Kelvin scale, is used by many scientists in their research. The Kelvin scale is similar to the Celsius scale, but it begins at absolute zero (the coldest possible temperature, which is around $-273°C$). To convert a Celsius temperature to a temperature on the Kelvin scale, we use the function $K(C) = C + 273$. Find $K[C(F)]$, which is the composite function that defines the temperature in Kelvins in terms of the temperature in degrees Fahrenheit.

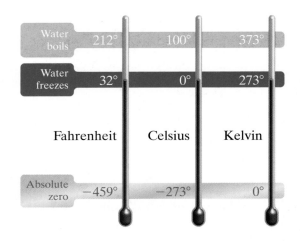

54. Suppose the dollar cost to produce n items in a factory is $c(n) = 5n + 4$. Furthermore, the number of items n produced in x hours is $n(x) = 3x$. Find $c[n(x)]$, which is the composite function that defines the dollar cost in terms of the number of hours of production x.

▲ **55.** The volume of polluted water emitted from a discharge pipe from a factory located on the ocean is shaped in a cone. The radius r of the cone of polluted water at the end of each day is given by the equation $r(h) = 3.5h$, where h is the number of hours the factory operated that day. The volume function that defines this cone is $v(r) = 31.4r^2$, where r is the radius of the cone measured in feet. Find $v[r(h)]$, which is the composite function that defines the volume of polluted water in terms of the number of hours h the factory has run. How large is the volume at the end of the day if the factory has been running for 8 hours?

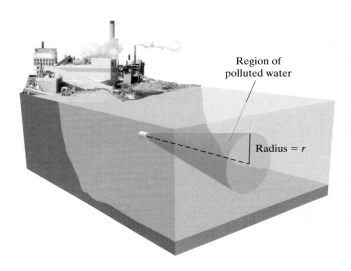

▲ **56.** An oil tanker with a ruptured hull is leaking oil off the coast of Africa. There is no wind or significant current, so the oil slick is spreading in a circle whose radius is defined by the function $r(t) = 3t$, where t is the time in minutes since the tanker began to leak. The area of the slick for any given radius is approximately determined by the function $a(r) = 3.14r^2$ where r is the radius of the circle measured in feet. Find $a[r(t)]$, which is the composite function that defines the area of the oil slick in terms of the minutes t since the beginning of the leak. How large is the area after 20 minutes?

Cumulative Review Problems

Factor each of the following:

57. $6ab - 3ac + 2b - c$

58. $36x^2 - 12x + 1$

59. $3x^2 - 7x + 2$

60. $x^4 - 10x^2 + 9$

61. Juanita has an hour-long radio show called "Talk of the Town." During the broadcast she has to break for twenty commercials. She has commercial tapes that are 30 seconds long and other commercial tapes that are 60 seconds long. How many of each type should she play in order to achieve the goal of 14 minutes of commercials?

62. The West Shore Community College Alumni Dinner was held for 470 people. A children's plate cost $4.50, and an adult's plate cost $7.50. The receipts for the dinner were $3180. How many children attended the dinner? How many adults attended the dinner?

Student Learning Objectives

After studying this section, you will be able to:

1 Determine whether a function is a one-to-one function.

2 Find the inverse function for a given function.

3 Graph a function and its inverse function.

SSM

PH TUTOR CENTER   CD & VIDEO   MATH PRO   WEB

Americans driving in Canada or Mexico need to be able to convert miles per hour to kilometers per hour and vice versa.

NOTRE SIGNALISATION ROUTIÈRE EST MÉTRIQUE
OUR TRAFFIC SIGNS ARE METRIC
MAXIMUM 55 → MAXIMUM 90 km/h

If someone is driving at 55 miles per hour, how fast is he or she going in kilometers per hour?

Approximate Value in Miles per Hour	Approximate Value in Kilometers per Hour
35	56
40	64
45	72
50	80
55	88
60	96
65	104

A function f that converts from miles per hour to an approximate value in kilometers per hour is $f(x) = 1.6x$.

For example, $f(40) = 1.6(40) = 64$.

This tells us that 40 miles per hour is approximately equivalent to 64 kilometers per hour.

We can come up with a function that does just the opposite—that is, that converts kilometers per hour to an approximate value in miles per hour. This function is $f^{-1}(x) = 0.625x$.

For example, $f^{-1}(64) = 0.625(64) = 40$.

This tells us that 64 kilometers per hour is approximately equivalent to 40 miles per hour.

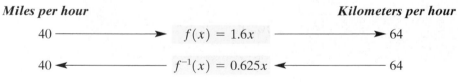

Miles per hour   *Kilometers per hour*

$40 \longrightarrow \quad f(x) = 1.6x \quad \longrightarrow 64$

$40 \longleftarrow \quad f^{-1}(x) = 0.625x \quad \longleftarrow 64$

We call a function f^{-1} that reverses the domain and range of a function f the **inverse function** f.

Most American cars have numbers showing kilometers per hour in smaller print on the car speedometer. Unfortunately, these numbers are usually hard to read. If we made a list of several function values of f and several inverse function values of f^{-1}, we could create a conversion scale like the one below that we could use if we should travel to Mexico or Canada with an American car.

Miles per Hour
0   10   20   30   **40**   50   60   70

0   10   20   30   40   50   **60**   70   80   90   100   110
Kilometers per Hour

The original function that we studied converts miles per hour to kilometers per hour. The corresponding inverse function converts kilometers per hour to miles per hour. How do we find inverse functions? Do all functions have inverse functions? These are questions we want to explore in this section.

1 Determining Whether a Function Is a One-to-One Function

First we state that not all functions have inverse functions. To have an inverse that is a function, a function must be one-to-one. This means that for every value of y, there is only one value of x. Or, in the language of ordered pairs, no ordered pairs have the same second coordinate.

Definition of a One-to-One Function

A **one-to-one function** is a function in which no ordered pairs have the same second coordinate.

To Think About   Why must a function be one-to-one in order to have an inverse that is a function?

EXAMPLE 1   Indicate whether the following functions are one-to-one.

(a) $M = \{(1, 3), (2, 7), (5, 8), (6, 12)\}$ **(b)** $P = \{(1, 4), (2, 9), (3, 4), (4, 18)\}$

(a) M is a function because no ordered pairs have the same first coordinate. M is also a one-to-one function because no ordered pairs have the same second coordinate.

(b) P is a function, but it is not one-to-one because the ordered pairs $(1, 4)$ and $(3, 4)$ have the same second coordinate.

Practice Problem 1

(a) Is the function $A = \{(-2, -6), (-3, -5), (-1, 2), (3, 5)\}$ one-to-one?

(b) Is the function $B = \{(0, 0), (1, 1), (2, 4), (3, 9), (-1, 1)\}$ one-to-one?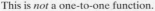

By examining the graph of a function, we can quickly tell whether it is one-to-one. If any horizontal line crosses the graph of a function in more than one place, the function is not one-to-one. If no such line exists, then the function is one-to-one.

Horizontal Line Test

If any horizontal line intersects the graph of a function more than once, the function is not one-to-one. If no such line exists, the function is one-to-one.

This is *not* a one-to-one function.

EXAMPLE 2 Determine whether the functions graphed are one-to-one functions.

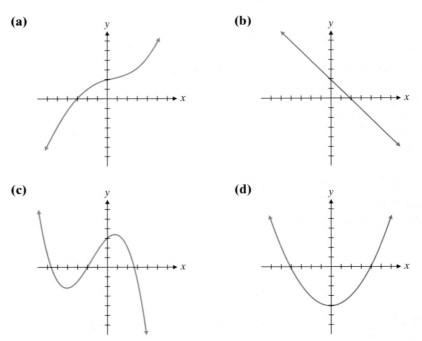

(a)

(b)

(c)

(d)

The graphs of **(a)** and **(b)** represent one-to-one functions. Horizontal lines cross the graphs at most once.

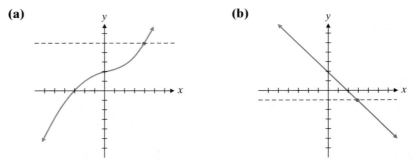

(a)

(b)

The graphs of **(c)** and **(d)** do not represent one-to-one functions. A horizontal line exists that crosses the graphs more than once.

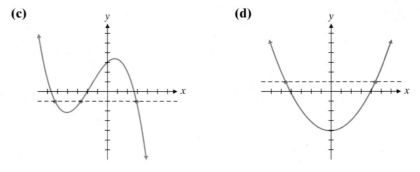

(c)

(d)

Practice Problem 2   Do the following graphs of functions represent one-to-one functions? Why or why not?

(a)

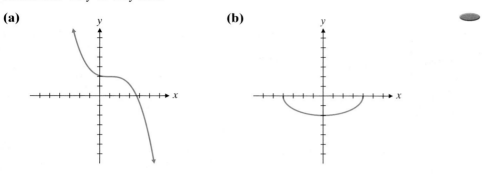

(b)

2 Finding the Inverse Function for a Given Function

How do we find the inverse of a function? If we have a list of ordered pairs, we simply interchange the coordinates of each ordered pair. In Example 1, we said that M has an inverse. What is it?

$$M = \{(1, 3), (2, 7), (5, 8), (6, 12)\}$$

The inverse of M, written M^{-1}, is

$$M^{-1} = \{(3, 1), (7, 2), (8, 5), (12, 6)\}.$$

Now do you see why a function must be one-to-one in order to have an inverse that is a function? Let's look at the function P from Example 1.

$$P = \{(1, 4), (2, 9), (3, 4), (4, 18)\}$$

If P had an inverse, it would be

$$P^{-1} = \{(4, 1), (9, 2), (4, 3), (18, 4)\}.$$

But we have two ordered pairs with the same first coordinate. Therefore, P^{-1} is not a function (in other words, the inverse function does not exist).

A number of real-world situations are described by functions that have inverses. Consider the function defined by the ordered pairs (year, U.S. budget in trillions of dollars). Some function values are

$$F = \{(2000, 1.83), (1995, 1.52), (1990, 1.25), (1985, 0.95), (1980, 0.59)\}.$$

In this case the inverse of the function is

$$F^{-1} = \{(1.83, 2000), (1.52, 1995), (1.25, 1990), (0.95, 1985), (0.59, 1980)\}.$$

⊘ **WARNING** F^{-1} does *not* mean $\dfrac{1}{F}$. Here the -1 simply means "inverse."

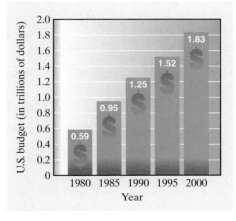

 EXAMPLE 3   Determine the inverse function of the function

$$F = \{(6, 1), (12, 2), (13, 5), (14, 6)\}.$$

The inverse function of F is $F^{-1} = \{(1, 6), (2, 12), (5, 13), (6, 14)\}.$

Practice Problem 3   Find the inverse of the one-to-one function $B = \{(1, 2), (7, 8),$ $(8, 7), (10, 12)\}$.

Suppose that a function is given in the form of an equation. How do we find the inverse? Since, by definition, we interchange the ordered pairs to find the inverse of a function, this means that the x-values of the function become the y-values of the inverse function and vice versa.

Four steps will help us find the inverse of a one-to-one function when we are given its equation.

Finding the Inverse of a One-to-One Function

1. Replace $f(x)$ with y.
2. Interchange x and y.
3. Solve for y in terms of x.
4. Replace y with $f^{-1}(x)$.

EXAMPLE 4   Find the inverse of $f(x) = 7x - 4$.

Step 1 $y = 7x - 4$ Replace $f(x)$ with y.

Step 2 $x = 7y - 4$ Interchange the variables x and y.

Step 3 $x + 4 = 7y$ Solve for y in terms of x.

$\dfrac{x + 4}{7} = y$

Step 4 $f^{-1}(x) = \dfrac{x + 4}{7}$ Replace y with $f^{-1}(x)$.

Practice Problem 4   Find the inverse of the function $g(x) = 4 - 6x$.

Recall the formula $f(x) = \frac{9}{5}x + 32$, which converts Celsius temperature (x) into the equivalent Fahrenheit temperature. Let's see if the above technique works to obtain the formula that converts Fahrenheit temperature to Celsius temperature.

EXAMPLE 5   Find the inverse function of $f(x) = \frac{9}{5}x + 32$.

Step 1 $y = \dfrac{9}{5}x + 32$ Replace $f(x)$ with y.

Step 2 $x = \dfrac{9}{5}y + 32$ Interchange x and y.

Step 3 $5(x) = 5\left(\dfrac{9}{5}\right)y + 5(32)$ Solve for y in terms of x.

$5x = 9y + 160$

$5x - 160 = 9y$

$\dfrac{5x - 160}{9} = \dfrac{9y}{9}$

$\dfrac{5x - 160}{9} = y$

Step 4 $f^{-1}(x) = \dfrac{5x - 160}{9}$ Replace y with $f^{-1}(x)$.

Note: Our inverse function $f^{-1}(x)$ will now convert Fahrenheit temperature to Celsius temperature.

$$f^{-1}(86) = \frac{5(86) - 160}{9} = \frac{270}{9} = 30$$

This tells us that a temperature of 86°F corresponds to a temperature of 30°C.

Practice Problem 5 Find the inverse function of $f(x) = 0.75 + 0.55(x - 1)$, which gives the cost of a telephone call for any call over 1 minute if the telephone company charges 75 cents for the first minute and 55 cents for each minute thereafter. Here $x =$ the number of minutes.

3 *Graphing a Function and Its Inverse Function*

The graph of a function and its inverse are symmetric about the line $y = x$. Why do you think that this is so?

EXAMPLE 6 If $f(x) = 3x - 2$, find $f^{-1}(x)$. Graph f and f^{-1} on the same set of axes. Draw the line $y = x$ as a dashed line for reference.

$$f(x) = 3x - 2$$
$$y = 3x - 2$$
$$x = 3y - 2$$
$$x + 2 = 3y$$
$$\frac{x + 2}{3} = y$$
$$f^{-1}(x) = \frac{x + 2}{3}$$

Now we graph each line.

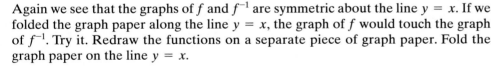

Again we see that the graphs of f and f^{-1} are symmetric about the line $y = x$. If we folded the graph paper along the line $y = x$, the graph of f would touch the graph of f^{-1}. Try it. Redraw the functions on a separate piece of graph paper. Fold the graph paper on the line $y = x$.

Practice Problem 6 If $f(x) = -\dfrac{1}{4}x + 1$, find $f^{-1}(x)$. Graph f and f^{-1} on the same coordinate plane. Draw the line $y = x$ as a dashed line for reference.

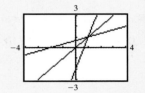
PRACTICE PROBLEM 6

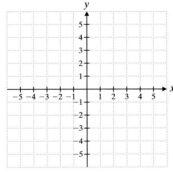

Exercises

Verbal and Writing Skills

Complete the following:

1. A one-to-one function is a function in which no ordered pairs _____

2. If any horizontal line intersects the graph of a function more than once, the function _____

3. The graphs of a function f and its inverse f^{-1} are symmetric about the line _____.

4. Do all functions have inverse functions? Why or why not?

Indicate whether each function is one-to-one.

5. $B = \{(0,1), (1,0), (10,0)\}$

6. $A = \{(-6,-2), (6,2), (3,4)\}$

7. $F = \{(\frac{2}{3}, 2), (3, -\frac{4}{5}), (-\frac{2}{3}, -2), (-3, \frac{4}{5})\}$

8. $C = \{(12,3), (-6,1), (6,3)\}$

9. $E = \{(1,3), (\frac{1}{2}, -5), (-1,-3), (-5, \frac{1}{2})\}$

10. $F = \{(5,0), (2,-7), (-2,7), (0,5)\}$

Indicate whether each graph represents a one-to-one function.

11.   **12.**   **13.**

14.   **15.**   **16.**

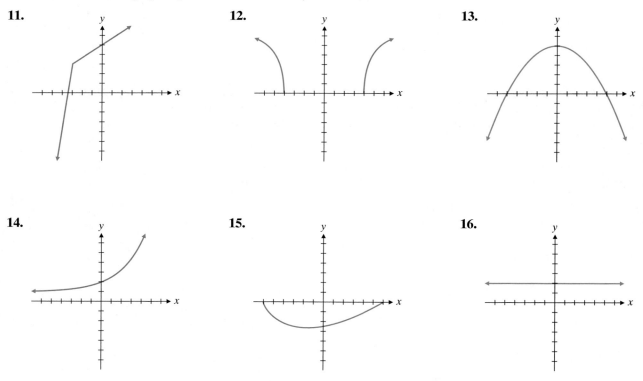

Verbal and Writing Skills

17. Does the graph of a horizontal line represent a function? Why or why not? Does it represent a one-to-one function? Explain.

18. Does the graph of a vertical line represent a function? Why or why not? Does it represent a one-to-one function? Explain.

Find the inverse of each one-to-one function. Graph the function and its inverse on one coordinate plane.

19. $J = \{(8, 2), (1, 1), (0, 0), (-8, -2)\}$

20. $K = \{(-7, 1), (6, 2), (3, -1), (2, 5)\}$

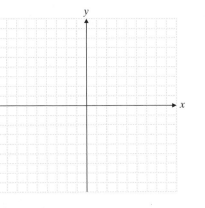

Find the inverse of each function.

21. $f(x) = 4x - 5$

22. $f(x) = \dfrac{3}{7}x + \dfrac{3}{2}$

23. $f(x) = x^3 - 2$

24. $f(x) = 5 + x^3$

25. $f(x) = -\dfrac{4}{x}$

26. $f(x) = \dfrac{3}{x}$

27. $f(x) = -\dfrac{3}{x - 2}$

28. $f(x) = \dfrac{3}{2x + 1}$

Verbal and Writing Skills

29. Can you find an inverse function for the function $f(x) = 2x^2 + 3$? Why or why not?

30. Can you find an inverse function for the function $f(x) = |3x + 4|$? Why or why not?

Find the inverse of each function. Graph the function and its inverse on one coordinate plane. Graph the line y = x as a dashed line.

31. $g(x) = 2x + 5$

32. $f(x) = 3x + 4$

33. $h(x) = \frac{1}{2}x - 2$

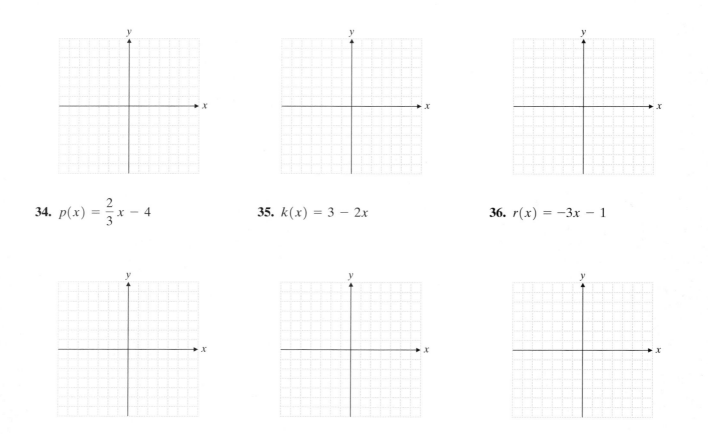

34. $p(x) = \frac{2}{3}x - 4$

35. $k(x) = 3 - 2x$

36. $r(x) = -3x - 1$

Applications

37. Manuela went to Spain. She brought Spanish pesetas back with her from Madrid and went to the bank to take advantage of the favorable exchange rate and convert her leftover Spanish pesetas to U.S. dollars. The bank teller informs Manuela that today's conversion rate is US $0.0063 for one Spanish peseta. The bank charges a fee of US $5.00 for each transaction. The function used to convert Spanish pesetas to U.S. dollars is given by $f(x) = 0.0063x - 5$, where x is the number of Spanish pesetas. Find the inverse function of f. What is the significance of the inverse function? If Manuela wanted to change U.S. dollars to Spanish pesetas, could she use this inverse function? Why or why not?

38. Sean went to Ireland. He brought Irish pounds back with him from Dublin and went to the bank to take advantage of the favorable exchange rate and convert his leftover Irish pounds to U.S. dollars. The bank teller informs Sean that today's conversion rate is US $1.437 for one Irish pound. The bank charges a fee of US $4.00 for each transaction. The function used to convert Irish pounds to U.S. dollars is given by $f(x) = 1.437x - 4$, where x is the number of Irish pounds. Find the inverse function of f. What is the significance of the inverse function? If Sean wanted to change U.S. dollars to Irish pounds, could he use this inverse function? Why or why not?

To Think About

For every function f and its inverse, f^{-1}, it is true that $f[f^{-1}(x)] = x$ and $f^{-1}[f(x)] = x$. Show that this is true for each pair of inverse functions.

39. $f(x) = 2x + \dfrac{3}{2}, f^{-1}(x) = \dfrac{1}{2}x - \dfrac{3}{4}$

40. $f(x) = -3x - 10, f^{-1}(x) = \dfrac{-x - 10}{3}$

Cumulative Review Problems

Solve for x.

41. $x^{2/3} + 7x^{1/3} + 12 = 0$

42. $x = \sqrt{15 - 2x}$

43. The average male human has more blood than the average female. In addition, each cubic centimeter of blood in males is usually richer in red blood cells. Each cubic centimeter of blood in men contains from 4.6 million to 6.2 million red blood cells, compared with 4.2 million to 5.4 million for women. Using the lower numbers, determine the ratio of red blood cells in men to red blood cells in women.

44. Catherine earns $17 per hour for a 40-hour week as an on-call nurse for City Hospital. She earns time and a half for every hour over 40 hours she works in one week. If Catherine made $1011.50 last week, how many overtime hours did she work?

45. Forests are a dominant feature of Canada. Ten percent of all the world's forests lie in Canada. One out of every sixteen people in the labor force in Canada works in a job that relates to forests. If the labor force in Canada in 2000 was 12,800,000 people, how many people worked in a job related to forests in that year?
Source: Canada Board of Tourism.

Math in the Media

Food-Borne Illness

The article, "A World of Choices and a World of Infectious Organisms," *The New York Times* late edition January 2001, reported findings from a study by the Center of Disease Control and Prevention in Atlanta.

The article notes that fewer than 5 percent of food-borne illnesses are reported. According to the article and study, it is estimated that 76 million illnesses, 323,914 hospitalizations, and 5,194 deaths each year in the United States are due to food-borne illness.

Use the skills you have learned in this chapter to find the equation that best describes the pattern in the questions that follow.

EXERCISES

Choose the equation form that best matches the description of each historical pattern of a food-borne illness. Assume each equation represents number of people affected, in thousands, as a function of time, in years:

a. The number of people falling ill varies up and down every 2–3 years.

b. The disease suddenly appeared 20 years ago, rose rather rapidly, then leveled off.

c. Occurrence of the disease is decreasing steadily, proportional to the passage of time.

d. The disease appeared 40 years ago, rose fairly rapidly, peaked 5 years ago, and is now diminishing at a rate that is a mirror-image of the previous growth rate.

1. $f(x) = Ax + B$

2. $f(x) = (Ax)^{1/2}$

3. $f(x) = Ax^2 + Bx + C$

4. $f(x) = Ax^3 + Bx^2 + Cx$

Chapter 11 Organizer

Topic	Procedure	Examples
Relations, functions, and one-to-one functions, pp. 630, and 647.	A relation is any set of ordered pairs. A function is a relation in which no ordered pairs have the same first coordinate. A one-to-one function is a function in which no ordered pairs have the same second coordinate.	Is $\{(3, 6), (2, 8), (9, 1), (4, 6)\}$ a one-to-one function? No, since $(3, 6)$ and $(4, 6)$ have the same second coordinate.
Vertical line test, p. 630.	If any vertical line intersects the graph of a relation more than once, the relation is not a function. If no such line exists, the relation is a function.	Does this graph represent a function? No, because a vertical line intersects the curve more than once.
Horizontal line test, p. 647.	If any horizontal line intersects the graph of a function more than once, the function is not one-to-one. If no such line exists, the function is one-to-one.	Does this graph represent a one-to-one function? Yes, any horizontal line will cross this function at most once.
Finding function values, p. 624.	Replace the variable by the quantity inside the parentheses. Simplify the result.	If $f(x) = 2x^2 + 3x - 4$, then we have the following: $$f(-2) = 2(-2)^2 + 3(-2) - 4$$ $$= 8 - 6 - 4 = -2$$ $$f(a) = 2a^2 + 3a - 4$$ $$f(a + 2) = 2(a + 2)^2 + 3(a + 2) - 4$$ $$= 2(a^2 + 4a + 4) + 3a + 6 - 4$$ $$= 2a^2 + 8a + 8 + 3a + 6 - 4$$ $$= 2a^2 + 11a + 10$$ $$f(3a) = 2(3a)^2 + 3(3a) - 4$$ $$= 2(9a^2) + 9a - 4$$ $$= 18a^2 + 9a - 4$$

Topic	Procedure	Examples
Vertical shifts of the graph of functions, p. 632.	If $k > 0$: **1.** The graph of $y = f(x) + k$ is shifted k units *upward* from the graph of $y = f(x)$.	Graph $f(x) = x^2$ and $g(x) = x^2 + 3$.
	2. The graph of $y = f(x) - k$ is shifted k units *downward* from the graph of $y = f(x)$.	Graph $f(x) = |x|$ and $g(x) = |x| - 2$.
Horizontal shifts of the graph of functions, p. 632.	If $h > 0$: **1.** The graph of $y = f(x - h)$ is shifted h units to the *right* of the graph of $y = f(x)$.	Graph $f(x) = x^2$ and $g(x) = (x - 3)^2$.
	2. The graph of $y = f(x + h)$ is shifted h units to the *left* of the graph of $y = f(x)$.	Graph $f(x) = x^3$ and $g(x) = (x + 4)^3$.
Sum, difference, product, and quotient of functions, p. 638.	**1.** $(f + g)(x) = f(x) + g(x)$ **2.** $(f - g)(x) = f(x) - g(x)$ **3.** $(f \cdot g)(x) = f(x) \cdot g(x)$ **4.** $\left(\dfrac{f}{g}\right)(x) = \dfrac{f(x)}{g(x)}, g(x) \neq 0$	If $f(x) = 2x + 3$ and $g(x) = 3x - 4$, then we have the following: **1.** $(f + g)(x) = (2x + 3) + (3x - 4)$ $\qquad = 5x - 1$ **2.** $(f - g)(x) = (2x + 3) - (3x - 4)$ $\qquad = 2x + 3 - 3x + 4$ $\qquad = -x + 7$ **3.** $(f \cdot g)(x) = (2x + 3)(3x - 4)$ $\qquad = 6x^2 + x - 12$ **4.** $\left(\dfrac{f}{g}\right)(x) = \dfrac{2x + 3}{3x - 4}, x \neq \dfrac{4}{3}$

Topic	Procedure	Examples
Composition of functions, p. 640.	The composition of functions f and g is written as $(f \circ g)(x) = f[g(x)]$. To find $f[g(x)]$ do the following: **1.** Replace $g(x)$ by its equation. **2.** Apply the formula for $f(x)$ to this expression. **3.** Simplify the results. Usually, $f[g(x)] \neq g[f(x)]$.	If $f(x) = x^2 - 5$ and $g(x) = -3x + 4$, find $f[g(x)]$ and $g[f(x)]$. $$f[g(x)] = f[-3x + 4]$$ $$= (-3x + 4)^2 - 5$$ $$= 9x^2 - 24x + 16 - 5$$ $$= 9x^2 - 24x + 11$$ $$g[f(x)] = g[x^2 - 5]$$ $$= -3(x^2 - 5) + 4$$ $$= -3x^2 + 15 + 4$$ $$= -3x^2 + 19$$
Finding the inverse of a function defined by a set of ordered pairs, p. 649.	Reverse the order of the coordinates of each ordered pair from (a, b) to (b, a).	Find the inverse of $A = \{(5, 6), (7, 8), (9, 10)\}$. $$A^{-1} = \{(6, 5), (8, 7), (10, 9)\}$$
Finding the inverse of a function defined by an equation, p. 650.	Any one-to-one function has an inverse function. To find the inverse f^{-1} of a one-to-one function f, do the following: **1.** Replace $f(x)$ with y. **2.** Interchange x and y. **3.** Solve for y in terms of x. **4.** Replace y with $f^{-1}(x)$.	Find the inverse of $f(x) = -\dfrac{2}{3}x + 4$. $$y = -\frac{2}{3}x + 4$$ $$x = -\frac{2}{3}y + 4$$ $$3x = -2y + 12$$ $$3x - 12 = -2y$$ $$\frac{3x - 12}{-2} = y$$ $$-\frac{3}{2}x + 6 = y$$ $$f^{-1}(x) = -\frac{3}{2}x + 6$$
Graphing the inverse of a function, p. 651.	Graph the line $y = x$ as a dashed line for reference. **1.** Graph $f(x)$. **2.** Graph $f^{-1}(x)$. The graphs of f and f^{-1} are symmetric about the line $y = x$.	$$f(x) = 2x + 3$$ $$f^{-1}(x) = \frac{x - 3}{2}$$ Graph f and f^{-1} on the same set of axes.

Putting Your Skills to Work

Using Polynomial Functions for Mathematical Modeling

We have studied linear functions (Chapter 3) and quadratic functions (Section 9.5). Now we turn to higher-degree polynomial functions. These types of functions are often useful models for studying phenomena in business and science.

A third-degree polynomial function is a function of the form

$$p(x) = ax^3 + bx^2 + cx + d, \quad \text{where } a \neq 0.$$

Now examine the graphs of the following two polynomial functions of degree three.

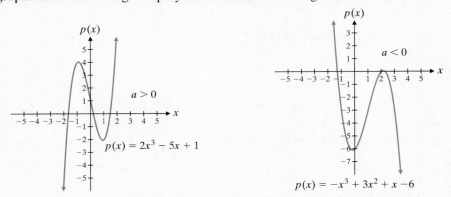

$a > 0$

$p(x) = 2x^3 - 5x + 1$

$a < 0$

$p(x) = -x^3 + 3x^2 + x - 6$

We observe that if the value of a is positive, the graph falls to the left and rises to the right.

Similarly we observe that if the value of a is negative, the graph rises to the left and falls to the right.

There are many additional properties of third-degree polynomials. We will not examine them now. They will be covered in higher-level mathematics courses.

Problems for Individual Investigation and Analysis

1. The number of cubic feet of lumber in thousands produced by the Kerr Sawmills of Fox River, Nova Scotia, Canada, can be approximated by the polynomial function

$$p(x) = x^3 - 80x^2 + 1900x + 2000, \text{ where } x \text{ is the number of years since 1940.}$$

Source: Loring Kerr, owner, Kerr Sawmills.
Evaluate $p(x)$ when $x = 0, 10, 20, 30, 40, 50,$ and 60.

2. Use the data obtained in Exercise 1 to graph the polynomial function. Connect the plotted points with a smooth curve. From your graph determine in what year between 1960 and 1990 the smallest number of cubic feet of lumber was produced by the Kerr Sawmills.

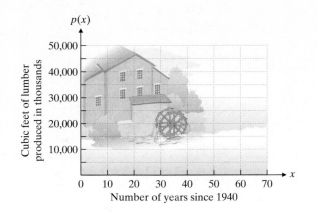

Problems for Group Investigation and Cooperative Learning

A fourth-degree polynomial function is a function of the form

$$p(x) = ax^4 + bx^3 + cx^2 + dx + e, \text{ where } a \neq 0.$$

Now examine the graphs of the following two polynomial functions of degree four.

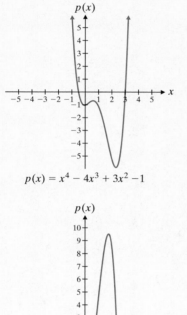

$$p(x) = x^4 - 4x^3 + 3x^2 - 1$$

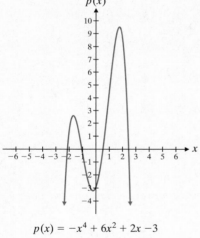

$$p(x) = -x^4 + 6x^2 + 2x - 3$$

We observe that if the value of a is positive, the graph rises to the left and also to the right.

Similarly, if the value of a is negative, the graph falls to the right and to the left.

3. The number of Big Macs served during a specific hour at the Hamilton Mall McDonald's can be approximated by the polynomial function

$$p(x) = -x^4 + 15x^3 - 66x^2 + 80x + 50,$$
where x is the number of hours since 11 A.M.

Using this equation, determine how many Big Macs will be served during the hour that begins at 12 noon. How many will be served during the hour that begins at 6 P.M. Evaluate $p(x)$ when $x = 0, 1, 2, 3, 4, 5, 6, 7,$ and 8.

4. Using the data obtained in Exercise 3, graph the polynomial function. Connect the plotted points with a smooth curve. From your graph determine at what hour the hourly demand for Big Macs is at its lowest.

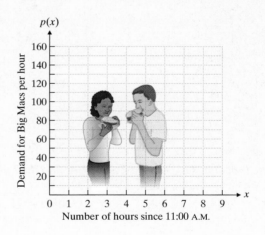

Internet Connections

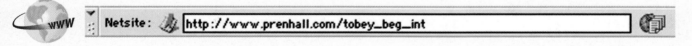

Site: U.S. National Oceanic and Atmospheric Administration

5. Search the site to find the exact number of tornadoes that occurred in the United States in each of the following years: 1987, 1988, 1989, 1990. A second-degree polynomial function that closely models this number is $p(x) = 57.75x^2 - 14.75x + 656.75$, where x is the number of years since 1987. Let us examine the equation for years 1987 to 1990. How accurately does the equation model the number of tornadoes that occurred in the United States in each of these years?

6. Now search the site to find the exact number of tornadoes that occurred in the United States in each of the following years: 1996, 1997, 1998, 1999. Is the equation used in Exercise 5 a reliable model for the number of tornadoes in each of these years? Why or why not? What does this tell us about constructing models with polynomial equations?

Chapter 11 Review Problems

For the function $f(x) = \dfrac{1}{2}x + 3$, *find the following:*

1. $f(a - 1)$

2. $f(a + 2)$

3. $f(a - 1) - f(a)$

4. $f(a + 2) - f(a)$

5. $f(2a + 3)$

6. $f(2a - 3)$

For the function $p(x) = -2x^2 + 3x - 1$, *find the following:*

7. $p(-3)$

8. $p(4)$

9. $p(2a) + p(-2)$

10. $p(3a) + p(3)$

11. $p(a + 2)$

12. $p(a - 3)$

For the function $h(x) = |2x - 1|$, *find the following:*

13. $h(8a)$

14. $h(7a)$

15. $h(\tfrac{1}{4}a)$

16. $h\left(\dfrac{3}{2}a\right)$

17. $h(a - 5)$

18. $h(a + 4)$

For the function $r(x) = \dfrac{3x}{x + 4}$, $x \neq -4$, *find the following. In each case, write your answers as one fraction.*

19. $r(5)$

20. $r(-6)$

21. $r(a + 3)$

22. $r(a - 2)$

23. $r(3) + r(a)$

24. $r(a) + r(-2)$

Find $\dfrac{f(x + h) - f(x)}{h}$ *for the following:*

25. $f(x) = 7x - 4$

26. $f(x) = 6x - 5$

27. $f(x) = 2x^2 - 5x$

28. $f(x) = 2x - 3x^2$

Examine each of the following graphs. **(a)** *Determine whether the graph is the graph of a function.* **(b)** *Determine whether the graph represents a one-to-one function.*

29.

30.

31.

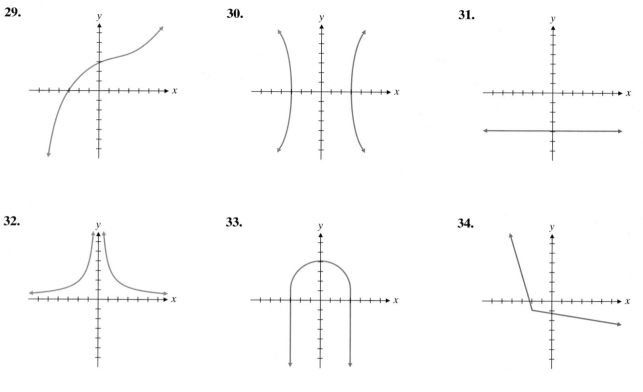

32.

33.

34.

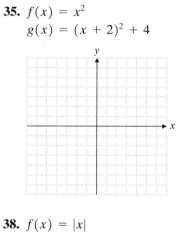

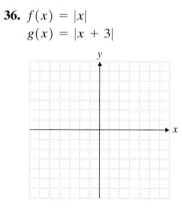

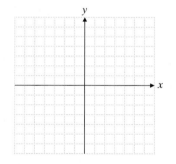

Graph each pair of functions on one set of axes.

35. $f(x) = x^2$
$\quad g(x) = (x + 2)^2 + 4$

36. $f(x) = |x|$
$\quad g(x) = |x + 3|$

37. $f(x) = |x|$
$\quad g(x) = |x - 4|$

38. $f(x) = |x|$
$\quad h(x) = |x| + 3$

39. $f(x) = |x|$
$\quad h(x) = |x| - 2$

40. $f(x) = x^3$
$\quad r(x) = (x + 3)^3 + 1$

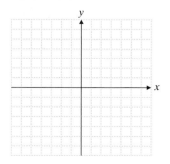

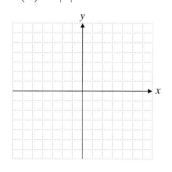

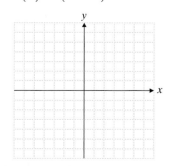

41. $f(x) = x^3$

$r(x) = (x - 1)^3 + 5$

42. $f(x) = \dfrac{2}{x}, x \neq 0$

$r(x) = \dfrac{2}{x + 3} - 2, x \neq -3$

43. $f(x) = \dfrac{4}{x}, x \neq 0$

$r(x) = \dfrac{4}{x + 2}, x \neq -2$

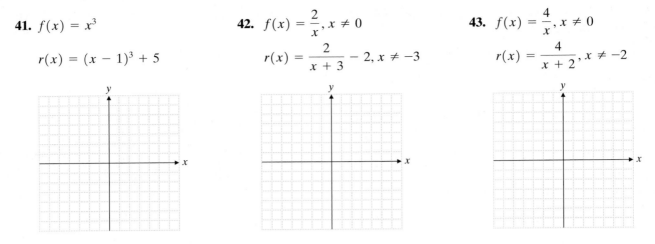

Given the functions,

$$f(x) = 3x + 5; \quad g(x) = \frac{2}{x}, \quad x \neq 0; \quad s(x) = \sqrt{x - 2}, \quad x \geq 2;$$

$$h(x) = \frac{x + 1}{x - 4}, \quad x \neq 4; \quad p(x) = 2x^2 - 3x + 4; \quad \text{and} \quad t(x) = -\frac{1}{2}x - 3,$$

find each of the following:

44. $(f + p)(x)$

45. $(f + t)(x)$

46. $(t - f)(x)$

47. $(p - f)(x)$

48. $(p - f)(2)$

49. $(t - f)(-3)$

50. $(fg)(x)$

51. $(tp)(x)$

52. $\left(\dfrac{g}{h}\right)(x)$

53. $\left(\dfrac{g}{f}\right)(x)$

54. $\left(\dfrac{g}{h}\right)(-2)$

55. $\left(\dfrac{g}{f}\right)(-3)$

56. $f\big[t(x)\big]$

57. $h\big[f(x)\big]$

58. $s\big[p(x)\big]$

59. $s\big[t(x)\big]$

60. $s\big[p(2)\big]$

61. $s\big[t(-18)\big]$

62. Show that
$f\big[g(x)\big] \neq g\big[f(x)\big].$

63. Show that
$p\big[g(x)\big] \neq g\big[p(x)\big].$

For each set, determine **(a)** *the domain,* **(b)** *the range,* **(c)** *whether the set defines a function, and* **(d)** *whether the set defines a one-to-one function.*

64. $B = \{(3, 7), (7, 3), (0, 8), (0, -8)\}$

65. $A = \{(100, 10), (200, 20), (300, 30), (400, 10)\}$

66. $D = \left\{\left(\frac{1}{2}, 2\right), \left(\frac{1}{4}, 4\right), \left(-\frac{1}{3}, -3\right), \left(4, \frac{1}{4}\right)\right\}$

67. $C = \{(12, 6), (0, 6), (0, -1), (-6, -12)\}$

68. $E = \{(0, 1), (1, 2), (2, 9), (-1, -2)\}$

69. $F = \{(3, 7), (2, 1), (0, -3), (1, 1)\}$

Find the inverse of each of the following functions.

70. $A = \left\{\left(3, \frac{1}{3}\right), \left(-2, -\frac{1}{2}\right), \left(-4, -\frac{1}{4}\right), \left(5, \frac{1}{5}\right)\right\}$

71. $B = \{(1, 10), (3, 7), (12, 15), (10, 1)\}$

72. $f(x) = -\frac{3}{4}x + 2$

73. $g(x) = -8 - 4x$

74. $h(x) = \frac{x + 2}{3}$

75. $j(x) = \frac{1}{x - 3}$

76. $p(x) = \sqrt[3]{x + 1}$

77. $r(x) = x^3 + 2$

Find the inverse of each function. Graph the function and its inverse on one coordinate plane. Then, on that same set of axes, graph the line $y = x$ *as a dashed line.*

78. $f(x) = \frac{-x - 2}{3}$

79. $f(x) = -\frac{3}{4}x + 1$

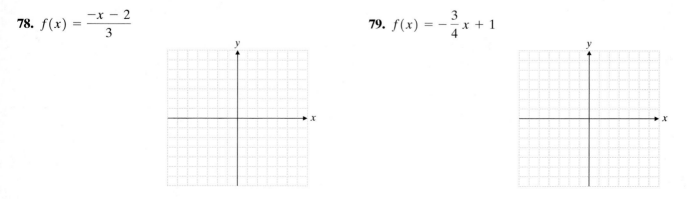

1. _____

2. _____

3. _____

4. _____

5. _____

6. _____

7. _____

8. (a) _____

(b) _____

9. (a) _____

(b) _____

10. _____

11. _____

12. (a) _____

(b) _____

(c) _____

For the function $f(x) = \dfrac{3}{4}x - 2$, *find the following:*

1. $f(-8)$ **2.** $f(2a)$ **3.** $f(a) - f(2)$

For the function $f(x) = 3x^2 - 2x + 4$, *find the following:*

4. $f(-6)$ **5.** $f(a + 1)$

6. $f(a) + f(1)$ **7.** $f(-2a) - 2$

Look at each graph below. **(a)** *Does the graph represent a function?* **(b)** *Does the graph represent a one-to-one function?*

8.

9.

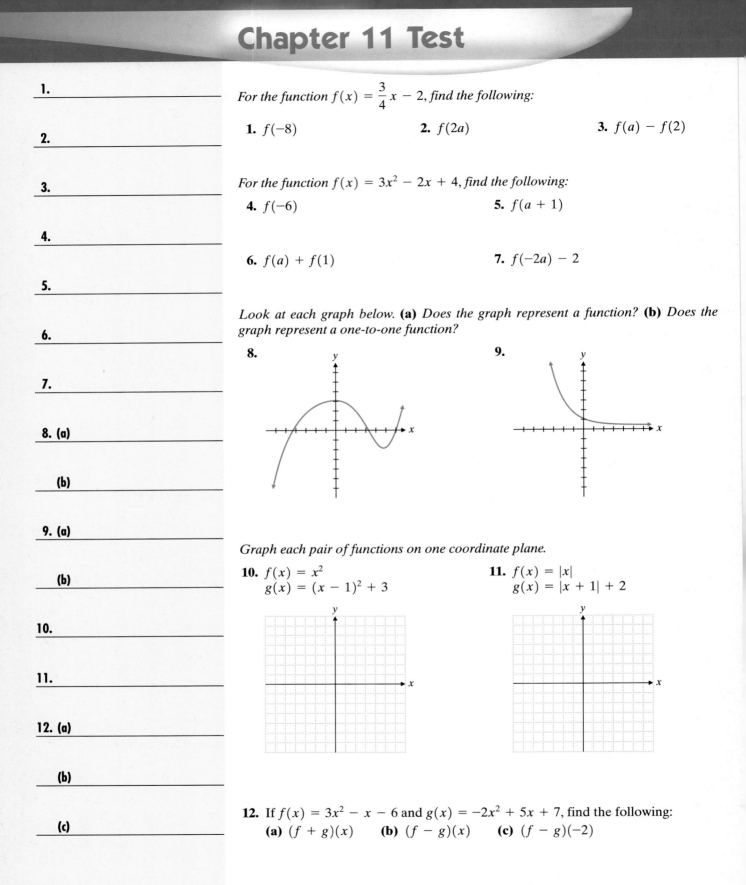

Graph each pair of functions on one coordinate plane.

10. $f(x) = x^2$
$g(x) = (x - 1)^2 + 3$

11. $f(x) = |x|$
$g(x) = |x + 1| + 2$

12. If $f(x) = 3x^2 - x - 6$ and $g(x) = -2x^2 + 5x + 7$, find the following:
(a) $(f + g)(x)$ **(b)** $(f - g)(x)$ **(c)** $(f - g)(-2)$

13. If $f(x) = \dfrac{3}{x}$, $x \neq 0$ and $g(x) = 2x - 1$, find the following:

    (a) $(fg)(x)$ **(b)** $\left(\dfrac{f}{g}\right)(x)$ **(c)** $g[f(x)]$

14. If $f(x) = \dfrac{1}{2}x - 3$ and $g(x) = 4x + 5$, find the following:

    (a) $(f \circ g)(x)$ **(b)** $(g \circ f)(x)$ **(c)** $f[f(x)]$

Look at the following functions. **(a)** *Is the function one-to-one?* **(b)** *If so, find the inverse of the function.*

15. $B = \{(1, 8), (8, 1), (9, 10), (-10, 9)\}$

16. $A = \{(1, 5), (2, 1), (4, -7), (0, 7)\}$

17. Determine the inverse of $f(x) = \dfrac{1}{2}x - \dfrac{1}{5}$.

18. Find f^{-1}. Graph f and its inverse f^{-1} on one coordinate plane. Graph $y = x$ as a dashed line for a reference.

$$f(x) = -3x + 2$$

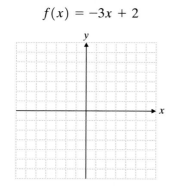

19. Given that $f(x) = \dfrac{3}{7}x + \dfrac{1}{2}$ and that $f^{-1}(x) = \dfrac{14x - 7}{6}$, find $f^{-1}[f(x)]$.

13. (a) _____

(b) _____

(c) _____

14. (a) _____

(b) _____

(c) _____

15. (a) _____

(b) _____

16. (a) _____

(b) _____

17. _____

18. _____

19. _____

1. _____

2. _____

3. _____

4. _____

5. _____

6. _____

7. _____

8. _____

9. _____

10. _____

11. _____

12. _____

13. _____

14. _____

15. (a) _____

(b) _____

(c) _____

Approximately one-half of this test covers the content of chapters 1–10. The remainder covers the content of Chapter 11.

1. Simplify. $3x\{2y - 3[x + 2(x + 2y)]\}$

2. Evaluate $2x^2 - 3xy + y^2$ when $x = 2$ and $y = -3$.

3. Solve for x. $\dfrac{1}{2}(x - 2) = \dfrac{1}{3}(x + 10) - 2x$

4. Factor completely. $16x^4 - 1$

5. Multiply. $(3x + 1)(2x - 3)(x + 4)$

6. Solve for x. $\dfrac{3x}{x^2 - 4} = \dfrac{2}{x + 2} + \dfrac{4}{2 - x}$

7. Find the equation of the line with slope $= -3$ that passes through the point $(2, -1)$. Write your answer in slope-intercept form.

8. Solve for (x, y).
$$3x + 2y = 5$$
$$7x + 5y = 11$$

9. Simplify. $\sqrt{18x^5y^6z^3}$

10. Multiply. $(\sqrt{2} + \sqrt{3})(2\sqrt{2} - 4\sqrt{3})$

11. Find the distance between $(6, -1)$ and $(-3, -4)$.

12. Factor $12x^2 - 11x + 2$.

13. Factor $x^4 - 10x^2 + 9$.

14. Write the standard form of the equation of a circle with radius 14 and center at $(-3, 6)$.

15. If $f(x) = 3x^2 - 2x + 1$, find the following:
(a) $f(-2)$
(b) $f(a - 2)$
(c) $f(a) + f(-2)$

16. Graph $f(x) = x^3$ and $g(x) = (x + 2)^3 + 4$ on one coordinate plane.

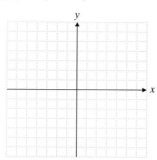

17. If $f(x) = 2x^2 - 5x - 6$ and $g(x) = 5x + 3$, find the following:
   (a) $(fg)(x)$
   (b) $\left(\dfrac{f}{g}\right)(x)$
   (c) $f[g(x)]$

18. $A = \{(3, 6), (1, 8), (2, 7), (4, 4)\}$
   (a) Is A a function?
   (b) Is A a one-to-one function?
   (c) Find A^{-1}.

19. Find the inverse function of $f(x) = 7x - 3$.

20. $f(x) = 5x^3 - 3x^2 - 6$
   (a) Find $f(5)$.
   (b) Find $f(-3)$.
   (c) Find $f(2a)$.

21. (a) Find the inverse function of $f(x) = -\dfrac{2}{3}x + 2$.

   (b) Graph f and f^{-1} on one coordinate plane. Graph $y = x$ as a reference.

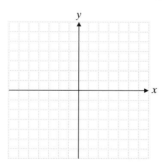

22. Find $f[f^{-1}(x)]$ using your results from Exercise 21.

16. _____

17. (a) _____

   (b) _____

   (c) _____

18. (a) _____

   (b) _____

   (c) _____

19. _____

20. (a) _____

   (b) _____

   (c) _____

21. (a) _____

   (b) _____

22. _____

669

Logarithmic and Exponential Functions

Have you ever noticed how some things cool quickly (like a cup of coffee in a ceramic mug) while other things cool more slowly (like a cup of coffee in a foam cup)? Have you wondered how that happens? If you know the temperature of a room, can you predict how long it will take a hot item in the room to cool? Turn to the Putting Your Skills to Work exercises on page 715.

If you are familiar with the topics in this chapter, take this test now. Check your answers with those in the back of the book. If an answer is wrong or you can't do an exercise, study the appropriate section of the chapter.

If you are not familiar with the topics in this chapter, don't take this test now. Instead, study the examples, work the practice exercises, and then take the test.

This test will help you identify those concepts that you have mastered and those that need more study.

Section 12.1

1. Sketch the graph of $f(x) = 2^{-x}$. Plot at least four points.

2. Solve for x. $3^{2x-1} = 27$

3. When a principal amount P is invested at interest rate r compounded annually, the amount of money A earned after t years is given by the equation $A = P(1 + r)^t$. How much money will Nancy have in 4 years if she invests $10,000 in a mutual fund that pays 12% interest compounded annually?

Section 12.2

4. Write in logarithmic form. $\dfrac{1}{49} = 7^{-2}$

5. Solve for x. $\log_5 x = 3$

6. Evaluate $\log_{10}(10{,}000)$.

Section 12.3

7. Write the logarithm in terms of $\log_5 x$, $\log_5 y$, and $\log_5 z$. $\log_5\left(\dfrac{x^2 y^5}{z^3}\right)$

8. Express as one logarithm. $\dfrac{1}{2}\log_4 x - 3\log_4 w$

9. Find x if $\log_3 x + \log_3 2 = 4$.

Section 12.4

Use a scientific calculator to evaluate each of the following. Round your answers to the nearest ten-thousandth.

10. Find x if $\log x = 3.9170$.

11. $\ln 4.79$

12. $\log_6 5.02$

13. $\log 0.7523$

14. Find the value of x in the equation $\ln x = 22.976$. (Express your answer in scientific notation.)

Section 12.5

15. Solve the following logarithmic equation and check your solution.

$$\log x - \log(x + 3) = -1$$

16. Solve the exponential equation $4^{2x+1} = 9$. (Round your answer to the nearest ten-thousandth.)

17. How long would it take for $2000 to grow to $7000 at 6% annual interest compounded yearly? Use the formula $A = P(1 + r)^t$. Round your answer to the nearest year.

1. _____

2. _____

3. _____

4. _____

5. _____

6. _____

7. _____

8. _____

9. _____

10. _____

11. _____

12. _____

13. _____

14. _____

15. _____

16. _____

17. _____

12.1 The Exponential Function

1 Graphing an Exponential Function

We have seen examples of exponential equations in previous chapters. For example,

$$2^{-2} = \frac{1}{4},$$

$$2^{1/2} = \sqrt{2}, \text{ and}$$

$$2^{1.7} = 2^{17/10} = \sqrt[10]{2^{17}}.$$

The above equations are in the form $a^x = b$ where x is a rational number. We can also define such equations when x is an irrational number, such as π or $\sqrt{2}$. However, we will leave this definition for a more-advanced course.

We define an **exponential function** for all real values of x as follows:

> **Definition of Exponential Function**
>
> The function $f(x) = b^x$, where $b > 0, b \neq 1$, and x is a real number, is called an **exponential function**. The number b is called the **base** of the function.

Now let's look at some graphs of exponential functions.

EXAMPLE 1 Graph $f(x) = 2^x$.

We make a table of values for x and $f(x)$.

$$f(-1) = 2^{-1} = \frac{1}{2}, \qquad f(0) = 2^0 = 1, \qquad f(1) = 2^1 = 2$$

Verify the other values in the table below. We then draw the graph.

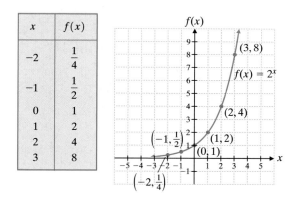

x	$f(x)$
-2	$\frac{1}{4}$
-1	$\frac{1}{2}$
0	1
1	2
2	4
3	8

Notice how the curve of $f(x) = 2^x$ comes *very close to* the x-axis but *never touches* it. The x-axis is an **asymptote** for every exponential function. You should also notice that $f(x)$ is always positive, so the range of f is the set of all positive real numbers (whereas the domain is the set of all real numbers). When the base is greater than one, as x increases, $f(x)$ increases faster and faster (that is, the curve gets steeper).

Practice Problem 1 Graph $f(x) = 3^x$.

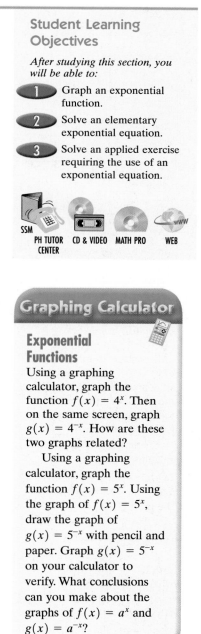

Graphing Calculator

Exponential Functions

Using a graphing calculator, graph the function $f(x) = 4^x$. Then on the same screen, graph $g(x) = 4^{-x}$. How are these two graphs related?

Using a graphing calculator, graph the function $f(x) = 5^x$. Using the graph of $f(x) = 5^x$, draw the graph of $g(x) = 5^{-x}$ with pencil and paper. Graph $g(x) = 5^{-x}$ on your calculator to verify. What conclusions can you make about the graphs of $f(x) = a^x$ and $g(x) = a^{-x}$?

PRACTICE PROBLEM 1

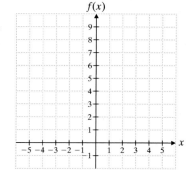

PRACTICE PROBLEM 2

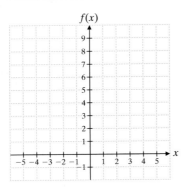

EXAMPLE 2 Graph $f(x) = \left(\dfrac{1}{2}\right)^x$.

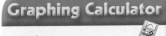

We can write $f(x) = \left(\dfrac{1}{2}\right)^x$ as $f(x) = \left(\dfrac{1}{2}\right)^x = (2^{-1})^x = 2^{-x}$ and evaluate it for a few values of x. We then draw the graph.

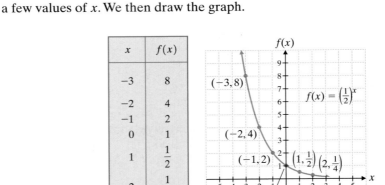

x	$f(x)$
-3	8
-2	4
-1	2
0	1
1	$\dfrac{1}{2}$
2	$\dfrac{1}{4}$

Note that as x increases, $f(x)$ decreases.

Practice Problem 2 Graph $f(x) = \left(\dfrac{1}{3}\right)^x$.

To Think About Look at the graph of $f(x) = 2^x$ in Example 1 and the graph of $f(x) = \left(\dfrac{1}{2}\right)^x = 2^{-x}$ in Example 2. How are the two graphs related?

EXAMPLE 3 Graph $f(x) = 3^{x-2}$.

We will make a table of values for a few values of x. Then we will graph the function.

$$f(0) = 3^{0-2} = 3^{-2} = \dfrac{1}{3^2} = \dfrac{1}{9}$$

$$f(1) = 3^{1-2} = 3^{-1} = \dfrac{1}{3}$$

$$f(2) = 3^{2-2} = 3^0 = 1$$
$$f(3) = 3^{3-2} = 3^1 = 3$$
$$f(4) = 3^{4-2} = 3^2 = 9$$

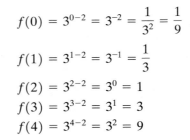

x	$f(x)$
0	$\dfrac{1}{9}$
1	$\dfrac{1}{3}$
2	1
3	3
4	9

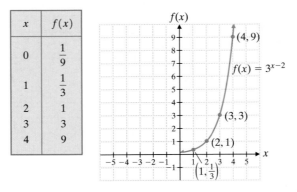

We observe that the curve is that of $f(x) = 3^x$ except that it has been shifted 2 units to the right.

Practice Problem 3 Graph $f(x) = 3^{x+2}$.

Graphing Calculator

Exploration
Using a graphing calculator, graph $f(x) = 2^x$. Then, on the same screen, graph $g(x) = 2^{x+3}$. Describe the shift that occurs. What will the graph of $g(x) = 2^{x+5}$ look like? Verify using the graphing calculator.

Using the graphing calculator, graph $f(x) = 2^{x-2}$. Describe the shift that occurs. What will the graph of $g(x) = 2^{x-3}$ look like? Verify using the graphing calculator.

Based on your experience with functions, what would the graph of $f(x) = 2^x + 3$ look like? How about $f(x) = 2^x - 4$? Verify using the graphing calculator.

PRACTICE PROBLEM 3

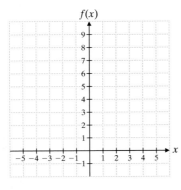

To Think About How is the graph of $f(x) = 3^{x+2}$ related to the graph of $f(x) = 3^x$? Without making a table of values, draw the graph of $f(x) = 3^{x+3}$. Draw the graph of $f(x) = 3^{x-3}$.

For the next example we need to discuss a special number that is denoted by the letter e. The letter e is a number like π. It is an irrational number. It occurs in many formulas that describe real-world phenomena, such as the growth of cells and radioactive decay. We need an approximate value for e to use this number in calculations: $e \approx \mathbf{2.7183}$.

An extremely useful function is the exponential function $f(x) = e^x$. We usually obtain values for e^x by using a calculator or a computer. If you have a scientific calculator, use the $\boxed{e^x}$ key. (Many scientific calculators require you to press $\boxed{\text{SHIFT}}$ $\boxed{\text{ln}}$ or $\boxed{\text{2nd F}}$ $\boxed{\text{ln}}$ or $\boxed{\text{INV}}$ $\boxed{\text{ln}}$ to obtain the operation e^x.) If you have a calculator that is not a scientific calculator, use $e \approx 2.7183$ as an approximate value. If you don't have any calculator, use Table A-2 in the appendix.

⬤ EXAMPLE 4 Graph $f(x) = e^x$.

We evaluate $f(x)$ for some negative and some positive values of x. We begin with $f(-2)$. Notice that the x-column in Table A-2 has only positive values. To find the value of $f(-2) = e^{-2}$, we must locate 2 in the x-column and then read across to the value in the column under e^{-x}. Thus, we see that $f(-2) \approx 0.1353$, or 0.14 rounded to the nearest hundredth.

To find $f(2) = e^2$ on a scientific calculator, we enter 2 $\boxed{e^x}$ and obtain 7.389056099 as an approximation. (On some scientific calculators you will need to use the keystrokes 2 $\boxed{\text{2nd F}}$ $\boxed{\text{ln}}$ or 2 $\boxed{\text{SHIFT}}$ $\boxed{\text{ln}}$ or 2 $\boxed{\text{INV}}$ $\boxed{\text{ln}}$.) Thus, $f(2) = e^2 \approx 7.39$ to the nearest hundredth.

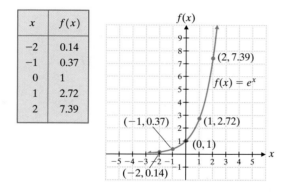

x	$f(x)$
-2	0.14
-1	0.37
0	1
1	2.72
2	7.39

Practice Problem 4 Graph $f(x) = e^{x-2}$.

⬤ **PRACTICE PROBLEM 4**

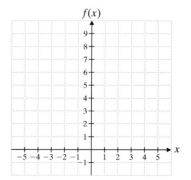

To Think About Look at the graphs of $f(x) = e^x$ and $f(x) = e^{x-2}$. Describe the shift that occurs. Without making a table of values, draw the graph of $f(x) = e^{x+3}$.

② Solving an Elementary Exponential Equation

All the usual laws of exponents are true for exponential functions. We also have the following important property to help us solve exponential equations.

Property of Exponential Equations

If $b^x = b^y$, then $x = y$ for $b > 0$ and $b \neq 1$.

EXAMPLE 5 Solve $2^x = \dfrac{1}{16}$.

To use the property of exponential equations, we must have the same base on both sides of the equation.

$$2^x = \frac{1}{16}$$

$$2^x = \frac{1}{2^4} \qquad \text{Because } 2^4 = 16.$$

$$2^x = 2^{-4} \qquad \text{Because } \frac{1}{2^4} = 2^{-4}.$$

$$x = -4 \qquad \text{Property of exponential equations.}$$

Practice Problem 5 Solve $2^x = \dfrac{1}{32}$.

3 Solving an Applied Exercise Requiring the Use of an Exponential Equation

An exponential function can be used to solve compound interest exercises. If a principal P is invested at an interest rate r compounded annually, the amount of money A accumulated after t years is $A = P(1 + r)^t$.

EXAMPLE 6 If a young married couple invests \$5000 in a mutual fund that pays 16% interest compounded annually, how much will they have in 3 years?

Here $P = 5000$, $r = 0.16$, and $t = 3$.

$$\begin{aligned}
A &= P(1 + r)^t \\
&= 5000(1 + 0.16)^3 \\
&= 5000(1.16)^3 \\
&= 5000(1.560896) \\
&= 7804.48
\end{aligned}$$

The couple will have \$7804.48.

If you have a scientific calculator, you can find the value of $5000(1.16)^3$ immediately by using the $\boxed{\times}$ key and the $\boxed{y^x}$ key. On most scientific calculators you can use the following keystrokes.

$$5000 \;\boxed{\times}\; 1.16 \;\boxed{y^x}\; 3 \;\boxed{=}\; 7804.48$$

Practice Problem 6 If Uncle Jose invests \$4000 in a mutual fund that pays 11% interest compounded annually, how much will he have in 2 years?

Interest is often compounded quarterly or monthly or even daily. Therefore, a more useful form of the interest formula that allows for variable compounding is needed. If a principal P is invested at an annual interest rate r that is compounded n times a year, then the amount of money A accumulated after t years is

$$A = P\left(1 + \frac{r}{n}\right)^{nt}.$$

EXAMPLE 7 If we invest $8000 in a fund that pays 15% annual interest compounded monthly, how much will we have after 6 years?

In this situation $P = 8000$, $r = 15\% = 0.15$, and $n = 12$. The interest is compounded monthly or twelve times per year. Finally, $t = 6$ since the interest will be compounded for 6 years.

$$A = 8000\left(1 + \frac{0.15}{12}\right)^{(12)(6)}$$
$$= 8000(1 + 0.0125)^{72}$$
$$= 8000(1.0125)^{72}$$
$$\approx 8000(2.445920268)$$
$$\approx 19{,}567.36215$$

Rounding to the nearest cent, we obtain the answer $19,567.36. Using a scientific calculator, we could have found the answer directly by using the following keystrokes.

$$8000 \boxed{\times} 1.0125 \boxed{y^x} 72 \boxed{=} 19{,}567.36215$$

Depending on your calculator, your answer may contain fewer or more digits.

Practice Problem 7 How much money would Collette have if she invested $1500 for 8 years at 8% annual interest if the interest is compounded quarterly?

An exponential function is used to describe radioactive decay. The equation $A = Ce^{kt}$ tells us how much of a radioactive element is left in a sample after a specified time.

EXAMPLE 8 The radioactive decay of the chemical element americium 241 can be described by the equation

$$A = Ce^{-0.0016008t},$$

where C is the original amount of the element in the sample; A is the amount of the element remaining after t years; and $k = -0.0016008$, the decay constant for americium. If 10 milligrams (mg) of americium 241 is sealed in a laboratory container today, how much will theoretically be present in 2000 years? Round your answer to the nearest hundredth.

Here $C = 10$ and $t = 2000$.

$$A = 10e^{-0.0016008(2000)} = 10e^{-3.2016}$$

Using a calculator or Table A-2, we have

$$A \approx 10(0.040697) = 0.40697 \approx 0.41 \text{ mg.}$$

The expression $10e^{-3.2016}$ can be found directly on some scientific calculators as follows:

$$10 \boxed{\times} 3.2016 \boxed{+/-} \boxed{e^x} \boxed{=} 0.406970366$$

(Scientific calculators with no $\boxed{e^x}$ key will require the keystrokes $\boxed{\text{INV}}$ $\boxed{\ln}$ or $\boxed{\text{2nd F}}$ $\boxed{\ln}$ or $\boxed{\text{SHIFT}}$ $\boxed{\ln}$ in place of the $\boxed{e^x}$.)

Thus, 0.41 milligrams of americium 241 would be present in 2000 years.

Practice Problem 8 If 20 milligrams of americium 241 is present in a sample now, how much will theoretically be present in 5000 years? Round your answer to the nearest thousandth.

Graphing Calculator

Exploration

Graph the function $f(t) = 10e^{-0.0016008t}$ from Example 8 for $t = 0$ to $t = 100$ years. Now graph the function for $t = 0$ to $t = 2000$ years. What significant change is there in the two graphs? From the graphs, estimate a value of t for which $f(t) = 5.0$. (Round your value of t to the nearest hundredth.)

12.1 Exercises

Verbal and Writing Skills

1. The exponential function is an equation of the form _____.

2. The irrational number *e* is a number that is approximately equal to _____. (Give your answer with four decimal places.)

Graph each function.

3. $f(x) = 3^x$

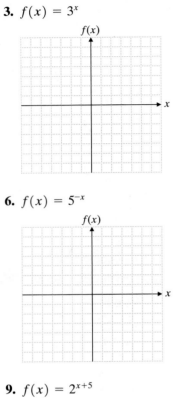

4. $f(x) = 2^x$

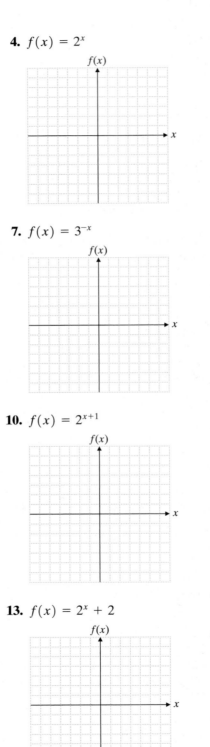

5. $f(x) = 2^{-x}$

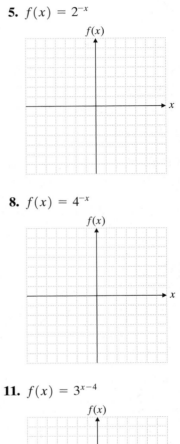

6. $f(x) = 5^{-x}$

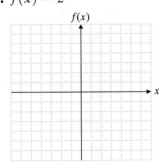

7. $f(x) = 3^{-x}$

8. $f(x) = 4^{-x}$

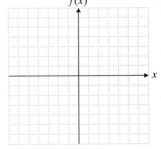

9. $f(x) = 2^{x+5}$

10. $f(x) = 2^{x+1}$

11. $f(x) = 3^{x-4}$

12. $f(x) = 3^{x-1}$

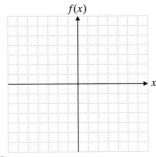

13. $f(x) = 2^x + 2$

14. $f(x) = 2^x - 2$

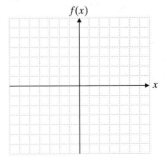

Graph each function. Use a calculator or Table A–2.

15. $f(x) = e^{x-1}$

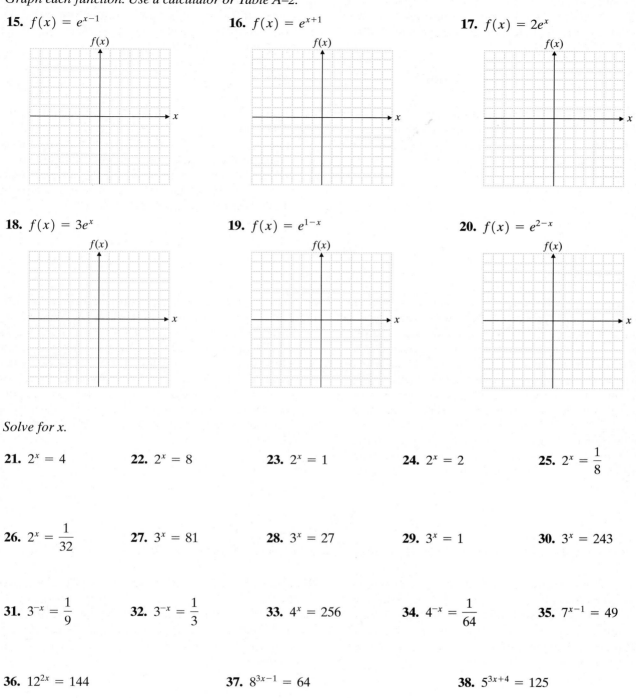

16. $f(x) = e^{x+1}$

17. $f(x) = 2e^x$

18. $f(x) = 3e^x$

19. $f(x) = e^{1-x}$

20. $f(x) = e^{2-x}$

Solve for x.

21. $2^x = 4$

22. $2^x = 8$

23. $2^x = 1$

24. $2^x = 2$

25. $2^x = \dfrac{1}{8}$

26. $2^x = \dfrac{1}{32}$

27. $3^x = 81$

28. $3^x = 27$

29. $3^x = 1$

30. $3^x = 243$

31. $3^{-x} = \dfrac{1}{9}$

32. $3^{-x} = \dfrac{1}{3}$

33. $4^x = 256$

34. $4^{-x} = \dfrac{1}{64}$

35. $7^{x-1} = 49$

36. $12^{2x} = 144$

37. $8^{3x-1} = 64$

38. $5^{3x+4} = 125$

To solve Exercises 39–42, use the interest formula $A = P\left(1 + \dfrac{r}{n}\right)^{nt}$. Round your answers to the nearest cent.

39. Alicia is investing $2000 at an annual rate of 6.3% compounded annually. How much money will Alicia have after 3 years?

40. Manza is investing $5000 at an annual rate of 7.1% compounded annually. How much money will Manza have after 4 years?

41. How much money will Suki have in 5 years if she invests $2000 at a 14% annual rate of interest compounded quarterly? How much will she have if it is compounded monthly?

42. How much money will Gina have in 2 years if she invests $8000 at a 17% annual rate of interest compounded quarterly? How much will she have if it is compounded monthly?

43. The number of bacteria in a culture is given by $B(t) = 4000(2^t)$, where t is the time in hours. How many bacteria will grow in the culture in the first 3 hours? In the first 9 hours?

44. Suppose that the cost of a college education is increasing 4% per year. The equation $C(t) = P(1.04)^t$ forecasts the tuition cost t years from now and is based on the present cost P in dollars. How much will a college now charging $3000 for tuition charge in 10 years? How much will a college now charging $12,000 for tuition charge in 15 years?

45. The city of Manchester just put in a municipal sewer to solve an underground water contamination problem, and many homeowners would like to have sewer lines connected to their homes. It is expected that each year the number of homeowners who use their own private septic tanks rather than the public sewer system will decrease by 8%. What percentage of people will still be using their private septic tanks in 5 years? The city feels that the underground water contamination problem will be solved when the number of homeowners still using septic tanks is less than 10%. Will that goal be achieved in the next 25 years?

46. U.S. Navy divers off the coast of Nantucket are searching for the wreckage of an old World War II–era submarine. They have found that if the water is relatively clear and the surface is calm, the ocean filters out 18% of the sunlight for each 4 feet they descend. How much sunlight is available at a depth of 20 feet? The divers need to use underwater spotlights when the amount of sunlight is less than 10%. Will they need spotlights when working at a depth of 48 feet?

Use an exponential equation to solve each problem. Round your answers to the nearest hundredth.

47. The radioactive decay of radium 226 can be described by the equation $A = Ce^{-0.0004279t}$, where C is the original amount of radium and A is the amount of radium remaining after t years. If 6 milligrams of radium are sealed in a container now, how much radium will be in the container after 1000 years?

48. The radioactive decay of radon 222 can be described by the equation $A = Ce^{-0.1813t}$, where C is the original amount of radon and A is the amount of radon after t days. If 1.5 milligrams are in a laboratory container today, how much was there in the container 10 days ago?

Use the following information for Exercises 49 and 50. The atmospheric pressure measured in pounds per square inch is given by the equation $P = 14.7e^{-0.21d}$, where d is the distance in miles above sea level. Round your answers to the nearest hundredth.

49. What is the pressure in pounds per square inch experienced by a man on a Colorado mountain that is 2 miles above sea level?

50. What is the pressure in pounds per square inch on an American Airlines jet plane flying 10 miles above sea level?

The total number of shares N (in millions) that were traded on the New York Stock Exchange in any given year between 1940 and 2000 can be approximated by the equation $N = 0.00472e^{0.11596t}$, where t is the number of years since 1900. Use this information for Exercises 51 and 52. Source: New York Stock Exchange.

51. Using the given equation, determine how many stocks were traded in 1980 and in 1990. What was the percent of increase from 1980 to 1990?

52. Using the given equation, determine how many stocks were traded in 1985 and in 2000. What was the percent of increase from 1985 to 2000?

The population of the world is growing exponentially. The following table and graph contain population data for selected years and show a pattern of significant increases.

Year	AD1	1650	1850	1930	1975	1995	2000
Approximate World Population in Billions	0.2	0.5	1	2	4	5.68	6.07

Source: Statistical Division of the United Nations.

53. Based on the graph, in approximately what year did the world's population reach three billion people?

54. Based on the graph, what was the approximate world population in 1900?

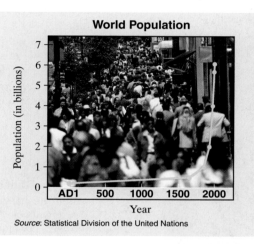

World Population

Population (in billions)

Year

Source: Statistical Division of the United Nations

55. The growth rate of the world's population during the period 1980–1990 was 1.7% per year. If that rate were to continue from 1995 to 2005, what would the world's population be in 2005?

56. The growth rate of the world's population during the period 1990–1997 was 1.4% per year. If that rate were to continue from 1995 to 2010, what would the world's population be in 2010?

Optional Graphing Calculator Problems

57. Let $f(x) = \dfrac{e^x + e^{-x}}{2}$. Evaluate $f(x)$ when $x = -1$, $-0.5, 0, 0.5, 1, 1.5,$ and 2. Now use these values to graph the function. (*f* defines a special function called *the hyperbolic cosine*. This function is used in advanced mathematics and science to study a variety of technical applications.)

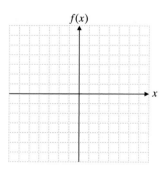

$f(x)$

x

58. Let $g(x) = \dfrac{e^x - e^{-x}}{2}$. Evaluate $g(x)$ when $x = -2$, $-1, -0.5, 0, 0.5, 1, 1.5,$ and 2. Now graph the function using these values. (*g* defines a special function called *the hyperbolic sine*.)

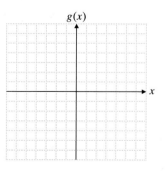

$g(x)$

x

Cumulative Review Problems *Solve for x.*

59. $5 - 2(3 - x) = 2(2x + 5) + 1$

60. $\dfrac{7}{12} + \dfrac{3}{4}x + \dfrac{5}{4} = -\dfrac{1}{6}x$

Student Learning Objectives

After studying this section, you will be able to:

1. Write exponential equations in logarithmic form.
2. Write logarithmic equations in exponential form.
3. Solve elementary logarithmic equations.
4. Graph a logarithmic function.

SSM
PH TUTOR CENTER   CD & VIDEO   MATH PRO   WEB

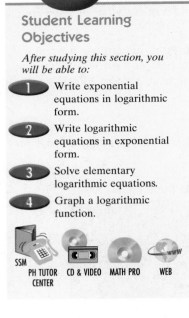

John Napier (1550–1617)

Logarithms were invented about 400 years ago by the Scottish mathematician John Napier. Napier's amazing invention reduced complicated exercises to simple subtraction and addition. Astronomers quickly saw the immense value of logarithms and began using them. The work of Johannes Kepler, Isaac Newton, and others would have been much more difficult without logarithms.

The most important thing to know for this chapter is that a logarithm is an exponent. In Section 12.1 we solved the equation $2^x = 8$. We found that $x = 3$. The question we faced was, "To what power do we raise 2 to get 8?" The answer was 3. Mathematicians have to solve this type of problem so often that we have invented a short-hand notation for asking the question. Instead of asking, "To what power do we raise 2 to get 8?" we say instead, "What is $\log_2 8$?" Both questions mean the same thing.

Now suppose we had a general equation $x = b^y$ and someone asked, "To what power do we raise b to get x?" We would abbreviate this question by asking, "What is $\log_b x$?" Thus, we see that $y = \log_b x$ is an equivalent form of the equation $x = b^y$.

The key concept you must remember is that a logarithm is an exponent. We write $\log_b x = y$ to mean that the logarithm of x to the base b is equal to y. y is the exponent.

Definition of Logarithm

The **logarithm**, base b, of a *positive* number x is the power (exponent) to which the base b must be raised to produce x. That is, $y = \log_b x$ is the same as $x = b^y$, where $b > 0$ and $b \neq 1$.

Often you will need to convert logarithmic expressions to exponential expressions, and vice versa, to solve equations.

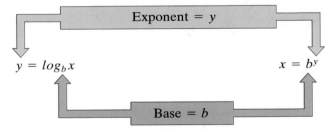

Exponent = y

$y = \log_b x$ $x = b^y$

Base = b

1 Writing Exponential Equations in Logarithmic Form

We begin by converting exponential expressions to logarithmic expressions.

EXAMPLE 1   Write in logarithmic form.

(a) $81 = 3^4$ **(b)** $\dfrac{1}{100} = 10^{-2}$

We use the fact that $x = b^y$ is equivalent to $\log_b x = y$.

(a) $81 = 3^4$
Here $x = 81$, $b = 3$, and $y = 4$. So $4 = \log_3 81$.

(b) $\dfrac{1}{100} = 10^{-2}$

Here $x = \dfrac{1}{100}$, $b = 10$, and $y = -2$. So $-2 = \log_{10}\left(\dfrac{1}{100}\right)$.

Practice Problem 1 Write in logarithmic form.

(a) $49 = 7^2$ **(b)** $\dfrac{1}{64} = 4^{-3}$

2 *Writing Logarithmic Equations in Exponential Form*

If we have an equation with a logarithm in it, we can write it in the form of an exponential equation. This is a very important skill. Carefully study the following example.

EXAMPLE 2 Write in exponential form.

(a) $2 = \log_5 25$ **(b)** $-4 = \log_{10}\left(\dfrac{1}{10,000}\right)$

(a) $2 = \log_5 25$
   Here $y = 2$, $b = 5$, and $x = 25$. Thus, since $x = b^y$, $25 = 5^2$.

(b) $-4 = \log_{10}\left(\dfrac{1}{10,000}\right)$

   Here $y = -4$, $b = 10$, and $x = \dfrac{1}{10,000}$. So $\dfrac{1}{10,000} = 10^{-4}$.

Practice Problem 2 Write in exponential form.

(a) $3 = \log_5 125$ **(b)** $-2 = \log_6\left(\dfrac{1}{36}\right)$

3 *Solving Elementary Logarithmic Equations*

Many logarithmic equations are fairly easy to solve if we first convert them to an equivalent exponential equation.

EXAMPLE 3 Solve for the variable.

(a) $\log_5 x = -3$ **(b)** $\log_a 16 = 4$

(a) $5^{-3} = x$

   $\dfrac{1}{5^3} = x$

   $\dfrac{1}{125} = x$

(b) $a^4 = 16$
   $a^4 = 2^4$
   $a = 2$

Practice Problem 3 Solve for the variable.

(a) $\log_b 125 = 3$ **(b)** $\log_{1/2} 32 = x$

With this knowledge we have the ability to solve an additional type of exercise.

EXAMPLE 4   Evaluate $\log_3 81$.

Now, what exactly is the exercise asking for? It is asking, "To what power must we raise 3 to get 81?" Since we do not know the power, we call it x. We have

$$\log_3 81 = x$$
$$81 = 3^x \quad \text{Write an equivalent exponential equation.}$$
$$3^4 = 3^x \quad \text{Write 81 as } 3^4.$$
$$x = 4 \quad \text{If } b^x = b^y, \text{ then } x = y \text{ for } b > 0 \text{ and } b \neq 1.$$

Thus, $\log_3 81 = 4$.

Practice Problem 4   Evaluate $\log_{10} 0.1$.

4   *Graphing a Logarithmic Function*

We found in Chapter 11 that the graphs of a function and its inverse have an interesting property. They are symmetric to one another with respect to the line $y = x$. We also found in Chapter 11 that the procedure for finding the inverse of a function is to interchange the x and y variables. For example, $y = 2x + 3$ and $x = 2y + 3$ are inverse functions. In similar fashion, $y = 2^x$ and $x = 2^y$ are inverse functions. Another way to write $x = 2^y$ is the logarithmic equation $y = \log_2 x$. Thus, the logarithmic function $y = \log_2 x$ is the **inverse** of the exponential function $y = 2^x$. If we graph the function $y = 2^x$ and $y = \log_2 x$ on the same set of axes, the graph of one is the reflection of the other about the line $y = x$.

EXAMPLE 5   Graph $y = \log_2 x$.

If we write $y = \log_2 x$ in exponential form, we have $x = 2^y$. We make a table of values and graph the function $x = 2^y$.

In each case, we pick a value of y as a first step.

$$\text{If} \quad y = -2, \qquad x = 2^y = 2^{-2} = \frac{1}{2^2} = \frac{1}{4}.$$

$$\text{If} \quad y = -1, \qquad x = 2^{-1} = \frac{1}{2}.$$

$$\text{If} \quad y = 0, \qquad x = 2^0 = 1.$$

$$\text{If} \quad y = 1, \qquad x = 2^1 = 2.$$

$$\text{If} \quad y = 2, \qquad x = 2^2 = 4.$$

x	y
$\frac{1}{4}$	-2
$\frac{1}{2}$	-1
1	0
2	1
4	2

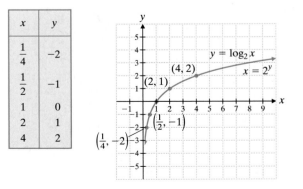

Practice Problem 5 Graph $y = \log_{1/2} x$.

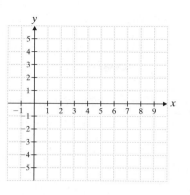

$f(x) = a^x$ and $f(x) = \log_a x$ are inverse functions. As such they have all the properties of inverse functions. We will review a few of these properties as we study the graphs of two inverse functions, $y = 2^x$ and $y = \log_2 x$.

EXAMPLE 6 Graph $y = \log_2 x$ and $y = 2^x$ on the same set of axes.

Make a table of values (ordered pairs) for each equation. Then draw each graph.

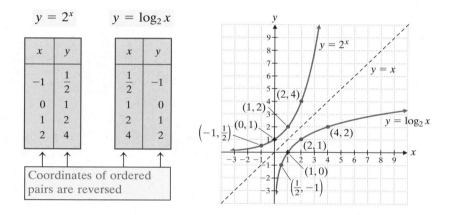

Note that $y = \log_2 x$ is the inverse of $y = 2^x$ because the ordered pairs (x, y) are reversed. The sketch of the two equations shows that they are inverses. If we reflect the graph of $y = 2^x$ about the line $y = x$, it will coincide with the graph of $y = \log_2 x$.

Recall that in function notation, f^{-1} means the inverse function of f. Thus, if we write $f(x) = \log_2 x$, then $f^{-1}(x) = 2^x$.

Practice Problem 6 Graph $y = \log_6 x$ and $y = 6^x$ on the same set of axes.

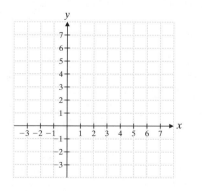

Exercises

Verbal and Writing Skills

1. A logarithm is an _____.

2. In the equation $y = \log_b x$, the value b is called the _____.

3. In the equation $y = \log_b x$, the domain (the set of permitted values of x) is _____.

4. In the equation $y = \log_b x$, the permitted values of b are _____.

Write in logarithmic form.

5. $81 = 3^4$

6. $343 = 7^3$

7. $36 = 6^2$

8. $100 = 10^2$

9. $\dfrac{1}{25} = 5^{-2}$

10. $0.01 = 10^{-2}$

11. $\dfrac{1}{32} = 2^{-5}$

12. $\dfrac{1}{64} = 2^{-6}$

13. $y = e^5$

14. $y = e^{-8}$

Write in exponential form.

15. $2 = \log_3 9$

16. $2 = \log_2 4$

17. $0 = \log_5 1$

18. $0 = \log_{13} 1$

19. $\dfrac{1}{2} = \log_{16} 4$

20. $\dfrac{1}{2} = \log_{100} 10$

21. $-2 = \log_{10}(0.01)$

22. $-3 = \log_{10}(0.001)$

23. $-4 = \log_3\left(\dfrac{1}{81}\right)$

24. $-5 = \log_2\left(\dfrac{1}{32}\right)$

25. $\dfrac{2}{3} = \log_e x$

26. $\dfrac{3}{7} = \log_e x$

Solve.

27. $\log_2 x = 4$

28. $\log_2 x = 6$

29. $\log_{10} x = -3$

30. $\log_{10} x = -2$

31. $\log_4 64 = y$

32. $\log_7 343 = y$

33. $\log_8\left(\dfrac{1}{64}\right) = y$

34. $\log_3\left(\dfrac{1}{243}\right) = y$

35. $\log_a 144 = 2$

36. $\log_a 625 = 4$

37. $\log_a 1000 = 3$

38. $\log_a 100 = 2$

39. $\log_{25} 5 = w$

40. $\log_8 2 = w$

41. $\log_3\left(\dfrac{1}{3}\right) = w$

42. $\log_{12} 1 = w$ **43.** $\log_{15} w = 0$ **44.** $\log_{10} w = -3$ **45.** $\log_w 81 = -2$ **46.** $\log_w 77 = -1$

Evaluate.

47. $\log_{10}(0.001)$ **48.** $\log_{10}(0.0001)$ **49.** $\log_2 128$ **50.** $\log_3 27$ **51.** $\log_{23} 1$

52. $\log_{18} \dfrac{1}{18}$ **53.** $\log_6 \sqrt{6}$ **54.** $\log_7 \sqrt{7}$ **55.** $\log_2 64$ **56.** $\log_3 \dfrac{1}{27}$

Graph.

57. $\log_4 x = y$

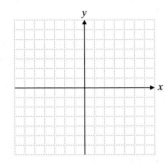

58. $\log_3 x = y$

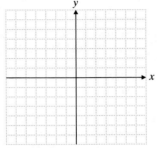

59. $\log_{1/3} x = y$

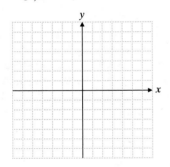

60. $\log_{1/4} x = y$

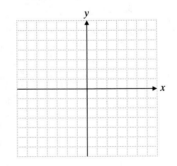

61. $\log_{10} x = y$

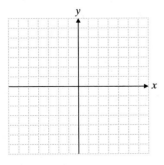

62. $\log_8 x = y$

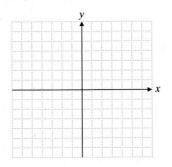

On one coordinate plane, graph the function f and the function f^{-1}. Then graph a dashed line for the equation $y = x$.

63. $f(x) = \log_3 x, f^{-1}(x) = 3^x$

64. $f(x) = \log_4 x, f^{-1}(x) = 4^x$

Applications

To determine whether a solution is an acid or a base, chemists check the solution's pH. A solution is an acid if its pH is less than 7 and a base if its pH is greater than 7. The pH is defined by $pH = -\log_{10}[H^+]$, where $[H^+]$ is the concentration of hydrogen ions in the solution.

65. The concentration of hydrogen ions in cranberries is approximately $10^{-2.5}$. What is the pH of cranberries?

66. The concentration of hydrogen ions in a lemon is approximately 10^{-2}. What is the pH of a lemon?

67. A scientist at the Center for Disease Control and Prevention has mixed a solution with a pH of 8. Find the concentration of hydrogen ions $[H^+]$ in the solution.

68. A construction team repairing the Cape Cod Canal has created a cement mixture with a pH of 9. Find the concentration of hydrogen ions $[H^+]$ in the solution.

69. The EPA is testing a batch of experimental industrial solvent with a pH of 9.25. Find the concentration of hydrogen ions $[H^+]$ in the solution. Give your answer in scientific notation rounded to three decimal places.

70. The chef of The Depot Restaurant has prepared a special balsamic vinaigrette salad dressing. What is the pH of the dressing if the concentration of hydrogen ions is 1.103×10^{-3}? The logarithm, base 10, of 0.001103 is approximately -2.957424488. Round your answer to three decimal places.

Maptech produces map software for people who want to use their home computers to locate any place in the United States and print a detailed map. In the first 2 years of sales of the software, they have found that for N sets of software to be sold, they need to invest d dollars for advertising according to the equation

$$N = 1200 + (2500)(\log_{10} d),$$

where d is always a positive number not less than 1.

71. How many sets of software were sold when they spent $100,000 on advertising?

72. How many sets of software were sold when they spent $10,000 on advertising?

73. They have a goal for 2 years from now of selling 18,700 sets of software. How much should they spend on advertising?

74. They have a goal for next year of selling 16,200 sets of software. How much should they spend on advertising?

75. Evaluate $5^{\log_5 4}$.

76. Evaluate $\log_2 \sqrt[4]{2}$.

Cumulative Review Problems

77. Graph $y = -\dfrac{2}{3}x + 5$.

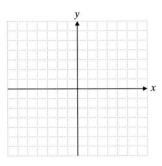

78. Graph $6x + 3y = -6$.

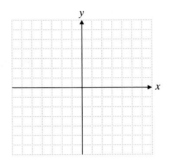

79. Find the slope of a straight line containing $(-6, 3)$ and $(-1, 2)$.

80. Find the equation of the line perpendicular to $y = -\dfrac{2}{3}x + 4$ that contains $(-4, 1)$.

81. Solve for x. $\dfrac{1}{4x} - \dfrac{3}{2x} = \dfrac{5}{8}$

82. Solve for x. $\sqrt{5x - 13} = \sqrt{4x - 9}$

83. Assume the cost of a college education is increasing at 4% per year. The equation $C(t) = P(1.04)^t$ forecasts the tuition cost t years from now, based on the present cost P in dollars.
  (a) How much will a college now charging $4400 for tuition charge in 5 years?
  (b) How much will a college now charging $16,500 for tuition charge in 10 years?

84. The number of viral cells in a laboratory culture in a biology research project is given by $A(t) = 9000(2^t)$, where t is measured in hours.
  (a) How many viral cells will grow in the culture in the first 2 hours?
  (b) How many viral cells will grow in the culture in the first 12 hours?

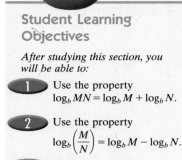
1 Using the Property $\log_b MN = \log_b M + \log_b N$

We can use logarithms to reduce complex expressions to addition and subtraction. The following properties show us how to use logarithms in this way.

> **Property 1   The Logarithm of a Product**
>
> For any positive real numbers M and N and any positive base $b \neq 1$,
> $$\log_b MN = \log_b M + \log_b N.$$

To see that this property is true, we let
$$\log_b M = x \quad \text{and} \quad \log_b N = y,$$
where x and y are any values. Now we write the expressions in exponential notation:
$$b^x = M \quad \text{and} \quad b^y = N.$$
Then
$$MN = b^x b^y = b^{x+y} \qquad \text{Laws of exponents.}$$
If we convert this equation to logarithmic form, then we have the following:
$$\log_b MN = x + y \qquad \text{Definition of logarithm.}$$
$$\log_b MN = \log_b M + \log_b N \qquad \text{By substitution.}$$
Note that the logarithms must have the same base.

EXAMPLE 1   Write $\log_3 XZ$ as a sum of logarithms.

By property 1, $\log_3 XZ = \log_3 X + \log_3 Z$.

Practice Problem 1   Write $\log_4 WXY$ as a sum of logarithms.

EXAMPLE 2   Write $\log_3 16 + \log_3 x + \log_3 y$ as a single logarithm.

If we extend our rule, we have $\log_b MNP = \log_b M + \log_b N + \log_b P$. Thus,
$$\log_3 16 + \log_3 x + \log_3 y = \log_3 16xy.$$

Practice Problem 2   Write $\log_7 w + \log_7 8 + \log_7 x$ as a single logarithm.

2 Using the Property $\log_b\left(\dfrac{M}{N}\right) = \log_b M - \log_b N$

Property 2 is similar to property 1 except that it involves two expressions that are divided, not multiplied.

> **Property 2   The Logarithm of a Quotient**
>
> For any positive real numbers M and N and any positive base $b \neq 1$,
> $$\log_b\left(\frac{M}{N}\right) = \log_b M - \log_b N.$$

Property 2 can be proved using a similar approach to the one used to prove property 1. The proof will be left as an exercise for you to try.

EXAMPLE 3 Write $\log_3\left(\dfrac{29}{7}\right)$ as the difference of two logarithms.

$$\log_3\left(\frac{29}{7}\right) = \log_3 29 - \log_3 7$$

Practice Problem 3 Write $\log_3\left(\dfrac{17}{5}\right)$ as the difference of two logarithms.

 EXAMPLE 4 Express $\log_b 36 - \log_b 9$ as a single logarithm.

$$\log_b 36 - \log_b 9 = \log_b\left(\frac{36}{9}\right) = \log_b 4$$

Practice Problem 4 Express $\log_b 132 - \log_b 4$ as a single logarithm.

Caution

Be sure you understand property 2!

$$\frac{\log_b M}{\log_b N} \neq \log_b M - \log_b N$$

Do you see why?

---

3 *Using the Property $\log_b M^p = p \log_b N$*

We now introduce the third property. We will not prove it now, but you will have a chance to verify this property as an exercise.

> **Property 3   The Logarithm of a Number Raised to a Power**
>
> For any positive real number M, any real number p, and any positive base $b \neq 1$,
>
> $$\log_b M^p = p \log_b M.$$

 EXAMPLE 5 Write $\dfrac{1}{3}\log_b x + 2 \log_b w - 3 \log_b z$ as a single logarithm.

First, we must eliminate the coefficients of the logarithm terms.

$$\log_b x^{1/3} + \log_b w^2 - \log_b z^3 \quad \text{By property 3.}$$

Now we can combine either the sum or difference of the logarithms. We'll do the sum.

$$\log_b x^{1/3}w^2 - \log_b z^3 \quad \text{By property 1.}$$

Now we combine the difference.

$$\log_b\left(\frac{x^{1/3}w^2}{z^3}\right) \quad \text{By property 2.}$$

Practice Problem 5 Write $\dfrac{1}{3}\log_7 x - 5 \log_7 y$ as one logarithm.

EXAMPLE 6 Write $\log_b\left(\dfrac{x^4 y^3}{z^2}\right)$ as three logarithms.

$$\log_b\left(\frac{x^4 y^3}{z^2}\right) = \log_b x^4 y^3 - \log_b z^2 \qquad \text{By property 2.}$$

$$= \log_b x^4 + \log_b y^3 - \log_b z^2 \qquad \text{By property 1.}$$

$$= 4\log_b x + 3\log_b y - 2\log_b z \qquad \text{By property 3.}$$

Practice Problem 6 Write $\log_3\left(\dfrac{x^4 y^5}{z}\right)$ as three logarithms.

Solving a Simple Logarithmic Equation

A major goal in solving many logarithmic equations is to obtain a logarithm on one side of the equation and no logarithm on the other side. In Example 7 we will use property 1 to combine two separate logarithms that are added.

EXAMPLE 7 Find x if $\log_2 x + \log_2 5 = 3$.

$$\log_2 5x = 3 \qquad \text{Use property 1.}$$

$$5x = 2^3 \qquad \text{Convert to exponential form.}$$

$$5x = 8 \qquad \text{Simplify.}$$

$$x = \frac{8}{5} \qquad \text{Divide both sides by 5.}$$

Practice Problem 7 Find x if $\log_4 x + \log_4 5 = 2$.

In Example 8, two logarithms are subtracted on the left side of the equation. We can use property 2 to combine these two logarithms. This will allow us to obtain the form of one logarithm on one side of the equation and no logarithm on the other side.

EXAMPLE 8 Find x if $\log_3(x + 4) - \log_3(x - 4) = 2$.

$$\log_3\left(\frac{x + 4}{x - 4}\right) = 2 \qquad \text{Use property 2.}$$

$$\frac{x + 4}{x - 4} = 3^2 \qquad \text{Convert to exponential form.}$$

$$x + 4 = 9(x - 4) \qquad \text{Multiply each side by } (x - 4).$$

$$x + 4 = 9x - 36 \qquad \text{Simplify.}$$

$$40 = 8x$$

$$5 = x$$

Practice Problem 8 Find x if $\log_{10} x - \log_{10}(x + 3) = -1$.

To solve some logarithmic equations, we need a few additional properties of logarithms. We state these properties now. The proofs of some of them will be left as exercises for you.

The following properties are true for all positive values of $b \neq 1$ and all positive values of x and y.

Property 4 $\log_b b = 1$

Property 5 $\log_b 1 = 0$

Property 6 If $\log_b x = \log_b y$, then $x = y$.

We now illustrate each property in Example 9.

🔵 EXAMPLE 9

(a) Evaluate $\log_7 7$.   **(b)** Evaluate $\log_5 1$.   **(c)** Find x if $\log_3 x = \log_3 17$.

(a) $\log_7 7 = 1$ because $\log_b b = 1$.   Property 4.

(b) $\log_5 1 = 0$ because $\log_b 1 = 0$.   Property 5.

(c) If $\log_3 x = \log_3 17$, then $x = 17$.   Property 6.

Practice Problem 9 Evaluate.

(a) $\log_7 1$ **(b)** $\log_8 8$ **(c)** Find y if $\log_{12} 13 = \log_{12}(y + 2)$.

We now have the mathematical tools needed to solve a variety of logarithmic equations.

🔵 EXAMPLE 10   Find x if $2 \log_7 3 - 4 \log_7 2 = \log_7 x$.

We can use property 3 in two cases.

$$2 \log_7 3 = \log_7 3^2 = \log_7 9$$
$$4 \log_7 2 = \log_7 2^4 = \log_7 16$$

By substituting these results, we have the following.

$$\log_7 9 - \log_7 16 = \log_7 x$$

$$\log_7 \frac{9}{16} = \log_7 x \quad \text{Property 2.}$$

$$\frac{9}{16} = x \qquad \text{Property 6.}$$

Practice Problem 10 Find x if $\log_3 2 - \log_3 5 = \log_3 6 + \log_3 x$.

Express as a sum of logarithms.

1. $\log_7 MN$

2. $\log_{12} CD$

3. $\log_5(7 \cdot 11)$

4. $\log_6(13 \cdot 5)$

5. $\log_b 9f$

6. $\log_b 5d$

Express as a difference of logarithms.

7. $\log_9\left(\dfrac{2}{7}\right)$

8. $\log_{11}\left(\dfrac{23}{17}\right)$

9. $\log_a\left(\dfrac{G}{7}\right)$

10. $\log_b\left(\dfrac{H}{10}\right)$

11. $\log_a\left(\dfrac{E}{F}\right)$

12. $\log_6\left(\dfrac{8}{M}\right)$

Express as a product.

13. $\log_8 a^7$

14. $\log_5 b^{10}$

15. $\log_b A^{-2}$

16. $\log_a B^{-5}$

17. $\log_5 \sqrt{w}$

18. $\log_6 \sqrt{z}$

Mixed Practice

Write each expression as the sum or difference of single logarithms of x, y, and z.

19. $\log_5 \sqrt{x}\, y^3$

20. $\log_3 x^4 \sqrt{y}$

21. $\log_{13}\left(\dfrac{5B}{A^2}\right)$

22. $\log_6\left(\dfrac{M^5 N}{3}\right)$

23. $\log_2\left(\dfrac{5xy^4}{\sqrt{z}}\right)$

24. $\log_5\left(\dfrac{3x^5 \sqrt[3]{y}}{z^4}\right)$

25. $\log_b \sqrt[3]{\dfrac{x}{y^2 z}}$

26. $\log_b \sqrt[4]{\dfrac{z}{x^2 y^3}}$

Write as a single logarithm.

27. $\log_4 13 + \log_4 y + \log_4 3$

28. $\log_8 15 + \log_8 a + \log_8 b$

29. $5 \log_3 x - \log_3 7$

30. $3 \log_8 5 - \log_8 z$

31. $\frac{2}{3} \log_b x + \frac{1}{2} \log_b y - 3 \log_b z$

32. $\frac{3}{4} \log_b x + 2 \log_b y - \frac{1}{2} \log_b z$

Use the properties of logarithms to simplify each of the following.

33. $\log_3 3$

34. $\log_7 7$

35. $\log_e e$

36. $\log_{10} 10$

37. $\log_9 1$

38. $\log_e 1$

39. $\log_5 5 + \log_5 1$

40. $\log_6 6 + \log_6 1$

Find x in each of the following.

41. $\log_8 x = \log_8 7$

42. $\log_9 x = \log_9 5$

43. $\log_5 (2x + 7) = \log_5 (29)$

44. $\log_{15} (26) = \log_{15} (3x - 1)$

45. $\log_3 1 = x$

46. $\log_8 1 = x$

47. $\log_7 7 = x$

48. $\log_5 5 = x$

49. $\log_{10} x + \log_{10} 25 = 2$

50. $\log_{10} x + \log_{10} 5 = 1$

51. $\log_2 7 = \log_2 x - \log_2 3$

52. $\log_5 1 = \log_5 x - \log_5 8$

53. $3 \log_5 x = \log_5 8$

54. $\frac{1}{2} \log_3 x = \log_3 4$

55. $\log_e x - \log_e 2 = 2$

56. $\log_e x + \log_e 3 = 1$

57. $\log_6 (5x + 21) - \log_6 (x + 3) = 1$

58. $\log_5 (15x + 4) - \log_5 (x - 1) = 2$

59. It can be shown that $y = b^{\log_b y}$. Use this property to evaluate $5^{\log_5 4} + 3^{\log_3 2}$.

60. It can be shown that $x = \log_b b^x$. Use this property to evaluate $\log_7 \sqrt[4]{7} + \log_6 \sqrt[12]{6}$.

To Think About

61. Prove that $\log_b\left(\dfrac{M}{N}\right) = \log_b M - \log_b N$ by using an argument similar to the proof of property 1.

62. Prove that $\log_b M^p = p \log_b M$ by using an argument similar to the proof of property 1.

Cumulative Review Problems

63. Find the area of a circle whose radius is 4 meters.

64. Find the volume of a cylinder with a radius of 2 meters and a height of 5 meters.

65. Solve for (x, y).
$$5x + 3y = 9$$
$$7x - 2y = 25$$

66. Solve for (x, y, z).
$$2x - y + z = 3$$
$$x + 2y + 2z = 1$$
$$4x + y + 2z = 0$$

67. Carbon dioxide emissions in China increased from 8.01×10^8 metric tons of carbon per year in 1996 to 9.30×10^8 metric tons in 2000. What is the percent of increase during this 4-year period? If an equal percent of increase occurs from 2000 to 2004, what will be the level of carbon dioxide emissions in China in 2004? *Source:* U.S. Energy Information Administration.

68. Carbon dioxide emissions in Japan decreased from 3.04×10^8 metric tons of carbon per year in 1996 to 2.73×10^8 metric tons in 2000. What is the percent of decrease during this 4-year period? If an equal percent of decrease occurs from 2000 to 2004, what will be the level of carbon dioxide emissions in Japan in 2004? *Source:* U.S. Energy Information Administration.

69. Natasha's biology professor gives a total of five tests during the semester. She received the following scores on the first 4 tests: 84, 80, 92, 95. What score must Natasha get on the fifth test to have an average of 88?

70. The area of a triangle is 56 square feet. Its altitude is 9 feet less than the length of the base. Find the dimensions of the triangle.

12.4 Finding Logarithmic Function Values on a Calculator

1 Finding Common Logarithms on a Scientific Calculator

Student Learning Objectives

After studying this section, you will be able to:

1 Find common logarithms.

2 Find the antilogarithm of a common logarithm.

3 Find natural logarithms.

4 Find the antilogarithm of a natural logarithm.

5 Evaluate a logarithm to a base other than 10 or e.

SSM PH TUTOR CD & VIDEO MATH PRO WEB
CENTER

Although we can find a logarithm of a number for any positive base except 1, the most frequently used bases are 10 and e. Base 10 logarithms are called *common logarithms* and are usually written with no subscript.

> **Definition**
>
> For all real numbers $x > 0$, the **common logarithm** of x is
> $$\log x = \log_{10} x.$$

Before the advent of calculators and computers, people used tables of common logarithms. Now most work with logarithms is done with the aid of a scientific calculator. We will take that approach in this section of the text. To find the common logarithm of a number on a scientific calculator, enter the number and then press the $\boxed{\log x}$ or $\boxed{\log}$ key.

EXAMPLE 1 On a scientific calculator, find a decimal approximation for each of the following.

(a) $\log 7.32$ **(b)** $\log 73.2$ **(c)** $\log 0.314$

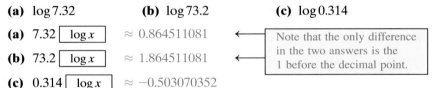

(a) $7.32 \boxed{\log x}$ ≈ 0.864511081 ←   Note that the only difference in the two answers is the 1 before the decimal point.

(b) $73.2 \boxed{\log x}$ ≈ 1.864511081 ←

(c) $0.314 \boxed{\log x}$ ≈ -0.503070352

Note: Your calculator may display fewer or more digits in the answer.

Practice Problem 1 On a scientific calculator, find a decimal approximation for each of the following.

(a) $\log 4.36$ **(b)** $\log 436$ **(c)** $\log 0.2418$

To Think About Why is the difference in the answers to Example 1 **(a)** and **(b)** equal to 1.00? Consider the following.

$$\begin{aligned} \log 73.2 &= \log(7.32 \times 10^1) && \text{Use scientific notation.} \\ &= \log 7.32 + \log 10^1 && \text{By property 1.} \\ &= \log 7.32 + 1 && \text{Because } \log_b b = 1. \\ &\approx 0.864511081 + 1 && \text{Use a calculator.} \\ &\approx 1.864511081 && \text{Add the decimals.} \end{aligned}$$

2 Finding Antilogarithms of a Common Logarithm on a Scientific Calculator

We have previously discussed the function $f(x) = \log x$ and the corresponding inverse function $f^{-1}(x) = 10^x$. The inverse of a logarithmic function is an exponential function. There is another name for this function. It is called an **antilogarithm**.

If $f(x) = \log x$ (here the base is understood to be 10), then $f^{-1}(x) = $ antilog $x = 10^x$.

EXAMPLE 2   Find an approximate value for x if $\log x = 4.326$.

Here we are given the value of the logarithm, and we want to find the number that has that logarithm. In other words, we want the antilogarithm. We know that $\log_{10} x = 4.326$ is equivalent to $10^{4.326} = x$. So to solve this problem, we want to find the value of 10 raised to the 4.326 power. Using a calculator, we have the following.

$$4.326 \boxed{10^x} \approx 21183.61135$$

Thus, $x \approx 21{,}183.61135$. (If your scientific calculator does not have a $\boxed{10^x}$ key, you can usually use $\boxed{\text{2nd Fn}}$ $\boxed{\log x}$ or $\boxed{\text{INV}}$ $\boxed{\log x}$ or $\boxed{\text{SHIFT}}$ $\boxed{\log x}$ to perform the operation.)

Practice Problem 2   Using a scientific calculator, find an approximate value for x if $\log x = 2.913$.

EXAMPLE 3   Evaluate antilog(-1.6784).

Asking what is antilog(-1.6784) is equivalent to asking what the value is of $10^{-1.6784}$. To determine this on a scientific calculator, it will be necessary to enter the numbers 1.6784 followed by the $\boxed{+/-}$ key.

$$1.6784 \boxed{+/-} \boxed{10^x} \approx 0.020970076$$

Thus, antilog$(-1.6784) \approx 0.020970076$.

Practice Problem 3   Evaluate antilog(-3.0705).

EXAMPLE 4   Using a scientific calculator, find an approximate value for x.

(a) $\log x = 0.07318$ **(b)** $\log x = -3.1621$

(a) $\log x = 0.07318$ is equivalent to $10^{0.07318} = x$.

$$0.07318 \boxed{10^x} \approx 1.183531987$$

Thus, $x \approx 1.183531987$.

(b) $\log x = -3.1621$ is equivalent to $10^{-3.1621} = x$.

$$3.1621 \boxed{+/-} \boxed{10^x} \approx 0.0006884937465.$$

Thus, $x \approx 0.0006884937465$.
(Some calculators may give the answer in scientific notation as $6.884937465 \times 10^{-4}$. This is often displayed on the calculator screen as $6.884937465 - 4$.)

Practice Problem 4   Using a scientific calculator, find an approximate value for x.

(a) $\log x = 0.06134$ **(b)** $\log x = -4.6218$

3   *Finding Natural Logarithms on a Scientific Calculator*

For most theoretical work in mathematics and other sciences, the most useful base for logarithms is e. Logarithms with base e are known as *natural logarithms* and are usually written $\ln x$.

> **Definition**
>
> For all real numbers $x > 0$, the **natural logarithm** of x is
>
> $$\ln x = \log_e x.$$

On a scientific calculator we can usually approximate natural logarithms with the $\boxed{\ln x}$ or $\boxed{\ln}$ key.

EXAMPLE 5 On a scientific calculator, approximate the following values.

(a) $\ln 7.21$ **(b)** $\ln 72.1$ **(c)** $\ln 0.0356$

(a) 7.21 | $\ln x$ | ≈ 1.975468951 **(b)** 72.1 | $\ln x$ | ≈ 4.278054044

(c) 0.0356 | $\ln x$ | ≈ -3.335409641

Note that there is no simple relationship between the answers to parts **(a)** and **(b)**. Do you see why these are different from common logarithms?

Practice Problem 5 On a scientific calculator, approximate the following values.

(a) $\ln 4.82$ **(b)** $\ln 48.2$ **(c)** $\ln 0.0793$

4 Finding Antilogarithms of a Natural Logarithm on a Scientific Calculator

EXAMPLE 6 On a scientific calculator, find an approximate value of x for each equation.

(a) $\ln x = 2.9836$ **(b)** $\ln x = -1.5619$

(a) If $\ln x = 2.9836$, then $e^{2.9836} = x$.

$$2.9836 \boxed{e^x} \approx 19.75882051$$

(b) If $\ln x = -1.5619$, then $e^{-1.5619} = x$.

$$1.5619 \boxed{+/-} \boxed{e^x} \approx 0.209737192$$

Practice Problem 6 On a scientific calculator, find an approximate value of x for each equation.

(a) $\ln x = 3.1628$ **(b)** $\ln x = -2.0573$

An alternative notation is sometimes used. This is $\text{antilog}_e(x)$.

5 Evaluating a Logarithm to a Base Other Than 10 or e

Although a scientific calculator has specific keys for finding common logarithms (base 10) and natural logarithms (base e), there are no keys for finding logarithms with other bases. What do we do in such cases? The logarithm of a number for a base other than 10 or e can be found with the following formula.

Change of Base Formula

$$\log_b x = \frac{\log_a x}{\log_a b},$$

where a, b, and $x > 0$, $a \neq 1$, and $b \neq 1$.

Let's see how this formula works. If we want to use common logarithms to find $\log_3 56$, we must first note that the value of b in the formula is 3. We then write

$$\log_3 56 = \frac{\log_{10} 56}{\log_{10} 3} = \frac{\log 56}{\log 3}.$$

Do you see why?

EXAMPLE 7 Evaluate using common logarithms. $\log_3 5.12$

$$\log_3 5.12 = \frac{\log 5.12}{\log 3}$$

On a calculator, we find the following.

$$5.12 \boxed{\log x} \div 3 \boxed{\log x} \boxed{=} \quad 1.486561234$$

Our answer is an approximate value with nine decimal places. Your answer may have more or fewer digits depending on your calculator.

Practice Problem 7 Evaluate using common logarithms. $\log_9 3.76$

If we desire to use base e, then the change of base formula is used with natural logarithms.

EXAMPLE 8 Obtain an approximate value for $\log_4 0.005739$ using natural logarithms.

Using the change of base formula, with $a = e$, $b = 4$, and $x = 0.005739$, we have the following.

$$\log_4 0.005739 = \frac{\log_e 0.005739}{\log_e 4} = \frac{\ln 0.005739}{\ln 4}$$

This is done on most scientific calculators as follows.

$$0.005739 \boxed{\ln} \div 4 \boxed{\ln} \boxed{=} \quad -3.722492455$$

Thus, we have $\log_4 0.005739 \approx -3.722492455$.

Check: To check our answer we want to know the following.

$$4^{-3.722492455} \overset{?}{=} 0.005739$$

Using a calculator, we can verify this with the $\boxed{y^x}$ key.

$$4 \boxed{y^x} 3.722492455 \boxed{+/-} \boxed{=} \quad 0.005739 \; \checkmark$$

Practice Problem 8 Obtain an approximate value for $\log_8 0.009312$ using natural logarithms.

EXAMPLE 9 Using a scientific calculator, graph $y = \log_2 x$.

If we use common logarithms ($\log_{10} x$), then for each value of x, we will need to calculate $\frac{\log x}{\log 2}$. Therefore, to find y when $x = 3$, we need to calculate $\frac{\log 3}{\log 2}$. On most scientific calculators, we would enter $3 \boxed{\log} \div 2 \boxed{\log} \boxed{=}$ and obtain 1.584962501. Rounded to the nearest tenth, we have $x = 3$ and $y = 1.6$. In a similar fashion we find other table values and then graph them.

<table>
<tr><td>x</td><td>$y = \log_2 x$</td></tr>
<tr><td>0.5</td><td>−1</td></tr>
<tr><td>1</td><td>0</td></tr>
<tr><td>2</td><td>1</td></tr>
<tr><td>3</td><td>1.6</td></tr>
<tr><td>4</td><td>2</td></tr>
<tr><td>6</td><td>2.6</td></tr>
<tr><td>8</td><td>3</td></tr>
</table>

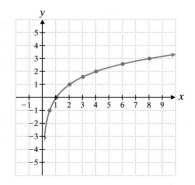

Graphing Calculator

Graphing Logarithmic Functions

You can use the change of base formula to graph logarithmic functions on a graphing calculator.

To graph $y = \log_2 x$ in Example 9 on a graphing calculator, enter the function $y = \dfrac{\log x}{\log 2}$ into the Y = editor of your calculator.

Display:

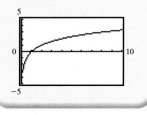

Practice Problem 9 Using a scientific calculator, graph $y = \log_5 x$.

Use a scientific calculator to approximate the following.

1. $\log 5.13$ **2.** $\log 4.95$ **3.** $\log 25.6$ **4.** $\log 83.8$ **5.** $\log 356$

6. $\log 896$ **7.** $\log 125,000$ **8.** $\log 78,500$ **9.** $\log 0.0123$ **10.** $\log 0.567$

Verbal and Writing Skills

11. Try to find $\log(-5.08)$. What happens? Why? **12.** Try to find $\log(-6.63)$. What happens? Why?

Find an approximate value of x using a scientific calculator.

13. $\log x = 2.016$ **14.** $\log x = 2.754$ **15.** $\log x = 1.7860$ **16.** $\log x = 1.7896$

17. $\log x = 3.9304$ **18.** $\log x = 3.9576$ **19.** $\log x = 6.4683$ **20.** $\log x = 5.6274$

21. $\log x = -3.3893$ **22.** $\log x = -4.0458$ **23.** $\log x = 2.0030$ **24.** $\log x = 2.1034$

Approximate the following with a scientific calculator.

25. antilog(7.6215) **26.** antilog(4.3894) **27.** antilog(-1.0826) **28.** antilog(-3.1145)

29. $\ln 5.62$ **30.** $\ln 8.81$ **31.** $\ln 107$ **32.** $\ln 463$

33. $\ln 136,000$ **34.** $\ln 129,000$ **35.** $\ln 0.00579$ **36.** $\ln 0.00134$

Find an approximate value of x using a scientific calculator.

37. $\ln x = 0.95$ **38.** $\ln x = 0.55$ **39.** $\ln x = 2.4$ **40.** $\ln x = 4.4$

41. $\ln x = -0.13$ **42.** $\ln x = -0.18$ **43.** $\ln x = -2.7$ **44.** $\ln x = -3.8$

Approximate the following with a scientific calculator.

45. antilog$_e(6.1582)$ **46.** antilog$_e(1.9047)$ **47.** antilog$_e(-2.1298)$ **48.** antilog$_e(-3.3712)$

*Use a scientific calculator and **common logarithms** to evaluate the following.*

49. $\log_3 9.2$　　**50.** $\log_2 6.13$　　**51.** $\log_7(7.35)$　　**52.** $\log_9(9.85)$

53. $\log_6 0.127$　　**54.** $\log_5 0.173$　　**55.** $\log_{11} 128$　　**56.** $\log_{12} 140$

*Use a scientific calculator and **natural logarithms** to evaluate the following.*

57. $\log_4 0.07733$　　**58.** $\log_7 0.004462$　　**59.** $\log_{18} 98{,}265$　　**60.** $\log_{20} 16{,}451$

Mixed Practice

Use a scientific calculator to find an approximate value for each of the following.

61. $\ln 1537$　　**62.** $\log 92.81$　　**63.** $\text{antilog}_e(-1.874)$　　**64.** $\log_6 0.5437$

Find an approximate value for x for each of the following.

65. $\log x = 8.5634$　　**66.** $\ln x = 7.9631$　　**67.** $\log_4 x = 0.8645$　　**68.** $\log_3 x = 0.5649$

Use a graphing calculator or a scientific calculator to graph the following.

69. $y = \log_6 x$　　　　**70.** $y = \log_4 x$　　　　**71.** $y = \log_{0.2} x$

Applications

The median age of people in the United States is slowly increasing as the population becomes older. In 1980 the median age of the population was 30.0 years. This means that approximately half the population of the country was under 30.0 years old and approximately half the population of the country was over 30.0 years old. By 1987 the median age had increased to 32.0 years. An equation that can be used to predict the median age N (in years) of the population of the United States is $N = 32.82 + 1.0249 \ln x$, where x is the number of years since 1990 and $x \geq 1$. Source: U.S. Census Bureau.

72. Use the equation to find the median age of the U.S. population in 1995 and in 2005. If this model is correct, by what percent will the median age increase from 1995 to 2005?

73. Use the equation to find the median age of the U.S. population in 2000 and in 2010. If this model is correct, by what percent will the median age increase from 2000 to 2010?

Suppose that we want to measure the magnitude of an earthquake. If an earthquake has a shock wave x times greater than the smallest shock wave that can be measured by a seismograph, then its magnitude R on the Richter scale is given by the equation $R = \log x$.

An earthquake that has a shock wave 25,000 times greater than the smallest shock wave that can be detected will have a magnitude of $R = \log 25{,}000 \approx 4.40$. (Usually we round the magnitude of an earthquake to the nearest hundredth.)

74. What is the magnitude of an earthquake that has a shock wave that is 56,000 times greater than the smallest shock wave that can be detected?

75. What is the magnitude of an earthquake that has a shock wave that is 184,000 times greater than the smallest shock wave that can be detected?

76. If the magnitude of an earthquake is $R = 6.6$ on the Richter scale, what can you say about the size of the earthquake's shock wave?

77. If the magnitude of an earthquake is $R = 5.4$ on the Richter scale, what can you say about the size of the earthquake's shock wave?

Cumulative Review Problems

Solve the quadratic equations. Simplify your answers.

78. $3x^2 - 11x - 5 = 0$

79. $2y^2 + 4y - 3 = 0$

80. Find the distance between the points $(-3, 6)$ and $(2, -4)$.

81. Write in standard form the equation of a circle with center at $(2, -1)$ and radius 3.

On a specific portion of Interstate 91, there are six exits. There is a distance of 12 miles between odd-numbered exits. There is a distance of 15 miles between even-numbered exits. The total distance between Exit 1 and Exit 6 is 36 miles.

82. Find the distance between Exit 1 and Exit 2. Find the distance between Exit 1 and Exit 3.

83. Find the distance between Exit 1 and Exit 4. Find the distance between Exit 1 and Exit 5.

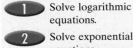

Student Learning Objectives

After studying this section, you will be able to:

1 Solve logarithmic equations.

2 Solve exponential equations.

3 Solve applied exercises using logarithmic or exponential equations.

SSM
PH TUTOR CENTER   CD & VIDEO   MATH PRO   WEB

Graphing Calculator

Solving Logarithmic Equations

Example 1 could be solved with a graphing calculator in the following way. First write the equation as

$\log 5 + \log(x + 3) - 2 = 0$

and then graph the function

$y = \log 5 + \log(x + 3) - 2$

to find an approximate value for x when $y = 0$. If you set the appropriate window and use your Zoom feature, you should be able to obtain $x = 17.0$, rounded to the nearest tenth. Now use your graphing calculator to solve the following.

1. $\log x + \log(x + 1) = 1$

2. $\log(x - 6) = 2 - \log(x + 15)$

3. $\ln(x + 2) = 12$

---

1 Solving Logarithmic Equations

In general, when solving logarithmic equations, we try to obtain all the logarithms on one side of the equation and all the numerical values on the other side. Then we seek to use the properties of logarithms to obtain a single logarithmic expression on one side.

We can describe a general procedure for solving logarithmic equations.

Step 1 If an equation contains some logarithms and some terms without logarithms, try to get one logarithm alone on one side and one numerical value on the other.

Step 2 Convert the logarithmic equation to an exponential equation using the definition of a logarithm.

Step 3 Solve the equation.

EXAMPLE 1 Solve $\log 5 = 2 - \log(x + 3)$.

$\log 5 + \log(x + 3) = 2$	Add $\log(x + 3)$ to each side.
$\log[5(x + 3)] = 2$	Property 1.
$\log(5x + 15) = 2$	Simplify.
$5x + 15 = 10^2$	Write the equation in exponential form.
$5x + 15 = 100$	Simplify.
$5x = 85$	Subtract 15 from each side.
$x = 17$	Divide each side by 5.

Check: $\log 5 \overset{?}{=} 2 - \log(17 + 3)$

$\log 5 \overset{?}{=} 2 - \log 20$

Since these are common logarithms (base 10), the easiest way to check the answer is to find decimal approximations for each logarithm on a calculator.

$0.698970004 \overset{?}{=} 2 - 1.301029996$

$0.698970004 = 0.698970004$ ✓

Practice Problem 1 Solve $\log(x + 5) = 2 - \log 5$.

EXAMPLE 2 Solve $\log_3(x + 6) - \log_3(x - 2) = 2$.

$\log_3\left(\dfrac{x + 6}{x - 2}\right) = 2$	Property 2.
$\dfrac{x + 6}{x - 2} = 3^2$	Write the equation in exponential form.
$\dfrac{x + 6}{x - 2} = 9$	Evaluate 3^2.
$x + 6 = 9(x - 2)$	Multiply each side by $(x - 2)$.
$x + 6 = 9x - 18$	Simplify.
$24 = 8x$	Add $18 - x$ to each side.
$3 = x$	Divide each side by 8.

Check: $\log_3(3 + 6) - \log_3(3 - 2) \overset{?}{=} 2$

$\log_3 9 - \log_3 1 \overset{?}{=} 2$

$2 - 0 \overset{?}{=} 2$

$2 = 2$ ✓

Practice Problem 2 Solve $\log(x + 3) - \log x = 1$.

Some equations consist of logarithmic terms only. In such cases we may be able to use property 6 to solve them. Recall that this rule states that if $b > 0, b \neq 1, x > 0,$ $y > 0,$ and $\log_b x = \log_b y,$ then $x = y.$

What if one of our possible solutions is the logarithm of a negative number? Can we evaluate the logarithm of a negative number? Look again at the graph of $y = \log_2 x$ on page 684. Note that the domain of this function is $x > 0.$ (The curve is located on the positive side of the x-axis.) Therefore, the logarithm of a negative number is *not defined*.

You should be able to see this by using the definition of logarithms. If $\log(-2)$ were valid, we could write the following.

$$y = \log_{10}(-2)$$
$$10^y = -2$$

Obviously, no value of y can make this equation true. Thus, we see that **it is not possible to take the logarithm of a negative number**.

Sometimes when we attempt to solve a logarithmic equation, we obtain a possible solution that leads to the logarithm of a negative number. We can immediately discard such a solution.

EXAMPLE 3 Solve $\log(x + 6) + \log(x + 2) = \log(x + 20).$

$$\log(x + 6)(x + 2) = \log(x + 20)$$
$$\log(x^2 + 8x + 12) = \log(x + 20)$$
$$x^2 + 8x + 12 = x + 20$$
$$x^2 + 7x - 8 = 0$$
$$(x + 8)(x - 1) = 0$$
$$x + 8 = 0 \qquad x - 1 = 0$$
$$x = -8 \qquad x = 1$$

Check: $\log(x + 6) + \log(x + 2) = \log(x + 20)$

$$x = 1\text{:} \quad \log(1 + 6) + \log(1 + 2) \overset{?}{=} \log(1 + 20)$$
$$\log(7) + \log(3) \overset{?}{=} \log(21)$$
$$\log(7 \cdot 3) \overset{?}{=} \log 21$$
$$\log 21 = \log 21 \quad \checkmark$$
$$x = -8\text{:} \quad \log(-8 + 6) + \log(-8 + 2) \overset{?}{=} \log(-8 + 20)$$
$$\log(-2) + \log(-6) \neq \log(12)$$

We can discard -8 because it leads to taking the logarithm of a negative number, which is not allowed. Only $x = 1$ is a solution. The only solution is 1.

Practice Problem 3 Solve $\log 5 - \log x = \log(6x - 7).$ Check your solution.

2 Solving Exponential Equations

You might expect that property 6 can be used in the reverse direction. It seems logical, for example, that if $x = 3,$ we should be able to state that $\log_4 x = \log_4 3.$ This is exactly the case, and we will formally state it as a property.

> **Property 7** If x and $y > 0$ and $x = y,$ then $\log_b x = \log_b y,$ where $b > 0$ and $b \neq 1.$

Property 7 is often referred to as "taking the logarithm of each side of the equation." Usually we will take the common logarithm of each side of the equation, but any base can be used.

EXAMPLE 4 Solve $2^x = 7$. Leave your answer in exact form.

$$\log 2^x = \log 7 \quad \text{Take the logarithm of each side (property 7).}$$
$$x \log 2 = \log 7 \quad \text{Property 3.}$$
$$x = \frac{\log 7}{\log 2} \quad \text{Divide each side by } \log 2.$$

Practice Problem 4 Solve $3^x = 5$. Leave your answer in exact form.

When we solve exponential equations, it will often be useful to find an approximate value for the answer.

EXAMPLE 5 Solve $3^x = 7^{x-1}$. Approximate your answer to the nearest thousandth.

$$\log 3^x = \log 7^{(x-1)}$$
$$x \log 3 = (x - 1) \log 7$$
$$x \log 3 = x \log 7 - \log 7$$
$$x \log 3 - x \log 7 = -\log 7$$
$$x(\log 3 - \log 7) = -\log 7$$
$$x = \frac{-\log 7}{\log 3 - \log 7}$$

We can approximate the value for x on most scientific calculators by using the following keystrokes.

$$7 \;\boxed{\log}\; \boxed{+/-}\; \boxed{\div}\; \boxed{(}\; 3 \;\boxed{\log}\; \boxed{-}\; 7 \;\boxed{\log}\; \boxed{)}\; \boxed{=}\; 2.296606943$$

Rounding to the nearest thousandth, we have $x \approx 2.297$.

Practice Problem 5 Solve $2^{3x+1} = 9^{x+1}$. Approximate your answer to the nearest thousandth.

If the exponential equation involves e raised to a power, it is best to take the natural logarithm of each side of the equation.

EXAMPLE 6 Solve $e^{2.5x} = 8.42$. Round your answer to the nearest ten-thousandth.

$$\ln e^{2.5x} = \ln 8.42 \quad \text{Take the natural logarithm of each side.}$$
$$(2.5x)(\ln e) = \ln 8.42 \quad \text{Property 3.}$$
$$2.5x = \ln 8.42 \quad \ln e = 1.$$
$$x = \frac{\ln 8.42}{2.5} \quad \text{Divide each side by 2.5.}$$

On most scientific calculators, the value of x can be approximated with the following keystrokes.

$$8.42 \;\boxed{\ln}\; \boxed{\div}\; 2.5 \;\boxed{=}\; 0.85224393$$

Rounding to the nearest ten-thousandth, we have $x \approx 0.8522$.

Practice Problem 6 Solve $20.98 = e^{3.6x}$. Round your answer to the nearest ten-thousandth.

3 **Solving Applied Exercises Using Logarithmic or Exponential Equations**

We now return to the compound interest formula and consider some other exercises that can be solved with it. For example, perhaps we would like to know how long it will take for a deposit to grow to a specified goal.

EXAMPLE 7 If P dollars are invested in an account that earns interest at 12% compounded annually, the amount available after t years is $A = P(1 + 0.12)^t$. How many years will it take for $300 in this account to grow to $1500? Round your answer to the nearest whole year.

$1500 = 300(1 + 0.12)^t$	Substitute $A = 1500$ and $P = 300$.
$1500 = 300(1.12)^t$	Simplify.
$\dfrac{1500}{300} = (1.12)^t$	Divide each side by 300.
$5 = (1.12)^t$	Simplify.
$\log 5 = \log(1.12)^t$	Take the common logarithm of each side.
$\log 5 = t(\log 1.12)$	Property 3.
$\dfrac{\log 5}{\log 1.12} = t$	Divide each side by $\log 1.12$.

On a scientific calculator we have the following.

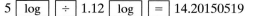

5 | log | ÷ | 1.12 | log | | = | 14.20150519

Thus, it would take approximately 14 years.

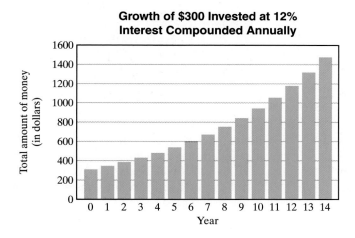

Growth of $300 Invested at 12% Interest Compounded Annually

Practice Problem 7 Mon Ling's father has an investment account that earns 8% interest compounded annually. How many years would it take for $4000 to grow to $10,000 in that account? Round your answer to the nearest whole year.

The growth equation for things that appear to be growing continuously is $A = A_0 e^{rt}$, where A is the final amount, A_0 is the original amount, r is the rate at which things are growing in a unit of time, and t is the total number of units of time.

For example, if a laboratory starts with 5000 cells and they reproduce at a rate of 35% per hour, the number of cells in 18 hours can be described by the following equation.

$$A = 5000e^{(0.35)(18)} = 5000e^{6.3} \approx 5000(544.57) = 2{,}722{,}850 \text{ cells}$$

EXAMPLE 8 If the world population is eight billion people and it continues to grow at the current rate of 2% per year, how many years will it take for the population to increase to fourteen billion people?

$$A = A_0 e^{rt}$$

To make our calculation easier, we write the values for the population in terms of billions. This will allow us to avoid writing numbers like 14,000,000,000 and 8,000,000,000. Do you see why we can do this?

$$14 = 8e^{0.02t} \qquad \text{Substitute known values.}$$

$$\frac{14}{8} = e^{0.02t} \qquad \text{Divide each side by 8.}$$

$$\ln\left(\frac{14}{8}\right) = \ln e^{0.02t} \qquad \text{Take the natural logarithm of each side.}$$

$$\ln(1.75) = (0.02t)\ln e \qquad \text{Property 3.}$$

$$\ln 1.75 = 0.02t \qquad \text{Since } \ln e = 1.$$

$$\frac{\ln 1.75}{0.02} = t \qquad \text{Divide each side by 0.02.}$$

Using our calculator, we obtain the following.

$$1.75 \; \boxed{\ln} \; \boxed{\div} \; 0.02 \; \boxed{=} \; 27.98078940$$

Rounding to the nearest whole year, we find that the population will grow from eight billion to fourteen billion in about 28 years if the growth continues at the same rate.

Practice Problem 8   The wildlife management team in one region of Alaska has determined that the black bear population is growing at the rate of 4.3% per year. This region currently has approximately 1300 black bears. A food shortage will develop if the population reaches 5000. If the growth rate remains unchanged, how many years will it take for this food shortage problem to occur?

EXAMPLE 9   The magnitude of an earthquake is measured by the formula $R = \log\left(\dfrac{I}{I_0}\right)$, where I is the intensity of the earthquake and I_0 is the minimum measurable intensity. The 1964 earthquake in Anchorage, Alaska, had a magnitude of 8.4. The 1906 earthquake in Taiwan had a magnitude of 7.1. How much more energy was released from the Anchorage earthquake than from the Taiwan earthquake? *Source:* National Oceanic and Atmospheric Administration.

Let I_A = intensity of the Alaska earthquake. Then

$$8.4 = \log\left(\frac{I_A}{I_0}\right) = \log I_A - \log I_0.$$

Solving for $\log I_0$ gives

$$\log I_0 = \log I_A - 8.4\,.$$

Let I_T = intensity of the Taiwan earthquake. Then

$$7.1 = \log\left(\frac{I_T}{I_0}\right) = \log I_T - \log I_0.$$

Solving for $\log I_0$ gives

$$\log I_0 = \log I_T - 7.1\,.$$

Therefore,

$$\log I_A - 8.4 = \log I_T - 7.1.$$

$$\log I_A - \log I_T = 8.4 - 7.1$$

$$\log \frac{I_A}{I_T} = 1.3$$

$$10^{1.3} = \frac{I_A}{I_T}$$

$$19.95262315 = \frac{I_A}{I_T} \qquad \text{Use a calculator.}$$

$$20 \approx \frac{I_A}{I_T} \qquad \text{Round to the nearest whole number.}$$

$$20 I_T \approx I_A$$

The Alaska earthquake had approximately twenty times the intensity of the Taiwan earthquake.

Practice Problem 9   The 1933 earthquake in Japan had a magnitude of 8.9. The 1989 earthquake in San Francisco had a magnitude of 7.1. How much more energy was released from the Japan earthquake than from the San Francisco earthquake? *Source:* National Oceanic and Atmospheric Administration.

Solve each logarithmic equation and check your solutions.

1. $\log_8(x + 12) + \log_8 4 = 2$

2. $\log_2(x + 1) + \log_2 3 = 3$

3. $\log_5(x - 2) + \log_5 3 = 2$

4. $\log_2 4 + \log_2(x - 1) = 5$

5. $\log_3(2x - 1) = 2 - \log_3 4$

6. $\log_5(3x + 1) = 1 - \log_5 2$

7. $\log(30x + 40) = 2 + \log(x - 1)$

8. $1 + \log x = \log(9x + 1)$

9. $2 + \log_6(x - 1) = \log_6(12x)$

10. $\log_2 x = \log_2(x + 5) - 1$

11. $\log(x + 20) - \log x = 2$

12. $\log_3(x + 6) + \log_3 x = 3$

13. $2\log_3 4 = \log_3 x - \log_3(x - 1)$

14. $\log_8 x + \log_8(x - 2) = 1$

15. $1 + \log(x - 2) = \log(6x)$

16. $\log_5(2x) - \log_5(x - 3) = 3\log_5 2$

17. $\log_2(x + 5) - 2 = \log_2 x$

18. $\log_3(x - 1) - 3 = \log_3(2x + 1)$

19. $2\log_7 x = \log_7(x + 4) + \log_7 2$

20. $\log x + \log(x - 1) = \log 12$

21. $\ln(10) - \ln x = \ln(x - 3)$ **22.** $\ln(2 + 2x) = 2\ln(x + 1)$

Solve each exponential equation. Leave your answers in exact form. Do not approximate.

23. $8^{x-1} = 11$ **24.** $3^{x+5} = 5$

25. $2^{3x+4} = 17$ **26.** $5^{2x-1} = 11$

Solve each exponential equation. Use your calculator to approximate your solutions to the nearest thousandth.

27. $6^{4x-1} = 225$ **28.** $12^{2x+1} = 150$ **29.** $5^x = 4^{x+1}$ **30.** $3^x = 2^{x+3}$

31. $e^{x-1} = 28$ **32.** $e^{x+1} = 17$ **33.** $88 = e^{2x+1}$ **34.** $3 = e^{1-x}$

Applications

When a principal P earns an annual interest rate r compounded yearly, the amount A after t years is $A = P(1 + r)^t$. Use this information to solve Exercises 35–38. Round all answers to the nearest whole year.

35. How long will it take $1500 to grow to $5000 at 8% compounded annually?

36. How long will it take $1000 to grow to $4500 at 7% compounded annually?

37. How long will it take for a principal to triple at 6% compounded annually?

38. How long will it take for a principal to double at 5% compounded annually?

39. What interest rate would be necessary to obtain $6500 in 6 years if $5000 is the amount of the original investment and the interest is compounded yearly? (Express the interest rate as a percent rounded to the nearest tenth.)

40. If $3000 is invested for 3 years with annual interest compounded yearly, what interest rate is needed to achieve an amount of $3600? (Express the interest rate as a percent rounded to the nearest tenth.)

The growth of the world's population can be described by the equation $A = A_0 e^{rt}$, where time t is measured in years; A_0 is the population of the world at time t = 0; r is the annual growth rate; and A is the population at time t. Assume that r = 2% per year. Use this information to solve Exercises 41–44. Round your answers to the nearest whole year.

41. How long will it take a population of six billion to increase to nine billion?

42. How long will it take a population of seven billion to increase to twelve billion?

43. How long will it take for the world's population to quadruple (become four times as large)?

44. How long will it take for the world's population to double?

The number N of employees in the cellular telephone industry in the United States can be approximated by the equation $N = 20,800(1.264)^x$, where x is the number of years since 1990. Source: Federal Communications Commission.

Use this equation to answer the following questions.

45. Approximately how many employees will there be in 2003?

46. Approximately how many employees will there be in 2006?

47. In what year will the number of employees reach 274,000?

48. In what year will the number of employees reach 699,000?

Use the equation $A = A_0 e^{rt}$ to solve Exercises 49–54. Round your answers to the nearest whole number.

49. The population of Melbourne, Australia, is approximately three million people. If the growth rate is 3% per year, in how many years will there be 3.5 million people?

50. The population of Bethel is 80,000 people, and it is growing at the rate of 1.5% per year. How many years will it take for the population to grow to 120,000 people?

51. The number of new skin cells on a revolutionary skin graft is growing at a rate of 4% per hour. How many hours will it take for 200 cells to become 1800 cells?

52. The workforce in a state is increasing at the rate of 1.5% per year. During the last measured year, the workforce was 3.5 million. If this rate continues, how many years will it be before the workforce reaches 4.5 million?

53. Unfortunately, U.S. deer carry ticks that spread Lyme disease. The number of people who are infected with the virus is increasing by 5% every year. If 24,500 people were confirmed to have Lyme disease in 1997, how many will be infected by the end of the year 2010?

54. In the city of Scranton, the number of videotape rentals is increasing by 7.5% per year. For the last year that data are available, 1.3 million videos were rented. How many years will it be before 2.0 million videos are rented per year?

To Think About

The magnitude of an earthquake (amount of energy released) is described by the formula $R = \log\left(\dfrac{I}{I_0}\right)$, where I is the intensity of the earthquake and I_0 is the minimum measurable intensity. Use this formula to solve Exercises 55–58. Round answers to the nearest tenth.

55. On January 17, 1993, in Northridge, California, residents experienced an earthquake that measured 6.8 on the Richter scale, killed sixty-one people, and undermined supposedly earthquake-proof steel-framed buildings. Exactly 1 year later near Kobe, Japan, an earthquake measuring 7.2 on the Richter scale killed more than 5300 people, injured more than 35,000, and destroyed nearly 200,000 homes, in spite of construction codes reputed to be the best in the world. How much more energy was released from the Japan earthquake than from the Northridge earthquake? *Source:* National Oceanic and Atmospheric Administration.

56. October 17, 1989, brought tragedy to the San Francisco/Oakland area when an earthquake measuring 7.1 on the Richter scale centered in the Loma Prieta area (Santa Cruz Mountains) and collapsed huge sections of freeway, killing sixty-three people. Almost 6 years later, an earthquake measuring 8.2 on the Richter scale killed 190 people in the Kurile Islands of Japan and Russia. How much more energy was released from the Kurile earthquake than from the Loma Prieta earthquake? *Source:* National Oceanic and Atmospheric Administration.

57. The 1906 earthquake in San Francisco had a magnitude of 8.3. In 1971 an earthquake in Japan measured 6.8. How much more energy was released from the San Francisco earthquake than from the Japan earthquake? *Source:* National Oceanic and Atmospheric Administration.

58. The 1933 Japan earthquake had a magnitude of 8.9. In Turkey a 1975 earthquake had a magnitude of 6.7. How much more energy was released from the Japan earthquake than from the Turkey earthquake? *Source:* National Oceanic and Atmospheric Administration.

Optional Graphing Calculator Problems

59. In Crystal Lake, north of Amherst, Nova Scotia, the fish population has been out of balance for several years because of an abundance of catfish. Environmentalists have taken a number of measures to increase the number of brook trout so that the populations of the two types of fish are at the same level. After several years of dealing with industrial pollution, the environmentalists have succeeded in cleaning the lake sufficiently enough so that the brook trout can reproduce more readily. The growth in the number of brook trout is now described by the equation $y = 300e^{0.12x}$, where x is the number of years from now. The growth in the number of catfish is given by the equation $y = 750 + 100x$, where x is the number of years from now. How many years will it take until the two fish populations are equal in number? Round your answer to the nearest tenth of a year.

60. Suppose that the population of wolves in one region of Alaska is growing according to the equation $y_1 = 34.572x + 850$, where x is the number of years from now. Suppose also that the food supply for wolves is growing according to the equation $y_2 = 1000e^{0.02x}$, where x is the number of years from now. In how many years (rounded to the nearest tenth of a year) will the food supply become inadequate for the number of wolves? (When will y_1 be greater than y_2 ?)

Cumulative Review Problems

Simplify. Assume that x and y are positive real numbers.

61. $\sqrt{98x^3y^2}$

62. $(\sqrt{3} + 2\sqrt{2})(\sqrt{6} - \sqrt{2})$

63. Nora just received a promotion at work along with an 8% salary increase. She now earns $40,500 per year. What was her old salary?

64. The London subway system has a staff of 16,000 and provides 2.5 million passenger journeys each day. An expansion of the system begun in 1993 will extend the system 16 kilometers and will cost 2.85 billion British pounds. How many dollars per mile will this extension cost? (Use 1 kilometer = 0.62 miles and 1 U.S. dollar = 0.63 British pounds.) *Source:* British Bureau of Tourism.

Solve for x.

65. $|3x - 5| = 7$

66. $|-2x + 6| \le 10$

Putting Your Skills to Work

Newton's Law of Cooling

Newton's law of cooling is an exponential formula that is used to determine how long it takes hot objects to cool to a certain temperature. It can also be used to find out what the temperature of a cooling object will be after a certain amount of time has passed. The formula is

$$T = C + (T_0 - C)e^{-kt},$$

where t is the time in minutes it takes an object to cool from an initial temperature T_0 to a temperature T; C is the temperature of the room that the cooling object is in; and k is a constant associated with the cooling object itself. All temperatures are measured on the Fahrenheit scale.

Problems for Individual Investigation and Analysis

1. A pie removed from an oven has a temperature of 220°F and is left to cool in a room that has a temperature of 72°F. After 30 minutes, the pie's temperature is 140°F. Use Newton's law of cooling to find the value of k for this pie. (Round your answer to the nearest thousandth.)

2. Now that you have the value of k, rewrite the formula for Newton's law of cooling by substituting the values of C and T_0 given in Exercise 1 and the value of k that you just found.

3. Use the formula from Exercise 2 to determine how many minutes it will take for the pie to cool to 90°F.

Problems for Cooperative Group Investigation

When you buy coffee at a convenience store or coffee shop, you are often given a choice of containers. Most of the time, you can have your coffee in a ceramic mug if you are sitting at the coffee shop, or you can have it poured into a Styrofoam or cardboard cup if you want it to go. Does it really make a difference what type of container your coffee is in? In which of these containers would your coffee cool the fastest? The slowest?

4. Suppose you have three cups of equal amounts of coffee. One cup is made of foam, one of cardboard, and the other is a ceramic mug. If the coffee has a temperature of 190°F when poured and the temperature of the room is 70°F, determine the temperatures of each cup of coffee after 20 minutes. (The values of k are 0.05 for foam, 0.08 for cardboard, and 0.13722 for ceramic.) Round your answers to the nearest tenth of a degree.

5. How long will it take each of the cups of coffee in Exercise 4 to cool to 125°F? Round your answers to the nearest tenth of a degree.

6. Interpret your results from the previous two exercises.

Internet Connections

http://www.prenhall.com/tobey_beg_int

Site: SOS Mathematics

More complex exercises involving Newton's law of cooling involve a subject known as differential equations and can be solved with more-advanced techniques. Study the web page and look at the example given there and its method of solution. Then try to solve the following two exercises.

7. A company executive is found murdered by the Monday morning cleaning staff at 7 A.M. The office had been at a constant temperature of 65°F all weekend. The first doctor who examined the body at 7:15 A.M. found the temperature of the body to be 84°F. When the medical forensics team had the body removed from the office building at 10:15 A.M., the temperature of the body had lowered to 76°F. Find the estimated time of death.

8. A wealthy socialite on vacation in the Cayman Islands is murdered. Her body is discovered at 10 P.M. by the maid who entered the room to turn down the bed. A doctor came with the police at 10:15 P.M. and found the temperature of the body to be 82°F. Two hours later when the police removed the body, the temperature had decreased to 74°F. The police noted that the temperature of the room was 68°F. Find the estimated time of death.

Math in the Media

Earth's Growing Population

In Steve Sternberg's article, "Earth Welcomes Six Billionth Baby," October 1999, *USA Today* on-line, it was reported that earth reached a population of 6 billion.

The article relayed that the population has increased by 1 billion in 12 years.

Use your knowledge from this chapter to investigate population growth and the growth rate in the questions that follow.

EXERCISES

The formula $t = \dfrac{\ln\left(\dfrac{E}{S}\right)}{r}$ *calculates the number of years, t, required for a population to grow from a starting number, S, to an end number, E, given a growth rate (per year) r, expressed as a decimal.*

1. Plot the number of years it takes a population to double from 6 billion to 12 billion as a function of growth rate for $r = 0.01, 0.011, 0.012 \ldots 0.02$. Would the curve look any different for a population doubling from 4 billion to 8 billion? From 8 billion to 16 billion?

2. What other types of growth-related data and projections would be needed to come to a judgment with regard to the sustainability of populations growing at a given rate, that is, to determine whether the planet can support a particular rate of population growth?

3. What growth rate would result in a population doubling time of 100 years? How might that rate come about?

Chapter 12 Organizer

Topic	Procedure	Examples
Exponential function, p. 673.	$f(x) = b^x$, where $b > 0$, $b \neq 1$, and x is a real number.	Graph $f(x) = \left(\dfrac{2}{3}\right)^x$.
Property of exponential equations, p. 675.	When $b > 0$ and $b \neq 1$, if $b^x = b^y$, then $x = y$.	Solve for x. $2^x = \dfrac{1}{32}$ $2^x = \dfrac{1}{2^5}$ $2^x = 2^{-5}$ $x = -5$
Definition of logarithm, p. 682.	$y = \log_b x$ is the same as $x = b^y$, where $x > 0$, $b > 0$, and $b \neq 1$.	Write $\log_3 17 = 2x$ in exponential form. $3^{2x} = 17$ Write $18 = 3^x$ in logarithmic form. $\log_3 18 = x$ Solve $\log_6\left(\dfrac{1}{36}\right) = x$ for x. $6^x = \dfrac{1}{36}$ $6^x = 6^{-2}$ $x = -2$
Properties of logarithms, pp. 690–693.	Suppose that $M > 0$, $N > 0$, $b > 0$, and $b \neq 1$. $\log_b MN = \log_b M + \log_b N$ $\log_b\left(\dfrac{M}{N}\right) = \log_b M - \log_b N$ $\log_b M^p = p\log_b M$ $\log_b b = 1$ $\log_b 1 = 0$ If $\log_b x = \log_b y$, then $x = y$. If $x = y$, then $\log_b x = \log_b y$,	Write as separate logarithms of x, y, and w. $\log_3\left(\dfrac{x^2 \sqrt[3]{y}}{w}\right)$ $= 2\log_3 x + \dfrac{1}{3}\log_3 y - \log_3 w$ Write as one logarithm. $5\log_6 x - 2\log_6 w - \dfrac{1}{4}\log_6 z$ $= \log_6\left(\dfrac{x^5}{w^2 \sqrt[4]{z}}\right)$ Simplify. $\log 10^5 + \log_3 3 + \log_5 1$ $= 5\log 10 + \log_3 3 + \log_5 1$ $= 5 + 1 + 0$ $= 6$
Finding logarithms, p. 697.	On a scientific calculator: $\log x = \log_{10} x$, $\quad$ for all $x > 0$ $\ln x = \log_e x$, $\quad$ for all $x > 0$	Find $\log 3.82$. 3.82 $\boxed{\log}$ $\log 3.82 \approx 0.5820634$ Find $\ln 52.8$. 52.8 $\boxed{\ln}$ $\ln 52.8 \approx 3.9665112$

Topic	Procedure	Examples
Finding antilogarithms, p. 698.	If $\log x = b$, then $10^b = x$. If $\ln x = b$, then $e^b = x$. Use a calculator or a table to solve.	Find x if $\log x = 2.1416$. $$10^{2.1416} = x$$ $2.1416 \boxed{10^x} \approx 138.54792$ Find x if $\ln x = 0.6218$. $$e^{0.6218} = x$$ $0.6218 \boxed{e^x} \approx 1.8622771$
Finding a logarithm to a different base, p. 699.	Change of base formula: $\log_b x = \dfrac{\log_a x}{\log_a b}$, where a, b, and $x > 0$, $a \neq 1$, and $b \neq 1$.	Evaluate $\log_7 1.86$. $$\frac{\log 1.86}{\log 7}$$ $1.86 \boxed{\log} \div 7 \boxed{\log} \boxed{=} 0.3189132$
Solving logarithmic equations, p. 704.	**1.** If some but not all of the terms of an equation have logarithms, try to rewrite the equation with one single logarithm on one side and one numerical value on the other. Then convert the equation to exponential form. **2.** If an equation contains logarithmic terms only, try to get only one logarithm on each side of the equation. Then use the property that if $\log_b x = \log_b y$, $x = y$. *Note:* Always check your solutions when solving logarithmic equations.	Solve for x. $\log_5 3x - \log_5(x^2 - 1) = \log_5 2$ $$\log_5 3x = \log_5 2 + \log_5(x^2 - 1)$$ $$\log_5 3x = \log_5[2(x^2 - 1)]$$ $$3x = 2x^2 - 2$$ $$0 = 2x^2 - 3x - 2$$ $$0 = (2x + 1)(x - 2)$$ $2x + 1 = 0 \qquad\qquad x - 2 = 0$ $$x = -\frac{1}{2} \qquad x = 2$$ *Check:* $x = 2$: $\log_5 3(2) - \log_5(2^2 - 1) \stackrel{?}{=} \log_5 2$ $\log_5 6 - \log_5 3 \stackrel{?}{=} \log_5 2$ $\log_5\left(\dfrac{6}{3}\right) \stackrel{?}{=} \log_5 2$ $\log_5 2 = \log_5 2$ ✓ $x = -\dfrac{1}{2}$: For the expression $\log_5(3x)$, we would obtain $\log_5(-1.5)$. You cannot take the logarithm of a negative number. $x = -\dfrac{1}{2}$ is not a solution. The solution is 2.
Solving exponential equations, p. 705.	**1.** See whether each expression can be written so that only one base appears on one side of the equation and the same base appears on the other side. Then use the property that if $b^x = b^y$, $x = y$. **2.** If you can't do step 1, take the logarithm of each side of the equation and use the properties of logarithms to solve for the variable.	Solve for x. $2^{x-1} = 7$ $$\log 2^{x-1} = \log 7$$ $$(x - 1)\log 2 = \log 7$$ $$x\log 2 - \log 2 = \log 7$$ $$x\log 2 = \log 7 + \log 2$$ $$x = \frac{\log 7 + \log 2}{\log 2}$$ (We can approximate the answer as $x \approx 3.8073549$.)

Chapter 12 Review Problems

Graph the functions in Exercises 1 and 2.

1. $f(x) = 4^{3+x}$

2. $f(x) = e^{x-3}$

3. Solve $5^{x+2} = 125$.

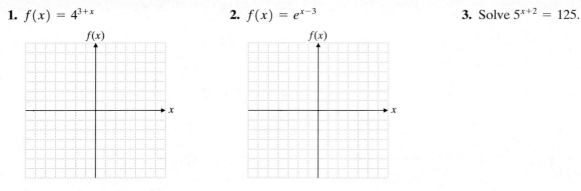

4. Write $-3 = \log_{10}(0.001)$ in exponential form.

5. Change $\dfrac{1}{32} = 2^{-5}$ to logarithmic form.

Solve.

6. $\log_w 16 = 4$

7. $\log_3 x = -2$

8. $\log_8 x = 0$

9. $\log_7 w = -1$

10. $\log_w 27 = 3$

11. $\log_{10} w = -3$

12. $\log_{10} 1000 = x$

13. $\log_2 64 = x$

14. $\log_2\left(\dfrac{1}{4}\right) = x$

15. $\log_5 125 = x$

16. Graph the equation $\log_3 x = y$.

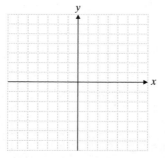

Write each expression as the sum or difference of $\log_2 x$, $\log_2 y$, and $\log_2 z$.

17. $\log_2\left(\dfrac{5x}{\sqrt{w}}\right)$

18. $\log_2 x^3 \sqrt{y}$

Write as a single logarithm.

19. $\log_3 x + \log_3 w^{1/2} - \log_3 2$

20. $4\log_8 w - \dfrac{1}{3}\log_8 z$

21. Evaluate $\log_e e^6$.

Solve.

22. $\log_5 100 - \log_5 x = \log_5 4$

23. $\log_8 x + \log_8 3 = \log_8 75$

Find the value with a scientific calculator.

24. $\log 23.8$

25. $\log 0.0817$

26. $\ln 3.92$

27. $\ln 803$

28. Find n if $\log n = 1.1367$.

29. Find n if $\ln n = 1.7$.

30. $\log_8 2.81$

Solve each equation and check your solutions.

31. $\log_7(x + 3) + \log_7(5) = 2$

32. $\log_3(2x + 3) = \log_3(2) - 3$

33. $\log_5(x + 1) - \log_5 8 = \log_5 x$

34. $2 \log_3(x + 3) - \log_3(x + 1) = 3 \log_3 2$

35. $\log_2(x - 2) + \log_2(x + 5) = 3$

36. $\log_5(x + 1) + \log_5(x - 3) = 1$

37. $\log(2t + 3) + \log(4t - 1) = 2 \log 3$

38. $\log(2t + 4) - \log(3t + 1) = \log 6$

Solve each equation. Leave your answers in exact form. Do not approximate.

39. $3^x = 14$

40. $5^x = 4^{x+2}$

41. $16e^{x+1} = 56$

42. $e^{2x} = 30.6$

Solve each equation. Round your answers to the nearest ten-thousandth.

43. $2^{3x+1} = 5^x$

44. $3^{x+1} = 7$

45. $e^{3x-4} = 20$

46. $(1.03)^x = 20$

For Exercises 47–50, use $A = P(1 + r)^t$, the formula for exercises involving interest that is compounded annually.

47. How long will it take Frances to double the money in her account if the interest rate is 8% compounded annually? (Round your answer to nearest year.)

48. How much money would Chou Lou have after 4 years if he invested $5000 at 6% compounded annually?

49. Melinda invested $12,000 at 7% compounded annually. How many years will it take for it to amount to $20,000? (Round your answer to the nearest year.)

50. Robert invested $3500 at 5% compounded annually. His brother invested $3500 at 6% compounded annually. How many years will it take for Robert's amount to be $500 less than his brother's amount? (Round your answer to the nearest year.)

The growth of the world's population can be described by the equation $A = A_0 e^{rt}$, where time t is measured in years; A_0 is the population of the world at time $t = 0$; r is the annual growth rate; and A is the population at time t. Use this information to solve Exercises 51–54. Round your answers to the nearest whole year.

51. How long will it take a population of six billion to increase to ten billion if $r = 2\%$ per year?

52. How long will it take a population of seven billion to increase to sixteen billion if $r = 2\%$ per year?

53. The number of moose in northern Maine is increasing at a rate of 3% per year. It is estimated in one county that there are now 2000 moose. If the growth rate remains unchanged, how many years will it be until there are 2600 moose in that county?

54. A town is growing at the rate of 8% per year. How long will it take the town to grow from 40,000 to 95,000 in population?

55. An earthquake's magnitude is given by $M = \log\left(\dfrac{I}{I_0}\right)$, where I is the intensity of the earthquake and I_0 is the minimum measurable intensity. The 1964 earthquake in Anchorage, Alaska, had a magnitude of 8.4. The 1975 earthquake in Turkey had a magnitude of 6.7. How much more energy was released from the Alaska earthquake than from the Turkey earthquake? *Source:* National Oceanic and Atmospheric Administration.

56. The work W done by a volume of gas expanding at a constant temperature from volume V_0 to volume V_1 is given by $W = p_0 V_0 \ln\left(\dfrac{V_1}{V_0}\right)$, where p_0 is the pressure at volume V_0.
   (a) Find W when $p_0 = 40$ pounds per cubic inch, $V_0 = 15$ cubic inches, and $V_1 = 24$ cubic inches.
   (b) If the amount of work is 100 pounds per cubic inch, $V_0 = 8$ cubic inches, and $V_1 = 40$ cubic inches, find p_0.

1. Graph $f(x) = 3^{4-x}$.

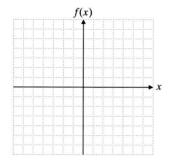

In Exercises 3 and 4, solve for the variable.

3. $\log_w 125 = 3$

4. $\log_8 x = -2$

5. Write $\log_8 x + \log_8 w - \dfrac{1}{4}\log_8 3$ as a single logarithm.

Evaluate using a calculator. Round your answers to the nearest ten-thousandth.

6. $\ln 5.99$ **7.** $\log 23.6$ **8.** $\log_3 1.62$

Use a scientific calculator to approximate x.

9. $\log x = 3.7284$ **10.** $\ln x = 0.14$

Solve the equation and check your solutions for Exercises 11 and 12.

11. $\log_8(x + 3) - \log_8 2x = \log_8 4$ **12.** $\log_8 2x + \log_8 6 = 2$

13. Find the exact value of x: $29 = 116e^{3x+1}$. Do not approximate.

14. Solve $5^{3x+6} = 17$. Approximate your answer to the nearest ten-thousandth.

15. How much money will Henry have if he invests $2000 for 5 years at 8% annual interest compounded annually?

16. How long will it take for Barb to double her money if she invests it at 5% compounded annually? Round to the nearest whole year.

2. Solve $4^{x+3} = 64$.

1. _____

2. _____

3. _____

4. _____

5. _____

6. _____

7. _____

8. _____

9. _____

10. _____

11. _____

12. _____

13. _____

14. _____

15. _____

16. _____

1. _____

2. _____

3. _____

4. _____

5. _____

6. _____

7. _____

8. _____

9. _____

10. _____

11. _____

Approximately one-half of this test covers the content of Chapters 1–11. The remainder covers the content of Chapter 12.

1. Evaluate $2(-3) + 12 \div (-2) + 3\sqrt{36}$.

2. Solve for x. $H = 3bx - 2ay$

3. Graph $y = -\dfrac{2}{3}x + 4$.

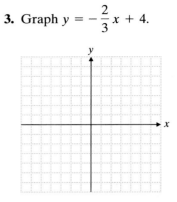

4. Factor $5ax + 5ay - 7wx - 7wy$.

5. Solve for (x, y, z).

$$3x - y + z = 6$$
$$2x - y + 2z = 7$$
$$x + y + z = 2$$

6. Simplify $\left(5\sqrt{2} + \sqrt{3}\right)\left(\sqrt{5} - 2\sqrt{6}\right)$.

7. Solve $x^4 - 5x^2 - 6 = 0$.
Express imaginary solutions in i notation.

8. Solve for x and y.

$$2x - y = 4$$
$$4x - y^2 = 0$$

9. Solve $2x - 3 = \sqrt{7x - 3}$.

10. Rationalize the denominator.

$$\frac{5}{\sqrt[3]{2xy^2}}$$

11. Graph $f(x) = 2^{3-2x}$.

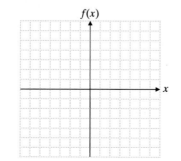

Solve for the variable.

12. $\log_x\left(\dfrac{1}{64}\right) = 3$

13. $5^{2x-1} = 25$

Evaluate using a calculator or a table.

14. $\log 7.67$

15. Find x if $\log x = 1.8209$.

16. $\log_3 7$

17. Find x if $\ln x = 1.9638$.

Solve the equation.

18. $\log_9 x = 1 - \log_9(x - 8)$

19. $\log_5 x = \log_5 2 + \log_5(x^2 - 3)$

Solve for x.

20. $3^{x+2} = 5$
   Approximate to the nearest thousandth.

21. $33 = 66e^{2x}$
   Leave your answer exact. Do not approximate.

22. How much money will Frank and Linda have in 4 years if they invest $3000 at 9% compounded annually? Round to the nearest cent.

12. _____

13. _____

14. _____

15. _____

16. _____

17. _____

18. _____

19. _____

20. _____

21. _____

22. _____

Practice Final Examination

1. _____

2. _____

3. _____

4. _____

5. _____

6. _____

7. _____

8. _____

9. _____

10. _____

11. _____

12. _____

13. _____

14. _____

Review the content areas of Chapters 1–12. Then try to solve the exercises in this Practice Final Examination.

Chapter 1

1. Evaluate $(4 - 3)^2 + \sqrt{9} \div (-3) + 4$.

2. Simplify. $5a - 2ab - 3a^2 - 6a - 8ab + 2a^2$

Chapter 2

3. Simplify $-2x + 3y\{7 - 2[x - (4x + y)]\}$.

4. Evaluate if $x = -2$ and $y = 3$. $2x^2 - 3xy - 4y$

5. Find the Fahrenheit temperature when the Celsius temperature is $-35°$. Use the formula $F = \frac{9}{5}C + 32$.

6. Solve for y. $\dfrac{1}{3}y - 4 = \dfrac{1}{2}y + 1$ **7.** Solve for b. $A = \dfrac{1}{2}a(b + c)$

8. Solve for x and graph the resulting inequality on a number line.
$5x + 3 - (4x - 2) \le 6x - 8$

▲ **9.** A piece of land is rectangular and has a perimeter of 1760 meters. The length is 200 meters less than twice the width. Find the dimensions of the land.

10. A man invested $4000, part at 12% interest and part at 14% interest. After 1 year he had earned $508 in interest. How much was invested at each interest rate?

Chapter 3

11. Find the intercepts and then graph the line $7x - 2y = -14$.

12. Graph the region $3x - 4y \le 6$.

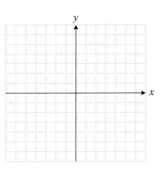

13. Find the slope of the line passing through $(1, 5)$ and $(-2, -3)$.

14. Write the equation in slope-intercept form of the line that is parallel to $3x + y = 8$ and passes through $(-1, 4)$.

Given the function defined by $f(x) = 3x^2 - 4x - 3$, find the following.

15. $f(3)$

16. $f(-2)$

17. Graph the function $f(x) = -\dfrac{1}{2}x^2$.

Chapter 4

18. Solve for x and y.

$$\frac{1}{2}x + \frac{2}{3}y = 1$$
$$\frac{1}{3}x + y = -1$$

19. Solve for x and y.

$$4x - 3y = 12$$
$$3x - 4y = 2$$

20. Solve for x, y, and z.

$$2x + 3y - z = 16$$
$$x - y + 3z = -9$$
$$5x + 2y - z = 15$$

21. Solve for x, y, and z.

$$y + z = 2$$
$$x + z = 5$$
$$x + y = 5$$

22. Graph the region.

$$3y \geq 8x - 12$$
$$2x + 3y \leq -6$$

23. Simplify. $(-3x^2 y)(-6x^3 y^4)$

Chapter 5

24. Multiply and simplify your answer.
$(3x - 2)(2x^2 - 4x + 3)$

25. Divide.
$(25x^3 + 9x + 2) \div (5x + 1)$

Chapter 6

Factor the following completely.

26. $9x^2 - 30x + 25$

27. $x^3 + 2x^2 - 4x - 8$

28. $2x^3 + 15x^2 - 8x$

29. Solve for x. $x^2 + 15x + 54 = 0$

15. _____

16. _____

17. _____

18. _____

19. _____

20. _____

21. _____

22. _____

23. _____

24. _____

25. _____

26. _____

27. _____

28. _____

29. _____

30.

31.

32.

33.

34.

35.

36.

37.

38.

39.

40.

41.

42.

43.

44.

45.

46.

47.

48.

49.

Chapter 7

Simplify the following.

30. $\dfrac{9x^3 - x}{3x^2 - 8x - 3}$

31. $\dfrac{x^2 - 9}{2x^2 + 7x + 3} \div \dfrac{x^2 - 3x}{2x^2 + 11x + 5}$

32. $\dfrac{3x}{x + 5} - \dfrac{2}{x^2 + 7x + 10}$

33. $\dfrac{\dfrac{3}{2x + 1} + 2}{1 - \dfrac{2}{4x^2 - 1}}$

34. Solve for x. $\dfrac{x - 1}{x^2 - 4} = \dfrac{2}{x + 2} + \dfrac{4}{x - 2}$

Chapter 8

35. Simplify $\dfrac{5x^{-4}y^{-2}}{15x^{-1/2}y^3}$.

36. Simplify $\sqrt[3]{40x^4y^7}$.

37. Combine like terms.

$5\sqrt{2} - 3\sqrt{50} + 4\sqrt{98}$

38. Rationalize the denominator.

$\dfrac{2\sqrt{3} + 1}{3\sqrt{3} - \sqrt{2}}$

39. Simplify and add together.
$i^3 + \sqrt{-25} + \sqrt{-16}$

40. Solve for x and check your solutions.
$\sqrt{x + 7} = x + 5$

41. If y varies directly with the square of x and $y = 15$ when $x = 2$, what will y be when $x = 3$?

Chapter 9

42. Solve for x. $5x(x + 1) = 1 + 6x$

43. Solve for x. $5x^2 - 9x = -12x$.

44. Solve for x. $x^{2/3} + 5x^{1/3} - 14 = 0$

45. Solve $3x^2 - 11x - 4 \geq 0$.

46. Graph the quadratic function $f(x) = -x^2 - 4x + 5$. Label the vertex and the intercepts.

▲ **47.** The area of a rectangle is 52 square centimeters. The length of the rectangle is 1 centimeter longer than 3 times its width. Find the dimensions of the rectangle.

48. Solve for x. $\left|\dfrac{2}{3}x - 4\right| = 2$

49. Solve the inequality. $|2x - 5| < 10$

Chapter 10

50. Place the equation of the circle in standard form. Find its center and radius.

$$x^2 + y^2 + 6x - 4y = -9$$

Identify and graph.

51. $\dfrac{x^2}{16} + \dfrac{y^2}{25} = 1$

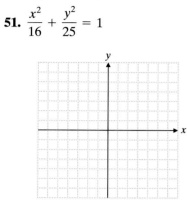

52. $\dfrac{x^2}{4} - \dfrac{y^2}{9} = 1$

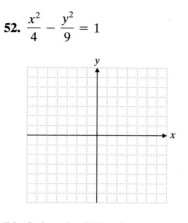

53. $x = (y - 3)^2 + 5$

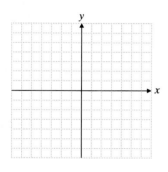

54. Solve the following system of equations.

$$x^2 + y^2 = 16$$
$$x^2 - y = 4$$

Chapter 11

55. Let $f(x) = 3x^2 - 2x + 5$.
   (a) Find $f(-1)$.
   (b) Find $f(a)$.
   (c) Find $f(a + 2)$.

56. If $f(x) = 5x^2 - 3$ and $g(x) = -4x - 2$, find $f[g(x)]$.

57. If $f(x) = \dfrac{1}{2}x - 7$, find $f^{-1}(x)$.

58. Graph on one set of axes: $f(x), f(x + 2),$ and $f(x) - 3$, if $f(x) = |x|$.

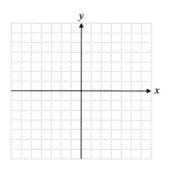

50. _____

51. _____

52. _____

53. _____

54. _____

55. _____

56. _____

57. _____

58. _____

Chapter 12

59. Graph $f(x) = 2^{1-x}$. Plot three points.

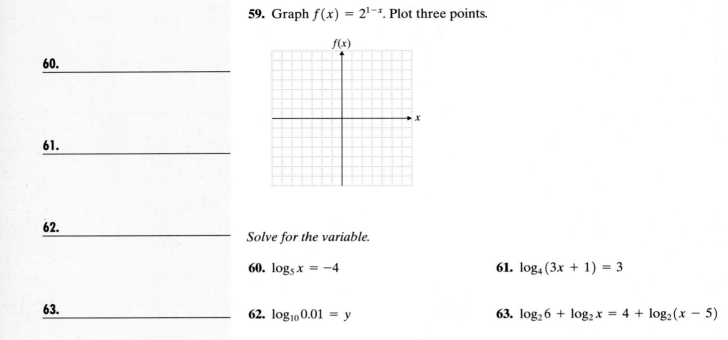

Solve for the variable.

60. $\log_5 x = -4$

61. $\log_4(3x + 1) = 3$

62. $\log_{10} 0.01 = y$

63. $\log_2 6 + \log_2 x = 4 + \log_2(x - 5)$

Glossary

Absolute value inequalities (9.8) Inequalities that contain at least one absolute value expression.

Absolute value of a number (1.1) The absolute value of a number x is the distance between 0 and the number x on the number line. It is written as $|x|$. $|x| = x$ if $x \geq 0$, but $|x| = -x$ if $x < 0$.

Algebraic fractions (7.1) The indicated quotient of two algebraic expressions.

$$\frac{x^2 + 3x + 2}{x - 4} \quad \text{and} \quad \frac{y - 6}{y + 8}$$

are algebraic fractions. In these fractions the value of the denominator cannot be zero.

Altitude of a geometric figure (1.8) The height of the geometric figure. In the three figures shown the altitude is labeled a.

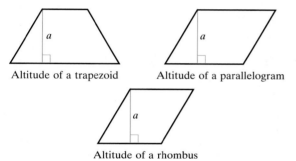

Altitude of a trapezoid   Altitude of a parallelogram

Altitude of a rhombus

Altitude of a triangle (1.8) The height of any given triangle. In the three triangles shown the altitude is labeled a.

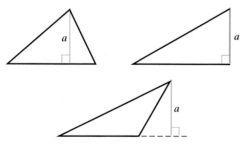

Associative property of addition (1.1) If a, b, and c are real numbers, then

$$a + (b + c) = (a + b) + c.$$

This property states that if three numbers are added it does not matter *which two numbers* are added first; the result will be the same.

Associative property of multiplication (1.3) If a, b, and c are real numbers, then

$$a \times (b \times c) = (a \times b) \times c.$$

This property states that if three numbers are multiplied it does not matter *which two numbers* are multiplied first; the result will be the same.

Asymptote (10.4) A line that a curve continues to approach but never actually touches. Often an asymptote is a helpful reference in making a sketch of a curve, such as a hyperbola.

Axis of symmetry of a parabola (10.2) A line passing through the focus and the vertex of a parabola, about which the two sides of the parabola are symmetric. See the sketch.

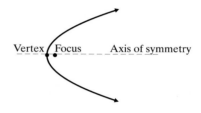

Vertex /Focus   Axis of symmetry

Base (1.4) The number or variable that is raised to a power. In the expression 2^6, the number 2 is the base.

Base of a triangle (1.8) The side of a triangle that is perpendicular to the altitude.

Base of an exponential function (12.1) The number b in the function $f(x) = b^x$.

Binomial (5.3) A polynomial of two terms. The expressions $a + 2b$, $6x^3 + 1$, and $5a^3b^2 + 6ab$ are all binomials.

Circumference of a circle (1.8) The distance around a circle. The circumference of a circle is given by the formula $C = \pi d$ or $C = 27\pi r$, where d is the diameter of the circle and r is the radius of the circle.

Coefficient (5.1) A coefficient is a factor or a group of factors in a product. In the term $4xy$ the coefficient of y is $4x$, but the coefficient of xy is 4. In the term $-5x^3y$ the coefficient of x^3y is -5.

Combined variation (8.7) When y varies directly with x and z and inversely with d^2, written $y = \dfrac{kxz}{d^2}$ where k is the constant of variation.

Common logarithm (12.4) The common logarithm of a number x is given by $\log x = \log_{10} x$ for all $x > 0$. A common logarithm is a logarithm using base 10.

Commutative property for addition (1.1) If a and b are any real numbers, then $a + b = b + a$.

G-1

Commutative property for multiplication (1.3) If a and b are any real numbers, then $ab = ba$.

Complex fraction (7.4) A fraction that contains at least one fraction in the numerator or in the denominator or both. These three fractions are complex fractions:

$$\frac{7 + \dfrac{1}{x}}{x^2 + 2}, \qquad \frac{1 + \dfrac{1}{5}}{2 - \dfrac{1}{7}}, \quad \text{and} \quad \frac{\dfrac{1}{3}}{\dfrac{3}{4}}$$

Complex number (8.6) A number that can be written in the form $a + bi$, where a and b are real numbers and $i = \sqrt{-1}$.

Conjugate of a binomial with radicals (8.4) The expressions $a\sqrt{x} + b\sqrt{y}$ and $a\sqrt{x} - b\sqrt{y}$. The conjugate of $2\sqrt{3} + 5\sqrt{2}$ is $2\sqrt{3} - 5\sqrt{2}$. The conjugate of $4 - \sqrt{x}$ is $4 + \sqrt{x}$.

Conjugate of a complex number (8.6) The expressions $a + bi$ and $a - bi$. The conjugate of $5 + 2i$ is $5 - 2i$. The conjugate of $7 - 3i$ is $7 + 3i$.

Constant (2.3) Symbol or letter that is used to represent exactly one single quantity during a particular problem or discussion.

Coordinates of a point (3.1) An ordered pair of numbers (x, y) that specifies the location of a point in a rectangular coordinate system.

Critical points of a quadratic inequality (9.6) In a quadratic inequality of the form $ax^2 + bx + c > 0$ or $ax^2 + bx + c < 0$, those points where $ax^2 + bx + c = 0$.

Degree of a polynomial (5.3) The degree of the highest-degree term of a polynomial. The degree of the polynomial $5x^3 + 2x^2 - 6x + 8$ is 3. The degree of the polynomial $5x^2y^2 + 3xy + 8$ is 4.

Degree of a term of a polynomial (5.3) The sum of the exponents of the variables in the term. The degree of $3x^3$ is 3. The degree of $4x^5y^2$ is 7.

Denominator (7.1) The bottom number or algebraic expression in a fraction. The denominator of

$$\frac{3x - 2}{x + 4}$$

is $x + 4$. The denominator of $\frac{3}{7}$ is 7. The denominator of a fraction may not be zero.

Dependent equations (4.1) Two equations are dependent if every value that satisfies one equation satisfies the other. A system of two dependent equations in two variables will not have a unique solution.

Determinant (Appendix E) A square array of numbers written between vertical lines. For example $\begin{vmatrix} 1 & 5 \\ 2 & 4 \end{vmatrix}$ is a

2×2 determinant. It is also called a *second-order determinant.* $\begin{vmatrix} 1 & 7 & 8 \\ 2 & -5 & -1 \\ -3 & 6 & 9 \end{vmatrix}$ is a 3×3 determinant. It is also called a *third-order determinant.*

Difference-of-two-squares polynomial (6.5) A polynomial of the form $a^2 - b^2$ that may be factored by using the formula

$$a^2 - b^2 = (a + b)(a - b).$$

Direct variation (8.7) When a variable y varies directly with x, written $y = kx$, where k represents some real number that will stay the same over a range of exercises. This value k is called the *constant of variation.*

Discriminant of a quadratic equation (9.2) In the equation $ax^2 + bx + c = 0$, where $a \neq 0$, the expression $b^2 - 4ac$. It can be used to determine the nature of the roots of the quadratic equation. If the discriminant is *positive*, there are two rational or irrational roots. The two roots will be rational only if the discriminant is a perfect square. If the discriminant is *zero*, there is only one rational root. If the discritninant is *negative*, there are two complex roots.

Distance between two points (10.1) The distance between point (x_1, y_1) and point (x_2, y_2) is given by the formula $d = \sqrt{(x_2 - x_1)^2 + (y_2 - y_1)^2}$.

Distributive property (1.6) For all real numbers a, b, and c, $a(b + c) = ab + ac$.

Domain of a relation (3.6) In any relation, the set of values that can be used for the independent variable is called its domain. This is the set of all the first coordinates of the ordered pairs that define the relation.

e (12.1) An irrational number that can be approximated by the value 2.7183.

Ellipse (10.3) The set of points in a plane such that for each point in the set, the sum of its distances to two fixed points is constant. Each of the fixed points is called a *focus.* Each of the following graphs is an ellipse.

Equilateral hyperbola (10.4) A hyperbola for which $a = b$ in the equation of the hyperbola.

Equilateral triangle (1.8) A triangle with three sides equal in length and three angles that measure 60°. Triangle ABC is an equilateral triangle.

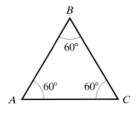

Even integers (1.3)   Integers that are exactly divisible by 2, such as $\ldots, -4, -2, 0, 2, 4, 6, \ldots$.

Exponent (1.4)   The number that indicates the power of a base. If the number is a positive integer it indicates how many times the base is multiplied. In the expression 2^6, the exponent is 6.

Exponential function (12.1) $f(x) = b^x$, where $b > 0$, $b \neq 1$, and x is any real number.

Expression (5.3)   A mathematic expression is any quantity using numbers and variables. Therefore, $2x$, $7x + 3$, and $5x^2 + 6x$ are all mathematical expressions.

Extraneous solution (7.5) and (8.5)   An obtained solution to an equation that when substituted back into the original equation, does *not* yield an identity. $x = 2$ is an extraneous solution to the equation

$$\frac{x}{x-2} - 4 = \frac{2}{x-2}$$

An extraneous solution is also called an extraneous root.

Factor (6.1)   When two or more numbers, variables, or algebraic expressions are multiplied, each is called a factor. If we write $3 \cdot 5 \cdot 2$, the factors are 3, 5, and 2. If we write $2xy$, the factors are $2, x$, and y. In the expression $(x - 6)(x + 2)$, the factors are $(x - 6)$ and $(x + 2)$.

Focus point of a parabola (10.2)   The focus point of a parabola has many properties. For example, the focus point of a parabolic mirror is the point to which all incoming light rays that are parallel to the axis of symmetry will collect. A parabola is a set of points that is the same distance from a fixed line called the *directrix* and a fixed point. This fixed point is the focus.

Function (3.6)   A relation in which no two different ordered pairs have the same first coordinate.

Graph of a function (11.2)   A graph in which a vertical line will never cross in more than one place. The following sketches represent the graphs of functions.

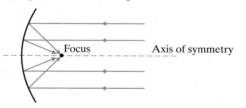

Graph of a one-to-one function (11.4)   A graph of a function with the additional property that a horizontal line will never cross the graph in more than one place. The following sketches represent the graphs of one-to-one functions.

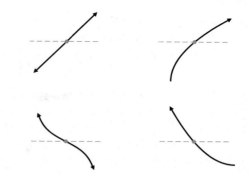

Higher-order equations (9.3)   Equations of degree 3 or higher. Examples of higher-order equations are

$$x^4 - 29x^2 + 100 = 0 \quad \text{and} \quad x^3 + 3x^2 - 4x - 12 = 0.$$

Higher-order roots (8.2)   Cube roots, fourth roots, and roots with an index greater than 2.

Horizontal parabolas (10.2)   Parabolas that open to the right or to the left. The following graphs represent horizontal parabolas.

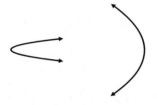

Hyperbola (10.4)   The set of points in a plane such that for each point in the set, the absolute value of the difference of its distances to two fixed points is constant. Each of these fixed points is called a *focus*. The following sketches represent graphs of hyperbolas.

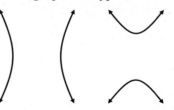

Hypotenuse of a right triangle (9.4)   The side opposite the right angle in any right triangle. The hypotenuse is always the longest side of a right triangle. In the following sketch the hypotenuse is side c.

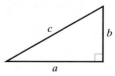

Imaginary number (8.6) i, defined as $i = \sqrt{-1}$ and $i^2 = -1$.

Inconsistent system of equations (4.1)   A system of equations that does not have a solution.

Independent equations (4.1)   Two equations that are not dependent are said to be independent.

Inequality (2.6), (2.7), and (3.5)   A mathematical relationship between quantities that are not equal. $x \leq -3$, $w > 5$, and $x < 2y + 1$ are mathematical inequalities.

Index of a radical (8.2)   Indicates what type of a root is being taken. The index of a cube root is $\sqrt[3]{x}$, the 3 is the index of the radical. In $\sqrt[4]{y}$, the index is 4. The index of a square root is 2, but the index is not written in the square root symbol, as shown: $\sqrt{x}$.

Integers (1.1)   The set of numbers $\ldots, -5, -4, -3, -2, -1, 0, 1, 2, 3, 4, 5, \ldots$.

Intercepts of an equation (3.2)   The point or points where the graph of the equation crosses the x-axis or the y-axis or both. (*See x-intercept or y-intercept.*)

Inverse function of a one-to-one function (11.4)   That function obtained by interchanging the first and second coordinates in each ordered pair of the function.

Inverse variation (8.7)   When a variable y varies inversely with x, written $y = \dfrac{k}{x}$, where k is the constant of variation.

Irrational number (1.1)   A real number that cannot be expressed in the form $\dfrac{a}{b}$, where a and b are integers and $b \neq 0$. $\sqrt{2}, \pi, 5 + 3\sqrt{2}$, and $-4\sqrt{7}$ are irrational numbers.

Isosceles triangle (1.8)   A triangle with two equal sides and two equal angles. Triangle ABC is an isosceles triangle. Angle BAC is equal to angle ACB. Side AB is equal in length to side BC.

Joint variation (8.7)   When a variable y varies jointly with x and z, written $y = kxz$, where k is the constant of variation.

Leg of a right triangle (9.4)   One of the two shorter sides of a right triangle. In the following sketch, sides a and b are the legs of the right triangle.

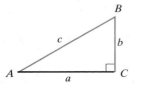

Like terms (1.7)   Terms that have identical variables and exponents. In the expression $5x^3 + 2xy^2 + 6x^2 - 3xy^2$, the term $2xy^2$ and the term $-3xy^2$ are like terms.

Linear equation in two variables (3.2)   An equation of the form $Ax + By = C$, where A, B, and C are real numbers. The graph of a linear equation in two variables is a straight line.

Minor of an element of a third-order determinant (Appendix B)   The second-order determinant that remains after we delete the row and column in which the element appears. The minor of the element 6 in the determinant $\begin{vmatrix} 1 & 2 & 3 \\ 7 & 6 & 8 \\ -3 & 5 & 9 \end{vmatrix}$ is the second-order determinant $\begin{vmatrix} 1 & 3 \\ -3 & 9 \end{vmatrix}$.

Logarithm (12.2)   For a positive number x, the power to which the base b must be raised to produce x. That is, $y = \log_b x$ is the same as $x = b^y$, where $b > 0$ and $b \neq 1$. A logarithm is an exponent.

Logarithmic equation (12.2)   An equation that contains at least one logarithm.

Magnitude of an earthquake (12.5)   The magnitude of an earthquake is measured by the formula $M = \log\left(\dfrac{I}{I_0}\right)$, where I is the intensity of the earthquake and I_0 is the minimum measurable intensity.

Natural logarithm (12.4)   For a number x, $\ln x = \log_e x$ for all $x > 0$. A natural logarithm is a logarithm using base e.

Nonlinear system of equations (10.5)   A system of equations in which at least one equation is not a linear equation.

Numerical coefficient (5.1)   The number that is multiplied by a variable or a group of variables. The numerical coefficient in $5x^3y^2$ is 5. The numerical coefficient in $-6abc$ is -6. The numerical coefficient in x^2y is 1. A numerical coefficient of 1 is not usually written.

Odd integers (1.3)   Integers that are not exactly divisible by 2, such as $\ldots, -3, -1, 1, 3, 5, 7, 9, \ldots$.

One-to-one function (11.4)   A function in which no two different ordered pairs have the same second coordinate.

Opposite of a number (1.1)   Two numbers that are the same distance from zero on the number line but lie on different sides of it are considered opposites. The opposite of -6 is 6. The opposite of $\dfrac{22}{7}$ is $-\dfrac{22}{7}$.

Ordered pair (3.1)   A pair of numbers presented in a specified order. An ordered pair is often used to specify a location on a graph. Every point in a rectangular coordinate system can be represented by an ordered pair (x, y).

Origin (3.1)   The point $(0, 0)$ in a rectangular coordinate system.

Parabola (10.2)   The set of points that is the same distance from some fixed line (called the *directrix*) and some fixed point (called the *focus*) that is not on the line. The graph of any equation of the form $y = ax^2 + bx + c$ or $x = ay^2 + by + c$, where a, b, and c are real numbers and $a \neq 0$, is a parabola. Some examples of the graphs of parabolas are shown.

Parallel lines (3.3) and (4.1)   Two straight lines that never intersect. The graph of an inconsistent system of two linear equations in two variables will result in parallel lines.

Parallelogram (1.8)   A four-sided figure with opposite sides parallel. Figure $ABCD$ is a parallelogram.

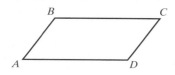

Perfect square number (6.5)   A number that is the square of an integer. The numbers 1, 4, 9, 16, 25, 36, 49, 64, 81, 100, 121, 144, ... are perfect square numbers.

Perfect-square trinomial (6.5)   A polynomial of the form $a^2 + 2ab + b^2$ or $a^2 - 2ab + b^2$ that may be factored using one of the following formulas:

$$a^2 + 2ab + b^2 = (a + b)^2$$

or

$$a^2 - 2ab + b^2 = (a - b)^2.$$

Perimeter (1.8)   The distance around any plane figure. The perimeter of this triangle is 13. The perimeter of this rectangle is 20.

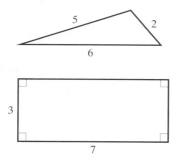

pH of a solution (12.2)   Defined by the equation $pH = -\log_{10}(H^+)$, where H^+ is the concentration of the hydrogen ion in the solution. The solution is an acid when the pH is less than 7 and a base when the pH is greater than 7.

Pi (1.8)   An irrational number, denoted by the symbol π, that is approximately equal to 3.141592654. In most cases 3.14 can be used as a sufficiently accurate approximation for π.

Polynomial (5.3)   Expressions that contain terms with nonnegative integer exponents. The expressions $5ab + 6$, $x^3 + 6x^2 + 3$, -12, and $x + 3y - 2$ *are all polynomials*. The expressions $x^{-2} + 2x^{-1}$, $2\sqrt{x} + 6$, and $\dfrac{5}{x} + 2x^2$ *are not polynomials*.

Prime polynomial (6.6)   A prime polynomial is a polynomial that cannot be factored by the methods of elementary algebra. $x^2 + x + 1$ is a prime polynomial.

Principal (2.9)   In monetary problems, the principal is the original amount of money invested or borrowed.

Principal square root (8.2)   The positive square root of a number. The symbol indicating the principal square root is $\sqrt{}$. Thus, $\sqrt{4}$ means to find the principal square root of 4, which is 2.

Proportion (7.6)   A proportion is an equation stating that two ratios are equal.

$$\frac{a}{b} = \frac{c}{d} \qquad \text{where } b, d \neq 0$$

is a proportion.

Pythagorean theorem (9.4)   In any right triangle, if c is the length of the hypotenuse and a and b are the lengths of the two legs, then $c^2 = a^2 + b^2$.

Quadratic equation in standard form (5.8, 9.1)   An equation of the form $ax^2 + bx + c = 0$, where a, b, and c are real numbers and $a \neq 0$. A quadratic equation is classified as a second-degree equation.

Quadratic formula (9.2)   If $ax^2 + bx + c = 0$ and $a \neq 0$, then the roots to the equation are found by the formula

$$x = \frac{-b \pm \sqrt{b^2 - 4ac}}{2a}.$$

Quadratic inequalities (10.5)   An inequality written in the form $ax^2 + bx + c > 0$, where $a \neq 0$ and a, b, and c are real numbers. The $>$ symbol may be replaced by a $<$, $\geq$, or $\leq$ symbol.

Radical equation (8.5)   An equation that contains one or more radicals. The following are examples of radical equations.

$$\sqrt{9x - 20} = x \quad \text{and} \quad 4 = \sqrt{x - 3} + \sqrt{x + 5}$$

Radical sign (8.2) The symbol $\sqrt{}$, which is used to indicate the root of a number.

Radicand (8.2) The expression beneath the radical sign. The radicand of $\sqrt{7x}$ is $7x$.

Range of a relation (3.6) In any relation, the set of values that represents the dependent variable is called its range. This is the set of all the second coordinates of the ordered pairs that define the relation.

Ratio (7.6) The ratio of one number a to another number b is the quotient $a \div b$ or $\frac{a}{b}$.

Rational numbers (1.1) and (8.1) A number that can be expressed in the form $\frac{a}{b}$, where a and b are integers and $b \neq 0$. $\frac{7}{3}, -\frac{2}{5}, \frac{7}{-8}, \frac{5}{1}$, 1.62, and 2.7156 are rational numbers.

Rationalizing the denominator (8.4) The process of transforming a fraction that contains one or more radicals in the denominator to an equivalent fraction that does not contain any radicals in the denominator. When we rationalize the denominator of $\frac{5}{\sqrt{3}}$, we obtain $\frac{5\sqrt{3}}{3}$. When we rationalize the denominator of $\frac{-2}{\sqrt{11} - \sqrt{7}}$, we obtain $-\frac{\sqrt{11} + \sqrt{7}}{2}$.

Rationalizing the numerator (8.4) The process of transforming a fraction that contains one or more radicals in the numerator to an equivalent fraction that does not contain any radicals in the numerator. When we rationalize the numerator of $\frac{\sqrt{5}}{x}$, we obtain $\frac{5}{x\sqrt{5}}$.

Real number (9.1) Any number that is rational or irrational. $2, 7, \sqrt{5}, \frac{3}{8}, \pi, -\frac{7}{5}$, and $-3\sqrt{5}$ are all real numbers.

Rectangle (1.8) A four-sided figure with opposite sides parallel and all interior angles measuring 90°. The opposite sides of a rectangle are equal.

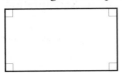

Relation (3.6) A relation is any set of ordered pairs.

Rhombus (1.8) A parallelogram with four equal sides. Figure $ABCD$ is a rhombus.

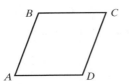

Right triangle (1.8) and (9.4) A triangle that contains one right angle (an angle that measures exactly 90 degrees). It is indicated by a small rectangle at the corner of the angle.

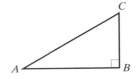

Root of an equation (2.1) and (6.7) A value of the variable that makes an equation into a true statement. The root of an equation is also called the solution of an equation.

Scientific notation (5.2) A positive number is written in scientific notation if it is in the form $a \times 10^n$, where $1 \leq a < 10$ and n is an integer.

Similar radicals (8.3) Two radicals that are simplified and have the same radicand and the same index. $2\sqrt[3]{7xy^2}$ and $-5\sqrt[3]{7xy^2}$ are similar radicals. Usually similar radicals are referred to as *like radicals*.

Simplifying a radical (8.3) To simplify a radical when the root cannot be found exactly, we use the product rule for radicals, $\sqrt[n]{ab} = \sqrt[n]{a}\sqrt[n]{b}$ for $a \geq 0$ and $b \geq 0$. To simplify $\sqrt{20}$, we have $= \sqrt{4}\sqrt{5} = 2\sqrt{5}$. To simplify $\sqrt[3]{16x^4}$, we have $= \sqrt[3]{8x^3}\sqrt[3]{2x} = 2x\sqrt[3]{2x}$.

Simplifying imaginary numbers (8.6) Using the property that states for all positive real numbers a, $\sqrt{-a} = \sqrt{-1}\sqrt{a} = i\sqrt{a}$. Thus, simplifying $\sqrt{-7}$, we have $\sqrt{-7} = \sqrt{-1}\sqrt{7} = i\sqrt{7}$.

Slope-intercept form (3.3) The equation of a line that has slope m and the y-intercept at $(0, b)$ is given by $y = mx + b$.

Slope of a line (3.3) The ratio of change in y over the change in x for any two different points on a nonvertical line. The slope m is determined by

$$m = \frac{y_2 - y_1}{x_2 - x_1},$$

where $x_2 \neq x_1$ for any two points (x_1, y_1) and (x_2, y_2) on a nonvertical line.

Solution of an equation (2.1) A number that, when substituted into a given equation, yields an identity. The solution of an equation is also called the root of an equation.

Solution of a linear inequality (2.7) The possible values that make a linear inequality true.

Solution of an inequality in two variables (3.5) The set of all possible ordered pairs that when substituted into the inequality will yield a true statement.

Square (1.8)   A rectangle with four equal sides.

Square root (8.2)   If x is a real number and a is positive real number such that $a = x^2$, then x is a square root of a. One square root of 16 is 4 since $4^2 = 16$. Another square root of 16 is -4 since $(-4)^2 = 16$.

Standard form of the equation of a circle (10.1)   For a circle with center at (h, k) and a radius of r,
$$(x - h)^2 + (y - k)^2 = r^2.$$

Standard form of the equation of an ellipse (10.3)   For an ellipse with center at the origin,
$$\frac{x^2}{a^2} + \frac{y^2}{b^2} = 1, \qquad \text{where } a \text{ and } b > 0.$$

This ellipse has intercepts at $(a, 0)$, $(-a, 0)$, $(0, b)$, and $(0, -b)$.

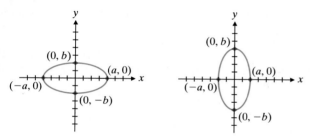

For an ellipse with center at (h, k),
$$\frac{(x - h)^2}{a^2} + \frac{(y - k)^2}{b^2} = 1, \qquad \text{where } a \text{ and } b > 0.$$

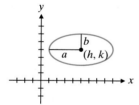

Standard form of the equation of a hyperbola with center at the origin (10.4)   For a horizontal hyperbola with center at the origin,
$$\frac{x^2}{a^2} - \frac{y^2}{b^2} = 1, \qquad \text{where } a \text{ and } b > 0.$$

The vertices are at $(-a, 0)$ and $(a, 0)$.

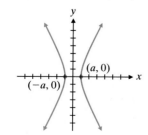

For a vertical hyperbola with center at the origin,

$$\frac{y^2}{b^2} - \frac{x^2}{a^2} = 1, \qquad \text{where } a \text{ and } b > 0.$$

The vertices are at $(0, b)$ and $(0, -b)$.

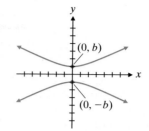

Standard form of the equation of a hyperbola with center at point (h, k) (10.4)   For a horizontal hyperbola with center at (h, k),
$$\frac{(x - h)^2}{a^2} - \frac{(y - k)^2}{b^2} = 1, \qquad \text{where } a \text{ and } b > 0.$$

The vertices are at $(h - a, k)$ and $(h + a, k)$.

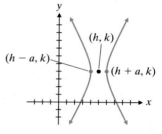

For a vertical hyperbola with center at (h, k),
$$\frac{(y - k)^2}{b^2} - \frac{(x - h)^2}{a^2} = 1, \qquad \text{where } a \text{ and } b > 0.$$

The vertices are at $(h, k + b)$ and $(h, k - b)$.

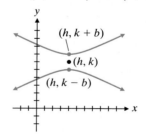

Standard form of the equation of a parabola (10.2)   For a vertical parabola with vertex at (h, k),
$$y = a(x - h)^2 + k, \text{ where } a \neq 0.$$

For a horizontal parabola with vertex at (h, k),
$$x = a(y - k)^2 + h, \text{ where } a \neq 0.$$

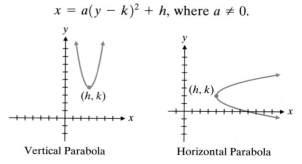

Vertical Parabola   Horizontal Parabola

Standard form of a quadratic equation (6.7)   A quadratic equation that is in the form $ax^2 + bx + c = 0$.

System of equations (4.1) A set of two or more equations that must be considered together. The solution is the value for each variable of the system that satisfies each equation.

$$x + 3y = -7 \qquad 4x + 3y = -1$$

is a system of two equations in two unknowns. The solution is $(2, -3)$, or the values $x = 2$, $y = -3$.

System of inequalities (4.4) Two or more inequalities in two variables that are considered at one time. The solution is the region that satisfies every inequality at one time. An example of a system of inequalities is

$$y > 2x + 1 \qquad y < \frac{1}{2}x + 2.$$

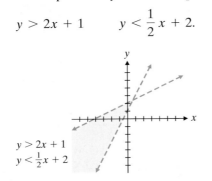

$$y > 2x + 1$$
$$y < \frac{1}{2}x + 2$$

Term (1.7) A number, a variable, or a product of numbers and variables. For example, in the expression $a^3 - 3a^2b + 4ab^2 + 6b^3 + 8$, there are five terms. They are a^3, $-3a^2b$, $4ab^2$, $6b^3$, and 8. The terms of a polynomial are separated by plus and minus signs.

Trapezoid (1.8) A four-sided figure with two sides parallel. The parallel sides are called the bases of the trapezoid. Figure $ABCD$ is a trapezoid.

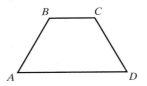

Trinomial (5.3) A polynomial of three terms. The expressions $x^2 + 6x - 8$ and $a + 2b - 3c$ are trinomials.

Value of a second-order determinant (Appendix E) For a second-order determinant $\begin{vmatrix} a & b \\ c & d \end{vmatrix}$, $ad - cb$.

Value of a third-order determinant (Appendix E) For third-order determinant $\begin{vmatrix} a_1 & b_1 & c_1 \\ a_2 & b_2 & c_2 \\ a_3 & b_3 & c_3 \end{vmatrix}$,

$a_1b_2c_3 + b_1c_2a_3 + c_1a_2b_3 - a_3b_2c_1 - b_3c_2a_1 - c_3a_2b_1.$

Variable (1.4) A letter that is used to represent a number or a set of numbers.

Variation (8.7) An equation relating values of one variable to those of other variables. An equation of the form $y = kx$, where k is a constant, indicates *direct variation*. An equation of the form $y = \dfrac{k}{x}$, where k is a

constant, indicates *inverse variation*. In both cases, k is called the *constant of variation*.

Vertex of a parabola (10.2) In a vertical parabola, the lowest point on a parabola opening upward or the highest point on a parabola opening downward.

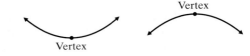

In a horizontal parabola, the leftmost point on a parabola opening to the right or the rightmost point on a parabola opening to the left.

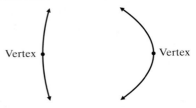

Vertical line test (3.6) If a vertical line can intersect the graph of a relation more than once, the relation is not a function.

Vertical parabolas (10.2) Parabolas that open upward or downward. The following graphs represent vertical parabolas.

x-intercept (3.2) The ordered pair $(a, 0)$ is the x-intercept of a line if the line crosses the x-axis at $(a, 0)$. The x-intercept of line l on the following graph is $(4, 0)$.

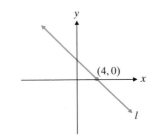

y-intercept (3.2) The ordered pair $(0, b)$ is the y-intercept of a line if the line crosses the y-axis at $(0, b)$. The y-intercept of line p on the following graph is $(0, 3)$.

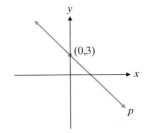

Appendix A
Tables

Table A-1: Table of Square Roots

Square root values ending in 000 are exact. All other values are approximate and are rounded to the nearest thousandth.

x	$\sqrt{x}$	x	$\sqrt{x}$	x	$\sqrt{x}$	x	$\sqrt{x}$	x	$\sqrt{x}$
1	1.000	41	6.403	81	9.000	121	11.000	161	12.689
2	1.414	42	6.481	82	9.055	122	11.045	162	12.728
3	1.732	43	6.557	83	9.110	123	11.091	163	12.767
4	2.000	44	6.633	84	9.165	124	11.136	164	12.806
5	2.236	45	6.708	85	9.220	125	11.180	165	12.845
6	2.449	46	6.782	86	9.274	126	11.225	166	12.884
7	2.646	47	6.856	87	9.327	127	11.269	167	12.923
8	2.828	48	6.928	88	9.381	128	11.314	168	12.961
9	3.000	49	7.000	89	9.434	129	11.358	169	13.000
10	3.162	50	7.071	90	9.487	130	11.402	170	13.038
11	3.317	51	7.141	91	9.539	131	11.446	171	13.077
12	3.464	52	7.211	92	9.592	132	11.489	172	13.115
13	3.606	53	7.280	93	9.644	133	11.533	173	13.153
14	3.742	54	7.348	94	9.695	134	11.576	174	13.191
15	3.873	55	7.416	95	9.747	135	11.619	175	13.229
16	4.000	56	7.483	96	9.798	136	11.662	176	13.266
17	4.123	57	7.550	97	9.849	137	11.705	177	13.304
18	4.243	58	7.616	98	9.899	138	11.747	178	13.342
19	4.359	59	7.681	99	9.950	139	11.790	179	13.379
20	4.472	60	7.746	100	10.000	140	11.832	180	13.416
21	4.583	61	7.810	101	10.050	141	11.874	181	13.454
22	4.690	62	7.874	102	10.100	142	11.916	182	13.491
23	4.796	63	7.937	103	10.149	143	11.958	183	13.528
24	4.899	64	8.000	104	10.198	144	12.000	184	13.565
25	5.000	65	8.062	105	10.247	145	12.042	185	13.601
26	5.099	66	8.124	106	10.296	146	12.083	186	13.638
27	5.196	67	8.185	107	10.344	147	12.124	187	13.675
28	5.292	68	8.246	108	10.392	148	12.166	188	13.711
29	5.385	69	8.307	109	10.440	149	12.207	189	13.748
30	5.477	70	8.367	110	10.488	150	12.247	190	13.784
31	5.568	71	8.426	111	10.536	151	12.288	191	13.820
32	5.657	72	8.485	112	10.583	152	12.329	192	13.856
33	5.745	73	8.544	113	10.630	153	12.369	193	13.892
34	5.831	74	8.602	114	10.677	154	12.410	194	13.928
35	5.916	75	8.660	115	10.724	155	12.450	195	13.964
36	6.000	76	8.718	116	10.770	156	12.490	196	14.000
37	6.083	77	8.775	117	10.817	157	12.530	197	14.036
38	6.164	78	8.832	118	10.863	158	12.570	198	14.071
39	6.245	79	8.888	119	10.909	159	12.610	199	14.107
40	6.325	80	8.944	120	10.954	160	12.649	200	14.142

Table A-2: Exponential Values

x	e^x	e^{-x}	x	e^x	e^{-x}
0.00	1.0000	1.0000	1.6	4.9530	0.2019
0.01	1.0101	0.9900	1.7	5.4739	0.1827
0.02	1.0202	0.9802	1.8	6.0496	0.1653
0.03	1.0305	0.9704	1.9	6.6859	0.1496
0.04	1.0408	0.9608	2.0	7.3891	0.1353
0.05	1.0513	0.9512	2.1	8.1662	0.1225
0.06	1.0618	0.9418	2.2	9.0250	0.1108
0.07	1.0725	0.9324	2.3	9.9742	0.1003
0.08	1.0833	0.9231	2.4	11.023	0.0907
0.09	1.0942	0.9139	2.5	12.182	0.0821
0.10	1.1052	0.9048	2.6	13.464	0.0743
0.11	1.1163	0.8958	2.7	14.880	0.0672
0.12	1.1275	0.8869	2.8	16.445	0.0608
0.13	1.1388	0.8781	2.9	18.174	0.0550
0.14	1.1503	0.8694	3.0	20.086	0.0498
0.15	1.1618	0.8607	3.1	22.198	0.0450
0.16	1.1735	0.8521	3.2	24.533	0.0408
0.17	1.1853	0.8437	3.3	27.113	0.0369
0.18	1.1972	0.8353	3.4	29.964	0.0334
0.19	1.2092	0.8270	3.5	33.115	0.0302
0.20	1.2214	0.8187	3.6	36.598	0.0273
0.21	1.2337	0.8106	3.7	40.447	0.0247
0.22	1.2461	0.8025	3.8	44.701	0.0224
0.23	1.2586	0.7945	3.9	49.402	0.0202
0.24	1.2712	0.7866	4.0	54.598	0.0183
0.25	1.2840	0.7788	4.1	60.340	0.0166
0.26	1.2969	0.7711	4.2	66.686	0.0150
0.27	1.3100	0.7634	4.3	73.700	0.0136
0.28	1.3231	0.7558	4.4	81.451	0.0123
0.29	1.3364	0.7483	4.5	90.017	0.0111
0.30	1.3499	0.7408	4.6	99.484	0.0101
0.35	1.4191	0.7047	4.7	109.95	0.0091
0.40	1.4918	0.6703	4.8	121.51	0.0082
0.45	1.5683	0.6376	4.9	134.29	0.0074
0.50	1.6487	0.6065	5.0	148.41	0.0067
0.55	1.7333	0.5769	5.5	244.69	0.0041
0.60	1.8221	0.5488	6.0	403.43	0.0025
0.65	1.9155	0.5220	6.5	665.14	0.0015
0.70	2.0138	0.4966	7.0	1,096.6	0.00091
0.75	2.1170	0.4724	7.5	1,808.0	0.00055
0.80	2.2255	0.4493	8.0	2,981.0	0.00034
0.85	2.3396	0.4274	8.5	4,914.8	0.00020
0.90	2.4596	0.4066	9.0	8,103.1	0.00012
0.95	2.5857	0.3867	9.5	13,360	0.000075
1.0	2.7183	0.3679	10	22,026	0.000045
1.1	3.0042	0.3329	11	59,874	0.000017
1.2	3.3201	0.3012	12	162,755	0.0000061
1.3	3.6693	0.2725	13	442,413	0.0000023
1.4	4.0552	0.2466	14	1,202,604	0.0000008
1.5	4.4817	0.2231	15	3,269,017	0.0000003

Appendix B
Mathematics Blueprint for Problem Solving

1 *Using the Mathematics Blueprint to Solve Real-life Problems*

Student Learning Objectives

After studying this section, you will be able to:

1 Use the Mathematics Blueprint to solve real-life problems.

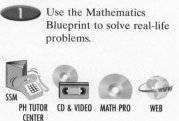

SSM · PH TUTOR CENTER · CD & VIDEO · MATH PRO · WEB

When a builder constructs a new home or office building, he or she often has a blueprint. This accurate drawing shows the basic form of the building. It also shows the dimensions of the structure to be built. This blueprint serves as a useful reference throughout the construction process.

Similarly, when solving real-life problems, it is helpful to have a "mathematics blueprint." This is a simple way to organize the information provided in the word problem, in a chart, or in a graph. You can record the facts you need to use. You can determine what it is you are trying to find and how you can go about actually finding it. You can record other information that you think will be helpful as you work through the problem.

As we solve real-life problems, we will use three steps.

Step 1 *Understand the problem.*   Here we will read through the problem. Draw a picture if it will help, and use the Mathematics Blueprint as a guide to assist us in thinking through the steps needed to solve the problem.

Step 2 *Solve and state the answer.*   We will use arithmetic or algebraic procedures along with problem-solving strategies to find a solution.

Step 3 *Check.*   We will use a variety of techniques to see if the answer in step 2 is the solution to the word problem. This will include estimating to see if the answer is reasonable, repeating our calculation, and working backward from the answer to see if we arrive at the original conditions of the problem.

▲ **EXAMPLE 1**   Nancy and John want to install wall-to-wall carpeting in their living room. The floor of the rectangular living room is $11\frac{2}{3}$ feet wide and $19\frac{1}{2}$ feet long. How much will it cost if the carpet is $18.00 per square yard?

1. *Understand the problem.*
   First, read the problem carefully. Drawing a sketch of the living room may help you see what is required. The carpet will cover the floor of the living room, so we need to find the area. Now we fill in the Mathematics Blueprint.

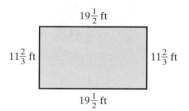

Mathematics Blueprint For Problem Solving

Gather the Facts	What Am I Solving For?	What Must I Calculate?	Key Points to Remember
The living room measures $11\frac{2}{3}$ ft by $19\frac{1}{2}$ ft. The carpet costs $18.00 per square yard.	**(a)** the area of the room in square feet **(b)** the area of the room in square yards **(c)** the cost of the carpet	**(a)** Multiply $11\frac{2}{3}$ ft by $19\frac{1}{2}$ ft to get area in square feet. **(b)** Divide the number of square feet by 9 to get the number of square yards. **(c)** Multiply the number of square yards by $18.00.	There are 9 square feet, 3 feet × 3 feet, in 1 square yard; therefore, we must divide the number of square feet by 9 to obtain square yards.

2. *Solve and state the answer.*

(a) To find the area of a rectangle, we multiply the length times the width.

$$11\frac{2}{3} \times 19\frac{1}{2} = \frac{35}{3} \times \frac{39}{2}$$

$$= \frac{455}{2} = 227\frac{1}{2} \text{ sq ft}$$

A minimum of $227\frac{1}{2}$ square feet of carpet will be needed. We say a minimum because some carpet may be wasted in cutting. Carpet is sold by the square yard. We will want to know the amount of carpet needed in square yards.

(b) To determine the area in square yards, we divide $227\frac{1}{2}$ by 9. (9 sq ft = 1 sq yd.)

$$227\frac{1}{2} \div 9 = \frac{455}{2} \div \frac{9}{1}$$

$$= \frac{455}{2} \times \frac{1}{9} = \frac{455}{18} = 25\frac{5}{18} \text{ sq yd}$$

A minimum of $25\frac{5}{18}$ square yards of carpet will be needed.

(c) Since the carpet costs $18.00 per square yard, we will multiply the number of square yards needed by $18.00.

$$25\frac{5}{18} \times 18 = \frac{455}{18} \times \frac{18}{1} = \$455$$

The carpet will cost a minimum of $455.00 for this room.

"Remember to estimate. It will save you time and money!"

3. *Check.*

We will estimate to see if our answers are reasonable.

(a) We will estimate by rounding each number to the nearest 10.

$$11\frac{2}{3} \times 19\frac{1}{2} \longrightarrow 10 \times 20 = 200 \text{ sq ft}$$

This is close to our answer of $227\frac{1}{2}$ sq ft. Our answer is reasonble. ✓

(b) We will estimate by rounding to one significant digit.

$$227\frac{1}{2} \div 9 \longrightarrow 200 \div 10 = 20 \text{ sq yd}$$

This is close to our answer of $25\frac{5}{18}$ sq yd. Our answer is reasonable. ✓

(c) We will estimate by rounding each number to the nearest 10.

$$25\frac{5}{18} \times 18 \longrightarrow 30 \times 20 = \$600$$

This is close to our answer of $455. Our answer seems reasonable. ✓

▲ **Practice Problem 1**   Jeff went to help Abby pick out wall-to-wall carpet for her new house. Her rectangular living room measures $16\frac{1}{2}$ feet by $10\frac{1}{2}$ feet. How much will it cost to carpet the room if the carpet costs $20 per square yard?

Mathematics Blueprint For Problem Solving

Gather the Facts	What Am I Solving For?	What Must I Calculate?	Key Points to Remember

To Think About   Assume that the carpet in Example 1 comes in a standard width of 12 feet. How much carpet will be wasted if it is laid out on the living room floor in one strip that is $19\frac{1}{2}$ feet long? How much carpet will be wasted if it is laid in two sections side by side that are each $11\frac{2}{3}$ feet long? Assuming you have to pay for wasted carpet, what is the minimum cost to carpet the room?

EXAMPLE 2   The following chart shows the 2000 sales of Micropower Computer Software for each of the four regions of the United States. Use the chart to answer the following questions (round all answers to the nearest whole percent):

(a) What percent of the sales personnel are assigned to the Northeast?

(b) What percent of the volume of sales is attributed to the Northeast?

(c) What percent of the sales personnel are assigned to the Southeast?

(d) What percent of the volume of sales is attributed to the Southeast?

(e) Which of these two regions of the country has sales personnel that appear to be more effective in terms of the volume of sales?

Region of the U.S.	Number of Sales Personnel	Dollar Volume of Sales
Northeast	12	1,560,000
Southeast	18	4,300,000
Northwest	10	3,660,000
Southwest	15	3,720,000
Total	55	13,240,000

1. ***Understand the problem.***

We will only need to deal with figures from the Northeast region and the Southeast region.

Mathematics Blueprint For Problem Solving

Gather the Facts	What Am I Solving For?	What Must I Calculate?	Key Points to Remember
Personnel: 12 Northeast 18 Southeast 55 total Sales Volume: $1,560,000 NE $4,300,000 SE $13,240,000 Total	**(a)** the percent of the total personnel that is in the Northeast **(b)** the percent of the total sales made in the Northeast **(c)** the percent of the total personnel that is in the Southeast **(d)** the percent of the total sales made in the Southeast **(e)** compare the percentages from the two regions	**(a)** 12 of 55 is what percent? Divide. $12 \div 55$ **(b)** 1,560,000 of 13,240,000 is what percent? 1,560,000 $\div$ 13,240,000 **(c)** $18 \div 55$ **(d)** 4,300,000 $\div$ 13,240,000	We do not need to use the numbers relating to the Northwest or the Southwest in this problem.

2. Solve and state the answer.

(a) $\dfrac{12}{55} = 0.21818\ldots$

$\approx 22\%$

(b) $\dfrac{1,560,000}{13,240,000} = \dfrac{156}{1324} \approx 0.1178$

$\approx 12\%$

(c) $\dfrac{18}{55} = 0.32727\ldots$

$\approx 33\%$

(d) $\dfrac{4,300,000}{13,240,000} = \dfrac{430}{1324} \approx 0.3248$

$\approx 32\%$

(e) We notice that 22% of the sales force in the Northeast made 12% of the sales. The percent of the sales compared to the percent of the sales force is about half (12% of 24% would be half), or 50%. 33% of the sales force in the Southeast made 32% of the sales. The percent of sales compared to the percent of the sales force is close to 100%. We must be cautious here. *If there are no other significant factors,* it would appear that the Southeast sales force is more effective. (There may be other significant factors affecting sales, such as a recession in the Northeast, new and inexperienced sales personnel, or fewer competing companies in the Southeast.)

3. Check.

You may want to use a calculator to check the division in step 2, or you may use estimation.

(a) $\dfrac{12}{55} \rightarrow \dfrac{10}{60} \approx 0.17$

$= 17\%$ ✓

(b) $\dfrac{1,560,000}{13,240,000} \rightarrow \dfrac{1,600,000}{13,000,000} \approx 0.12$

$= 12\%$ ✓

(c) $\dfrac{18}{55} \rightarrow \dfrac{20}{60} \approx 0.33$

$= 33\%$ ✓

(d) $\dfrac{4,300,000}{13,240,000} \rightarrow \dfrac{4,300,000}{13,000,000} \approx 0.33$

$= 33\%$ ✓

Practice Problem 2   Using the chart for Example 2, answer the following questions. (Round all answers to the nearest whole percent.)

(a) What percent of the sales personnel are assigned to the Northwest?

(b) What percent of the sales volume is attributed to the Northwest?

(c) What percent of the sales personnel are assigned to the Southwest?

(d) What percent of the sales volume is attributed to the Southwest?

(e) Which of these two regions of the country has sales personnel that appear to be more effective in terms of volume of sales?

Mathematics Blueprint For Problem Solving

Gather the Facts	What Am I Solving For?	What Must I Calculate?	Key Points to Remember

To Think About   Suppose in 2001 the number of sales personnel (55) increases by 60%. What would the new number of sales personnel be? Suppose in 2002 that the number of sales personnel decreases by 60% from the number of sales personnel in 2001. What would the new number be? Why is this number not 55, since we have increased the number by 60% and then decreased the result by 60%? Explain.

Appendix B Exercises

Use the Mathematics Blueprint for Problem Solving to help you solve each of the following exercises.

▲ **1.** Jocelyn wants to put new vinyl flooring in her kitchen. The kitchen measures $12\frac{3}{4}$ feet long by $9\frac{1}{2}$ feet wide. If the vinyl flooring she chose costs $20.00 per square yard, how much will the new flooring cost her? (Round your answer to the nearest cent.)

▲ **2.** The Carters need to replace the deck floor off their kitchen door. The deck is $11\frac{1}{2}$ feet by $20\frac{1}{2}$ feet. If the new decking costs $4.50 per square foot, how much will it cost them to replace the deck? (Round your answer to the nearest cent.)

▲ **3.** In order to put in a new lawn, the landscaper told Mr. Lopez to add new loam to a depth of $\frac{1}{2}$ foot. Mr. Lopez's lawn is $85\frac{1}{2}$ feet by 60 feet. How many cubic yards of loam does he need? (There are 27 cubic feet in 1 cubic yard.)

▲ **4.** The Brock family removed a built-in swimming pool and have decided to fill in the hole with dirt and seed the area. The pool hole is 30 feet by 12 feet by 9 feet deep. How many cubic yards of dirt are needed to fill in the hole? (There are 27 cubic feet in 1 cubic yard.)

The following directions are posted on the wall at the gym.

Beginning exercise training schedule

On day 1, each athlete will begin the morning as follows:

Jog.................. $1\frac{1}{2}$ miles

Walk.............. $1\frac{3}{4}$ miles

Rest............... $2\frac{1}{2}$ minutes

Walk............. 1 mile

5. Betty's athletic trainer told her to follow the beginning exercise training schedule on day 1. On day 2, she is to increase all distances and times by $\frac{1}{3}$ that of day 1. On day 3, she is to increase all distances and times by $\frac{1}{3}$ that of day 2. What will be her training schedule on day 3?

6. Melinda's athletic trainer told her to follow the beginning exercise training schedule on day 1. On day 2, she is to increase all distances and times by $\frac{1}{3}$ that of day 1. On day 3, she is to once again increase all distances and times by $\frac{1}{3}$ that of day 1. What will be her training schedule on day 3?

To Think About

Refer to exercises 5 and 6 in working exercises 7–10.

7. Who will have a more demanding schedule on day 3, Betty or Melinda? Why?

8. If Betty kept up the same type of increase day after day, how many miles would she be jogging on day 5?

9. If Melinda kept up the same type of increase day after day, how many miles would she be jogging on day 7?

10. Which athletic trainer would appear to have the best plan for training athletes if they used this plan for 14 days? Why?

11. In 1985, the average selling price of an existing single-family home in Atlanta, Georgia, was $66,200. Between 1985 and 1990, the average price increased by 30%. Between 1990 and 2000, the average price increased again, this time by 15%. What was the median house price in Atlanta in 2000?

12. Chicken eggs are classified by weight per dozen eggs. Large eggs weigh 24 ounces per dozen and medium eggs weigh 21 ounces per dozen.

   (a) If you do not include the shell, which is 12% of the total weight of an egg, how many ounces of eggs do you get from a dozen large eggs? From a dozen medium eggs?

   (b) At a local market, large eggs sell for $1.79 a dozen, and medium eggs for $1.39 a dozen. If you do not include the shell, which is a better buy, large or medium eggs?

North Shore Community College students recently conducted a survey of 1000 passengers at Logan Airport. The passengers were classified as shown in the following table.

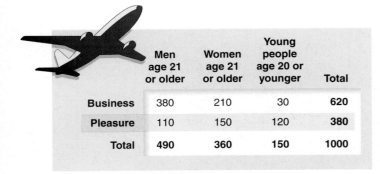

	Men age 21 or older	Women age 21 or older	Young people age 20 or younger	Total
Business	380	210	30	**620**
Pleasure	110	150	120	**380**
Total	**490**	**360**	**150**	**1000**

13. What percent of the total travelers were men age 21 or older? What percent of the business travelers were men age 21 or older? Which percentage is greater? Why do you suppose this is the case?

14. What percent of the business travelers were men age 21 or older? What percent of the business travelers were women age 21 or older? What percent of the business travelers were young people age 20 or younger? Why do you suppose this percentage is so small?

Use the following information from a paycheck stub to solve exercises 15–18.

TOBEY & SLATER INC. 5000 Stillwell Avenue Queens, NY 10001		Check Number	Payroll Period		Pay Date
			From Date	To Date	
		495885	10-30-99	11-30-99	12-01-99

Name	Social Security No.	I.D. Number	File Number	Rate/Salary	Department	MS	DEP	Res
Fred J. Gilliani	012-34-5678	01	1379	1150.00	0100	M	5	NY

	Current	Year to Date		Current	Year to Date
GROSS	1,150.00	6,670.00	STATE	67.76	388.45
FEDERAL	138.97	781.07	LOCAL	5.18	30.04
FICA	87.98	510.28	DIS-SUI	.00	.00
W-2 GROSS		6,670.00	NET	790.47	4,960.16

Earnings						Deductions/Specials		
No.	Type	Hours	Rate	Amount	Dept/Job No.	No.	Description	Amount
96	REGULAR			1,150.00	0100	82	Retirement	12.56
						75	Medical	36.28
						56	Union Dues	10.80

Gross pay is the pay an employee receives for his or her services before deductions. Net pay is the pay the employee actually gets to take home. You may round each amount to the nearest whole percent for exercises 15–18.

15. What percent of Fred's gross pay is deducted for federal, state, and local taxes?

16. What percent of Fred's gross pay is deducted for retirement and medical?

17. What percent of Fred's gross pay does he actually get to take home?

18. What percent of Fred's deductions are special deductions?

Appendix C
Interpreting Data from Tables, Charts, and Graphs

 **Tables**

A table is a device used to organize information into categories. Using it you can readily find details about each category.

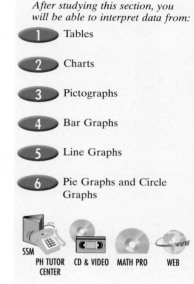

After studying this section, you will be able to interpret data from:

1 Tables

2 Charts

3 Pictographs

4 Bar Graphs

5 Line Graphs

6 Pie Graphs and Circle Graphs

SSM   PH TUTOR   CD & VIDEO   MATH PRO   WEB
     CENTER

EXAMPLE 1

Table of Nutritive Values of Certain Popular "Fast Foods"

Type of Sandwich	Calories	Protein (g)	Fat (g)	Cholesterol (g)	Sodium (mg)
Burger King Whopper	630	27	38	90	880
McDonald's Big Mac	500	25	26	100	890
Wendy's Bacon Cheeseburger	440	22	25	65	870
Burger King BK Broiler (chicken)	280	20	10	50	770
McDonald's McChicken	415	19	20	50	830
Wendy's Grilled Chicken	290	24	7	60	670

Source: U.S. Government Agencies and Food Manufacturers

(a) Which food item has the least amount of fat per serving?

(b) Which beef item has the least amount of calories per serving?

(c) How much more protein does a Burger King Whopper have than a McDonald's Big Mac?

(a) The least amount of fat, 7 grams, is in the Wendy's Grilled Chicken Sandwich.

(b) The least amount of calories, 440, for a beef sandwich is the Wendy's Bacon Cheeseburger.

(c) The Burger King Whopper has 2 grams of protein more than the McDonald's Big Mac.

Practice Problem 1

(a) Which sandwich has the lowest level of sodium?

(b) Which sandwich has the highest level of cholesterol?

2 Charts

A chart is a device used to organize information in which not every category is the same. Example 2 illustrates a chart containing different types of data.

● **EXAMPLE 2** The following chart shows how people in Topsfield indicated they spent their free time.

Survey of Use of Leisure Time

Category	Activity	Hours spent per week
Single men	Gym	6
	Outdoor sports	4
	Dating	7
	Watching pro sports	12
	Reading & TV	3
Single women	Gym	4
	Outdoor sports	2
	Dating	7
	Time with friends	10
	Reading & TV	9
Couples	Time with family	21
	Time as a couple	8
	Time with friends	4
	Reading & TV	9
Children	Watching TV	28
	Playing outside	8
	Reading	1

Use the chart to answer the following questions about people in Topsfield.

(a) What is the average amount of time a couple spends together as a family during the week?

(b) How much more time do children spend watching TV than playing outside?

(c) What activity do single men spend most of their time doing?

(a) The average amount of time a couple spends together as a family is 21 hours per week.

(b) Children spend 20 more hours per week watching TV than playing outside.

(c) Single men spend more time per week watching pro sports (12 hr) than any other activity.

Practice Problem 2

(a) What two categories do single women spend the most time doing?

(b) What do couples spend most of their time doing?

(c) What is the most significant numerical difference in terms of number of hours per week spent by single women versus single men?

3 *Pictographs*

A pictograph uses a visually appropriate symbol to represent an amount of items. A pictograph is used in Example 3.

EXAMPLE 3 Consider the following pictograph.

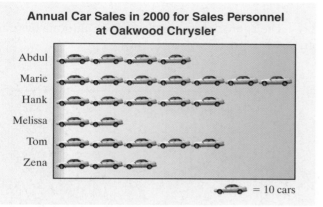

(a) How many cars did Melissa sell in 2000?

(b) Who sold the greatest number of cars?

(c) How many more cars did Tom sell than Zena?

(a) Melissa sold $2 \times 10 = 20$ cars.

(b) Marie sold the greatest number of cars.

(c) Tom sold $5 \times 10 = 50$ cars. Zena sold $3 \times 10 = 30$ cars. Now $50 - 30 = 20$. Therefore, Tom sold 20 cars more than Zena.

Practice Problem 3

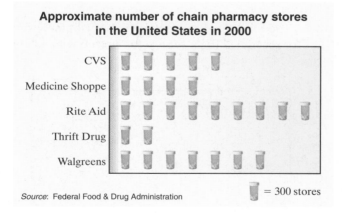

(a) Approximately how many stores does Walgreens have?

(b) Approximately how many more stores does Rite Aid have than CVS?

(c) What is the combined number of Thrift Drug and Medicine Shoppe stores?

4   *Bar Graphs*

A bar graph is helpful for making comparisons and noting changes or trends. A scale is provided so that the height of the bar graph indicates a specific number. A bar graph is displayed in Example 4. A bar graph may be represented horizontally or vertically. In either case the basic concepts of interpreting a bar graph are the same.

◼ EXAMPLE 4   The approximate population of California by year is given in the following bar graph.

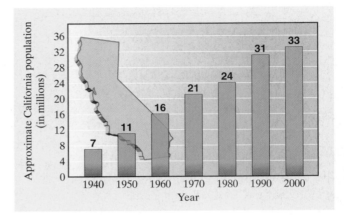

(a) What was the approximate population of California in 2000?

(b) How much greater was the population of California in 1980 than in 1970?

(a) The approximate population of California in 2000 was about 33 million.

(b) In 1980 it was 24 million. In 1970 it was 21 million. The population was approximately 3 million people more in California in 1980 than in 1970.

Practice Problem 4   The following bar graph depicts the number of fatal accidents for U.S. air carriers for scheduled flight service for aircraft with 30 seats or more.

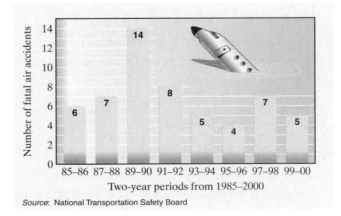

Source: National Transportation Safety Board

(a) What two-year period had the greatest number of fatal accidents?

(b) What was the increase in the number of fatal accidents from the 1987–1988 period to the 1989–1990 period?

(c) What was the decrease in the number of fatal accidents from the 1991–1992 period to the 1993–1994 period?

5 **Line Graphs**

A line graph is often used to display data when significant changes or trends are present. In a line graph, only a few points are actually plotted from measured values. The points are then connected by straight lines in order to show a trend. The intervening values between points may not exactly lie on the line. A line graph is displayed in Example 5.

EXAMPLE 5   The following line graph shows the number of customers per month coming to a restaurant in a tourist vacation community.

(a) What month had the greatest number of customers?

(b) How many customers came to the restaurant in April?

(c) How many more customers came in July than in August?

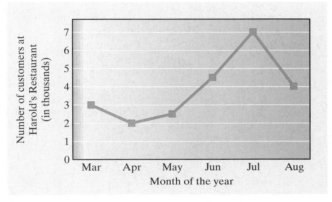

(a) The greatest number of customers came during the month of July.

(b) Approximately 2000 customers came in April.

(c) In July there were 7000 customers, while in August there were 4000 customers. Thus, there were 3000 more customers in July than in August.

Practice Problem 5   The quality of the air is measured by the Pollutant Standards Index (PSI). To meet the national air quality standards set by the U.S. government, the air in a city cannot have a PSI greater than 100. The following line graph indicates the number of days that the PSI was greater than 100 in the city of Baltimore during an 11-year period.

(a) What was the number of days the PSI exceeded 100 in Baltimore in 1993?

(b) In what year did Baltimore have the fewest days in which the PSI exceeded 100?

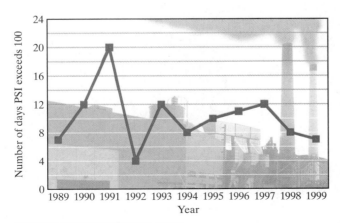

Source: U.S. Environmental Protective Agency

 **Pie Graphs and Circle Graphs**

A pie graph or a circle graph indicates how a whole quantity is divided into parts. These graphs help you to visualize the size of the relative proportions of parts. Each piece of the pie or circle is called a sector. Example 6 uses a pie graph.

EXAMPLE 6   Together, the Great Lakes form the largest body of fresh water in the world. The total area of these five lakes is about 290,000 square miles, almost all of which is suitable for boating. The percentage of this total area taken up by each of the Great Lakes is shown in the pie graph.

(a) What percentage of the area is taken up by Lake Michigan?

(b) What lake takes up the largest percentage of area?

(c) How many square miles are taken up by Lake Huron and Lake Michigan together?

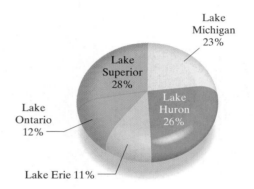

(a) Lake Michigan takes up 23% of the area.

(b) Lake Superior takes up the largest percentage.

(c) If we add 26 + 23, we get 49. Thus, Lake Huron and Lake Michigan together take up 49% of the total area. 49% of 290,000 = (0.49)(290,000) = 142,100 square miles.

Practice Problem 6   Seattle receives on average about 37 inches of rain per year. However, the amount of rainfall per month varies significantly. The percent of rainfall that occurs during each quarter of the year is shown by the circle graph.

(a) What percent of the rain in Seattle falls between April and June?

(b) Forty percent of the rainfall occurs in what three-month period?

(c) How many inches of rainfall in Seattle from January to March?

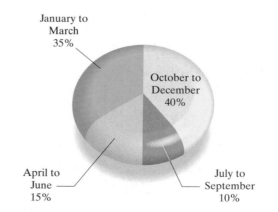

Appendix C Exercises

In each case, study carefully the appropriate visual display, then answer the questions.

Consider the following table in answering exercises 1–10.

Table of Facts of the Rocky Mountain States

State	Area in Square Miles	Date Admitted To the Union	Estimated 1990 Population	Number of Representatives in U.S. Congress	Popular Name
Colorado	104,247	1876	3,500,000	5	Centennial State
Idaho	83,557	1890	1,100,000	2	Gem State
Montana	147,138	1889	800,000	2	Treasure State
Nevada	110,540	1864	1,100,000	1	Silver State
Utah	84,916	1896	1,700,000	2	Beehive State
Wyoming	97,914	1890	500,000	1	Equality State

1. What is the area of Utah in square miles?

2. What is the area of Montana in square miles?

3. What is the 1990 estimated population of Colorado?

4. What is the 1990 estimated population of Nevada?

5. How many representatives in the U.S. Congress come from Idaho?

6. How many representatives in the U.S. Congress come from Wyoming?

7. What is the popular name for Montana?

8. What is the popular name for Utah?

9. Which of these six states was the first one to be admitted to the Union?

10. In what year did two of these six states both get admitted to the Union?

**Number of new homes built in 2000
in each of five counties**

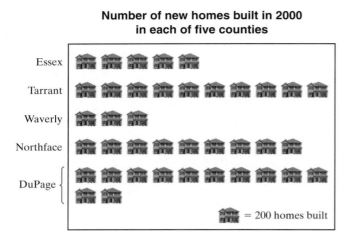

Essex

Tarrant

Waverly

Northface

DuPage

= 200 homes built

Approximate number of apartments in U.S. in 2000

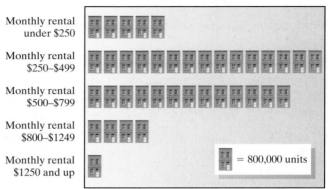

Monthly rental
under $250

Monthly rental
$250–$499

Monthly rental
$500–$799

Monthly rental
$800–$1249

Monthly rental
$1250 and up

= 800,000 units

Source: U.S. Bureau of the Census

Use this pictograph to answer exercises 11–16.

11. How many homes were built in Tarrant County in the year 2000?

12. How many homes were built in Essex County in the year 2000?

13. In what county were the most homes built?

14. How many more homes were built in Tarrant County than Waverly County?

15. How many homes were built in Essex County and Northface County combined?

16. How many homes were built in DuPage County and Waverly County combined?

Use this pictograph to answer exercises 17–20.

17. How many apartment units are rented for under $250 per month?

18. How many apartment units are rented for $800–$1249 per month?

19. How many more apartment units are available in the $500–$799 range than in the $800–$1249 range?

20. How many more apartment units are available in the $800–$1249 range than in the $1250 and up range?

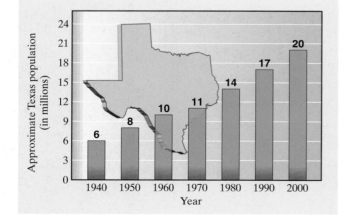

Use this bar graph to answer exercises 21–26.

Number of people engaged in various activities at least once in the last twelve months

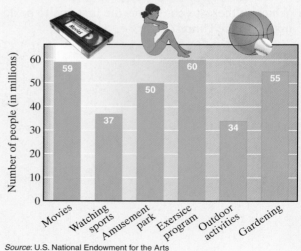

Source: U.S. National Endowment for the Arts

Use this bar graph to answer Questions 27–30.

21. What was the population in Texas in 1950?

22. What was the population in Texas in 1990?

23. Between what two years was the increase in population the greatest?

24. Between what two years was the increase in population the smallest?

25. How many more people lived in Texas in 1970 than in 1950?

26. How many more people lived in Texas in 1980 than in 1960?

27. According to the bar graph, how many people watched a sports event at least once in the last 12 months?

28. According to the bar graph, how many people were involved in gardening at least once in the last 12 months?

29. What two activities were the most common?

30. What two activities were the least common?

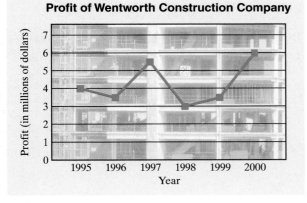

Profit of Wentworth Construction Company

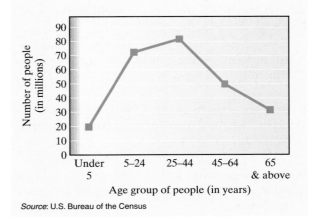

Source: U.S. Bureau of the Census

Use this line graph to answer exercises 31–36.

31. What was the profit in 1997?

32. What was the profit in 1996?

33. How much greater was the profit in 2000 than 1999?

34. In what year did the smallest profit occur?

35. Between what two years did the profit decrease the most?

36. Between what two years did the profit increase the most?

Use this line graph to answer exercises 37–40.

37. How many people in the U.S. are in the age group of 5 to 24 years?

38. How many people in the U.S. are in the age group of 25 to 44 years?

39. Thirty-three million people are in what age group?

40. Fifty-one million people are in what age group?

Religious faith distribution in the world

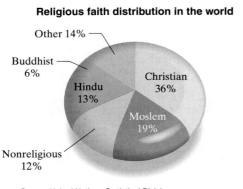

Source: United Nations Statistical Division

Distribution of spending by "average" two-income American family in 2000

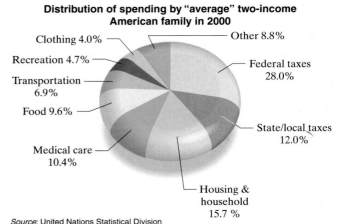

Source: United Nations Statistical Division

Use this circle graph to answer exercises 41–46.

41. What percent of the world's population is either Hindu or Buddhist?

42. What percent of the world's population is either Moslem or nonreligious?

43. What percent of the world's population is *not* Moslem?

44. What percent of the world's population is *not* Christian?

45. If there are approximately 6.5 billion people in the world in 2005, how many of them would we expect to be Hindu?

46. If there are approximately 6.9 billion people in the world in 2010, how many of them would we expect to be Moslem?

Use this circle graph to answer exercises 47–50.

47. What percent of the family income is spent for federal, state, and local taxes?

48. What percent of the family income is spent for food, medical care, and housing?

49. If the average two-income family earns $52,000 per year, how much is spent on transportation?

50. If the average two-income family earns $52,000 per year, how much is spent on recreation?

Appendix D
Inductive and Deductive Reasoning

1 Using Inductive Reasoning to Reach a Conclusion

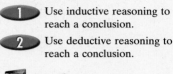

When we reach a conclusion based on specific observations, we are using **inductive reasoning**. Much of our early learning is based on simple cases of inductive reasoning. If a child touches a hot stove or other appliance several times and each time he gets burned, he is likely to conclude, "If I touch something that is hot, I will get burned." This is inductive reasoning. The child has thought about several actions and their outcomes and has made a conclusion or generalization.

The next few examples show how inductive reasoning can be used in mathematics.

EXAMPLE 1   Find the next number in the sequence 10, 13, 16, 19, 22, 25, 28, ... We observe a pattern that each number is 3 more than the preceding number: $10 + 3 = 13$; $13 + 3 = 16$, and so on. Therefore, if we add 3 to 28, we conclude that the next number in the sequence is 31.

Practice Problem 1   Find the next number in the sequence 24, 31, 38, 45, 52, 59, 66, ...

EXAMPLE 2   Find the next number in the sequence 1, 8, 27, 64, 125, ... The sequence can be written as $1^3, 2^3, 3^3, 4^3, 5^3, \ldots$. Each successive integer is cubed. The next number would be 6^3 or 216.

Practice Problem 2   Find the next number in the sequence.

$$3, 8, 15, 24, 35, 48, 63, 80, \ldots.$$

EXAMPLE 3   Guess the next seven digits in the following irrational number:

$$5.636336333633336\ldots$$

Between 6's there are the digits 3, 33, 333, 3333, and so on. The pattern is that the number of 3's keeps increasing by 1 each time. Thus the next seven digits are 3333363.

Practice Problem 3   Guess the next seven digits in the following irrational number:

$$6.1213314441\ldots$$

EXAMPLE 4   Find the next two figures that would appear in the sequence.

We notice an alternating pattern: square, square, circle, circle, square .... We would next expect a square followed by a circle.

We notice a shading pattern of horizontal, vertical, horizontal, vertical, horizontal .... We would next expect vertical, then horizontal. Thus, the next two figures are

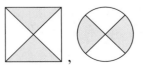

A-19

Practice Problem 4   Find the next two figures that would appear in the sequence.

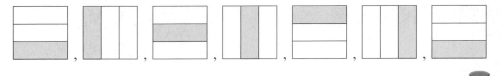

How accurate is inductive reasoning? Do we always come to the right conclusion? Conclusions arrived at by inductive reasoning are always tentative. They may require further investigation. When we use inductive reasoning we are using specific data to reach a general conclusion. However, we may have reached the wrong conclusion. To illustrate:

Take the sequence of numbers 20, 10, 5. What is the next number? You might say 2.5. In the sequence 20, 10, 5, each number appears to be one-half of the preceding number. Thus we would predict 20, 10, 5, 2.5, ...

But wait! There is another possibility. Maybe the sequence is 20, 10, 5, −5, −10, −20, −25, −35, −40, ...

To know for sure which answer is correct, we would need more information, such as more numbers in the sequence to verify the pattern. **You should always treat inductive reasoning conclusions as tentative, requiring further verification**.

2 Using Deductive Reasoning to Reach a Conclusion

Deductive reasoning requires us to take general facts, postulates, or accepted truths and use them to reach a specific conclusion. Suppose we know the following rules or "facts" of algebra:

For all numbers a, b, and c, and for all numbers $d \neq 0$:

1. *Addition principle of equations:* If $a = b$, then $a + c = b + c$.

2. *Division principle of equations:* If $a = b$, then $\dfrac{a}{d} = \dfrac{b}{d}$.

3. *Multiplication principle of equations:* If $a = b$, then $ac = bc$.

4. *Distributive property:* $a(b + c) = ab + ac$.

EXAMPLE 5   Use deductive reasoning and the four properties listed in the preceding box to justify each step in solving the equation.

$$2(7x - 2) = 38$$

STATEMENT	REASON
1. $14x - 4 = 38$	**1.** Distributive property: $a(b + c) = ab + ac$. Here we distributed the 2.
2. $14x = 42$	**2.** Addition principle of equations: If $a = b$ then $a + c = b + c$. Here we added 4 to each side of the equation.
3. $\dfrac{14x}{14} = \dfrac{42}{14}$	**3.** Division principle of equations: If $a = b$ then $\dfrac{a}{d} = \dfrac{b}{d}$.
$x = 3$	Here we divided each side by 14.

Practice Problem 5   Use deductive reasoning and the four properties listed in the preceding box to justify each step in solving the equation.

$$\frac{1}{6}x = \frac{1}{3}x + 4$$

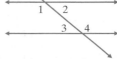

Sometimes we need to make conclusions about angles and lines in geometry. The following properties are useful. We will refer to angles 1, 2, 3, 4.

1. If two lines intersect, the opposite angles are equal. Here, $\angle 1 = \angle 2$ and $\angle 4 = \angle 3$.

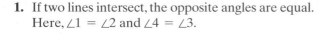

2. If two lines intersect, the adjacent angles are supplementary (they add up to 180°). Here, $\angle 4 + \angle 1 = 180°$, and $\angle 1 + \angle 3 = 180°$, also $\angle 3 + \angle 2 = 180°$, and $\angle 4 + \angle 2 = 180°$.

3. If a transversal (intersecting line) crosses two parallel lines, the alternate interior angles are equal. In the figure at right, line P is a transversal. If line M is parallel to line N, then $\angle 3 = \angle 4$ and $\angle 1 = \angle 2$.

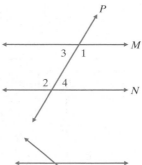

4. If two alternate interior angles on each side of a transversal cutting two straight lines are equal, the two straight lines are parallel. If $\angle 3 = \angle 4$, then line M is parallel to line N.

5. Supplements of equal angles are equal. In the figure at right, if $\angle 1$ and $\angle 2$ are supplementary, and $\angle 3$ and $\angle 4$ are supplementary, then if $\angle 1 = \angle 4$, then $\angle 2 = \angle 3$.

6. Transitive property of equality:

$$\text{If } a = b \text{ and } b = c, \text{ then } a = c.$$

We will now use these facts to prove some geometric conclusions.

▲ **EXAMPLE 6**   Prove that line M is parallel to line N if $\angle 1 = \angle 3$.

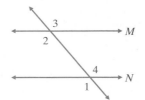

STATEMENT	*REASON*
1. $\angle 3 = \angle 2$ and $\angle 4 = \angle 1$	**1.** If two lines intersect, the opposite angles are equal.
2. $\angle 2 = \angle 4$	**2.** Transitive property of equality.
3. Therefore, line M is parallel to line N. (We write this as $M \parallel N$.)	**3.** If two alternate interior angles on each side of a transversal cutting two straight lines are equal, the straight lines are parallel.

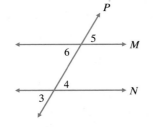

▲ **Practice Problem 6**   Refer to the diagram in the margin. Line M is parallel to line N. $\angle 5 = 26°$. Prove $\angle 4 = 26°$.

Now let us see if we can use deductive reasoning to solve the following problem.

EXAMPLE 7   For next semester, four professors from the psychology department have expressed the following desires for freshman courses. Each professor will teach only *one* freshman course next semester. The four freshman courses are General Psychology, Social Psychology, Psychology of Adjustment, and Educational Psychology.

   1. Professors A and B don't want to teach General Psychology.
   2. Professor C wants to teach Social Psychology.
   3. Professor D will be happy to teach any course.
   4. Professor B wants to teach Psychology of Adjustment.

Which professor will teach Educational Psychology, if all professors are given the courses they desire?

Let us organize the facts by listing the four freshman courses and each professor: A, B, C, and D.

General Psychology	Social Psychology	Psychology of Adjustment	Educational Psychology
A	A	A	A
B	B	B	B
C	C	C	C
D	D	D	D

STEP	*REASON*
1. We cross off Professors A and B from the General Psychology list.	**1.** Professors A and B don't want to teach General Psychology.

General Psychology			
$\cancel{A}$	A	A	A
$\cancel{B}$	B	B	B
C	C	C	C
D	D	D	D

2. We cross Professor C off every list (except Social Psychology) and mark that he will teach it.	**2.** Professor C wants to teach Social Psychology.

	Social Psychology		
$\cancel{A}$			
$\cancel{B}$			
$\cancel{C}$	$\boxed{C}$	$\cancel{C}$	$\cancel{C}$
D			

STEP	*REASON*
3. Professor *D* is thus the only person who can teach General Psychology. We cross him off every other list and mark that he will teach General Psychology.	**3.** Professor *D* is happy to teach any course.

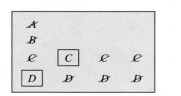

STEP	*REASON*
4. We cross out all courses for Professor *B* except Psychology of Adjustment.	**4.** Professor *B* wants to teach Psychology of Adjustment.

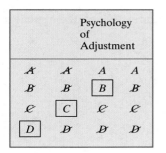

STEP	*REASON*
5. Professor *A* will teach Educational Psychology.	**5.** He is the only professor left. All others are assigned.

Educational Psychology

A

Practice Problem 7   A Honda, Toyota, Mustang, and Camaro are parked side by side, but not in that order.

1. The Camaro is parked on the right end.
2. The Mustang is between the Honda and the Toyota.
3. The Honda is not next to the Camaro.

Which car is parked on the left end?

Appendix D Exercises

Find the next number in the sequence.

1. 2, 4, 6, 8, 10, 12, …

2. 0, 5, 10, 15, 20, 25, …

3. 7, 16, 25, 34, 43, …

4. 12, 25, 38, 51, 64, …

5. 1, 16, 81, 256, 625, …

6. 1, 6, 13, 22, 33, 46, …

7. −7, 3, −6, 4, −5, 5, −4, 6, …

8. 2, −4, 8, −16, 32, −64, 128, …

9. $5x, 6x − 1, 7x − 2, 8x − 3, 9x − 4, …$

10. $60x, 30x, 15x, 7.5x, 3.75x, 1.875x, …$

In exercises 11–16, guess the next seven digits in each irrational number.

11. 8.181181118…

12. 3.043004300043…

13. 12.98987987698765…

14. 7.6574839201102938…

15. 2.14916253649…

16. 6.112223333…

17. Find the next row in this triangular pattern.

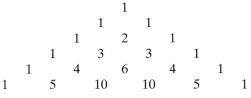

18. Find the next two figures that would appear in the sequence.

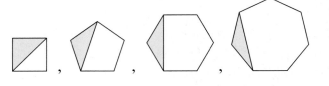

Use deductive reasoning and the four properties discussed in Example 5 to justify each step in solving each equation.

19. $12x − 30 = 6$

20. $3x + 7 = 4$

21. $4x − 3 = 3x − 5$

22. $2x − 9 = 4x + 5$

23. $8x − 3(x − 5) = 30$

24. $3x − 5(x − 1) = −5$

25. $\frac{1}{2}x + 6 = \frac{3}{2} + 6x$

26. $\frac{4}{5}x − 3 = \frac{1}{10}x + \frac{3}{5}$

In exercises 27–28, use deductive reasoning and the properties of geometry discussed in Example 6 to prove each statement.

▲ **27.** If $\angle 1 = \angle 5$, prove that line P is parallel to line S.

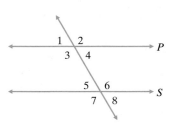

▲ **28.** If line R is parallel to line S, prove that $\angle 6 = \angle 3$.

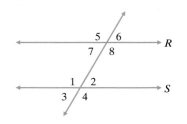

29. William, Brent, Charlie, and Dave competed in the Olympics. These four divers placed first, second, third, and fourth in the competition.
   1. James ranked between Brent and Dave.
   2. William did better than Brent.
   3. Brent did better than Dave.
Who finished in each of the four places?

30. The four floors of Hotel Royale are color-coded.
   1. The blue floor is directly below the green floor.
   2. The red floor is next to the yellow floor.
   3. The green floor is above the red floor.
   4. The blue floor is above the yellow floor.
   5. There is no floor below the red floor.
What is the order of the colors from top to bottom of Hotel Royale?

31. There are four people named Peter, Michael, Linda and Judy. Each of them has one occupation. They are a teacher, a butcher, a baker, and a candlestick maker.
   1. Linda is the baker.
   2. The teacher plans to marry the baker in October.
   3. Judy and Peter were married last year.
Who is the teacher?

32. A president of a company, a lawyer, a salesman, and a doctor are seated at the head table at a banquet.
   1. The lawyer prosecuted the doctor in a malpractice lawsuit and they should not sit together.
   2. The salesman should be at one end of the table.
   3. The lawyer should be at the far right.
From left to right how should they be seated?

33. John is considering purchasing a new car when he starts his new full-time job. The car with the best gas mileage is a Honda Civic, the car with the most conveniently located dealer is a Toyota Corolla, the car with the lowest price is the Ford Escort, and the car that has best handling is the Dodge Neon. His next-door neighbor has a Honda and he does not want to copy his neighbor. If he does not buy a Toyota then he will definitely not go to graduate school. He told his best friend that he probably will not purchase a car based on its handling. He has definitely decided to go to graduate school. What car is he most likely to purchase?

34. Detective Smith recently found a stolen Corvette abandoned near Skull Rocks. A small empty boat was anchored 1 mile offshore with no person in the boat. Divers later found the body of a 35-year-old man $\frac{1}{2}$ mile further out to sea than the boat. If the man entered the water after 10:00 A.M., then the current of the outgoing tide was strong enough to take a dead body a distance of $\frac{1}{2}$ mile further out to sea. The coroner determined when the body was found at 4:00 P.M. that the man had been dead for seven hours. The detective determined that he died instantly when his head hit the rocks directly beneath the anchored boat. Could the man have acted alone based on the facts revealed so far?

35. Dr. Parad is deciding whether he should do a root canal in order to save Fred's tooth or just remove the tooth and put in a false tooth. If Dr. Parad works alone on the tooth the procedure will take 2 hours and 15 minutes. Fred wants to save the tooth but he does not want the dental bill for Tuesday's work to be more than $200. Dr. Parad charges $100 per hour for dental procedures. If the bill is under $150 it is because Dr. Parad's efficient dental assistant is helping with the oral surgery. The dental procedure is scheduled for Tuesday, when Dr. Parad does not have any dental assistants available. Will Fred's tooth be saved or will he get a false tooth?

36. If Marcia passes the Bilingual Teacher's exam, she will stay in the Chicago area. If she devotes at least 50 hours of study time to practice her Spanish, she is confident that she can pass the exam. Either Marcia will stay in the Chicago area or she will move back to Massachusetts. If she did not pass the Bilingual Teacher's exam it is because she needed 12 hours of study time per month. Marcia started studying for the exam in April. She will be moving to Massachusetts. When was the Bilingual Teacher's exam given, if we know that Marcia took the exam?

Appendix E
Determinants and Cramer's Rule

Student Learning Objectives

After studying this section, you will be able to:

 1 Evaluate a second-order determinant.

2 Evaluate a third-order determinant.

3 Solve a system of two linear equations with two unknowns using Cramer's rule.

4 Solve a system of three linear equations with three unknowns using Cramer's rule.

SSM

PH TUTOR CENTER   CD & VIDEO   MATH PRO   WEB

 1 *Evaluating a Second-Order Determinant*

Mathematicians have developed techniques to solve systems of linear equations by focusing on the coefficients of the variables and the constants in the equations. The computational techniques can be easily carried out by computers or calculators. We will learn to do them by hand so that you will have a better understanding of what is involved.

To begin, we need to define a matrix and a determinant. A **matrix** is any rectangular array of numbers that is arranged in rows and columns. We use the symbol [ ] to indicate a matrix.

$$\begin{bmatrix} 3 & 2 & 4 \\ -1 & 4 & 0 \end{bmatrix}, \qquad \begin{bmatrix} 4 & -3 \\ 2 & \frac{1}{2} \\ 1 & 5 \end{bmatrix}, \qquad [-4 \quad 1 \quad 6], \quad \text{and} \quad \begin{bmatrix} \frac{1}{4} \\ 3 \\ -2 \end{bmatrix}$$

are matrices. If you have a graphing calculator, you can enter the elements of a matrix and store them for future use. Let's examine two systems of equations.

$$\begin{aligned} 3x + 2y &= 16 \\ x + 4y &= 22 \end{aligned} \quad \text{and} \quad \begin{aligned} -6x \quad &= 18 \\ x + 3y &= 9 \end{aligned}$$

We could write the coefficients of the variables in each of these systems as a matrix.

$$\begin{aligned} 3x + 2y \\ x + 4y \end{aligned} \Rightarrow \begin{bmatrix} 3 & 2 \\ 1 & 4 \end{bmatrix} \quad \text{and} \quad \begin{aligned} -6x \\ x + 3y \end{aligned} \Rightarrow \begin{bmatrix} -6 & 0 \\ 1 & 3 \end{bmatrix}$$

Now we define a determinant. A **determinant** is a *square* arrangement of numbers. We use the symbol | | to indicate a determinant.

$$\begin{vmatrix} 3 & 2 \\ 1 & 4 \end{vmatrix} \quad \text{and} \quad \begin{vmatrix} -6 & 0 \\ 1 & 3 \end{vmatrix}$$

are determinants. The value of a determinant is a *real number* and is defined as follows:

Definition

The value of the second-order determinant $\begin{vmatrix} a & c \\ b & d \end{vmatrix}$ is $ad - bc$.

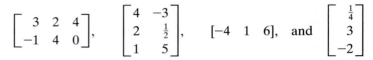 **EXAMPLE 1** Find the value of each determinant.

(a) $\begin{vmatrix} -6 & 2 \\ -1 & 4 \end{vmatrix}$ **(b)** $\begin{vmatrix} 0 & -3 \\ -2 & 6 \end{vmatrix}$

(a) $\begin{vmatrix} -6 & 2 \\ -1 & 4 \end{vmatrix} = (-6)(4) - (-1)(2) = -24 - (-2) = -24 + 2 = -22$

(b) $\begin{vmatrix} 0 & -3 \\ -2 & 6 \end{vmatrix} = (0)(6) - (-2)(-3) = 0 - (+6) = -6$

Practice Problem 1 Find the value of each determinant.

(a) $\begin{vmatrix} -7 & 3 \\ -4 & -2 \end{vmatrix}$
(b) $\begin{vmatrix} 5 & 6 \\ 0 & -5 \end{vmatrix}$

2 Evaluating a Third-Order Determinant

Third-order determinants have three rows and three columns. Again, each determinant has exactly one value.

Definition

The value of the third-order determinant

$$\begin{vmatrix} a_1 & b_1 & c_1 \\ a_2 & b_2 & c_2 \\ a_3 & b_3 & c_3 \end{vmatrix}$$

is

$$a_1 b_2 c_3 + b_1 c_2 a_3 + c_1 a_2 b_3 - a_3 b_2 c_1 - b_3 c_2 a_1 - c_3 a_2 b_1.$$

Because this definition is difficult to memorize and cumbersome to use, we evaluate third-order determinants by a simpler method called **expansion by minors**. The **minor** of an element (number or variable) of a third-order determinant is the second-order determinant that remains after we delete the row and column in which the element appears.

EXAMPLE 2 Find **(a)** the minor of 6 and **(b)** the minor of -3 in the determinant.

$$\begin{vmatrix} 6 & 1 & 2 \\ -3 & 4 & 5 \\ -2 & 7 & 8 \end{vmatrix}$$

(a) Since the element 6 appears in the first row and the first column, we delete them.

$$\begin{vmatrix} 6 & 1 & 2 \\ -3 & 4 & 5 \\ -2 & 7 & 8 \end{vmatrix}$$

Therefore, the minor of 6 is

$$\begin{vmatrix} 4 & 5 \\ 7 & 8 \end{vmatrix}$$

(b) Since -3 appears in the first column and the second row, we delete them.

$$\begin{vmatrix} 6 & 1 & 2 \\ -3 & 4 & 5 \\ -2 & 7 & 8 \end{vmatrix}$$

The minor of -3 is

$$\begin{vmatrix} 1 & 2 \\ 7 & 8 \end{vmatrix}.$$

Practice Problem 2 Find **(a)** the minor of 3 and **(b)** the minor of -6 in the determinant.

$$\begin{vmatrix} 1 & 2 & 7 \\ -4 & -5 & -6 \\ 3 & 4 & -9 \end{vmatrix}$$

To evaluate a third-order determinant, we use expansion by minors of elements in the first column; for example, we have

$$\begin{vmatrix} a_1 & b_1 & c_1 \\ a_2 & b_2 & c_2 \\ a_3 & b_3 & c_3 \end{vmatrix} = a_1 \begin{vmatrix} b_2 & c_2 \\ b_3 & c_3 \end{vmatrix} - a_2 \begin{vmatrix} b_1 & c_1 \\ b_3 & c_3 \end{vmatrix} + a_3 \begin{vmatrix} b_1 & c_1 \\ b_2 & c_2 \end{vmatrix}$$

Note that the signs alternate. We then evaluate the second-order determinant according to our definition.

EXAMPLE 3   Evaluate the determinant $\begin{vmatrix} 2 & 3 & 6 \\ 4 & -2 & 0 \\ 1 & -5 & -3 \end{vmatrix}$ by expanding it by minors of elements in the first column.

$$\begin{vmatrix} 2 & 3 & 6 \\ 4 & -2 & 0 \\ 1 & -5 & -3 \end{vmatrix} = 2 \begin{vmatrix} -2 & 0 \\ -5 & -3 \end{vmatrix} - 4 \begin{vmatrix} 3 & 6 \\ -5 & -3 \end{vmatrix} + 1 \begin{vmatrix} 3 & 6 \\ -2 & 0 \end{vmatrix}$$

$$= 2[(-2)(-3) - (-5)(0)] - 4[(3)(-3) - (-5)(6)] + 1[(3)(0) - (-2)(6)]$$

$$= 2[6 - 0] - 4[-9 - (-30)] + 1[0 - (-12)]$$

$$= 2(6) - 4(21) + 1(12)$$

$$= 12 - 84 + 12$$

$$= -60$$

Practice Problem 3   Evaluate the determinant $\begin{vmatrix} 1 & 2 & -3 \\ 2 & -1 & 2 \\ 3 & 1 & 4 \end{vmatrix}$.

3   *Solving a System of Two Linear Equations with Two Unknowns Using Cramer's Rule*

We can solve a linear system of two equations with two unknowns by Cramer's rule. The rule is named for Gabriel Cramer, a Swiss mathematician who lived from 1704 to 1752. Cramer's rule expresses the solution of each variable in a linear system as the quotient of two determinants. Computer programs are available to solve a system of equations by Cramer's rule.

Cramer's Rule

The solution to
$$a_1 x + b_1 y = c_1$$
$$a_2 x + b_2 y = c_2$$

is
$$x = \frac{D_x}{D} \quad \text{and} \quad y = \frac{D_y}{D}, \quad D \neq 0,$$

where
$$D_x = \begin{vmatrix} c_1 & b_1 \\ c_2 & b_2 \end{vmatrix}, \quad D_y = \begin{vmatrix} a_1 & c_1 \\ a_2 & c_2 \end{vmatrix}, \quad \text{and} \quad D = \begin{vmatrix} a_1 & b_1 \\ a_2 & b_2 \end{vmatrix}.$$

Cramer's rule may look complicated, but it soon becomes easy. Notice that D is just the determinant of the coefficients of the variables.

EXAMPLE 4   Solve by Cramer's rule.

$$-3x + y = 7$$
$$-4x - 3y = 5$$

$$D = \begin{vmatrix} -3 & 1 \\ -4 & -3 \end{vmatrix} \qquad D_x = \begin{vmatrix} 7 & 1 \\ 5 & -3 \end{vmatrix} \qquad D_y = \begin{vmatrix} -3 & 7 \\ -4 & 5 \end{vmatrix}$$

$$= (-3)(-3) - (-4)(1) \qquad = (7)(-3) - (5)(1) \qquad = (-3)(5) - (-4)(7)$$

$$= 9 - (-4) \qquad\qquad = -21 - 5 \qquad\qquad = -15 - (-28)$$

$$= 9 + 4 \qquad\qquad\quad = -26 \qquad\qquad\quad = -15 + 28$$

$$= 13 \qquad\qquad\qquad\qquad\qquad\qquad\qquad = 13$$

Hence,

$$x = \frac{D_x}{D} = \frac{-26}{13} = -2$$

$$y = \frac{D_y}{D} = \frac{13}{13} = 1.$$

The solution to the system is $x = -2$ and $y = 1$. Verify this.

Practice Problem 4   Solve by Cramer's rule.

$$5x + 3y = 17$$
$$2x - 5y = 13$$

4   Solving a System of Three Linear Equations with Three Unknowns Using Cramer's Rule

It is quite easy to extend Cramer's rule to three linear equations.

Cramer's Rule

The solution to the system $a_1x + b_1y + c_1z = d_1$

$$a_2x + b_2y + c_2z = d_2$$

$$a_3x + b_3y + c_3z = d_3$$

is $x = \dfrac{D_x}{D}, \qquad y = \dfrac{D_y}{D}, \quad \text{and} \quad z = \dfrac{D_z}{D}, \qquad D \ne 0,$

where $D = \begin{vmatrix} a_1 & b_1 & c_1 \\ a_2 & b_2 & c_2 \\ a_3 & b_3 & c_3 \end{vmatrix}, \qquad D_x = \begin{vmatrix} d_1 & b_1 & c_1 \\ d_2 & b_2 & c_2 \\ d_3 & b_3 & c_3 \end{vmatrix},$

$$D_y = \begin{vmatrix} a_1 & d_1 & c_1 \\ a_2 & d_2 & c_2 \\ a_3 & d_3 & c_3 \end{vmatrix}, \quad \text{and} \quad D_z = \begin{vmatrix} a_1 & b_1 & d_1 \\ a_2 & b_2 & d_2 \\ a_3 & b_3 & d_3 \end{vmatrix}.$$

● **EXAMPLE 5**   Use Cramer's rule to solve the system.

$$2x - y + z = 6$$
$$3x + 2y - z = 5$$
$$2x + 3y - 2z = 1$$

We will expand each determinant by the first column.

$$D = \begin{vmatrix} 2 & -1 & 1 \\ 3 & 2 & -1 \\ 2 & 3 & -2 \end{vmatrix}$$

$$= 2\begin{vmatrix} 2 & -1 \\ 3 & -2 \end{vmatrix} - 3\begin{vmatrix} -1 & 1 \\ 3 & -2 \end{vmatrix} + 2\begin{vmatrix} -1 & 1 \\ 2 & -1 \end{vmatrix}$$

$$= 2[-4 - (-3)] - 3[2 - 3] + 2[1 - 2]$$

$$= 2[-1] - 3[-1] + 2[-1]$$

$$= -2 + 3 - 2$$

$$= -1$$

$$D_x = \begin{vmatrix} 6 & -1 & 1 \\ 5 & 2 & -1 \\ 1 & 3 & -2 \end{vmatrix}$$

$$= 6\begin{vmatrix} 2 & -1 \\ 3 & -2 \end{vmatrix} - 5\begin{vmatrix} -1 & 1 \\ 3 & -2 \end{vmatrix} + 1\begin{vmatrix} -1 & 1 \\ 2 & -1 \end{vmatrix}$$

$$= 6[-4 - (-3)] - 5[2 - 3] + 1[1 - 2]$$

$$= 6[-1] - 5[-1] + 1[-1]$$

$$= -6 + 5 - 1$$

$$= -2$$

$$D_y = \begin{vmatrix} 2 & 6 & 1 \\ 3 & 5 & -1 \\ 2 & 1 & -2 \end{vmatrix}$$

$$= 2\begin{vmatrix} 5 & -1 \\ 1 & -2 \end{vmatrix} - 3\begin{vmatrix} 6 & 1 \\ 1 & -2 \end{vmatrix} + 2\begin{vmatrix} 6 & 1 \\ 5 & -1 \end{vmatrix}$$

$$= 2[-10 - (-1)] - 3[-12 - 1] + 2[-6 - 5]$$

$$= 2[-9] - 3[-13] + 2[-11]$$

$$= -18 + 39 - 22$$

$$= -1$$

$$D_z = \begin{vmatrix} 2 & -1 & 6 \\ 3 & 2 & 5 \\ 2 & 3 & 1 \end{vmatrix}$$

$$= 2\begin{vmatrix} 2 & 5 \\ 3 & 1 \end{vmatrix} - 3\begin{vmatrix} -1 & 6 \\ 3 & 1 \end{vmatrix} + 2\begin{vmatrix} -1 & 6 \\ 2 & 5 \end{vmatrix}$$

$$= 2[2 - 15] - 3[-1 - 18] + 2[-5 - 12]$$

$$= 2[-13] - 3[-19] + 2[-17]$$

$$= -26 + 57 - 34$$

$$= -3$$

$$x = \frac{D_x}{D} = \frac{-2}{-1} = 2; \qquad y = \frac{D_y}{D} = \frac{-1}{-1} = 1; \qquad z = \frac{D_z}{D} = \frac{-3}{-1} = 3$$

Check: Verify this solution.

Graphing Calculator

Copying Matrices
If you are using a graphing calculator to evaluate the four determinants in Example 5 or similar exercises, first enter matrix D into the calculator. Then copy the matrix using the copy function to three additional locations. Usually we store matrix D as matrix A. Then store a copy of it as matrix B, C, and D. Finally, use the Edit function and modify one column of each of matrices B, C, and D so that they become D_x, D_y, and D_z. This allows you to evaluate all four determinants in a minimum amount of time.

Practice Problem 5   Find the solution to the system by Cramer's rule.

$$2x + 3y - z = -1$$
$$3x + 5y - 2z = -3$$
$$x + 2y + 3z = 2$$

Cramer's rule cannot be used for every system of linear equations. If the equations are dependent or if the system of equations is inconsistent, the determinant of coefficients will be zero. Division by zero is not defined. In such a situation the system will not have a unique answer.

If $D = 0$, then the following are true:

1. If $D_x = 0$ and $D_y = 0$ (and $D_z = 0$, if there are three equations), then the equations are *dependent*. Such a system will have an infinite number of solutions.

2. If at least one of D_x or D_y (or D_z if there are three equations) is nonzero, then the system of equations is *inconsistent*. Such a system will have no solution.

Appendix E Exercises

Evaluate each determinant.

1. $\begin{vmatrix} 5 & 6 \\ 2 & 1 \end{vmatrix}$

2. $\begin{vmatrix} 3 & 4 \\ 1 & 8 \end{vmatrix}$

3. $\begin{vmatrix} 2 & -1 \\ 3 & 6 \end{vmatrix}$

4. $\begin{vmatrix} -4 & 2 \\ 1 & 5 \end{vmatrix}$

5. $\begin{vmatrix} -\frac{1}{2} & -\frac{2}{3} \\ 9 & 8 \end{vmatrix}$

6. $\begin{vmatrix} 10 & 4 \\ -\frac{3}{2} & -\frac{2}{5} \end{vmatrix}$

7. $\begin{vmatrix} -5 & 3 \\ -4 & -7 \end{vmatrix}$

8. $\begin{vmatrix} 2 & -3 \\ -4 & -6 \end{vmatrix}$

9. $\begin{vmatrix} 0 & -6 \\ 3 & -4 \end{vmatrix}$

10. $\begin{vmatrix} -5 & 0 \\ 2 & -7 \end{vmatrix}$

11. $\begin{vmatrix} 2 & -5 \\ -4 & 10 \end{vmatrix}$

12. $\begin{vmatrix} -3 & 6 \\ 7 & -14 \end{vmatrix}$

13. $\begin{vmatrix} 0 & 0 \\ -2 & 6 \end{vmatrix}$

14. $\begin{vmatrix} -4 & 0 \\ -3 & 0 \end{vmatrix}$

15. $\begin{vmatrix} 0.3 & 0.6 \\ 1.2 & 0.4 \end{vmatrix}$

16. $\begin{vmatrix} 0.1 & 0.7 \\ 0.5 & 0.8 \end{vmatrix}$

17. $\begin{vmatrix} 7 & 4 \\ b & -a \end{vmatrix}$

18. $\begin{vmatrix} \frac{1}{4} & \frac{3}{5} \\ \frac{2}{3} & \frac{1}{5} \end{vmatrix}$

19. $\begin{vmatrix} \frac{3}{7} & -\frac{1}{3} \\ -\frac{1}{4} & \frac{1}{2} \end{vmatrix}$

20. $\begin{vmatrix} -3 & y \\ -2 & x \end{vmatrix}$

In the following determinant $\begin{vmatrix} 3 & -4 & 7 \\ -2 & 6 & 10 \\ 1 & -5 & 9 \end{vmatrix}$,

21. Find the minor of 3.

22. Find the minor of -2.

23. Find the minor of 10.

24. Find the minor of 9.

Evaluate each of the following determinants.

25. $\begin{vmatrix} 4 & 1 & 2 \\ 3 & -1 & 0 \\ 1 & 2 & 3 \end{vmatrix}$

26. $\begin{vmatrix} 2 & 3 & 1 \\ -3 & 1 & 0 \\ 2 & 1 & 4 \end{vmatrix}$

27. $\begin{vmatrix} -4 & 0 & -1 \\ 2 & 1 & -1 \\ 0 & 3 & 2 \end{vmatrix}$

28. $\begin{vmatrix} 3 & -4 & -1 \\ -2 & 1 & 3 \\ 0 & 1 & 4 \end{vmatrix}$

29. $\begin{vmatrix} \frac{1}{2} & 1 & -1 \\ \frac{3}{2} & 1 & 2 \\ 3 & 0 & -2 \end{vmatrix}$

30. $\begin{vmatrix} 1 & 2 & 3 \\ 4 & -2 & -1 \\ 5 & -3 & 2 \end{vmatrix}$

31. $\begin{vmatrix} 4 & 1 & 2 \\ -1 & -2 & -3 \\ 4 & -1 & 3 \end{vmatrix}$

32. $\begin{vmatrix} -\frac{1}{2} & 2 & 3 \\ \frac{5}{2} & -2 & -1 \\ \frac{3}{4} & -3 & 2 \end{vmatrix}$

33. $\begin{vmatrix} 2 & 0 & -2 \\ -1 & 0 & 2 \\ 3 & 4 & 3 \end{vmatrix}$

34. $\begin{vmatrix} 7 & 0 & 2 \\ 1 & 0 & -5 \\ 3 & 0 & 6 \end{vmatrix}$

35. $\begin{vmatrix} 6 & -4 & 3 \\ 1 & 2 & 4 \\ 0 & 0 & 0 \end{vmatrix}$

36. $\begin{vmatrix} 7 & 0 & 3 \\ 1 & 2 & 4 \\ 3 & 0 & -7 \end{vmatrix}$

Optional Graphing Calculator Problems

If you have a graphing calculator, use the determinant function to evaluate the following:

37. $\begin{vmatrix} 1.3 & 1.8 & 2.5 \\ 7.9 & 5.3 & 6.0 \\ 1.7 & 1.8 & 2.8 \end{vmatrix}$

38. $\begin{vmatrix} 0.7 & 5.3 & 0.4 \\ 1.6 & 0.3 & 3.7 \\ 0.8 & 6.7 & 4.2 \end{vmatrix}$

39. $\begin{vmatrix} -55 & 17 & 19 \\ -62 & 23 & 31 \\ 81 & 51 & 74 \end{vmatrix}$

40. $\begin{vmatrix} 82 & -20 & 56 \\ 93 & -18 & 39 \\ 65 & -27 & 72 \end{vmatrix}$

Solve each system by Cramer's rule.

41. $\begin{aligned} x + 2y &= 8 \\ 2x + y &= 7 \end{aligned}$

42. $\begin{aligned} x + 3y &= 6 \\ 2x + y &= 7 \end{aligned}$

43. $\begin{aligned} 5x + 4y &= 10 \\ -x + 2y &= 12 \end{aligned}$

44. $\begin{aligned} 3x + 5y &= 11 \\ 2x + y &= -2 \end{aligned}$

45. $\begin{aligned} x - 5y &= 0 \\ x + 6y &= 22 \end{aligned}$

46. $\begin{aligned} x - 3y &= 4 \\ -3x + 4y &= -12 \end{aligned}$

47. $\begin{aligned} 0.3x + 0.5y &= 0.2 \\ 0.1x + 0.2y &= 0.0 \end{aligned}$

48. $\begin{aligned} 0.5x + 0.3y &= -0.7 \\ 0.4x + 0.5y &= -0.3 \end{aligned}$

Solve by Cramer's rule. Round your answers to four decimal places.

49. $52.9634x - 27.3715y = 86.1239$
$31.9872x + 61.4598y = 44.9812$

50. $0.0076x + 0.0092y = 0.01237$
$-0.5628x - 0.2374y = -0.7635$

Solve each system by Cramer's rule.

51. $2x + y + z = 4$
$x - y - 2z = -2$
$x + y - z = 1$

52. $x + 2y - z = -4$
$x + 4y - 2z = -6$
$2x + 3y + z = 3$

53. $2x + 2y + 3z = 6$
$x - y + z = 1$
$3x + y + z = 1$

54. $4x + y + 2z = 6$
$x + y + z = 1$
$-x + 3y - z = -5$

55. $x + 2y + z = 1$
$3x - 4z = 8$
$3y + 5z = -1$

56. $3x + y + z = 2$
$2y + 3z = -6$
$2x - y = -1$

Optional Graphing Calculator Problems

Solve. Round your answers to the nearest thousandth.

57. $10x + 20y + 10z = -2$
$-24x - 31y - 11z = -12$
$61x + 39y + 28z = -45$

58. $121x + 134y + 101z = 146$
$315x - 112y - 108z = 426$
$148x + 503y + 516z = -127$

59. $28w + 35x - 18y + 40z = 60$
$60w + 32x + 28y = 400$
$30w + 15x + 18y + 66z = 720$
$26w - 18x - 15y + 75z = 125$

Appendix F
Sets

After studying this section, you will be able to:

1 Write a set in roster form.

2 Write a set in set-builder notation.

3 Find the union and intersection of sets.

4 Identify subsets.

SSM
PH TUTOR   CD & VIDEO   MATH PRO   WEB
CENTER

1 Roster Form

Set theory is the basis of several mathematical topics. Sorting and classifying objects into categories is something we do every day. You may organize your closet so all your sweaters are together. When you go through your mail, you may separate bills from junk mail.

A **set** is a collection of objects called **elements**. Numbers can be classified into several different sets. The natural numbers, for example, are the set of whole numbers excluding 0. Prime numbers make up another set. They are the set of natural numbers greater than 1 whose only natural number factors are 1 and itself. We can write these sets the following way.

$$N = \{1, 2, 3, 4, 5, \dots\}$$
$$P = \{2, 3, 5, 7, 11, 13, \dots\}$$

There are several things to notice. Capital letters are usually used to represent sets. When elements of a set are listed, they are separated by commas and enclosed by braces. When we list the elements of a set this way, we say the set is in **roster form**. The three dots in sets N and P indicate that the pattern of numbers continues.

To indicate an element is part of a set, we use the symbol $\in$. Since 17 is a prime number, we can write $17 \in P$. This is read "17 is an element of set P".

EXAMPLE 1   Write in roster form.
(a) Set D is the set of Beatles.
(b) Set X is the set of natural numbers between 2 and 7.
(c) Set Y is the set of natural numbers between 2 and 7, inclusive.

(a) Writing set D in roster form, we have

$D = \{$*John Lennon, Paul McCartney, George Harrison, Ringo Starr*$\}$.

(b) $X = \{3, 4, 5, 6\}$.
(c) The word inclusive means the numbers 2 and 7 are included.

$$Y = \{2, 3, 4, 5, 6, 7\}$$

Practice Problem 1   Write in roster form.
(a) Set A is the set of continents on Earth.
(b) Set C is the set of natural numbers between 35 and 42.
(c) Set D is the set of natural numbers between 35 and 42, inclusive.

The sets in Example 1 are **finite sets**; we can count the number of elements. Set D in part (a) contains four elements and set Y in part (c) has six elements. The set of natural numbers $N = \{1, 2, 3, \dots\}$ is an example of an **infinite set**. The list of numbers continues without bound; there are infinitely many elements. Some sets contain no elements and are called **empty sets**. The empty set is denoted by the symbol $\{ \ \}$ or $\varnothing$. The set of students in your class that are 11 feet tall is an empty set.

2 Set-Builder Notation

All the sets we have seen so far have been in roster form or have been described in words. Another way to write a set is **set-builder notation**, used often in higher mathematics. An example of a set written in set-builder notation is

$$A = \{x \,|\, x \text{ is a natural number greater than } 10\}.$$

We read this as "Set A is the set of all elements x such that x is a natural number greater than 10." We also could have written $A = \{x \,|\, x \in N \text{ and } x > 10\}$. $x \in N$ means x is an element of the natural numbers.

Let's look at each part of set-builder notation and its meaning.

$\{$	x	$|$	criteria	$\}$
$\downarrow$	$\downarrow$	$\downarrow$	$\downarrow$	
The set of	all elements x	such that	x meets these criteria	

EXAMPLE 2   Write set B in set-builder notation. $B = \{a, e, i, o, u\}$

For an element to be in set B, it must be a vowel of the alphabet. We write $B = \{x \,|\, x \text{ is a vowel}\}$. Notice what is written to the right of the bar. We don't describe the set in words here. We simply indicate what criteria an element must meet to be in the set. $B = \{x \,|\, x \text{ is the set of vowels}\}$ is *not* correct.

Practice Problem 2   Write set C in set-builder notation. $C = \{4, 6, 8, 10, 12\}$ (*Hint:* There is more than one acceptable answer.)

3 The Union and Intersection of Sets

Addition, subtraction, multiplication, and division are operations used on numbers. There are other operations used on sets. The two most common operations are union and intersection.

The **union** of two sets A and B, written $A \cup B$, is the set of elements that are in set A, *or* set B.

EXAMPLE 3   Find $A \cup B$ if $A = \{a, b, c, d, e\}$ and $B = \{a, c, d, g\}$.

To find the set $A \cup B$, we combine the elements of A with those of B. We have $A \cup B = \{a, b, c, d, e, g\}$.

Practice Problem 3   Find $G \cup H$ if $G = \{!, *, \%, \$\}$ and $H = \{\$, ?, \wedge, +\}$.

The **intersection** of sets A and B, written $A \cap B$, is the set of elements in set A *and* set B.

EXAMPLE 4   Find $A \cap B$ if $A = \{a, b, c, d, e\}$ and $B = \{a, c, d, g\}$.

The elements that are common to sets A and B are $a, c,$ and d. Thus $A \cap B = \{a, c, d\}$.

Practice Problem 4   Find $G \cap H$ if $G = \{!, *, \%, \$\}$ and $H = \{\$, ?, \wedge, +\}$.

We have talked about one important set of numbers, the natural numbers. There are several other sets of numbers that we summarize below.

Sets of Numbers	
Real number	$\{x \,|\, x \text{ can be placed on the number line}\}$
Natural numbers	$\{1, 2, 3, 4, 5, \dots\}$
Whole numbers	$\{0, 1, 2, 3, 4, 5, \dots\}$
Integers	$\{\dots, -2, -1, 0, 1, 2, \dots\}$
Rational numbers	$\{x \,|\, x \text{ can be written as } \dfrac{p}{q} \text{ where } p \text{ and } q \text{ are integers, and } q \neq 0\}$
Irrational numbers	$\{x \,|\, x \text{ is a real number that is not rational}\}$

You are probably more familiar with the terms natural numbers, whole numbers, and integer than the others. Let's look at rational, irrational, and real numbers in more detail.

A **rational number** is a number that can be written as a fraction (with a denominator not equal to 0). Here are some examples of rational numbers:

$$-5, \quad 3.54, \quad \sqrt{9}, \quad \frac{1}{4}$$

The first three numbers can be written as $\frac{-5}{1}$, $\frac{354}{100}$, and $\frac{3}{1}$, respectively, and so are considered rational numbers. Every integer is rational since it can be written with 1 in the denominator. When a number is written in decimal form, we can easily determine whether or not it is a rational number. If the decimal repeats or terminates, it is a rational number.

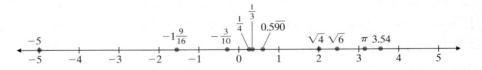

$$\frac{1}{3} = 0.3333\ldots = 0.\overline{3} \quad \text{and} \qquad \frac{13}{22} = 0.5909090\ldots = 0.5\overline{90} \text{ are repeating decimals}$$

rational numbers

$$\frac{3}{10} = 0.3 \qquad\qquad \text{and} \quad -1\frac{9}{16} = -0.5625 \qquad\qquad \text{are terminating decimals}$$

There are some numbers whose decimal representation is not a repeating or terminating decimal. For example, if we looked at the decimal forms of $\sqrt{2}$, $\sqrt{6}$, and $\sqrt{7}$, we would see that the decimal does not end and does not contain digits that repeat. These are **irrational numbers**. Pi (π) is another irrational number. When these numbers are used in calculations, we use approximations: $\sqrt{6} \approx 2.449$ and $\pi \approx 3.14$.

All the numbers we have discussed above can be placed on the number line. Some of them are shown below.

Any number that can be placed on the number line is a **real number**. The set of real numbers is the union of the rational and irrational numbers.

4 ▶ Subsets

We have seen that all integers are rational numbers and all rational numbers are real numbers. When all the elements of one set are contained in another set, we say that it is a **subset**. Here is a more formal definition.

> Set A is a subset of set B, written $A \subseteq B$, if all elements in A are also in B.

Consider the sets $A = \{$Amy, Jack, Ron$\}$ and $B = \{$Amy, Harry, Lena, Jack, Ron$\}$. All three elements of set A are also in set B. Therefore, set A is a subset of set B, and we can write $A \subseteq B$.

EXAMPLE 5   Determine if the statement is true or false. If false, state the reason.

(a) $A = \{t, v\}$ and $B = \{r, s, t, u, v, w\}$, so $A \subseteq B$.

(b) The set of integers is a subset of the natural numbers.

(a) True. All the elements of A are also in B.

(b) False. To see why, consider the number -3. -3 is an element of the set of integers, but -3 is not a natural number. So the integers is not a subset of the natural numbers.

The natural numbers, however, is a subset of the set of integers. Do you see why?

Practice Problem 5   Determine if the statement is true or false. If false, state the reason.

(a) $C = \{a, b, c, d, e, f\}$ and $D = \{c, f\}$ so $C \subseteq D$.

(b) The set of whole numbers is a subset of the rational numbers.

The table below shows the relationship among the sets of numbers we have discussed.

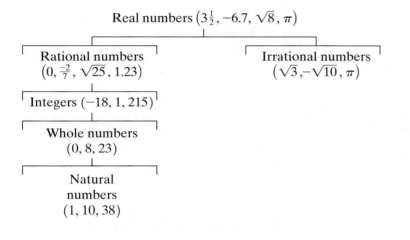

Appendix F Exercises

Fill in the blank with appropriate word or words.

1. The objects of a set are called _____.

2. When the elements of a set are listed in braces, the set is in _____.

3. The _____ of two sets is the elements the sets have in common.

4. The symbol _____ means 'intersection'.

5. A set that contains no elements is called the _____.

6. If a set is _____, we can count the number of elements it contains.

In exercises 7–10, write the set in roster form.

7. The set of states in the United States that begin with the letter C.

8. $A = \{x \mid x \in N \text{ and } 4 < x < 10\}$

9. $C = \{x \mid x \text{ is odd}\}$

10. The set of the last five months of the year

Exercises 11–14, write the set in set-builder notation.

11. $O = \{\text{Atlantic, Pacific, Indian, Arctic}\}$

12. $B = \{3, 6, 9, 12, 15\}$

13. $T = \{\text{scalene, isosceles, equilateral}\}$

14. C is the set of planets in our solar system

15. If $A = \{-1, 2, 3, 5, 8\}$ and $B = \{-2, -1, 3, 5\}$, find
    (a) $A \cup B$ **(b)** $A \cap B$

16. If $C = \{d, e, f, g, h, i, j\}$ and $D = \{x, z\}$, find
    (a) $C \cup D$ **(b)** $C \cap D$

Given sets A, B, and C, decide if the statements in exercises 17–24 are true or false. If it is false, give the reason.
$A = \{1, 2, 3, 4, 5, \ldots\}$, $B = \{10, 20, 30, 40, \ldots\}$, $C = \{1, 2, 3, 4, 5\}$

17. B is a finite set

18. $A \subseteq C$

19. $68 \in A$

20. $120 \in B$

21. $B \subseteq A$

22. $B \cap C = \{10, 20, 30, 40\}$

23. $A \cap C = \{\ \}$

24. $A \cup B = \{1, 2, 3, 4, 5, \ldots\}$

25. Give an example of a subset of *A*. *A* = {Joe, Ann, Nina, Doug}

26. Give an example of a set of which *B* is a subset. *B* = {poodle, Irish setter, dachshund}

Below is a table of the top 10 most popular boy and girl names for 1980 and 2000. Use the table to answer exercises 27 and 28. Source: www.cherishedmoments.com

	1980		2000	
	Boy	Girl	Boy	Girl
1.	Michael	Jennifer	Jacob	Emily
2.	Jason	Jessica	Michael	Hannah
3.	Christopher	Amanda	Matthew	Madison
4.	David	Melissa	Joshua	Ashley
5.	James	Sarah	Christopher	Sarah
6.	Matthew	Nicole	Nicholas	Alexis
7.	John	Heather	Andrew	Samantha
8.	Joshua	Amy	Joseph	Jessica
9.	Robert	Michelle	Daniel	Taylor
10.	Daniel	Elizabeth	Tyler	Elizabeth

27. (a) Write set *B* in roster form. *B* is the set of the most popular boys' names in 1980 or 2000.

(b) Your answer to part (a) represents a(n) _____ of two sets.

28.. (a) Write set *G* in roster form. *G* is the set of the most popular girls' names in 1980 and 2000.

(b) Your answer to part (a) represents a(n) _____ of two sets.

29. Decide which elements of the following set are whole numbers, natural numbers, integers, rational numbers, irrational numbers, or real numbers.

$$\left\{3.62, \sqrt{20}, \frac{-3}{11}, 15, \frac{22}{3}, 0, \sqrt{81}, -17\right\}$$

30. Is the set of whole numbers a subset of the real numbers? Explain.

31. Is the set of integers a subset of the whole numbers? Explain.

32. Which sets of numbers are subsets of the rational numbers?

33. List all sets of numbers of which the whole numbers are a subset.

Solutions to Practice Problems

Chapter 1

1.1 Practice Problems

1.

	Number	Integer	Rational Number	Irrational Number	Real Number
(a)	$-\frac{2}{5}$		X		X
(b)	$1.515151\ldots$		X		X
(c)	-8	X	X		X
(d)	π			X	X

2. (a) Population growth of 1,259 is +1,259.
  (b) Depreciation of $763 is −$763.00.
  (c) Windchill factor of minus 10 is −10.

3. (a) The additive inverse of $+\frac{2}{5}$ is $-\frac{2}{5}$.
  (b) The additive inverse of −1.92 is +1.92.
  (c) The opposite of a loss of 12 yards on a football play is a gain of 12 yards on the play.

4. (a) $|-7.34| = 7.34$

  (b) $\left|\dfrac{5}{8}\right| = \dfrac{5}{8}$ **(c)** $\left|\dfrac{0}{2}\right| = \dfrac{0}{2} = 0$

5. $-23 + (-35)$
    $23 + 35 = 58$
    $-23 + (-35) = -58$

6. $-\dfrac{3}{5} + \left(-\dfrac{4}{7}\right)$

    $-\dfrac{21}{35} + \left(-\dfrac{20}{35}\right)$

    $-\dfrac{21}{35} + \left(-\dfrac{20}{35}\right) = -\dfrac{41}{35}$

7. $-12.7 + (-9.38)$
    $12.7 + 9.38 = 22.08$
    $-12.7 + (-9.38) = -22.08$

8. $-7 + (-11) + (-33)$
    $= -18 + (-33)$
    $= -51$

9. $-9 + 15$
    $15 - 9 = 6$
    $-9 + 15 = 6$

10. $-\dfrac{5}{12} + \dfrac{7}{12} + \left(-\dfrac{11}{12}\right)$

    $= \dfrac{2}{12} + \left(-\dfrac{11}{12}\right) = -\dfrac{9}{12} = -\dfrac{3}{4}$

11. $-6.3 + (-8.0) + 3.5$
    $= -14.3 + 3.5$
    $= -10.8$

12. $-6 + 5 + (-7) + (-2) + 5 + 3$
    $-6 + 5$
    $-7 + 5$
    $\underline{-2 + 3}$
    $-15 \quad 13$
    $-15 + 13 = -2$

13. (a) $-2.9 + (-5.7) = -8.6$
  (c) $-10 + (-3) + 15 + 4$
    $= -13 + 15 + 4$
    $= 2 + 4 = 6$

  (b) $\dfrac{2}{3} + \left(-\dfrac{1}{4}\right)$
    $= \dfrac{8}{12} + \left(-\dfrac{3}{12}\right) = \dfrac{5}{12}$

1.2 Practice Problems

1. $9 - (-3) = 9 + (+3)$
    $= 12$

2. $-12 - (-5)$
    $= -12 + (+5)$
    $= -7$

3. $-\dfrac{1}{5} - \dfrac{1}{4} = -\dfrac{1}{5} + \left(-\dfrac{1}{4}\right) = -\dfrac{4}{20} + \left(-\dfrac{5}{20}\right) = -\dfrac{9}{20}$

4. $-17.3 - (-17.3)$
    $= -17.3 + 17.3$
    $= 0$

5. (a) $-21 - 9$
    $= -21 + (-9)$
    $= -30$

  (b) $17 - 36$
    $= 17 + (-36)$
    $= -19$

6. $350 - (-186)$
    $= 350 + 186$
    $= 536$ The helicopter is 536 feet from the sunken vessel.

1.3 Practice Problems

1. (a) $(-6)(-2) = 12$ **(b)** $(7)(9) = 63$

  (c) $\left(-\dfrac{3}{5}\right)\left(\dfrac{2}{7}\right) = -\dfrac{6}{35}$ **(d)** $40(-20) = -800$

2. $(-5)(-2)(-6)$
    $= (+10)(-6) = -60$

3. (a) positive; $-2(-3)(-4)(-1)$
               $= 6(-4)(-1)$
               $= -24(-1)$
               $= +24$ or 24

  (b) negative; $(-1)(-3)(-2)$
               $= 3(-2)$
               $= -6$

  (c) positive; $-4(-\frac{1}{4})(-2)(-6)$
               $= 1(-2)(-6)$
               $= -2(-6)$
               $= +12$ or 12

4. (a) $-36 \div (-2) = 18$ **(b)** $\dfrac{50}{-10} = -5$

  (c) $-49 \div 7 = -7$

5. (a) $-1.242 \div (-1.8)$

    $\begin{array}{r} .69 \\ 1.8_\wedge \overline{)1.2_\wedge 42} \\ \underline{10\ 8} \\ 1\ 62 \\ \underline{1\ 62} \\ 0 \end{array}$ Thus $-1.242 \div (-1.8) = 0.69$

  (b) $0.235 \div (-0.0025)$

    $\begin{array}{r} 94\ . \\ 0.0025_\wedge \overline{)2350_\wedge} \\ \underline{225} \\ 100 \\ \underline{100} \\ 0 \end{array}$ Thus $0.235 \div (-0.0025) = -94$

6. $-\dfrac{5}{16} \div \left(-\dfrac{10}{13}\right) = \left(-\dfrac{5}{16}\right)\left(-\dfrac{13}{10}\right) = \left(-\dfrac{\overset{1}{\cancel{5}}}{16}\right)\left(-\dfrac{13}{\underset{2}{\cancel{10}}}\right) = \dfrac{13}{32}$

7. (a) $\dfrac{-12}{-\dfrac{4}{5}} = -12 \div \left(-\dfrac{4}{5}\right) = -12\left(-\dfrac{5}{4}\right) = \left(-\dfrac{\overset{3}{\cancel{12}}}{1}\right)\left(-\dfrac{5}{\cancel{4}}\right) = 15$

(b) $\dfrac{-\dfrac{2}{9}}{\dfrac{8}{13}} = -\dfrac{2}{9} \div \dfrac{8}{13} = -\dfrac{\overset{1}{\cancel{2}}}{9}\left(\dfrac{13}{\underset{4}{\cancel{8}}}\right) = -\dfrac{13}{36}$

8. (a) $6(-10) = -60$ yards

(b) $7(15) = 105$ yards

(c) $-60 + 105 = $ a gain of 45 yards

1.4 Practice Problems

1. (a) $6(6)(6)(6) = 6^4$

(b) $(-2)(-2)(-2)(-2)(-2) = (-2)^5$

(c) $108(108)(108) = 108^3$

(d) $(-11)(-11)(-11)(-11)(-11)(-11) = (-11)^6$

(e) $(w)(w)(w) = w^3$

(f) $(z)(z)(z)(z) = z^4$

2. (a) $3^5 = (3)(3)(3)(3)(3) = 243$

(b) $2^2 + 3^3$

$2^2 = (2)(2) = 4$
$3^3 = (3)(3)(3) = 27$
$4 + 27 = 31$

3. (a) $(-3)^3 = -27$

(b) $(-2)^6 = 64$

(c) $-2^4 = -(2^4) = -16$

(d) $-(3^6) = -729$

4. (a) $\left(\frac{1}{3}\right)^3 = \left(\frac{1}{3}\right)\left(\frac{1}{3}\right)\left(\frac{1}{3}\right) = \frac{1}{27}$

(b) $(0.3)^4 = (0.3)(0.3)(0.3)(0.3) = 0.0081$

(c) $\left(\frac{3}{2}\right)^4 = \left(\frac{3}{2}\right)\left(\frac{3}{2}\right)\left(\frac{3}{2}\right)\left(\frac{3}{2}\right) = \frac{81}{16}$

(d) $(3)^4(4)^2$

$3^4 = (3)(3)(3)(3) = 81$
$4^2 = (4)(4) = 16$
$(81)(16) = 1296$

(e) $4^2 - 2^4 = 16 - 16 = 0$

1.5 Practice Problems

1. $25 \div 5 \cdot 6 + 2^3$

$= 25 \div 5 \cdot 6 + 8$
$= 5 \cdot 6 + 8$
$= 30 + 8$
$= 38$

2. $(-4)^3 - 2^6$

$= -64 - 64$
$= -128$

3. $6 - (8 - 12)^2 + 8 \div 2$

$= 6 - (-4)^2 + 8 \div 2$
$= 6 - (16) + 8 \div 2$
$= 6 - 16 + 4$
$= -10 + 4$
$= -6$

4. $\left(-\dfrac{1}{7}\right)\left(-\dfrac{14}{5}\right) + \left(-\dfrac{1}{2}\right) \div \left(\dfrac{3}{4}\right)$

$= \left(-\dfrac{1}{7}\right)\left(-\dfrac{14}{5}\right) + \left(-\dfrac{1}{2}\right) \times \left(\dfrac{4}{3}\right)$

$= \dfrac{2}{5} + \left(-\dfrac{2}{3}\right)$

$= \dfrac{2 \cdot 3}{5 \cdot 3} + \left(-\dfrac{2 \cdot 5}{3 \cdot 5}\right)$

$= \dfrac{6}{15} + \left(-\dfrac{10}{15}\right) = -\dfrac{4}{15}$

1.6 Practice Problems

1. (a) $-3(x + 2y) = -3(x) + (-3)(2y) = -3x - 6y$

(b) $-a(a - 3b) = -a(a) + (-a)(-3b) = -a^2 + 3ab$

2. (a) $-(-3x + y) = (-1)(-3x + y) = (-1)(-3x) + (-1)(y)$
$= 3x - y$

3. (a) $\dfrac{3}{5}(a^2 - 5a + 25) = \left(\dfrac{3}{5}\right)(a^2) + \left(\dfrac{3}{5}\right)(-5a) + \left(\dfrac{3}{5}\right)(25)$

$= \dfrac{3}{5}a^2 - 3a + 15$

(b) $2.5(x^2 - 3.5x + 1.2)$
$= (2.5)(x^2) + (2.5)(-3.5x) + (2.5)(1.2)$
$= 2.5x^2 - 8.75x + 3$

4. $-4x(x - 2y + 3) = (-4)(x)(x) - (-4)(x)(2)(y) + (-4)(x)(3)$
$= -4x^2 + 8xy - 12x$

5. $(3x^2 - 2x)(-4) = (3x^2)(-4) - (2x)(-4) = -12x^2 + 8x$

6. $400(6x + 9y) = 400(6x) + 400(9y)$
$= 2400x + 3600y$

1.7 Practice Problems

1. (a) $5a$ and $8a$ are like terms.

$2b$ and $-4b$ are like terms.

(b) y^2 and $-7y^2$ are like terms. These are the only like terms.

2. (a) $5a + 7a + 4a = (5 + 7 + 4)a = 16a$

(b) $16y^3 + 9y^3 = (16 + 9)y^3 = 25y^3$

3. $-8y^2 - 9y^2 + 4y^2 = (-8 - 9 + 4)y^2 = -13y^2$

4. (a) $-x + 3a - 9x + 2a = -x - 9x + 3a + 2a = -10x + 5a$

(b) $5ab - 2ab^2 - 3a^2b + 6ab = 5ab + 6ab - 2ab^2 - 3a^2b$
$= 11ab - 2ab^2 - 3a^2b$

(c) $7x^2y - 2xy^2 - 3x^2y - 4xy^2 + 5x^2y$
$= 7x^2y - 3x^2y + 5x^2y - 2xy^2 - 4xy^2 = 9x^2y - 6xy^2$

5. $5xy - 2x^2y + 6xy^2 - xy - 3xy^2 - 7x^2y$
$= 5xy - xy - 2x^2y - 7x^2y + 6xy^2 - 3xy^2$
$= 4xy - 9x^2y + 3xy^2$

6. $5a(2 - 3b) - 4(6a + 2ab) = 10a - 15ab - 24a - 8ab$
$= -14a - 23ab$

7. $\dfrac{1}{7}a^2 + 2a^2 = \dfrac{1}{7}a^2 + \dfrac{2}{1}a^2 = \dfrac{1}{7}a^2 + \dfrac{2 \cdot 7}{1 \cdot 7}a^2$

$= \dfrac{1}{7}a^2 + \dfrac{14}{7}a^2 = \dfrac{15}{7}a^2$

$-\dfrac{5}{12}b - \dfrac{1}{3}b = -\dfrac{5}{12}b - \dfrac{1 \cdot 4}{3 \cdot 4}b = -\dfrac{5}{12}b - \dfrac{4}{12}b$

$= -\dfrac{9}{12}b = -\dfrac{3}{4}b$

Thus, our solution is $\dfrac{15}{7}a^2 - \dfrac{3}{4}b$.

1.8 Practice Problems

1. $4 - \dfrac{1}{2}x = 4 - \dfrac{1}{2}(-8)$

$= 4 + 4$

$= 8$

2. **(a)** $-x^4 = -(-3)^4$

$= -(81) = -81$

(b) $(-x)^4 = \left[-(-3)\right]^4$

$= (3)^4 = 81$

3. $(5x)^3 + 2x = \left[5(-2)\right]^3 + 2(-2)$

$= (-10)^3 + (-4)$

$= (-1000) + (-4)$

$= -1004$

4. Area of a triangle is

$A = \frac{1}{2}ba$

altitude $= 3$ meters (m)

base $= 7$ meters (m)

$A = \dfrac{1}{2}(7\text{m})(3\text{m})$

$= \dfrac{1}{2}(7)(3)(m)(m)$

$= \left(\dfrac{7}{2}\right)(3)(m)^2$

$= \dfrac{21}{2}(m)^2$

$= 10.5$ square meters

5. Area of a circle is

$A = \pi r^2$

$r = 3$ meters

$A = 3.14(3 \text{ m})^2$

$= 3.14(9)(m)^2$

$= 28.26$ square meters

6. Formula $\qquad C = \dfrac{5}{9}(F - 32)$

$= \dfrac{5}{9}(68 - 32)$

$= \dfrac{5}{9}(36)$

$= 5(4)$

$= 20°$ Celsius

7. Use the formula.

$k = 1.61$ (r) $\qquad$ Replace r by 35.

$k = 1.61$ (35)

$k = 56.35$ $\qquad$ The truck is violating the minimum law.

1.9 Practice Problems

1. $5\left[4x - 3(y - 2)\right]$

$= 5\left[4x - 3y + 6\right]$

$= 20x - 15y + 30$

2. $3ab - \left[2ab - (2 - a)\right]$

$= 3ab - \left[2ab - 2 + a\right]$

$= 3ab - 2ab + 2 - a$

$= ab + 2 - a$

3. $3\left[4x - 2(1 - x)\right] - \left[3x + (x - 2)\right]$

$= 3\left[4x - 2 + 2x\right] - \left[3x + x - 2\right]$

$= 12x - 6 + 6x - 3x - x + 2$

$= 14x - 4$

4. $-2\{5x - 3x[2x - (x^2 - 4x)]\}$

$= -2\{5x - 3x[2x - x^2 + 4x]\}$

$= -2\{5x - 3x[6x - x^2]\}$

$= -2\{5x - 18x^2 + 3x^3\}$

$= -10x + 36x^2 - 6x^3$

$= -6x^3 + 36x^2 - 10x$

Chapter 2

2.1 Practice Problems

1. $x + 0.3 = 1.2$

$\dfrac{-0.3 \quad -0.3}{x \qquad = 0.9}$ $\qquad$ Check. $\qquad 0.9 + 0.3 \overset{?}{=} 1.2$

$1.2 = 1.2$

2. $17 = x - 5$

$\dfrac{+ 5 \qquad + 5}{22 = x}$ $\qquad$ Check. $\qquad 17 \overset{?}{=} 22 - 5$

$17 = 17$

3. $5 - 12 = x - 3$

$-7 = x - 3$

$\dfrac{+3 \qquad + 3}{-4 = x}$ $\qquad$ Check. $\qquad 5 - 12 \overset{?}{=} -4 - 3$

$-7 = -7$

4. $x + 8 = -22 + 6$

$x = -2$

Check. $\qquad -2 + 8 \overset{?}{=} -22 + 6$

$6 \neq -16$ $\qquad$ This is not true.

Thus $x = -2$ is not a solution. Solve to find the solution.

$x + 8 = -22 + 6 = -16$

$x = -16 - 8$

$x = -24$

5. $\dfrac{1}{20} - \dfrac{1}{2} = x + \dfrac{3}{5}$ $\qquad$ Check.

$\dfrac{1}{20} - \dfrac{1 \cdot 10}{2 \cdot 10} = x + \dfrac{3 \cdot 4}{5 \cdot 4}$

$\dfrac{1}{20} - \dfrac{10}{20} = x + \dfrac{12}{20}$

$-\dfrac{9}{20} = x + \dfrac{12}{20}$

$-\dfrac{9}{20} - \dfrac{12}{20} = x + \dfrac{12}{20} - \dfrac{12}{20}$

$-\dfrac{21}{20} = -1\dfrac{1}{20} = x$

$\dfrac{1}{20} - \dfrac{1}{2} \overset{?}{=} -1\dfrac{1}{20} + \dfrac{3}{5}$

$\dfrac{1}{20} - \dfrac{10}{20} \overset{?}{=} -\dfrac{21}{20} + \dfrac{12}{20}$

$-\dfrac{9}{20} = -\dfrac{9}{20}$ ✓

2.2 Practice Problems

1. $(8)\dfrac{1}{8}x = -2(8)$

$x = -16$

2. $\dfrac{9x}{9} = \dfrac{72}{9}$

$x = 8$

3. $\dfrac{6x}{6} = \dfrac{50}{6}$

$x = \dfrac{25}{3}$

4. $\dfrac{-27x}{-27} = \dfrac{54}{-27}$

$x = -2$

5. $\dfrac{-x}{-1} = \dfrac{36}{-1}$

$x = -36$

6. $\dfrac{-51}{6} = \dfrac{6x}{6}$

$-\dfrac{17}{2} = x$

7. $\dfrac{21}{4.2} = \dfrac{4.2x}{4.2}$

$5 = x$

2.3 Practice Problems

1. $9x + 2 = 38$ $\qquad$ Check. $\qquad 9(4) + 2 \overset{?}{=} 38$

$\dfrac{- 2 \qquad -2}{\dfrac{9x}{9} = \dfrac{36}{9}}$

$x = 4$

$36 + 2 \overset{?}{=} 38$

$38 = 38$ ✓

2. $13x = 2x - 66$ Check. $13(-6) \stackrel{?}{=} 2(-6) - 66$

$$\frac{-2x \quad -2x}{\frac{11x}{11} = \frac{-66}{11}}$$

$$x = -6$$

$-78 \stackrel{?}{=} -12 - 66$

$-78 = -78$ ✓

3. $3x + 2 = 5x + 2$ Check. $3(0) + 2 \stackrel{?}{=} 5(0) + 2$

$$\frac{-2 \qquad -2}{3x \quad = 5x}$$

$$\frac{-5x \qquad -5x}{-2x \quad = 0}$$

$$x = 0$$

$2 = 2$ ✓

4. $-z + 8 - z = 3z + 10 - 3$

$-2z + 8 = 3z + 7$

$$\frac{-3z \qquad -3z}{-5z + 8 = 7}$$

$$\frac{-8 \quad = -8}{\frac{-5z}{-5} = \frac{-1}{-5}}$$

$$z = \frac{1}{5}$$

5. $2x^2 - 6x + 3 = -4x - 7 + 2x^2$

$$\frac{-2x^2 \qquad\qquad\qquad - 2x^2}{-6x + 3 = -4x - 7}$$

$$\frac{+4x \qquad +4x}{-2x + 3 = -7}$$

$$\frac{-3 = -3}{\frac{-2x}{-2} = \frac{-10}{-2}}$$

$$x = 5$$

6. $4x - (x + 3) = 12 - 3(x - 2)$

$4x - x - 3 = 12 - 3x + 6$

$3x - 3 = -3x + 18$

$$\frac{+3x \qquad\qquad + 3x}{6x - 3 = 18}$$

$$\frac{+3 \qquad\qquad +3}{\frac{6x}{6} = \frac{21}{6}}$$

$$x = \frac{21}{6} = \frac{7}{2}$$

Check. $4\left(\frac{7}{2}\right) - \left(\frac{7}{2} + 3\right) \stackrel{?}{=} 12 - 3\left(\frac{7}{2} - 2\right)$

$$14 - \frac{13}{2} \stackrel{?}{=} 12 - 3\left(\frac{3}{2}\right)$$

$$\frac{28}{2} - \frac{13}{2} \stackrel{?}{=} \frac{24}{2} - \frac{9}{2}; \qquad \frac{15}{2} = \frac{15}{2} \; ✓$$

7. $4(-2x - 3) = -5(x - 2) + 2$

$-8x - 12 = -5x + 10 + 2$

$-8x - 12 = -5x + 12$

$$\frac{+5x \qquad\qquad +5x}{-3x - 12 = 12}$$

$$\frac{+12 = +12}{\frac{-3x}{-3} = \frac{24}{-3}}$$

$$x = -8$$

8. $0.3x - 2(x + 0.1) = 0.4(x - 3) - 1.1$

$0.3x - 2x - 0.2 = 0.4x - 1.2 - 1.1$

$-1.7x - 0.2 = 0.4x - 2.3$

$$\frac{-0.4x \qquad\qquad -0.4x}{-2.1x - 0.2 = -2.3}$$

$$\frac{+0.2 \qquad +0.2}{\frac{-2.1x}{-2.1} = \frac{-2.1}{-2.1}}$$

$$x = 1$$

9. $5(2z - 1) + 7 = 7z - 4(z + 3)$

$10z - 5 + 7 = 7z - 4z - 12$

$10z + 2 = 3z - 12$

$$\frac{-3z \qquad\qquad - 3z}{7z + 2 = -12}$$

$$\frac{-2 \qquad\qquad - 2}{\frac{7z}{7} = \frac{-14}{7}}$$

$$z = -2$$

Check. $5[2(-2) - 1] + 7 \stackrel{?}{=} 7(-2) - 4[-2 + 3]$

$$5(-5) + 7 \stackrel{?}{=} -14 - 4(1)$$

$$-25 + 7 \stackrel{?}{=} -18$$

$$-18 = -18 \; ✓$$

2.4 Practice Problems

1. $\dfrac{3}{8}x - \dfrac{3}{2} = \dfrac{1}{4}x$

$3x - 12 = 2x$

$$\frac{-2x \qquad -2x}{x - 12 = 0}$$

$$\frac{+12 \quad +12}{x = 12}$$

2. $\dfrac{5x}{4} - 1 = \dfrac{3x}{4} + \dfrac{1}{2}$

$5x - 4 = 3x + 2$

$$\frac{-3x \qquad -3x}{2x - 4 = +2}$$

$$\frac{+4 = +4}{\frac{2x}{2} = \frac{6}{2}}$$

$$x = 3$$

Check. $\dfrac{5(3)}{4} - 1 \stackrel{?}{=} \dfrac{3(3)}{4} + \dfrac{1}{2}$

$$\frac{15}{4} - 1 \stackrel{?}{=} \frac{9}{4} + \frac{1}{2}$$

$$\frac{15}{4} - \frac{4}{4} \stackrel{?}{=} \frac{9}{4} + \frac{2}{4}$$

$$\frac{11}{4} = \frac{11}{4} \; ✓$$

3. $\dfrac{5x}{6} - \dfrac{5}{8} = \dfrac{3x}{4} - \dfrac{1}{3}$

$20x - 15 = 18x - 8$

$$\frac{-18x \qquad -18x}{2x - 15 = -8}$$

$$\frac{+15 = +15}{\frac{2x}{2} = \frac{7}{2}}$$

$$x = \frac{7}{2}$$

4. $\frac{1}{3}(x-2) = \frac{1}{4}(x+5) - \frac{5}{3}$ Check. $\frac{1}{3}(3-2) \stackrel{?}{=} \frac{1}{4}(3+5) - \frac{5}{3}$

$$\frac{1}{3}x - \frac{2}{3} = \frac{1}{4}x + \frac{5}{4} - \frac{5}{3}$$

$$4x - 8 = 3x + 15 - 20$$

$$4x - 8 = 3x - 5$$

$$\underline{-3x \quad\quad = -3x}$$

$$\underline{x - 8 = \quad\quad -5}$$

$$\underline{+8 \quad\quad\quad +8}$$

$$x \quad = \quad\quad 3$$

$\frac{1}{3}(1) \stackrel{?}{=} \frac{1}{4}(8) - \frac{5}{3}$

$\frac{1}{3} \stackrel{?}{=} 2 - \frac{5}{3}$

$\frac{1}{3} \stackrel{?}{=} \frac{6}{3} - \frac{5}{3}$

$\frac{1}{3} = \frac{1}{3}$ ✓

5. $2.8 = 0.3(x-2) + 2(0.1x - 0.3)$

$$2.8 = 0.3x - 0.6 + 0.2x - 0.6$$

$$10(2.8) = 10(0.3x) - 10(0.6) + 10(0.2x) - 10(0.6)$$

$$28 = 3x - 6 + 2x - 6$$

$$28 = 5x - 12$$

$$\underline{+12 \quad\quad + 12}$$

$$40 = 5x$$

$$8 = x$$

2.5 Practice Problems

1. $d = rt$

$$3525 = r(2.5)$$

$$\frac{3525}{2.5} = r$$

$$1410 \text{ mph} = r$$

2. Solve for m. $\dfrac{E}{c^2} = \dfrac{mc^2}{c^2}$ $\dfrac{E}{c^2} = m$

3. Solve for y. $8 - 2y + 3x = 0$

$$\underline{-3x = \quad - 3x}$$

$$\underline{8 - 2y \quad\quad = \quad - 3x}$$

$$\underline{-8 \quad\quad\quad\quad - 8}$$

$$\frac{-2y}{-2} = \frac{-3x - 8}{-2}$$

$$y = \frac{3x + 8}{2} \text{ or } y = \frac{3}{2}x + 4$$

4. Solve for d. $C = \pi d$

$$\frac{C}{\pi} = \frac{\pi d}{\pi}$$

$$\frac{C}{\pi} = d$$

2.6 Practice Problems

1. (a) $7 > 2$ **(b)** $-3 > -4$ **(c)** $-1 < 2$

(d) $-8 < -5$ **(e)** $0 > -2$ **(f)** $\dfrac{2}{5} > \dfrac{3}{8}$

2. (a) $x > 5$; x is greater than 5.

(b) $x \le -2$; x is less than or equal to -2.

(c) $3 > x$; 3 is greater than x (or x is less than 3).

(d) $x \ge -\dfrac{3}{2}$; x is greater than or equal to $-\dfrac{3}{2}$.

3. (a) $t \le 180$ **(b)** $d < 15,000$

2.7 Practice Problems

1. (a) $9 + 4 > 6 + 4$

$$13 > 10$$

(b) $-2 - 3 < 5 - 3$

$$-5 < 2$$

(c) $1(2) > -3(2)$

$$2 > -6$$

(d) $\dfrac{10}{5} < \dfrac{15}{5}$

$$2 < 3$$

2. (a) $7 > 2$ **(b)** $-3 < -1$

$\quad\quad -14 < -4$ $\quad\quad 3 > 1$

(c) $-10 \ge -20$ **(d)** $-15 \le -5$

$\quad\quad 1 \le 2$ $\quad\quad 3 \ge 1$

3. $8x - 2 < 3$

$$\underline{\quad +2 \quad +2}$$

$$\frac{8x}{8} < \frac{5}{8}$$

$$x < \frac{5}{8}$$

4. $4 - 5x > 7$

$$\underline{-4 \quad\quad\quad -4}$$

$$\frac{-5x}{-5} < \frac{3}{-5}$$

$$x < -\frac{3}{5}$$

5. $\dfrac{1}{2}x + 3 < \dfrac{2}{3}x$

$$3x + 18 < 4x$$

$$\underline{-4x \quad\quad\quad -4x}$$

$$\underline{-x + 18 < \quad 0}$$

$$\underline{-18 \quad\quad -18}$$

$$\frac{-x}{-1} > \frac{-18}{-1}$$

$$x > 18$$

6. $\dfrac{1}{2}(3 - x) \le 2x + 5$

$$\frac{3}{2} - \frac{1}{2}x \le 2x + 5$$

$$3 - x \le 4x + 10$$

$$-5x \le 7$$

$$x \ge -\frac{7}{5}$$

7. $2000n - 700,000 \ge 2,500,000$

$$2000n \ge 3,200,000$$

$$n \ge 1600$$

2.8 Practice Problems

1. short piece = x
long piece = $x + 17$
$$x + (x + 17) = 89$$
$$2x + 17 = 89$$
$$2x = 72$$
$$x = 36 \text{ feet}$$

Therefore   Short piece   36 feet
            Long piece $36 + 17 = 53$ feet

2. Family 1 = $x + 360$
Family 2 = x
Family 3 = $2x - 200$
$$(x) + (x + 360) + (2x - 200) = 3960$$
$$4x + 160 = 3960$$
$$x = 950$$

Therefore   Family #2 = 950.00
            Family #3 = $2(950) - 200 = \$1700.00$
            Family #1 = $950 + 360 = \$1310.00$

3. Let the second side = x Therefore the
      first side = $x - 30$ second side = 300 meters
      third side = $\frac{1}{2}x$ first side = 270 meters
$x + x - 30 + \frac{1}{2}x = 720$ third side = 150 meters
$$5x - 60 = 1440$$
$$5x = 1500$$
$$x = 300$$

2.9 Practice Problems

1. $3(25) + (0.20)m = 350$
$$75 + 0.20m = 350$$
$$0.20m = 275$$
$$m = 1375 \text{ miles}$$

2. $0.38x = 4560$
$$x = \$12,000.00$$

3. $x + 0.07x = 13,910$
$$1.07x = 13,910$$
$$x = \$13,000.00$$

4. $x = 7000(0.12)(1)$
$$x = \$840.00$$

5. $0.09x + 0.07(8000 - x) = 630$
$$0.09x + 560 - 0.07x = 630$$
$$0.02x + 560 = 630$$
$$0.02x = 70$$
$$x = \$3500$$

Therefore, she invested $3500.00 at 9%.
$(8000 - 3500) = \$4500.00$ at 7%

6. x = Dimes
$x + 5$ = Quarters
$$0.10x + 0.25(x + 5) = 5.10$$
$$0.10x + 0.25x + 1.25 = 5.10$$
$$0.35x + 1.25 = 5.10$$
$$0.35x = 3.85$$
$$x = 11$$

Therefore, she has 11 dimes and 16 quarters.

7.   Nickels = $2x$
     Dimes = x
Quarters = $x + 4$
$$0.05(2x) + 0.10(x) + 0.25(x + 4) = 2.35$$
$$0.1x + 0.10x + 0.25x + 1 = 2.35$$
$$0.45x + 1 = 2.35$$
$$0.45x = 1.35$$
$$x = 3$$

Therefore, the boy has 3 dimes, 6 nickels, and 7 quarters.

Chapter 3

3.1 Practice Problems

1. Point B is 3 units to the right on the x-axis and 4 units up from the point where we stopped on the x-axis.

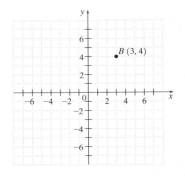

2. (a) Begin by counting 2 squares to the left starting at the origin. Since the y-coordinate is negative, count 4 units down from the point where we stopped on the x-axis. Label the point I.
(b) Begin by counting 4 squares to the left of the origin. Then count 5 units up because the y-coordinate is positive. Label the point J.
(c) Begin by counting 4 units to the right of the origin. Then count 2 units down because the y-coordinate is negative. Label the point K.

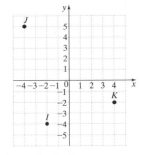

3. The points are plotted in the figure.

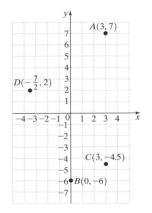

4. Move along the x-axis to get as close as possible to B. We end up at 2. Thus the first number of the ordered pair is 2. Then count 7 units upward on a line parallel to the y-axis to reach B. So the second number of the ordered pair is 7. Thus, $B = (2, 7)$.

5. $A = (-2, -1); B = (-1, 3); C = (0, 0); D = (2, -1); E = (3, 1)$

6. (a) Replace x by 0 in the equation. $3(0) - 4y = 12$
$$0 - 4y = 12$$
$$y = -3 \qquad (0, -3)$$

(b) Replace the variable y by 3. $3x - 4(3) = 12$
$$3x - 12 = 12$$
$$3x = 24$$
$$x = 8 \qquad (8, 3)$$

(c) Replace the variable y by -6.
$$3x - 4(-6) = 12$$
$$3x + 24 = 12$$
$$3x = -12$$
$$x = -4 \qquad (-4, -6)$$

7. (a)

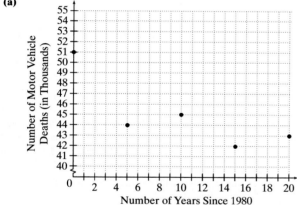

(b) 1980 was a significant high in motor vehicle deaths. During 1985–2000, the number of motor vehicle deaths was relatively stable.

3.2 Practice Problems

1. Graph $x + y = 10$.

Let $x = 0$.
$$0 + y = 10$$
$$y = 10$$

Let $x = 5$.
$$5 + y = 10$$
$$y = 5$$

Let $x = 2$.
$$2 + y = 10$$
$$y = 8$$

Plot the ordered pairs
$(0, 10), (5, 5),$ and $(2, 8)$.

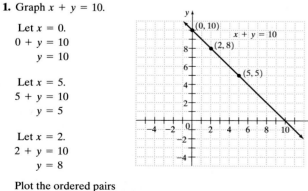

2.
$$7x + 3 = -2y + 3$$
$$7x + 3 - 3 = -2y + 3 - 3$$
$$7x = -2y$$
$$7x + 2y = -2y + 2y$$
$$7x + 2y = 0$$

Let $x = 0$.
$$7(0) + 2y = 0$$
$$2y = 0$$
$$y = 0$$

Let $x = -2$.
$$7(-2) + 2y = 0$$
$$-14 + 2y = 0$$
$$2y = 14$$
$$y = 7$$

Let $x = 2$.
$$7(2) + 2y = 0$$
$$14 + 2y = 0$$
$$2y = -14$$
$$y = -7$$

Graph the ordered pairs $(0, 0), (2, -7),$ and $(-2, 7)$.

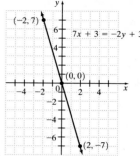

3. $2y - x = 6$

Find the two intercepts.

Let $x = 0$.
$$2y - 0 = 6$$
$$2y = 6$$
$$y = 3$$

Let $y = 0$.
$$2(0) - x = 6$$
$$-x = 6$$
$$x = -6$$

Find a third point.

Let $y = 1$.
$$2(1) - x = 6$$
$$2 - x = 6$$
$$-x = 4$$
$$x = -4$$

Graph the ordered pairs
$(0, 3), (-6, 0),$ and $(-4, 1)$.

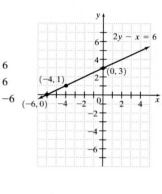

4. $2y - 3 = 0$

Solve for y.
$$2y = 3$$
$$y = \frac{3}{2}$$

This line is parallel to the x-axis. It is a horizontal line $1\frac{1}{2}$ units above the x-axis.

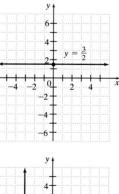

5. $x + 3 = 0$

Solve for x.
$$x = -3$$

This line is parallel to the y-axis. It is a vertical line 3 units to the left of the y-axis.

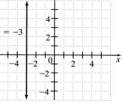

3.3 Practice Problems

1. $m = \dfrac{y_2 - y_1}{x_2 - x_1} = \dfrac{-1 - 1}{-4 - 6} = \dfrac{-2}{-10} = \dfrac{1}{5}$

2. $m = \dfrac{y_2 - y_1}{x_2 - x_1} = \dfrac{1 - 0}{-1 - 2} = \dfrac{1}{-3} = -\dfrac{1}{3}$

3. (a) $m = \dfrac{6 - 3}{-5 - (-5)} = \dfrac{3}{0}$

$\dfrac{3}{0}$ is undefined. Therefore there is no slope and the line is a vertical line through $x = -5$.

(b) $m = \dfrac{-11 - (-11)}{3 - (-7)} = \dfrac{0}{10} = 0$

$m = 0$. The line is a horizontal line through $y = -11$.

4. Solve for y.

$4x - 2y = -5$

$-2y = -4x - 5$

$y = \dfrac{-4x - 5}{-2}$

$y = 2x + \dfrac{5}{2}$ Slope $= 2$ y-intercept $= \left(0, \dfrac{5}{2}\right)$

5. (a) $y = mx + b$

$m = -\dfrac{3}{7}$ y-intercept $= \left(0, \dfrac{2}{7}\right)$

$y = -\dfrac{3}{7}x + \dfrac{2}{7}$

(b) $y = -\dfrac{3}{7}x + \dfrac{2}{7}$

$7(y) = 7\left(-\dfrac{3}{7}x\right) + 7\left(\dfrac{2}{7}\right)$

$7y = -3x + 2$

$3x + 7y = 2$

6. y-intercept $= (0, -1)$. Thus the coordinates of the y-intercept for this line are $(0, -1)$. Plot the point. Slope is $\dfrac{\text{rise}}{\text{run}}$. Since the slope for this line is $\dfrac{3}{4}$, we will go up (rise) 3 units and go over (run) 4 units to the right from the point $(0, -1)$. This is the point $(4, 2)$.

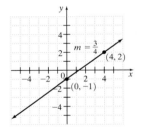

7. $y = -\dfrac{2}{3}x + 5$

The y-intercept is $(0, 5)$ since $b = 5$. Plot the point $(0, 5)$. The slope is $-\dfrac{2}{3} = \dfrac{-2}{3}$. Begin at $(0, 5)$, go down 2 units and to the right 3 units. This is the point $(3, 3)$. Draw a line that connects the points $(0, 5)$ and $(3, 3)$.

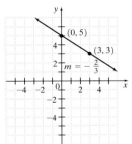

8. (a) Parallel lines have the same slope. Line j has a slope of $\dfrac{1}{4}$.

(b) Perpendicular lines have slopes whose product is -1.

$m_1 m_2 = -1$

$\dfrac{1}{4}m_2 = -1$

$4\left(\dfrac{1}{4}\right)m_2 = -1(4)$

$m_2 = -4$

Thus line k has a slope of -4.

9. (a) The slope of line n is $\dfrac{1}{4}$. The slope of a line that is parallel to line n is $\dfrac{1}{4}$.

(b) $m_1 m_2 = -1$

$\dfrac{1}{4}m_2 = -1$

$m_2 = -4$

The slope of a line that is perpendicular to n is -4.

3.4 Practice Problems

1. $y = mx + b$

$12 = -\dfrac{3}{4}(-8) + b$

$12 = 6 + b$

$6 = b$

The equation of the line is $y = -\dfrac{3}{4}x + 6$.

2. Find the slope.

$m = \dfrac{y_2 - y_1}{x_2 - x_1} = \dfrac{1 - 5}{-1 - 3} = \dfrac{-4}{-4} = 1$

Using either of the two points given, substitute x and y values into the equation $y = mx + b$.

$m = 1$ $x = 3$ and $y = 5$

$y = mx + b$

$5 = 1(3) + b$

$5 = 3 + b$

$2 = b$

The equation of the line is $y = x + 2$.

3. The y-intercept is $(0, 1)$. Thus $b = 1$. Look for another point on the line. We choose $(6, 2)$. Count the number of vertical units from 1 to 2 (rise). Count the number of horizontal units from 0 to 6 (run).

$m = \dfrac{1}{6}$ Now we can write the equation of the line.

$y = mx + b$

$y = \dfrac{1}{6}x + 1$

3.5 Practice Problems

1. Graph $x - y \geq -10$.

Begin by graphing the line $x - y = -10$. You may use any method discussed previously. Since there is an equal sign in the inequality, we will draw a solid line to indicate that the line is part of the solution set. The easiest test point is $(0, 0)$. Substitute $x = 0$, $y = 0$ in the inequality.

$x - y \geq -10$

$0 - 0 \geq -10$

$0 \geq -10$ True

Therefore shade the side of the line that includes the point $(0, 0)$.

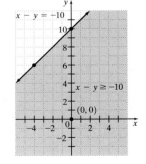

2. Step 1 Graph $y = \frac{1}{2}x$. Since $>$ is used, the line should be a dashed line.

Step 2 We see the line passes through $(0, 0)$.

Step 3 Choose another test point. We will choose $(-1, 1)$.

$$y > \frac{1}{2}x$$

$$1 > \frac{1}{2}(-1)$$

$$1 > -\frac{1}{2} \quad \text{true}$$

Shade the region that includes $(-1, 1)$, that is, the region above the line.

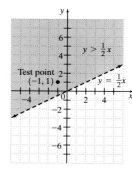

3. Step 1 Graph $y = -3$. Since $\geq$ is used, the line should be solid.

Step 2 Test $(0, 0)$ in the inequality.

$$y \geq -3$$

$$0 \geq -3 \quad \text{true}$$

Shade the region that includes $(0, 0)$, that is, the region above the line $y = -3$.

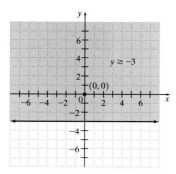

3.6 Practice Problems

1. The domain is $\{-3, 3, 0, 20\}$. The range is $\{-5, 5\}$.

2. (a) Look at the ordered pairs. No two ordered pairs have the same first coordinate. Thus this set of ordered pairs defines a function.

(b) Look at the ordered pairs. Two different ordered pairs, $(60, 30)$ and $(60, 120)$, have the same first coordinate. Thus this relation is not a function.

3. (a) Looking at the table, we see that no two different ordered pairs have the same first coordinate. The cost of gasoline is a function of the distance traveled.

Note that cost depends on distance. Thus distance is the independent variable. Since a negative distance does not make sense, the domain is {all nonnegative real numbers}.

The range is {all nonnegative real numbers}.

(b) Looking at the table, we see two ordered pairs, $(5, 20)$ and $(5, 30)$, have the same first coordinate. Thus this relation is not a function.

4. Construct a table, plot the ordered pairs and connect the points.

x	$y = x^2 - 2$	y
-2	$y = (-2)^2 - 2 = 2$	2
-1	$y = (-1)^2 - 2 = -1$	-1
0	$y = 0 - 2 = -2$	-2
1	$y = (1)^2 - 2 = -1$	-1
2	$y = (2)^2 - 2 = 2$	2

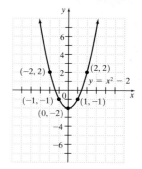

5. Select values of y and then substitute them into the equation to obtain the x values.

y	$x = y^2 - 1$	x	y
-2	$x = (-2)^2 - 1 = 3$	3	-2
-1	$x = (-1)^2 - 1 = 0$	0	-1
0	$x = (0)^2 - 1 = -1$	-1	0
1	$x = (1)^2 - 1 = 0$	0	1
2	$x = (2)^2 - 1 = 3$	3	2

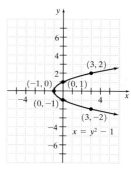

6. $y = \frac{6}{x}$

x	$y = \frac{6}{x}$	y
-3	$y = \frac{6}{-3} = -2$	-2
-2	$y = \frac{6}{-2} = -3$	-3
-1	$y = \frac{6}{-1} = -6$	-6
0	We cannot divide by 0.	
1	$y = \frac{6}{1} = 6$	6
2	$y = \frac{6}{2} = 3$	3
3	$y = \frac{6}{3} = 2$	2

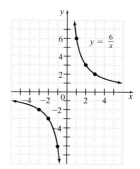

7. (a) The graph of a vertical line is not a function.

(b) This curve is a function. Any vertical line will cross the curve in only one location.

(c) This curve is not the graph of a function. There exist vertical lines that will cross the curve in more than one place.

8. $f(x) = -2x^2 + 3x - 8$

(a) $f(2) = -2(2)^2 + 3(2) - 8$
$f(2) = -2(4) + 3(2) - 8$
$f(2) = -8 + 6 - 8$
$f(2) = -10$

(b) $f(-3) = -2(-3)^2 + 3(-3) - 8$
$f(-3) = -2(9) + 3(-3) - 8$
$f(-3) = -18 - 9 - 8$
$f(-3) = -35$

(c) $f(0) = -2(0)^2 + 3(0) - 8$
$f(0) = -2(0) + 3(0) - 8$
$f(0) = 0 + 0 - 8$
$f(0) = -8$

Chapter 4

4.1 Practice Problems

1. Substitute $(-3, 4)$ into the first equation to see if the ordered pair is a solution.

$$2x + 3y = 6$$
$$2(-3) + 3(4) \overset{?}{=} 6$$
$$-6 + 12 \overset{?}{=} 6$$
$$6 = 6 \checkmark$$

Likewise, we will determine if $(-3, 4)$ is a solution to the second equation.

$$3x - 4y = 7$$
$$3(-3) - 4(4) \overset{?}{=} 7$$
$$-9 - 16 \overset{?}{=} 7$$
$$-25 \neq 7$$

Since $(-3, 4)$ is not a solution to each equation in the system, it is not a solution to the system itself.

2. You can use any method we developed in Chapter 3 to graph each line. We will change each equation to slope-intercept form to graph.

$$3x + 2y = 10$$
$$2y = -3x + 10$$
$$y = -\frac{3}{2}x + 5$$

$$x - y = 5$$
$$y = x - 5$$

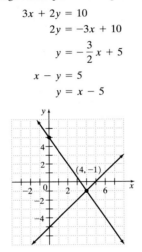

The lines intersect at the point $(4, -1)$. Thus $(4, -1)$ is a solution. We verify this by substituting $x = 4$ and $y = -1$ into the system of equations.

$$3x + 2y = 10 \qquad x - y = 5$$
$$3(4) + 2(-1) \overset{?}{=} 10 \qquad 4 - (-1) \overset{?}{=} 5$$
$$12 - 2 \overset{?}{=} 10 \qquad 5 = 5 \checkmark$$
$$10 = 10 \checkmark$$

3. $2x - y = 7$ [1]
$3x + 4y = -6$ [2]

Solve equation [1] for y.

$$-y = 7 - 2x$$
$$y = -7 + 2x \quad [3]$$

Substitute $-7 + 2x$ for y in equation [2].

$$3x + 4(-7 + 2x) = -6$$
$$3x - 28 + 8x = -6$$
$$11x - 28 = -6$$
$$11x = 22$$
$$x = 2$$

Substitute $x = 2$ into equation [3].

$$y = -7 + 2(2)$$
$$y = -7 + 4$$
$$y = -3$$

The solution is $(2, -3)$.

4. $\dfrac{x}{2} + \dfrac{2y}{3} = 1$

$\dfrac{x}{3} + y = -1$

First we clear both equations of fractions.

$$3x + 4y = 6 \quad [1]$$
$$x + 3y = -3 \quad [2]$$

Step 1 Solve for x in equation [2].

$$x + 3y = -3$$
$$x = -3y - 3$$

Steps 2 and 3 Substitute this expression for x into equation [1] and solve for y.

$$3(-3y - 3) + 4y = 6$$
$$-9y - 9 + 4y = 6$$
$$-5y = 15$$
$$y = -3$$

Step 4 Substitute $y = -3$ into equation [1] or [2].

$$x + 3y = -3$$
$$x + 3(-3) = -3$$
$$x - 9 = -3$$
$$x = 6$$

So our solution is $(6, -3)$.

5. $-3x + y = 5$ [1]
$2x + 3y = 4$ [2]

Multiply equation [1] by -3 and add to equation [2].

$$9x - 3y = -15$$
$$\underline{2x + 3y = 4}$$
$$11x = -11$$
$$x = -1$$

Now substitute $x = -1$ into equation [1].

$$-3(-1) + y = 5$$
$$3 + y = 5$$
$$y = 2$$

The solution is $(-1, 2)$.

6. $5x + 4y = 23$ [1]
$7x - 3y = 15$ [2]

Multiply equation [1] by 3 and equation [2] by 4.

$$15x + 12y = 69$$
$$\underline{28x - 12y = 60}$$
$$43x = 129$$
$$x = 3$$

Now substitute $x = 3$ into equation [1] or [2].

$$5(3) + 4y = 23$$
$$15 + 4y = 23$$
$$4y = 8$$
$$y = 2$$

The solution is $(3, 2)$.

7. $4x - 2y = 6$ [1]
$-6x + 3y = 9$ [2]

Multiply equation [1] by 3 and equation [2] by 2 and add together.

$$12x - 6y = 18$$
$$\underline{-12x + 6y = 18}$$
$$0 = 36$$

This statement is of course false. Thus, we conclude that this system of equations is inconsistent, so there is **no solution**.

8. $0.3x - 0.9y = 1.8$ [1]
$-0.4x + 1.2y = -2.4$ [2]

Multiply both equations by 10 to obtain a more convenient form.

$$3x - 9y = 18 \quad [3]$$
$$-4x + 12y = -24 \quad [4]$$

Multiply equation [3] by 4 and equation [4] by 3.

$$12x - 36y = 72$$
$$\underline{-12x + 36y = -72}$$
$$0 = 0$$

This statement is always true. Hence these are dependent equations. There are an infinite number of solutions.

4.2 Practice Problems

1. Substitute $x = 3$, $y = -2$, $z = 2$ into each equation.

$$2(3) + 4(-2) + (2) \overset{?}{=} 0$$
$$6 - 8 + 2 \overset{?}{=} 0$$
$$-2 + 2 \overset{?}{=} 0$$
$$0 = 0 \checkmark$$

$$(3) - 2(-2) + 5(2) \overset{?}{=} 17$$
$$3 + 4 + 10 \overset{?}{=} 17$$
$$17 = 17 \checkmark$$

$$3(3) - 4(-2) + 2 \overset{?}{=} 19$$
$$9 + 8 + 2 \overset{?}{=} 19$$
$$19 = 19 \checkmark$$

Since we obtained three true statements, the ordered triple $(3, -2, 2)$ is a solution to the system.

2. $x + 2y + 3z = 4$ [1]
$2x + y - 2z = 3$ [2]
$3x + 3y + 4z = 10$ [3]

We eliminate x by multiplying equation [1] by -2 and adding it to equation [2].

$$-2x - 4y - 6z = -8 \quad [4]$$
$$\underline{2x + y - 2z = 3 \quad [2]}$$
$$-3y - 8z = -5 \quad [5]$$

Now we eliminate x by multiplying equation [1] by -3 and adding it to equation [3].

$$-3x - 6y - 9z = -12 \quad [6]$$
$$\underline{3x + 3y + 4z = 10 \quad [3]}$$
$$-3y - 5z = -2 \quad [7]$$

We now eliminate y and solve for z in the system formed by equation [5] and equation [7].

$$-3y - 8z = -5 \quad [5]$$
$$-3y - 5z = -2 \quad [7]$$

To do this we multiply equation [5] by -1 and add it to equation [7].

$$3y + 8z = 5 \quad [8]$$
$$\underline{-3y - 5z = -2 \quad [7]}$$
$$3z = 3$$
$$z = 1$$

Substitute $z = 1$ into equation [8] and solve for y.

$$3y + 8(1) = 5$$
$$3y + 8 = 5$$
$$3y = -3$$
$$y = -1$$

Substitute $z = 1$, $y = -1$ into equation [1] and solve for x.

$$x + 2(-1) + 3(1) = 4$$
$$x + 1 = 4$$
$$x = 3$$

The solution is $(3, -1, 1)$.

3. $2x + y + z = 11$ [1]
$4y + 3z = -8$ [2]
$x - 5y = 2$ [3]

Multiply equation [1] by -3 and add the results to equation [2], thus eliminating the z terms.

$$-6x - 3y - 3z = -33 \quad [4]$$
$$\underline{4y + 3z = -8 \quad [2]}$$
$$-6x + y = -41 \quad [5]$$

We can solve the system formed by equation [3] and equation [5].

$$x - 5y = 2 \quad [3]$$
$$-6x + y = -41 \quad [5]$$

Multiply equation [3] by 6 and add the results to equation [5].

$$6x - 30y = 12 \quad [6]$$
$$\underline{-6x + y = -41 \quad [5]}$$
$$-29y = -29$$
$$y = 1$$

Now substitute $y = 1$ into equation [2] and solve for z.

$$4(1) + 3z = -8$$
$$4 + 3z = -8$$
$$3z = -12$$
$$z = -4$$

Now substitute $y = 1$, $z = -4$ into equation [1] and solve for x.

$$2x + 1 + (-4) = 11$$
$$2x - 3 = 11$$
$$2x = 14$$
$$x = 7$$

The solution is $(7, 1, -4)$.

4.3 Practice Problems

1. Let x = the number of baseballs purchased and y = the number of bats purchased.

Last week: $6x + 21y = 318$ [1]

This week: $5x + 17y = 259$ [2]

Multiply equation [1] by 5 and equation [2] by -6.

$$30x + 105y = 1590 \quad [3]$$
$$\underline{-30x - 102y = -1554 \quad [4]}$$
$$3y = 36 \quad \text{Add equations [3] and [4].}$$
$$y = 12$$

Substitute $y = 12$ into equation [2].

$$5x + 17(12) = 259$$
$$5x + 204 = 259$$
$$5x = 55$$
$$x = 11$$

Thus 11 baseballs and 12 bats were purchased.

2. Let x = the number of small chairs and y = the number of large chairs. (HINT: change all hours to minutes)

$$30x + 40y = 1560 \quad [1]$$
$$75x + 80y = 3420 \quad [2]$$

Multiply equation [1] by -2 and add the results to equation [2].

$$\begin{aligned} -60x - 80y &= -3120 \quad [3] \\ 75x + 80y &= 3420 \quad [2] \\ \hline 15x &= 300 \\ x &= 20 \end{aligned}$$

Substitute $x = 20$ in either equation [1] or [2] and solve for y.

$$30(20) + 40y = 1560$$
$$600 + 40y = 1560$$
$$40y = 960$$
$$y = 24$$

Therefore, the company can make 20 small chairs and 24 large chairs each day.

3. Let a = the speed of the airplane in still air in kilometers per hour and w = the speed of the wind in kilometers per hour.

	R	$\cdot$	T	$=$	D
Against the wind	$a - w$		3		1950
With the wind	$a + w$		2		1600

We obtain a system of equations from the chart.

$$(a - w)3 = 1950$$
$$(a + w)2 = 1600$$

We remove the parentheses.

$$3a - 3w = 1950 \quad [1]$$
$$2a + 2w = 1600 \quad [2]$$

Multiply equation [1] by 2 and equation [2] by 3 and add the resulting equations.

$$\begin{aligned} 6a - 6w &= 3900 \\ 6a + 6w &= 4800 \\ \hline 12a &= 8700 \\ a &= 725 \end{aligned}$$

Substituting $a = 725$ into equation [2] we have

$$2(725) + 2w = 1600$$
$$1450 + 2w = 1600$$
$$2w = 150$$
$$w = 75$$

Thus, the speed of the plane in still air is 725 kilometers per hour and the speed of the wind is 75 kilometers per hour.

4. $\quad A + B + C = 260 \quad [1]$
$\quad 3A + 2B = 390 \quad [2]$
$\quad 3B + 4C = 655 \quad [3]$

Multiply equation [1] by -3 and add it to equation [2].

$$\begin{aligned} -3A - 3B - 3C &= -780 \quad [4] \\ 3A + 2B &= 390 \quad [2] \\ \hline -B - 3C &= -390 \quad [5] \end{aligned}$$

Now multiply equation [5] by 3 and add it to equation [3].

$$\begin{aligned} -3B - 9C &= -1170 \quad [6] \\ 3B + 4C &= 655 \quad [3] \\ \hline -5C &= -515 \\ C &= 103 \end{aligned}$$

Substitute $C = 103$ into equation [3] and solve for B.

$$3B + 4(103) = 655$$
$$3B + 412 = 655$$
$$3B = 243$$
$$B = 81$$

Now substitute $B = 81$ into equation [2] and solve for A.

$$3A + 2(81) = 390$$
$$3A + 162 = 390$$
$$3A = 228$$
$$A = 76$$

Machine A wraps 76 boxes per hour, machine B wraps 81 boxes per hour, and machine C wraps 103 boxes per hour.

4.4 Practice Problems

1.

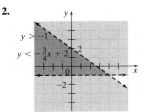

2.

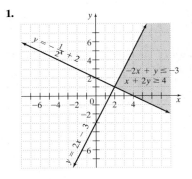

3.

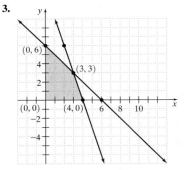

The vertices of the solution are $(3, 3)$, $(0, 6)$, $(4, 0)$, and $(0, 0)$.

Chapter 5

5.1 Practice Problems

1. (a) $a^7 \cdot a^5 = a^{7+5} = a^{12}$ **(b)** $w^{10} \cdot w = w^{10+1} = w^{11}$

2. (a) $x^3 \cdot x^9 = x^{3+9} = x^{12}$ **(b)** $3^7 \cdot 3^4 = 3^{7+4} = 3^{11}$

    (c) $a^3 \cdot b^2 = a^3 \cdot b^2$ (cannot be simplified)

3. (a) $(-a^8)(a^4) = (-1 \cdot 1)(a^8 \cdot a^4)$
$$= -1(a^8 \cdot a^4)$$
$$= -1a^{12} = -a^{12}$$

    (b) $(3y^2)(-2y^3) = (3)(-2)(y^2 \cdot y^3) = -6y^5$

    (c) $(-4x^3)(-5x^2) = (-4)(-5)(x^3 \cdot x^2) = 20x^5$

4. $(2xy)\left(-\dfrac{1}{4}x^2y\right)(6xy^3) = (2)\left(-\dfrac{1}{4}\right)(6)(x \cdot x^2 \cdot x)(y \cdot y \cdot y^3)$

$$= -3x^4y^5$$

5. (a) $\dfrac{10^{13}}{10^7} = 10^{13-7} = 10^6$ **(b)** $\dfrac{x^{11}}{x} = x^{11-1} = x^{10}$

6. (a) $\dfrac{c^3}{c^4} = \dfrac{1}{c^{4-3}} = \dfrac{1}{c}$ **(b)** $\dfrac{10^{31}}{10^{56}} = \dfrac{1}{10^{56-31}} = \dfrac{1}{10^{25}}$

7. (a) $\dfrac{-7x^7}{-21x^9} = \dfrac{1}{3x^{9-7}} = \dfrac{1}{3x^2}$ **(b)** $\dfrac{15x^{11}}{-3x^4} = -5x^{11-4} = -5x^7$

8. (a) $\dfrac{x^7y^9}{y^{10}} = \dfrac{x^7}{y}$ **(b)** $\dfrac{12x^5y^6}{-24x^3y^8} = -\dfrac{x^2}{2y^2}$

9. (a) $\dfrac{10^7}{10^7} = 1$

    (b) $\dfrac{12a^4}{15a^4} = \dfrac{4}{5}\left(\dfrac{a^4}{a^4}\right) = \dfrac{4}{5}(1) = \dfrac{4}{5}$

10. $\dfrac{-20a^3b^8c^4}{28a^3b^7c^5} = -\dfrac{5a^0b}{7c} = -\dfrac{5(1)b}{7c} = -\dfrac{5b}{7c}$

11. $\dfrac{(-6ab^5)(3a^2b^4)}{16a^5b^7} = \dfrac{-18a^3b^9}{16a^5b^7} = -\dfrac{9b^2}{8a^2}$

12. (a) $(a^4)^3 = a^{4 \cdot 3} = a^{12}$

    (b) $(10^5)^2 = 10^{5 \cdot 2} = 10^{10}$ **(c)** $(-1)^{15} = -1$

13. (a) $(3xy)^3 = (3)^3x^3y^3 = 27x^3y^3$ **(b)** $(yz)^{37} = y^{37}z^{37}$

14. $\left(\dfrac{4a}{b}\right)^6 = \dfrac{4^6a^6}{b^6} = \dfrac{4096a^6}{b^6}$

15. $\left(\dfrac{-2x^3y^0z}{4xz^2}\right)^5 = \left(\dfrac{-x^2}{2z}\right)^5 = \dfrac{(-1)^5(x^2)^5}{2^5z^5} = -\dfrac{x^{10}}{32z^5}$

5.2 Practice Problems

1. (a) $x^{-12} = \dfrac{1}{x^{12}}$ **(b)** $w^{-5} = \dfrac{1}{w^5}$ **(c)** $z^{-2} = \dfrac{1}{z^2}$

2. $4^{-3} = \dfrac{1}{4^3} = \dfrac{1}{64}$

3. (a) $\dfrac{3}{w^{-4}} = 3w^4$ **(b)** $\dfrac{x^{-6}y^4}{z^{-2}} = \dfrac{y^4z^2}{x^6}$ **(c)** $x^{-6}y^{-5} = \dfrac{1}{x^6y^5}$

4. (a) $(2x^4y^{-5})^{-2} = 2^{-2}x^{-8}y^{10} = \dfrac{y^{10}}{2^2x^8} = \dfrac{y^{10}}{4x^8}$

    (b) $\dfrac{y^{-3}z^{-4}}{y^2z^{-6}} = \dfrac{z^6}{y^2y^3z^4} = \dfrac{z^6}{y^5z^4} = \dfrac{z^2}{y^5}$

5. (a) $78{,}200 = 7.82 \times 10{,}000 = 7.82 \times 10^4$

    (b) $4{,}786{,}000 = 4.786 \times 1{,}000{,}000 = 4.786 \times 10^6$

6. (a) $0.98 = 9.8 \times 10^{-1}$ **(b)** $0.000092 = 9.2 \times 10^{-5}$

7. (a) $2.96 \times 10^3 = 2.96 \times 1000 = 2960$

    (b) $1.93 \times 10^6 = 1.93 \times 1{,}000{,}000 = 1{,}930{,}000$

    (c) $5.43 \times 10^{-2} = 5.43 \times \dfrac{1}{100} = 0.0543$

    (d) $8.562 \times 10^{-5} = 8.562 \times \dfrac{1}{100{,}000} = 0.00008562$

8. $30{,}900{,}000{,}000{,}000{,}000 = 3.09 \times 10^{16}$ meters

9. (a) $(56{,}000)(1{,}400{,}000{,}000) = (5.6 \times 10^4)(1.4 \times 10^9)$

$$= (5.6)(1.4)(10^4)(10^9)$$

$$= 7.84 \times 10^{13}$$

    (b) $\dfrac{0.000111}{0.00000037} = \dfrac{1.11 \times 10^{-4}}{3.7 \times 10^{-7}} = \dfrac{1.11}{3.7} \times \dfrac{10^{-4}}{10^{-7}}$

$$= \dfrac{1.11}{3.7} \times \dfrac{10^7}{10^4} = 0.3 \times 10^3 = 3.0 \times 10^2$$

---

10. 159 parsecs $= (159 \text{ parsecs}) \dfrac{(3.09 \times 10^{13} \text{ kilometers})}{1 \text{ parsec}}$

$$= 491.31 \times 10^{13} \text{ kilometers}$$

$d = r \times t$

$$491.31 \times 10^{13} \text{ km} = \dfrac{50{,}000 \text{ km}}{1 \text{ hr}} \times t$$

$$4.9131 \times 10^{15} \text{ km} = \dfrac{5 \times 10^4 \text{ km}}{1 \text{ hr}} \times t$$

$$\dfrac{4.9131 \times 10^{15} \text{ km}}{\dfrac{5 \times 10^4 \text{ km}}{1 \text{ hr}}} = t$$

$$\dfrac{4.9131 \times 10^{15} \text{ km } (1 \text{ hr})}{5.0 \times 10^4 \text{ km}} = t$$

$$0.98262 \times 10^{11} \text{ hr} = t$$

$$9.83 \times 10^{10} \text{ hours}$$

5.3 Practice Problems

1. (a) This polynomial is of degree 5. It has two terms, so it is a binomial.

    (b) This polynomial is of degree 7 since the sum of the exponents is $3 + 4 = 7$. It has one term, so it is a monomial.

    (c) This polynomial is of degree 3. It has three terms, so it is a trinomial.

2. $(-8x^3 + 3x^2 + 6) + (2x^3 - 7x^2 - 3)$

$$= [-8x^3 + 2x^3] + [3x^2 - 7x^2] + [6 - 3]$$

$$= [(-8 + 2)x^3] + [(3 - 7)x^2] + [6 - 3]$$

$$= -6x^3 - 4x^2 + 3$$

3. $\left(-\dfrac{1}{3}x^2 - 6x - \dfrac{1}{12}\right) + \left(\dfrac{1}{4}x^2 + 5x - \dfrac{1}{3}\right)$

$$= \left[-\dfrac{1}{3}x^2 + \dfrac{1}{4}x^2\right] + [-6x + 5x] + \left[-\dfrac{1}{12} - \dfrac{1}{3}\right]$$

$$= \left[\left(-\dfrac{1}{3} + \dfrac{1}{4}\right)x^2\right] + [(-6 + 5)x] + \left[-\dfrac{1}{12} - \dfrac{1}{3}\right]$$

$$= \left[\left(-\dfrac{4}{12} + \dfrac{3}{12}\right)x^2\right] + [-x] + \left[-\dfrac{1}{12} - \dfrac{4}{12}\right]$$

$$= -\dfrac{1}{12}x^2 - x - \dfrac{5}{12}$$

4. $(3.5x^3 - 0.02x^2 + 1.56x - 3.5) + (-0.08x^2 - 1.98x + 4)$

$$= 3.5x^3 + [-0.02 - 0.08]x^2 + [1.56 - 1.98]x + [-3.5 + 4]$$

$$= 3.5x^3 - 0.1x^2 - 0.42x + 0.5$$

5. $(5x^3 - 15x^2 + 6x - 3) - (-4x^3 - 10x^2 + 5x + 13)$

$$= (5x^3 - 15x^2 + 6x - 3) + (4x^3 + 10x^2 - 5x - 13)$$

$$= [5 + 4]x^3 + [-15 + 10]x^2 + [6 - 5]x + [-3 - 13]$$

$$= 9x^3 - 5x^2 + x - 16$$

6. $(x^3 - 7x^2y + 3xy^2 - 2y^3) - (2x^3 + 4xy - 6y^3)$

$$= [1 - 2]x^3 - 7x^2y + 3xy^2 - 4xy + [-2 + 6]y^3$$

$$= -x^3 - 7x^2y + 3xy^2 - 4xy + 4y^3$$

7. (a) 1974 is 4 years later than 1970, so $x = 4$.

$$0.03(4) + 5.4 = 0.12 + 5.4$$

$$= 5.52$$

We estimate that the average truck in 1974 obtained 5.52 miles per gallon.

    (b) 2006 is 36 years later than 1970, so $x = 36$.

$$0.03(36) + 5.4 = 1.08 + 5.4$$

$$= 6.48$$

We predict that the average truck in 2006 will obtain 6.48 miles per gallon.

5.4 Practice Problems

1. $4x^3(-2x^2 + 3x) = 4x^3(-2x^2) + 4x^3(3x)$
$$= 4(-2)(x^3)(x^2) + 4(3)(x^3)(x)$$
$$= -8x^5 + 12x^4$$

2. (a) $-3x(x^2 + 2x - 4) = -3x^3 - 6x^2 + 12x$

   (b) $6xy(x^3 + 2x^2y - y^2) = 6x^4y + 12x^3y^2 - 6xy^3$

3. $(-6x^3 + 4x^2 - 2x)(-3xy) = 18x^4y - 12x^3y + 6x^2y$

4. $(5x - 1)(x - 2) = 5x^2 - 10x - x + 2 = 5x^2 - 11x + 2$

5. $(8a - 5b)(3a - b) = 24a^2 - 8ab - 15ab + 5b^2$
$$= 24a^2 - 23ab + 5b^2$$

6. $(3a + 2b)(2a - 3c) = 6a^2 - 9ac + 4ab - 6bc$

7. $(3x - 2y)(3x - 2y) = 9x^2 - 6xy - 6xy + 4y^2$
$$= 9x^2 - 12xy + 4y^2$$

8. $(2x^2 + 3y^2)(5x^2 + 6y^2) = 10x^4 + 12x^2y^2 + 15x^2y^2 + 18y^4$
$$= 10x^4 + 27x^2y^2 + 18y^4$$

9. $A = (\text{length})(\text{width}) = (7x + 3)(2x - 1)$
$$= 14x^2 - 7x + 6x - 3$$
$$= 14x^2 - x - 3$$

There are $(14x^2 - x - 3)$ square feet in the room.

5.5 Practice Problems

1. $(6x + 7)(6x - 7) = (6x)^2 - (7)^2 = 36x^2 - 49$

2. $(3x + 5y)(3x - 5y) = (3x)^2 - (5y)^2 = 9x^2 - 25y^2$

3. (a) $(5x + 4)^2 = (5x)^2 + 2(5x)(4) + (4)^2 = 25x^2 + 40x + 16$

   (b) $(4a - 9b)^2 = (4a)^2 - 2(4a)(9b) + (9b)^2$
$$= 16a^2 - 72ab + 81b^2$$

4.
$$
\begin{array}{r}
3x^2 - 2xy + 4y^2 \\
\underline{x - 2y} \\
-6x^2y + 4xy^2 - 8y^3 \\
\underline{3x^3 - 2x^2y + 4xy^2} \\
3x^3 - 8x^2y + 8xy^2 - 8y^3
\end{array}
$$

5. $(2x^2 - 5x + 3)(x^2 + 3x - 4)$
$$= 2x^2(x^2 + 3x - 4) - 5x(x^2 + 3x - 4) + 3(x^2 + 3x - 4)$$
$$= 2x^4 + 6x^3 - 8x^2 - 5x^3 - 15x^2 + 20x + 3x^2 + 9x - 12$$
$$= 2x^4 + x^3 - 20x^2 + 29x - 12$$

6. $(3x - 2)(2x + 3)(3x + 2) = (3x - 2)(3x + 2)(2x + 3)$
$$= [(3x)^2 - 2^2](2x + 3)$$
$$= (9x^2 - 4)(2x + 3)$$

$$
\begin{array}{r}
9x^2 - 4 \\
\underline{2x + 3} \\
27x^2 + 0x - 12 \\
\underline{18x^3 + 0x^2 - 8x} \\
18x^3 + 27x^2 - 8x - 12
\end{array}
$$

Thus we have
$(3x - 2)(2x + 3)(3x + 2) = 18x^3 + 27x^2 - 8x - 12.$

5.6 Practice Problems

1. $\dfrac{15y^4 - 27y^3 - 21y^2}{3y^2} = \dfrac{15y^4}{3y^2} - \dfrac{27y^3}{3y^2} - \dfrac{21y^2}{3y^2} = 5y^2 - 9y - 7$

2.
$$
\begin{array}{r}
x^2 + 6x + 7 \\
x + 4{\overline{\smash{\big)}\,x^3 + 10x^2 + 31x + 35}} \\
\underline{x^3 + 4x^2} \\
6x^2 + 31x \\
\underline{6x^2 + 24x} \\
7x + 35 \\
\underline{7x + 28} \\
7
\end{array}
$$

Ans.: $x^2 + 6x + 7 + \dfrac{7}{x + 4}$

3.
$$
\begin{array}{r}
2x^2 + x + 1 \\
x - 1{\overline{\smash{\big)}\,2x^3 - x^2 + 0x + 1}} \\
\underline{2x^3 - 2x^2} \\
x^2 + 0x \\
\underline{x^2 - x} \\
x + 1 \\
\underline{x - 1} \\
2
\end{array}
$$

Ans.: $2x^2 + x + 1 + \dfrac{2}{x - 1}$

4.
$$
\begin{array}{r}
5x^2 + x - 2 \\
4x - 3{\overline{\smash{\big)}\,20x^3 - 11x^2 - 11x + 6}} \\
\underline{20x^3 - 15x^2} \\
4x^2 - 11x \\
\underline{4x^2 - 3x} \\
-8x + 6 \\
\underline{-8x + 6} \\
0
\end{array}
$$

Ans.: $5x^2 + x - 2$

Check. $(4x - 3)(5x^2 + x - 2)$
$$= 20x^3 + 4x^2 - 8x - 15x^2 - 3x + 6$$
$$= 20x^3 - 11x^2 - 11x + 6$$

Chapter 6

6.1 Practice Problems

1. (a) $7(3a - b)$ **(b)** $p(1 + rt)$

2. $4a(3a + 4b^2 - 3ab)$

3. (a) $8(2a^3 - 3b^3)$ **(b)** $r^3(s^2 - 4rs + 7r^2)$

4. $9ab^2c(2a^2 - 3bc - 5ac)$

5. $6xy^2(5x^2 - 4x + 1) = 30x^3y^2 - 24x^2y^2 + 6xy^2$

6. $(a + b)(3 + x)$ **7.** $(8y - 1)(9y^2 - 2)$

8. $\pi b^2 - \pi a^2 = \pi(b^2 - a^2)$

6.2 Practice Problems

1. $(2x - 7)(3y - 8)$

2. $3x(2x - 5) + 2(2x - 5)$
$$= (2x - 5)(3x + 2)$$

3. $a(x + 2) + 4b(x + 2)$
$$= (x + 2)(a + 4b)$$

4. $6a^2 + 3ac + 10ab + 5bc$
$$= 3a(2a + c) + 5b(2a + c)$$
$$= (2a + c)(3a + 5b)$$

5. $6xy + 14x - 15y - 35$
$$= 2x(3y + 7) - 5(3y + 7)$$
$$= (3y + 7)(2x - 5)$$

6. $3x + 6y - 5ax - 10ay$
$$= 3(x + 2y) - 5a(x + 2y)$$
$$= (3 - 5a)(x + 2y)$$

7. $10ad + 27bc - 6bd - 45ac$
$$= 10ad - 6bd - 45ac + 27bc$$
$$= 2d(5a - 3b) - 9c(5a - 3b)$$
$$= (2d - 9c)(5a - 3b)$$

Check: $10ad - 6bd - 45ac + 27bc$

6.3 Practice Problems

1. $(x + 6)(x + 2)$ **2.** $(x + 2)(x + 15)$

3. $(x - 9)(x - 2)$ **4.** $(y - 3)(y - 8)$

5. $(a + 3)(a - 8)$

6. $(x + 20)(x - 3)$

Check: $x^2 - 3x + 20x - 60$
$$= x^2 + 17x - 60$$

7. $(x + 5)(x - 12)$ **8.** $(a^2 + 7)(a^2 - 6)$

9. $3x^2 + 45x + 150$
$= 3(x^2 + 15x + 50)$
$= 3(x + 5)(x + 10)$
10. $4x^2 - 8x - 140$
$= 4(x^2 - 2x - 35)$
$= 4(x + 5)(x - 7)$
11. $x(x + 1) - 4(5)$
$= x^2 + x - 20$
$= (x + 5)(x - 4)$

6.4 Practice Problems

1. $(2x - 5)(x - 1)$ **2.** $(9x - 1)(x - 7)$
3. $(3x - 7)(x + 2)$
4. $2x^2 - 2x - 5x + 5$
$= 2x(x - 1) - 5(x - 1)$
$= (2x - 5)(x - 1)$
5. $9x^2 - 63x - x + 7$
$= 9x(x - 7) - 1(x - 7)$
$= (x - 7)(9x - 1)$
6. $3x^2 + 6x - 2x - 4$
$= 3x(x + 2) - 2(x + 2)$
$= (x + 2)(3x - 2)$
7. $8x^2 - 8x - 6$
$= 2(4x^2 - 4x - 3)$
$= 2(2x + 1)(2x - 3)$
8. $24x^2 - 38x + 10$
$= 2(12x^2 - 19x + 5)$
$= 2(12x^2 - 4x - 15x + 5)$
$= 2[4x(3x - 1) - 5(3x - 1)]$
$= 2(4x - 5)(3x - 1)$

6.5 Practice Problems

1. $(1 + 8x)(1 - 8x)$ **2.** $(6x + 7)(6x - 7)$
3. $(10x + 9y)(10x - 9y)$
4. $(x^4 + 1)(x^4 - 1)$
$= (x^4 + 1)(x^2 + 1)(x^2 - 1)$
$= (x^4 + 1)(x^2 + 1)(x + 1)(x - 1)$
5. $(4x + 1)(4x + 1) = (4x + 1)^2$
6. $(5x - 3)(5x - 3) = (5x - 3)^2$
7. (a) $(5x - 6y)(5x - 6y) = (5x - 6y)^2$
   (b) $(8x^3 - 3)(8x^3 - 3) = (8x^3 - 3)^2$
8. $9x^2 - 12x - 3x + 4$
$= 3x(3x - 4) - 1(3x - 4)$
$= (3x - 4)(3x - 1)$
9. $20x^2 - 45$
$= 5(4x^2 - 9)$
$= 5(2x - 3)(2x + 3)$
10. $3(25x^2 - 20x + 4)$
$= 3(5x - 2)(5x - 2)$
$= 3(5x - 2)^2$

6.6 Practice Problems

1. (a) $9y^2(x^4 - 1)$
$= 9y^2(x^2 + 1)(x^2 - 1)$
$= 9y^2(x^2 + 1)(x + 1)(x - 1)$
   (b) $-(4x^2 - 12x + 9)$
$= -(2x - 3)(2x - 3)$
$= -(2x - 3)^2$
   (c) $3(x^2 - 12x + 36)$
$= 3(x - 6)(x - 6)$
$= 3(x - 6)^2$
   (d) $5x(x^2 - 3xy + 2x - 6y)$
$= 5x[x(x - 3y) + 2(x - 3y)]$
$= 5x(x + 2)(x - 3y)$

2. $x^2 - 9x - 8$
The factors of -8 are $\quad (-2)(4) = -8$
$\qquad\qquad\qquad\qquad (2)(-4) = -8$
$\qquad\qquad\qquad\qquad (-8)(1) = -8$
$\qquad\qquad\qquad\qquad (-1)(8) = -8.$
None of these pairs will add up to be the coefficient of the middle term. Thus the polynomial cannot be factored. It is prime.
3. $25x^2 + 82x + 4$
Check to see if this is a perfect square trinomial.
$2[(5)(2)] = 2(10) = 20$
This is not the coefficient of the middle term. The grouping number equals 100. No factors add to 82. It is prime.

6.7 Practice Problems

1. $10x^2 - x - 2 = 0$
$(5x + 2)(2x - 1) = 0 \qquad 5x + 2 = 0 \qquad\quad 2x - 1 = 0$
$\qquad\qquad\qquad\qquad\qquad\quad 5x = -2 \qquad\qquad 2x = 1$
$\qquad\qquad\qquad\qquad\qquad\quad x = -\dfrac{2}{5} \qquad\qquad x = \dfrac{1}{2}$

Check: $\quad 10\left(-\dfrac{2}{5}\right)^2 - \left(-\dfrac{2}{5}\right) - 2 \overset{?}{=} 0 \qquad 10\left(\dfrac{1}{2}\right)^2 - \dfrac{1}{2} - 2 \overset{?}{=} 0$

$\qquad\qquad 10\left(\dfrac{4}{25}\right) + \dfrac{2}{5} - 2 \overset{?}{=} 0 \qquad 10\left(\dfrac{1}{4}\right) - \dfrac{1}{2} - 2 \overset{?}{=} 0$

$\qquad\qquad\qquad \dfrac{8}{5} + \dfrac{2}{5} - 2 \overset{?}{=} 0 \qquad\qquad \dfrac{5}{2} - \dfrac{1}{2} - 2 \overset{?}{=} 0$

$\qquad\qquad\qquad \dfrac{10}{5} - \dfrac{10}{5} \overset{?}{=} 0 \qquad\qquad\quad \dfrac{4}{2} - 2 \overset{?}{=} 0$

$\qquad\qquad\qquad\qquad 0 = 0 \qquad\qquad\qquad 2 - 2 \overset{?}{=} 0$

$\qquad\qquad\qquad\qquad\qquad\qquad\qquad\qquad\quad 0 = 0$

Thus $-\dfrac{2}{5}$ and $\dfrac{1}{2}$ are both roots for the equation.

2. $\quad 3x^2 - 5x + 2 = 0 \qquad 3x - 2 = 0 \qquad x - 1 = 0$
$\quad (3x - 2)(x - 1) = 0 \qquad\quad 3x = 2 \qquad\quad x = 1$
$\qquad\qquad\qquad\qquad\qquad\qquad x = \dfrac{2}{3}$

3. $\quad 7x^2 + 11x = 0 \qquad\quad x = 0 \qquad\quad 7x + 11 = 0$
$\quad x(7x + 11) = 0 \qquad\qquad\qquad\qquad\quad 7x = -11$
$\qquad\qquad\qquad\qquad\qquad\qquad\qquad\qquad x = \dfrac{-11}{7}$

4. $\quad x^2 - 6x + 4 = -8 + x \qquad x - 3 = 0 \qquad x - 4 = 0$
$\qquad\quad x^2 - 7x + 12 = 0 \qquad\qquad x = 3 \qquad\qquad x = 4$
$\quad (x - 3)(x - 4) = 0$

5. $\quad \dfrac{2x^2 - 7x}{3} = 5$
$\qquad\quad 2x^2 - 7x = 15 \qquad 2x + 3 = 0 \qquad x - 5 = 0$
$\quad 2x^2 - 7x - 15 = 0 \qquad\qquad x = -\dfrac{3}{2} \qquad x = 5$
$\quad (2x + 3)(x - 5) = 0$

6. Let w = width, then
$3w + 2$ = length.
$\qquad (3w + 2)w = 85 \qquad 3w + 17 = 0 \qquad w - 5 = 0$
$\qquad\qquad 3w^2 + 2w = 85 \qquad\qquad w = -\dfrac{17}{3} \qquad w = 5$
$\qquad 3w^2 + 2w - 85 = 0$
$\qquad (3w + 17)(w - 5) = 0$
The only valid answer is width = 5 meters.
$\qquad\qquad\qquad\qquad\qquad$ length = $3(5) + 2 = 17$ meters
7. Let b = base, then
$b - 3$ = altitude.
$\qquad \dfrac{b(b - 3)}{2} = 35$
$\qquad\quad b^2 - 3b = 70$
$\quad b^2 - 3b - 70 = 0$
$\quad (b + 7)(b - 10) = 0$
$\qquad b + 7 = 0 \qquad\qquad b - 10 = 0$
$\qquad\quad b = -7 \qquad\qquad\quad b = 10$
This is not a $\qquad$ Thus the base = 10 centimeters.
valid answer. $\qquad$ altitude = $10 - 3 = 7$ centimeters

8.
$$-5t^2 + 45 = 0$$
$$-5(t^2 - 9) = 0$$
$$t^2 - 9 = 0$$
$$(t + 3)(t - 3) = 0$$
$$t = 3 \qquad t = -3 \qquad t = -3 \text{ is not a valid answer.}$$
Thus it will be 3 seconds before he breaks the water's surface.

Chapter 7

7.1 Practice Problems

1. (a) $\dfrac{7}{3} = \dfrac{?}{18}$

$$\dfrac{7 \cdot 6}{3 \cdot 6} = \dfrac{42}{18}$$

(b) $\dfrac{28}{63} = \dfrac{7 \times 2 \times 2}{7 \times 3 \times 3} = \dfrac{4}{9}$

2. $\dfrac{12x - 6}{14x - 7} = \dfrac{6(2x - 1)}{7(2x - 1)} = \dfrac{6}{7}$

3. $\dfrac{4x - 6}{2x^2 - x - 3} = \dfrac{2(2x - 3)}{(2x - 3)(x + 1)} = \dfrac{2}{x + 1}$

4. $\dfrac{x^3 + 11x^2 + 30x}{3x^3 + 17x^2 - 6x} = \dfrac{x(x^2 + 11x + 30)}{x(3x^2 + 17x - 6)} = \dfrac{(x + 5)(x + 6)}{(3x - 1)(x + 6)}$

$$= \dfrac{x + 5}{3x - 1}$$

5. $\dfrac{2x - 5}{5 - 2x} = \dfrac{-1(-2x + 5)}{(5 - 2x)} = \dfrac{-1(5 - 2x)}{(5 - 2x)} = -1$

6. $\dfrac{4x^2 + 3x - 10}{25 - 16x^2} = \dfrac{(4x - 5)(x + 2)}{(5 + 4x)(5 - 4x)} = \dfrac{(4x - 5)(x + 2)}{-1(4x - 5)(5 + 4x)}$

$$= \dfrac{x + 2}{-1(5 + 4x)} = -\dfrac{x + 2}{5 + 4x}$$

7. $\dfrac{4x^2 - 9y^2}{4x^2 + 12xy + 9y^2} = \dfrac{(2x + 3y)(2x - 3y)}{(2x + 3y)(2x + 3y)} = \dfrac{2x - 3y}{2x + 3y}$

8. $\dfrac{12x^3 - 48x}{6x - 3x^2} = \dfrac{12x(x^2 - 4)}{3x(2 - x)} = \dfrac{12x(x + 2)(x - 2)}{3x(2 - x)} = -4(x + 2)$

7.2 Practice Problems

1. $\dfrac{10x}{x^2 - 7x + 10} \cdot \dfrac{x^2 + 3x - 10}{25x}$

$$= \dfrac{10x}{(x - 5)(x - 2)} \cdot \dfrac{(x - 2)(x + 5)}{25x}$$

$$= \dfrac{2(x + 5)}{5(x - 5)}$$

2. $\dfrac{2y^2 - 6y - 8}{y^2 - y - 2} \cdot \dfrac{y^2 - 5y + 6}{2y^2 - 4y - 6}$

$$= \dfrac{2(y + 1)(y - 4)}{(y + 1)(y - 2)} \cdot \dfrac{(y - 2)(y - 3)}{2(y + 1)(y - 3)} = \dfrac{y - 4}{y + 1}$$

3. $\dfrac{x^2 + 5x + 6}{x^2 + 8x} \div \dfrac{2x^2 + 5x + 2}{2x^2 + x}$

$$= \dfrac{(x + 2)(x + 3)}{x(x + 8)} \cdot \dfrac{x(2x + 1)}{(2x + 1)(x + 2)} = \dfrac{x + 3}{x + 8}$$

4. $\dfrac{x + 3}{x - 3} \div (9 - x^2) = \dfrac{x + 3}{x - 3} \cdot \dfrac{1}{(3 + x)(3 - x)}$

$$= \dfrac{1}{(x - 3)(3 - x)}$$

7.3 Practice Problems

1. $\dfrac{2s + t}{2s - t} + \dfrac{s - t}{2s - t} = \dfrac{2s + t + s - t}{2s - t} = \dfrac{3s}{2s - t}$

2. $\dfrac{b}{(a - 2b)(a + b)} - \dfrac{2b}{(a - 2b)(a + b)} = \dfrac{b - 2b}{(a - 2b)(a + b)}$

$$= \dfrac{-b}{(a - 2b)(a + b)}$$

3. $\dfrac{7}{6x + 21}, \quad \dfrac{13}{10x + 35}$

$$6x + 21 = 3(2x + 7)$$
$$10x + 35 = 5(2x + 7)$$
$$\text{LCD} = 3 \cdot 5 \cdot (2x + 7) = 15(2x + 7)$$

4. (a) $\dfrac{3}{50xy^2z}, \quad \dfrac{19}{40x^3yz}$

$$50xy^2z = 2 \cdot 5^2 \cdot x \cdot y^2 \cdot z$$
$$40x^3yz = 2^3 \cdot 5 \cdot x^3 \cdot y \cdot z$$
$$\text{LCD} = 2^3 \cdot 5^2 \cdot x^3 \cdot y^2 \cdot z$$
$$\text{LCD} = 200x^3y^2z$$

(b) $\dfrac{2}{x^2 + 5x + 6}, \quad \dfrac{6}{3x^2 + 5x - 2}$

$$x^2 + 5x + 6 = (x + 3)(x + 2)$$
$$3x^2 + 5x - 2 = (3x - 1)(x + 2)$$
$$\text{LCD} = (x + 2)(x + 3)(3x - 1)$$

5. $\dfrac{7}{a} + \dfrac{3}{abc} = \dfrac{7bc + 3}{abc} \qquad \text{LCD} = abc$

6. $\dfrac{2a - b}{(a + 2b)(a - 2b)} + \dfrac{2}{(a + 2b)} \qquad \text{LCD} = (a + 2b)(a - 2b)$

$$= \dfrac{2a - b}{(a + 2b)(a - 2b)} + \dfrac{2(a - 2b)}{(a + 2b)(a - 2b)}$$

$$= \dfrac{2a - b + 2a - 4b}{(a + 2b)(a - 2b)} = \dfrac{4a - 5b}{(a + 2b)(a - 2b)}$$

7. $\dfrac{7a}{a^2 + 2ab + b^2} + \dfrac{4}{a^2 + ab}$

$$= \dfrac{7a}{(a + b)(a + b)} + \dfrac{4}{a(a + b)} \qquad \text{LCD} = a(a + b)^2$$

$$= \dfrac{7a^2}{a(a + b)(a + b)} + \dfrac{4(a + b)}{a(a + b)(a + b)} = \dfrac{7a^2 + 4a + 4b}{a(a + b)^2}$$

8. $\dfrac{x + 7}{3x - 9} - \dfrac{x - 6}{x - 3} = \dfrac{x + 7}{3(x - 3)} - \dfrac{x - 6}{x - 3} \qquad \text{LCD} = 3(x - 3)$

$$= \dfrac{x + 7 - 3(x - 6)}{3(x - 3)} = \dfrac{x + 7 - 3x + 18}{3(x - 3)} = \dfrac{-2x + 25}{3(x - 3)}$$

9. $\dfrac{x - 2}{x^2 - 4} - \dfrac{x + 1}{2x^2 + 4x} = \dfrac{x - 2}{(x + 2)(x - 2)} - \dfrac{x + 1}{2x(x + 2)}$

$$\text{LCD} = 2x(x + 2)(x - 2)$$

$$= \dfrac{2x(x - 2)}{2x(x + 2)(x - 2)} - \dfrac{(x - 2)(x + 1)}{2x(x + 2)(x - 2)}$$

$$= \dfrac{2x^2 - 4x - (x^2 - x - 2)}{2x(x + 2)(x - 2)} = \dfrac{2x^2 - 4x - x^2 + x + 2}{2x(x + 2)(x - 2)}$$

$$= \dfrac{x^2 - 3x + 2}{2x(x + 2)(x - 2)} = \dfrac{(x - 2)(x - 1)}{2x(x + 2)(x - 2)} = \dfrac{x - 1}{2x(x + 2)}$$

7.4 Practice Problems

1. $\dfrac{\dfrac{1}{a} + \dfrac{1}{b}}{\dfrac{2}{ab^2}} = \dfrac{\dfrac{b + a}{ab}}{\dfrac{2}{ab^2}}$

$$= \dfrac{b + a}{ab} \div \dfrac{2}{ab^2} = \dfrac{b + a}{ab} \cdot \dfrac{ab^2}{2} = \dfrac{ab^2(a + b)}{2\,ab} = \dfrac{b(a + b)}{2}$$

2. $\dfrac{\dfrac{1}{a}+\dfrac{1}{b}}{\dfrac{1}{a}-\dfrac{1}{b}} = \dfrac{\dfrac{b+a}{ab}}{\dfrac{b-a}{ab}} = \dfrac{b+a}{ab} \cdot \dfrac{ab}{b-a} = \dfrac{b+a}{b-a}$

3. $\dfrac{\dfrac{x}{x^2+4x+3}+\dfrac{2}{x+1}}{x+1} = \dfrac{\dfrac{x}{(x+1)(x+3)}+\dfrac{2}{x+1}}{x+1}$

$= \dfrac{\dfrac{x+2(x+3)}{(x+1)(x+3)}}{x+1} = \dfrac{\dfrac{x+2x+6}{(x+1)(x+3)}}{(x+1)}$

$= \dfrac{3x+6}{(x+1)(x+3)} \cdot \dfrac{1}{(x+1)}$

$= \dfrac{3(x+2)}{(x+1)^2(x+3)}$

4. $\dfrac{\dfrac{6}{x^2-y^2}}{\dfrac{1}{x-y}+\dfrac{3}{x+y}} = \dfrac{\dfrac{6}{(x+y)(x-y)}}{\dfrac{(x+y)+3(x-y)}{(x+y)(x-y)}} = \dfrac{6}{(x+y)(x-y)}$

$\div \dfrac{x+y+3x-3y}{(x+y)(x-y)} = \dfrac{6}{(x+y)(x-y)} \cdot \dfrac{(x+y)(x-y)}{4x-2y}$

$= \dfrac{6}{\cancel{(x+y)(x-y)}} \cdot \dfrac{\cancel{(x+y)(x-y)}}{2(2x-y)} = \dfrac{3}{2x-y}$

5. $\dfrac{\dfrac{2}{3x^2}-\dfrac{3}{y}}{\dfrac{5}{xy}-4}$ LCD $= 3x^2y$

$\dfrac{3x^2y\left(\dfrac{2}{3x^2}\right)-3x^2y\left(\dfrac{3}{y}\right)}{3x^2y\left(\dfrac{5}{xy}\right)-3x^2y(4)} = \dfrac{2y-3x^2(3)}{3x(5)-12x^2y} = \dfrac{2y-9x^2}{15x-12x^2y}$

6. $\dfrac{\dfrac{6}{x^2-y^2}}{\dfrac{7}{x-y}+\dfrac{3}{x+y}} = \dfrac{\dfrac{6}{(x+y)(x-y)}}{\dfrac{7}{x-y}+\dfrac{3}{x+y}}$

LCD $= (x+y)(x-y)$

$\dfrac{(x+y)(x-y)\left(\dfrac{6}{(x+y)(x-y)}\right)}{(x+y)(x-y)\left(\dfrac{7}{x-y}\right)+(x+y)(x-y)\left(\dfrac{3}{x+y}\right)}$

$= \dfrac{6}{7(x+y)+3(x-y)} = \dfrac{6}{7x+7y+3x-3y} = \dfrac{6}{10x+4y}$

$= \dfrac{3}{5x+2y}$

7.5 Practice Problems

1. $\dfrac{2}{x}+\dfrac{4}{x} = 3-\dfrac{1}{x}-\dfrac{17}{8}$

LCD $= 8x$

$16+32 = 24x-8-17x$ Check. $\dfrac{2}{8}+\dfrac{4}{8} \overset{?}{=} 3-\dfrac{1}{8}-\dfrac{17}{8}$

$48 = 7x-8$ $\dfrac{6}{8} \overset{?}{=} 3-\dfrac{18}{8}$

$56 = 7x$ $\dfrac{6}{8} \overset{?}{=} \dfrac{24}{8}-\dfrac{18}{8}$

$x = 8$ $\dfrac{6}{8} = \dfrac{6}{8}$ ✓

2. $\dfrac{4}{2x+1} = \dfrac{6}{2x-1}$ LCD $= (2x+1)(2x-1)$

$(2x+1)(2x-1)\left[\dfrac{4}{2x+1}\right] = (2x+1)(2x-1)\left[\dfrac{6}{2x-1}\right]$

$4(2x-1) = 6(2x+1)$

$8x-4 = 12x+6$

$-4x = 10$

$x = -\dfrac{5}{2}$

Check. $\dfrac{4}{2\left(-\dfrac{5}{2}\right)+1} \overset{?}{=} \dfrac{6}{2\left(-\dfrac{5}{2}\right)-1}$

$\dfrac{4}{-5+1} \overset{?}{=} \dfrac{6}{-5-1}$

$\dfrac{4}{-4} \overset{?}{=} \dfrac{6}{-6}$

$-1 = -1$ ✓

3. $\dfrac{x-1}{x^2-4} = \dfrac{2}{x+2}+\dfrac{4}{x-2}$

$\dfrac{x-1}{(x+2)(x-2)} = \dfrac{2}{x+2}+\dfrac{4}{x-2}$ LCD $= (x+2)(x-2)$

$(x+2)(x-2)\left[\dfrac{x-1}{(x+2)(x-2)}\right]$

$= (x+2)(x-2)\left[\dfrac{2}{x+2}\right]+(x+2)(x-2)\left[\dfrac{4}{x-2}\right]$

$x-1 = 2(x-2)+4(x+2)$

$x-1 = 2x-4+4x+8$

$x-1 = 6x+4$

$-5x = 5$

$x = -1$

Check. $\dfrac{-1-1}{(-1)^2-4} \overset{?}{=} \dfrac{2}{-1+2}+\dfrac{4}{-1-2}$

$\dfrac{-2}{-3} \overset{?}{=} \dfrac{2}{1}+\dfrac{4}{-3}$

$\dfrac{2}{3} \overset{?}{=} \dfrac{6}{3}-\dfrac{4}{3}$

$\dfrac{2}{3} = \dfrac{2}{3}$ ✓

4. $\dfrac{2x}{x+1} = \dfrac{-2}{x+1}+1$ LCD $= (x+1)$

$(x+1)\left[\dfrac{2x}{x+1}\right] = (x+1)\left[\dfrac{-2}{x+1}\right]+(x+1)[1]$

$2x = -2+x+1$

$2x = x-1$

$x = -1$

Check. $\dfrac{2(-1)}{-1+1} \overset{?}{=} \dfrac{-2}{-1+1}+1$

$\dfrac{-2}{0} \overset{?}{=} \dfrac{-2}{0}+1$

These expressions are not defined; therefore, there is no solution to this problem.

7.6 Practice Problems

1. $\dfrac{8}{420} = \dfrac{x}{315}$

$8(315) = 420x$

$2520 = 420x$

$x = 6$

It would take Brenda 6 hours to drive 315 miles.

2. $\dfrac{\frac{5}{8}}{30} = \dfrac{2\frac{1}{2}}{x}$

$\dfrac{5}{8}x = 30\left(2\dfrac{1}{2}\right)$

$\dfrac{5}{8}x = 75$

$x = 120$

Therefore $2\frac{1}{2}$ inches would represent 120 miles.

3. $\dfrac{13}{x} = \dfrac{16}{18}$

$13(18) = 16x$

$234 = 16x$

$x = 14\dfrac{5}{8}$ cm

4. $\dfrac{6}{7} = \dfrac{x}{38.5}$

$6(38.5) = 7x$

$231 = 7x$

$x = 33$ feet

5. Train A time $= \dfrac{180}{x + 10}$ Train B time $= \dfrac{150}{x}$

$\dfrac{180}{x + 10} = \dfrac{150}{x}$

$180x = 150(x + 10)$

$180x = 150x + 1500$

$30x = 1500$

$x = 50$

Train B travels 50 kilometers per hour. Train A travels $50 + 10 = 60$ kilometers per hour.

6.

	Number of Hours	Part of the Job Done in One Hour
John	6 hours	$\dfrac{1}{6}$
Dave	7 hours	$\dfrac{1}{7}$
John & Dave Together	x	$\dfrac{1}{x}$

$\dfrac{1}{6} + \dfrac{1}{7} = \dfrac{1}{x}$ LCD $= 42x$

$7x + 6x = 42$

$13x = 42$

$x = 3\dfrac{3}{13}$ $\dfrac{3}{13}$ hour $\times \dfrac{60 \text{ min}}{1 \text{ hour}} = \dfrac{180}{13}$ min ≈ 13.846 min

Thus, doing the job together will take about 3 hours and 14 minutes.

Chapter 8

8.1 Practice Problems

1. $\left(\dfrac{3x^{-2}y^4}{2x^{-5}y^2}\right)^{-3} = \dfrac{(3x^{-2}y^4)^{-3}}{(2x^{-5}y^2)^{-3}}$

$= \dfrac{3^{-3}(x^{-2})^{-3}(y^4)^{-3}}{2^{-3}(x^{-5})^{-3}(y^2)^{-3}}$

$= \dfrac{3^{-3}x^6y^{-12}}{2^{-3}x^{15}y^{-6}}$

$= \dfrac{3^{-3}}{2^{-3}} \cdot \dfrac{x^6}{x^{15}} \cdot \dfrac{y^{-12}}{y^{-6}}$

$= \dfrac{2^3}{3^3} \cdot x^{6-15} \cdot y^{-12+6}$

$= \dfrac{8}{27}x^{-9}y^{-6}$ or $\dfrac{8}{27x^9y^6}$

2. (a) $\left(x^4\right)^{3/8} = x^{(4/1)(3/8)} = x^{3/2}$

(b) $\dfrac{x^{3/7}}{x^{2/7}} = x^{3/7-2/7} = x^{1/7}$

(c) $x^{-7/5} \cdot x^{4/5} = x^{-7/5+4/5} = x^{-3/5}$

3. (a) $\left(-3x^{1/4}\right)\left(2x^{1/2}\right) = -6x^{1/4+1/2} = -6x^{1/4+2/4} = -6x^{3/4}$

(b) $\dfrac{13x^{1/12}y^{-1/4}}{26x^{-1/3}y^{1/2}} = \dfrac{x^{1/12-(-1/3)}y^{-1/4-1/2}}{2}$

$= \dfrac{x^{1/12+4/12}y^{-1/4-2/4}}{2}$

$= \dfrac{x^{5/12}y^{-3/4}}{2}$

$= \dfrac{x^{5/12}}{2y^{3/4}}$

4. $-3x^{1/2}\left(2x^{1/4} + 3x^{-1/2}\right) = -6x^{1/2+1/4} - 9x^{1/2-1/2}$

$= -6x^{2/4+1/4} - 9x^0$

$= -6x^{3/4} - 9$

5. (a) $(4)^{5/2} = \left(2^2\right)^{5/2} = 2^{2/1 \cdot 5/2} = 2^5 = 32$

(b) $(27)^{4/3} = \left(3^3\right)^{4/3} = 3^{3/1 \cdot 4/3} = 3^4 = 81$

6. $3x^{1/3} + x^{-1/3} = 3x^{1/3} + \dfrac{1}{x^{1/3}}$

$= \dfrac{x^{1/3}}{x^{1/3}}\left(3x^{1/3}\right) + \dfrac{1}{x^{1/3}}$

$= \dfrac{3x^{2/3} + 1}{x^{1/3}}$

7. $4y^{3/2} - 8y^{5/2} = 4y^{2/2+1/2} - 8y^{2/2+3/2}$

$= 4\left(y^{2/2}\right)\left(y^{1/2}\right) - 8\left(y^{2/2}\right)\left(y^{3/2}\right)$

$= 4y\left(y^{1/2} - 2y^{3/2}\right)$

8.2 Practice Problems

1. (a) $\sqrt[3]{216} = \sqrt[3]{(6)^3} = 6$

(b) $\sqrt[5]{32} = \sqrt[5]{(2)^5} = 2$

(c) $\sqrt[3]{-8} = \sqrt[3]{(-2)^3} = -2$

(d) $\sqrt[4]{-81}$ is not a real number.

2. (a) $f(3) = \sqrt{4(3) - 3} = \sqrt{12 - 3} = \sqrt{9} = 3$

(b) $f(4) = \sqrt{4(4) - 3} = \sqrt{16 - 3} = \sqrt{13} \approx 3.6$

(c) $f(7) = \sqrt{4(7) - 3} = \sqrt{28 - 3} = \sqrt{25} = 5$

3. $0.5x + 2 \geq 0$

$$0.5x \geq -2$$

$$x \geq -4$$

The domain is all real numbers x where $x \geq -4$.

4.

x	$f(x)$
3	0
4	1.7
5	2.4
6	3
15	6

$f(x) = \sqrt{3x - 9}$

5. (a) $\sqrt[3]{x^3} = (x^3)^{1/3} = x^{3/3} = x^1 = x$

(b) $\sqrt[4]{y^4} = (y^4)^{1/4} = y^{4/4} = y^1 = y$

6. (a) $\sqrt[4]{x^3} = (x^3)^{1/4} = x^{3/4}$

(b) $\sqrt[5]{(xy)^7} = [(xy)^7]^{1/5} = (xy)^{7/5}$

7. (a) $\sqrt[4]{81x^{12}} = (3^4 x^{12})^{1/4} = 3x^3$

(b) $\sqrt[3]{27x^6} = [(3)^3 x^6]^{1/3} = 3x^2$

(c) $(32x^5)^{3/5} = (2^5 x^5)^{3/5} = 2^3 x^3 = 8x^3$

8. (a) $x^{3/4} = \sqrt[4]{x^3}$

(b) $y^{-1/3} = \dfrac{1}{y^{1/3}} = \dfrac{1}{\sqrt[3]{y}}$

(c) $(2x)^{4/5} = \sqrt[5]{(2x)^4} = \sqrt[5]{16x^4}$

(d) $2x^{4/5} = 2\sqrt[5]{x^4}$

9. (a) $8^{2/3} = (\sqrt[3]{8})^2 = 2^2 = 4$

(b) $(-8)^{4/3} = (\sqrt[3]{-8})^4 = (-2)^4 = 16$

(c) $100^{-3/2} = \dfrac{1}{100^{3/2}} = \dfrac{1}{(\sqrt{100})^3} = \dfrac{1}{10^3} = \dfrac{1}{1000}$

10. (a) $\sqrt[5]{(-3)^5} = -3$

(b) $\sqrt[4]{(-5)^4} = |-5| = 5$

(c) $\sqrt[4]{w^4} = |w|$

(d) $\sqrt[7]{y^7} = y$

11. (a) $\sqrt{36x^2} = 6|x|$

(b) $\sqrt[3]{125x^3 y^6} = \sqrt[3]{(5)^3 (x)^3 (y^2)^3} = 5xy^2$

(c) $\sqrt[4]{16y^8} = 2|y^2| = 2y^2$

8.3 Practice Problems

1. $\sqrt{20} = \sqrt{4 \cdot 5} = \sqrt{4} \cdot \sqrt{5} = 2\sqrt{5}$

2. $\sqrt{27} = \sqrt{9 \cdot 3} = \sqrt{9} \cdot \sqrt{3} = 3\sqrt{3}$

3. (a) $\sqrt[3]{24} = \sqrt[3]{8} \cdot \sqrt[3]{3} = 2\sqrt[3]{3}$

(b) $\sqrt[3]{-108} = \sqrt[3]{-27} \cdot \sqrt[3]{4} = -3\sqrt[3]{4}$

4. $\sqrt[4]{64} = \sqrt[4]{16} \cdot \sqrt[4]{4} = 2\sqrt[4]{4}$

5. (a) $\sqrt{45x^6 y^7} = \sqrt{9 \cdot 5 \cdot x^6 \cdot y^6 \cdot y} = \sqrt{9x^6 y^6} \cdot \sqrt{5y}$

$$= 3x^3 y^3 \sqrt{5y}$$

(b) $\sqrt{27a^7 b^8 c^9} = \sqrt{9 \cdot 3a^6 \cdot a \cdot b^8 \cdot c^8 \cdot c}$

$$= \sqrt{9a^6 b^8 c^8} \cdot \sqrt{3ac}$$

$$= 3a^3 b^4 c^4 \sqrt{3ac}$$

6. $19\sqrt{xy} + 5\sqrt{xy} - 10\sqrt{xy} = (19 + 5 - 10)\sqrt{xy} = 14\sqrt{xy}$

7. $4\sqrt{2} - 5\sqrt{50} - 3\sqrt{98}$

$$= 4\sqrt{2} - 5\sqrt{25} \cdot \sqrt{2} - 3\sqrt{49} \cdot \sqrt{2}$$

$$= 4\sqrt{2} - 5(5)\sqrt{2} - 3(7)\sqrt{2}$$

$$= 4\sqrt{2} - 25\sqrt{2} - 21\sqrt{2}$$

$$= (4 - 25 - 21)\sqrt{2}$$

$$= -42\sqrt{2}$$

8. $4\sqrt{2x} + \sqrt{18x} - 2\sqrt{125x} - 6\sqrt{20x}$

$$= 4\sqrt{2x} + \sqrt{9} \cdot \sqrt{2x} - 2\sqrt{25} \cdot \sqrt{5x} - 6\sqrt{4} \cdot \sqrt{5x}$$

$$= 4\sqrt{2x} + 3\sqrt{2x} - 2(5)\sqrt{5x} - 6(2)\sqrt{5x}$$

$$= 4\sqrt{2x} + 3\sqrt{2x} - 10\sqrt{5x} - 12\sqrt{5x}$$

$$= 7\sqrt{2x} - 22\sqrt{5x}$$

9. $3x\sqrt[3]{54x^4} - 3\sqrt[3]{16x^7}$

$$= 3x\sqrt[3]{27x^3} \cdot \sqrt[3]{2x} - 3\sqrt[3]{8x^6} \cdot \sqrt[3]{2x}$$

$$= 3x(3x)\sqrt[3]{2x} - 3(2x^2)\sqrt[3]{2x}$$

$$= 9x^2 \sqrt[3]{2x} - 6x^2 \sqrt[3]{2x}$$

$$= 3x^2 \sqrt[3]{2x}$$

8.4 Practice Problems

1. $(-4\sqrt{2})(-3\sqrt{13x}) = (-4)(-3)\sqrt{2 \cdot 13x} = 12\sqrt{26x}$

2. $\sqrt{2x}(\sqrt{5} + 2\sqrt{3x} + \sqrt{8})$

$$= (\sqrt{2x})(\sqrt{5}) + (\sqrt{2x})(2\sqrt{3x}) + (\sqrt{2x})(\sqrt{8})$$

$$= \sqrt{10x} + 2\sqrt{6x^2} + \sqrt{16x}$$

$$= \sqrt{10x} + 2\sqrt{x^2}\sqrt{6} + \sqrt{16}\sqrt{x}$$

$$= \sqrt{10x} + 2x\sqrt{6} + 4\sqrt{x}$$

3. $(\sqrt{7} + 4\sqrt{2})(2\sqrt{7} - 3\sqrt{2})$

$$= 2\sqrt{49} - 3\sqrt{14} + 8\sqrt{14} - 12\sqrt{4}$$

$$= 2(7) + 5\sqrt{14} - 12(2)$$

$$= 14 + 5\sqrt{14} - 24$$

$$= -10 + 5\sqrt{14}$$

4. $(2 - 5\sqrt{5})(3 - 2\sqrt{2}) = 6 - 4\sqrt{2} - 15\sqrt{5} + 10\sqrt{10}$

5. $(\sqrt{5x} + \sqrt{10})^2 = (\sqrt{5x} + \sqrt{10})(\sqrt{5x} + \sqrt{10})$

$$= \sqrt{25x^2} + \sqrt{50x} + \sqrt{50x} + \sqrt{100}$$

$$= 5x + 2\sqrt{25}\sqrt{2x} + 10$$

$$= 5x + 2(5)\sqrt{2x} + 10$$

$$= 5x + 10\sqrt{2x} + 10$$

6. (a) $\sqrt[3]{2x}(\sqrt[3]{4x^2} + 3\sqrt[3]{y})$

$$= (\sqrt[3]{2x})(\sqrt[3]{4x^2}) + (\sqrt[3]{2x})(3\sqrt[3]{y})$$

$$= \sqrt[3]{8x^3} + 3\sqrt[3]{2xy}$$

$$= 2x + 3\sqrt[3]{2xy}$$

(b) $(\sqrt[3]{7} + \sqrt[3]{x^2})(2\sqrt[3]{49} - \sqrt[3]{x})$

$$= 2\sqrt[3]{343} - \sqrt[3]{7x} + 2\sqrt[3]{49x^2} - \sqrt[3]{x^3}$$

$$= 2\sqrt[3]{7^3} - \sqrt[3]{7x} + 2\sqrt[3]{49x^2} - x$$

$$= 2(7) - \sqrt[3]{7x} + 2\sqrt[3]{49x^2} - x$$

$$= 14 - \sqrt[3]{7x} + 2\sqrt[3]{49x^2} - x$$

7. (a) $\dfrac{\sqrt{75}}{\sqrt{3}} = \sqrt{\dfrac{75}{3}} = \sqrt{25} = 5$

(b) $\sqrt[3]{\dfrac{27}{64}} = \dfrac{\sqrt[3]{27}}{\sqrt[3]{64}} = \dfrac{3}{4}$

(c) $\dfrac{\sqrt{54a^3b^7}}{\sqrt{6b^5}} = \sqrt{\dfrac{54a^3b^7}{6b^5}} = \sqrt{9a^3b^2} = 3ab\sqrt{a}$

8. $\dfrac{7}{\sqrt{3}} = \dfrac{7}{\sqrt{3}} \cdot \dfrac{\sqrt{3}}{\sqrt{3}} = \dfrac{7\sqrt{3}}{\sqrt{9}} = \dfrac{7\sqrt{3}}{3}$

9. $\dfrac{8}{\sqrt{20x}} = \dfrac{8}{\sqrt{4}\sqrt{5x}} = \dfrac{8}{2\sqrt{5x}} \cdot \dfrac{\sqrt{5x}}{\sqrt{5x}} = \dfrac{8\sqrt{5x}}{10x} = \dfrac{4\sqrt{5x}}{5x}$

10. $\sqrt[3]{\dfrac{6}{5x}} = \dfrac{\sqrt[3]{6}}{\sqrt[3]{5x}} = \dfrac{\sqrt[3]{6}}{\sqrt[3]{5x}} \cdot \dfrac{\sqrt[3]{25x^2}}{\sqrt[3]{25x^2}} = \dfrac{\sqrt[3]{150x^2}}{\sqrt[3]{125x^3}} = \dfrac{\sqrt[3]{150x^2}}{5x}$

11. $\dfrac{4}{2 + \sqrt{5}} = \dfrac{4}{2 + \sqrt{5}} \cdot \dfrac{2 - \sqrt{5}}{2 - \sqrt{5}}$

$$= \dfrac{4(2 - \sqrt{5})}{2^2 - (\sqrt{5})^2}$$

$$= \dfrac{4(2 - \sqrt{5})}{4 - 5}$$

$$= \dfrac{4(2 - \sqrt{5})}{-1}$$

$$= -(8 - 4\sqrt{5})$$

$$= -8 + 4\sqrt{5}$$

12. $\dfrac{\sqrt{11} + \sqrt{2}}{\sqrt{11} - \sqrt{2}} \cdot \dfrac{\sqrt{11} + \sqrt{2}}{\sqrt{11} + \sqrt{2}}$

$$= \dfrac{\sqrt{121} + \sqrt{22} + \sqrt{22} + \sqrt{4}}{(\sqrt{11})^2 - (\sqrt{2})^2}$$

$$= \dfrac{11 + 2\sqrt{22} + 2}{11 - 2} = \dfrac{13 + 2\sqrt{22}}{9}$$

8.5 Practice Problems

1. $\sqrt{3x - 8} = x - 2$

$$(\sqrt{3x - 8})^2 = (x - 2)^2$$

$$3x - 8 = x^2 - 4x + 4$$

$$0 = x^2 - 7x + 12$$

$$0 = (x - 3)(x - 4)$$

$$x - 3 = 0 \quad \text{or} \quad x - 4 = 0$$

$$x = 3 \qquad\qquad x = 4$$

Check:

For $x = 3$: $\sqrt{3(3) - 8} \overset{?}{=} 3 - 2$

$$\sqrt{1} \overset{?}{=} 1$$

$$1 = 1 \checkmark$$

For $x = 4$: $\sqrt{3(4) - 8} \overset{?}{=} 4 - 2$

$$\sqrt{4} \overset{?}{=} 2$$

$$2 = 2 \checkmark$$

The solutions are 3 and 4.

2. $\sqrt{x + 4} = x + 4$

$$(\sqrt{x + 4})^2 = (x + 4)^2$$

$$x + 4 = x^2 + 8x + 16$$

$$0 = x^2 + 7x + 12$$

$$0 = (x + 3)(x + 4)$$

$$x + 3 = 0 \quad \text{or} \quad x + 4 = 0$$

$$x = -3 \qquad\qquad x = -4$$

Check:

For $x = -3$: $\sqrt{-3 + 4} \overset{?}{=} -3 + 4$

$$\sqrt{1} \overset{?}{=} 1$$

$$1 = 1 \checkmark$$

For $x = -4$: $\sqrt{-4 + 4} \overset{?}{=} -4 + 4$

$$\sqrt{0} \overset{?}{=} 0$$

$$0 = 0 \checkmark$$

The solutions are -4 and -3.

3. $\sqrt{2x + 5} - 2\sqrt{2x} = 1$

$$\sqrt{2x + 5} = 2\sqrt{2x} + 1$$

$$(\sqrt{2x + 5})^2 = (2\sqrt{2x} + 1)^2$$

$$2x + 5 = (2\sqrt{2x} + 1)(2\sqrt{2x} + 1)$$

$$2x + 5 = 8x + 4\sqrt{2x} + 1$$

$$-6x + 4 = 4\sqrt{2x}$$

$$-3x + 2 = 2\sqrt{2x}$$

$$(-3x + 2)^2 = (2\sqrt{2x})^2$$

$$9x^2 - 12x + 4 = 8x$$

$$9x^2 - 20x + 4 = 0$$

$$(9x - 2)(x - 2) = 0$$

$$9x - 2 = 0 \quad \text{or} \quad x - 2 = 0$$

$$x = \dfrac{2}{9} \qquad\qquad x = 2$$

Check: For $x = \dfrac{2}{9}$:

$$\sqrt{2\left(\dfrac{2}{9}\right) + 5} - 2\sqrt{2\left(\dfrac{2}{9}\right)} \overset{?}{=} 1$$

$$\sqrt{\dfrac{4}{9} + 5} - 2\sqrt{\dfrac{4}{9}} \overset{?}{=} 1$$

$$\sqrt{\dfrac{49}{9}} - 2\sqrt{\dfrac{4}{9}} \overset{?}{=} 1$$

$$\dfrac{7}{3} - \dfrac{4}{3} \overset{?}{=} 1$$

$$\dfrac{3}{3} \overset{?}{=} 1$$

$$1 = 1 \ \checkmark$$

For $x = 2$:

$$\sqrt{2(2) + 5} - 2\sqrt{2(2)} \overset{?}{=} 1$$

$$\sqrt{9} - 2\sqrt{4} \overset{?}{=} 1$$

$$3 - 4 \overset{?}{=} 1$$

$$-1 \neq 1$$

The only solution is $\dfrac{2}{9}$.

4.
$$\sqrt{y - 1} + \sqrt{y - 4} = \sqrt{4y - 11}$$
$$\left(\sqrt{y - 1} + \sqrt{y - 4}\right)^2 = \left(\sqrt{4y - 11}\right)^2$$
$$\left(\sqrt{y - 1} + \sqrt{y - 4}\right)\left(\sqrt{y - 1} + \sqrt{y - 4}\right) = 4y - 11$$
$$y - 1 + 2\left(\sqrt{y - 1}\right)\left(\sqrt{y - 4}\right) + y - 4 = 4y - 11$$
$$2y - 5 + 2\left(\sqrt{y - 1}\right)\left(\sqrt{y - 4}\right) = 4y - 11$$
$$2\left(\sqrt{y - 1}\right)\left(\sqrt{y - 4}\right) = 2y - 6$$
$$\left(\sqrt{y - 1}\right)\left(\sqrt{y - 4}\right) = y - 3$$
$$\left(\sqrt{y^2 - 5y + 4}\right)^2 = (y - 3)^2$$
$$y^2 - 5y + 4 = y^2 - 6y + 9$$
$$y - 5 = 0$$
$$y = 5$$

Check: $\sqrt{5 - 1} + \sqrt{5 - 4} \overset{?}{=} \sqrt{4(5) - 11}$

$$2 + 1 \overset{?}{=} 3$$

$$3 = 3 \ \checkmark$$

The solution is 5.

8.6 Practice Problems

1. (a) $\sqrt{-49} = \sqrt{-1}\,\sqrt{49} = (i)(7) = 7i$

(b) $\sqrt{-31} = \sqrt{-1}\,\sqrt{31} = i\sqrt{31}$

2. $\sqrt{-98} = \sqrt{-1}\,\sqrt{98} = i\sqrt{98} = i\sqrt{49}\,\sqrt{2} = 7i\sqrt{2}$

3. $\sqrt{-8} \cdot \sqrt{-2} = \sqrt{-1}\,\sqrt{8} \cdot \sqrt{-1}\,\sqrt{2}$

$$= i\sqrt{8} \cdot i\sqrt{2}$$

$$= i^2\sqrt{16}$$

$$= -1(4) = -4$$

4. $-7 + 2yi\sqrt{3} = x + 6i\sqrt{3}$

$$x = -7 \qquad 2y\sqrt{3} = 6\sqrt{3}$$

$$y = 3$$

5. $(3 - 4i) - (-2 - 18i)$

$$= [3 - (-2)] + [-4 - (-18)]i$$

$$= (3 + 2) + (-4 + 18)i$$

$$= 5 + 14i$$

6. $(4 - 2i)(3 - 7i)$

$$= (4)(3) + (4)(-7i) + (-2i)(3) + (-2i)(-7i)$$

$$= 12 - 28i - 6i + 14i^2$$

$$= 12 - 28i - 6i + 14(-1)$$

$$= 12 - 28i - 6i - 14 = -2 - 34i$$

7. $-2i(5 + 6i)$

$$= (-2)(5)i + (-2)(6)i^2$$

$$= -10i - 12i^2$$

$$= -10i - 12(-1) = 12 - 10i$$

8. (a) $\sqrt{-50} \cdot \sqrt{-4}$

$$= \sqrt{-1}\,\sqrt{50} \cdot \sqrt{-1}\,\sqrt{4}$$

$$= i\sqrt{50} \cdot i\sqrt{4}$$

$$= i^2\sqrt{200} = 10i^2\sqrt{2} = 10(-1)\sqrt{2} = -10\sqrt{2}$$

9. (a) $i^{42} = \left(i^{40+2}\right) = \left(i^{40}\right)\left(i^2\right) = \left(i^4\right)^{10}\left(i^2\right) = (1)^{10}(-1) = -1$

(b) $i^{53} = \left(i^{52+1}\right) = \left(i^{52}\right)(i) = \left(i^4\right)^{13}(i) = (1)^{13}(i) = i$

10. $\dfrac{4 + 2i}{3 + 4i} \cdot \dfrac{3 - 4i}{3 - 4i} = \dfrac{12 - 16i + 6i - 8i^2}{9 - 16i^2}$

$$= \dfrac{12 - 10i - 8(-1)}{9 - 16(-1)}$$

$$= \dfrac{12 - 10i + 8}{9 + 16} = \dfrac{20 - 10i}{25}$$

$$= \dfrac{5(4 - 2i)}{25} = \dfrac{4 - 2i}{5}$$

11. $\dfrac{5 - 6i}{-2i} \cdot \dfrac{2i}{2i}$

$$= \dfrac{10i - 12i^2}{-4i^2} = \dfrac{10i - 12(-1)}{-4(-1)}$$

$$= \dfrac{10i + 12}{4} = \dfrac{2(5i + 6)}{4} = \dfrac{6 + 5i}{2}$$

8.7 Practice Problems

1. Let $s = $ speed,

$h = $ horsepower.

$s = k\sqrt{h}$

Substitute $s = 128$ and $h = 256$.

$128 = k\sqrt{256}$

$128 = 16k$

$8 = k$

Now we know the value of k so

$s = 8\sqrt{h}$.

when $h = 225$

$s = 8\left(\sqrt{225}\right)$

$s = 8(15)$

$s = 120$ miles per hour

2. $y = \dfrac{k}{x}$

Substitute $y = 45$ and $x = 16$.

$45 = \dfrac{k}{16}$

$720 = k$

We now write the equation $y = \dfrac{720}{x}$.

Find the value of y when $x = 36$.

$y = \dfrac{720}{36}$

$y = 20$

3. Let r = resistance,

c = amount of current.

$r = \dfrac{k}{c^2}$

Find the value of k when $r = 800$ ohms and $c = 0.01$ amps.

$800 = \dfrac{k}{(0.01)^2}$

$0.08 = k$

We now write the equation $r = \dfrac{0.08}{c^2}$.

Now substitute $c = 0.04$ and solve for r.

$r = \dfrac{0.08}{(0.04)^2}$

$r = \dfrac{0.08}{0.0016}$

$r = 50$ ohms

4. $y = \dfrac{kzw^2}{x}$

To find the value of k substitute $y = 20$, $z = 3$, $w = 5$, and $x = 4$.
Solve for k.

$20 = \dfrac{k(3)(5)^2}{4}$

$20 = \dfrac{75k}{4}$

$\dfrac{80}{75} = k$

$\dfrac{16}{15} = k$

We now substitute $\dfrac{16}{15}$ for k.

$y = \dfrac{16zw^2}{15x}$

We use this equation to find y when $z = 4$, $w = 6$, and $x = 2$.

$y = \dfrac{16(4)(6)^2}{15(2)} = \dfrac{2304}{30}$

$y = \dfrac{384}{5}$

Chapter 9

9.1 Practice Problems

1. $x^2 - 121 = 0$

$x^2 = 121$

$x = \pm 11$

Check:

$(11)^2 - 121 \overset{?}{=} 0 \qquad\qquad (-11)^2 - 121 \overset{?}{=} 0$

$121 - 121 \overset{?}{=} 0 \qquad\qquad 121 - 121 \overset{?}{=} 0$

$0 = 0 \checkmark \qquad\qquad\qquad 0 = 0 \checkmark$

2. $x^2 = 18$

$x = \pm\sqrt{18}$

$x = \pm 3\sqrt{2}$

3. $5x^2 + 1 = 46$

$5x^2 = 45$

$x = \pm\sqrt{9}$

$x = \pm 3$

4. $3x^2 = -27$

$x = \pm\sqrt{-9}$

$x = \pm 3i$

Check:

$3(3i)^2 \overset{?}{=} -27 \qquad\qquad 3(-3i)^2 \overset{?}{=} -27$

$3(9)(-1) \overset{?}{=} -27 \qquad\qquad 3(9)(-1) \overset{?}{=} -27$

$-27 = -27 \checkmark \qquad\qquad -27 = -27 \checkmark$

5. $(2x + 3)^2 = 7$

$(2x + 3) = \pm\sqrt{7}$

$2x + 3 = \pm\sqrt{7}$

$2x = -3 \pm \sqrt{7}$

$x = \dfrac{-3 \pm \sqrt{7}}{2}$

6. $x^2 + 8x + 3 = 0$

$x^2 + 8x = -3$

$x^2 + 8x + (4)^2 = -3 + (4)^2$

$(x + 4)^2 = 13$

$x + 4 = \pm\sqrt{13}$

$x = -4 \pm \sqrt{13}$

7. $2x^2 + 4x + 1 = 0$

$x^2 + 2x = \dfrac{-1}{2}$

$x^2 + 2x + (1)^2 = \dfrac{-1}{2} + 1$

$(x + 1)^2 = \dfrac{1}{2}$

$(x + 1) = \pm\sqrt{\dfrac{1}{2}}$

$x + 1 = \pm\dfrac{1}{\sqrt{2}}$

$x = -1 \pm \dfrac{\sqrt{2}}{2} \quad \text{or} \quad \dfrac{-2 \pm \sqrt{2}}{2}$

9.2 Practice Problems

1. $x^2 + 5x = -1 + 2x$

$x^2 + 3x + 1 = 0$

$a = 1, b = 3, c = 1$

$x = \dfrac{-3 \pm \sqrt{3^2 - 4(1)(1)}}{2(1)}$

$x = \dfrac{-3 \pm \sqrt{5}}{2}$

2. $2x^2 + 7x + 6 = 0$

$a = 2, b = 7, c = 6$

$x = \dfrac{-7 \pm \sqrt{7^2 - 4(2)(6)}}{2(2)}$

$x = \dfrac{-7 \pm \sqrt{49 - 48}}{4}$

$x = \dfrac{-7 \pm \sqrt{1}}{4}$

$x = \dfrac{-7 + 1}{4} \quad \text{or} \quad x = \dfrac{-7 - 1}{4}$

$x = -\dfrac{6}{4} = -\dfrac{3}{2} \qquad\qquad x = -2$

3. $2x^2 - 26 = 0$

$a = 2, b = 0, c = -26$

$$x = \frac{-0 \pm \sqrt{0^2 - 4(2)(-26)}}{2(2)}$$

$$x = \frac{\pm\sqrt{208}}{4} = \frac{\pm 4\sqrt{13}}{4} = \pm\sqrt{13}$$

4. $0 = -100x^2 + 4800x - 52{,}559$

$a = -100, b = 4800, c = -52{,}559$

$$x = \frac{-4800 \pm \sqrt{(4800)^2 - 4(-100)(-52{,}559)}}{2(-100)}$$

$$x = \frac{-4800 \pm \sqrt{23{,}040{,}000 - 21{,}023{,}600}}{-200}$$

$$x = \frac{-4800 \pm \sqrt{2{,}016{,}400}}{-200}$$

$$x = \frac{-4800 \pm 1420}{-200}$$

$$x = \frac{-4800 + 1420}{-200} = 16.9 \approx 17$$

or

$$x = \frac{-4800 - 1420}{-200} = 31.1 \approx 31$$

5. $\dfrac{1}{x} + \dfrac{1}{x-1} = \dfrac{5}{6}$ LCD is $6x(x-1)$.

$$6x(x-1)\left[\frac{1}{x}\right] + 6x(x-1)\left[\frac{1}{x-1}\right] = 6x(x-1)\left[\frac{5}{6}\right]$$

$$6(x-1) + 6x = 5(x^2 - x)$$

$$6x - 6 + 6x = 5x^2 - 5x$$

$$0 = 5x^2 - 17x + 6$$

$a = 5, b = -17, c = 6$

$$x = \frac{-(-17) \pm \sqrt{(-17)^2 - 4(5)(6)}}{2(5)}$$

$$x = \frac{17 \pm \sqrt{289 - 120}}{10}$$

$$x = \frac{17 \pm \sqrt{169}}{10}$$

$$x = \frac{17 \pm 13}{10}$$

$$x = \frac{17 + 13}{10} = \frac{30}{10} = 3 \qquad x = \frac{17 - 13}{10} = \frac{2}{5}$$

6. $2x^2 - 4x + 5 = 0$

$a = 2, b = -4, c = 5$

$$x = \frac{-(-4) \pm \sqrt{(-4)^2 - 4(2)(5)}}{2(2)}$$

$$x = \frac{4 \pm \sqrt{-24}}{4}$$

$$x = \frac{4 \pm 2i\sqrt{6}}{4} = \frac{2 \pm i\sqrt{6}}{2}$$

7. $9x^2 + 12x + 4 = 0$

$a = 9, b = 12, c = 4$

$b^2 - 4ac = 12^2 - 4(9)(4) = 144 - 144 = 0$

Since the discriminant is 0, there is one rational solution.

8. (a) $x^2 - 4x + 13 = 0$

$a = 1, b = -4, c = 13$

$b^2 - 4ac = (-4)^2 - 4(1)(13) = 16 - 52 = -36$

Since the discriminant is negative, there are two complex solutions containing i.

(b) $9x^2 + 6x + 7 = 0$

$a = 9, b = 6, c = 7$

$b^2 - 4ac = 6^2 - 4(9)(7)$

$\qquad\qquad = 36 - 252 = -216$

Since the discriminant is negative, there are two complex solutions containing i.

9. $x = -10$ $\qquad\qquad$ $x = -6$

$x + 10 = 0$ $\qquad\qquad$ $x + 6 = 0$

$(x + 10)(x + 6) = 0$

$x^2 + 6x + 10x + 60 = 0$

$x^2 + 16x + 60 = 0$

10. $x = 2i\sqrt{3}$ $\qquad\qquad$ $x = -2i\sqrt{3}$

$x - 2i\sqrt{3} = 0$ $\qquad\qquad$ $x + 2i\sqrt{3} = 0$

$(x - 2i\sqrt{3})(x + 2i\sqrt{3}) = 0$

$x^2 - 4i^2(\sqrt{9}) = 0$

$x^2 - 4(-1)(3) = 0$

$x^2 + 12 = 0$

9.3 Practice Problems

1. $x^4 - 5x^2 - 36 = 0$

Let $y = x^2$. Then $y^2 = x^4$.

Thus, our new equation is

$\qquad y^2 - 5y - 36 = 0.$

$(y - 9)(y + 4) = 0$

$y - 9 = 0 \qquad\qquad y + 4 = 0$

$y = 9 \qquad\qquad y = -4$

$x^2 = 9 \qquad\qquad x^2 = -4$

$x = \pm\sqrt{9} \qquad\qquad x = \pm\sqrt{-4}$

$x = \pm 3 \qquad\qquad x = \pm 2i$

2. $x^6 - 5x^3 + 4 = 0$

Let $y = x^3$. Then $y^2 = x^6$.

$y^2 - 5y + 4 = 0$

$(y - 1)(y - 4) = 0$

$y - 1 = 0 \qquad\qquad y - 4 = 0$

$y = 1 \qquad\qquad y = 4$

$x^3 = 1 \qquad\qquad x^3 = 4$

$x = 1 \qquad\qquad x = \sqrt[3]{4}$

3. $3x^{4/3} - 5x^{2/3} + 2 = 0$

Let $y = x^{2/3}$ and $y^2 = x^{4/3}$.

$3y^2 - 5y + 2 = 0$

$(3y - 2)(y - 1) = 0$

$3y - 2 = 0 \qquad\qquad y - 1 = 0$

$y = \dfrac{2}{3} \qquad\qquad y = 1$

$x^{2/3} = \dfrac{2}{3} \qquad\qquad x^{2/3} = 1$

$(x^{2/3})^3 = \left(\dfrac{2}{3}\right)^3 \qquad\qquad (x^{2/3})^3 = 1^3$

$x^2 = \dfrac{8}{27} \qquad\qquad x^2 = 1$

$x = \pm\sqrt{\dfrac{8}{27}} \qquad\qquad x = \pm\sqrt{1}$

$x = \pm\dfrac{2\sqrt{2}}{3\sqrt{3}} \qquad\qquad x = \pm 1$

$x = \pm\dfrac{2\sqrt{2}}{3\sqrt{3}} \cdot \dfrac{\sqrt{3}}{\sqrt{3}}$

$x = \pm\dfrac{2\sqrt{6}}{9}$

Check: for $x = \dfrac{2\sqrt{6}}{9}$

$$3\left(\frac{2\sqrt{6}}{9}\right)^{4/3} - 5\left(\frac{2\sqrt{6}}{9}\right)^{2/3} + 2 \stackrel{?}{=} 0$$

$$3\left(\frac{4}{9}\right) - 5\left(\frac{2}{3}\right) + 2 \stackrel{?}{=} 0$$

$$\frac{4}{3} - \frac{10}{3} + 2 \stackrel{?}{=} 0$$

$$0 = 0 \checkmark$$

for $x = -\dfrac{2\sqrt{6}}{9}$

$$3\left(-\frac{2\sqrt{6}}{9}\right)^{4/3} - 5\left(-\frac{2\sqrt{6}}{9}\right)^{2/3} + 2 \stackrel{?}{=} 0$$

$$3\left(\frac{4}{9}\right) - 5\left(\frac{2}{3}\right) + 2 \stackrel{?}{=} 0$$

$$\frac{4}{3} - \frac{10}{3} + 2 \stackrel{?}{=} 0$$

$$0 = 0 \checkmark$$

for $x = 1$

$$3(1)^{4/3} - 5(1)^{2/3} + 2 \stackrel{?}{=} 0$$

$$3(1) - 5(1) + 2 \stackrel{?}{=} 0$$

$$0 = 0 \checkmark$$

for $x = -1$

$$3(-1)^{4/3} - 5(-1)^{2/3} + 2 \stackrel{?}{=} 0$$

$$3(1) - 5(1) + 2 \stackrel{?}{=} 0$$

$$0 = 0 \checkmark$$

4. $\quad 3x^{1/2} = 8x^{1/4} - 4$

$$3x^{1/2} - 8x^{1/4} + 4 = 0$$

Let $y = x^{1/4}$ and $y^2 = x^{1/2}$.

$$3y^2 - 8y + 4 = 0$$

$$(3y - 2)(y - 2) = 0$$

$3y - 2 = 0$	$y - 2 = 0$
$y = \dfrac{2}{3}$	$y = 2$
$x^{1/4} = \dfrac{2}{3}$	$x^{1/4} = 2$
$(x^{1/4})^4 = \left(\dfrac{2}{3}\right)^4$	$(x^{1/4})^4 = (2)^4$
$x = \dfrac{16}{81}$	$x = 16$

Check: for $x = \dfrac{16}{81}$ $\qquad$ for $x = 16$

$$3\left(\frac{16}{81}\right)^{1/2} \stackrel{?}{=} 8\left(\frac{16}{81}\right)^{1/4} - 4 \qquad 3(16)^{1/2} \stackrel{?}{=} 8(16)^{1/4} - 4$$

$$3\left(\frac{4}{9}\right) \stackrel{?}{=} 8\left(\frac{2}{3}\right) - 4 \qquad\qquad 3(4) \stackrel{?}{=} 8(2) - 4$$

$$\frac{4}{3} = \frac{4}{3} \checkmark \qquad\qquad\qquad 12 = 12 \checkmark$$

9.4 Practice Problems

1. $\qquad V = \dfrac{1}{3}\pi r^2 h \qquad$ Solve for r.

$$\frac{3V}{\pi h} = r^2$$

$$\pm\sqrt{\frac{3V}{\pi h}} = r \qquad$$ Since we are solving for the length of the radius, we use the positive answer.

$$r = \sqrt{\frac{3V}{\pi h}}$$

2. $\quad 2y^2 + 9wy + 7w^2 = 0 \qquad$ Solve for y.

$$(2y + 7w)(y + w) = 0$$

$2y + 7w = 0$	$y + w = 0$
$2y = -7w$	$y = -w$
$y = -\dfrac{7}{2}w$	

3. $3y^2 + 2fy - 7g = 0 \qquad$ Solve for y.

Use the quadratic formula.

$$a = 3, b = 2f, c = -7g$$

$$y = \frac{-2f \pm \sqrt{(2f)^2 - 4(3)(-7g)}}{2(3)}$$

$$y = \frac{-2f \pm \sqrt{4f^2 + 84g}}{6}$$

$$y = \frac{-2f \pm \sqrt{4(f^2 + 21g)}}{6}$$

$$y = \frac{-2f + 2\sqrt{(f^2 + 21g)}}{6}$$

$$y = \frac{-f \pm \sqrt{f^2 + 21g}}{3}$$

4. $d = \dfrac{n^2 - 3n}{2} \qquad$ Solve for n.

Multiply each term by 2.

$$2d = n^2 - 3n$$

$$0 = n^2 - 3n - 2d$$

Use the quadratic formula.

$$a = 1, b = -3, c = -2d$$

$$n = \frac{-(-3) \pm \sqrt{(-3)^2 - 4(1)(-2d)}}{2}$$

$$n = \frac{3 \pm \sqrt{9 + 8d}}{2}$$

5. (a) $a^2 + b^2 = c^2 \qquad\qquad$ Solve for b.

$$b^2 = c^2 - a^2$$

$$b = \sqrt{c^2 - a^2}$$

(b) $b = \sqrt{c^2 - a^2}$

$$b = \sqrt{(26)^2 - (24)^2}$$

$$b = \sqrt{676 - 576}$$

$$b = \sqrt{100}$$

$$b = 10$$

6. $x + x - 7 + c = 30$

$$2x - 7 + c = 30$$

$$c = -2x + 37$$

$$a = x, b = x - 7, c = -2x + 37$$

By the Pythagorean theorem,

$$x^2 + (x - 7)^2 = (-2x + 37)^2$$

$$x^2 + x^2 - 14x + 49 = 4x^2 - 148x + 1369$$

$$-2x^2 + 134x - 1320 = 0$$

$$x^2 - 67x + 660 = 0$$

By the quadratic formula,

$$a = 1, b = -67, c = 660$$

$$x = \frac{67 \pm \sqrt{(67)^2 - 4(1)(660)}}{2}$$

$$x = \frac{67 \pm \sqrt{4489 - 2640}}{2}$$

$$x = \frac{67 \pm \sqrt{1849}}{2}$$

$$x = \frac{67 \pm 43}{2}$$

$$x = \frac{67 + 43}{2} = 55 \quad \text{or} \quad x = \frac{67 - 43}{2} = 12$$

The only answer that makes sense is $x = 12$; therefore,

$$x = 12$$

$$x - 7 = 5$$

$$-2x + 37 = 13$$

The legs are 5 miles and 12 miles long. The hypotenuse of the triangle is 13 miles long.

7. $A = \pi r^2$

$$A = \pi(6)^2$$

$$= 36\pi$$

Let $x =$ the radius of the new pipe.

(area of new pipe) minus (area of old pipe) $= 45\pi$

$$\pi x^2 - 36\pi = 45\pi$$

$$\pi x^2 = 45\pi + 36\pi$$

$$x^2 = 81$$

$$x = \pm 9$$

Since the radius must be positive, we select $x = 9$. The radius of the new pipe is 9 inches. The radius of the new pipe has been increased by 3 inches.

8. Let $x =$ width. Then $2x - 3 =$ the length.

$$x(2x - 3) = 54$$

$$2x^2 - 3x = 54$$

$$2x^2 - 3x - 54 = 0$$

$$(2x + 9)(x - 6) = 0$$

$$2x + 9 = 0 \qquad x - 6 = 6$$

$$x = -\frac{9}{2} \qquad x = 6$$

We do not use the negative value.

Thus, width = 6 feet

length = $2x - 3 = 2(6) - 3 = 9$ feet

9.

	Distance	Rate	Time
Secondary Road	150	x	$\dfrac{150}{x}$
Better Road	240	$x + 10$	$\dfrac{240}{x + 10}$
TOTAL	390	(not used)	7

$$\frac{150}{x} + \frac{240}{x + 10} = 7$$

The LCD of this equation is $x(x + 10)$. Multiply each term by the LCD.

$$x(x + 10)\left[\frac{150}{x}\right] + x(x + 10)\left[\frac{240}{x + 10}\right] = x(x + 10)[7]$$

$$150(x + 10) + 240x = 7x(x + 10)$$

$$150x + 1500 + 240x = 7x^2 + 70x$$

$$7x^2 - 320x - 1500 = 0$$

$$(x - 50)(7x + 30) = 0$$

$$x - 50 = 0 \qquad 7x + 30 = 0$$

$$x = 50 \qquad x = \frac{-30}{7}$$

We disregard the negative answer. Thus, $x = 50$ mph, so Carlos drove 50 mph on the secondary road and 60 mph on the better road.

9.5 Practice Problems

1. $f(x) = x^2 - 6x + 5$

$a = 1, b = -6, c = 15$

Step 1 The vertex occurs at $x = \dfrac{-b}{2a}$. Thus, $x = \dfrac{-(-6)}{2(1)} = 3$.

The vertex has an x-coordinate of 3.

To find the y-coordinate, we evaluate $f(3)$.

$$f(3) = 3^2 - 6(3) + 5$$

$$= 9 - 18 + 5$$

$$= -4$$

Thus, the vertex is $(3, -4)$.

Step 2 The y-intercept is at $f(0)$.

$$f(0) = 0^2 - 6(0) + 5$$

$$= 5$$

The y-intercept is $(0, 5)$.

Step 3 The x-intercept is at $f(x) = 0$.

$$x^2 - 6x + 5 = 0$$

$$(x - 5)(x - 1) = 0$$

$$x - 5 = 0 \qquad x - 1 = 0$$

$$x = 5 \qquad x = 1$$

Thus, the x-intercepts are $(5, 0)$ and $(1, 0)$.

2. $g(x) = x^2 - 2x - 2$

$a = 1, b = -2, c = -2$

Step 1 The vertex occurs at

$$x = \frac{-b}{2a}.$$

$$x = \frac{-(-2)}{2(1)} = \frac{2}{2} = 1$$

The vertex has an x-coordinate of 1. To find the y-coordinate, we evaluate $f(1)$.

$$g(1) = 1^2 - 2(1) - 2$$

$$= 1 - 2 - 2$$

$$= -3$$

Thus, the vertex is $(1, -3)$.

Step 2 The y-intercept is at $g(0)$.
$$g(0) = 0^2 - 2(0) - 2$$
$$= -2$$
The y-intercept is $(0, -2)$.

Step 3 The x-intercepts occur when $g(x) = 0$.

We set $x^2 - 2x - 2 = 0$ and solve for x. The equation does not factor, so we use the quadratic formula.
$$x = \frac{-(-2) \pm \sqrt{12}}{2} = \frac{2 \pm 2\sqrt{3}}{2} = 1 \pm \sqrt{3}$$
The x-intercepts are approximately $(2.7, 0)$ and $(-0.7, 0)$.

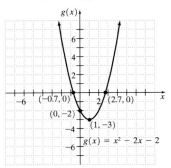

3. $g(x) = -2x^2 - 8x - 6$

$a = -2, b = -8, c = -6$

Since $a < 0$, the parabola opens downward.

The vertex occurs at
$$x = \frac{-b}{2a} = \frac{-(-8)}{2(-2)} = -2.$$
To find the y-coordinate, evaluate $g(-2)$.
$$g(-2) = -2(-2)^2 - 8(-2) - 6$$
$$= -8 + 16 - 6$$
$$= 2$$
Thus, the vertex is $(-2, 2)$.

The y-intercept is at $g(0)$.
$$g(0) = -2(0)^2 - 8(0) - 6$$
$$= -6$$
The y-intercept is $(0, -6)$.

The x-intercepts occur when $g(x) = 0$.

Using the quadratic formula.
$$x = \frac{-(-8) \pm \sqrt{64 - 4(-2)(-6)}}{2(-2)}$$
$$= \frac{8 \pm \sqrt{16}}{-4} = -2 \pm -1$$
$$x = -3, \quad x = -1$$
The x-intercepts are $(-3, 0)$ and $(-1, 0)$.

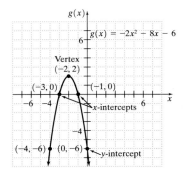

9.6 Practice Problems

1. $x^2 - 2x - 8 < 0$

Replace the inequality by an equal sign and solve.

$(x - 4)(x + 2) = 0$

$x = 4 \qquad x = -2$

Region I	$x < -2$	$(-3)^2 - 2(-3) - 8$
	$x = -3$	$= 9 + 6 - 8 = 7 > 0$
Region II	$-2 < x < 4$	$0^2 - 2(0) - 8 = -8 < 0$
	$x = 0$	
Region III	$x > 4$	$(5)^2 - 2(5) - 8 = 7 > 0$
	$x = 5$	

Thus, $x^2 - 2x - 8 < 0$ when $-2 < x < 4$.

2. $3x^2 - x - 2 \geq 0$

$(3x + 2)(x - 1) = 0$

$3x + 2 = 0 \qquad x - 1 = 0$

$x = -\frac{2}{3} \qquad x = 1$

Region I	$x < -\dfrac{2}{3}$
	$x = -1 \qquad 3(-1)^2 + 1 - 2 = 2 > 0$
Region II	$-\dfrac{2}{3} < x < 1$
	$x = 0 \qquad 3(0) - 0 - 2 = -2 < 0$
Region III	$x > 1$
	$x = 2 \qquad 3(2)^2 - 2 - 2 = 8 > 0$

Thus, $3x^2 - x - 2 \geq 0$ when $x \leq -\dfrac{2}{3}$ or when $x \geq 1$.

3. $x^2 + 2x < 7$

$x^2 + 2x - 7 < 0$

$x^2 + 2x - 7 = 0$

$$x = \frac{-2 \pm \sqrt{4 + 28}}{2}$$

$x = -1 \pm 2\sqrt{2}$

$x \approx 1.8 \quad \text{and} \quad x \approx -3.8$

Region I	$x < -3.8$	
	$x = -5$	$(-5)^2 + 2(-5) - 7 = 8 > 0$
Region II	$-3.8 < x < 1.8$	
	$x = 0$	$(0)^2 + 2(0) - 7 = -7 < 0$
Region III	$x > 1.8$	
	$x = 3$	$(3)^2 + 2(3) - 7 = 8 > 0$

Thus, $x^2 + 2x - 7 < 0$ when $-1 - 2\sqrt{2} < x < -1 + 2\sqrt{2}$.

Approximately $-3.8 < x < 1.8$

9.7 Practice Problems

1. Solve $|3x - 4| = 23$ and check.

We have two equations.

$$3x - 4 = 23 \qquad \text{or} \qquad 3x - 4 = -23$$
$$3x = 27 \qquad\qquad 3x = -19$$
$$x = 9 \qquad\qquad x = -\frac{19}{3}$$

Check: if $x = 9$

$$|3x - 4| = 23$$
$$|3(9) - 4| \stackrel{?}{=} 23$$
$$|27 - 4| \stackrel{?}{=} 23$$
$$|23| \stackrel{?}{=} 23$$
$$23 = 23 \checkmark$$

if $x = -\dfrac{19}{3}$

$$|3x - 4| = 23$$
$$\left|3\left(-\frac{19}{3}\right) - 4\right| \stackrel{?}{=} 23$$
$$|-19 - 4| \stackrel{?}{=} 23$$
$$|-23| \stackrel{?}{=} 23$$
$$23 = 23 \checkmark$$

2. Solve. $\left|\dfrac{2}{3}x + 4\right| = 2$

We have two equations.

$$\frac{2}{3}x + 4 = 2 \quad \text{or} \quad \frac{2}{3}x + 4 = -2$$
$$2x + 12 = 6 \qquad\qquad 2x + 12 = -6$$
$$2x = -6 \qquad\qquad\quad 2x = -18$$
$$x = -3 \qquad\qquad\quad x = -9$$

Check: if $x = -3$ $\qquad$ if $x = -9$

$$\left|\frac{2}{3}(-3) + 4\right| \stackrel{?}{=} 2 \quad \left|\frac{2}{3}(-9) + 4\right| \stackrel{?}{=} 2$$
$$|-2 + 4| \stackrel{?}{=} 2 \qquad |-6 + 4| \stackrel{?}{=} 2$$
$$|2| \stackrel{?}{=} 2 \qquad\qquad |-2| \stackrel{?}{=} 2$$
$$2 = 2 \checkmark \qquad\qquad 2 = 2 \checkmark$$

3. Solve $|2x + 1| + 3 = 8$.
First change the equation so that the absolute value expression is alone on one side of the equation.

$$|2x + 1| + 3 = 8$$
$$|2x + 1| + 3 - 3 = 8 - 3$$
$$|2x + 1| = 5$$

We have the two equations

$$2x + 1 = 5 \quad \text{or} \quad 2x + 1 = -5$$
$$2x = 4 \qquad\qquad 2x = -6$$
$$x = 2 \qquad\qquad x = -3$$

Check:

if $x = 2$ $\qquad\qquad$ if $x = -3$

$$|2(2) + 1| + 3 \stackrel{?}{=} 8 \quad |2(-3) + 1| + 3 \stackrel{?}{=} 8$$
$$|5| + 3 \stackrel{?}{=} 8 \qquad |-5| + 3 \stackrel{?}{=} 8$$
$$8 = 8 \checkmark \qquad\qquad 8 = 8 \checkmark$$

4. Solve $|x - 6| = |5x + 8|$.
We write the two possible equations and solve each equation.

$$x - 6 = 5x + 8 \quad \text{or} \quad x - 6 = -(5x + 8)$$
$$-4x = 14 \qquad\qquad x - 6 = -5x - 8$$
$$x = \frac{14}{-4} = -\frac{7}{2} \qquad 6x = -2$$
$$x = \frac{-2}{6} = -\frac{1}{3}$$

Check: if $x = -\dfrac{7}{2}$

$$\left|-\frac{7}{2} - 6\right| \stackrel{?}{=} \left|5\left(-\frac{7}{2}\right) + 8\right|$$
$$\left|-\frac{7}{2} - 6\right| \stackrel{?}{=} \left|-\frac{35}{2} + 8\right|$$
$$\left|-\frac{7}{2} - \frac{12}{2}\right| \stackrel{?}{=} \left|-\frac{35}{2} + \frac{16}{2}\right|$$
$$\left|-\frac{19}{2}\right| \stackrel{?}{=} \left|-\frac{19}{2}\right|$$
$$\frac{19}{2} = \frac{19}{2} \checkmark$$

if $x = -\dfrac{1}{3}$

$$\left|-\frac{1}{3} - 6\right| \stackrel{?}{=} \left|5\left(-\frac{1}{3}\right) + 8\right|$$
$$\left|-\frac{1}{3} - 6\right| \stackrel{?}{=} \left|-\frac{5}{3} + 8\right|$$
$$\left|-\frac{19}{3}\right| \stackrel{?}{=} \left|\frac{19}{3}\right|$$
$$\frac{19}{3} = \frac{19}{3} \checkmark$$

9.8 Practice Problems

1. $|x| < 2$

$$-2 < x < 2$$

$$-2 < x < 2$$

2. $|x - 6| < 15$

$$-15 < x - 6 < 15$$
$$-15 + 6 < x - 6 + 6 < 15 + 6$$
$$-9 < x < 21$$

$$-9 < x < 21$$

3. $\left|x + \dfrac{3}{4}\right| \leq \dfrac{7}{6}$

$$-\frac{7}{6} \leq x + \frac{3}{4} \leq \frac{7}{6}$$
$$-14 \leq 12x + 9 \leq 14$$
$$-14 - 9 \leq 12x + 9 - 9 \leq 14 - 9$$
$$-23 \leq 12x \leq 5$$
$$\frac{-23}{12} \leq \frac{12x}{12} \leq \frac{5}{12}$$
$$-1\frac{11}{12} \leq x \leq \frac{5}{12}$$

$$-1\tfrac{11}{12} \leq x \leq \tfrac{5}{12}$$

4. $|2 + 3(x - 1)| < 20$

$|2 + 3x - 3| < 20$

$|-1 + 3x| < 20$

$-20 < -1 + 3x < 20$

$-20 + 1 < 1 - 1 + 3x < 20 + 1$

$-19 < 3x < 21$

$\dfrac{-19}{3} < \dfrac{3x}{3} < \dfrac{21}{3}$

$\dfrac{-19}{3} < x < 7$

$-6\dfrac{1}{3} < x < 7$

```
  -6⅓                                          7
——+—o—+——+——+——+——+——+——+——+——+——+—o——
 -7  -6  -5  -4  -3  -2  -1   0   1   2   3   4   5   6   7
              -6⅓ < x < 7
```

5. $|x| > 2.5$

$x > 2.5 \quad \text{or} \quad x < -2.5$

```
←——+—o—+——+——+——+——+—o—+——→
   -3  -2  -1   0   1   2   3
```

6. $|x + 6| > 2$

$x + 6 > 2 \quad \text{or} \quad x + 6 < -2$

$x > -4 \quad \text{or} \quad \quad x < -8$

```
←—+——+—⊕—+——+——+—⊕—+——+——→
 -10 -9  -8  -7  -6  -5  -4  -3  -2
         x < -8 or x > -4
```

7. $|-5x - 2| > 13$

$-5x - 2 > 13 \quad \text{or} \quad -5x - 2 < -13$

$-5x > 15 \quad \text{or} \quad \quad -5x < -11$

$x < -3 \quad \text{or} \quad \quad x > \dfrac{11}{5}$

$x > 2\dfrac{1}{5}$

```
                    2⅕
←——+—⊕—+——+——+——+—⊕—+——+——→
 -4  -3  -2  -1   0   1   2   3   4
```

8. $\left|4 - \dfrac{3}{4}x\right| \geq 5$

$4 - \dfrac{3}{4}x \geq 5 \quad \text{or} \quad 4 - \dfrac{3}{4}x \leq -5$

$16 - 3x \geq 20 \quad \quad 16 - 3x \leq -20$

$-3x \geq 4 \quad \quad \quad -3x \leq -36$

$x \leq \dfrac{-4}{3} \quad \quad \quad x \geq 12$

$x \leq -1\dfrac{1}{3}$

```
←—+—●—+——+——+——+——+——+——+——+——+—●—+——→
  -1⅓ 0                                  12
```

9.

$|d - s| \leq 0.37$

$|d - 276.53| \leq 0.37$

$-0.37 \leq d - 276.53 \leq 0.37$

$-0.37 + 276.53 \leq d - 276.53 + 276.53 \leq 0.37 + 276.53$

$276.16 \leq d \leq 276.90$

Thus the diameter of the transmission must be at least 276.16 millimeters, but not greater than 276.90 millimeters.

Chapter 10

10.1 Practice Problems

1. Let $(x_1, y_1) = (-6, -2)$ and $(x_2, y_2) = (3, 1)$.

$$d = \sqrt{(x_2 - x_1)^2 + (y_2 - y_1)^2}$$

$$= \sqrt{[3 - (-6)]^2 + [1 - (-2)]^2}$$

$$= \sqrt{(3 + 6)^2 + (1 + 2)^2}$$

$$= \sqrt{(9)^2 + (3)^2}$$

$$= \sqrt{81 + 9} = \sqrt{90} = 3\sqrt{10}$$

2. $(x + 1)^2 + (y + 2)^2 = 9$

If we compare this to $(x - h)^2 + (y - k)^2 = r^2$, we can write it in the form

$$[x - (-1)]^2 + [y - (-2)]^2 = 3^2.$$

Thus, we see the center is $(h, k) = (-1, -2)$ and the radius is $r = 3$.

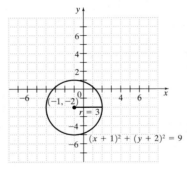

3. We are given that $(h, k) = (-5, 0)$ and $r = \sqrt{3}$. Thus, $(x - h)^2 + (y - k)^2 = r^2$ becomes

$$[x - (-5)]^2 + (y - 0)^2 = (\sqrt{3})^2$$

$$(x + 5)^2 + y^2 = 3.$$

4. To write $x^2 + 4x + y^2 + 2y - 20 = 0$ in standard form, we complete the square.

$$x^2 + 4x + \underline{\quad\quad} + y^2 + 2y + \underline{\quad\quad} = 20$$

$$x^2 + 4x + 4 + y^2 + 2y + 1 = 20 + 4 + 1$$

$$x^2 + 4x + 4 + y^2 + 2y + 1 = 25$$

$$(x + 2)^2 + (y + 1)^2 = 25$$

The circle has its center at $(-2, -1)$ and the radius is 5.

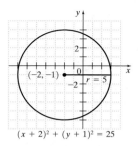

$(x + 2)^2 + (y + 1)^2 = 25$

10.2 Practice Problems

1. Make a table of values. Begin with $x = -3$ in the middle of the table because $(-3 + 3) = 0$. Plot the points and draw the graph.

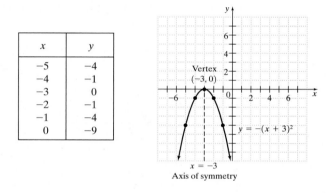

x	y
-5	-4
-4	-1
-3	0
-2	-1
-1	-4
0	-9

The vertex is $(-3, 0)$, and the axis of symmetry is the line $x = -3$.

2. This graph looks like the graph of $y = x^2$, except that it is shifted 6 units to the right and 4 units up.
The vertex is $(6, 4)$. The axis of symmetry is $x = 6$.
If $x = 0$, $y = (0 - 6)^2 + 4 = 36 + 4 = 40$, so the y-intercept is $(0, 40)$.

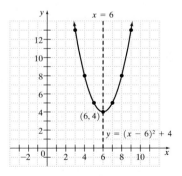

3. Step 1 The equation has the form $y = a(x - h)^2 + k$, where $a = \frac{1}{4}$, $h = 2$, and $k = 3$, so it is a vertical parabola.

Step 2 $a > 0$; so the parabola opens upward.

Step 3 We have $h = 2$ and $k = 3$. Therefore, the vertex is $(2, 3)$.

Step 4 The axis of symmetry is the line $x = 2$. Plot a few points on either side of the axis of symmetry. At $x = 4$, $y = 4$. Thus, the point is $(4, 4)$. The image from symmetry is $(0, 4)$. At $x = 6$, $y = 7$. Thus the point is $(6, 7)$. The image from symmetry is $(-2, 7)$.

Step 5 At $x = 0$,

$$y = \frac{1}{4}(0 - 2)^2 + 3$$
$$= \frac{1}{4}(4) + 3$$
$$= 1 + 3 = 4.$$

Thus, the y-intercept is $(0, 4)$.

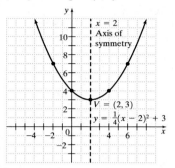

4. Make a table, plot points and draw the graph. Choose values of y and find x. Begin with $y = 0$.

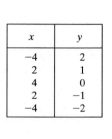 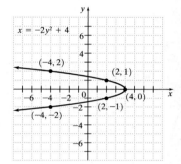

x	y
-4	2
2	1
4	0
2	-1
-4	-2

The vertex is at $(4, 0)$. The axis of symmetry is the x-axis.

5. Step 1 The equation has the form $x = a(y - k)^2 + h$ where $a = -1, k = -1$, and $h = -3$, so it is a horizontal parabola.

Step 2 $a < 0$; so the parabola opens to the left.

Step 3 We have $k = -1$ and $h = -3$. Therefore, the vertex is $(-3, -1)$.

Step 4 The line $y = -1$ is the axis of symmetry. At $y = 0$, $x = -4$. Thus, we have the point $(-4, 0)$ and $(-4, -2)$ from symmetry. At $y = 1$, $x = -7$. Thus, we have the point $(-7, 1)$ and $(-7, -3)$ from symmetry.

Step 5 At $y = 0$,

$$x = -(0 + 1)^2 - 3$$
$$= -(1) - 3$$
$$= -1 - 3 = -4$$

Thus, the x-intercept is $(-4, 0)$.

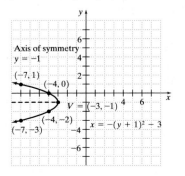

6. Since the y-term is squared, we have a horizontal parabola. The standard form is $x = a(y - k)^2 + h$.

$$x = y^2 - 6y + 13$$
$$= y^2 - 6y + \left(\frac{6}{2}\right)^2 - \left(\frac{6}{2}\right)^2 + 13 \quad \text{Complete the square.}$$
$$= \left(y^2 - 6y + 9\right) + 4$$
$$= (y - 3)^2 + 4$$

Therefore, we know that $a = 1, k = 3$, and $h = 4$. The vertex is at $(4, 3)$. The axis of symmetry is $y = 3$. If $y = 0$, $x = (-3)^2 + 4 = 9 + 4 = 13$. So the x-intercept is $(13, 0)$.

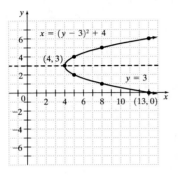

7. $y = 2x^2 + 8x + 9$

Since the x-term is squared, we have a vertical parabola. The standard form is

$$y = a(x - h)^2 + k$$
$$y = 2x^2 + 8x + 9$$
$$= 2\left(x^2 + 4x + \underline{\quad}\right) - \underline{\quad} + 9 \quad \text{Complete the square.}$$
$$= 2\left(x^2 + 4x + 4\right) - 2(4) + 9$$
$$= 2(x + 2)^2 + 1$$

The parabola opens upward. The vertex is $(-2, 1)$, and the y-intercept is $(0, 9)$. The axis of symmetry is $x = -2$.

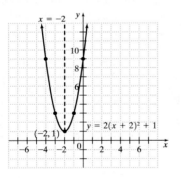

10.3 Practice Problems

1. Write the equation in standard form.

$$4x^2 + y^2 = 16$$
$$\frac{4x^2}{16} + \frac{y^2}{16} = \frac{16}{16}$$
$$\frac{x^2}{4} + \frac{y^2}{16} = 1$$

Thus, we have:

$$a^2 = 4 \quad \text{so} \quad a = 2$$
$$b^2 = 16 \quad \text{so} \quad b = 4$$

The x-intercepts are $(2, 0)$ and $(-2, 0)$ and the y-intercepts are $(0, 4)$ and $(0, -4)$.

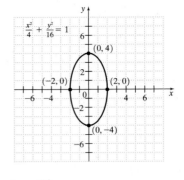

2. $\dfrac{(x - 2)^2}{16} + \dfrac{(y + 3)^2}{9} = 1$

The center is $(h, k) = (2, -3)$, $a = 4$, and $b = 3$. We start at $(2, -3)$ and measure to the right and to the left 4 units, and up and down 3 units.

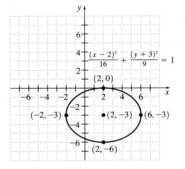

10.4 Practice Problems

1. $\dfrac{x^2}{16} - \dfrac{y^2}{25} = 1$

This is the equation of a horizontal hyperbola with center $(0, 0)$, where $a = 4$ and $b = 5$. The vertices are $(-4, 0)$ and $(4, 0)$.

Construct a fundamental rectangle with corners at $(4, 5)$, $(4, -5)$, $(-4, 5)$, and $(-4, -5)$. Draw extended diagonals as the asymptotes.

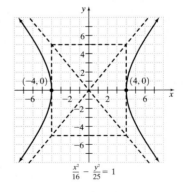

2. Write the equation in standard form.

$$y^2 - 4x^2 = 4$$
$$\frac{y^2}{4} - \frac{4x^2}{4} = \frac{4}{4}$$
$$\frac{y^2}{4} - \frac{x^2}{1} = 1$$

This is the equation of a vertical hyperbola with center $(0, 0)$, where $a = 1$ and $b = 2$. The vertices are $(0, 2)$ and $(0, -2)$.

The fundamental rectangle has corners at $(1, 2)$, $(1, -2)$, $(-1, 2)$, and $(-1, -2)$.

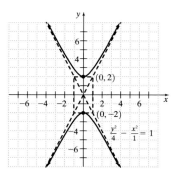

3. $\dfrac{(y + 2)^2}{9} - \dfrac{(x - 3)^2}{16} = 1$

This is a vertical hyperbola with center at $(3, -2)$, where $a = 4$ and $b = 3$. The vertices are $(3, 1)$ and $(3, -5)$. The fundamental rectangle has corners at $(7, 1)$, $(7, -5)$, $(-1, 1)$, and $(-1, -5)$.

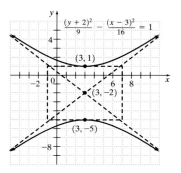

10.5 Practice Problems

1. $\dfrac{x^2}{4} - \dfrac{y^2}{4} = 1 \qquad\qquad x + y + 1 = 0$

$x^2 - y^2 = 4$ (1) $\qquad\qquad y = -x - 1$ (2)

Substitute (2) into (1).

$$x^2 - (-x - 1)^2 = 4$$
$$x^2 - (x^2 + 2x + 1) = 4$$
$$x^2 - x^2 - 2x - 1 = 4$$
$$-2x - 1 = 4$$
$$-2x = 5$$
$$x = \frac{5}{-2} = -2.5$$

Now substitute the value for x in the equation $y = -x - 1$.

For $x = -2.5$: $y = -(-2.5) - 1 = 2.5 - 1 = 1.5$.

The solution is $(-2.5, 1.5)$.

2. $2x - 9 = y$ (1)

$\qquad xy = -4$ (2) $\qquad$ Solve equation (2) for y.

$\qquad y = \dfrac{-4}{x}$ (3)

Substitute equation (3) into (1) and solve for x.

$$2x - 9 = \frac{-4}{x}$$
$$2x^2 - 9x + 4 = 0$$
$$(2x - 1)(x - 4) = 0$$
$$2x - 1 = 0 \qquad x - 4 = 0$$
$$x = \frac{1}{2} \quad \text{or} \qquad x = 4$$

$x = \dfrac{1}{2}$ and $x = 4$

For $x = \dfrac{1}{2}$:

$$y = \frac{-4}{\frac{1}{2}} = -8.$$

For $x = 4$:

$$y = \frac{-4}{4} = -1.$$

The solutions are $(4, -1)$ and $\left(\dfrac{1}{2}, -8\right)$.

The graph of $y = \dfrac{-4}{x}$ is the graph of $y = \dfrac{1}{x}$ reflected across the x-axis and stretched by a factor of 4. The graph of $y = 2x - 9$ is a line with slope 2 passing through the point $(0, -9)$, The sketch show that the points $(4, -1)$ and $\left(\dfrac{1}{2}, -8\right)$ seem reasonable.

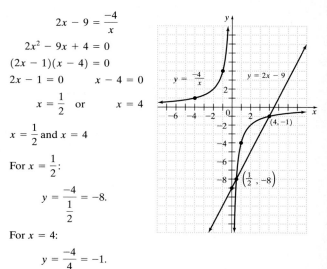

3. (1) $\quad x^2 + y^2 = 12 \quad 4x^2 + 4y^2 = 48 \qquad$ Multiply (1) by 4.

$\quad$ (2) $\quad 3x^2 - 4y^2 = 8 \quad \dfrac{3x^2 - 4y^2 = 8}{7x^2 = 56} \qquad$ Add the equations.

$$x^2 = 8$$
$$x = \pm\sqrt{8}$$
$$x = \pm 2\sqrt{2}$$

If $x = 2\sqrt{2}$, then $x^2 = 8$. Substituting this value into equation (1) gives

$$8 + y^2 = 12$$
$$y^2 = 4$$
$$y = \pm\sqrt{4}$$
$$y = \pm 2$$

Similarly, if $x = -2\sqrt{2}$, then $y = \pm 2$.

Thus, the four solutions are $(2\sqrt{2}, 2)$, $(2\sqrt{2}, -2)$, $(-2\sqrt{2}, 2)$, and $(-2\sqrt{2}, -2)$.

Chapter 11

11.1 Practice Problems

1. (a) $g(a) = \dfrac{1}{2}a - 3$

(b) $g(a + 4) = \dfrac{1}{2}(a + 4) - 3$

$$= \frac{1}{2}a + 2 - 3$$

$$= \frac{1}{2}a - 1$$

(c) $g(a) = \dfrac{1}{2}a - 3$

$$g(4) = \dfrac{1}{2}(4) - 3 = 2 - 3 = -1$$

Thus, $g(a) + g(4) = \left(\dfrac{1}{2}a - 3\right) + (-1)$

$$= \dfrac{1}{2}a - 3 - 1$$

$$= \dfrac{1}{2}a - 4$$

2. (a) $p(-3) = -3(-3)^2 + 2(-3) + 4$

$$= -3(9) + 2(-3) + 4$$

$$= -27 - 6 + 4$$

$$= -29$$

(b) $p(a) = -3(a)^2 + 2(a) + 4$

$$= -3a^2 + 2a + 4$$

(c) $p(2a) = -3(2a)^2 + 2(2a) + 4$

$$= -3(4a^2) + 2(2a) + 4$$

$$= -12a^2 + 4a + 4$$

(d) $p(a - 3) = -3(a - 3)^2 + 2(a - 3) + 4$

$$= -3(a - 3)(a - 3) + 2(a - 3) + 4$$

$$= -3(a^2 - 6a + 9) + 2(a - 3) + 4$$

$$= -3a^2 + 18a - 27 + 2a - 6 + 4$$

$$= -3a^2 + 20a - 29$$

3. (a) $r(a + 2) = \dfrac{-3}{(a + 2) + 1}$

$$= \dfrac{-3}{a + 3}$$

(b) $r(a) = \dfrac{-3}{a + 1}$

(c) $r(a + 2) - r(a) = \dfrac{-3}{a + 3} - \left(\dfrac{-3}{a + 1}\right)$

$$= \dfrac{-3}{a + 3} + \dfrac{3}{a + 1}$$

$$= \dfrac{(a + 1)(-3)}{(a + 1)(a + 3)} + \dfrac{3(a + 3)}{(a + 1)(a + 3)}$$

$$= \dfrac{-3a - 3}{(a + 1)(a + 3)} + \dfrac{3a + 9}{(a + 1)(a + 3)}$$

$$= \dfrac{-3a - 3 + 3a + 9}{(a + 1)(a + 3)}$$

$$= \dfrac{6}{(a + 1)(a + 3)}$$

4. $g(x + h) = 2 - 5(x + h) = 2 - 5x - 5h$

$g(x) = 2 - 5x$

$g(x + h) - g(x) = (2 - 5x - 5h) - (2 - 5x)$

$$= 2 - 5x - 5h - 2 + 5x$$

$$= -5h$$

Therefore, $\dfrac{g(x + h) - g(x)}{h} = \dfrac{-5h}{h} = -5.$

5. (a) $S(r) = 16(3.14)r + 2(3.14)r^2 = 50.24r + 6.28r^2$

(b) $S(2) = 50.24(2) + 6.28(2)^2$

$$= 50.24(2) + 6.28(4)$$

$$= 125.6 \text{ square meters}$$

(c) $S(2 + e) = 50.24(2 + e) + 6.28(2 + e)^2$

$$= 50.24(2 + e) + 6.28(2 + e)(2 + e)$$

$$= 50.24(2 + e) + 6.28(4 + 4e + e^2)$$

$$= 100.48 + 50.24e + 25.12 + 25.12e + 6.28e^2$$

$$= 125.6 + 75.36e + 6.28e^2$$

(d) $S = 125.6 + 75.36e + 6.28e^2$

$$= 125.6 + 75.36(0.3) + 6.28(0.3)^2$$

$$= 125.6 + 22.608 + 0.5652$$

$$= 148.77 \text{ square meters}$$

Thus, if the radius of 2 square meters was incorrectly measured as 2.3 meters, the surface area would be approximately 23.17 square meters too large.

11.2 Practice Problems

1. By the vertical line test, this relation is not a function.

2. $f(x) = x^2$ \hspace{2cm} $h(x) = x^2 - 5$

x	$f(x)$
-2	4
-1	1
0	0
1	1
2	4

x	$h(x)$
-2	-1
-1	-4
0	-5
1	-4
2	-1

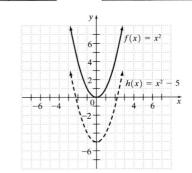

3. $f(x) = |x|$ \hspace{2cm} $p(x) = |x + 2|$

x	$f(x)$
-2	2
-1	1
0	0
1	1
2	2

x	$p(x)$
-4	2
-3	1
-2	0
-1	1
0	2
1	3
2	4

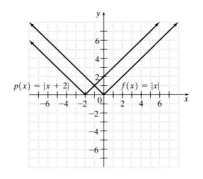

4. $f(x) = x^3$

x	$f(x)$
-2	-8
-1	-1
0	0
1	1
2	8

We recognize that $h(x)$ will have a similar shape, but the curve will be shifted 4 units to the left and 3 units upward.

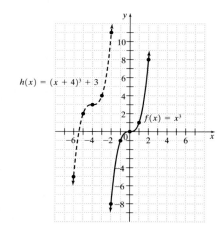

$h(x) = (x + 4)^3 + 3$

$f(x) = x^3$

5. $f(x) = \dfrac{2}{x}$

x	$f(x)$
-4	$-\dfrac{1}{2}$
-2	-1
-1	-2
$-\dfrac{1}{2}$	-4
0	undefined
$\dfrac{1}{2}$	4
1	2
2	1
4	$\dfrac{1}{2}$

The graph of $g(x)$ is 1 unit to the left and 2 units below $f(x)$. We use each point on $f(x)$ to guide us in graphing $g(x)$. *Note:* $x = -1$ is undefined on $g(x)$.

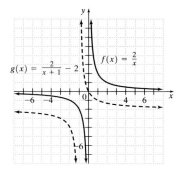

$g(x) = \dfrac{2}{x+1} - 2$

$f(x) = \dfrac{2}{x}$

11.3 Practice Problems

1. (a)
$$\begin{aligned}
(f + g)(x) &= f(x) + g(x) \\
&= (4x + 5) + (2x^2 + 7x - 8) \\
&= 4x + 5 + 2x^2 + 7x - 8 \\
&= 2x^2 + 11x - 3
\end{aligned}$$

(b) Using the formula obtained in **(a)**,
$$\begin{aligned}
(f + g)(x) &= 2x^2 + 11x - 3 \\
(f + g)(4) &= 2(4)^2 + 11(4) - 3 \\
&= 2(16) + 11(4) - 3 \\
&= 32 + 44 - 3 \\
&= 73
\end{aligned}$$

2. (a)
$$\begin{aligned}
(fg)(x) &= f(x) \cdot g(x) \\
&= (3x + 2)(x^2 - 3x - 4) \\
&= 3x^3 - 9x^2 - 12x + 2x^2 - 6x - 8 \\
&= 3x^3 - 7x^2 - 18x - 8
\end{aligned}$$

(b) Using the formula obtained in **(a)**,
$$\begin{aligned}
(fg)(x) &= 3x^3 - 7x^2 - 18x - 8 \\
(fg)(2) &= 3(2)^3 - 7(2)^2 - 18(2) - 8 \\
&= 3(8) - 7(4) - 18(2) - 8 \\
&= 24 - 28 - 36 - 8 \\
&= -48
\end{aligned}$$

3. (a) $\left(\dfrac{g}{h}\right)(x) = \dfrac{5x + 1}{3x - 2}$, where $x \neq \dfrac{2}{3}$

(b) $\left(\dfrac{g}{p}\right)(x) = \dfrac{5x + 1}{5x^2 + 6x + 1} = \dfrac{5x + 1}{(5x + 1)(x + 1)} = \dfrac{1}{(x + 1)}$,

where $x \neq -\dfrac{1}{5}, x \neq -1$

(c) $\left(\dfrac{g}{h}\right)(x) = \dfrac{5x + 1}{3x - 2}$

$\left(\dfrac{g}{h}\right)(3) = \dfrac{5(3) + 1}{3(3) - 2} = \dfrac{15 + 1}{9 - 2} = \dfrac{16}{7}$

4.
$$\begin{aligned}
f[g(x)] &= f(3x - 4) \\
&= 2(3x - 4) - 1 \\
&= 6x - 8 - 1 \\
&= 6x - 9
\end{aligned}$$

5. (a)
$$\begin{aligned}
f[g(x)] &= f(x + 2) \\
&= 2(x + 2)^2 - 3(x + 2) + 1 \\
&= 2(x^2 + 4x + 4) - 3x - 6 + 1 \\
&= 2x^2 + 8x + 8 - 3x - 6 + 1 \\
&= 2x^2 + 5x + 3
\end{aligned}$$

(b)
$$\begin{aligned}
g[f(x)] &= g(2x^2 - 3x + 1) \\
&= (2x^2 - 3x + 1) + 2 \\
&= 2x^2 - 3x + 1 + 2 \\
&= 2x^2 - 3x + 3
\end{aligned}$$

6. (a)
$$\begin{aligned}
(g \circ f)(x) &= g[f(x)] \\
&= g(3x + 1) \\
&= \dfrac{2}{(3x + 1) - 3} \\
&= \dfrac{2}{3x + 1 - 3} \\
&= \dfrac{2}{3x - 2}
\end{aligned}$$

(b) $(g \circ f)(-3) = \dfrac{2}{3(-3) - 2} = \dfrac{2}{-9 - 2} = \dfrac{2}{-11} = -\dfrac{2}{11}$

11.4 Practice Problems

1. (a) A is a function. No two pairs have the same second coordinate. Thus, A is a one-to-one function.

(b) B is a function. The pair $(1, 1)$ and $(-1, 1)$ share a common second coordinate. Therefore, B is not a one-to-one function.

2. A horizontal line through the graphs of **(a)** and **(b)** intersects the graphs more than once. The graphs do not represent one-to-one functions.

(a)

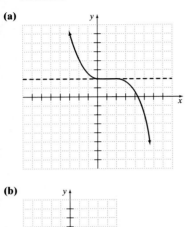

(b)

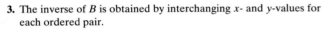

3. The inverse of B is obtained by interchanging x- and y-values for each ordered pair.

$$B^{-1} = \{(2, 1), (8, 7), (7, 8), (12, 10)\}$$

4.
$$y = 4 - 6x$$
$$x = 4 - 6y$$
$$x - 4 = -6y$$
$$-x + 4 = 6y$$
$$\frac{-x + 4}{6} = y$$
$$g^{-1}(x) = \frac{-x + 4}{6}$$

5.
$$y = 0.75 + 0.55(x - 1)$$
$$x = 0.75 + 0.55(y - 1)$$
$$x = 0.75 + 0.55y - 0.55$$
$$x = 0.55y + 0.2$$
$$x - 0.2 = 0.55y$$
$$\frac{x - 0.2}{0.55} = y$$
$$f^{-1}(x) = \frac{x - 0.2}{0.55}$$

6.
$$f(x) = -\frac{1}{4}x + 1$$
$$y = -\frac{1}{4}x + 1$$
$$x = -\frac{1}{4}y + 1$$
$$x - 1 = -\frac{1}{4}y$$
$$-4x + 4 = y$$
$$f^{-1}(x) = -4x + 4$$

Now graph each line.

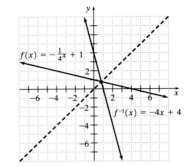

$f(x) = -\frac{1}{4}x + 1$

$f^{-1}(x) = -4x + 4$

Chapter 12

12.1 Practice Problems

1. Graph $f(x) = 3^x$.

$$f(x) = 3^x$$
$$f(-2) = 3^{-2} = \left(\frac{1}{3}\right)^2 = \frac{1}{9}$$
$$f(-1) = 3^{-1} = \left(\frac{1}{3}\right) = \frac{1}{3}$$
$$f(0) = 3^0 = 1$$
$$f(1) = 3^1 = 3$$
$$f(2) = 3^2 = 9$$

x	$f(x)$
-2	$\frac{1}{9}$
-1	$\frac{1}{3}$
0	1
1	3
2	9

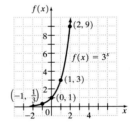

$f(x) = 3^x$

2. Graph $f(x) = \left(\frac{1}{3}\right)^x$.

$$f(x) = \left(\frac{1}{3}\right)^x$$
$$f(-2) = \left(\frac{1}{3}\right)^{-2} = 3^2 = 9$$
$$f(-1) = \left(\frac{1}{3}\right)^{-1} = 3^1 = 3$$
$$f(0) = \left(\frac{1}{3}\right)^0 = 1$$
$$f(1) = \left(\frac{1}{3}\right)^1 = \frac{1}{3}$$
$$f(2) = \left(\frac{1}{3}\right)^2 = \frac{1}{9}$$

x	$f(x)$
-2	9
-1	3
0	1
1	$\frac{1}{3}$
2	$\frac{1}{9}$

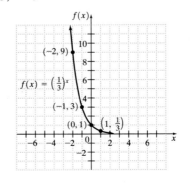

$f(x) = \left(\frac{1}{3}\right)^x$

3. Graph $f(x) = 3^{x+2}$.

$f(-4) = 3^{-4+2} = 3^{-2} = \dfrac{1}{3^2} = \dfrac{1}{9}$

$f(-3) = 3^{-3+2} = 3^{-1} = \dfrac{1}{3}$

$f(-2) = 3^{-2+2} = 3^0 = 1$

$f(-1) = 3^{-1+2} = 3^1 = 3$

$f(0) = 3^{0+2} = 3^2 = 9$

x	$f(x)$
-4	$\dfrac{1}{9}$
-3	$\dfrac{1}{3}$
-2	1
-1	3
0	9

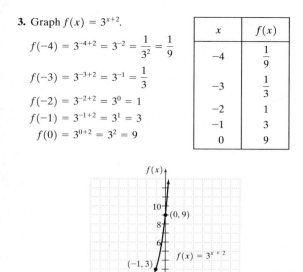

4. Graph $f(x) = e^{x-2}$.

$f(x) = e^{x-2}$ (values rounded to nearest hundredth)

$f(4) = e^{4-2} = e^2 = 7.39$

$f(3) = e^{3-2} = e^1 = 2.72$

$f(2) = e^{2-2} = e^0 = 1$

$f(1) = e^{1-2} = e^{-1} = 0.37$

$f(0) = e^{0-2} = e^{-2} = 0.14$

x	$f(x)$
4	7.39
3	2.72
2	1
1	0.37
0	0.14

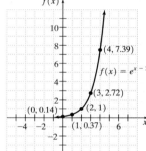

5. Solve $2^x = \dfrac{1}{32}$.

$2^x = \dfrac{1}{32}$

$2^x = \dfrac{1}{2^5}$ Because $2^5 = 32$.

$2^x = 2^{-5}$ Because $\dfrac{1}{2^5} = 2^{-5}$.

$x = -5$ Property of exponential equations.

6. Here $P = 4000$, $r = 0.11$, and $t = 2$.

$A = P(1 + r)^t$

$= 4000(1 + 0.11)^2$

$= 4000(1.11)^2$

$= 4000(1.2321)$

$= 4928.4$

Uncle Jose will have $4928.40.

7. How much money would Collette have if she invested $1500 for 8 years at 8% annual interest if the interest is compounded quarterly (four times a year)?

Here $P = 1500$, $r = 0.08$, $t = 8$, and $n = 4$.

$A = P\left(1 + \dfrac{r}{n}\right)^{nt}$

$= 1500\left(1 + \dfrac{.08}{4}\right)^{[4(8)]}$

$= 1500(1 + .02)^{32}$

$= 1500(1.02)^{32}$

$\approx 1500(1.884540592)$

≈ 2826.81089

Collette will have approximately $2826.81.

8. Here $C = 20$ and $t = 5000$.

$A = 20e^{-0.0016008(5000)}$

$A = 20e^{-8.004}$

$A \approx 20(0.0003341) = 0.006682$

Thus, 0.007 milligrams of americium 241 would be present in 5000 years.

12.2 Practice Problems

1. Use the fact that $x = b^y$ is equivalent to $\log_b x = y$.

  (a) Here $x = 49$, $b = 7$, and $y = 2$. So $2 = \log_7 49$.

  (b) Here $x = \dfrac{1}{64}$, $b = 4$, and $y = -3$. So $-3 = \log_4\left(\dfrac{1}{64}\right)$.

2. (a) Here $y = 3$, $b = 5$ and $x = 125$. Thus, since $x = b^y$, $125 = 5^3$.

  (b) Here $y = -2$, $b = 6$, and $x = \dfrac{1}{36}$. So $\dfrac{1}{36} = 6^{-2}$.

3. (a) $\log_b 125 = 3$; then $125 = b^3$

$5^3 = b^3$

$b = 5$

  (b) $\log_{1/2} 32 = x$; then $32 = \left(\dfrac{1}{2}\right)^x$

$2^5 = \left(\dfrac{1}{2}\right)^x$

$\dfrac{1}{2^{-5}} = \left(\dfrac{1}{2}\right)^x$

$\left(\dfrac{1}{2}\right)^{-5} = \left(\dfrac{1}{2}\right)^x$

$x = -5$

4. $\log_{10} 0.1 = x$

$0.1 = 10^x$

$10^{-1} = 10^x$

$-1 = x$

Thus, $\log_{10} 0.1 = -1$.

5. Graph $y = \log_{1/2} x$.

To graph $y = \log_{1/2} x$, we first write $x = \left(\dfrac{1}{2}\right)^y$. We make a table of values.

x	y
$\dfrac{1}{4}$	2
$\dfrac{1}{2}$	1
1	0
2	-1
4	-2

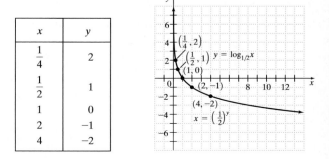

6. Make a table of values for each equation.

$$y = \log_6 x \qquad\qquad y = 6^x$$

x	y
$\frac{1}{6}$	-1
1	0
6	1
36	2

x	y
-1	$\frac{1}{6}$
0	1
1	6
2	36

Ordered pairs reversed

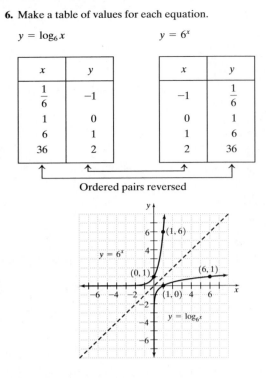

12.3 Practice Problems

1. Property 1 can be extended to three logarithms.
$$\log_b MNP = \log_b M + \log_b N + \log_b P$$
Thus,
$$\log_4 WXY = \log_4 W + \log_4 X + \log_4 Y.$$

2. $\log_b M + \log_b N + \log_b P = \log_b MNP$

Therefore,
$$\log_7 w + \log_7 8 + \log_7 x = \log_7 (w \cdot 8 \cdot x) = \log_7 8wx.$$

3. $\log_3 \left(\dfrac{17}{5}\right) = \log_3 17 - \log_3 5$

4. $\log_b 132 - \log_b 4 = \log_b \left(\dfrac{132}{4}\right) = \log_b 33$

5. $\dfrac{1}{3}\log_7 x - 5\log_7 y = \log_7 x^{1/3} - \log_7 y^5$ By property 3.

$$\phantom{\dfrac{1}{3}\log_7 x - 5\log_7 y} = \log_7 \left(\dfrac{x^{1/3}}{y^5}\right) \quad \text{By property 2.}$$

6. $\log_3 \left(\dfrac{x^4 y^5}{z}\right) = \log_3 x^4 y^5 - \log_3 z$ By property 2.

$$\phantom{\log_3 \left(\dfrac{x^4 y^5}{z}\right)} = \log_3 x^4 + \log_3 y^5 - \log_3 z \quad \text{By property 1.}$$

$$\phantom{\log_3 \left(\dfrac{x^4 y^5}{z}\right)} = 4\log_3 x + 5\log_3 y - \log_3 z \quad \text{By property 3.}$$

7. $\log_4 x + \log_4 5 = 2$

$\quad \log_4 5x = 2 \qquad$ By property 1.

Converting to exponential form, we have

$4^2 = 5x$

$16 = 5x$

$\dfrac{16}{5} = x$

$x = \dfrac{16}{5}$

8. $\log_{10} x - \log_{10}(x + 3) = -1$

$\log_{10}\left[\dfrac{x}{(x + 3)}\right] = -1 \qquad$ By property 2.

$\left[\dfrac{x}{x + 3}\right] = 10^{-1} \qquad$ Convert to exponential form.

$x = \dfrac{1}{10}(x + 3) \qquad$ Multiply each side by $(x + 3)$.

$x = \dfrac{1}{10}x + \dfrac{3}{10} \qquad$ Simplify.

$\dfrac{9}{10}x = \dfrac{3}{10}$

$x = \dfrac{1}{3}$

9. (a) $\log_7 1 = 0 \qquad$ By property 5.

(b) $\log_8 8 = 1 \qquad$ By property 4.

(c) $\log_{12} 13 = \log_{12}(y + 2)$

$\quad 13 = y + 2 \qquad$ By property 6.

$\quad y = 11$

10. $\log_3 2 - \log_3 5 = \log_3 6 + \log_3 x$

$\log_3 \left(\dfrac{2}{5}\right) = \log_3 6x \qquad$ By property 2 and property 3.

$\dfrac{2}{5} = 6x \qquad$ By property 6.

$x = \dfrac{1}{15}$

12.4 Practice Problems

1. (a) $4.36 \boxed{\log x} \approx 0.639486489$

(b) $436 \boxed{\log x} \approx 2.639486489$

(c) $0.2418 \boxed{\log x} \approx -0.616543703$

2. We know that $\log x = 2.913$ is equivalent to $10^{2.913} = x$. Using a calculator we have

$2.913 \boxed{10^x} \approx 818.46479.$

Thus, $x \approx 818.46479.$

3. To evaluate antilog(-3.0705) using a scientific calculator, we have

$3.0705 \boxed{+/-} \boxed{10^x} \approx 8.5015869 \times 10^{-4}.$

Thus, antilog$(-3.0705) \approx 0.00085015869.$

4. (a) $\log x = 0.06134$ is equivalent to $10^{0.06134} = x$.

$0.06134 \boxed{10^x} \approx 1.1517017$

Thus, $x \approx 1.1517017.$

(b) $\log x = -4.6218$ is equivalent to $10^{-4.6218} = x$.

$4.6218 \boxed{+/-} \boxed{10^x} \approx 2.3889112 \times 10^{-5}$

Thus, $x \approx 0.000023889112.$

5. (a) $4.82 \boxed{\ln x} \approx 1.572773928$

(b) $48.2 \boxed{\ln x} \approx 3.875359021$

(c) $0.0793 \boxed{\ln x} \approx -2.53451715$

6. (a) If $\ln x = 3.1628$, then $e^{3.1628} = x$.

$3.1628 \boxed{e^x} \approx 23.636686$

Thus, $x \approx 23.636686.$

(b) If $\ln x = -2.0573$, then $e^{-2.0573} = x$.

$$2.0573 \;\boxed{+/-}\; \boxed{e^x} \approx 0.1277986$$

Thus, $x \approx 0.1277986$.

7. To evaluate $\log_9 3.76$, we use the change of base formula.

$$\log_9 3.76 = \frac{\log 3.76}{\log 9}$$

On a calculator, find the following.

$$3.76 \;\boxed{\log x}\; \boxed{\div}\; 9 \;\boxed{\log x}\; \boxed{=}\; 0.602769044.$$

Thus, $\log_9 3.76 \approx 0.602769$.

8. By the change of base formula,

$$\log_8 0.009312 = \frac{\log_e 0.009312}{\log_e 8} = \frac{\ln 0.009312}{\ln 8}$$

On a calculator, find the following.

$$0.009312 \;\boxed{\ln}\; \boxed{\div}\; 8 \;\boxed{\ln}\; \boxed{=}\; -2.24889774$$

Thus, $\log_8 0.009312 \approx -2.24889774$.

9. Make a table of values.

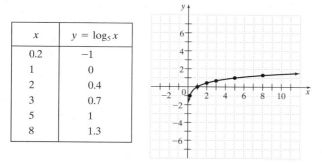

x	$y = \log_5 x$
0.2	-1
1	0
2	0.4
3	0.7
5	1
8	1.3

12.5 Practice Problems

1.
$$\log(x + 5) = 2 - \log 5$$
$$\log(x + 5) + \log 5 = 2$$
$$\log[5(x + 5)] = 2$$
$$\log(5x + 25) = 2$$
$$5x + 25 = 10^2$$
$$5x + 25 = 100$$
$$5x = 75$$
$$x = 15$$

Check:
$$\log(15 + 5) \overset{?}{=} 2 - \log 5$$
$$\log 20 \overset{?}{=} 2 - \log 5$$
$$1.301029996 \overset{?}{=} 2 - 0.698970004$$
$$1.301029996 = 1.301029996 \;\checkmark$$

2. $\log(x + 3) - \log x = 1$

$$\log\left(\frac{x + 3}{x}\right) = 1$$
$$\frac{x + 3}{x} = 10^1$$
$$\frac{x + 3}{x} = 10$$
$$x + 3 = 10x$$
$$3 = 9x$$
$$\frac{1}{3} = x$$

Check:
$$\log\left(\frac{1}{3} + 3\right) - \log\frac{1}{3} \overset{?}{=} 1$$
$$\log\frac{10}{3} - \log\frac{1}{3} \overset{?}{=} 1$$
$$\log\left(\frac{\frac{10}{3}}{\frac{1}{3}}\right) \overset{?}{=} 1$$
$$\log 10 \overset{?}{=} 1$$
$$1 = 1 \;\checkmark$$

3. $\log 5 - \log x = \log(6x - 7)$

$$\log\left(\frac{5}{x}\right) = \log(6x - 7)$$
$$\frac{5}{x} = 6x - 7$$
$$5 = 6x^2 - 7x$$
$$0 = 6x^2 - 7x - 5$$
$$0 = (3x - 5)(2x + 1)$$

$3x - 5 = 0 \quad$ or $\quad 2x + 1 = 0$
$3x = 5 \qquad\qquad 2x = -1$
$x = \dfrac{5}{3} \qquad\qquad x = -\dfrac{1}{2}$

Check: $\log 5 - \log x = \log(6x - 7)$

$x = \dfrac{5}{3}$: $\quad \log 5 - \log\left(\dfrac{5}{3}\right) \overset{?}{=} \log\left[6\left(\dfrac{5}{3}\right) - 7\right]$
$$\log 5 - \log\frac{5}{3} \overset{?}{=} \log(10 - 7)$$
$$\log 5 - \log\frac{5}{3} \overset{?}{=} \log 3$$
$$\log\left[\frac{5}{\frac{5}{3}}\right] \overset{?}{=} \log 3$$
$$\log 3 = \log 3 \;\checkmark$$

$x = -\dfrac{1}{2}$: $\quad \log 5 - \log\left(-\dfrac{1}{2}\right) \overset{?}{=} \log\left[6\left(-\dfrac{1}{2}\right) - 7\right]$

Since logarithms of negative numbers do not exist, $x = -\dfrac{1}{2}$ is not a valid solution.

The only solution is $\dfrac{5}{3}$.

4. Take the logarithm of each side.

$$\log 3^x = \log 5$$
$$x \log 3 = \log 5$$
$$x = \frac{\log 5}{\log 3}$$

5. Take the logarithm of each side.

$$\log 2^{3x+1} = \log 9^{x+1}$$
$$(3x + 1)\log 2 = (x + 1)\log 9$$
$$3x \log 2 + \log 2 = x \log 9 + \log 9$$
$$3x \log 2 - x \log 9 = \log 9 - \log 2$$
$$x(3 \log 2 - \log 9) = \log 9 - \log 2$$
$$x = \frac{\log 9 - \log 2}{3 \log 2 - \log 9}$$

Use the following keystrokes.

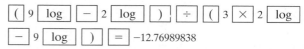

$$\boxed{(}\; 9 \;\boxed{\log}\; \boxed{-}\; 2 \;\boxed{\log}\; \boxed{)}\; \boxed{\div}\; \boxed{(}\; 3 \;\boxed{\times}\; 2 \;\boxed{\log}$$
$$\boxed{-}\; 9 \;\boxed{\log}\; \boxed{)}\; \boxed{=}\; -12.76989838$$

Rounding to the nearest thousandth, we have $x \approx -12.770$.

6. Take the natural logarithm of each side.

$\ln 20.98 = \ln e^{3.6x}$

$\ln 20.98 = (3.6x)(\ln e)$

$\ln 20.98 = 3.6x$

$\dfrac{\ln 20.98}{3.6} = x$

On a scientific calculator, find the following.

20.98 | ln | ÷ | 3.6 | = | 0.845436001

Rounding to the nearest ten-thousandth, we have $x \approx 0.8454$.

7. We use the formula $A = P(1 + r)^t$, where

$A = \$10{,}000$, $P = \$4000$, and $r = 0.08$.

$10{,}000 = 4000(1 + 0.08)^t$

$10{,}000 = 4000(1.08)^t$

$\dfrac{10{,}000}{4000} = (1.08)^t$

$2.5 = (1.08)^t$

$\log(2.5) = \log(1.08)^t$

$\log 2.5 = t(\log 1.08)$

$\dfrac{\log 2.5}{\log 1.08} = t$

On a scientific calculator,

2.5 | log | ÷ | 1.08 | log | = | 11.905904.

Thus, it would take approximately 12 years.

8.

$A = A_0 e^{rt}$

$5000 = 1300 e^{0.043t}$ Substitute known values.

$\dfrac{5000}{1300} = e^{0.043t}$ Divide each side by 1300.

$\ln\left(\dfrac{5000}{1300}\right) = \ln e^{0.043t}$ Take the natural logarithm of each side.

$\ln\left(\dfrac{50}{13}\right) = 0.043t \ln e$

$\ln 50 - \ln 13 = 0.043t$

$\dfrac{\ln 50 - \ln 13}{0.043} = t$

Using a scientific calculator,

(| 50 | ln | − | 13 | ln |) | ÷ | 0.043 | = | 31.32729414

Rounding to the nearest whole years, the food shortage will develope in about 31 years.

9. Let I_J = intensity of the Japan earthquake.

Let I_S = intensity of the San Francisco earthquake.

$8.9 = \log\left(\dfrac{I_J}{I_{10}}\right) = \log I_J - \log I_0$

Solving for $\log I_0$ gives $\log I_0 = \log I_J - 8.9$

$7.1 = \log\left(\dfrac{I_S}{I_0}\right) = \log I_S - \log I_0$

Solving for $\log I_0$ gives $\log I_0 = \log I_S - 7.1$

Therefore, $\log I_J - 8.9 = \log I_S - 7.1$

$\log I_J - \log I_S = 1.8$

$\log\left(\dfrac{I_J}{I_S}\right) = 1.8$

$\dfrac{I_J}{I_S} = 10^{1.8}$

$\dfrac{I_J}{I_S} = 63.09573445$

$\dfrac{I_J}{I_S} \approx 63$

$I_J \approx 63 I_S$

The earthquake in Japan was about sixty-three times as intense as the San Francisco earthquake.

Answers to Selected Exercises

Chapter 1

Pretest Chapter 1

1. 3   **2.** −15   **3.** −18   **4.** 20   **5.** −42   **6.** $\frac{8}{3}$ **7.** 9.2   **8.** −4   **9.** 16   **10.** 64   **11.** $-\frac{8}{27}$ **12.** −256   **13.** 28   **14.** 197   **15.** $\frac{107}{12}$ **16.** 6.37   **17.** $-6x^2 + 4x^2y - 2xz$ **18.** $-3x + 4y + 12$ **19.** $5x^2 - 6x^2y - 11xy$ **20.** $7x - 2y + 7$ **21.** $5x - 7y$ **22.** $-x^2y + 17y^2$ **23.** 18   **24.** 29   **25.** 25°C   **26.** $-9x - 3y$ **27.** $-x^2 + 6xy$

1.1 Exercises

1. Whole number, rational number, real number   **3.** Irrational number, real number   **5.** Rational number, real number   **7.** Rational number, real number   **9.** Irrational number, real number   **11.** −20,000   **13.** $-2\frac{3}{8}$ **15.** +7   **17.** $-\frac{3}{4}$ **19.** 2.73   **21.** 1.3   **23.** $\frac{5}{6}$ **25.** −11   **27.** $\frac{1}{3}$ **29.** −31   **31.** $-\frac{7}{13}$ **33.** −3.8   **35.** 0.4   **37.** −25   **39.** $\frac{1}{35}$ **41.** −6   **43.** 3   **45.** 17   **47.** −28   **49.** $\frac{1}{15}$ **51.** $-\frac{11}{12}$ **53.** −3.4   **55.** 12   **57.** 1   **59.** 16.39   **61.** −5°F   **63.** −265 feet   **65.** no; $16.65 short   **67.** $140   **69.** $22,000,000   **71.** 28   **73.** 41%; 120 more kittens

1.2 Exercises

1. First change subtracting −3 to adding a positive three. Then use the rules for addition of two real numbers with different signs. Thus $-8 - (-3) = -8 + 3 = -5$.   **3.** −26   **5.** −11   **7.** 8   **9.** 5   **11.** −5   **13.** 0   **15.** −3   **17.** −0.9   **19.** 4.47   **21.** 3.7   **23.** $1\frac{7}{20}$ **25.** $-1\frac{7}{12}$ **27.** −53   **29.** −99   **31.** 7.1   **33.** $8\frac{3}{4}$ **35.** $8\frac{3}{5}$ or $\frac{43}{5}$ **37.** $-\frac{21}{20}$ or $-1\frac{1}{20}$ **39.** −8.5   **41.** 0.0499   **43.** $-5\frac{4}{5}$ **45.** −6.0023   **47.** 7   **49.** −48   **51.** −2   **53.** 11   **55.** −62   **57.** 7   **59.** 38   **61.** $149   **63.** −16°F   **65.** −21   **66.** −51   **67.** −19   **68.** −8°C   **69.** 23°F

1.3 Exercises

1. To multiply two real numbers, multiply the absolute values. The sign of the result is positive if both numbers have the same sign, but negative if the two numbers have opposite signs.   **3.** −24   **5.** 0   **7.** 24   **9.** 0.264   **11.** −1.75   **13.** −24   **15.** $\frac{4}{15}$ **17.** −3   **19.** 0   **21.** −5   **23.** −16   **25.** −2   **27.** −9   **29.** −0.9   **31.** $-\frac{3}{10}$ **33.** $-\frac{20}{3}$ **35.** −30   **37.** $\frac{9}{16}$ **39.** $-\frac{7}{9}$ **41.** −24   **43.** 24   **45.** −16   **47.** −18   **49.** $-\frac{8}{35}$ **51.** $\frac{2}{27}$ **53.** −4   **55.** −2   **57.** 17   **59.** −72   **61.** −1   **63.** $30   **65.** $328.50   **67.** 20 yards.   **69.** 70 yards   **71.** The Panthers gained 20 yards.   **73.** The Panthers would have gained 105 fewer yards.   **74.** −6.69   **75.** −72   **76.** 4.63   **77.** −88   **78.** 266 square yards

1.4 Exercises

1. The base is 4 and the exponent is 4. Thus you multiply $(4)(4)(4)(4) = 256$.   **3.** The answer is negative. When you raise a negative number to an odd power the result is always negative.   **5.** If you have parentheses surrounding the −2, then the base is −2 and the exponent is 4. The result is 16. If you do not have parentheses, then the base is 2. You evaluate to obtain 16 and then take the negative of 16, which is −16. Thus $(-2)^4 = 16$ but $-2^4 = -16$.   **7.** 27   **9.** 81   **11.** 343   **13.** −27   **15.** 64   **17.** −125   **19.** $\frac{1}{16}$ **21.** $\frac{8}{125}$ **23.** 0.81   **25.** 0.0016   **27.** 256   **29.** −256   **31.** 6^5 **33.** w^2 **35.** x^4 **37.** $(3q)^3$ **39.** 161   **41.** −91   **43.** 152   **45.** −576   **47.** −512   **49.** 16,777,216   **51.** 2   **53.** −19   **54.** $-\frac{5}{3}$ **55.** −8   **56.** 2.52   **57.** $\frac{3}{44}$

1.5 Exercises

1. $3(4) + 6(5)$ **3.** (a) 90 (b) 42   **5.** 12   **7.** −29   **9.** 14   **11.** 13   **13.** −6   **15.** 42   **17.** $\frac{9}{4}$ **19.** 0.848   **21.** $\frac{3}{10}$ **23.** 7.56   **25.** $\frac{1}{4}$ **27.** 0.125   **28.** $-\frac{19}{12}$ **29.** −1   **30.** 6   **31.** 45 ounces

1.6 Exercises

1. variable   **3.** Here we are multiplying 4 by x by x. Since we know from the definition of exponents that x multiplied by x is x^2, this gives us an answer of $4x^2$.
5. Yes, $a(b - c)$ can be written as $a[b + (-c)]$
$$3(10 - 2) = (3 \times 10) - (3 \times 2)$$
$$3 \times 8 = 30 - 6$$
$$24 = 24$$

7. $-x - 4y$ **9.** $-6a + 15b$ **11.** $8x + 2y - 4$ **13.** $6x^2 - 9xy + 3xz$ **15.** $8x^2 - 2xy - 12x$ **17.** $-15x - 45 + 35y$ **19.** $\dfrac{x^2}{4} + \dfrac{x}{2} - 2$

21. $-18a^4 + 6a^2 - 14$ **23.** $y^2 - \dfrac{4xy}{3} - 2y$ **25.** $6a^2 + 3ab - 3ac - 12a$ **27.** $-20x - 4$ **29.** $12x^2 - 15xy - 18x$

31. $-4a^2b + 2ab^2 + ab$ **33.** $-8x^2y + 4xy^2 - 12xy$ **35.** $3.75a^2 - 8.75a + 5$ **37.** $-1.89q^2 + 0.18qr + 0.72qs$

39. $8400x + 5600y$ square feet   **41.** $4500x - 12xy$ square feet   **43.** -16 **44.** 64 **45.** 14 **46.** 0 **47.** -13

1.7 Exercises

1. A term is a number, a variable, or a product of numbers and variables.   **3.** The two terms $5x$ and $-8x$ are like terms because they both have the variable x with the exponent of one.   **5.** The only like terms are $7xy$ and $-14xy$ because the other two have different exponents even though they have the same variables.   **7.** $-25b^2$ **9.** $18x^4 + 7x^2$ **11.** $-4ab - 7$ **13.** $7.1x - 3.5y$ **15.** $-2x - 8.7y$ **17.** $\dfrac{3}{4}x^2 - \dfrac{10}{3}y$

19. $-\dfrac{1}{15}x - \dfrac{2}{21}y$ **21.** $5p + q - 18$ **23.** $5x^2y - 10xy^6 - xy^2$ **25.** $5bc - 6ac$ **27.** $x^2 - 10x + 3$ **29.** $-10y^2 - 16y + 12$

31. $5a - 3ab - 8b$ **33.** $-12x + 13y$ **35.** $14x^2 + 2xy$ **37.** $-17xy + 27y^2$ **39.** $71x - 27$ **41.** $20x + 4$ meters   **43.** $32a + 10$ feet

44. $-\dfrac{2}{15}$ **45.** $-\dfrac{5}{6}$ **46.** $\dfrac{23}{50}$ **47.** $-\dfrac{15}{98}$ **48.** 0.2 liter

1.8 Exercises

1. -5 **3.** -11 **5.** $\dfrac{25}{2}$ **7.** -26 **9.** 10 **11.** 3 **13.** -24 **15.** $\dfrac{25}{4}$ **17.** 9 **19.** 39 **21.** -2 **23.** 15 **25.** -9 **27.** 29

29. 42 **31.** 29 **33.** 4968 square feet   **35.** 129 square millimeters   **37.** 166.5 square inches   **39.** 133 square feet   **41.** 78.5 square meters

43. $14°\text{F}$ **45.** $\$300.26$ **47.** $-76°\text{F}$ to $-22°\text{F}$ **49.** 12.4 miles   **50.** 16 **51.** $-x^2 + 2x - 4y$ **52.** 6.2 minutes/song

Putting Your Skills to Work

1. 21.3 **2.** yes; 25.6

1.9 Exercises

1. $-(3x + 2y)$ **3.** distributive   **5.** $3x + 6y$ **7.** $5a + 3b$ **9.** $-245 + 25x$ **11.** $8x^3 - 4x^2 + 12x$ **13.** $4x - 6y - 3$ **15.** $15a - 60ab$

17. $-x^3 + 2x^2 - 3x - 12$ **19.** $3a^2 + 16b + 12b^2$ **21.** $-7a + 8b$ **23.** $12a^2 - 8b$ **25.** 219 successful attempts; 8541 unsuccessful attempts

27. $97.52°\text{F}$ **28.** $453,416$ square feet   **29.** $300,000$ square feet; $\$16,500,000$ **30.** 11.375 square feet; $\$1387.75$

Putting Your Skills to Work

1. $\$335.20$ **2.** $\$22,050.00$ **3.** $\$598.00$ **4.** $\$507.00$

Chapter 1 Review Problems

1. -8 **2.** -4.2 **3.** -9 **4.** 1.9 **5.** $-\dfrac{1}{3}$ **6.** $-\dfrac{7}{22}$ **7.** $\dfrac{1}{6}$ **8.** $\dfrac{22}{15}$ **9.** 8 **10.** 13 **11.** -33 **12.** 9.2 **13.** $-\dfrac{13}{8}$ **14.** $\dfrac{1}{2}$

15. -22.7 **16.** -88 **17.** -4 **18.** 16 **19.** -29 **20.** 1 **21.** -3 **22.** 18 **23.** 32 **24.** $-\dfrac{2}{3}$ **25.** $-\dfrac{25}{7}$ **26.** -72 **27.** 30

28. -30 **29.** $-\dfrac{1}{2}$ **30.** $-\dfrac{4}{7}$ **31.** -30 **32.** -5 **33.** -9.1 **34.** 0.9 **35.** 10.1 **36.** -1.2 **37.** 1.9 **38.** -1.3 **39.** 24 yards

40. $-22°\text{F}$ **41.** 7177 feet   **42.** $2\dfrac{1}{4}$ point loss   **43.** -243 **44.** -128 **45.** 625 **46.** $\dfrac{8}{27}$ **47.** -81 **48.** 0.36 **49.** $\dfrac{25}{36}$ **50.** $\dfrac{27}{64}$

51. -44 **52.** 30 **53.** 1 **54.** $15x - 35y$ **55.** $6x^2 - 14xy + 8x$ **56.** $-7x^2 + 3x - 11$ **57.** $-6xy^2 - 3xy + 3y^2$ **58.** $-5a^2b + 3bc$

59. $-3x - 4y$ **60.** $-5x^2 - 35x - 9$ **61.** $10x^2 - 8x - \dfrac{1}{2}$ **62.** -55 **63.** 1 **64.** -4 **65.** -15 **66.** 10 **67.** -16 **68.** $\dfrac{32}{5}$

69. $\$810$ **70.** $86°\text{F}$ **71.** $\$2119.50$ **72.** $\$8580.00$ **73.** $100,000$ square feet; $\$200,000$ **74.** 10.45 square feet; $\$689.70$ **75.** $-2x + 42$

76. $-17x - 18$ **77.** $-2 + 10x$ **78.** $-12x^2 + 63x$ **79.** $5xy^3 - 6x^3y - 13x^2y^2 - 6x^2y$ **80.** $x - 10y + 35 - 15xy$ **81.** $10x - 22y - 36$

82. $-10a + 25ab - 15b^2 - 10ab^2$ **83.** $-3x - 9xy + 18y^2$ **84.** $10x + 8xy - 32y$

Chapter 1 Test

1. 4 **2.** 0.2 **3.** 96 **4.** -70 **5.** 4 **6.** -3 **7.** -64 **8.** 1.69 **9.** $\dfrac{16}{81}$ **10.** 8 **11.** -25 **12.** $-5x^2 - 10xy + 35x$

13. $6a^2b^2 + 4ab^3 - 14a^2b^3$ **14.** $2a^2b + \dfrac{15}{2}ab$ **15.** $8a^2 + 20ab$ **16.** $5a + 30$ **17.** $14x - 16y$ **18.** 122 **19.** 37 **20.** $\dfrac{13}{6}$

21. 96.6 kilometers/hour   **22.** $22,800$ square feet   **23.** $\$23.12$ **24.** 452.16 square feet   **25.** $-3a - 9ab + 3b^2 - 3ab^2$ **26.** $-69x + 90y - 51$

Chapter 2

Pretest Chapter 2

1. $x = -16$ **2.** $x = 48$ **3.** $x = -4$ **4.** $x = 4$ **5.** $x = -\dfrac{7}{2}$ **6.** $x = -\dfrac{1}{7}$ **7.** $x = 1$ **8.** $x = -4$ **9.** $x = \dfrac{5}{2}$ **10.** $x = \dfrac{17}{10}$

11. $x = -\dfrac{2}{3}$ **12.** $x = 17$ **13. (a)** $F = \dfrac{9C + 160}{5}$ **(b)** $F = 5°$ **14. (a)** $r = \dfrac{I}{Pt}$ **(b)** 6% **15.** $<$ **16.** $>$ **17.** $>$ **18.** $<$

19. (number line with point at -2) x **20.** (number line with open circle at 6) x **21.** one side $= 18$ feet, second side $= 7$ feet, third side $= 13$ feet

22. 1st package $= 7.5$ pounds, 2nd package $= 4$ pounds, 3rd package $= 5.5$ pounds   **23.** width $= 11$ feet, length $= 33$ feet   **24.** \$400 at 12%, \$600 at 9%.   **25.** 22,000 people   **26.** 3 nickels, 7 dimes, 10 quarters

2.1 Exercises

1. equal, equal   **3.** solution   **5.** Answers may vary. A sample answer is to isolate the variable.   **7.** $x = 6$ **9.** $6 = x$ **11.** $x = 16$ **13.** $x = 7$ **15.** $x = 1$ **17.** $x = 0$ **19.** $x = -2$ **21.** $x = 66$ **23.** $x = 27$ **25.** $x = -11$ **27.** no; $x = 26$ **29.** yes   **31.** no; $x = 8$

33. yes   **35.** 4   **37.** 22   **39.** 3.5   **41.** $x = -2.9$ **43.** $\frac{1}{3}$ **45.** $x = -\frac{1}{5}$ **47.** $\frac{7}{6}$ or $1\frac{1}{6}$ **49.** $27\frac{1}{8}$ **51.** 7.2   **53.** $x = -3.783$

54. $-2x - 4y$ **55.** $-2y^2 - 4y + 4$ **56.** 117 feet   **57.** \$14.92

2.2 Exercises

1. 6   **3.** 7   **5.** $x = 35$ **7.** $x = -27$ **9.** $x = 80$ **11.** $x = -15$ **13.** $x = 4$ **15.** $x = -\frac{8}{3}$ **17.** $x = 50$ **19.** $x = 15$ **21.** $x = -7$

23. $x = 0.2$ or $\frac{1}{5}$ **25.** no, $x = -7$ **27.** yes   **29.** $y = -0.8$ **31.** $t = \frac{8}{3}$ **33.** $y = -0.7$ **35.** $x = 3$ **37.** $x = -4$ **39.** $x = \frac{7}{9}$ **41.** $x = \frac{7}{6}$

43. $x = -5.26$ **45.** To solve an equation, we are performing steps to get an equivalent equation that has the same solution. Now $a = b$ and $a(0) = b(0)$ are not equivalent equations because they do not have the same solution. So we must have the requirement that when we multiply both sides of the equation by c, it is absolutely essential that c is nonzero.   **47.** 42   **48.** -37 **49.** 21   **50.** 104 calves   **51.** \$632

2.3 Exercises

1. 2   **3.** 6   **5.** -17.5 **7.** 13   **9.** 15   **11.** 40   **13.** -36 **15.** 8   **17.** 3   **19.** 7   **21.** -9 **23.** yes

25. no; $x = 12$ **27.** 7; 7   **29.** 0   **31.** 5   **33.** -4 **35.** 6   **37.** 5   **39.** $\frac{1}{2}$ **41.** 1   **43.** 3.1   **45.** -4 **47.** 0   **49.** 2

51. 1.2   **53.** -1.2 **55.** 2   **57.** $-\frac{3}{5}$ **59.** 1   **61.** 1   **63.** -7 **65.** 42   **67.** -4 **69.** 2.78   **70.** $14x^2 - 14xy$ **71.** $-10x - 60$

72. \$1216.20   **73.** $1\frac{3}{4}x + 8$ feet

2.4 Exercises

1. 2   **3.** $2\frac{1}{2}$ or $\frac{5}{2}$ **5.** 1   **7.** 24   **9.** 20   **11.** 7   **13.** 18   **15.** -3.5 **17.** yes   **19.** no   **21.** 1   **23.** 0   **25.** 2   **27.** -3 **29.** 4

31. -22 **33.** 2   **35.** -12 **37.** $-\frac{13}{12}$ or $-1\frac{1}{12}$ **39.** $\frac{4}{113}$ **41.** $\frac{18}{17}$ **43.** no solution   **45.** infinite number of solutions

47. $\frac{27}{14}$ or $1\frac{13}{14}$ **48.** $-\frac{69}{8}$ or $-8\frac{5}{8}$ **49.** $\frac{66}{5}$ or $13\frac{1}{5}$ **50.** 12   **51.** 18.6 miles   **52.** \$108   **53.** \$226.08   **54.** \$273.60

2.5 Exercises

1. Multiply each term by 5. Then add -160 to each side. Then divide each side by 9. We would obtain $\frac{5F - 160}{9} = C$.

3. (a) 10 meters   **(b)** 16 meters   **5. (a)** $y = \frac{3}{5}x - 3$ or $\frac{15 - 3x}{-5}$ **(b)** -6 **7.** $b = \frac{2A}{h}$ **9.** $P = \frac{I}{rt}$ **11.** $m = \frac{y - b}{x}$ **13.** $t = \frac{A - P}{Pr}$

15. $y = \frac{5}{6}x - 1$ **17.** $x = -\frac{4}{3}y + 12$ **19.** $y = \frac{c - ax}{b}$ **21.** $r^2 = \frac{A}{\pi}$ **23.** $g = \frac{2S}{t^2}$ **25.** $h = \frac{S - 2\pi r^2}{2\pi r}$ **27.** $h = \frac{V}{\pi r^2}$ **29.** $L = \frac{V}{WH}$

31. $r^2 = \frac{3V}{\pi h}$ **33.** $W = \frac{P - 2L}{2}$ **35.** $a^2 = c^2 - b^2$ **37.** $C = \frac{5(F - 32)}{9}$ or $C = \frac{5F - 160}{9}$ **39.** $T = \frac{PV}{k}$ **41.** $S = \frac{360A}{\pi r^2}$

43. The width is 0.8 miles.   **45.** The length is 30 feet.   **47. (a)** $\frac{V - 7050}{1100} = x$ **(b)** 2002   **49.** I doubles.   **51.** A increases four times.

53. 10   **54.** $\frac{3}{10}$ **55.** 30   **56.** 65   **57.** 39,000 square feet   **58.** $10\frac{7}{12}$ hours

Putting Your Skills to Work

1. 3.3 million tons   **2.** 1.9 million tons

2.6 Exercises

1. Yes, both statements imply 5 is to the right of -6 on the number line.   **3.** $>$ **5.** $<$ **7.** $>$ **9.** $<$ **11.** $<$ **13.** $<$ **15.** 0.0247

17. (number line: -7 -6 -5 -4 -3, point at -6)   **19.** (number line: 5 6 7 8 9, open circle at 7)   **21.** (number line: $\frac{1}{4}$ $\frac{1}{2}$ $\frac{3}{4}$ 1 $\frac{5}{4}$, open circle at $\frac{3}{4}$)   **23.** (number line: -3.8 -3.6 -3.4)

25. (number line: 10 15 20 25 30, open circle at 25)   **27.** $x \geq -\frac{2}{3}$ **29.** $x < -20$ **31.** $x > 6.5$ **33.** $V > 580$ **35.** $h \geq 37$ **37.** $h \geq 48$

39. (number line: -3 -2 -1 0 1 2, point at $-\frac{5}{2}$)   **41.** $\frac{9}{16}$ **42.** $-10x + 2$ **43.** $x = 78$ **44.** yes   **45.** \$1575   **46.** 495 feet

2.7 Exercises

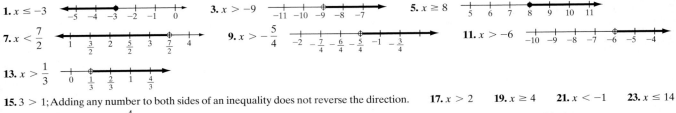

1. $x \le -3$ **3.** $x > -9$ **5.** $x \ge 8$

7. $x < \dfrac{7}{2}$ **9.** $x > -\dfrac{5}{4}$ **11.** $x > -6$

13. $x > \dfrac{1}{3}$

15. $3 > 1$; Adding any number to both sides of an inequality does not reverse the direction.   **17.** $x > 2$ **19.** $x \ge 4$ **21.** $x < -1$ **23.** $x \le 14$
25. $x < -3$ **27.** $x > -\dfrac{4}{11}$ **29.** $x > -1.46$ **31.** 76 or greater   **33.** 8 days or more   **35.** 260 feet   **36.** 4.5 inches
37. 63.6 square inches   **38.** 15,600 square feet

2.8 Exercises

1. Long piece is 32 meters long; short piece is 15 meters long   **3.** David worked 30 hours; Sarah worked 45 hours; Kate worked 25 hours
5. Mt. McKinley is 20,320 feet high; Mt. Whitney is 14,494 feet high; Mt. Oxford is 14,153 feet high   **7.** width = 100 meters; length = 250 meters
9. width = 32.5 cm; length = 62.5 cm   **11.** The cheetah can run 70 mph; the jackal can run 35 mph; the elk can run 45 mph.   **13.** longest
side = 18 in.; shortest side = 13 in.; third side = 15 in.   **15. (a)** 38 mph   **(b)** 57 mph   **(c)** 19 mph   **17.** Original square was 11 m × 11 m.
19. $-8x^3 + 12x^2 - 32x$ **20.** $5a^2b + 30ab - 10a^2$ **21.** $-19x + 2y - 2$ **22.** $9x^2y - 6xy^2 + 7xy$ **23.** $x = -\dfrac{1}{2}$ **24.** $h = \dfrac{by + 2a}{3}$

2.9 Exercises

1. 11 bags   **3.** 7 hours   **5.** $360   **7.** Walter's profit = $4500; Jim's profit = $5040   **9.** $6000   **11.** $3000 at 7%; $2000 at 5%   **13.** $250,000
in the conservative fund; $150,000 in the growth fund   **15.** $12,000   **17.** 13 quarters; 9 nickels   **19.** 18 nickels; 6 dimes; 9 quarters   **21.** eight
$10 bills; sixteen $20 bills; eleven $100 bills   **23.** first angle measures 70°; second angle measures 35°; third angle measures 75°   **25.** $1328 at
7%, $1641 at 11%   **27.** more than 225 miles   **29.** 12   **30.** 15   **31.** -28 **32.** -25 **33.** $95.20   **34.** $19,040

Putting Your Skills to Work

1. $248.90; $1947.20   **2.** $141.21; $389.04

Chapter 2 Review Problems

1. -10 **2.** -1 **3.** 2   **4.** -3 **5.** 2.75   **6.** -13 **7.** 40.4   **8.** -7 **9.** -2 **10.** -64 **11.** -11 **12.** $\dfrac{22}{3}$ or $7\dfrac{1}{3}$ **13.** -3 **14.** -1

15. 4   **16.** 3   **17.** 3   **18.** $-\dfrac{7}{2}$ or $-3\dfrac{1}{2}$ or -3.5 **19.** -3 **20.** 2.1   **21.** $-\dfrac{7}{3}$ or $-2\dfrac{1}{3}$ **22.** 5   **23.** 0   **24.** 1   **25.** $\dfrac{2}{3}$ **26.** 5   **27.** $\dfrac{35}{11}$

28. 4   **29.** -17 **30.** $\dfrac{2}{5}$ or 0.4   **31.** 32   **32.** $\dfrac{26}{7}$ **33.** -1 **34.** $-\dfrac{5}{2}$ or $-2\dfrac{1}{2}$ or -2.5 **35.** 4   **36.** -17 **37.** 4   **38.** 4   **39.** -5

40. 23   **41.** -32 **42.** $-\dfrac{17}{5}$ or -3.4 **43.** 0   **44.** 9.75 or $9\dfrac{3}{4}$ or $\dfrac{39}{4}$ **45.** $y = 3x - 10$ **46.** $y = \dfrac{-5x - 7}{2}$ **47.** $r = \dfrac{A - P}{Pt}$

48. $h = \dfrac{A - 4\pi r^2}{2\pi r}$ **49.** $p = \dfrac{3H - a - 3}{2}$ **50.** $y = \dfrac{c - ax}{b}$ **51.** $d = \dfrac{6c - H}{5}$ **52.** $b = \dfrac{4H - 3c}{2}$

53. (a) $T = \dfrac{1000C}{WR}$ **(b)** $T = 6000$ **54. (a)** $y = \dfrac{5}{3}x - 4$ **(b)** $y = 11$ **55. (a)** $R = \dfrac{E}{I}$ **(b)** $R = 5$

56. $x \ge 1$ **57.** $x < -4$ **58.** $x > -3$

59. $x \ge 3$ **60.** $x < 2$ **61.** $x \le -8$

62. $x \ge 3$ **63.** $x < 10$ **64.** $x > \dfrac{17}{2}$

65. $x \ge \dfrac{19}{7}$ **66.** $x \ge -15$ **67.** $x > \dfrac{7}{5}$

68. $n \le 46$ **69.** $n \le 17$ **70.** 1st side = 8 yd; 2nd side = 15 yd; 3rd side = 17 yd   **71.** 1st angle = 32°; 2nd angle = 96°; 3rd angle = 52°
72. 31.25 yd and 18.75 yd   **73.** Jon = $30,000; Lauren = $18,000   **74.** 310 kilowatt-hours   **75.** 280 miles   **76.** $22,500   **77.** $200
78. $7000 at 12% and $2000 at 8%   **79.** $2000 at 4.5% and $3000 at 6%   **80.** 18 nickels; 6 dimes; 9 quarters   **81.** 7 nickels; 8 dimes;
10 quarters

Chapter 2 Test

1. $x = 2$ **2.** $x = \dfrac{1}{3}$ **3.** $y = -\dfrac{7}{2}$ or $-3\dfrac{1}{2}$ or -3.5 **4.** $y = 8.4$ or $8\dfrac{2}{5}$ or $\dfrac{42}{5}$ **5.** $x = 8$ **6.** $x = -1.2$ **7.** $y = 7$ **8.** $y = \dfrac{7}{3}$ **9.** $x = 13$

10. $x = 125$ **11.** $x = 10$ **12.** $x = -4$ **13.** $x = 12$ **14.** $x = -\dfrac{1}{5}$ or -0.2 **15.** $x = 3$ **16.** $x = 2$ **17.** $x = 18$ **18.** $w = \dfrac{A - 2P}{3}$

19. $w = \dfrac{6 - 3x}{4}$ **20.** $a = \dfrac{2A - hb}{h}$ **21.** $y = \dfrac{10ax - 5}{8ax}$ **22.** $\dfrac{P - 2L}{2} = W$ **23.** 18 feet

24. $x \le -\dfrac{1}{2}$ **25.** $x > -\dfrac{5}{4}$ **26.** $x < 2$ **27.** $x \ge \dfrac{1}{2}$

28. first side $= 20$ m; second side $= 30$ m; third side $= 16$ m **29.** width $= 20$ m; length $= 47$ m **30.** 1st pollutant $= 8$ ppm; 2nd pollutant $= 4$ ppm; 3rd pollutant $= 3$ ppm **31.** 15 months **32.** \$1400 at 14%; \$2600 at 11% **33.** 16 nickels, 7 dimes, 8 quarters

Cumulative Test for Chapters 1–2

1. $-\dfrac{20}{21}$ **2.** 30 **3.** 0.41 **4.** $12ab - 28ab^2$ **5.** $25x^2$ **6.** 5 **7.** $2x + 6y - 4xy + 16xy^2$ **8.** $x = \dfrac{40}{11}$ or $3\dfrac{7}{11}$

9. $x = 4$ **10.** $y = 1$ **11.** $b = \dfrac{3H - 8a}{2}$ **12.** $t = \dfrac{I}{Pr}$ **13.** $a = \dfrac{2A}{h} - b$

14. $x \le 3$ **15.** $x > -15$ **16.** $x \le -1$ **17.** $x \ge -12$

18. 92 **19.** width $= 7$ cm; length $= 32$ cm **20.** \$3000 at 15%; \$4000 at 7% **21.** 7 nickels, 12 dimes, 4 quarters

Chapter 3

Pretest Chapter 3

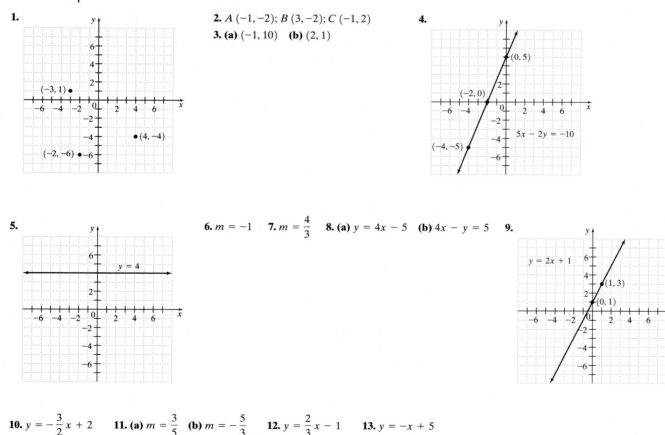

2. $A\,(-1, -2);\ B\,(3, -2);\ C\,(-1, 2)$
3. (a) $(-1, 10)$ **(b)** $(2, 1)$

6. $m = -1$ **7.** $m = \dfrac{4}{3}$ **8. (a)** $y = 4x - 5$ **(b)** $4x - y = 5$

10. $y = -\dfrac{3}{2}x + 2$ **11. (a)** $m = \dfrac{3}{5}$ **(b)** $m = -\dfrac{5}{3}$ **12.** $y = \dfrac{2}{3}x - 1$ **13.** $y = -x + 5$

14.

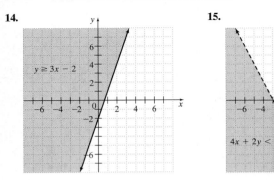

$y \geq 3x - 2$

15.

4x + 2y < -12

16. function   **17.** not a function

18. (a) −13   **(b)** 14

19. (a) 72   **(b)** $\frac{1}{2}$

3.1 Exercises

1.

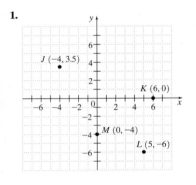

J (−4, 3.5)
K (6, 0)
M (0, −4)
L (5, −6)

3. R: $(-3, -5)$; S: $\left(-4\frac{1}{2}, 0\right)$; X: $(3, -5)$; Y: $\left(2\frac{1}{2}, 6\right)$ **5.** 0

7. The order in which you write the numbers matters. The graph of $(5, 1)$ is not the same as the graph of $(1, 5)$.

9. $(-5, -3), (-4, -4), (-4, -5), (-3, -3), (-2, -1), (-2, -2), (-1, -3), (0, -4), (0, -5)$, and $(1, -3)$

11. (a) $(0, 8)$ **(b)** $(4, 20)$ **13. (a)** $(-6, 15)$ **(b)** $(3, -3)$ **15. (a)** $(7, 13)$ **(b)** $(-1, -7)$

17. (a) $(10, 4)$ **(b)** $(-5, -8)$ **19. (a)** $(3, -12)$ **(b)** $\left(\frac{3}{2}, 6\right)$ **21.** D5   **23.** D1   **25.** C3

27. (a)

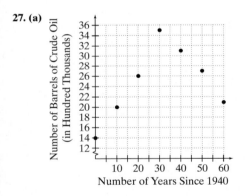

Number of Barrels of Crude Oil (in Hundred Thousands)

Number of Years Since 1940

(b) The number of barrels of crude oil increased significantly from 1940 to 1970. From 1970 to 2000, it has decreased significantly.

28. $x = 0$ **29.** $x \leq \frac{11}{2}$

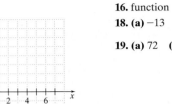

30. 1133.54 square yards   **31.** 922   **32.** 95.7%

33. (a) $55 per square foot   **(b)** $6160

3.2 Exercises

1. No; replacing x by -2 and y by 5 in the equation does not result in a true statement.   **3.** x-axis

5. $(0, 1), (-2, 5), (1, -1)$

7. $(0, -4), (2, -2), (4, 0)$

9.

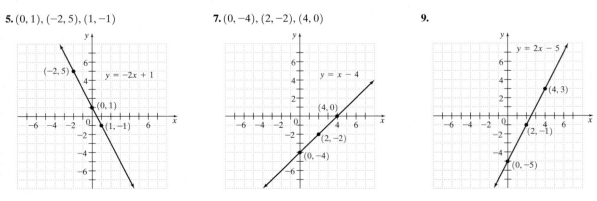

(−2, 5)   y = −2x + 1   (0, 1)   (1, −1)

y = x − 4   (4, 0)   (2, −2)   (0, −4)

y = 2x − 5   (4, 3)   (2, −1)   (0, −5)

11.
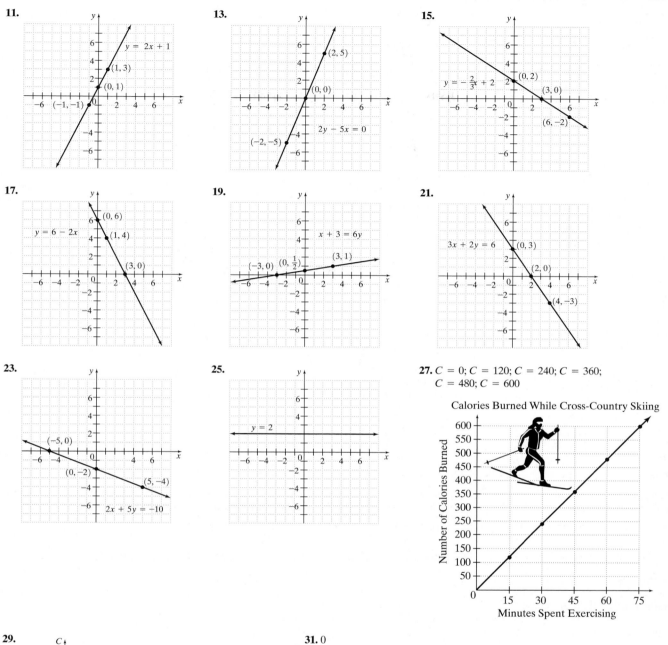
$y = 2x + 1$; points $(1, 3)$, $(0, 1)$, $(-1, -1)$

13. $2y - 5x = 0$; points $(2, 5)$, $(0, 0)$, $(-2, -5)$

15. $y = -\frac{2}{3}x + 2$; points $(0, 2)$, $(3, 0)$, $(6, -2)$

17. $y = 6 - 2x$; points $(0, 6)$, $(1, 4)$, $(3, 0)$

19. $x + 3 = 6y$; points $(-3, 0)$, $(0, \frac{1}{2})$, $(3, 1)$

21. $3x + 2y = 6$; points $(0, 3)$, $(2, 0)$, $(4, -3)$

23. $2x + 5y = -10$; points $(-5, 0)$, $(0, -2)$, $(5, -4)$

25. $y = 2$

27. $C = 0$; $C = 120$; $C = 240$; $C = 360$; $C = 480$; $C = 600$

Calories Burned While Cross-Country Skiing
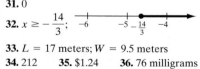

29.
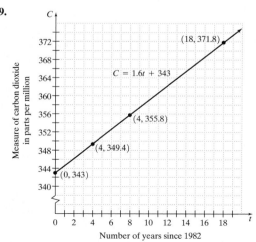
$C = 1.6t + 343$; points $(0, 343)$, $(4, 349.4)$, $(4, 355.8)$, $(18, 371.8)$

31. 0

32. $x \geq -\frac{14}{3}$;

33. $L = 17$ meters; $W = 9.5$ meters

34. 212   **35.** $1.24   **36.** 76 milligrams

3.3 Exercises

1. -1 **3.** $\dfrac{3}{5}$ **5.** $-\dfrac{2}{5}$ **7.** $-\dfrac{4}{3}$ **9.** $-\dfrac{16}{5}$ **11.** No; division by zero is impossible, so the slope is undefined. **13.** $m = 8; (0, 9)$

15. $m = -3; (0, 4)$ **17.** $m = \dfrac{5}{6}; \left(0, -\dfrac{2}{9}\right)$ **19.** $m = -6; (0, 0)$ **21.** $m = -6; \left(0, \dfrac{4}{5}\right)$ **23.** $m = -\dfrac{5}{2}; \left(0, \dfrac{3}{2}\right)$ **25.** $m = \dfrac{7}{3}; \left(0, -\dfrac{4}{3}\right)$

27. (a) $y = \dfrac{3}{4}x + 2$ **(b)** $3x - 4y = -8$ **29. (a)** $y = 6x - 3$ **(b)** $6x - y = 3$ **31. (a)** $y = -\dfrac{5}{4}x - \dfrac{3}{4}$ **(b)** $5x + 4y = -3$

33. **35.** **37.**

39. **41.**

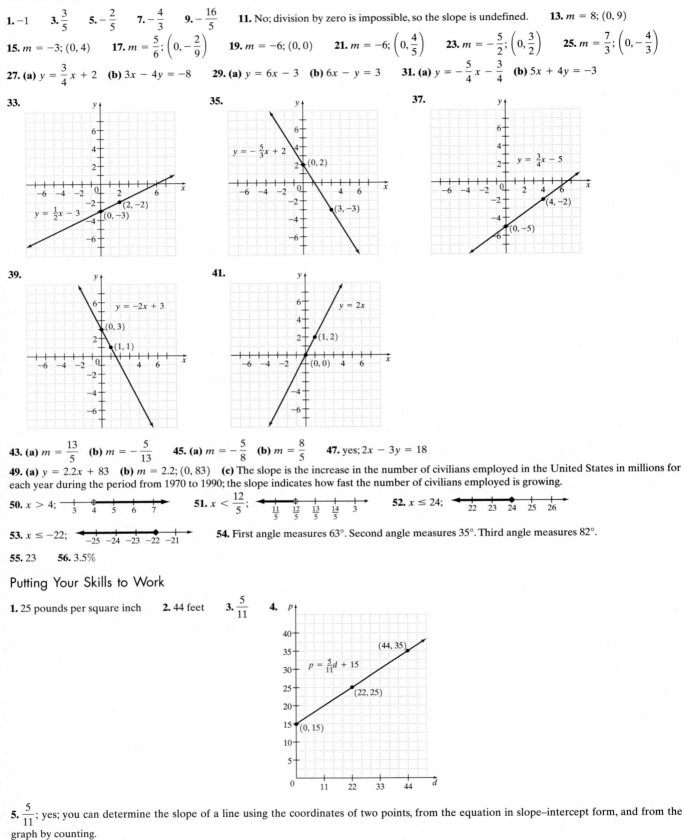

43. (a) $m = \dfrac{13}{5}$ **(b)** $m = -\dfrac{5}{13}$ **45. (a)** $m = -\dfrac{5}{8}$ **(b)** $m = \dfrac{8}{5}$ **47.** yes; $2x - 3y = 18$

49. (a) $y = 2.2x + 83$ **(b)** $m = 2.2; (0, 83)$ **(c)** The slope is the increase in the number of civilians employed in the United States in millions for each year during the period from 1970 to 1990; the slope indicates how fast the number of civilians employed is growing.

50. $x > 4$; **51.** $x < \dfrac{12}{5}$; **52.** $x \le 24$;

53. $x \le -22$; **54.** First angle measures $63°$. Second angle measures $35°$. Third angle measures $82°$.

55. 23 **56.** 3.5%

Putting Your Skills to Work

1. 25 pounds per square inch **2.** 44 feet **3.** $\dfrac{5}{11}$ **4.**

5. $\dfrac{5}{11}$; yes; you can determine the slope of a line using the coordinates of two points, from the equation in slope–intercept form, and from the graph by counting.

6.

d	0	11	22	33	44
p	15	20	25	30	35

(a) 5 **(b)** 11 **(c)** The slope is the ratio of these two numbers; slope is the difference between the p values over the difference between the d values.

3.4 Exercises

1. $y = 4x + 12$ **3.** $y = -2x + 11$ **5.** $y = -3x + \dfrac{7}{2}$ **7.** $y = -\dfrac{2}{5}x - 1$ **9.** $y = -2x - 6$ **11.** $y = -3x + 3$ **13.** $y = 5x - 10$

15. $y = \dfrac{1}{3}x + \dfrac{1}{2}$ **17.** $y = -\dfrac{2}{3}x + 1$ **19.** $y = \dfrac{2}{3}x - 4$ **21.** $y = 3x$ **23.** $y = -2$ **25.** $y = -2$ **27.** $y = 2$ **29.** $y = \dfrac{3}{4}x - 4$

31. $y = -\dfrac{1}{2}x + 4$ **33.** $y = 2.4x + 227$ **35.** Yes   **36.** $x = -\dfrac{14}{11}$ or $x = -1\dfrac{3}{11}$ **37.** \$61.20   **38.** 290 minutes

39. 105 entertainment centers   **40.** \$12,640

3.5 Exercises

1. No, all points in one region will be solutions to the inequality while all points in the other region will not be solutions. Thus testing any point will give the same result, as long as the point is not on the boundary line.

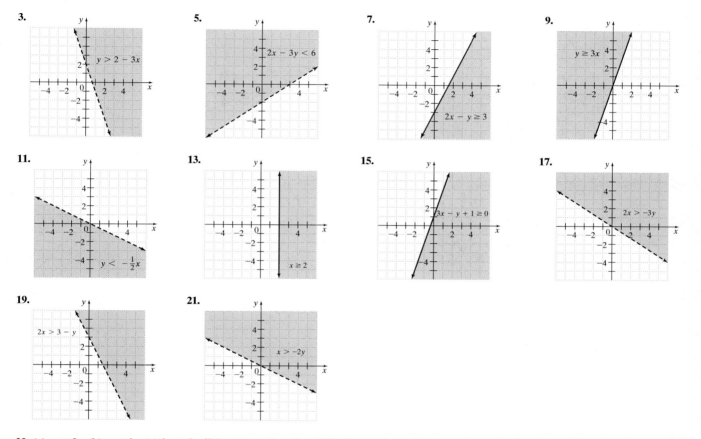

23. (a) $x < 3$ (b) $x < 3$ (c) $3x + 3$ will be greater than $5x - 3$ for those values of x where the graph of $y = 3x + 3$ lies *above* the graph of $y = 5x - 3$.   **24.** $-24a^2 + 6ab - 3a^2b$ **25.** $y = -\dfrac{7}{3}x + 7$ **26.** 9200 miles   **27.** 16

3.6 Exercises

1. You can describe a function using a table of values, an algebraic equation, or a graph.   **3.** Possible values; independent   **5.** If a vertical line can intersect the graph more than once, the relation is not a function. If no such line exists, then the relation is a function.

7. (a) Domain $= \left\{\dfrac{1}{4}, \dfrac{1}{2}, \dfrac{3}{4}\right\}$; Range $= \{5, 6, 10\}$ **(b)** not a function   **9. (a)** Domain $= \{0, 2, 7.3\}$; Range $= \{1, 8\}$ **(b)** function

11. (a) Domain $= \{5, 5.6, 5.8, 6\}$; Range $= \{5.8, 6, 8\}$ **(b)** function   **13. (a)** Domain $= \{85, 95, 110\}$; Range $= \{3, 11, 15, 20\}$ **(b)** not a function

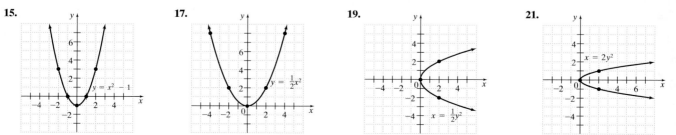

23.

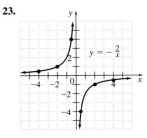

25.

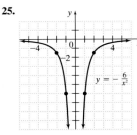

27.

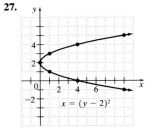

29.

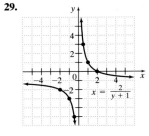

31. function   **33.** not a function   **35.** function   **37.** not a function   **39. (a)** 8   **(b)** -2 **(c)** -7 **41. (a)** -4 **(b)** -3 **(c)** 12   **43. (a)** -8
(b) 7   **(c)** 4   **45.** 30%   **47.** between 1985 and 1990

49.

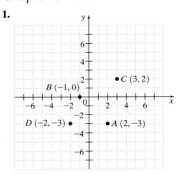

$f(0) = 411; f(4) = 489.4; f(10) = 571; f(20) = 611$
The curve slopes upward more steeply for smaller values of x.
Growth rate is decreasing as x gets larger.

50. $-2x^2 - 14x + 6$ **51.** $-3x^2 - 7x + 11$ **52.** $4x^3 + 6x^2 - x + 6$ **53.** $2x + 4 + \dfrac{-1}{3x - 1}$

Putting Your Skills to Work

1. 213,408 acres; 78,000,000 acres   **2.** 4337 million acres or 4,337,000,000 acres   **3.** 4181 million acres; 3947 million acres

Chapter 3 Review Problems

1.

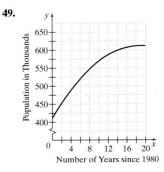

2.

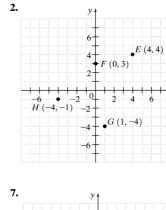

3. (a) $(0, -5)$ **(b)** $(3, 4)$ **4. (a)** $(1, 2)$ **(b)** $(-4, 4)$
5. (a) $(6, -1)$ **(b)** $(6, 3)$

6.

7.

8.

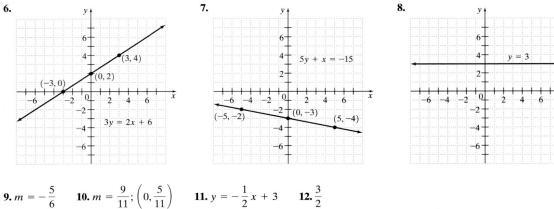

9. $m = -\dfrac{5}{6}$ **10.** $m = \dfrac{9}{11}; \left(0, \dfrac{5}{11}\right)$ **11.** $y = -\dfrac{1}{2}x + 3$ **12.** $\dfrac{3}{2}$

13.

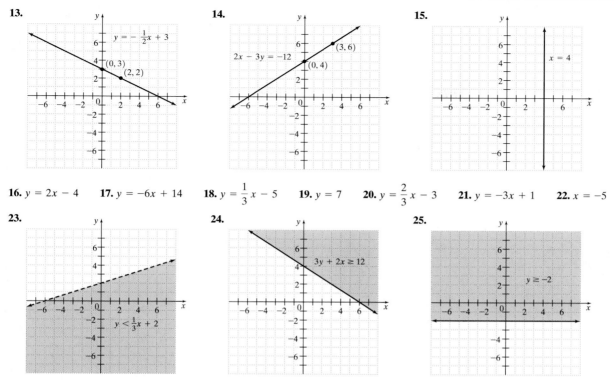

$y = -\frac{1}{2}x + 3$

(0, 3)
(2, 2)

14.

$2x - 3y = -12$

(3, 6)
(0, 4)

15.

$x = 4$

16. $y = 2x - 4$ **17.** $y = -6x + 14$ **18.** $y = \frac{1}{3}x - 5$ **19.** $y = 7$ **20.** $y = \frac{2}{3}x - 3$ **21.** $y = -3x + 1$ **22.** $x = -5$

23.

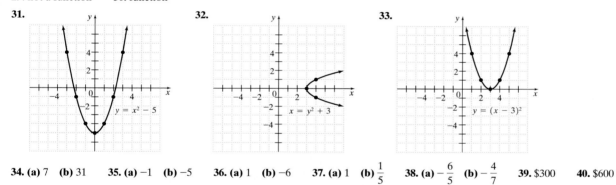

$y < \frac{1}{3}x + 2$

24.

$3y + 2x \geq 12$

25.

$y \geq -2$

26. Domain: $\{-6, -5, 5\}$; Range: $\{-6, 5\}$; not a function   **27.** Domain: $\{-7, -3, 3, 7\}$; Range: $\{-7, -3, 3, 7\}$; function   **28.** function
29. not a function   **30.** function

31.

$y = x^2 - 5$

32.

$x = y^2 + 3$

33.

$y = (x - 3)^2$

34. (a) 7 **(b)** 31   **35. (a)** -1 **(b)** -5 **36. (a)** 1 **(b)** -6 **37. (a)** 1 **(b)** $\frac{1}{5}$ **38. (a)** $-\frac{6}{5}$ **(b)** $-\frac{4}{7}$ **39.** \$300   **40.** \$600
41. $y = 0.15x + 150$; 0.15   **42.** The cost of the trip increases \$0.15 for each mile.   **43.** 2600 miles   **44.** 4000 miles   **45.** \$210   **46.** \$174
47. $y = 0.09x + 30$ (0, 30); it tells us that if Russ and Norma use no electricity, the minimum cost is \$30.   **48.** $m = 0.09$; the electric bill increases
\$0.09 for each kilowatt-hour of use.   **49.** 1300 kilowatt-hours   **50.** 2400 kilowatt-hours

Chapter 3 Test

1.

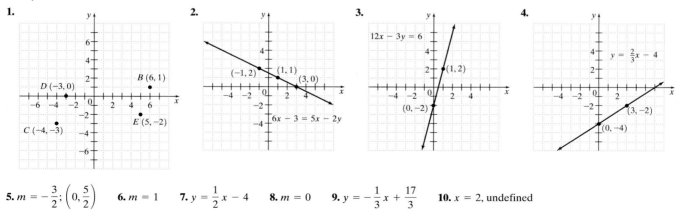

$D (-3, 0)$
$B (6, 1)$
$E (5, -2)$
$C (-4, -3)$

2.

$(-1, 2)$
$(1, 1)$
$(3, 0)$
$6x - 3 = 5x - 2y$

3.

$12x - 3y = 6$
$(1, 2)$
$(0, -2)$

4.

$y = \frac{2}{3}x - 4$
$(3, -2)$
$(0, -4)$

5. $m = -\frac{3}{2}; \left(0, \frac{5}{2}\right)$ **6.** $m = 1$ **7.** $y = \frac{1}{2}x - 4$ **8.** $m = 0$ **9.** $y = -\frac{1}{3}x + \frac{17}{3}$ **10.** $x = 2$, undefined

11.

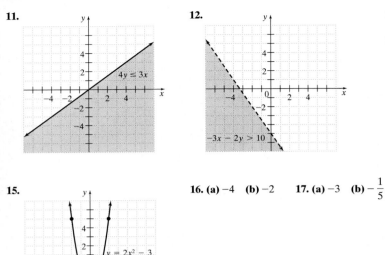

12.

13. Yes; no two different ordered pairs have the same first coordinate.

14. Yes; it passes the vertical line test.

15.

16. (a) -4 **(b)** -2 **17. (a)** -3 **(b)** $-\dfrac{1}{5}$

Cumulative Test for Chapters 1–3

1. $-\dfrac{22}{75}$ **2.** $x = 5$ **3.** $-25x^2 + 60x$ **4.** \$1225   **5.** $x = 7$ **6.** $x = 6$ **7.** $w = \dfrac{A - lh}{l}$ or $w = \dfrac{A}{l} - h$

8. $x \geq 1$ **9.** height $= 12$ feet; length $= 33$ feet   **10.** $y = 3x - 10$ **11.** $x = 7$ **12.** $y = \dfrac{1}{3}x + \dfrac{11}{3}$

13. $m = 0$ **14.** $m = \dfrac{3}{7}$

15.

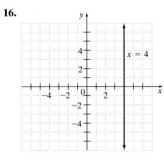

16.

17.

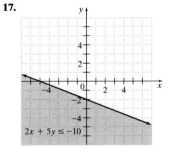

18. no

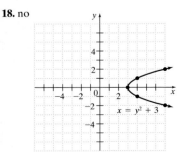

19. yes   **20. (a)** -39 **(b)** 3

Chapter 4

Pretest Chapter 4

1. $x = 9, y = 9$ **2.** $x = 3, y = -2$ **3.** Infinite number of solutions; dependent equations   **4.** $x = 2, y = -2$ **5.** $x = 1, y = 3, z = 0$
6. $x = -2, y = 3, z = 4$ **7.** $x = -1, y = 3, z = 2$ **8.** \$12 shirts, \$17 pants   **9.** Packet $A = 3$; packet $B = 2$; packet $C = 4$
10. 4 difficult exercises; 6 easy exercises

11.   **12.**

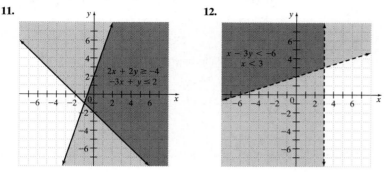

4.1 Exercises

1. There is no solution. There is no point (x, y) that satisfies both equations. The graph of such a system yields two parallel lines.
3. $\left(\dfrac{3}{2}, -1\right)$ is a solution to the system.

5.   **7.**   **9.**

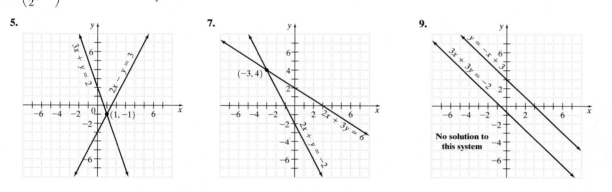

11. $x = 23, y = -43$ **13.** $x = 2, y = -2$ **15.** $x = 15, y = 4$ **17.** $x = 3, y = -2$ **19.** $x = 0, y = 1$ **21.** $s = 1, t = \dfrac{5}{3}$

23. $x = 1, y = -3$ **25.** $x = 3, y = -2$ **27.** $x = 6, y = -8$ **29.** No solution; inconsistent system of equations
31. Infinite number of solutions; dependent equations   **33.** $x = 5, y = -3$ **35.** No solution; inconsistent system of equations

37. $x = 16, y = 8$ **39. (a)** $y = 300 + 30x$, $y = 200 + 50x$ **(b)**

x	$y = 300 + 30x$		x	$y = 200 + 50x$
0	300		0	200
4	420		4	400
8	540		8	600

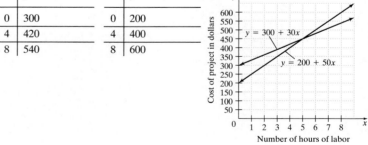

(c) The cost will be the same for 5 hours of installing new tile.   **(d)** The cost will be less for Modern Bathroom Headquarters.
41. $(2.46, -0.38)$ **43.** $(-2.45, 6.11)$ **45.** \$27   **46.** 341,889 cars

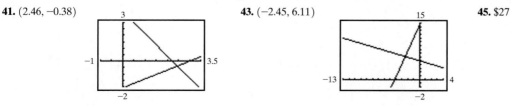

Putting Your Skills to Work

1. 880 cheetahs lost per year; $c(t) = 100,000 - 880t$ **2.** 1980 black rhinos lost per year; $r(t) = 70,000 - 1980t$ **3.** $c(t) = 20,800 - 880t$
4. $r(t) = 30,400 - 1980t$

4.2 Exercises

1. $(2, 1, -4)$ is a solution.   **3.** $x = 1, y = 3, z = -2$ **5.** $x = 3, y = -1, z = 4$ **7.** $x = -1, y = -2, z = 0$ **9.** $x = 1, y = -1, z = 2$
11. $x = 2, y = 1, z = -4$ **13.** $a = 4, b = 0, c = 2$ **15.** $a = 3, b = -1, c = -2$ **17.** $x = 1, y = 2, z = -3$

19. $x = 1.10551, y = 2.93991, z = 1.73307$ **21.** $x = 3, y = -2, z = 1$ **23.** $x = \frac{1}{2}, y = \frac{2}{3}, z = \frac{5}{6}$ **25.** $x = 1, y = 3, z = 5$

27. $a = -2, b = -5, c = 4$ **29.** Infinite number of solutions; dependent equations   **31.** No solution; inconsistent system of equations
33. $y = \frac{1}{3}x + \frac{11}{3}$ **34.** $y = \frac{3}{2}x + 8$

35. He will buy 57 sheep and 22 cattle. After the purchase he will have 346 horses, 602 sheep, and 623 cattle.   **36.** 12.5 miles per hour

4.3 Exercises

1. 16 heavy equipment operators; 19 general laborers   **3.** 51 tickets for regular coach seats; 47 tickets for sleeper car seats   **5.** 45 managers with computer experience; 20 managers without computer experience   **7.** 30 packages of old fertilizer; 25 packages of new fertilizer
9. One doughnut costs \$0.45; one large coffee costs \$0.89   **11.** Speed of plane in still air = 216 mph; speed of wind = 36 mph   **13.** Speed of plane in still air = 195 mph; speed of the wind = 15 mph   **15.** He scored 10 free throws and 11 2-point baskets.   **17.** He drove 192 miles on the highway and 240 miles in city driving.   **19.** The department pays \$10,258 for a car and \$17,300 for a truck.   **21.** The lodge has 2 large vans, 4 Dodge minivans, and 8 Ford Explorers.   **23.** A total of 80 adults, 170 high school students, and 50 children not yet in high school attended.   **25.** A total of 800 senior citizens, 10,000 adults, and 1200 children under 12 ride during the rush hour.   **27.** 7 medium sandwiches, 8 large sandwiches, 9 extra large sandwiches   **29.** She can prepare 2 of Box A, 3 of Box B, and 4 of Box C.   **31.** First angle measures 20°, second angle measures 40°, third angle measures 120°   **33.** The scientist should use 2 of packet A, 1 of packet B, 5 of packet C, and 3 of packet D.   **34.** $x = \frac{26}{7}$ **35.** $x = \frac{7}{18}$ **36.** $y = \frac{5}{3}$

4.4 Exercises

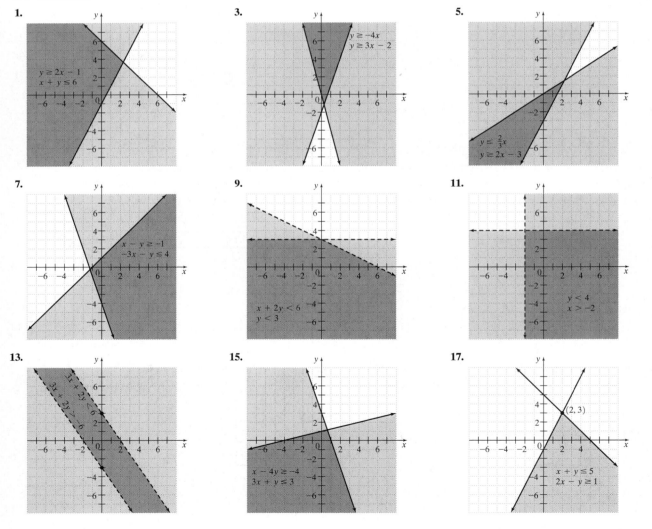

19.

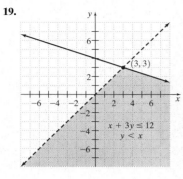

$x + 3y \le 12$
$y < x$

21.

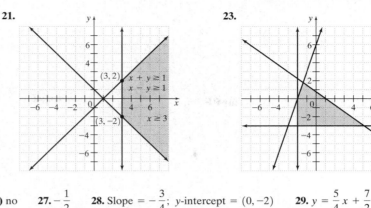

$(3, 2)$
$(3, -2)$
$x + y \ge 1$
$x - y \ge 1$
$x \ge 3$

23.

(graph)

25. (a)

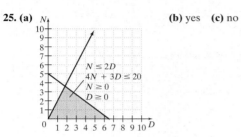

$N \le 2D$
$4N + 3D \le 20$
$N \ge 0$
$D \ge 0$

(b) yes **(c)** no

27. $-\dfrac{1}{2}$ **28.** Slope $= -\dfrac{3}{4}$; y-intercept $= (0, -2)$ **29.** $y = \dfrac{5}{4}x + \dfrac{7}{2}$

30. $y = -2x + 10$ **31.** The Cinema takes in \$4200 on a rainy day and \$3000 on a sunny day. **32.** The volunteers establish 43 ft of bicycle trails each day. The professionals establish 65 ft of bicycle trails each day. **33.** They had each worked for 12 weeks and had each sold \$100,000 worth of goods. **34.** One roast beef sandwich costs \$2.50. One order of french fries costs \$1.75. One soda costs \$0.95.

Chapter 4 Review Problems

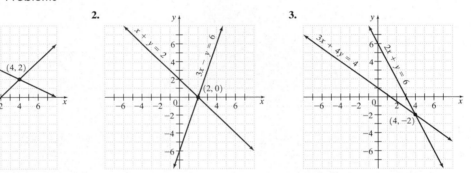

1.
$x + 2y = 8$
$(4, 2)$
$x - y = 2$

2.
$x + y = 2$
$3x - y = 6$
$(2, 0)$

3.
$3x + 4y = 4$
$2x + y = 6$
$(4, -2)$

4. $x = -1, y = 3$ **5.** $x = 1, y = -7$ **6.** $x = 1, y = 2$ **7.** $x = 1, y = 3$ **8.** $x = 1, y = -2$ **9.** $x = 3, y = -2$ **10.** $x = 2, y = 3$
11. $x = 2, y = -3$ **12.** No solution; inconsistent system of equations **13.** Infinite number of solutions; dependent equations **14.** $x = 2, y = -4$
15. $x = -\dfrac{1}{3}, y = \dfrac{1}{2}$ **16.** $x = 0, y = 3$ **17.** $x = -2, y = 6$ **18.** $a = \dfrac{4}{3}, b = -\dfrac{1}{2}$ **19.** $a = \dfrac{17}{2}, b = -\dfrac{7}{2}$ **20.** $x = 0, y = \dfrac{2}{3}$
21. No solution; inconsistent system of equations **22.** No solution; inconsistent system of equations **23.** $x = 5, y = 2$
24. $x = 1, y = 1, z = -2$ **25.** $x = 1, y = -2, z = 3$ **26.** $x = 5, y = -3, z = 8$ **27.** $x = 7, y = \dfrac{1}{2}, z = -3$ **28.** $x = 3, y = 0, z = -2$
29. $x = 1, y = -2, z = 3$ **30.** $x = 1, y = 2, z = -4$ **31.** $x = -2, y = -4, z = -8$ **32.** Speed of plane in still air $= 264$ mph; speed of wind $= 24$ mph **33.** New employees $= 7$; laid-off employees $= 10$ **34.** Laborers $= 15$; mechanics $= 10$ **35.** Children's tickets $= 340$; adult tickets $= 250$ **36.** Hats $= \$3$; shirts $= \$15$; pants $= \$12$ **37.** $A = 2; B = 3; C = 4$ **38.** One jar of jelly $= \$0.70$; one jar of peanut butter $= \$1.00$; one jar of honey $= \$0.80$ **39.** Buses $= 2$; station wagons $= 4$; sedans $= 3$ **40.** $x = 0, z = 1$ **41.** $x = \dfrac{20}{3}, y = \dfrac{10}{3}$ **42.** $x = -3, y = 2$
43. $x = 0, y = 2$ **44.** $x = 10, y = 0$ **45.** $x = -1, y = -2$ **46.** $x = -3, y = -4$ **47.** $x = 2, y = 5$ **48.** $x = 11, y = 6$
49. $x = -4, y = -7$ **50.** $x = -3, y = -2, z = 2$ **51.** $x = 1, y = 0, z = -1$ **52.** $x = 3, y = -1, z = -2$ **53.** $x = 5, y = -5, z = -20$

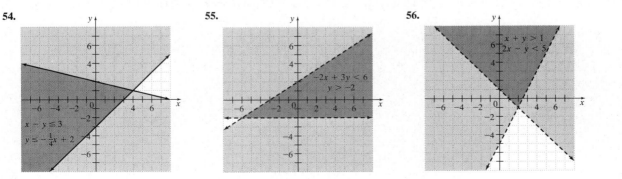

54.
$x - y \le 3$
$y \le -\dfrac{1}{4}x + 2$

55.
$-2x + 3y < 6$
$y > -2$

56.
$x + y > 1$
$2x - y < 5$

57.

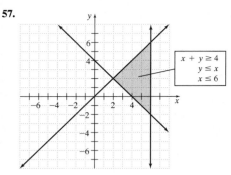

$$x + y \geq 4$$
$$y \leq x$$
$$x \leq 6$$

Chapter 4 Test

1. $x = -2, y = 1$ **2.** $x = \dfrac{2}{3}, y = 3$ **3.** $a = \dfrac{1}{2}, b = \dfrac{3}{2}$ **4.** No solution; inconsistent system of equations   **5.** $x = 3, y = 4$

6. $x = 1, y = 2$ **7.** $x = 2, y = -1, z = 3$ **8.** $x = -2, y = 3, z = 5$ **9.** $x = -4, y = 1, z = -1$ **10.** Speed of plane in still air = 450 mph; speed of wind = 50 mph   **11.** Station wagons = 4, 2-door sedans = 6, 4-door sedans = 6   **12.** They charge \$30 per day and \$0.20 per mile.

13.   **14.**

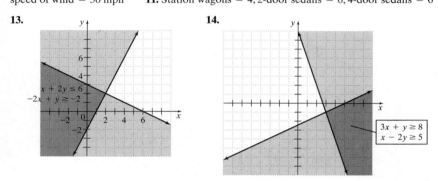

$$x + 2y \leq 6$$
$$-2x + y \geq -2$$

$$3x + y \geq 8$$
$$x - 2y \geq 5$$

Cumulative Test for Chapters 1–4

1. $1\dfrac{5}{8}$ **2.** $-1\dfrac{2}{5}$ **3.** -23 **4.** $22x + 12$ **5.** $P = \dfrac{A}{3 + 4rt}$ **6.** $x = 68$

7.

x	y
0	-1.25
4	0.75
7	2.25

8. $m = \dfrac{1}{10}$ **9.** $x > -10$ **10.** $5 \leq x \leq 11$ **11.** $y = 2x - 7$

12. 1st side = 17 m, 2nd side = 24 m, 3rd side = 28 m   **13.** \$1500 at 7%; \$4500 at 9%   **14.** $x = 2, y = -4$ **15.** $x = 2, y = -1, z = -1$
16. Shirts = \$21, slacks = \$30   **17.** $x = 5, y = 3$ **18.** $x = 1, y = 2, z = -2$ **19.** Infinite number of solutions; dependent equations

20.

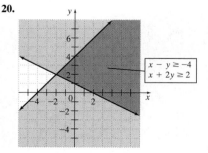

$$x - y \geq -4$$
$$x + 2y \geq 2$$

Chapter 5

Pretest Chapter 5

1. 3^{16} **2.** $-10x^7$ **3.** $-24a^5b^4$ **4.** x^{23} **5.** $-\dfrac{2}{x^2y^2}$ **6.** $\dfrac{5bc}{3a}$ **7.** x^{50} **8.** $-8x^6y^3$ **9.** $\dfrac{64a^9b^3}{c^6}$ **10.** $\dfrac{3x^2}{y^3z^4}$ **11.** $-\dfrac{2b^2}{a^3c^4}$ **12.** $\dfrac{1}{9}$

13. 6.38×10^{-4} **14.** $1{,}894{,}000{,}000{,}000$ **15.** $-2x^2 + 7x - 29$ **16.** $x^3 - 5x^2 + 11x - 12$ **17.** $-6x + 21x^2$ **18.** $2x^4 + 8x^3 - 2x^2$

19. $5x^4y^2 - 15x^2y^3 + 5xy^4$ **20.** $x^2 + 14x + 45$ **21.** $12x^2 + xy - 6y^2$ **22.** $2x^4 - 7x^2y^2 + 3y^4$ **23.** $64x^2 - 121y^2$

24. $25x^2 - 30xy + 9y^2$ **25.** $25a^2b^2 - 60ab + 36$ **26.** $4x^3 - 13x^2 + 11x - 2$ **27.** $16x^4 - 81$ **28.** $7x^3 - 6x^2 + 11x$

29. $5x^2 - 6x + 2 - \dfrac{3}{3x - 2}$

5.1 Exercises

1. When you multiply exponential expressions with the same base, keep the base the same and add the exponents.

3. A sample example is $\dfrac{2^2}{2^3} = \dfrac{\cancel{2} \cdot \cancel{2}}{\cancel{2} \cdot \cancel{2} \cdot 2} = \dfrac{1}{2} = \dfrac{1}{2^{3-2}}$.

5. $6; x, y; 11$ and 1 **7.** 2^2a^3b **9.** $-3a^2b^2c^3$ **11.** 3^{15} **13.** 5^{26} **15.** $3^5 \cdot 8^2$ **17.** $-54x^5$ **19.** $12x^5$ **21.** $-12x^{11}$ **23.** $\dfrac{3}{4}x^7y^5$

25. $-8.05wx^5y^4$ **27.** 0 **29.** $80x^3y^7$ **31.** $-28a^5b^6$ **33.** 0 **35.** 0 **37.** $-24w^5xyz^6$ **39.** $\dfrac{1}{y^3}$ **41.** y^7 **43.** $\dfrac{1}{13^{10}}$ **45.** 3^4

47. $\dfrac{a^8}{4}$ **49.** $\dfrac{x^7}{y^9}$ **51.** $\dfrac{x^4}{2}$ **53.** 1 **55.** $-\dfrac{2x^4}{y^5}$ **57.** $\dfrac{t^2}{20s^3}$ **59.** $6x^2$ **61.** $\dfrac{y^2}{16x^3}$ **63.** $\dfrac{3a}{4}$ **65.** $\dfrac{5x^6}{7y^8}$ **67.** $3x^2$ **69.** $-\dfrac{2}{3a^3}$

71. $-7ab^5$ **73.** Answers will vary. **75.** x^{3a} **77.** c^{y+2} **79.** w^{40} **81.** $a^{12}b^4$ **83.** $m^{15}n^{10}p^5$ **85.** $3^8x^4y^8$ **87.** $16a^{20}$

89. $\dfrac{12^5x^5}{y^{10}}$ **91.** $\dfrac{16a^{16}}{81b^{12}}$ **93.** $-32a^{25}b^{10}c^5$ **95.** $-64x^3z^{12}$ **97.** $\dfrac{a}{4b^4}$ **99.** $-8a^7b^{11}$ **101.** $\dfrac{64}{x^{18}}$ **103.** $\dfrac{a^{15}b^5}{c^{25}d^5}$ **105.** $16x^9y^3$

107. $\pm 2x^5y^4z^7$ **108.** -11 **109.** -46 **110.** $-\dfrac{7}{4}$ **111.** 5 **112.** $x = 2, y = 3$ **113.** $x = -1, y = 4, z = 0$

5.2 Exercises

1. $\dfrac{3}{x^2}$ **3.** $\dfrac{1}{16x^4y^2}$ **5.** $\dfrac{3xz^3}{y^2}$ **7.** $3x$ **9.** $\dfrac{wy^3}{x^5z^2}$ **11.** $\dfrac{1}{8}$ **13.** $\dfrac{z^8}{9x^2y^4}$ **15.** $\dfrac{1}{x^6y}$ **17.** 1.2378×10^5 **19.** 7.42×10^{-4} **21.** 7.652×10^9

23. $302{,}000$ **25.** 0.000033 **27.** $983{,}000$ **29.** 1.67×10^{-27} kilogram **31.** 0.000007 meter **33.** 1.0×10^1 **35.** 3.2×10^{-19}

37. 4.5×10^5 **39.** 2.03×10^4 dollars **41.** 1.90×10^{11} hours **43.** 1.15×10^{11} hours **45.** 4.71744×10^{32} joules

47. 46.3% **49.** -0.8 **50.** -1 **51.** $-\dfrac{1}{28}$ **52.** Candidate #1 shook 2524 hands; candidate #2 shook 1016 hands.

53. Mario earns $\$36{,}094$; Alfonso earns $\$27{,}352$; Gina earns $\$48{,}554$.

54. **55.**

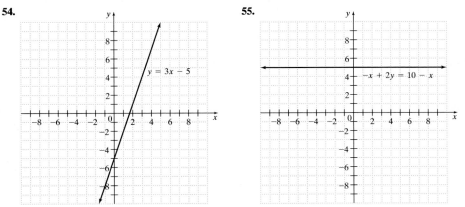

Putting Your Skills to Work

1. 3.02×10^8 acres **2.** 2.203×10^9 acres

5.3 Exercises

1. A polynomial in x is the sum of a finite number of terms of the form ax^n, where a is any real number and n is a whole number. An example is $3x^2 - 5x - 9$. **3.** The degree of a polynomial in x is the largest exponent of x in any of the terms of the polynomial. **5.** degree 4; monomial

7. degree 5; trinomial **9.** degree 5; binomial **11.** $5x - 28$ **13.** $-3x^2 - 10x + 10$ **15.** $\dfrac{5}{6}x^2 + \dfrac{1}{2}x - 9$ **17.** $1.7x^3 - 3.4x^2 - 13.2x - 5.4$

19. $5x - 24$ **21.** $-\dfrac{1}{6}x^2 + 7x - \dfrac{3}{10}$ **23.** $-2x^3 + 5x - 10$ **25.** $-4.7x^4 - 0.7x^2 - 1.6x + 0.4$ **27.** $10x + 10$ **29.** $-3x^2y + 6xy^2 - 4$

31. $x^4 - 3x^3 - 4x^2 - 24$ **33.** 5.4 **35.** 2020 **37.** $727{,}000$ **39.** $434{,}400$ **41.** $3x^2 + 12x$ **43.** $x = \dfrac{3y - 2}{8}$ **44.** $d = \dfrac{5xy}{B}$

45. approximately 696.8 billion dollars **46.** $m = -\dfrac{2}{3}$ **47.** $y = -4x - 3$

5.4 Exercises

1. $-12x^4 + 2x^2$ **3.** $-15x^3 + 10x^2 - 25x$ **5.** $6x^7 - 4x^6 + 10x^4 - 2x^3$ **7.** $x + \frac{3}{2}x^2 + \frac{5}{2}x^3$ **9.** $-15x^4y^2 + 6x^3y^2 - 18x^2y^2$

11. $4b^4 + 6b^3 - 8b^2$ **13.** $3x^4 - 9x^3 + 15x^2 - 6x$ **15.** $-2x^3y^3 + 12x^2y^2 - 16xy$ **17.** $-28x^5y + 12x^4y + 8x^3y - 4x^2y$

19. $-6c^2d^5 + 8c^2d^3 - 12c^2d$ **21.** $12x^7 - 6x^5 + 18x^4 + 54x^3$ **23.** $-24x^7 + 12x^5 - 8x^3$ **25.** $x^2 + 13x + 30$ **27.** $x^2 + 8x + 12$

29. $x^2 - 3x - 18$ **31.** $x^2 - 11x + 30$ **33.** $-14x^2 - 23x - 3$ **35.** $3x^2 + 4xy - 27x - 36y$ **37.** $12y^2 - y - 6$ **39.** $20y^2 - 22y + 6$

41. The last term is incorrect. The result should be $-3x + 7$. **43.** $25x^2 - 1$; it is the difference of two perfect squares.

45. $6b^4 - 31b^2c + 35c^2$ **47.** $64x^2 - 32x + 4$ **49.** $25a^4 - 30a^2b^2 + 9b^4$ **51.** $0.8x^2 + 11.94x - 0.9$ **53.** $30x^2 + 43xy - 8y^2$

55. $\frac{1}{9}x^2 - \frac{1}{10}x - \frac{1}{10}$ **57.** $5bx - 10b^2 - 7cx + 14bc$ **59.** $10x^2 - 11x - 6$ **61.** $x = -10$ **62.** $w = -\frac{25}{7}$ **63.** 8 dimes; 11 quarters

64. -5 **65.** $y = -\frac{1}{2}x + 1$ **66.** 19 hours **67.** 22 hours **68.** 31 hours **69.** 32.5 hours

5.5 Exercises

1. binomial **3.** The middle term is missing. The answer should be $16x^2 - 56x + 49$. **5.** $y^2 - 49$ **7.** $x^2 - 81$ **9.** $64x^2 - 9$ **11.** $4x^2 - 49$

13. $25x^2 - 9y^2$ **15.** $0.36x^2 - 9$ **17.** $9y^2 + 6y + 1$ **19.** $25x^2 - 40x + 16$ **21.** $81x^2 + 90x + 25$ **23.** $9x^2 - 42x + 49$

25. $\frac{4}{9}x^2 + \frac{1}{3}x + \frac{1}{16}$ **27.** $36w^2 + 60wz + 25z^2$ **29.** $49x^2 - 9y^2$ **31.** $49c^6 - 84c^3d + 36d^2$ **33.** $x^3 - 11x + 6$

35. $4x^4 - 7x^3 + 2x^2 - 3x - 1$ **37.** $3x^3 - 2x^2 - 25x + 24$ **39.** $3x^3 - 13x^2 - 6x + 40$ **41.** $2x^3 - 7x^2 - 32x + 112$

43. $a^4 + a^3 - 13a^2 + 17a - 6$ **45.** $24x^3 + 14x^2 - 11x - 6$ **47.** \$11,000 at 7%; \$7,000 at 11% **48.** Width is 7 meters; length is 10 meters.

49. $-28.12°F$ **50.** 2.0625×10^5 meters

Putting Your Skills to Work

1. 1.53×10^6 nanometers **2.** 1.53 millimeters

5.6 Exercises

1. $5x^3 - 3x + 4$ **3.** $2y^2 - 3y - 1$ **5.** $7x^4 - 3x^2 + 8$ **7.** $8x^4 - 9x + 6$ **9.** $3x + 5$ **11.** $x - 2 - \frac{20}{x-7}$ **13.** $3x^2 - 4x + 8 - \frac{10}{x+1}$

15. $2x^2 - 3x - 2 - \frac{5}{2x+5}$ **17.** $2x^2 + x - 2$ **19.** $6y^2 + 3y - 8 + \frac{7}{2y-3}$ **21.** $y^2 - 4y - 1 - \frac{9}{y+3}$

23. $y^3 + 2y^2 - 5y - 10 - \frac{25}{y-2}$ **25.** $2y^2 + \frac{1}{2}y + \frac{7}{8} - \frac{49}{32y-8}$ **27.** 110,000 gallons **28.** approximately 3.4 million

29. 200 cats **30.** 519 and 520

Putting Your Skills to Work

1. 4.1 million **2.** 700,000

Chapter 5 Review Problems

1. $-18a^7$ **2.** 5^{23} **3.** $6x^4y^6$ **4.** 8^{17} **5.** $\frac{1}{7^{12}}$ **6.** $\frac{1}{x^5}$ **7.** y^{14} **8.** $\frac{x^4}{3}$ **9.** $-\frac{3}{5x^5y^4}$ **10.** $-\frac{2a}{3b^6}$ **11.** x^{24} **12.** $125x^3y^6$ **13.** $9a^6b^4$

14. $\frac{2x^4}{3y^2}$ **15.** $\frac{25a^2b^4}{c^6}$ **16.** $\frac{y^9}{64w^{15}z^6}$ **17.** $\frac{1}{x^3}$ **18.** $\frac{1}{x^5y^{11}}$ **19.** $\frac{2y^3}{x^6}$ **20.** $\frac{x^5}{2y^6}$ **21.** $\frac{1}{4x^6}$ **22.** $\frac{3y^2}{x^3}$ **23.** $\frac{4w^2}{x^5y^6z^8}$ **24.** $\frac{b^5c^3d^4}{27a^2}$

25. 1.563402×10^{11} **26.** 1.79632×10^5 **27.** 7.8×10^{-3} **28.** 6.173×10^{-5} **29.** 120,000 **30.** 83,670,000,000 **31.** 3,000,000 **32.** 0.25

33. 0.00000005708 **34.** 0.000000006 **35.** 2×10^{13} **36.** 9.36×10^{19} **37.** 9.6×10^{-10} **38.** 7.8×10^{-11} **39.** 3.504×10^8 kilometers

40. 7.94×10^{14} cycles **41.** 6×10^9 **42.** $-5x^2 - 11x - 18$ **43.** $6.7x^2 - 11x + 3$ **44.** $-x^3 + 2x^2 - x + 8$ **45.** $7x^3 - 3x^2 - 6x + 4$

46. $5x^3y^3 - 2x^2y^2 + 10xy - 4$ **47.** $\frac{1}{4}x^2 - \frac{1}{4}x + \frac{1}{10}$ **48.** $-x^2 - 2x + 10$ **49.** $-6x^2 - 6x - 3$ **50.** $15x^2 + 2x - 1$

51. $28x^2 - 29x + 6$ **52.** $20x^2 + 48x + 27$ **53.** $10x^3 - 30x^2 + 15x$ **54.** $-6x^3y^3 + 10x^2y^2 - 12xy$ **55.** $4x^4 - 12x^3 + 20x^2 - 8x$

56. $5a^2 - 8ab - 21b^2$ **57.** $8x^4 - 10x^2y - 12x^2 + 15y$ **58.** $-15x^6y^2 - 9x^4y + 6x^2y$ **59.** $9x^2 - 12x + 4$ **60.** $25x^2 - 9$

61. $49x^2 - 36y^2$ **62.** $25a^2 - 20ab + 4b^2$ **63.** $64x^2 + 144xy + 81y^2$ **64.** $4x^3 + 27x^2 + 5x - 3$ **65.** $2x^3 - 7x^2 - 42x + 72$

66. $2y^2 + 3y + 4$ **67.** $6x^3 + 7x^2 - 18x$ **68.** $4x^2y - 6x + 8y$ **69.** $53x^3 - 12x^2 + 19x + 13$ **70.** $4x + 1$ **71.** $3x + 7$

72. $3x^2 + 2x + 4 + \frac{9}{2x-1}$ **73.** $2x^2 - 5x + 13 - \frac{27}{x+2}$ **74.** $4x + 1$ **75.** $4x - 5 + \frac{11}{2x+1}$ **76.** $x^2 + 3x + 8$

77. $2x^2 + 4x + 5 + \frac{11}{x-2}$ **78.** 1.43×10^9 people **79.** 1.147×10^9 people **80.** 2.733×10^{-23} gram **81.** 3.3696×10^{31} joules

82. $3xy + 2x$ **83.** $2x^2 - 4y^2$

Chapter 5 Test

1. 3^{34} **2.** 3^{28} **3.** 8^{24} **4.** $12x^4y^{10}$ **5.** $-\frac{7x^3}{5}$ **6.** $-125x^3y^{18}$ **7.** $\frac{49a^{14}b^4}{9}$ **8.** $\frac{1}{4a^5b^6}$ **9.** $\frac{1}{64}$ **10.** $\frac{6c^5}{a^4b^3}$ **11.** $\frac{2w^6}{x^3y^4z^8}$

12. 5.482×10^{-4} **13.** 582,000,000 **14.** 2.4×10^{-6} **15.** $-2x^2 + 5x$ **16.** $3x^2 + 3xy - 6y - 7y^2$ **17.** $-21x^5 + 28x^4 - 42x^3 + 14x^2$

18. $15x^4y^3 - 18x^3y^2 + 6x^2y$ **19.** $10a^2 + 7ab - 12b^2$ **20.** $6x^3 - 11x^2 - 19x - 6$ **21.** $49x^4 + 28x^2y^2 + 4y^4$ **22.** $81x^2 - 4y^2$

23. $12x^4 - 14x^3 + 25x^2 - 29x + 10$ **24.** $3x^4 + 4x^3y - 15x^2y^2$ **25.** $3x^3 - x + 5$ **26.** $2x^2 - 7x + 4$ **27.** $2x^2 + 6x + 12$

28. 7.85×10^1 people **29.** 4.18×10^6 miles

Cumulative Test for Chapters 1–5

1. $2x^2 - 13x$ **2.** 35 **3.** $x = -\dfrac{9}{2}$ **4.** $x = 44$ **5.** $x > -1$ **6.** $f = \dfrac{2B}{3a} - \dfrac{c}{3}$ **7.** 12,400 employees

8. **9.** $y = -\dfrac{1}{3}x + \dfrac{11}{3}$ **10.**

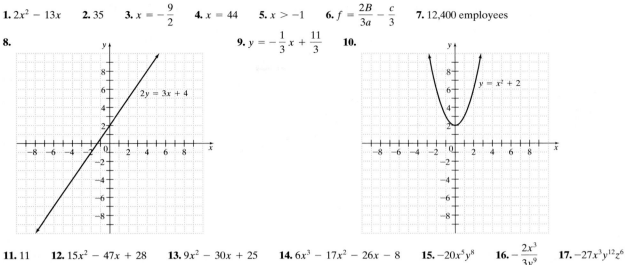

11. 11 **12.** $15x^2 - 47x + 28$ **13.** $9x^2 - 30x + 25$ **14.** $6x^3 - 17x^2 - 26x - 8$ **15.** $-20x^5y^8$ **16.** $-\dfrac{2x^3}{3y^9}$ **17.** $-27x^3y^{12}z^6$

18. $\dfrac{9z^8}{w^2x^3y^4}$ **19.** 1.36×10^{15} **20.** 5.6×10^{-4} **21.** 4.0×10^{-35} **22.** $5x^3 + 7x^2 - 6x + 50$ **23.** $-36x^3y^2 + 18x^2y^3 - 48xy^4$

24. $2x^4 - 15x^3 + 28x^2 - 33x + 12$ **25.** $x + 5 + \dfrac{3}{x - 3}$

Chapter 6

Pretest Chapter 6

1. $2x(x - 3y + 6y^2)$ **2.** $(3x + 4y)(a - 2b)$ **3.** $18ab(2b - 1)$ **4.** $(a - 2b)(5 - 3x)$ **5.** $(3x - 4)(x + y)$ **6.** $(7x - 3)(3x - 2)$
7. $(x - 24)(x + 2)$ **8.** $(x - 5)(x - 3)$ **9.** $(x + 8)(x + 1)$ **10.** $2(x + 6)(x - 2)$ **11.** $3(x - 9)(x + 7)$ **12.** $(5x - 2)(3x - 2)$
13. $(3y - 2z)(2y + 3z)$ **14.** $4(3x + 5)(x + 2)$ **15.** $(9x^2 + 4)(3x + 2)(3x - 2)$ **16.** $(7x - 2y)^2$ **17.** $(5x + 8)^2$
18. $3x(2x - 1)(x + 3)$ **19.** $2y^2(4x - 3)^2$ **20.** cannot be factored **21.** $x = 1, x = -\dfrac{3}{2}$ **22.** $x = 5, x = 6$ **23.** $x = -\dfrac{2}{3}, x = 3$
24. altitude $= 8$ cm, base $= 15$ cm

6.1 Exercises

1. factors **3.** No; $6a^3 + 3a^2 - 9a$ has a common factor of 3a. **5.** $2c(c + 1)$ **7.** $9wz(2 - 3w)$ **9.** $2x(4x^2 - 5x - 7)$
11. $6(2xy - 3yz - 6xz)$ **13.** $b^2(2ab + 3x - 5b^2 + 2)$ **15.** $2x^5(3x^4 - 4x^2 + 2)$ **17.** $9ab(ab - 4)$ **19.** $8a(5a - 2b - 3)$
21. $8y(6x - 3y + 5)$ **23.** $(3a + b)(6 - z)$ **25.** $(x - 7)(5x + 3)$ **27.** $(3y + 5z)(7x - 6t)$ **29.** $(bc - 1)(5a + b + c)$
31. $(bc - 3a)(3c - 2 - 6b)$ **33.** $(x - 2y)(3x^2 - 1)$ **35.** $(5x - 3)(d - 1)$ **37.** $C = 29.95(a + b + c + d)$ **38.** $17, 19, 21$

39. $26{,}000$ **40.** $-27y^3z^9$ **41.** $\dfrac{4b^4}{9a^2}$ **42.** 345 people **43.** 650 people

6.2 Exercises

1. $(a + 4)(b - 3)$ **3.** $(a + 3b)(2x - y)$ **5.** $(x - 4)(x^2 + 3)$ **7.** $(3a + b)(x - 2)$ **9.** $(a + 2b)(5 + 6c)$ **11.** $(a - b)(5 - 2x)$
13. $(y - 2)(y - 3)$ **15.** $(7 + y)(2 - y)$ **17.** $(3x + y)(2a - 1)$ **19.** $(2x - 3)(x + 4)$ **21.** $(4x + 3w)(7x + 2y^2)$ **23.** We must rearrange the terms in a different order so that the expression in the parentheses is the same in each case. We use the order $6a^2 - 8ad + 9ab - 12bd$
to factor $2a(3a - 4d) + 3b(3a - 4d) = (3a - 4d)(2a + 3b)$. **24.** $y = 2x + \dfrac{1}{2}$ **25.** $x \geq 6$

26. $126.9 million **27.** $170.4 million

6.3 Exercises

1. product; sum **3.** $(x + 1)^2$ **5.** $(x + 7)(x + 5)$ **7.** $(x - 3)(x - 1)$ **9.** $(x - 7)(x - 4)$ **11.** $(x + 4)(x - 3)$
13. $(x - 14)(x + 1)$ **15.** $(x + 7)(x - 5)$ **17.** $(x - 6)(x + 4)$ **19.** $(x + 7)(x - 2)$ **21.** $(x - 6)(x - 4)$ **23.** $(x + 3)(x + 10)$
25. $(y - 5)(y + 1)$ **27.** $(a + 8)(a - 2)$ **29.** $(x - 4)(x - 8)$ **31.** $(x + 7)(x - 3)$ **33.** $(x + 5)(x + 8)$ **35.** $(x + 2)(x - 23)$
37. $(x + 12)(x - 3)$ **39.** $(x + 3y)(x - 5y)$ **41.** $(x - 7y)(x - 9y)$ **43.** $2(x - 2)(x - 4)$ **45.** $3(x + 4)(x - 6)$
47. $4(x + 5)(x + 1)$ **49.** $7(x + 5)(x - 2)$ **51.** $6(x + 1)(x + 2)$ **53.** $3(x - 1)(x - 5)$ **55.** $4(\pi x^2 - 9)$

57. first angle $= 62°$; second angle $= 30°$; third angle $= 88°$ **58.** $130,000 **59.** $y = \dfrac{1}{2}x + 4$ **60.** $y = -x + 2$ **61.** 1:30 A.M.
62. 9:00 P.M. **63.** $25°C$ **64.** June

6.4 Exercises

1. $(4x + 1)(x + 3)$ **3.** $(2x - 1)(x - 2)$ **5.** $(3x - 7)(x + 1)$ **7.** $(2x + 1)(x - 3)$ **9.** $(5x - 2)(x + 1)$ **11.** $(3x - 5)(5x - 3)$
13. $(2x - 5)(x + 4)$ **15.** $(3x + 1)(3x + 2)$ **17.** $(3x + 2)(2x - 3)$ **19.** $(3x - 2)(2x - 5)$ **21.** $(x - 2)(7x + 9)$
23. $(9y - 4)(y - 1)$ **25.** $(5a + 2)(a - 3)$ **27.** $(6x - 1)(2x - 3)$ **29.** $(5x - 2)(3x + 2)$ **31.** $(6x + 5)(2x + 3)$

33. $(6x + 1)(2x - 3)$ **35.** $(2x^2 - 1)(x^2 + 8)$ **37.** $(2x - y)(2x + 5y)$ **39.** $(5x - 4y)(x + 4y)$ **41.** $2(5x + 6)(x + 1)$

43. $3(2x - 1)(2x - 3)$ **45.** $5(2x + 1)(x - 3)$ **47.** $2x(3x + 1)(x - 3)$ **49.** $(2x + 5)(6x - 7)$ **51.** $(4x - 3)(5x - 3)$ **53.** $x = \dfrac{1}{10}$

54. 18.8 million children   **55.** 36.7 million children

6.5 Exercises

1. $(9x - 4)(9x + 4)$ **3.** $(4 - 3x)(4 + 3x)$ **5.** $(3x + 5)(3x - 5)$ **7.** $(2x - 5)(2x + 5)$ **9.** $(6x - 5)(6x + 5)$ **11.** $(1 - 7x)(1 + 7x)$
13. $(4x - 7y)(4x + 7y)$ **15.** $(5 + 11x)(5 - 11x)$ **17.** $(9x + 10y)(9x - 10y)$ **19.** $(5a + 7)(5a - 7)$ **21.** $(3x + 1)^2$ **23.** $(y - 3)^2$
25. $(3x - 4)^2$ **27.** $(7x + 2)^2$ **29.** $(x + 7)^2$ **31.** $(5x - 4)^2$ **33.** $(9x + 2y)^2$ **35.** $(5x - 3y)^2$ **37.** $(4a + 9b)^2$ **39.** $(3x^2 - y)^2$
41. $(7x + 1)(7x + 9)$ **43.** $(4x^2 + 1)(2x + 1)(2x - 1)$ **45.** $(x^5 + 6y^5)(x^5 - 6y^5)$ **47.** $(3x^5 - 2)^2$ **49.** Because no matter what combination you try, you cannot multiply two binomials to equal $9x^2 + 1$.   **51.** 49; one answer   **53.** $4(2x - 3)(2x + 3)$ **55.** $3(7x - y)(7x + y)$
57. $3(2x - 3)^2$ **59.** $2(7x + 3)^2$ **61.** $(x - 2)(x - 7)$ **63.** $(2x - 1)(x + 3)$ **65.** $(4x - 11)(4x + 11)$ **67.** $(3x + 7)^2$

69. $3(x + 5)(x - 3)$ **71.** $5(x - 4)(x + 4)$ **73.** $5(x + 2)^2$ **75.** $2(x - 9)(x - 7)$ **77.** $x^2 + 3x + 4 + \dfrac{-3}{x - 2}$

78. $2x^2 + x - 5$ **79.** 1.2 ounces of greens, 1.05 ounces of bulk vegetables, 0.75 ounce of fruit   **80.** 120 ounces   **81.** 3838 ft above sea level
82. 5 miles

6.6 Exercises

1. $a(6a + 2b - 3)$ **3.** $9(2x - y)(2x + y)$ **5.** $(3x - 2y)^2$ **7.** $(x + 5)(x + 3)$ **9.** $(3x + 2)(5x - 1)$ **11.** $(x - 3y)(a + 2b)$
13. $3(x^2 + 2)(x^2 - 2)$ **15.** $(2x - 3)^2$ **17.** $(2x - 3)(x - 4)$ **19.** $(x - 10y)(x + 7y)$ **21.** $(a + 3)(x - 5)$ **23.** $5x(3 - x)(3 + x)$
25. $5xy^3(x - 1)^2$ **27.** $3xy(3z + 2)(3z - 2)$ **29.** $3(x + 7)(x - 5)$ **31.** $5(x - 2)(x - 4)$ **33.** $-1(2x^2 + 1)(x + 2)(x - 2)$ **35.** prime
37. $5(x^2 + 2xy - 6y)$ **39.** $3x(2x + y)(5x - 2y)$ **41.** $4(2x - 1)(x + 4)$ **43.** prime   **45.** \$28,000   **46.** \$159,091   **47.** $x = 4, y = -5$
48. $x = -1, y = 2, z = -3$ **49.** 57 hardcover books, 94 softcover books, 47 magazines   **50.** 55 hardcover books, 92 softcover books,
46 magazines

6.7 Exercises

1. $-3, 7$ **3.** $-\dfrac{1}{2}, 3$ **5.** $\dfrac{3}{2}, 2$ **7.** $\dfrac{2}{3}, \dfrac{3}{2}$ **9.** $0, -13$ **11.** $3, -3$ **13.** $0, 1$ **15.** $-3, 2$ **17.** $-\dfrac{1}{2}$ **19.** $0, -2$ **21.** $-3, -4$ **23.** $-\dfrac{5}{3}, 3$

25. You can always factor out x.   **27.** $L = 14$ m; $W = 10$ m   **29.** 182 games   **31.** 15 teams   **33.** It will hit the ground after 3 seconds.
12 meters above ground after 2 seconds   **35.** 5 additional helicopters   **37.** 2415 telephone calls   **38.** 3160 telephone calls   **39.** 18 people

41. $-12x^3y^7$ **42.** $12a^{10}b^{13}$ **43.** $-\dfrac{3a^4}{2b^2}$ **44.** $\dfrac{1}{3x^5y^4}$

Putting Your Skills to Work

1. \$765,000   **2.** \$3,160,000

Chapter 6 Review Problems

1. $3x^2y(5x - 3y)$ **2.** $8x(5x - 4)$ **3.** $7xy(x - 2y - 3x^2y^2)$ **4.** $25a^4b^4(2b - 1 + 3ab)$ **5.** $9x^2(3x - 1)$ **6.** $2(x - 2y + 3z + 6)$
7. $(a + 3b)(2a - 5)$ **8.** $3xy(5x^2 + 2y + 1)$ **9.** $(3x - 7)(a - 2)$ **10.** $(a + 5b)(a - 4)$ **11.** $(x^2 + 3)(y - 2)$ **12.** $(4a - 5)(d + 5c)$
13. $(5x - 1)(3x + 2)$ **14.** $(5w - 3)(6w + z)$ **15.** $(x - 7)(x + 5)$ **16.** $(x - 4)(x - 6)$ **17.** $(x + 6)(x + 8)$
18. $(x + 3y)(x + 5y)$ **19.** $(x^2 + 7)(x^2 + 6)$ **20.** $(x - 17)(x + 3)$ **21.** $5(x + 1)(x + 3)$ **22.** $3(x + 1)(x + 12)$
23. $2(x - 6)(x - 8)$ **24.** $4(x - 5)(x - 6)$ **25.** $(4x - 5)(x + 3)$ **26.** $(3x - 1)(4x + 5)$ **27.** $(5x + 4)(3x - 1)$
28. $(3x - 2)(2x - 3)$ **29.** $(2x - 3)(x + 1)$ **30.** $(3x - 4)(x + 2)$ **31.** $(10x - 1)(2x + 5)$ **32.** $(5x - 1)(4x + 5)$
33. $(4a + 1)(a - 3)$ **34.** $(4a + 3)(a - 1)$ **35.** $2(x - 1)(3x + 5)$ **36.** $2(x + 1)(3x - 5)$ **37.** $2(2x - 3)(x - 5)$
38. $4(x - 9)(x + 4)$ **39.** $(3x - 1)(4x + 3)$ **40.** $2(x - 1)(8x + 15)$ **41.** $(3x - 2y)(2x - 5y)$ **42.** $2(3x - y)(x - 5y)$
43. $(7x + y)(7x - y)$ **44.** $4(2x - 3y)(2x + 3y)$ **45.** $(3x - 2)^2$ **46.** $(8x + 1)(8x - 1)$ **47.** $(5x - 6)(5x + 6)$
48. $(10x - 3)(10x + 3)$ **49.** $(1 - 7x)(1 + 7x)$ **50.** $(2 - 7x)(2 + 7x)$ **51.** $(6x + 1)^2$ **52.** $(5x - 2)^2$ **53.** $(4x - 3y)^2$
54. $(7x - 2y)^2$ **55.** $2(x - 3)(x + 3)$ **56.** $3(x - 5)(x + 5)$ **57.** $2(2x + 5)^2$ **58.** $2(5x - 6)^2$ **59.** $(2x + 3y)(2x - 3y)$
60. $(x + 3)^2$ **61.** $(x - 3)(x - 6)$ **62.** $(x + 15)(x - 2)$ **63.** $(x - 1)(6x + 7)$ **64.** $(5x - 2)(2x + 1)$ **65.** $4(3x + 4)$
66. $4xy(2xy - 1)$ **67.** $10x^2y^2(5x + 2)$ **68.** $13ab(2a^2 - b^2 + 4ab^3)$ **69.** $x(x - 8)^2$ **70.** $2(x + 10)^2$ **71.** $3(x - 3)^2$
72. $x(5x - 6)^2$ **73.** $(7x + 5)(x - 2)$ **74.** $(4x + 3)(x - 4)$ **75.** $xy(3x + 2y)(3x - 2y)$ **76.** $x^3a(3a + 4x)(a - 5x)$
77. $2(3a + 5b)(2a - b)$ **78.** $(4a - 5b)^2$ **79.** $(a - 1)(7 - b)$ **80.** $(3d - 4)(1 - c)$ **81.** $(2x - 1)(1 + b)$ **82.** $(b - 7)(5x + 4y)$
83. $x(2a - 1)(a - 7)$ **84.** $x(x + 4)(x - 4)(x + 1)(x - 1)$ **85.** $(x^2 + 9y^6)(x + 3y^3)(x - 3y^3)$ **86.** $(3x^2 - 5)(2x^2 + 3)$
87. $yz(14 - x)(2 - x)$ **88.** $x(3x + 2)(4x + 3)$ **89.** $(2w + 1)(8w - 5)$ **90.** $3(2w - 1)^2$ **91.** $2y(2y - 1)(y + 3)$
92. $(5y - 1)(2y + 7)$ **93.** $8y^8(y^2 - 2)$ **94.** $49(x^2 + 1)(x + 1)(x - 1)$ **95.** prime   **96.** prime   **97.** $4y(2y^2 - 5)(y^2 + 3)$
98. $3x(3y + 7)(y - 2)$ **99.** $(4x^2y - 7)^2$ **100.** $2xy(8x + 1)(8x - 1)$ **101.** $(2x + 5)(a - 2b)$ **102.** $(2x + 1)(x + 3)(x - 3)$

103. $-3, 6$ **104.** $3, -9$ **105.** $0, \dfrac{1}{6}$ **106.** $0, -\dfrac{11}{6}$ **107.** $-5, \dfrac{1}{2}$ **108.** $-8, -3$ **109.** $-5, -9$ **110.** $-\dfrac{3}{5}, 2$ **111.** -3 **112.** $-3, \dfrac{3}{4}$

113. $\dfrac{1}{5}, 2$ **114.** base $= 10$ cm, altitude $= 7$ cm   **115.** width $= 7$ feet, length $= 15$ feet   **116.** 6 seconds   **117.** 8 amperes, 12 amperes

Putting Your Skills to Work

1. 3.0 times greater   **2.** 5.0 times greater

Chapter 6 Test

1. $(x + 14)(x - 2)$ **2.** $(5x + 7y)(5x - 7y)$ **3.** $(5x + 1)(2x + 5)$ **4.** $(3a - 5b)^2$ **5.** $x(7 - 9x + 14y)$ **6.** $(x + 2y)(3x - 2w)$
7. $2x(3x - 4)(x - 2)$ **8.** $c(5a - b)(a - 2b)$ **9.** $4(5x^2 + 2y^2)(5x^2 - 2y^2)$ **10.** $(3x - y)(3x - 4y)$ **11.** $7x(x - 6)$ **12.** prime
13. prime   **14.** $-5y(2x - 3y)^2$ **15.** $(9x + 1)(9x - 1)$ **16.** $(x^8 + 1)(x^4 + 1)(x^2 + 1)(x + 1)(x - 1)$ **17.** $(x + 3)(2a - 5)$
18. $(a + 2b)(w + 2)(w - 2)$ **19.** $3(x - 6)(x + 5)$ **20.** $x(2x + 5)(x - 3)$ **21.** $-5, -9$ **22.** $-\dfrac{7}{3}, -2$ **23.** $-\dfrac{5}{2}, 2$
24. width $= 7$ miles, length $= 13$ miles

Cumulative Test for Chapters 1–6

1. -10.36 **2.** $8x^4y^{10}$ **3.** 81 **4.** $27x^2 + 6x - 8$ **5.** $2x^3 - 12x^2 + 19x - 3$ **6.** $x \le -3$ **7.** 2 **8.** -15 **9.** $t = \dfrac{2s - 2a}{3}$
10. $m = 1$ **11.** $-\dfrac{1}{2}$ **12.** $x = -5, y = 3$ **13.** $(3x - 1)(2x - 1)$ **14.** $(3x + 4)(2x - 1)$ **15.** $(3x + 2)(3x - 1)$
16. $(11x + 8y)(11x - 8y)$ **17.** $-4(5x + 6)(4x - 5)$ **18.** prime   **19.** $x(4x + 5)^2$ **20.** $(9x^2 + 4b^2)(3x + 2b)(3x - 2b)$
21. $(2x + 3)(a - 2b)$ **22.** $(x^2 + 5)(x^2 + 3)$ **23.** $-8, 3$ **24.** $\dfrac{5}{3}, 2$ **25.** base $= 6$ miles, altitude $= 19$ miles

Chapter 7

Pretest Chapter 7

1. -2 **2.** $\dfrac{x - 3}{2x + 3}$ **3.** $\dfrac{b}{2a - b}$ **4.** $\dfrac{2a - b}{6}$ **5.** $\dfrac{x}{x + 5}$ **6.** 1 **7.** $\dfrac{y}{x - 1}$ **8.** 3 **9.** $\dfrac{-2y - 7}{(y - 1)(2y + 3)}$ **10.** $\dfrac{1}{x + 5}$
11. $\dfrac{5x^2 + 6x - 9}{3x(x - 3)^2}$ **12.** $\dfrac{2a - 3}{5a^2 + a}$ **13.** $\dfrac{a^2 - 2a - 2}{3a^2(a + 1)}$ **14.** $\dfrac{-x^2 - y^2}{x^2 + 2xy - y^2}$ **15.** $x = -5$ **16.** $x = 7$ **17.** 9.3 **18.** $\$263.50$

7.1 Exercises

1. 3 **3.** $\dfrac{6}{x}$ **5.** $\dfrac{2}{x - 4}$ **7.** $\dfrac{x + y^2}{xy}$ **9.** $\dfrac{x + 2}{x}$ **11.** $\dfrac{x - 5}{3x - 1}$ **13.** $\dfrac{x(x - 4)}{x + 6}$ **15.** $\dfrac{3x - 2}{x + 4}$ **17.** $\dfrac{3x - 5}{4x - 1}$ **19.** $\dfrac{x - 5}{x + 1}$ **21.** $\dfrac{-3}{2}$
23. $\dfrac{-2x - 3}{x + 5}$ **25.** $\dfrac{4x + 5}{2x - 1}$ **27.** $\dfrac{2x - 3}{-x + 5}$ **29.** $\dfrac{a - b}{2a - b}$ **31.** $\dfrac{3x - 2y}{3x + 2y}$ **33.** $\dfrac{x(x - 2)}{2(x + 3)}$ **35.** $9x^2 - 42x + 49$ **36.** $2x^3 - 9x^2 - 2x + 24$
37. $-8a^3c^9$ **38.** $\dfrac{5y}{6xz^3}$ **39.** $885,000,000$ people   **40.** $332,000,000$ people

7.2 Exercises

1. Factor the numerator and denominator completely and divide out any common factors.
3. $\dfrac{2(x + 4)}{(x - 4)}$ **5.** $\dfrac{x^2}{2(x - 2)}$ **7.** $\dfrac{x - 2}{x + 3}$ **9.** $x + 2$ **11.** $\dfrac{3(x + 2y)}{4(x + 3y)}$ **13.** $\dfrac{-3(x - 2)}{x + 5}$ **15.** $\dfrac{(x + 5)(x - 2)}{3x - 1}$ **17.** 1
19. By definition, the denominator of a rational expression cannot have the value zero. So the original expression cannot have a replacement value of 2 (or the first denominator would be zero) or 6 (or the second denominator would be zero). If we multiply the first fraction by the reciprocal of the second fraction, then the denominator $x + 7$ cannot be zero. Thus x also cannot have a replacement value of -7. The domain of the given expression is all real numbers except 2, -7, and 6.
21. $x = \dfrac{3}{2}$ **22.** $7x^3 - 22x^2 + 2x + 3$ **23.** $\$713,989.5x$ **24.** $\$2.38$

7.3 Exercises

1. The LCD would be a product that contains each factor. However, any repeated factor in any one denominator must be repeated the greatest number of times it occurs in any one denominator. So the LCD would be $(x + 5)(x + 3)^2$.
3. $\dfrac{3x + 1}{x + 5}$ **5.** $\dfrac{-1}{x - 4}$ **7.** $\dfrac{2x - 7}{5x + 7}$ **9.** $5a^3$ **11.** $112x^2y^4$ **13.** $x^2 - 16$ **15.** $(x + 2)(4x - 1)^2$ **17.** $\dfrac{3x + 2}{(x - 1)(x - 1)}$
19. $\dfrac{4y}{(y + 1)(y - 1)}$ **21.** $\dfrac{10a + 5b + 2ab}{2ab(2a + b)}$ **23.** $\dfrac{x^2 - 3x + 24}{4x^2}$ **25.** $\dfrac{5x + 21}{(x - 4)(x + 3)(x + 5)}$ **27.** $\dfrac{2x}{(x + 1)(x - 1)}$ **29.** $\dfrac{a + 5}{6}$
31. $\dfrac{9x + 2}{(3x - 4)(4x - 3)}$ **33.** $\dfrac{-4x - 2}{x(x - 2)^2}$ **35.** $\dfrac{x^2 - 5x - 2}{(x + 2)(x + 3)(x - 1)}$ **37.** $\dfrac{9y}{y - 3}$ **39.** $\dfrac{3}{y + 4}$ **41.** $\dfrac{x^2 + 9x - 10}{(x - 4)(x + 4)}$
43. $\dfrac{(4y + 1)(y - 1)}{2y(3y + 1)(y - 3)}$ **45.** $\dfrac{1}{10(3x + 2)}$ **46.** $x = -7$ **47.** $y = \dfrac{5ax + 6bc}{2a}$ **48.** $m = -1$ **49.** $m = -\dfrac{1}{2}; b = -3$ **50.** at least 17 days
51. $566,100$ more people   **52.** approximately 7.8%

Putting Your Skills to Work

1. $883 billion   **2.** $423 billion

7.4 Exercises

1. $\dfrac{5}{7x}$ **3.** $y + x$ **5.** $\dfrac{y}{x}$ **7.** $\dfrac{x - 3}{x}$ **9.** $\dfrac{-2x}{3(x + 1)}$ **11.** $\dfrac{3a^2 + 9}{a^2 + 2}$ **13.** $\dfrac{(2y + 1)(y - 2)}{2y}$ **15.** $\dfrac{x^2 - 2x}{4 + 5x}$ **17.** $\dfrac{2x - 5}{3(x + 3)}$ **19.** $\dfrac{y + 1}{-y + 1}$

21. No expression in any denominator can be allowed to be zero since division by zero is undefined. So -3, 5, and 0 are not allowable replacements for the variable x. The domain of the given expression is all real numbers except $-3, 5$, and 0.   **23.** $w = \dfrac{P - 2l}{2}$

24. $x > -1$ **25.**   **26.** $x = -3; \; y = \dfrac{1}{2}$

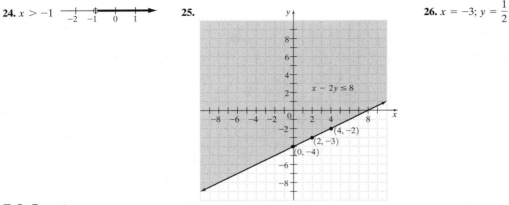

$x - 2y \le 8$

7.5 Exercises

1. $x = 3$ **3.** $x = -8$ **5.** $x = -\dfrac{3}{4}$ **7.** $x = -2$ **9.** $x = 4$ **11.** $x = \dfrac{-14}{3}$ **13.** There is no solution.   **15.** $x = -3$ **17.** $y = -5$

19. $x = 12$ **21.** There is no solution.   **23.** There is no solution.   **25.** $x = -11$ **27.** $x = 2, x = 4$ **29.** width $= 7$ m; length $= 20$ m

30. Domain $= \{7, 2, -2\}$; Range $= \{3, 2, 0, -2, -3\}$; not a function   **31.**

32. $(3x + 4)(2x - 3)$ **33.** $(2y + 11z)(2y - 11z)$ **34.** $x = -3, x = 7$

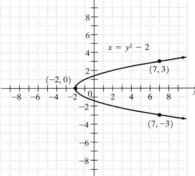

$x = y^2 - 2$

7.6 Exercises

1. $x = 18$ **3.** $x = 40\dfrac{4}{5}$ **5.** $x = \dfrac{40}{3}$ **7.** $x = 22.75$ **9.** 110 miles   **11. (a)** 522.88 British pounds   **(b)** $214.20   **13.** 56 miles per hour

15. 29 miles   **17.** $n = 18\dfrac{17}{20}$ inches   **19.** $d = 9\dfrac{1}{3}$ in.   **21.** $k = 38\dfrac{2}{5}$ m   **23.** 48 inches   **25.** 86 meters   **27.** 61.5 miles per hour

29. commuter airline, 250 kilometers per hour; helicopter, 210 kilometers per hour   **31. (a)** $22.06   **(b)** $24.71   **(c)** $27

33. $2\dfrac{2}{9}$ hours or 2 hours, 13 minutes   **35.** $3\dfrac{3}{7}$ hours or 3 hours, 26 minutes   **37.** 6.316×10^{-7} **38.** 582,000,000   **39.** $\dfrac{w^8}{x^3 y^2 z^4}$ **40.** $\dfrac{27}{8}$ or $3\dfrac{3}{8}$

Putting Your Skills to Work

1. 462 days   **2.** 21 orbits

Chapter 7 Review Problems

1. $-\dfrac{4}{5}$ **2.** $\dfrac{x}{x - y}$ **3.** $\dfrac{x + 3}{x - 4}$ **4.** $\dfrac{x + 2}{x + 4}$ **5.** $\dfrac{x + 3}{x - 7}$ **6.** $\dfrac{2(x + 4)}{3}$ **7.** $\dfrac{2x + 3}{2x}$ **8.** $\dfrac{x}{x - 5}$ **9.** $\dfrac{2(x - 4y)}{2x - y}$ **10.** $\dfrac{2 - y}{3y - 1}$

11. $\dfrac{x - 2}{(5x + 6)(x - 1)}$ **12.** $4x + 2y$ **13.** $\dfrac{x - 25}{x + 5}$ **14.** $\dfrac{2(y - 3)}{3y}$ **15.** $\dfrac{(y + 2)(2y + 1)(y - 2)}{(2y - 1)(y + 1)^2}$ **16.** $\dfrac{4y(3y - 1)}{(3y + 1)(2y + 5)}$

17. $\dfrac{3y(4x + 3)}{2(x - 5)}$ **18.** $\dfrac{x + 2}{2}$ **19.** $\dfrac{3}{16}$ **20.** $\dfrac{2(x + 3y)}{x - 4y}$ **21.** $\dfrac{9x + 2}{x(x + 1)}$ **22.** $\dfrac{5x^2 + 7x + 1}{x(x + 1)}$ **23.** $\dfrac{(x - 1)(x - 2)}{(x + 3)(x - 3)}$

24. $\dfrac{10x - 22}{(x + 2)(x - 4)}$ **25.** $\dfrac{2xy + 4x + 5y + 6}{2y(y + 2)}$ **26.** $\dfrac{(2a + b)(a + 4b)}{ab(a + b)}$ **27.** $\dfrac{3x - 2}{3x}$ **28.** $\dfrac{2x^2 + 7x - 2}{2x(x + 2)}$ **29.** $\dfrac{1 - 2x - x^2}{(x + 5)(x + 2)}$

30. $\dfrac{3}{2(x - 9)}$ **31.** $\dfrac{1}{11}$ **32.** $\dfrac{5}{3x^2}$ **33.** $w - 2$ **34.** 1 **35.** $-\dfrac{y^2}{2}$ **36.** $\dfrac{x + 2y}{y(x + y + 2)}$ **37.** $\dfrac{-1}{a(a + b)}$ **38.** $\dfrac{-3a - b}{b}$ **39.** $\dfrac{-3y}{2(x + 2y)}$

40. $\dfrac{5y(x + 5y)^2}{x(x - 6y)}$ **41.** $a = 15$ **42.** $a = 2$ **43.** $x = \dfrac{2}{7}$ **44.** $x = -8$ **45.** $y = -4$ **46.** $x = -2$ **47.** $x = \dfrac{1}{2}$ **48.** $x = \dfrac{1}{5}$

49. no solution **50.** $x = -4, x = 6$ **51.** $y = 2.0$ **52.** $y = 6$ **53.** $y = -2$ **54.** no solution **55.** $y = 9$ **56.** $y = 0$ **57.** $x = 1.3$
58. $x = 2.8$ **59.** $x = 26.4$ **60.** $x = 2$ **61.** $x = 16$ **62.** $x = 9.1$ **63.** 8.3 gallons **64.** 167 cookies **65.** 46 gallons **66.** 91.5 miles
67. train, 60 miles per hour; car, 40 miles per hour **68.** 3 hours, 5 minutes **69.** 1200 feet **70.** 182 feet

71. $3\dfrac{1}{3}$ hours or 3 hours, 20 minutes **72.** 12 hours

Chapter 7 Test

1. $\dfrac{2}{3a}$ **2.** $\dfrac{2x^2(2 - y)}{(y + 2)}$ **3.** $\dfrac{5}{12}$ **4.** $\dfrac{1}{3y(x - y)}$ **5.** $\dfrac{(a - 2)(4a + 7)}{(a + 2)^2}$ **6.** $\dfrac{4x + 18}{(x + 5)(x - 2)(x + 3)}$ **7.** $\dfrac{x - a}{ax}$ **8.** $-\dfrac{x + 2}{x + 3}$ **9.** $\dfrac{x}{4}$

10. $\dfrac{x + 1}{x(x + 3)^2}$ **11.** $\dfrac{2x - 3y}{4x + y}$ **12.** $\dfrac{x}{(x + 2)(x + 4)}$ **13.** $x = 3$ **14.** $x = 4$ **15.** no solution **16.** $x = \dfrac{47}{6}$ **17.** $x = \dfrac{45}{13}$

18. $x = 37.2$ **19.** 6 hours, 25 minutes **20.** $368 **21.** 102 feet

Cumulative Test for Chapters 1–7

1. $x = 6$ **2.** $h = \dfrac{A}{\pi r^2}$ **3.** $x > 1.25$ **4.** $x > 5$ **5.**

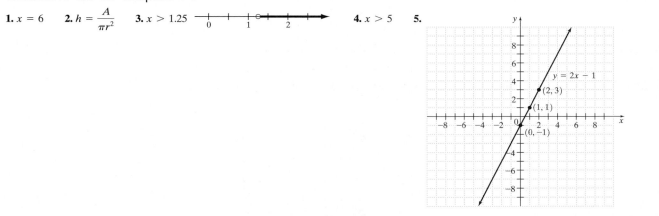

6. $y = x + 3$ **7.** $x = 1, y = -2, z = -5$ **8.** $-12x^5y^7$ **9.** $4x^2 - 28x + 49$ **10.** $(a + b)(3x - 2y)$ **11.** $\dfrac{2x + 5}{x + 7}$ **12.** $\dfrac{(x - 2)(3x + 1)}{3x(x + 5)}$

13. $\dfrac{1}{2}$ **14.** $\dfrac{11x - 3}{2(x + 2)(x - 3)}$ **15.** $\dfrac{x^2 + 2x + 6}{(x + 4)(x - 3)(x + 2)}$ **16.** $x = -2$ **17.** $x = -\dfrac{9}{2}$ **18.** $\dfrac{x + 8}{6x(x - 3)}$ **19.** $\dfrac{3ab^2 + 2a^2b}{5b^2 - 2a^2}$

20. $\dfrac{98}{5}$ or $19\dfrac{3}{5}$ **21.** 208 miles **22.** 484 phone calls

Chapter 8

Pretest Chapter 8

1. $\dfrac{6y^{5/6}}{x^{1/4}}$ **2.** $-\dfrac{64y}{x^{3/4}}$ **3.** $6x^3y^{5/3}$ **4.** $\dfrac{9x^4}{y^6}$ **5.** $\dfrac{1}{81}$ **6.** 9 **7.** $3a^4b^2c^5$ **8.** $2x^2y^3\sqrt[4]{2y^3}$ **9.** $2y\sqrt{3y} + 5\sqrt[3]{2}$ **10.** $108 - 57\sqrt{2}$

11. $\dfrac{2\sqrt[3]{3x^2}}{x}$ **12.** $-5 - 2\sqrt{6}$ **13.** $x = 7$ **14.** $x = 3, x = 11$ **15.** $4 - 5i$ **16.** $4i$ **17.** $-16 + 30i$ **18.** $\dfrac{12 - 5i}{13}$ **19.** $y = 50$

20. $y = 7.2$

8.1 Exercises

1. $\dfrac{16x^2}{y^6}$ **3.** $\dfrac{8b^3}{-27a^3}$ **5.** $x^{3/2}$ **7.** y^8 **9.** $x^{1/2}$ **11.** $x^{5/2}$ **13.** $x^{4/7}$ **15.** $y^{1/2}$ **17.** $\dfrac{1}{x^{3/4}}$ **19.** $\dfrac{b^{1/3}}{a^{5/6}}$ **21.** $\dfrac{1}{6^{1/2}}$ **23.** $\dfrac{2}{a^{1/4}}$ **25.** $x^{5/6}y$

27. $-14x^{7/12}y^{1/12}$ **29.** $6^{4/3}$ **31.** $2x^{7/10}$ **33.** $-\dfrac{4x^{5/2}}{y^{6/5}}$ **35.** $2ab$ **37.** $9x^{4/5}y^3z^{2/3}$ **39.** $x^2 - x^{13/15}$ **41.** $m^{3/8} + 2m^{15/8}$ **43.** $\dfrac{1}{x^{1/4}}$ **45.** 9

47. 8 **49.** 32 **51.** $\dfrac{2y + 1}{y^{2/3}}$ **53.** $\dfrac{x^{1/2} + 5^{1/4}}{5^{1/4}x^{1/2}}$ **55.** $2a(3a^{1/3} - 4a^{1/2})$ **57.** $b = \dfrac{1}{4}$ **59.** radius = 2.48 meters **61.** radius = 10 feet

62. $x = -\dfrac{3}{2}$ **63.** $b = \dfrac{2A - ah}{h}$ or $b = \dfrac{2A}{h} - a$ **64.** 147 milligrams **65.** 5 years old

8.2 Exercises

1. A square root of a number is a value that when multiplied by itself is equal to the original number.　　**3.** $\sqrt[3]{-8} = -2$ because $(-2)(-2)(-2) = -8$.

5. 8　　**7.** 12　　**9.** $-\dfrac{1}{3}$　　**11.** 1　　**13.** 0.2　　**15.** 2.2, 3.9, 5, 5.9; $x \geq -0.5$　　**17.** 0, 1, 2, 2.2; $x \geq 6$

19.　　　　　　　　**21.**　　　　　　　　　　**23.** 6　　**25.** 4　　**27.** -2　　**29.** 3　　**31.** 8　　**33.** 5　　**35.** $-\dfrac{1}{4}$　　**37.** $\dfrac{3}{5}$

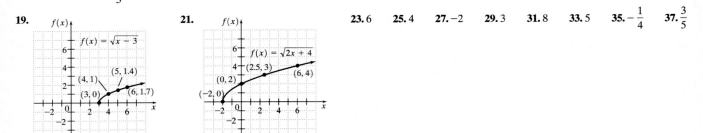

39. $a^{1/2}$　　**41.** $(3y)^{1/4}$　　**43.** $(a-b)^{5/9}$　　**45.** $\left(y^{1/2}\right)^{1/5} = y^{1/10}$　　**47.** $(2x)^{3/5}$　　**49.** -11　　**51.** a^2b　　**53.** $7xy^4$　　**55.** $3a^3b^5$

57. $2xy^3$　　**59.** $\left(\sqrt[6]{x}\right)^5$　　**61.** $\dfrac{1}{\sqrt[5]{125}}$　　**63.** $\left(\sqrt[7]{x+3y}\right)^4$　　**65.** $\left(\sqrt[7]{-y}\right)^5$　　**67.** $\sqrt[3]{9a^2b^2}$　　**69.** 9　　**71.** $\dfrac{2}{5}$　　**73.** 25　　**75.** $\dfrac{1}{6y^4}$　　**77.** $7x^4$

79. $5a^7b^9$　　**81.** $10x^5y^6z$　　**83.** $-5a^2b^5c^7$　　**85.** $10|x|$　　**87.** $-3x^3$　　**89.** x^4y^{10}　　**91.** $|ab^5|$　　**93.** $7a^6b^2$　　**95.** \$1095

96. 2.0135×10^{16} Btu　　**97.** 24.47%　　**98.** $2(x+3y)^2$　　**99.** $(2x-7)(x+3)$　　**100.** $\dfrac{x+3}{4y}$　　**101.** $\dfrac{37x+10}{4(x-2)(3x+1)}$

8.3 Exercises

1. $2\sqrt{2}$　　**3.** $3\sqrt{2}$　　**5.** $2\sqrt{30}$　　**7.** $2\sqrt{11}$　　**9.** $3x\sqrt{x}$　　**11.** $2a^2b^2\sqrt{15b}$　　**13.** $7x^2y^3\sqrt{2xz}$　　**15.** 2　　**17.** $3\sqrt[3]{4}$　　**19.** $2\sqrt[3]{7y}$　　**21.** $2ab^2\sqrt[3]{b^2}$

23. $2x^2y^3\sqrt[3]{3y^2}$　　**25.** $3p^5\sqrt[4]{kp^3}$　　**27.** $-2xy\sqrt[5]{y}$　　**29.** $a=4$　　**31.** 17　　**33.** $6\sqrt{3}$　　**35.** $8\sqrt{2}$　　**37.** $-12\sqrt{2}$　　**39.** 0　　**41.** $2\sqrt{11} - \sqrt{7x}$

43. $6x\sqrt{2x}$　　**45.** $11\sqrt[3]{2}$　　**47.** $-10xy\sqrt[3]{y} + 6xy^2$　　**49.** $20.78460969 = 20.78460969$ ✓　　**51.** 7.071 amps　　**53.** 3.14 seconds

55.　　　　　　　　　　　　**56.**　　　　　　　　　　　　**57.** $\dfrac{9y^4}{4x^2}$　　**58.** $-27a^6c^{12}$

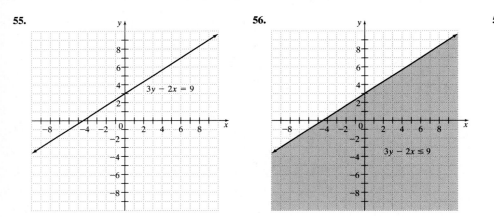

8.4 Exercises

1. $-12\sqrt{3}$　　**3.** $-3\sqrt{5xy}$　　**5.** $14\sqrt{3x} - 35x$　　**7.** $22 - 5\sqrt{2}$　　**9.** $4 - 6\sqrt{6}$　　**11.** $14 + 11\sqrt{35x} + 60x$　　**13.** $\sqrt{15} + 3 + 2\sqrt{10} + 2\sqrt{6}$

15. $-11x$　　**17.** $29 - 4\sqrt{30}$　　**19.** $3x + 13 + 6\sqrt{3x+4}$　　**21.** $36 - 60\sqrt{a} + 25a$　　**23.** $3x\sqrt[3]{4} - 4x^2\sqrt[3]{x}$　　**25.** $\dfrac{7}{5}$　　**27.** $\dfrac{2\sqrt{3x}}{7y^3}$

29. $\dfrac{2xy^2\sqrt[3]{x^2}}{3}$　　**31.** $2xy$　　**33.** $\dfrac{3\sqrt{2}}{2}$　　**35.** $\dfrac{2\sqrt{3}}{3}$　　**37.** $\dfrac{\sqrt{5y}}{5y}$　　**39.** $\dfrac{x(\sqrt{5}+\sqrt{2})}{3}$　　**41.** $\sqrt{15} + 2\sqrt{3}$　　**43.** $\dfrac{x\sqrt{3}-\sqrt{2x}}{3x-2}$　　**45.** $4 + \sqrt{15}$

47. $\dfrac{3x - 3\sqrt{3xy} + 2y}{3x - y}$　　**49.** $5x - 2x\sqrt{5} + \sqrt{5} - 2$　　**51.** $-5(\sqrt{2} + \sqrt{3})$　　**53.** $\dfrac{\sqrt[3]{49}}{7}$　　**55.** 1.194938299, 1.194938299; yes; yes

57. $\dfrac{-25}{8(\sqrt{3} - 2\sqrt{7})}$　　**59.** \$2.92　　**61.** $(x + 8\sqrt{x} + 15)$ square millimeters　　**63.** $x = 2, y = 3$　　**64.** $x = 1, y = -5, z = 3$

65. January 11　　**66.** 5 cups of coffee and 6 cups of tea; on January 22

8.5 Exercises

1. Isolate one of the radicals on one side of the equation.　　**3.** $x = 3$　　**5.** $x = 1$　　**7.** $y = 7$　　**9.** $y = 0, y = -1$　　**11.** $x = 3, x = 7$

13. $x = 0, x = \dfrac{1}{2}$　　**15.** $x = \dfrac{5}{2}$　　**17.** $x = 7$　　**19.** $x = 12$　　**21.** $x = \dfrac{1}{4}$　　**23.** $x = 5$　　**25.** $x = 0, x = 8$　　**27.** $x = -1$　　**29.** $x = 9$

31. $x = 4.9232, x = 0.4028$ **33. (a)** $S = \dfrac{V^2}{12}$ **(b)** 27 feet **35.** $x = 0.055y^2 + 1.25y - 10$ **37.** $c = 9$ **39.** $16x^4$ **40.** $\dfrac{1}{2x^2}$

41. $-6x^2y^3$ **42.** $2x^3y^4\sqrt[4]{4}$ **43.** $(8x^3 + 16x^2 + 24x + 27)$ cubic centimeters **44.** $(4r^3 + 18r^2 + 26r + 12)$ boxes **45.** 3 miles per hour

46. 15 miles per hour

8.6 Exercises

1. No. There is no real number that, when squared, will equal -9. **3.** No. To be equal, the real number parts must be equal, and the imaginary parts

must be equal. $2 \neq 3$ and $3i \neq 2i$. **5.** $6i$ **7.** $5i\sqrt{2}$ **9.** $\dfrac{1}{2}i$ **11.** $-9i$ **13.** $2 + i\sqrt{3}$ **15.** $-3 + 2i\sqrt{6}$ **17.** $x = 5, y = -3$

19. $x = 1.3, y = 2$ **21.** $x = -6, y = 3$ **23.** $1 - i$ **25.** $1.2 + 2.1i$ **27.** $7 + 4i$ **29.** $8 + 3i$ **31.** $-10 - 12i$ **33.** $-\dfrac{3}{4} + i$

35. $-\sqrt{21}$ **37.** $-\sqrt{6}$ **39.** -12 **41.** $12 - \sqrt{10} + 4i\sqrt{2} + 3i\sqrt{5}$ **43.** i **45.** 1 **47.** -1 **49.** 0 **51.** $\dfrac{1 + i}{2}$ **53.** $\dfrac{3 + 6i}{10}$

55. $-\dfrac{2 + 5i}{6}$ **57.** $\dfrac{35 + 42i}{61}$ **59.** $\dfrac{11 - 6i}{13}$ **61.** $\dfrac{1 - 8i}{5}$ **63.** $-2299.95 + 3293.32i$ **65.** $Z = \dfrac{2 - 3i}{3}$ **67.** 18 hours producing juice in

glass bottles, 25 hours producing juice in cans, 62 hours producing juice in plastic bottles **68.** 57 small bowls, 63 large bowls

8.7 Exercises

1. Answers will vary. A person's weekly paycheck varies as the number of hours worked. $y = kx$, y is the weekly salary, k is the hourly salary, x is

the number of hours. **3.** $y = \dfrac{k}{x}$ **5.** $y = 24$ **7.** 71.4 pounds per square inch **9.** 16 feet; 64 feet **11.** $y = 160$

13. 1333.3 Btu per hour **15.** 30 miles per hour **17.** 400 pounds **19.** approximately 2.3 oersteds **21.** approximately 62.7 miles per hour

23. $x = \dfrac{2}{3}, x = 2$ **24.** $x = 1, x = -8$ **25.** \$460 **26.** 55 gallons **27.** 65 gold leaf frames, 45 silver frames

28. first side is 16 centimeters; second side is 20 centimeters; third side is 14 centimeters

Putting Your Skills to Work

1. 3.9 miles, 8.7 miles, 12.2 miles **2.** 15 miles, 19.4 miles

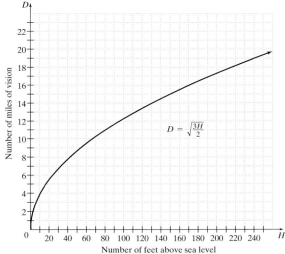

Chapter 8 Review Problems

1. $\dfrac{15x^3}{y^{5/2}}$ **2.** $\dfrac{1}{2}x^{1/2}$ **3.** $5a^{3/2}b^2$ **4.** $5^{3/4}$ **5.** $-6a^{5/6}b^{3/4}$ **6.** $\dfrac{x^{1/2}y^{3/10}}{2}$ **7.** $\dfrac{x}{32y^{1/2}z^4}$ **8.** $7a^5b$ **9.** $x^{1/2}y^{1/5}$ **10.** $3x^{n+1}$ **11.** $5^{12/7}$

12. $\dfrac{2x + 1}{x^{2/3}}$ **13.** $3x(2x^{1/2} - 3x^{-1/2})$ **14.** $(2x)^{1/10}$ **15.** $\left(\sqrt[9]{2x + 3y}\right)^4$ **16.** 8 **17.** $\sqrt[6]{-64}$ is not a real number; $-\sqrt[6]{64} = -2$ **18.** $\dfrac{1}{81}$

19. $\dfrac{8}{27}$ **20.** $3xy^3z^5\sqrt{11x}$ **21.** $-2a^2b^3c^4\sqrt[3]{7a^2b}$ **22.** $12x^5y^6$ **23.** $5a^3b^2c^{100}$ **24.** y **25.** $|y|$ **26.** $|xy|$ **27.** x^2 **28.** x^7 **29.** x^4

30. $11\sqrt{2}$ **31.** $13\sqrt{7}$ **32.** $2 - 6\sqrt[3]{2}$ **33.** $11\sqrt{2x} - x\sqrt{2}$ **34.** $90\sqrt{2}$ **35.** $12x\sqrt{2} - 36\sqrt{3x}$ **36.** $4 - 9\sqrt{6}$ **37.** $34 - 14\sqrt{3}$

38. $74 - 12\sqrt{30}$ **39.** $2x - \sqrt[3]{2xy} + 2\sqrt[3]{3x^2} - \sqrt[3]{6y}$ **40. (a)** $f(16) = 10$ **(b)** all real numbers x where $x \geq -4$

41. (a) $f(5) = 4$ **(b)** all real numbers x where $x \leq 9$ **42. (a)** $f(1) = \dfrac{1}{2}$ **(b)** all real numbers x where $x \geq \dfrac{2}{3}$ **43.** $\dfrac{x\sqrt{3y}}{y}$ **44.** $\dfrac{2\sqrt{3y}}{3y}$

45. $\sqrt{3}$ **46.** $2\sqrt{6} + 2\sqrt{5}$ **47.** $\dfrac{3x - \sqrt{xy}}{9x - y}$ **48.** $\dfrac{-(\sqrt{35} + 3\sqrt{5})}{2}$ **49.** $\dfrac{2 + 3\sqrt{2}}{7}$ **50.** $\dfrac{10\sqrt{3} - 3\sqrt{2} + 5\sqrt{6} - 3}{3}$

51. $\dfrac{3x + 4\sqrt{xy} + y}{x - y}$ **52.** $\dfrac{\sqrt[3]{4x^2y}}{2y}$ **53.** $4i + 3i\sqrt{5}$ **54.** $x = \dfrac{-7 + \sqrt{6}}{2}$; $y = -3$ **55.** $-9 - 11i$ **56.** $-10 + 2i$ **57.** $29 - 29i$

58. $48 - 64i$ **59.** $-8 + 6i$ **60.** $-5 - 4i$ **61.** -1 **62.** i **63.** $\dfrac{13 - 34i}{25}$ **64.** $\dfrac{11 + 13i}{10}$ **65.** $-\dfrac{3 + 4i}{5}$ **66.** $\dfrac{18 + 30i}{17}$ **67.** $-2i$

68. $x = 4$ **69.** $x = -1$ **70.** $x = 4$ **71.** $x = 5$ **72.** $x = 5, x = 1$ **73.** $x = 1, x = \dfrac{3}{2}$ **74.** $y = 9.6$ **75.** $y = 12.5$

76. 168.1 feet   **77.** 3.5 seconds   **78.** $y = 0.5$ **79.** 16.8 pounds per square inch   **80.** $y = 1.3$ **81.** 160 cubic centimeters

Chapter 8 Test

1. $-6x^{5/6}y^{1/2}$ **2.** $\dfrac{7x^{9/4}}{4}$ **3.** $8^{3/2}x^{1/2}$ or $16(2x)^{1/2}$ **4.** $6^{4/5}$ **5.** $\dfrac{1}{4}$ **6.** 32 **7.** $5a^2b^4\sqrt{3b}$ **8.** $8x^3y^2\sqrt{y}$ **9.** $5xy^2\sqrt[3]{2x}$

10. $18\sqrt{3} + x\sqrt[3]{2x^2}$ **11.** $10\sqrt{2}$ **12.** $18\sqrt{2} - 10\sqrt{6}$ **13.** $12 + 39\sqrt{2}$ **14.** $\dfrac{4\sqrt{5x}}{5x}$ **15.** $\dfrac{\sqrt{3xy}}{3}$ **16.** $2 + \sqrt{3}$ **17.** $x = 2, x = 1$

18. $x = 10$ **19.** $x = 6$ **20.** $2 + 14i$ **21.** $-1 + 4i$ **22.** $18 + i$ **23.** $\dfrac{-13 + 11i}{10}$ **24.** $27 + 36i$ **25.** $-i$ **26.** $y = 3$ **27.** $y = \dfrac{5}{6}$

28. about 83.3 feet

Cumulative Test for Chapters 1–8

1. $3\dfrac{7}{16}$ **2.** $6a^4 - 3a^3 + 15a^2 - 8a$ **3.** -64 **4.** $x = \dfrac{-4y + 8}{3}$ **5.**

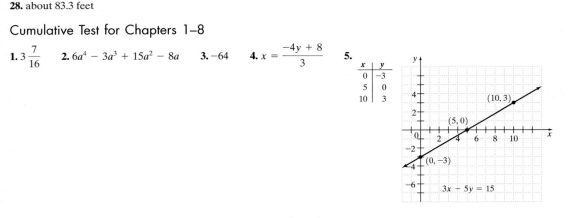

x	y
0	−3
5	0
10	3

$3x - 5y = 15$

6. $8(2x - 1)(x + 2)$ **7.** $x = 2; y = 1; z = -4$ **8.** $\dfrac{5x - 6}{(x - 5)(x + 3)}$ **9.** width = 7 meters; length = 17 meters   **10.** $b = \dfrac{26x + 1}{4}$

11. $\dfrac{1}{2x^{1/2}y^{15/2}}$ **12.** $\dfrac{x^{1/6}}{3^{1/3}y^{2/3}}$ **13.** $\dfrac{1}{4}$ **14.** $2xy^3\sqrt[3]{5x^2}$ **15.** $4\sqrt{5x}$ **16.** $-34 + 3\sqrt{6}$ **17.** $-\dfrac{16 + 9\sqrt{3}}{13}$ **18.** $12i$ **19.** $-7 - 24i$

20. $\dfrac{13 + i}{10}$ **21.** $x = 8$ **22.** $x = -1$ **23.** $y = 75$ **24.** about 53.3 lumens

Chapter 9

Pretest Chapter 9

1. $x = \pm 3\sqrt{2}$ **2.** $x = \dfrac{2 \pm \sqrt{10}}{2}$ **3.** $x = \dfrac{1 \pm \sqrt{57}}{8}$ **4.** $x = -2 \pm \sqrt{11}$ **5.** $x = \dfrac{-3 \pm 2i\sqrt{2}}{2}$ **6.** $x = -2, \dfrac{6}{5}$ **7.** $x = 0, \dfrac{12}{7}$

8. $x = -\dfrac{2}{3}, 3$ **9.** $x = 2, -1$ **10.** $w = \pm 8, \pm 2\sqrt{2}$ **11.** $x = \dfrac{-w \pm \sqrt{w^2 - 24w}}{3}$ **12.** $W = 4$ m; $L = 13$ m

13. Vertex $(-1, -12)$; y-intercept $(0, -9)$; x-intercepts $(1, 0), (-3, 0)$ **14.**

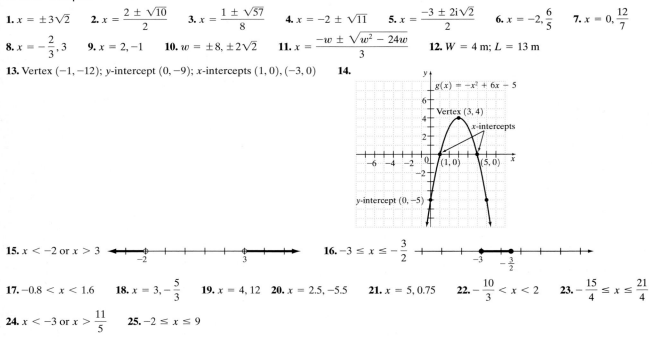

15. $x < -2$ or $x > 3$ **16.** $-3 \le x \le -\dfrac{3}{2}$

17. $-0.8 < x < 1.6$ **18.** $x = 3, -\dfrac{5}{3}$ **19.** $x = 4, 12$ **20.** $x = 2.5, -5.5$ **21.** $x = 5, 0.75$ **22.** $-\dfrac{10}{3} < x < 2$ **23.** $-\dfrac{15}{4} \le x \le \dfrac{21}{4}$

24. $x < -3$ or $x > \dfrac{11}{5}$ **25.** $-2 \le x \le 9$

9.1 Exercises

1. $x = \pm 10$ **3.** $x = \pm 9i$ **5.** $x = \pm\sqrt{15}$ **7.** $x = \pm\sqrt{2}$ **9.** $x = \pm 9i$ **11.** $x = \pm i\sqrt{2}$ **13.** $x = -2 \pm 3\sqrt{2}$ **15.** $x = \dfrac{-2 \pm \sqrt{5}}{3}$

17. $x = \dfrac{7}{5}, -\dfrac{3}{5}$ **19.** $x = 3 \pm 9\sqrt{5}$ **21.** $x = -3 \pm \sqrt{7}$ **23.** $x = 6 \pm 2\sqrt{10}$ **25.** $x = \dfrac{1 \pm \sqrt{37}}{2}$ **27.** $x = \dfrac{-2 \pm \sqrt{39}}{7}$

29. $x = \dfrac{-2 \pm \sqrt{19}}{5}$ **31.** $y = 3, y = -\dfrac{5}{2}$ **33.** $x = \dfrac{3 \pm i\sqrt{7}}{4}$

35. $(-1 + \sqrt{6})^2 + 2(-1 + \sqrt{6}) - 5 \overset{?}{=} 0; 1 - 2\sqrt{6} + 6 - 2 + 2\sqrt{6} - 5 \overset{?}{=} 0; 0 = 0$ ✓   **37.** $x = 16$ feet   **39.** Approximately 0.88 second

41. 15 seconds   **43.** 8   **44.** 7   **45.** 40   **46.** 4

9.2 Exercises

1. Place the quadratic equation in standard form. Find a, b, and c. Substitute these values into the quadratic formula.   **3.** one real

5. $x = \dfrac{1 \pm \sqrt{13}}{2}$ **7.** $x = \dfrac{-1 \pm \sqrt{33}}{4}$ **9.** $x = 0, \dfrac{2}{3}$ **11.** $x = \dfrac{1}{2}, -\dfrac{1}{3}$ **13.** $x = \dfrac{-3 \pm \sqrt{41}}{8}$ **15.** $x = \dfrac{\pm\sqrt{21}}{3}$

17. $x = \dfrac{-1 \pm \sqrt{3}}{2}$ **19.** $x = \pm 1$ **21.** $x = \dfrac{2 \pm 3\sqrt{2}}{2}$ **23.** $x = \dfrac{4 \pm \sqrt{40}}{2} = 2 \pm \sqrt{10}$ **25.** $y = \dfrac{-5 \pm \sqrt{61}}{6}$ **27.** $y = 6, -8$

29. $x = \dfrac{2 \pm \sqrt{-12}}{2} = 1 \pm i\sqrt{3}$ **31.** $x = \dfrac{\pm\sqrt{-60}}{10} = \dfrac{\pm i\sqrt{15}}{5}$ **33.** Two irrational roots   **35.** Two rational roots

37. One rational root   **39.** $x^2 - 11x - 26 = 0$ **41.** $x^2 + 17x + 60 = 0$ **43.** $x^2 + 16 = 0$ **45.** $2x^2 - x - 15 = 0$

47. $x = -2.7554, 1.0888$ **49.** $x = 2.8515, 0.7116$ **51.** Fourteen parachutes or twenty-eight parachutes per day

53. The profit is \$4624 per day. 21 is the average of 14 and 28.   **55.** $-3x^2 - 10x + 11$ **56.** $-y^2 + 3y$

57. The width is 9 feet and length is 16 feet.   **58.** The suits cost \$95 and the goggles cost \$29 last year.

9.3 Exercises

1. $x = \pm\sqrt{5}, x = \pm 2$ **3.** $x = \pm\sqrt{3}, x = \pm 2i$ **5.** $x = \pm\dfrac{i\sqrt{6}}{3}, x = \pm 2$ **7.** $x = 2, x = -1$ **9.** $x = \sqrt[3]{7}, x = -\sqrt[3]{2}$

11. $x = \pm\sqrt[4]{2}, x = \pm 1$ **13.** $x = \pm\dfrac{\sqrt[4]{54}}{3}$; these are the only real roots.   **15.** $x = 8, x = -64$ **17.** $x = -\dfrac{1}{8}; x = 64$ **19.** $x = 625$

21. $x = 16$ **23.** $x = 1, x = -32$ **25.** $x = -2, x = 1, x = \dfrac{-1 \pm \sqrt{13}}{2}$ **27.** $x = 9, x = 4$ **29.** $x = -5, x = -2$ **31.** $x = \dfrac{5}{6}, x = \dfrac{3}{2}$

32. $x = \sqrt[3]{\dfrac{-13 \pm \sqrt{41}}{2}}$ **33.** $-15\sqrt{2x}$ **34.** $\sqrt{3x}$ **35.** $3\sqrt{10} - 12\sqrt{3}$ **36.** $6 - 2\sqrt{10} + 6\sqrt{3} - 2\sqrt{30}$

9.4 Exercises

1. $t = \pm\dfrac{\sqrt{S}}{4}$ **3.** $d = \pm\sqrt{\dfrac{4A}{\pi}}$ **5.** $x = \pm\sqrt{\dfrac{6H}{a}}$ **7.** $y = \pm\dfrac{\sqrt{7R - 4w + 5}}{2}$ **9.** $M = \pm\sqrt{\dfrac{2cQ}{3mw}}$ **11.** $r = \pm\sqrt{\dfrac{V - \pi R^2 h}{\pi h}}$

13. $x = -5b, x = 2b$ **15.** $I = \dfrac{E \pm \sqrt{E^2 - 4RP}}{2R}$ **17.** $w = \dfrac{3q \pm \sqrt{9q^2 + 160}}{20}$ **19.** $r = \dfrac{-\pi h \pm \sqrt{\pi^2 h^2 + S\pi}}{\pi}$

21. $x = \dfrac{-5 \pm \sqrt{25 - 8aw - 8w}}{2a + 2}$ **23.** $c = 2\sqrt{13}$ **25.** $\sqrt{15} = a$ **27.** $b = \dfrac{24\sqrt{5}}{5}, a = \dfrac{12\sqrt{5}}{5}$

29. The hypotenuse is 15 miles long. The shorter leg is 9 miles long.   **31.** The shorter leg was approximately 6.13 miles long. The final leg was approximately 9.13 miles long.   **33.** Width is 7 feet; length is 18 feet.   **35.** Base is 8 cm; altitude is 18 cm.   **37.** The rate of the car in rain was 45 mph; the rate of the car without rain was 50 mph.   **39.** 30 miles   **41.** 1,659,250   **43.** In the year 2004

45. $w = \dfrac{-7b - 21 \pm \sqrt{529b^2 + 294b + 441}}{10}$ **47.** $\dfrac{4\sqrt{3x}}{3x}$ **48.** $\dfrac{\sqrt{30}}{2}$ **49.** $\dfrac{3(\sqrt{x} - \sqrt{y})}{x - y}$ **50.** $-2 - 2\sqrt{2}$ **51.** $\dfrac{3\sqrt[3]{a^2 b}}{2}$

9.5 Exercises

1. $V(1, -9); I(0, -8); (4, 0); (-2, 0)$ **3.** $V(-2, 16); I(0, 12); (-6, 0); (2, 0)$ **5.** $V(-2, -9); I(0, 3); (-0.3, 0); (-3.7, 0)$

7. $V\left(-\dfrac{1}{3}, -\dfrac{17}{3}\right)$; no x-intercepts; y-intercept $(0, -6)$ **9.** $V\left(-\dfrac{1}{2}, -\dfrac{9}{2}\right); I(0, -4); (1, 0); (-2, 0)$

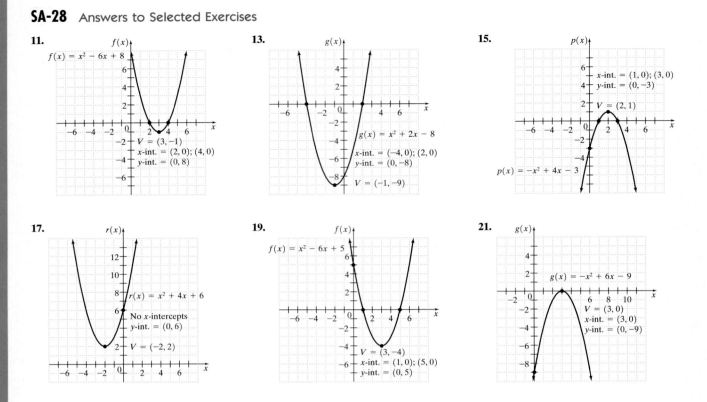

11. $f(x) = x^2 - 6x + 8$; $V = (3, -1)$; x-int. $= (2, 0); (4, 0)$; y-int. $= (0, 8)$

13. $g(x) = x^2 + 2x - 8$; x-int. $= (-4, 0); (2, 0)$; y-int. $= (0, -8)$; $V = (-1, -9)$

15. $p(x) = -x^2 + 4x - 3$; x-int. $= (1, 0); (3, 0)$; y-int. $= (0, -3)$; $V = (2, 1)$

17. $r(x) = x^2 + 4x + 6$; No x-intercepts; y-int. $= (0, 6)$; $V = (-2, 2)$

19. $f(x) = x^2 - 6x + 5$; $V = (3, -4)$; x-int. $= (1, 0); (5, 0)$; y-int. $= (0, 5)$

21. $g(x) = -x^2 + 6x - 9$; $V = (3, 0)$; x-int. $= (3, 0)$; y-int. $= (0, -9)$

23. $N(20) = 110{,}650$; $N(40) = 263{,}050$; $N(60) = 559{,}450$; $N(80) = 999{,}550$; $N(100) = 1{,}584{,}250$; **25.** $N(70) \approx 750{,}000$. Approximately 750,000 people scuba dive and have a mean income of \$70,000. **27.** Approximately 50. This means that 390,000 people who scuba dive have a mean income of \$50,000.

29.

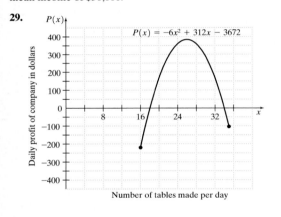

$P(x) = -6x^2 + 312x - 3672$

Daily profit of company in dollars

Number of tables made per day

31. Twenty-four tables per day or twenty-eight tables per day. The parabola is symmetric.

33. The maximum height is 56 feet. It will take about 2.9 seconds.

35. Vertex $(2.2, 2.8)$; y-intercept $(0, 7.6)$; no x-intercepts

37. x-intercepts $(-0.3, 0)$ and $(2.6, 0)$

39. $(-1, 3)$ **40.** $(11, 5)$ **41.** $(3, 1, 2)$ **42.** $(3, -5, 8)$

9.6 Exercises

1. The critical points divide the number line into regions. All values of x in a given region produce results that are greater than zero, or else all the values of x in a given region produce results that are less than zero.

3. $-4 < x < 3$

5. $-\dfrac{3}{2} < x < 1$

7. $x \le -2$ or $x \ge 2$

9. $-\dfrac{1}{5} \le x \le 1$ **11.** $-5 < x < 4$ **13.** $x < -\dfrac{2}{3}$ or $x > \dfrac{3}{2}$

15. $-10 \le x \le 3$ **17.** $x = 2$ **19.** Approximately $x < -1.2$ or $x > 3.2$ **21.** Approximately $1.6 < x < 4.4$ **23.** All real numbers satisfy this inequality. **25.** For time t greater than fifteen seconds but less than twenty-five seconds **27.** (a) Approximately $11.5 < x < 208.5$ (b) \$122,000 (c) \$144,000 **29.** She must score a combined total of 167 points on the two tests. Any two test scores that total 167 will be sufficient to participate in synchronized swimming. **30.** 52 ounces of potato chips, 122 ounces of peanuts, 62 ounces of popcorn, and 124 ounces of pretzels.

31. **32.** **33.** $(4y + 1)(4y - 1)$
34. $-3(x + 3y)^2$

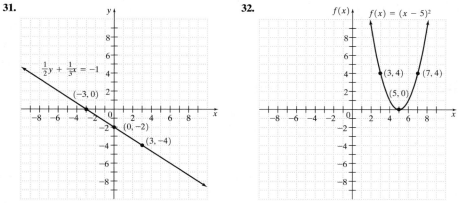

Putting Your Skills to Work

1. 39,388,160 cubic feet of rock   **2.** 7,988,160 cubic feet of concrete

9.7 Exercises

1. $x = 30, -30$ **3.** $x = 9, -4$ **5.** $x = -\dfrac{3}{2}, 4$ **7.** $x = 10, 2$ **9.** $x = \dfrac{13}{3}, 11$ **11.** $x = 6, -10$ **13.** $x = 10, -8$ **15.** $x = -\dfrac{8}{3}, \dfrac{16}{3}$

17. $x = 9, -1$ **19.** $x = -\dfrac{7}{3}, -1$ **21.** $x = 3, 1$ **23.** $x = 0, 12$ **25.** $x = -18.41, -3.92$ **27.** $x = 6.5, -2.5$ **29.** $x = -12$

31. No solution   **33.** $x = \dfrac{1}{6}, -\dfrac{5}{6}$ **35.** $x = 30.5, -1$ **37.** $x = 11, -5$ **38.** $x = \dfrac{3}{2}$ **39.** $\dfrac{8y^3 z^3}{x^6}$

40. Each beaker was $30. Each Bunsen burner was $75.   **41.** 2.4 hours

9.8 Exercises

1. $-8 \le x \le 8$ **3.** $x > 5$ or $x < -5$

5. $-9.5 < x < 0.5$ **7.** $-2 \le x \le 8$ **9.** $-1 \le x \le 6$ **11.** $2 \le x \le 5$ **13.** $1 < x < 2$

15. $-32 < x < 16$ **17.** $-7 < x < 9$ **19.** $-\dfrac{10}{3} < x < \dfrac{14}{3}$ **21.** $x > 3$ or $x < -7$ **23.** $x \ge 3$ or $x \le -1$ **25.** $x \ge 5$ or $x \le \dfrac{1}{3}$

27. $x < -8$ or $x > 16$ **29.** $x < -\dfrac{19}{2}$ or $x > \dfrac{21}{2}$ **31.** $-13 < x < 17$ **33.** $18.53 \le m \le 18.77$ **35.** $9.63 \le n \le 9.73$

37. The statement says $12 < -12$, which is a false statement.   **39.** 2   **40.** -11 **41.** $\dfrac{3x - 19}{2(x + 5)(x - 5)}$ **42.** $x = -1$

Chapter 9 Review Problems

1. $x = \pm 2\sqrt{5}$ **2.** $x = 1, -17$ **3.** $x = -4 \pm \sqrt{3}$ **4.** $x = \dfrac{2 \pm \sqrt{3}}{2}$ or $1 \pm \dfrac{\sqrt{3}}{2}$ **5.** $x = \dfrac{5 \pm \sqrt{7}}{3}$ **6.** $x = 3 \pm \sqrt{13}$ **7.** $x = \dfrac{3}{2}$

8. $x = \dfrac{4 \pm i\sqrt{2}}{3}$ **9.** $x = 0, \dfrac{9}{2}$ **10.** $x = \dfrac{3}{4}, \dfrac{5}{3}$ **11.** $x = 7, -4$ **12.** $x = 0, -\dfrac{7}{3}$ **13.** $x = \dfrac{2}{3}, 5$ **14.** $x = -1 \pm \sqrt{5}$ **15.** $x = \dfrac{3 \pm i\sqrt{23}}{8}$

16. $x = \dfrac{-5 \pm \sqrt{13}}{6}$ **17.** $x = -\dfrac{2}{3}, \dfrac{1}{3}$ **18.** $x = \dfrac{11 \pm \sqrt{21}}{10}$ **19.** $x = \dfrac{3 \pm i}{5}$ **20.** $x = -\dfrac{1}{4}, -1$ **21.** $y = -\dfrac{5}{6}, -2$ **22.** $y = -\dfrac{5}{2}, \dfrac{3}{5}$

23. $y = -5, 3$ **24.** $y = 9$ **25.** $y = -3, 1$ **26.** $y = 0, -\dfrac{5}{2}$ **27.** $x = -2, 3$ **28.** $x = -3, 2$ **29.** Two rational solutions   **30.** Two irra-

tional solutions   **31.** Two complex solutions   **32.** One rational solution   **33.** $x^2 - 25 = 0$ **34.** $x^2 + 9 = 0$ **35.** $x^2 - 32 = 0$

36. $8x^2 + 10x + 3 = 0$ **37.** $x = \pm 2, \pm \sqrt{2}$ **38.** $x = \dfrac{-\sqrt[3]{4}}{2}, \sqrt[3]{3}$ **39.** $x = -512, -1$ **40.** $x = 256$ **41.** $x = 1, 2$ **42.** $x = \pm 1, \pm \sqrt{2}$

43. $B = \pm \sqrt{\dfrac{3AH}{2C}}$ **44.** $b = \pm \sqrt{\dfrac{2H}{3g} - a^2}$ **45.** $d = \dfrac{x}{4}, \dfrac{-x}{5}$ **46.** $x = \dfrac{3 \pm \sqrt{9 + 28y}}{2y}$ **47.** $y = \dfrac{2a \pm \sqrt{4a^2 - 6a}}{3}$

48. $x = \dfrac{-1 \pm \sqrt{1 - 15y^2 + 5PV}}{5}$ **49.** $a = 4\sqrt{15}$ **50.** $c = \sqrt{22}$ **51.** The car is approximately 3.3 miles from the observer.

52. Width is 7 m; length is 29 m   **53.** Base is 7 cm; altitude is 20 cm   **54.** 50 mph during first part; 45 mph during rain   **55.** 20 mph cruising; 5 mph trolling   **56.** The walkway should be approximately 2.4 feet wide.   **57.** The walkway should be 2 feet wide.
58. Vertex $(3, -2)$; y-intercept $(0, -11)$; no x-intercepts   **59.** Vertex $(-5, 0)$; one x-intercept $(-5, 0)$; y-intercept $(0, 25)$

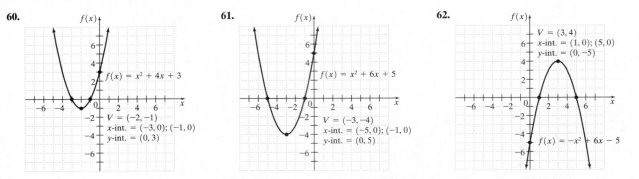

60. $f(x) = x^2 + 4x + 3$; $V = (-2, -1)$; x-int. $= (-3, 0); (-1, 0)$; y-int. $= (0, 3)$

61. $f(x) = x^2 + 6x + 5$; $V = (-3, -4)$; x-int. $= (-5, 0); (-1, 0)$; y-int. $= (0, 5)$

62. $V = (3, 4)$; x-int. $= (1, 0); (5, 0)$; y-int. $= (0, -5)$; $f(x) = -x^2 + 6x - 5$

63. The maximum height is 2540 feet. The amount of time for the complete flight is 25.1 seconds.   **64.** $R(x) = x(1200 - x)$; the maximum revenue will occur if the price is \$600 for each unit.   **65.** $-9 < x < 2$

66. $-7 < x < 3$

67. $x < 4$ or $x > 5$ **68.** $x < 4$ or $x > 7$

69. $-\dfrac{1}{2} \le x \le 3$ **70.** $-\dfrac{1}{3} \le x \le 2$ **71.** $x < -\dfrac{5}{4}$ or $x > \dfrac{5}{4}$ **72.** $x < -\dfrac{2}{3}$ or $x > \dfrac{2}{3}$

73. $x < \left(1 - \sqrt{5}\right)$ or $x > \left(1 + \sqrt{5}\right)$; approximately $x < -1.2$ or $x > 3.2$ **74.** $x \le -6$ or $x \ge -2$ **75.** $x < -8$ or $x > 2$

76. $x < 1.4$ or $x > 2.6$ **77.** No real solution   **78.** No real solution   **79.** $x = 7, -9$ **80.** $x = 3, -\dfrac{1}{2}$ **81.** $x = 6, -\dfrac{22}{3}$

82. $x = 2, \dfrac{8}{3}$ **83.** $x = 44, -20$ **84.** $x = \dfrac{13}{2}, \dfrac{3}{2}$ **85.** $-22 < x < 8$ **86.** $-27 < x < 9$ **87.** $-\dfrac{15}{2} < x < -\dfrac{1}{2}$ **88.** $x \ge 5$ or $x \le -4$

89. $x \ge 1$ or $x \le -\dfrac{1}{3}$ **90.** $x \ge \dfrac{8}{3}$ or $x \le -\dfrac{2}{3}$ **91.** $x < -4$ or $2 < x < 3$ **92.** $-4 < x < -1$ or $x > 2$

Chapter 9 Test

1. $x = 0, -\dfrac{9}{8}$ **2.** $x = \dfrac{1}{4}, -\dfrac{3}{2}$ **3.** $x = 2, -\dfrac{2}{9}$ **4.** $x = -2, 10$ **5.** $x = \pm 2\sqrt{2}$ **6.** $x = \dfrac{7}{2}, -1$ **7.** $x = \dfrac{3 \pm i}{2}$ **8.** $x = \dfrac{3 \pm \sqrt{3}}{2}$

9. $x = \pm\sqrt{7}, \pm\sqrt{2}$ **10.** $x = \dfrac{1}{5}, -\dfrac{3}{4}$ **11.** $x = \left(1 \pm \sqrt{13}\right)^3$ **12.** $z = \pm\sqrt{\dfrac{xyw}{B}}$ **13.** $y = \dfrac{-b \pm \sqrt{b^2 - 30w}}{5}$

14. Width is 5 miles; length is 16 miles.   **15.** $c = 4\sqrt{3}$ **16.** 2 mph during first part; 3 mph after lunch

17. $V = (-3, 4)$; y-int. $(0, -5)$; x-int. $(-5, 0)$; $(-1, 0)$

$f(x) = -x^2 - 6x - 5$

18. $x \le -\dfrac{9}{2}$ or $x \ge 3$ **19.** $-\dfrac{2}{3} \le x \le 4$

20. $x < -4.5$ or $x > 1.5$ **21.** $x = -7, \dfrac{39}{5}$ **22.** $x = 6, -18$ **23.** $-\dfrac{15}{7} \le x \le 3$ **24.** $x < -\dfrac{8}{3}$ or $x > 2$

Cumulative Test for Chapters 1–9

1. $\dfrac{81y^{12}}{x^8}$ **2.** $\dfrac{1}{4}a^3 - a^2 - 3a$ **3.** $y = \dfrac{ab + 4}{a}$ **4.**

x	y
0	4
-2	0

5. $x + 2y = 4$ **6.** $\dfrac{32\pi}{3}$ cubic inches

7. $(5x - 3y)(25x^2 + 15xy + 9y^2)$ **8.** $6xy^3\sqrt{2x}$ **9.** $4\sqrt{6} + 5\sqrt{3}$ **10.** $\dfrac{3\sqrt{11}}{11}$ **11.** $x = 0, \dfrac{14}{3}$ **12.** $x = \dfrac{2}{3}, \dfrac{1}{4}$ **13.** $x = \dfrac{3 \pm 2\sqrt{3}}{2}$

14. $x = \dfrac{2 \pm i\sqrt{11}}{3}$ **15.** $x = 16$ **16.** $x = -27, -216$ **17.** $y = \dfrac{-5w \pm \sqrt{25w^2 + 56z}}{4}$ **18.** $y = \dfrac{\pm\sqrt{15w - 48z^2}}{3}$

19. $x = \sqrt{15}$ **20.** Base is 5 meters; altitude is 18 meters. **21.** Vertex $(4, 4)$; y-intercept $(0, -12)$; x-intercepts $(2, 0)$; $(6, 0)$

22.

23. $-\dfrac{1}{2} \le x \le \dfrac{2}{3}$ **24.** $x < -5 \text{ or } x > 3$ **25.** $x = 5 \text{ or } x = -\dfrac{17}{3}$ **26.** $-20 \le x \le 12$

27. $x < -\dfrac{7}{3} \text{ or } x > 5$

Chapter 10

Pretest Chapter 10

1. $(x - 8)^2 + (y + 2)^2 = 7$ **2.** $3\sqrt{5}$

3. center $= (1, 2)$; radius$=2$;
$(x - 1)^2 + (y - 2)^2 = 4$

4. $x = (y + 1)^2 + 2$

5. $y = (x + 2)^2 - 3$

6. $\dfrac{x^2}{9} + \dfrac{y^2}{36} = 1$

7. $\dfrac{(x + 3)^2}{25} + \dfrac{(y - 1)^2}{16} = 1$

8. $\dfrac{y^2}{9} - \dfrac{x^2}{25} = 1$

9. $\dfrac{(x - 2)^2}{4} - \dfrac{(y + 1)^2}{9} = 1$

10. $(4, -3), (-4, 3)$ **11.** $(0, 1)$

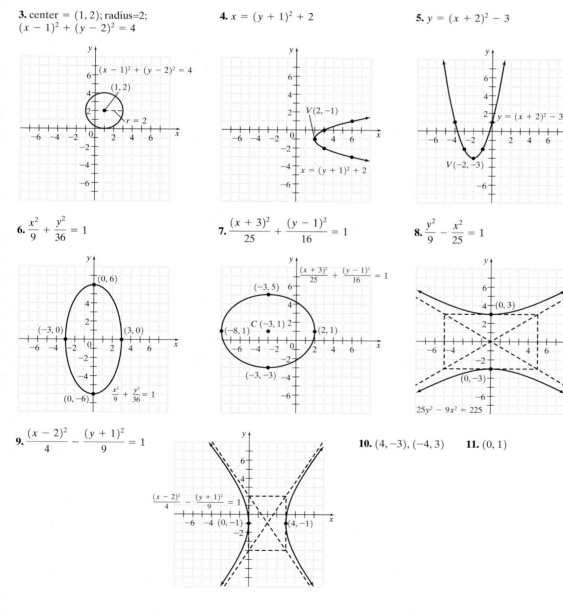

10.1 Exercises

1. Subtract the value of the points and use the absolute value: $|-2 - 4| = 6$ **3.** Since the equation is given in standard form, we determine the values of h, k, and r to find the center and radius. $h = 1$, $k = -2$, and $r = 3$. Thus, the center is $(1, -2)$ and the radius is 3.

5. $\sqrt{5}$ **7.** $\dfrac{\sqrt{17}}{4}$ **9.** 13   **11.** $4\sqrt{2}$ **13.** $\dfrac{2\sqrt{26}}{5}$ **15.** $5\sqrt{2}$ **17.** $y = 10$, $y = -6$ **19.** $y = 0$, $y = 4$ **21.** $x = 6$, $x = 8$

23. 9.5 miles   **25.** $(x + 1)^2 + (y + 7)^2 = 5$ **27.** $(x + 3.5)^2 + y^2 = 36$ **29.** $\left(x - \dfrac{7}{4}\right)^2 + y^2 = \dfrac{1}{9}$

31.   **33.**   **35.**

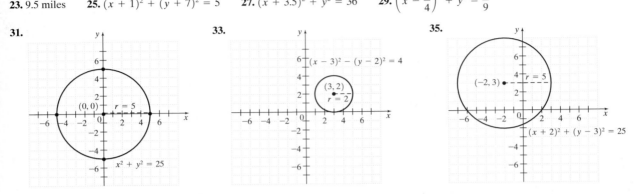

37. $(x + 3)^2 + (y - 2)^2 = 16$; center $(-3, 2)$, $r = 4$ **39.** $(x - 6)^2 + (y + 1)^2 = 49$; center $(6, -1)$, $r = 7$

41. $\left(x + \dfrac{3}{2}\right)^2 + y^2 = \dfrac{17}{4}$; center $\left(-\dfrac{3}{2}, 0\right)$, $r = \dfrac{\sqrt{17}}{2}$ **43.** $(x - 42.7)^2 + (y - 29.7)^2 = 630.01$ **45.**

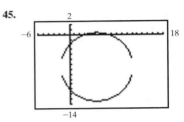

47. $x = -\dfrac{2}{3}$, $x = \dfrac{1}{3}$ **48.** $x = \dfrac{2}{3}$, $x = 1$ **49.** $x = \dfrac{3 \pm 2\sqrt{11}}{5}$ **50.** $x = \dfrac{-1 \pm \sqrt{5}}{4}$ **51.** approximately 8.364×10^{10} cubic feet

52. approximately 81 seconds

10.2 Exercises

1. y-axis, x-axis   **3.** If it is in the standard form $y = a(x - h)^2 + k$, the vertex is (h, k). So in this case the vertex is $(3, 4)$.

5.   **7.**   **9.**

11.   **13.**   **15.**

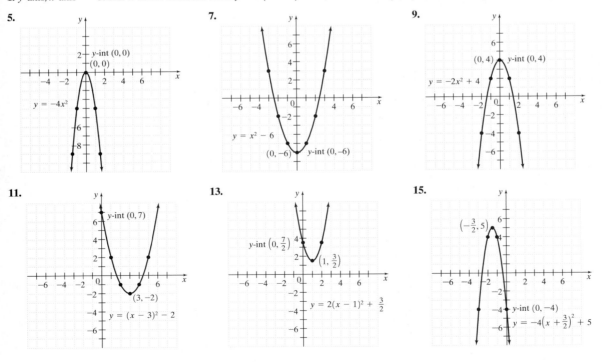

17.

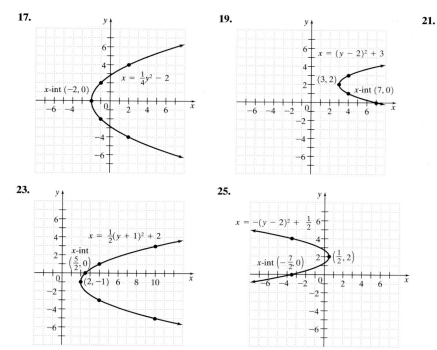

19.

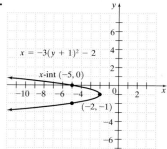

21.

23.

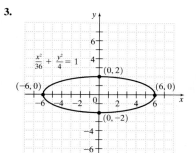

25.

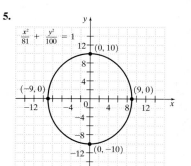

27. $y = (x + 6)^2 - 11$ **(a)** vertical **(b)** opens upward **(c)** vertex $(-6, -11)$ **29.** $y = -2(x - 1)^2 - 1$ **(a)** vertical **(b)** opens downward **(c)** vertex $(1, -1)$ **31.** $x = (y + 4)^2 - 7$ **(a)** horizontal **(b)** opens right **(c)** vertex $(-7, -4)$ **33.** $y = \dfrac{1}{32} x^2$ **35.** 8 inches

37. vertex $(-1.62, -5.38)$; y-intercept $(0, -0.1312)$; x-intercepts $(0.020121947, 0)$ and $(-3.260121947, 0)$

39. maximum profit $= \$52{,}000$; number of items produced $= 50$ **41.** maximum sensitivity $= 52{,}812.5$; dosage $= 162.5$ milligrams

43. $5x\sqrt{2x}$ **44.** $2xy\sqrt[3]{5y}$ **45.** $2x\sqrt{2} - 8\sqrt{2x}$ **46.** $2x\sqrt[3]{2x} - 20x\sqrt[3]{2}$ **47.** $27\dfrac{1}{3}$ miles **48.** approximately 42.2 miles **49.** 7392 blooms

50. approximately 1098 buds

10.3 Exercises

1. In the ellipse $\dfrac{(x - h)^2}{a^2} + \dfrac{(y - k)^2}{b^2} = 1$ the center is (h, k). In this case the center of the ellipse is $(-2, 3)$.

3.

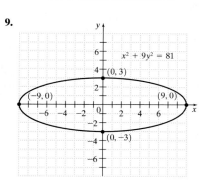

5.

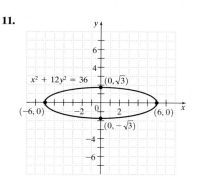

7.

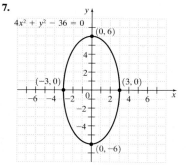

9.

11.

13.

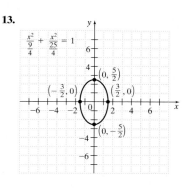

15.

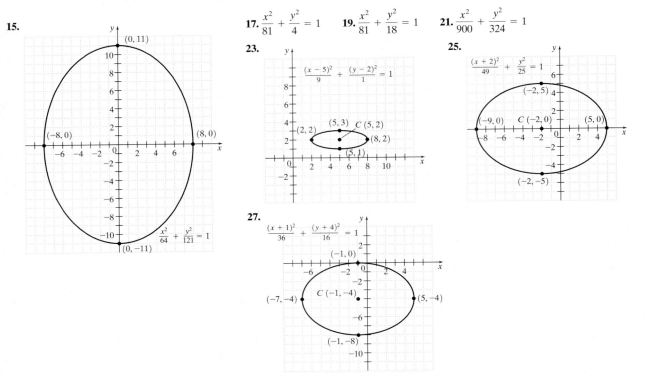

17. $\dfrac{x^2}{81} + \dfrac{y^2}{4} = 1$ **19.** $\dfrac{x^2}{81} + \dfrac{y^2}{18} = 1$ **21.** $\dfrac{x^2}{900} + \dfrac{y^2}{324} = 1$

23.

25.

27.

29. $\dfrac{(x-4)^2}{4} + \dfrac{(y-3)^2}{16} = 1$ **31.** $a = \dfrac{-8 \pm 3\sqrt{3}}{2}$ **33.** $(0, 7.2768), (0, 3.3232), (4.2783, 0), (2.9217, 0)$ **35.** $22{,}376.0$ square meters

36. $30\sqrt{2} + 40\sqrt{3} - 2\sqrt{6} - 8$ **37.** $\dfrac{5(\sqrt{2x} + \sqrt{y})}{2x - y}$ **38.** $y = \dfrac{1}{3}x + 6$ **39.** $y = 5x - 2$ **40.** $-5 - 12i$ **41.** $\dfrac{-5i + 3}{2}$

10.4 Exercises

1. The standard form of a horizontal hyperbola centered at the origin is $\dfrac{x^2}{a^2} - \dfrac{y^2}{b^2} = 1$ with a and b being positive real numbers.

3. This is a horizontal hyperbola, centered at the origin, with vertices at $(4, 0)$ and $(-4, 0)$. Draw a fundamental rectangle with corners at $(4, 2)$, $(4, -2), (-4, 2)$, and $(-4, -2)$. Extend the diagonals through the rectangle as asymptotes of the hyperbola. Construct each branch of the hyperbola passing through the vertex and approaching the asymptotes.

5.

7.

9.

11.

13.

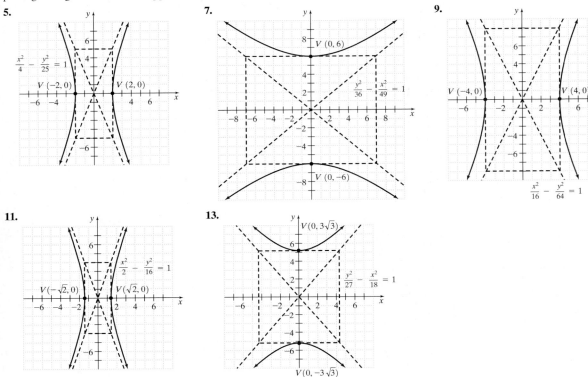

15. $\dfrac{x^2}{9} - \dfrac{y^2}{16} = 1$ **17.** $\dfrac{y^2}{49} - \dfrac{x^2}{9} = 1$ **19.** $\dfrac{x^2}{14,400} - \dfrac{y^2}{129,600} = 1$, where x and y are measured in millions of miles

21. $\dfrac{(x-6)^2}{25} - \dfrac{(y-4)^2}{49} = 1$ **23.**

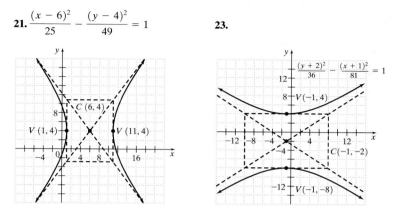

25. center $(-6, 0)$; vertices $(-6 + \sqrt{7}, 0)$, $(-6 - \sqrt{7}, 0)$ **27.** $\dfrac{(y-7)^2}{49} - \dfrac{(x-5)^2}{25} = 1$ **29.** $y = \pm 9.055385138$ **31.** $(4x + 3)(3x - 2)$

32. $2(2x - 3y)^2$ **33.** $\dfrac{5x}{(x-3)(x-2)(x+2)}$ **34.** $\dfrac{-x-6}{(5x-1)(x+2)}$ or $-\dfrac{x+6}{(5x-1)(x+2)}$ **35.** 31.5 feet

36. The width is 25 feet, the length along the building is 50 feet.   **37.** 640,000,000 pencils   **38.** approximately 38.7%.

10.5 Exercises

1. $(1, 2)$ **3.** $(1, 4)$, $(-1, -4)$

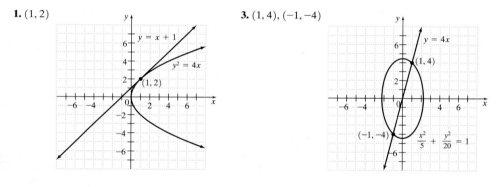

5. $(-2, 3)$, $(1, 0)$ **7.** $(-5, 0)$, $(4, 3)$ **9.** $\left(\dfrac{2}{3}, \dfrac{4}{3}\right)$, $(2, 0)$ **11.** $\left(\dfrac{5}{2}, \dfrac{3}{2}\right)$ **13.** $(3, 2)$, $(-3, 2)$, $(3, -2)$, $(-3, -2)$

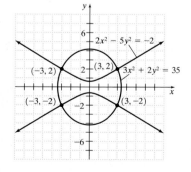

15. $(1, 2\sqrt{2})$, $(-1, 2\sqrt{2})$, $(1, -2\sqrt{2})$, $(-1, -2\sqrt{2})$ **17.** $\left(\dfrac{\sqrt{30}}{3}, \dfrac{\sqrt{21}}{3}\right)$, $\left(-\dfrac{\sqrt{30}}{3}, \dfrac{\sqrt{21}}{3}\right)$, $\left(\dfrac{\sqrt{30}}{3}, -\dfrac{\sqrt{21}}{3}\right)$, $\left(-\dfrac{\sqrt{30}}{3}, -\dfrac{\sqrt{21}}{3}\right)$

19. $(2, \sqrt{3})$, $(-2, \sqrt{3})$, $(2, -\sqrt{3})$, $(-2, -\sqrt{3})$ **21.** $(1, 3)$, $(-3, -1)$ **23.** $(-4, -2)$, $(2, 4)$ **25.** no real solution

27. 15 meters by 36 meters.   **29.** Yes, the hyperbola intersects the circle; $(3290, 2270)$

31. $\dfrac{2x(x+1)}{x-2}$ **32.** $x^2 - 3x - 10$ **33.** 25 miles per hour   **34.** 128,500 CD-ROMs

Putting Your Skills to Work

1. The epicenter is approximately 30.4 miles east and 26.0 miles north of station A or approximately 37.2 miles east and 14.6 miles north of station A.
2. The epicenter is approximately 34.7 miles east and 36.0 miles north of station A or approximately 49.3 miles east and 80 miles south of station A.

Chapter 10 Review Problems

1. $\sqrt{13}$ **2.** $\sqrt{41}$ **3.** $(x + 6)^2 + (y - 3)^2 = 15$ **4.** $x^2 + (y + 7)^2 = 25$ **5.** $(x - 3)^2 + (y - 4)^2 = 22$; center $(3, 4)$, $r = \sqrt{22}$

6. $(x - 5)^2 + (y + 6)^2 = 9$; center $(5, -6)$, $r = 3$

7. 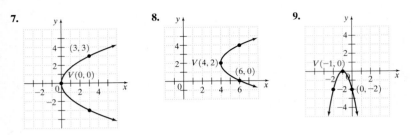   **8.**   **9.**

10. $x = (y - 4)^2 - 6$; vertex at $(-6, 4)$; opens to the right   **11.** $y = (x + 3)^2 - 5$; vertex at $(-3, -5)$; opens upward

12.   **13.**   **14.** center $(-5, -3)$; vertices are $(-3, -3)$, $(-7, -3)$, $(-5, 2)$, $(-5, -8)$

Scale: each unit is $\sqrt{2}$ units

15. center $(-1, 2)$; vertices are $(2, 2)$, $(-4, 2)$, $(-1, 6)$, $(-1, -2)$ **16.** 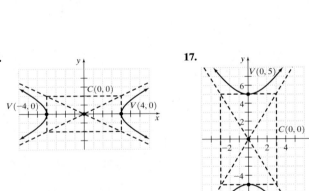   **17.**

18. vertices are $(0, -3)$, $(4, -3)$; center $(2, -3)$ **19.** vertices are $(-5, 3)$, $(-5, 1)$; center $(-5, 2)$ **20.** $(-3, 0)$, $(2, 5)$

21. $\left(\dfrac{1 + 2\sqrt{14}}{5}, \dfrac{-2 + \sqrt{14}}{5}\right)$, $\left(\dfrac{1 - 2\sqrt{14}}{5}, \dfrac{-2 - \sqrt{14}}{5}\right)$ **22.** $(2, 3)$, $(-2, 3)$, $(2, -3)$, $(-2, -3)$ **23.** $(2, -1)$, $(-2, 1)$, $(1, -2)$, $(-1, 2)$

24. no real solution   **25.** $(0, 1)$, $(\sqrt{5}, 6)$, $(-\sqrt{5}, 6)$ **26.** $(1, 4)$, $(-1, -4)$, $(2\sqrt{2}, \sqrt{2})$, $(-2\sqrt{2}, -\sqrt{2})$ **27.** $(1, 2)$, $(-1, 2)$, $(1, -2)$, $(-1, -2)$

28. $\left(\dfrac{1}{2}, -\dfrac{1}{2}\right)$, $(2, 1)$ **29.** $(2, 2)$ **30.** 1.56 feet   **31.** 0.78 feet

Chapter 10 Test

1. $\sqrt{185}$ **2.** $x = (y - 3)^2 + 4$; vertex $(4, 3)$; parabola   **3.** $(x + 3)^2 + (y - 2)^2 = 4$; center $(-3, 2)$; circle   **4.** ellipse; center $(0, 0)$

5. hyperbola; center $(0, 0)$ **6.** parabola; vertex $(-3, 4)$ **7.** ellipse; center $(-2, 5)$

8. hyperbola; center $(0, 0)$ **9.** $(x - 3)^2 + (y + 5)^2 = 8$ **10.** $\dfrac{(x + 4)^2}{1} + \dfrac{(y + 2)^2}{9} = 1$

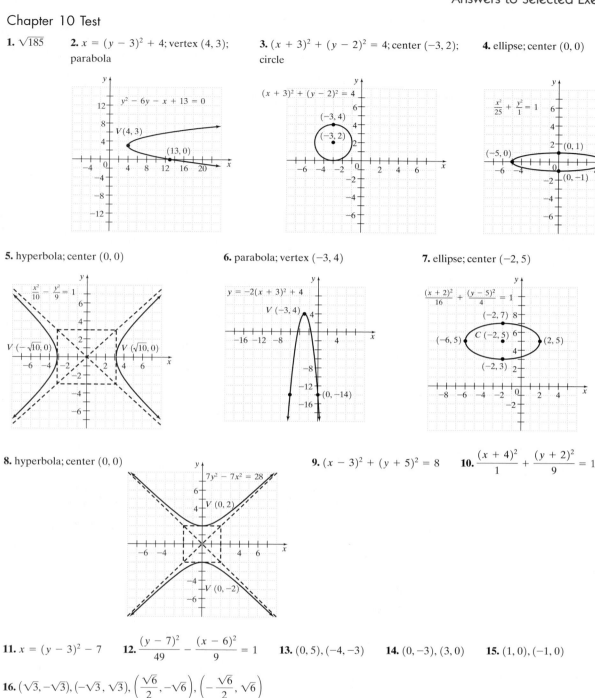

11. $x = (y - 3)^2 - 7$ **12.** $\dfrac{(y - 7)^2}{49} - \dfrac{(x - 6)^2}{9} = 1$ **13.** $(0, 5), (-4, -3)$ **14.** $(0, -3), (3, 0)$ **15.** $(1, 0), (-1, 0)$

16. $\left(\sqrt{3}, -\sqrt{3}\right), \left(-\sqrt{3}, \sqrt{3}\right), \left(\dfrac{\sqrt{6}}{2}, -\sqrt{6}\right), \left(-\dfrac{\sqrt{6}}{2}, \sqrt{6}\right)$

Cumulative Test for Chapters 1–10

1. $8x + 12$ **2.** -19 **3.** $p = \dfrac{A - 3bt}{rt}$ **4.** undefined slope or no slope   **5.** $(5x - 4y)^2$ **6.** $\dfrac{3x + 18}{(x + 4)(x - 4)}$

7. $x = \dfrac{7}{2}$ **8.** $x = 4, y = -3, z = 1$ **9.** $4\sqrt{3} + 6\sqrt{2} - \sqrt{6} - 3$ **10.** $2\sqrt{2x} - x\sqrt{2}$ **11.** $x > 1$ **12.** $x \geq 14$ **13.** $3\sqrt{10}$

14. parabola

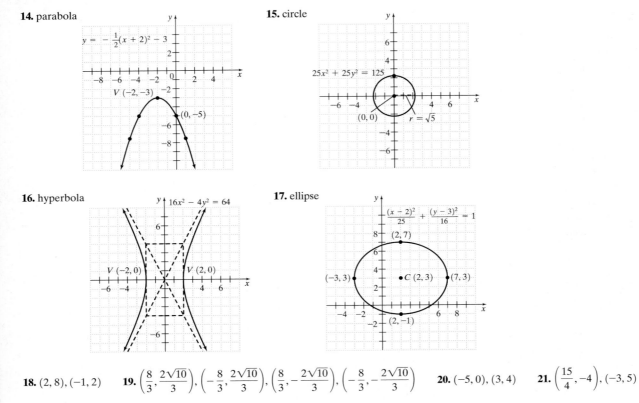

15. circle

16. hyperbola

17. ellipse

18. $(2, 8), (-1, 2)$ **19.** $\left(\dfrac{8}{3}, \dfrac{2\sqrt{10}}{3}\right), \left(-\dfrac{8}{3}, \dfrac{2\sqrt{10}}{3}\right), \left(\dfrac{8}{3}, -\dfrac{2\sqrt{10}}{3}\right), \left(-\dfrac{8}{3}, -\dfrac{2\sqrt{10}}{3}\right)$ **20.** $(-5, 0), (3, 4)$ **21.** $\left(\dfrac{15}{4}, -4\right), (-3, 5)$

Chapter 11

Pretest Chapter 11

1. (a) -12 **(b)** $2a - 6$ **(c)** $4a - 6$ **(d)** $2a - 2$ **2. (a)** 13 **(b)** $5a^2 + 2a - 3$ **(c)** $5a^2 + 12a + 4$ **3. (a)** $\dfrac{6(a^2 - 2)}{a(a + 2)}$ **(b)** $\dfrac{18(a - 1)}{5(3a + 2)}$

4. function **5.** not a function **6.** **7.**

8. (a) $\dfrac{-6x + 2}{x + 6}$ **(b)** 13 **(c)** $\dfrac{2}{-3x + 7}, x \neq \dfrac{7}{3}$ **9. (a)** $-2x^3 - 3x - 1$ **(b)** -23 **(c)** $-6x^3 - 18x + 5$ **10. (a)** $2x + 1, x \neq \dfrac{4}{3}$ **(b)** -1

(c) $54x^2 - 159x + 112$ **(d)** $18x^2 - 15x - 16$ **11.** one-to-one **12.** not one-to-one **13.** Yes, A is one-to-one

14. $F^{-1} = \{(1, 7), (3, 6), (-1, 2), (5, -1)\}$ **15.** $g^{-1}(x) = \dfrac{3 - x}{5}$

11.1 Exercises

1. -7 **3.** $3a - 17$ **5.** $\dfrac{1}{2}a - 4$ **7.** $a - 3$ **9.** $a - 5$ **11.** $\dfrac{1}{2}a^2 - \dfrac{1}{5}$ **13.** 2 **15.** $\dfrac{3}{4}$ **17.** $3a^2 + 10a + 5$ **19.** $\dfrac{3a^4}{4} - 2a^2 - 2$

21. 2 **23.** $2\sqrt{2}$ **25.** $\sqrt{a^2 + 4}$ **27.** $\sqrt{3a + 5}$ **29.** $2\sqrt{a + 1}$ **31.** $\sqrt{b^2 + b + 5}$ **33.** $\dfrac{7}{4}$ **35.** $-\dfrac{14}{3}$ **37.** $\dfrac{7}{a^2 - 3}$

39. $\dfrac{7}{a - 1}$ **41.** $-\dfrac{7}{5}$ **43.** -2 **45.** $4x + 2h$ **47.** **(a)** $P(w) = 2.5w^2$ **(b)** 1000 kilowatts **(c)** $P(e) = 2.5e^2 + 100e + 1000$

(d) 1210 kilowatts **49.** The percent would decrease by $13; 26$ **51.** $38.021a^2 - 16.376a + 1.23$ **53.** $3a^2 - 5.512a + 1.999$

55. $A(x) = \dfrac{x^2 - 20x + 200}{8}$; $A(2) = 20.5$; $A(5) = 15.625$; $A(8) = 13$ **57.** $x = -3$ **58.** $x = 5$ **59.** approximately 17.9 times greater

60. approximately 26.3 times greater

11.2 Exercises

1. No, $f(x + 2)$ means to substitute $x + 2$ for x in the function $f(x)$. $f(x) + f(2)$ means to evaluate $f(x)$ and to evaluate $f(2)$ and then to add the two solutions. **3.** up **5.** not a function **7.** function **9.** function **11.** not a function **13.** not a function

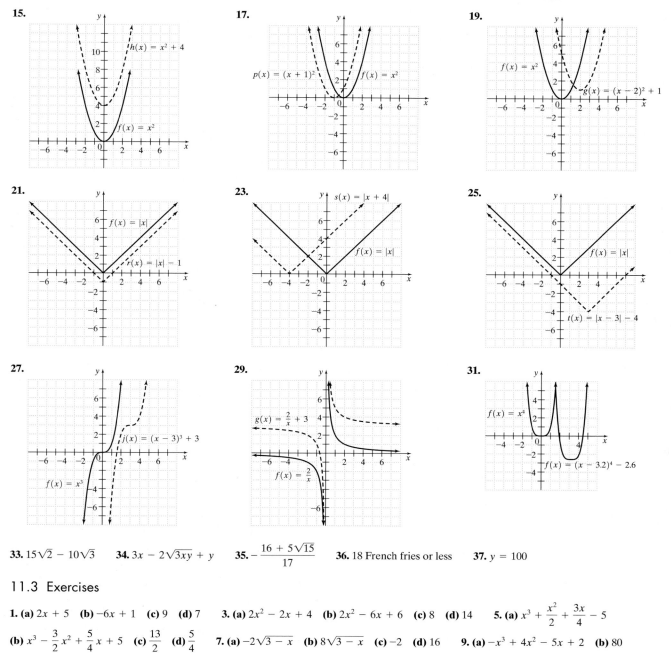

33. $15\sqrt{2} - 10\sqrt{3}$ **34.** $3x - 2\sqrt{3xy} + y$ **35.** $-\dfrac{16 + 5\sqrt{15}}{17}$ **36.** 18 French fries or less **37.** $y = 100$

11.3 Exercises

1. **(a)** $2x + 5$ **(b)** $-6x + 1$ **(c)** 9 **(d)** 7 **3.** **(a)** $2x^2 - 2x + 4$ **(b)** $2x^2 - 6x + 6$ **(c)** 8 **(d)** 14 **5.** **(a)** $x^3 + \dfrac{x^2}{2} + \dfrac{3x}{4} - 5$

(b) $x^3 - \dfrac{3}{2}x^2 + \dfrac{5}{4}x + 5$ **(c)** $\dfrac{13}{2}$ **(d)** $\dfrac{5}{4}$ **7.** **(a)** $-2\sqrt{3 - x}$ **(b)** $8\sqrt{3 - x}$ **(c)** -2 **(d)** 16 **9.** **(a)** $-x^3 + 4x^2 - 5x + 2$ **(b)** 80

11. (a) $\dfrac{2(x-1)}{x}$ **(b)** $\dfrac{8}{3}$ **13. (a)** $-3x\sqrt{-2x+1}$ **(b)** $9\sqrt{7}$ **15. (a)** $\dfrac{3x}{4x-1}, x \neq \dfrac{1}{4}$ **(b)** $\dfrac{6}{7}$ **17. (a)** $x+1, x \neq 1$ **(b)** 3

19. (a) $x+5, x \neq -5$ **(b)** 7 **21. (a)** $\dfrac{1}{x+2}, x \neq -2, x \neq \dfrac{1}{4}$ **(b)** $\dfrac{1}{4}$ **23.** $-x^2 + 5x + 2$ **25.** $3x^3 - 4x^2 - 4x$ **27.** -3

29. $\dfrac{9x+6}{x-2}, x \neq 2$ **31.** $-6x - 13$ **33.** $3x^2 - 24x + 48$ **35.** $-6x^2 + 7$ **37.** $\dfrac{3}{2x}, x \neq 0$ **39.** $|2x + 2|$ **41.** $9x^2 + 30x + 27$

43. $3x^2 + 11$ **45.** 38 **47.** $\sqrt{x^2 + 1}$ **49.** $\dfrac{3\sqrt{2}}{2} + 5$ **51.** $\sqrt{26}$ **53.** $K[C(F)] = \dfrac{5F + 2297}{9}$

55. $v[r(h)] = 384.65h^2$; 24,617.6 cubic feet **57.** $(3a + 1)(2b - c)$ **58.** $(6x - 1)^2$ **59.** $(3x - 1)(x - 2)$
60. $(x + 3)(x - 3)(x + 1)(x - 1)$ **61.** 8 commercials that are 60 seconds long and 12 commercials that are 30 seconds long
62. 115 children; 335 adults

11.4 Exercises

1. have the same second coordinate. **3.** $y = x$ **5.** not one-to-one **7.** one-to-one **9.** one-to-one **11.** one-to-one
13. not one-to-one **15.** not one-to-one **17.** Yes, it passes the vertical line test. No, it does not pass the horizontal line test.

19. $J^{-1} = \{(2, 8), (1, 1), (0, 0), (-2, -8)\}$

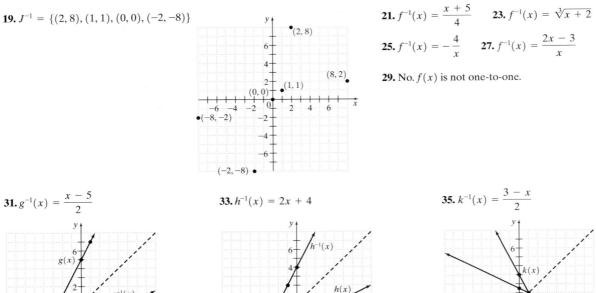

21. $f^{-1}(x) = \dfrac{x + 5}{4}$ **23.** $f^{-1}(x) = \sqrt[3]{x + 2}$

25. $f^{-1}(x) = -\dfrac{4}{x}$ **27.** $f^{-1}(x) = \dfrac{2x - 3}{x}$

29. No. $f(x)$ is not one-to-one.

31. $g^{-1}(x) = \dfrac{x - 5}{2}$

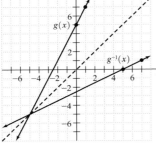

33. $h^{-1}(x) = 2x + 4$

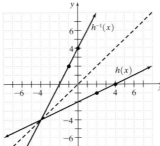

35. $k^{-1}(x) = \dfrac{3 - x}{2}$

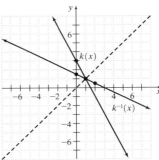

37. $f^{-1}(x) = \dfrac{x + 5}{0.0063}$; The inverse function would tell you how many Spanish pesetas were given by the bank for x dollars. No. Because of the bank

fee, it would not work for Manuela's transaction. **39.** $f[f^{-1}(x)] = 2\left[\dfrac{1}{2}x - \dfrac{3}{4}\right] + \dfrac{3}{2} = x; f^{-1}[f(x)] = \dfrac{1}{2}\left[2x + \dfrac{3}{2}\right] - \dfrac{3}{4} = x$

41. $x = -64; x = -27$ **42.** $x = 3$ **43.** $23:21$ **44.** 13 overtime hours **45.** 800,000 people

Putting Your Skills to Work

1.

x	0	10	20	30	40	50	60
$p(x)$	2000	14,000	16,000	14,000	14,000	22,000	44,000

2. 1976

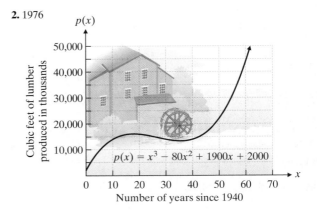

Chapter 11 Review Problems

1. $\frac{1}{2}a + \frac{5}{2}$ **2.** $\frac{1}{2}a + 4$ **3.** $-\frac{1}{2}$ **4.** 1 **5.** $a + \frac{9}{2}$ **6.** $a + \frac{3}{2}$ **7.** -28 **8.** -21 **9.** $-8a^2 + 6a - 16$ **10.** $-18a^2 + 9a - 11$

11. $-2a^2 - 5a - 3$ **12.** $-2a^2 + 15a - 28$ **13.** $|16a - 1|$ **14.** $|14a - 1|$ **15.** $\left|\frac{1}{2}a - 1\right|$ **16.** $|3a - 1|$ **17.** $|2a - 11|$

18. $|2a + 7|$ **19.** $\frac{5}{3}$ **20.** 9 **21.** $\frac{3a + 9}{a + 7}$ **22.** $\frac{3a - 6}{a + 2}$ **23.** $\frac{30a + 36}{7a + 28}$ **24.** $-\frac{12}{a + 4}$ **25.** 7 **26.** 6 **27.** $4x + 2h - 5$

28. $-6x - 3h + 2$ **29. (a)** function **(b)** one-to-one **30. (a)** not a function **(b)** not one-to-one **31. (a)** function **(b)** not one-to-one

32. (a) function **(b)** not one-to-one **33. (a)** not a function **(b)** not one-to-one **34. (a)** function **(b)** one-to-one

35. $g(x) = (x + 2)^2 + 4$

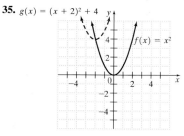

36.

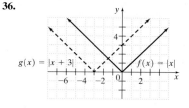

37.

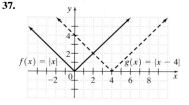

38.

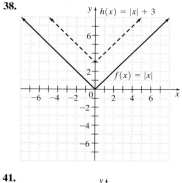

39.

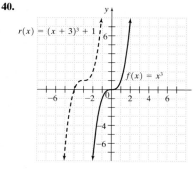

40.

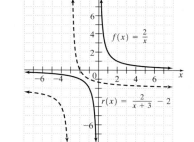

41.

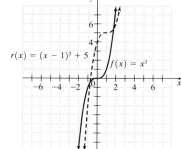

42.

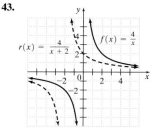

43.

44. $2x^2 + 9$ **45.** $\frac{5}{2}x + 2$ **46.** $-\frac{7}{2}x - 8$ **47.** $2x^2 - 6x - 1$ **48.** -5 **49.** $\frac{5}{2}$ **50.** $\frac{6x + 10}{x}$, $x \neq 0$ **51.** $-x^3 - \frac{9}{2}x^2 + 7x - 12$

52. $\frac{2x - 8}{x^2 + x}$, $x \neq 0, x \neq -1, x \neq 4$ **53.** $\frac{2}{3x^2 + 5x}$, $x \neq 0, x \neq -\frac{5}{3}$ **54.** -6 **55.** $\frac{1}{6}$ **56.** $-\frac{3}{2}x - 4$ **57.** $\frac{3x + 6}{3x + 1}$, $x \neq -\frac{1}{3}$

58. $\sqrt{2x^2 - 3x + 2}$ **59.** $\sqrt{-\frac{1}{2}x - 5}$, $x \leq -10$ **60.** 2 **61.** 2 **62.** $f[g(x)] = \frac{6}{x} + 5 = \frac{6 + 5x}{x}$; $g[f(x)] = \frac{2}{3x + 5}$; $f[g(x)] \neq g[f(x)]$

63. $p[g(x)] = \frac{8}{x^2} - \frac{6}{x} + 4 = \frac{8 - 6x + 4x^2}{x^2}$; $g[p(x)] = \frac{2}{2x^2 - 3x + 4}$; $p[g(x)] \neq g[p(x)]$ **64. (a)** domain $= \{0, 3, 7\}$

(b) range $= \{-8, 3, 7, 8\}$ **(c)** not a function **(d)** not one-to-one **65. (a)** domain $= \{100, 200, 300, 400\}$ **(b)** range $= \{10, 20, 30\}$

(c) function **(d)** not one-to-one **66. (a)** domain $= \left\{-\frac{1}{3}, \frac{1}{4}, \frac{1}{2}, 4\right\}$ **(b)** range $= \left\{-3, \frac{1}{4}, 2, 4\right\}$ **(c)** function **(d)** one-to-one

67. (a) domain $= \{-6, 0, 12\}$ **(b)** range $= \{-12, -1, 6\}$ **(c)** not a function **(d)** not one-to-one **68. (a)** domain $= \{-1, 0, 1, 2\}$

(b) range $= \{-2, 1, 2, 9\}$ **(c)** function **(d)** one-to-one **69. (a)** domain $= \{0, 1, 2, 3\}$ **(b)** range $= \{-3, 1, 7\}$ **(c)** function

(d) not one-to-one **70.** $A^{-1} = \left\{\left(\frac{1}{3}, 3\right), \left(-\frac{1}{2}, -2\right), \left(-\frac{1}{4}, -4\right), \left(\frac{1}{5}, 5\right)\right\}$ **71.** $B^{-1} = \{(10, 1), (7, 3), (15, 12), (1, 10)\}$

72. $f^{-1}(x) = -\dfrac{4}{3}x + \dfrac{8}{3}$ **73.** $g^{-1}(x) = -\dfrac{1}{4}x - 2$ **74.** $h^{-1}(x) = 3x - 2$ **75.** $j^{-1}(x) = \dfrac{1}{x} + 3$ **76.** $p^{-1}(x) = x^3 - 1$

77. $r^{-1}(x) = \sqrt[3]{x - 2}$

78. $f^{-1}(x) = -3x - 2$

79. $f^{-1}(x) = -\dfrac{4}{3}x + \dfrac{4}{3}$

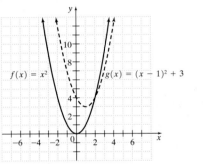

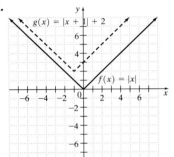

Chapter 11 Test

1. -8 **2.** $\dfrac{3}{2}a - 2$ **3.** $\dfrac{3}{4}a - \dfrac{3}{2}$ **4.** 124 **5.** $3a^2 + 4a + 5$ **6.** $3a^2 - 2a + 9$ **7.** $12a^2 + 4a + 2$ **8. (a)** function **(b)** not one-to-one

9. (a) function **(b)** one-to-one **10.**

11.

12. (a) $x^2 + 4x + 1$ **(b)** $5x^2 - 6x - 13$ **(c)** 19 **13. (a)** $\dfrac{6x - 3}{x}, x \neq 0$ **(b)** $\dfrac{3}{2x^2 - x}, x \neq 0, x \neq \dfrac{1}{2}$ **(c)** $\dfrac{6}{x} - 1, x \neq 0$

14. (a) $2x - \dfrac{1}{2}$ **(b)** $2x - 7$ **(c)** $\dfrac{1}{4}x - \dfrac{9}{2}$ **15. (a)** one-to-one **(b)** $B^{-1} = \{(8, 1), (1, 8), (10, 9), (9, -10)\}$

16. (a) one-to-one **(b)** $A^{-1} = \{(5, 1), (1, 2), (-7, 4), (7, 0)\}$ **17.** $f^{-1}(x) = 2x + \dfrac{2}{5}$ **18.** $f^{-1}(x) = -\dfrac{1}{3}x + \dfrac{2}{3}$

19. $f^{-1}[f(x)] = x$

Cumulative Test for Chapters 1–11

1. $-27x^2 - 30xy$ **2.** 35 **3.** 2 **4.** $(2x - 1)(2x + 1)(4x^2 + 1)$ **5.** $6x^3 + 17x^2 - 31x - 12$ **6.** $x = -\dfrac{12}{5}$

7. $y = -3x + 5$ **8.** $(3, -2)$ **9.** $3x^2y^3z\sqrt{2xz}$ **10.** $-2\sqrt{6} - 8$ **11.** $3\sqrt{10}$ **12.** $(4x - 1)(3x - 2)$

13. $(x + 3)(x - 3)(x + 1)(x - 1)$ **14.** $(x + 3)^2 + (y - 6)^2 = 196$ **15. (a)** 17 **(b)** $3a^2 - 14a + 17$ **(c)** $3a^2 - 2a + 18$

16.

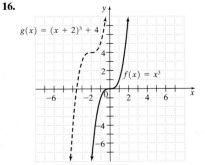

17. (a) $10x^3 - 19x^2 - 45x - 18$ **(b)** $\dfrac{2x^2 - 5x - 6}{5x + 3}, x \neq -\dfrac{3}{5}$ **(c)** $50x^2 + 35x - 3$

18. (a) yes **(b)** yes **(c)** $A^{-1} = \{(6, 3), (8, 1), (7, 2), (4, 4)\}$ **19.** $f^{-1}(x) = \dfrac{x + 3}{7}$

20. (a) 544 **(b)** -168 **(c)** $40a^3 - 12a^2 - 6$

21. (a) $f^{-1}(x) = -\dfrac{3}{2}x + 3$ **(b)**

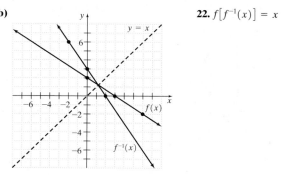

22. $f[f^{-1}(x)] = x$

Chapter 12

Pretest Chapter 12

1.

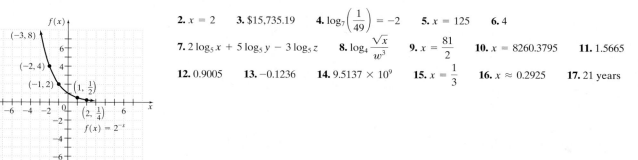

2. $x = 2$ **3.** \$15,735.19 **4.** $\log_7\left(\dfrac{1}{49}\right) = -2$ **5.** $x = 125$ **6.** 4

7. $2\log_5 x + 5\log_5 y - 3\log_5 z$ **8.** $\log_4 \dfrac{\sqrt{x}}{w^3}$ **9.** $x = \dfrac{81}{2}$ **10.** $x = 8260.3795$ **11.** 1.5665

12. 0.9005 **13.** -0.1236 **14.** 9.5137×10^9 **15.** $x = \dfrac{1}{3}$ **16.** $x \approx 0.2925$ **17.** 21 years

12.1 Exercises

1. $f(x) = b^x$, where $b > 0$, $b \neq 1$, and x is a real number. **3.** $f(x) = 3^x$ **5.** $f(x) = 2^{-x}$

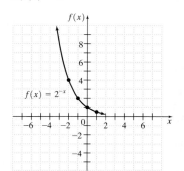

7. $f(x) = 3^{-x}$ **9.** $f(x) = 2^{x+5}$ **11.** $f(x) = 3^{x-4}$

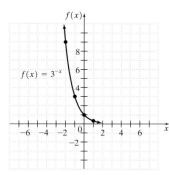

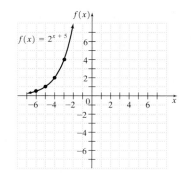

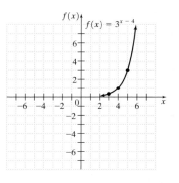

13. $f(x) = 2^x + 2$

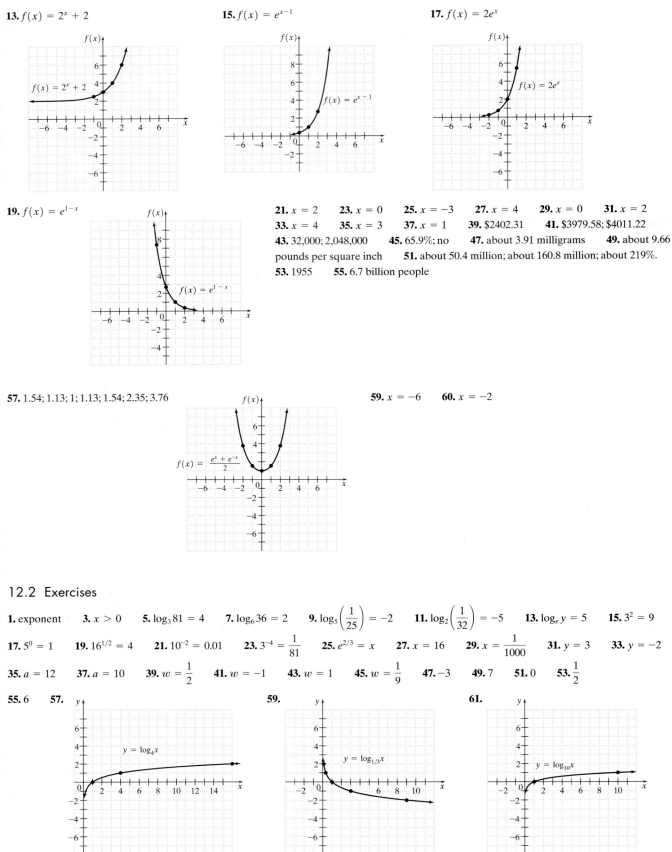

15. $f(x) = e^{x-1}$

17. $f(x) = 2e^x$

19. $f(x) = e^{1-x}$

21. $x = 2$ **23.** $x = 0$ **25.** $x = -3$ **27.** $x = 4$ **29.** $x = 0$ **31.** $x = 2$
33. $x = 4$ **35.** $x = 3$ **37.** $x = 1$ **39.** \$2402.31 **41.** \$3979.58; \$4011.22
43. 32,000; 2,048,000 **45.** 65.9%; no **47.** about 3.91 milligrams **49.** about 9.66
pounds per square inch **51.** about 50.4 million; about 160.8 million; about 219%.
53. 1955 **55.** 6.7 billion people

57. 1.54; 1.13; 1; 1.13; 1.54; 2.35; 3.76

59. $x = -6$ **60.** $x = -2$

12.2 Exercises

1. exponent **3.** $x > 0$ **5.** $\log_3 81 = 4$ **7.** $\log_6 36 = 2$ **9.** $\log_5\left(\dfrac{1}{25}\right) = -2$ **11.** $\log_2\left(\dfrac{1}{32}\right) = -5$ **13.** $\log_e y = 5$ **15.** $3^2 = 9$

17. $5^0 = 1$ **19.** $16^{1/2} = 4$ **21.** $10^{-2} = 0.01$ **23.** $3^{-4} = \dfrac{1}{81}$ **25.** $e^{2/3} = x$ **27.** $x = 16$ **29.** $x = \dfrac{1}{1000}$ **31.** $y = 3$ **33.** $y = -2$

35. $a = 12$ **37.** $a = 10$ **39.** $w = \dfrac{1}{2}$ **41.** $w = -1$ **43.** $w = 1$ **45.** $w = \dfrac{1}{9}$ **47.** -3 **49.** 7 **51.** 0 **53.** $\dfrac{1}{2}$

55. 6 **57.** **59.** **61.**

63.

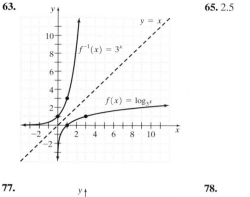

65. 2.5   **67.** 10^{-8} **69.** 5.623×10^{-10} **71.** 13,700 sets   **73.** $10,000,000   **75.** 4

77.

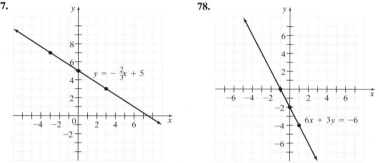

78.

79. $m = -\dfrac{1}{5}$ **80.** $y = \dfrac{3}{2}x + 7$ **81.** $x = -2$ **82.** $x = 4$ **83. (a)** $5353.27 **(b)** $24,424.03   **84. (a)** 36,000 cells **(b)** 36,864,000 cells

12.3 Exercises

1. $\log_7 M + \log_7 N$ **3.** $\log_5 7 + \log_5 11$ **5.** $\log_b 9 + \log_b f$ **7.** $\log_9 2 - \log_9 7$ **9.** $\log_a G - \log_a 7$ **11.** $\log_a E - \log_a F$ **13.** $7 \log_8 a$

15. $-2 \log_b A$ **17.** $\dfrac{1}{2} \log_5 w$ **19.** $\dfrac{1}{2} \log_5 x + 3 \log_5 y$ **21.** $\log_{13} 5 + \log_{13} B - 2 \log_{13} A$ **23.** $\log_2 5 + \log_2 x + 4 \log_2 y - \dfrac{1}{2} \log_2 z$

25. $\dfrac{1}{3} \log_b x - \dfrac{2}{3} \log_b y - \dfrac{1}{3} \log_b z$ **27.** $\log_4 39y$ **29.** $\log_3 \left(\dfrac{x^5}{7}\right)$ **31.** $\log_b \left(\dfrac{\sqrt[3]{x^2}\sqrt{y}}{z^3}\right)$ **33.** 1   **35.** 1   **37.** 0   **39.** 1   **41.** $x = 7$

43. $x = 11$ **45.** $x = 0$ **47.** $x = 1$ **49.** $x = 4$ **51.** $x = 21$ **53.** $x = 2$ **55.** $x = 2e^2$ **57.** $x = 3$ **59.** 6   **63.** approximately 50.27 square meters   **64.** approximately 62.83 cubic meters   **65.** $(3, -2)$ **66.** $(-1, -2, 3)$ **67.** about 16.1%; about 1.08×10^9 metric tons   **68.** about 10.2%; about 2.45×10^8 metric tons   **69.** 89   **70.** base = 16 feet; altitude = 7 feet

12.4 Exercises

1. 0.710117365   **3.** 1.408239965   **5.** 2.551449998   **7.** 5.096910013   **9.** -1.910094889 **11.** Error. You cannot take the log of a negative number.   **13.** 103.752842   **15.** 61.09420249   **17.** 8519.223264   **19.** 2939679.609   **21.** 0.000408037   **23.** 100.6931669   **25.** 41831168.87   **27.** 0.082679911   **29.** 1.726331664   **31.** 4.67282883   **33.** 11.82041016   **35.** -5.15162299 **37.** 2.585709659   **39.** 11.02317638   **41.** 0.878095431   **43.** 0.067205513   **45.** 472.576671   **47.** 0.1188610637   **49.** 2.020006063   **51.** 1.02507318   **53.** -1.151699337 **55.** 2.02345378   **57.** -1.846414 **59.** 3.97714348   **61.** 7.3375877   **63.** 0.1535084   **65.** 3.6593167×10^8 **67.** 3.3149796

69.

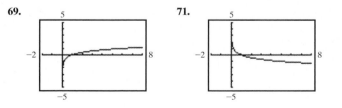

71.

73. 35.18; 35.89; approximately 2.0%   **75.** $R \approx 5.26$
77. The shock wave is about 251,000 times greater than the smallest detectable shock wave.

78. $x = \dfrac{11 \pm \sqrt{181}}{6}$ **79.** $y = \dfrac{-2 \pm \sqrt{10}}{2}$ **80.** $5\sqrt{5}$ **81.** $(x - 2)^2 + (y + 1)^2 = 9$ **82.** 6 miles; 12 miles   **83.** 21 miles; 24 miles

12.5 Exercises

1. $x = 4$ **3.** $x = \dfrac{31}{3}$ **5.** $x = \dfrac{13}{8}$ **7.** $x = 2$ **9.** $x = \dfrac{3}{2}$ **11.** $x = \dfrac{20}{99}$ **13.** $x = \dfrac{16}{15}$ **15.** $x = 5$ **17.** $x = \dfrac{5}{3}$ **19.** $x = 4$ **21.** $x = 5$

23. $x = \dfrac{\log 11 + \log 8}{\log 8}$ **25.** $x = \dfrac{\log 17 - 4\log 2}{3\log 2}$ **27.** $x \approx 1.006$ **29.** $x \approx 6.213$ **31.** $x \approx 4.332$ **33.** $x \approx 1.739$ **35.** $t \approx 16$ years

37. $t \approx 19$ years   **39.** 4.5%   **41.** 20 years   **43.** 69 years   **45.** 437,000 employees   **47.** 2001   **49.** 5 years   **51.** 55 hours

53. 46,931 people   **55.** about 2.5 times greater   **57.** about 31.6 times greater   **59.** 17.8 years   **61.** $7xy\sqrt{2x}$ **62.** $3\sqrt{2} + 4\sqrt{3} - \sqrt{6} - 4$

63. \$37,500   **64.** approximately \$456 million per mile   **65.** $x = 4$ or $x = -\dfrac{2}{3}$ **66.** $-2 \le x \le 8$

Putting Your Skills to Work

1. $k \approx 0.026$ **2.** $T = 72 + (220 - 72)e^{-0.026t}$ **3.** 81 minutes

Chapter 12 Review Problems

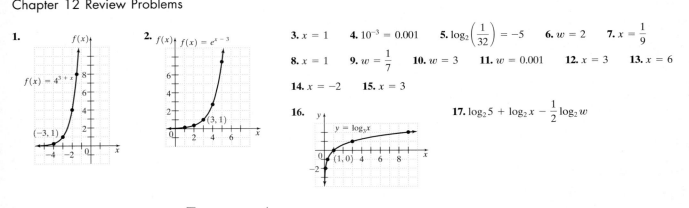

1.

2. $f(x) = e^{x-3}$

$f(x) = 4^{3+x}$ $(-3, 1)$

$(3, 1)$

3. $x = 1$ **4.** $10^{-3} = 0.001$ **5.** $\log_2\left(\dfrac{1}{32}\right) = -5$ **6.** $w = 2$ **7.** $x = \dfrac{1}{9}$

8. $x = 1$ **9.** $w = \dfrac{1}{7}$ **10.** $w = 3$ **11.** $w = 0.001$ **12.** $x = 3$ **13.** $x = 6$

14. $x = -2$ **15.** $x = 3$

16. $y = \log_3 x$ $(1, 0)$ **17.** $\log_2 5 + \log_2 x - \dfrac{1}{2}\log_2 w$

18. $3\log_2 x + \dfrac{1}{2}\log_2 y$ **19.** $\log_3 \dfrac{x\sqrt{w}}{2}$ **20.** $\log_8 \dfrac{w^4}{\sqrt[3]{z}}$ **21.** 6   **22.** $x = 25$ **23.** $x = 25$ **24.** 1.376576957   **25.** -1.087777943

26. 1.366091654   **27.** 6.688354714   **28.** $n = 13.69935122$ **29.** $n = 5.473947392$ **30.** 0.49685671   **31.** $x = \dfrac{34}{5}$ **32.** $x = -\dfrac{79}{54}$

33. $x = \dfrac{1}{7}$ **34.** $x = 1$ **35.** $x = 3$ **36.** $x = 4$ **37.** $t = \dfrac{3}{4}$ **38.** $t = -\dfrac{1}{8}$ **39.** $x = \dfrac{\log 14}{\log 3}$ **40.** $x = \dfrac{2\log 4}{\log 5 - \log 4}$

41. $x = -1 + \ln 3.5$ **42.** $x = \dfrac{\ln 30.6}{2}$ **43.** $x \approx -1.4748$ **44.** $x \approx 0.7712$ **45.** $x \approx 2.3319$ **46.** $x \approx 101.3482$ **47.** 9 yrs

48. \$6312.38   **49.** 8 years   **50.** 9 years   **51.** 26 years   **52.** 41 years   **53.** 9 years   **54.** 11 years   **55.** 50.1 times more
56. (a) approximately 282 pounds   **(b)** 7.77 pounds per cubic inch

Chapter 12 Test

1.

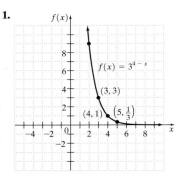

$f(x) = 3^{4-x}$ $(3, 3)$ $(4, 1)$ $\left(5, \dfrac{1}{3}\right)$

2. $x = 0$ **3.** $w = 5$ **4.** $x = \dfrac{1}{64}$ **5.** $\log_8 \dfrac{xw}{\sqrt[4]{3}}$ **6.** 1.7901   **7.** 1.3729   **8.** 0.4391

9. $x \approx 5350.569382$ **10.** $x \approx 1.150273799$ **11.** $x = \dfrac{3}{7}$ **12.** $x = \dfrac{16}{3}$ **13.** $x = \dfrac{-1 + \ln 0.25}{3}$

14. $x \approx -1.4132$ **15.** \$2938.66   **16.** 14 years

Cumulative Test for Chapters 1–12

1. 6   **2.** $x = \dfrac{H + 2ay}{3b}$ **3.**

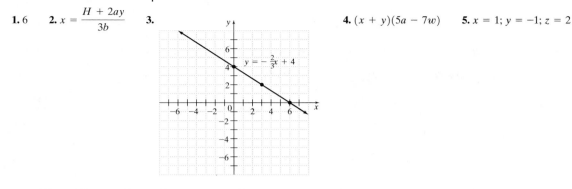
$y = -\dfrac{2}{3}x + 4$

4. $(x + y)(5a - 7w)$ **5.** $x = 1; y = -1; z = 2$

6. $5\sqrt{10} + \sqrt{15} - 20\sqrt{3} - 6\sqrt{2}$ **7.** $x = \pm\sqrt{6}; x = \pm i$ **8.** $x = 4, y = 4, x = 1, y = -2$ **9.** $x = 4$

10. $\dfrac{5\sqrt[3]{4x^2y}}{2xy}$ **11.**

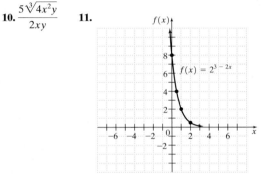
$f(x) = 2^{3 - 2x}$

12. $x = \dfrac{1}{4}$ **13.** $x = \dfrac{3}{2}$ **14.** 0.884795364   **15.** $x \approx 66.20640403$

16. 1.771243749   **17.** $x \approx 7.1263558$ **18.** $x = 9$ **19.** $x = 2$

20. $x \approx -0.535$ **21.** $x = \dfrac{\ln 0.5}{2}$ **22.** \$4234.74

Practice Final Examination

1. 4   **2.** $-a^2 - 10ab - a$ **3.** $-2x + 21y + 18xy + 6y^2$ **4.** 14   **5.** $F = -31°$ **6.** $y = -30$ **7.** $b = \dfrac{2A - ac}{a}$

8. $x \geq 2.6$ 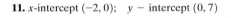   **9.** Length $= 520$ m; width $= 360$ m   **10.** \$2600 at 12%; \$1400 at 14%

11. x-intercept $(-2, 0)$; $y -$ intercept $(0, 7)$

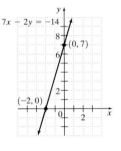
$7x - 2y = -14$
$(0, 7)$
$(-2, 0)$

12.

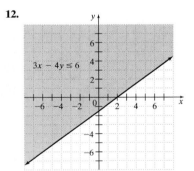
$3x - 4y \leq 6$

13. $m = \dfrac{8}{3}$ **14.** $y = -3x + 1$ **15.** $f(3) = 12$ **16.** $f(-2) = 17$ **17.**

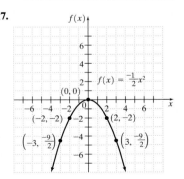
$f(x) = \dfrac{-1}{2}x^2$
$(0, 0)$
$(-2, -2)$ $(2, -2)$
$\left(-3, \dfrac{-9}{2}\right)$ $\left(3, \dfrac{-9}{2}\right)$

18. $x = 6$; $y = -3$ **19.** $x = 6$; $y = 4$ **20.** $x = 1$; $y = 4$; $z = -2$ **21.** $x = 4$; $y = 1$; $z = 1$ **23.** $18x^5y^5$

22.

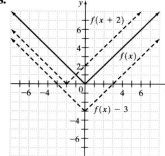

24. $6x^3 - 16x^2 + 17x - 6$ **25.** $5x^2 - x + 2$ **26.** $(3x - 5)^2$ **27.** $(x + 2)(x + 2)(x - 2)$

28. $x(2x - 1)(x + 8)$ **29.** $x = -6$, $x = -9$ **30.** $\dfrac{x(3x - 1)}{x - 3}$ **31.** $\dfrac{x + 5}{x}$

32. $\dfrac{3x^2 + 6x - 2}{(x + 5)(x + 2)}$ **33.** $\dfrac{8x^2 + 6x - 5}{4x^2 - 3}$ **34.** $x = -1$ **35.** $\dfrac{1}{3x^{7/2}y^5}$ **36.** $2xy^2\sqrt[3]{5xy}$

37. $18\sqrt{2}$ **38.** $\dfrac{18 + 2\sqrt{6} + 3\sqrt{3} + \sqrt{2}}{25}$ **39.** $8i$ **40.** $x = -3$ **41.** $y = 33.75$

42. $x = \dfrac{1 \pm \sqrt{21}}{10}$ **43.** $x = 0$, $x = -\dfrac{3}{5}$ **44.** $x = 8$, $x = -343$ **45.** $x \le -\dfrac{1}{3}$ or $x \ge 4$

46.

47. Width = 4 cm; length = 13 cm **48.** $x = 3$, $x = 9$ **49.** $-\dfrac{5}{2} < x < \dfrac{15}{2}$

50. $(x + 3)^2 + (y - 2)^2 = 4$; center at $(-3, 2)$; radius = 2

51. $\dfrac{x^2}{16} + \dfrac{y^2}{25} = 1$; ellipse

52. $\dfrac{x^2}{4} - \dfrac{y^2}{9} = 1$; hyperbola

53. Parabola opening right

54. $(0, -4)$, $(\sqrt{7}, 3)$, $(-\sqrt{7}, 3)$

55. $f(-1) = 10$; $f(a) = 3a^2 - 2a + 5$; $f(a + 2) = 3a^2 + 10a + 13$

56. $f[g(x)] = 80x^2 + 80x + 17$ **57.** $f^{-1}(x) = 2x + 14$

58.

59. $f(x) = 2^{1-x}$

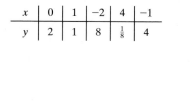

x	0	1	-2	4	-1
y	2	1	8	$\frac{1}{8}$	4

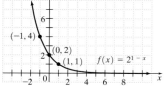

60. $0.0016 = x$ or $\dfrac{1}{625} = x$ **61.** $x = 21$ **62.** $y = -2$ **63.** $x = 8$

Appendix B Exercises

1. $269.17   **3.** 95 cubic yards   **5.** Jog $2\frac{2}{3}$ miles; walk $3\frac{1}{9}$ miles; rest $4\frac{4}{9}$ minutes; walk $1\frac{7}{9}$ miles.   **7.** Betty; Melinda increases each activity by $\frac{2}{3}$ by day 3 and Betty increases each activity by $\frac{7}{9}$ by day 3.   **9.** $4\frac{1}{2}$ miles   **11.** $98,969.00   **13.** 49%: 61%; business travelers; answers may vary. A sample is: Men 21 or older spend more of their travel time on business than for pleasure.   **15.** 18%   **17.** 69%

Appendix C Exercises

1. 84,916 square miles   **3.** 3,500,000   **5.** 2   **7.** Treasure State   **9.** Nevada   **11.** 2000   **13.** Dupage County   **15.** 2800   **17.** 4,000,000   **19.** 7,200,000   **21.** 8,000,000   **23.** It is a three-way tie. The increase was 3 million in three cases. It occurred between 1970 and 1980, again between 1980 and 1990, and finally again between 1990 and 2000.   **25.** 3 million more   **27.** 37 million   **29.** movies and an exercise program   **31.** $5.6 million   **33.** $2.5 million greater   **35.** Between 1997 and 1998   **37.** 74 million   **39.** 65 years and above   **41.** 19%   **43.** 81%   **45.** 845 million people   **47.** 40%   **49.** $3588

Appendix D Exercises

1. 14   **3.** 52   **5.** 1296   **7.** -3 **9.** $10x - 5$ **11.** 1111811   **13.** 9876549   **15.** 6481100   **17.** 1 6 15 20 15 6 1

19. $x = 3$, steps will vary   **21.** $x = -2$, steps will vary   **23.** $x = 3$, steps will vary   **25.** $x = \frac{9}{11}$, steps will vary   **27.** Answers will vary.

29. William was first, Brent was second, James was third, and Dave was fourth.   **31.** Michael   **33.** Toyota Corolla   **35.** Unless Fred is willing to pay over $200 for Tuesday's procedure, he will probably get a false tooth.

Appendix E Exercises

1. -7 **3.** 15   **5.** 2   **7.** 47   **9.** 18   **11.** 0   **13.** 0   **15.** -0.6 **17.** $-7a - 4b$ **19.** $\frac{11}{84}$ **21.** $\begin{vmatrix} 6 & 10 \\ -5 & 9 \end{vmatrix}$ **23.** $\begin{vmatrix} 3 & -4 \\ 1 & -5 \end{vmatrix}$ **25.** -7 **27.** -26 **29.** 11   **31.** -27 **33.** -8 **35.** 0   **37.** -3.179 **39.** 18,553   **41.** $x = 2$; $y = 3$ **43.** $x = -2$; $y = 5$ **45.** $x = 10$; $y = 2$ **47.** $x = 4$; $y = -2$ **49.** $x = 1.5795$; $y = -0.0902$ **51.** $x = 1$; $y = 1$; $z = 1$ **53.** $x = -\frac{1}{2}$; $y = \frac{1}{2}$; $z = 2$ **55.** $x = 4$; $y = -2$; $z = 1$ **57.** $x = -0.219$; $y = 1.893$; $z = -3.768$ **59.** $w = -3.105$; $x = 4.402$; $y = 15.909$; $z = 6.981$

Appendix F Exercises

1. elements   **3.** intersection   **5.** empty set   **7.** {California, Colorado, Connecticut}   **9.** $\{1, 3, 5, 7, \ldots\}$ **11.** $O = \{x \mid x \text{ is an ocean}\}$ **13.** $T = \{x \mid x \text{ is a type of triangle}\}$ **15. (a)** $\{-2, -1, 2, 3, 5, 8\}$ **(b)** $\{-1, 3, 5\}$ **17.** False. B is an infinite set.   **19.** True   **21.** True   **23.** False; $A \cap C = \{1, 2, 3, 4, 5\}$ **25.** Answers will vary. One possible answer is $\{\text{Ann, Nina}\}$.   **27. (a)** {Andrew, Christopher, Daniel, David, Jacob, James, Jason, John, Joseph, Joshua, Matthew, Michael, Nicholas, Robert, Tyler}   **(b)** union   **29.** whole numbers: 15, 0, $\sqrt{81}$; natural numbers: 15, $\sqrt{81}$; integers: 15, 0, $\sqrt{81}$, -17; rational numbers: 3.62, $\frac{-3}{11}$, 15, 0, $\sqrt{81}$, -17; irrational numbers: $\sqrt{20}$; real numbers: 3.62, $\sqrt{20}$, $\frac{-3}{11}$, 15, $\frac{22}{3}$, 0, $\sqrt{81}$, -17

31. No. The set of integers contains negative numbers. Negative numbers are not part of the set of whole numbers.   **33.** The set of whole numbers is a subset of the integers, the rational numbers, and the real numbers.

Applications Index

Subject Index

Photo Credits

Chapter 1 CO Charles Gupton/Stock Boston **p. 4** Peter Skinner/Photo Researchers, Inc. **p. 47** John Coletti/Stock Boston **p. 50** Wojnarowicz/The Image Works

Chapter 2 CO Mark Richards/PhotoEdit **p. 78** Jerry Wachter/Photo Researchers, Inc. **p. 108** Jeffrey Dunn/Stock Boston **p. 111** Stephen J. Krasemann/Photo Researchers, Inc. **p. 112** Bob Daemmrich/Stock Boston **p. 116** Dave Bartruff/Corbis **p. 128** Dana White/PhotoEdit

Chapter 3 CO Kaz Mori/The Image Bank **p. 176** Rondi/Tani/Science Studios/Photo Researchers, Inc. **p. 199** Jacques Jangoux/Stone

Chapter 4 CO Tim Davis/Stone **p. 234** B. Daemmrich/The Image Works **p. 235** Mitch Wojnarowicz/The Image Works

Chapter 5 CO D. Young-Wolff/PhotoEdit **p. 299** Bsip Laurent/Photo Researchers, Inc.

Chapter 6 CO Joseph Nettis/Stock Boston **p. 323** Malcolm Fielding, Johnson Matthey PLC/Science Photo Library/Photo Researchers, Inc. **p. 347** Alok Kavan/Photo Researchers, Inc. **p. 360** Laima Druskis/Stock Boston **p. 366** Alan Schein/Corbis Stock Market

Chapter 7 CO Owen Franken/Stock Boston

Chapter 8 CO Stone **p. 469** Stuart Westmorland/Stone **p. 474** Glen Allison/Stone **p. 479** Alan Carey/The Image Works

Chapter 9 CO Robert Cameron/Stone **p. 498** Kevin Miller/Stone **p. 514** Ellis Herwig/Stock Boston **p. 516** Woodfin Camp & Associates **p. 536** Hulton Getty/Hulton/Archive **p. 546** Jay Freis/The Image Bank

Chaper 10 CO Andrew Rafkind/Stone

Chapter 11 CO John Blaustein/Woodfin Camp & Associates **p. 640** Mark C. Burnett/Photo Researchers, Inc. **p. 646** John Coletti/Stock Boston

Chapter 12 CO Lee Snider/The Image Works **p. 676** Ford Smith/Corbis **p. 681** © 2000 Richard Laird/FPG International LLC **p. 709** Index Stock Imagery, Inc. **p. 715** © Adam Woolfitt/Corbis **p. 716** NASA